Potential Invasive Pests of Agricultural Crops

CABI Invasive Species Series

Invasive species are plants, animals or microorganisms not native to an ecosystem, whose introduction has threatened biodiversity, food security, health or economic development. Many ecosystems are affected by invasive species and they pose one of the biggest threats to biodiversity worldwide. Globalization through increased trade, transport, travel and tourism will inevitably increase the intentional or accidental introduction of organisms to new environments, and it is widely predicted that climate change will further increase the threat posed by invasive species. To help control and mitigate the effects of invasive species, scientists need access to information that not only provides an overview of and background to the field, but also keeps them up to date with the latest research findings.

This series addresses all topics relating to invasive species, including biosecurity surveillance, mapping and modelling, economics of invasive species and species interactions in plant invasions. Aimed at researchers, upper-level students and policy makers, titles in the series provide international coverage of topics related to invasive species, including both a synthesis of facts and discussions of future research perspectives and possible solutions.

Titles Available

1. Invasive Alien Plants : An Ecological Appraisal for the Indian Subcontinent
   Edited by J.R. BHATT, J.S. SINGH, S.P. SINGH, R.S. TRIPATHI and R.K. KOHLI

2. Invasive Plant Ecology and Management : Linking Processes to Practice
   Edited by T.A. MONACO and R.L. SHELEY

3. Potential Invasive Pests of Agricultural Crops
   Edited by J.E. PEÑA

Potential Invasive Pests of Agricultural Crops

Edited by

Jorge E. Peña

University of Florida, USA

www.cabi.org

CABI is a trading name of CAB International

CABI
Nosworthy Way
Wallingford
Oxfordshire OX10 8DE
UK

CABI
38 Chauncey Street
Suite 1002
Boston, MA 02111
USA

Tel: +44 (0)1491 832111
Fax: +44 (0)1491 833508
E-mail: info@cabi.org
Website: www.cabi.org

Tel: +1 800 552 3083 (toll free)
Tel: +1 (0)617 395 4051
E-mail: cabi-nao@cabi.org

© CAB International 2013. All rights reserved. No part of this publication may be reproduced in any form or by any means, electronically, mechanically, by photocopying, recording or otherwise, without the prior permission of the copyright owners.

A catalogue record for this book is available from the British Library, London, UK.

Library of Congress Cataloging-in-Publication Data

Potential invasive pests of agricultural crops / [editor, Jorge E. Peña].
   p. ; cm. -- (CAB invasive ; [3])
   ISBN 978-1-84593-829-1 (hbk)
   1. Agricultural pests. 2. Introduced organisms. I. Peña, Jorge E., 1948-II. Series: CABI invasive species series ; 3.

   SB601.P68 2012
   632'.9--dc23
                                       2012029378

ISBN-13: 978-1-84593-829-1

Commissioning editor: Dave Hemming
Editorial assistant: Chris Shire
Production editor: Simon Hill

Typeset by SPi, Pondicherry, India.
Printed and bound in the UK by CPI Group (UK) Ltd, Croydon, CR0 4YY.

Contents

Acknowledgments ix

Contributors xi

Preface xv
JORGE E. PEÑA

1. **Biology and Management of the Red Palm Weevil,**
   Rhynchophorus ferrugineus 1
   Robin M. Giblin-Davis, Jose Romeno Faleiro, Josep A. Jacas, Jorge E. Peña, and P.S.P.V. Vidyasagar

2. **Avocado Weevils of the Genus *Heilipus*** 35
   Alvaro Castañeda-Vildózola, Armando Equihua-Martinez and Jorge E. Peña

3. **Exotic Bark and Ambrosia Beetles in the USA: Potential and Current Invaders** 48
   Robert A. Haack and Robert J. Rabaglia

4. ***Diabrotica speciosa*: an Important Soil Pest in South America** 75
   Crébio José Ávila and Alexa Gabriela Santana

5. **Potential Lepidopteran Pests Associated with Avocado Fruit in Parts of the Home Range of *Persea americana*** 86
   Mark S. Hoddle and J.R.P. Parra

6. **Biology, Ecology and Management of the South American Tomato Pinworm, *Tuta absoluta*** 98
   Alberto Urbaneja, Nicolas Desneux, Rosa Gabarra, Judit Arnó, Joel González-Cabrera, Agenor Mafra-Neto, Lyndsie Stoltman, Alexandre de Sene Pinto and José R.P. Parra

7. ***Tecia solanivora* Povolny (Lepidoptera: Gelechiidae), an Invasive Pest of Potatoes *Solanum tuberosum* L. in the Northern Andes** 126
   Daniel Carrillo and Edison Torrado-Leon

8. **The Tomato Fruit Borer, *Neoleucinodes elegantalis* (Guenée) (Lepidoptera: Crambidae), an Insect Pest of Neotropical Solanaceous Fruits** 137
   Ana Elizabeth Diaz Montilla, Maria Alma Solis and Takumasa Kondo

9. ***Copitarsia* spp.: Biology and Risk Posed by Potentially Invasive Lepidoptera from South and Central America** 160
   Juli Gould, Rebecca Simmons and Robert Venette

10. **Host Range of the Nettle Caterpillar *Darna pallivitta* (Moore) (Lepidoptera: Limacodidae) in Hawai'i** 183
    Arnold H. Hara, Christopher M. Kishimoto and Ruth Y. Niino-DuPonte

11. **Fruit flies *Anastrepha ludens* (Loew), *A. obliqua* (Macquart) and *A. grandis* (Macquart) (Diptera: Tephritidae): Three Pestiferous Tropical Fruit Flies That Could Potentially Expand Their Range to Temperate Areas** 192
    Andrea Birke, Larissa Guillén, David Midgarden and Martin Aluja

12. ***Bactrocera* Species that Pose a Threat to Florida: *B. carambolae* and *B. invadens*** 214
    Aldo Malavasi, David Midgarden and Marc de Meyer

13. **Signature Chemicals for Detection of *Citrus* Infestation by Fruit Fly Larvae (Diptera: Tephritidae)** 228
    Paul E. Kendra, Amy L. Roda, Wayne S. Montgomery, Elena Q. Schnell, Jerome Niogret, Nancy D. Epsky and Robert R. Heath

14. **Gall Midges (Cecidomyiidae) attacking Horticultural Crops in the Caribbean Region and South America** 240
    Juliet Goldsmith, Jorge Castillo and Dionne Clarke-Harris

15. **Recent Mite Invasions in South America** 251
    Denise Navia, Alberto Luiz Marsaro Júnior, Manoel Guedes Correa Gondim Jr, Renata Santos de Mendonça and Paulo Roberto Valle da Silva Pereira

16. ***Planococcus minor* (Hemiptera: Pseudococcidae): Bioecology, Survey and Mitigation Strategies** 288
    Amy Roda, Antonio Francis, Moses T.K. Kairo and Mark Culik

17. **The Citrus Orthezia *Praelongorthezia praelonga* (Douglas) (Hemiptera: Ortheziidae), a Potential Invasive Species** 301
    Takumasa Kondo, Ana Lucia Peronti, Ferenc Kozár and Éva Szita

18. **Potential Invasive Species of Scale Insects for the USA and Caribbean Basin** 320
    Gregory A. Evans and John W. Dooley

19. **Recent Adventive Scale Insects (Hemiptera: Coccoidea) and Whiteflies (Hemiptera: Aleyrodidae) in Florida and the Caribbean Region** 342
    Ian Stocks

20. **Biology, Ecology and Control of the Ficus Whitefly, *Singhiella simplex* (Hemiptera: Aleyrodidae)** 363
    Jesusa Crisostomo Legaspi, Catharine Mannion, Divina Amalin and Benjamin C. Legaspi Jr

21. **Invasion of Exotic Arthropods in South America's Biodiversity Hotspots and Agro-Production Systems** 373
    K.A.G. Wyckhuys, T. Kondo, B.V. Herrera, D.R. Miller, N. Naranjo and G. Hyman

22. **Likelihood of Dispersal of the Armored Scale, *Aonidiella orientalis* (Hemiptera: Diaspididae), to Avocado Trees from Infested Fruit Discarded on the Ground, and Observations on Spread by Handlers** 401
    M.K. Hennessey, J.E. Peña, M. Zlotina and K. Santos

23. **Insect Life Cycle Modelling (ILCYM) Software – A New Tool for Regional and Global Insect Pest Risk Assessments under Current and Future Climate Change Scenarios** 412
    Marc Sporleder, Henri E.Z. Tonnang, Pablo Carhuapoma, Juan C. Gonzales, Henry Juarez and Jürgen Kroschel

Index 429

Dedication

To A.C. Bellotti and the Peña-Rojas and Litz Families

Acknowledgments

Extensive thanks to the authors for their willingness to participate in this endeavour. Dr. J.L. Capinera, Ms. Rita E. Duncan and Ms. H. Glenn for their help. This Book was the result of the Potential Invasive Pest Workshop supported by a grant from TSTAR and by the University of Florida Center for Tropical Agriculture under the direction of R.E. Litz.

Contributors

Martin Aluja *Red de Manejo Biorracional de Plagas y Vectores, Instituto de Ecología A.C., Xalapa, Veracruz, México.*
Divina Amalin *Tropical Research and Education Center, University of Florida, 18905 SW 280th Street, Homestead, Florida 33031, USA.*
Judit Arnó *Entomology IRTA, Ctra. de Cabrils km 2, 08348 Cabrils (Barcelona), Spain.*
Crébio José Ávila *Embrapa Western Agriculture, Caixa Postal 661, Dourados 79824-100, Brazil; E-mail: crebio@cpao.embrapa.br*
Andrea Birke *Red de Manejo Biorracional de Plagas y Vectores, Instituto de Ecología A.C., Xalapa, Veracruz, México.*
Pablo Carhuapoma *International Potato Center (CIP), Apartado 1558, Lima 12, Peru.*
Daniel Carrillo *Department of Entomology and Nematology, Tropical Research and Education Center, University of Florida, Homestead, Florida 33031, USA; E-mail: dancar@ufl.edu*
Alvaro Castañeda-Vild zola *Facultad de Ciencias Agricolas, Universidad Autónoma del Estado de México, Toluca, Estado de México, México.*
Jorge Castillo *Universidad Nacional Agraria La Molina, Lima, Peru.*
Dionne Clarke-Harris *Caribbean Agricultural Research and Development Institute, Jamaica.*
Manoel Guedes Correa Gondim Jr *Universidade Federal Rural de Pernambuco – UFRPE, Avenida Dom Manoel de Medeiros s/n, Dois Irmãos, 52171-900, Recife, Pernambuco, Brazil.*
Jesusa Crisostomo Legaspi *U.S. Department of Agriculture, Agricultural Research Service, CMAVE/Florida A&M University – Center for Biological Control, 6383 Mahan Dr., Tallahassee, Florida 32308, USA.*
Mark Culik *Instituto Capixaba de Pesquisa, Assistência Técnica e Extensão Rural – INCAPER, Vitória, Espírito Santo, Brazil.*
Marc De Meyer *Royal Museum for Central Africa, Tervuren, Belgium; E-mail: marc.de.meyer@africamuseum.be*
Alexandre de Sene Pinto *Centro Universitário Moura Lacerda, Av. Dr. Oscar de Moura Lacerda, 1520, 14076-510, Ribeirão Preto, Brazil.*
Nicolas Desneux *INRA (French National Institute for Agricultural Research), UR 880, 400 route des chappes, 06903 Sophia-Antipolis, France.*
Ana Elizabeth Diaz Montilla *Corporación Colombiana de Investigación Agropecuaria, Corpoica, Colombia; E-mail: aediaz@corpoica.org.co*

John W. Dooley USDA/APHIS/PPQ 389 Oyster Point Blvd, Suite 2A, South San Francisco, California 94080, USA; E-mail: john.w.dooley@aphis.usda.gov

Nancy D. Epsky USDA-ARS, Subtropical Horticulture Research Station, Miami, Florida 33158, USA.

Armando Equihua-Martinez Instituto de Fitosanidad, Colegio de Posgraduados, Montecillo, Texcoco, Estado de México, México.

Gregory A. Evans USDA/APHIS/PPQ/National Identification Service, 10300 Baltimore Avenue, BARC-West, Bldg. 005, Rm 09A, Beltsville, Maryland 20705, USA; E-mail: Gregory.A.Evans@usda.gov

Jose Romeno Faleiro Mariella, Arlem-Raia, Salcette, Goa 403 720, India; E-mail: jrfaleiro@yahoo.co.in

Antonio Francis Center for Biological Control, College of Engineering Sciences, Technology and Agriculture, Florida Agricultural and Mechanical University, Tallahassee, Florida, USA.

Rosa Gabarra Entomology IRTA, Ctra. de Cabrils km 2, 08348 Cabrils (Barcelona), Spain.

Robin M. Giblin-Davis Fort Lauderdale Research and Education Center, University of Florida/IFAS, Davie, Florida, USA.

Juliet Goldsmith Plant Quarantine, Produce Inspection Branch, Ministry of Agriculture and Fisheries, Jamaica.

Juan C. Gonzales International Potato Center (CIP), Apartado 1558, Lima 12, Peru.

Joel González-Cabrera Unidad de Entomología, Centro de Protección Vegetal y Biotecnología, Instituto Valenciano de Investigaciones Agrarias (IVIA), Carretera Moncada-Náquera km 4,5, 46113 Moncada, Valencia, Spain.

Juli Gould USDA Animal and Plant Health Inspection Service, Center for Plant Health Science and Technology, Buzzards Bay, Massachusetts, USA.

Larissa Guillén Red de Manejo Biorracional de Plagas y Vectores, Instituto de Ecología A.C., Xalapa, Veracruz, México.

Robert A. Haack USDA Forest Service, Northern Research Station, 1407 S Harrison Road, East Lansing, Michigan 48823, USA; E-mail: rhaack@fs.fed.us

Arnold H. Hara Department of Plant and Environmental Protection Sciences, Komohana Research and Extension Center, University of Hawai'i at Manoa, 875 Komohana Street, Hilo, Hawai'i 96720, USA.

Robert R. Heath USDA-ARS, Subtropical Horticulture Research Station, Miami, Florida 33158, USA.

M.K. Hennessey USDA-APHIS-PPQ, 1730 Varsity Dr., Suite 300, Raleigh, North Carolina 27606, USA.

B.V. Herrera International Center for Tropical Agriculture CIAT, Recta Palmira-Cali, Cali, Valle del Cauca, Colombia.

Mark S. Hoddle Department of Entomology, and Center for Invasive Species Research, University of California, Riverside, California 92521, USA.

G. Hyman International Center for Tropical Agriculture CIAT, Recta Palmira-Cali, Cali, Valle del Cauca, Colombia.

Josep A. Jacas Universitat Jaume I, Campus del Riu Sec, Castelló de la Plana, Spain.

Henry Juarez International Potato Center (CIP), Apartado 1558, Lima 12, Peru.

Moses T.K. Kairo Center for Biological Control, College of Engineering Sciences, Technology and Agriculture, Florida Agricultural and Mechanical University, Tallahassee, Florida, USA.

Paul E. Kendra USDA-ARS, Subtropical Horticulture Research Station, Miami, Florida 33158, USA; E-mail: paul.kendra@ars.usda.gov

Christopher M. Kishimoto 1255 Nu'uanu Ave, # E-3110, Honolulu, Hawai'i 96817, USA.

Takumasa Kondo Corporación Colombiana de Investigación Agropecuaria, Corpoica, Colombia; E-mail: takumasa.kondo@gmail.com

Ferenc Kozár Plant Protection Institute, Hungarian Academy of Sciences, Budapest, Hungary; E-mail: h2405koz@ella.hu

Jürgen Kroschel International Potato Center (CIP), Apartado 1558, Lima 12, Peru; E-mail: j.kroschel@cgiar.org

Benjamin C. Legaspi, Jr Tropical Research and Education Center, University of Florida, 18905 SW 280th Street, Homestead, Florida 33031, USA

Agenor Mafra-Neto *ISCA Technologies, Inc., 1230 W. Spring Street, Riverside, California 92507, USA.*

Aldo Malavasi *Medfly Rearing Facility – Moscamed Brasil, Juazeiro, Bahia, Brazil; E-mail: malavasi@moscamed.org.br*

Catharine Mannion *Tropical Research and Education Center, University of Florida, 18905 SW 280th Street, Homestead, Florida 33031, USA.*

Alberto Luiz Marsaro Júnior *Embrapa Trigo, Rodovia BR 285, Km 294, Cx. Postal 451, 99001-970, Passo Fundo, Rio Grande do Sul, Brazil.*

David Midgarden *USDA APHIS Medfly Program, Guatemala City, Guatemala.*

D.R. Miller *Systematic Entomology Laboratory, Agricultural Research Service, U.S. Department of Agriculture, Beltsville, Maryland 20705, USA.*

Wayne S. Montgomery *USDA-ARS, Subtropical Horticulture Research Station, Miami, Florida 33158, USA.*

N. Naranjo *Horticulture Research Center CIAA, Universidad Jorge Tadeo Lozano, Bogota, Colombia.*

Denise Navia *Embrapa Recursos Genéticos e Biotecnologia, Parque Estação Biológica, W5 Norte Final, Cx. Postal 02372, 70770-917 Brasília, Distrito Federal, Brazil; E-mail: navia@cenargen.embrapa.br*

Ruth Y. Niino-DuPonte *Department of Plant and Environmental Protection Sciences, Komohana Research and Extension Center, University of Hawai'i at Manoa, 875 Komohana Street, Hilo, Hawai'i 96720, USA.*

Jerome Niogret *USDA-ARS, Subtropical Horticulture Research Station, Miami, Florida 33158, USA.*

Jose R.P. Parra *Departamento de Entomologia, Fitopatologia e Zoologia Agrícola, Escola Superior de Agricultura 'Luiz de Queiroz', Universidade de São Paulo, 13418-900, Brazil.*

Jorge E. Peña *Tropical Research and Education Center, University of Florida/IFAS, Homestead, Florida 33031, USA; E-mail: jepe@ifas.ufl.edu*

Lucia Peronti *Departamento de Ecologia e Biologia Evolutiva, Universidade Federal de São Carlos (UFSCar), São Carlos/SP, Brazil; E-mail: anaperonti@hotmail.com*

Robert J. Rabaglia *USDA Forest Service, Forest Health Protection, 1601 N Kent Street, RPC–7, Arlington, Virginia 22209, USA; E-mail: brabaglia@fs.fed.us*

Amy L. Roda *USDA-APHIS-PPQ, Center for Plant Health Science and Technology, Miami, Florida 33158, USA.*

Alexa Gabriela Santana *Embrapa Western Agriculture, Caixa Postal 661, Dourados 79824-100, Brazil; E-mail: Alexagsantana27@hotmail.com*

Katia Santos *Tropical Research and Education Center, University of Florida/IFAS, Homestead, Florida 33031, USA.*

Renata Santos de Mendonça *Embrapa Recursos Genéticos e Biotecnologia, Parque Estação Biológica, W5 Norte Final, Cx. Postal 02372, 70770-917 Brasília, Distrito Federal, Brazil.*

Elena Q. Schnell *USDA-ARS, Subtropical Horticulture Research Station, Miami, Florida 33158, USA.*

Rebecca Simmons *University of North Dakota, Grand Forks, North Dakota, USA.*

Maria Alma Solis *USDA/ARS, SEL, Room 133, Building 005, BARC-West, 10300 Baltimore Ave., Beltsville, Maryland 20705, USA; E-mail: alma.solis@ars.usda.gov*

Marc Sporleder *International Potato Center (CIP), Apartado 1558, Lima 12, Peru.*

Ian Stocks *Division of Plant Industry, Florida Department of Agriculture & Consumer Services, 1911 SW 34th Street, PO Box 147100,Gainesville, Florida 32614-7100, USA.*

Lyndsie Stoltman *ISCA Technologies, Inc., 1230 W. Spring Street, Riverside, California 92507, USA.*

Éva Szita *Plant Protection Institute, Hungarian Academy of Sciences, Budapest, Hungary; E-mail: szita@julia-nki.hu*

Henri E.Z. Tonnang *International Potato Center (CIP), Apartado 1558, Lima 12, Peru.*

Edison Torrado-Leon *Facultad de Agronomia, Universidad Nacional de Colombia, Bogota D.C., Colombia.*

Alberto Urbaneja *Unidad de Entomología, Centro de Protección Vegetal y Biotecnología, Instituto Valenciano de Investigaciones Agrarias (IVIA), Carretera Moncada-Náquera km 4,5, 46113 Moncada, Valencia, Spain.*

Paulo Roberto Valle da Silva Pereira *Embrapa Trigo, Rodovia BR 285, Km 294, Cx. Postal 451, 99001-970, Passo Fundo, Rio Grande do Sul, Brazil.*
Robert Venette *USDA Forest Service, Northern Research, St. Paul, Minnesota, USA.*
P.S.P.V. Vidyasagar *King Saud University, Riyadh, Kingdom of Saudi Arabia.*
K.A.G. Wyckhuys *International Center for Tropical Agriculture CIAT, Recta Palmira-Cali, Cali, Valle del Cauca, Colombia; E-mail: k.wyckhuys@cgiar.org*
M. Zlotina *USDA-APHIS-PPQ, 1730 Varsity Dr., Suite 300, Raleigh, North Carolina 27606, USA.*

Preface

Numbers of exotic pests and the damage that they inflict to agricultural crops and ecosystems are increasing all over the world because of expanding international transport, tourism, global climate change and trade in agricultural products (Kiritani, 2001; Yan et al., 2001). While inherent arthropod pest dispersion has resulted in invasion of foreign habitats, we humans have been responsible for trade and transportation of alien species for millennia (Hulme, 2009).

According to Pimentel et al. (2005) c. 50,000 alien-invasive species are estimated to have been introduced into the USA. Of these, 4500 species are arthropods (Pimentel et al., 2005). According to CSIR (2009), California acquires on average six invasive species each year. In Florida, 271 insect species were reported as new arrivals between 1971 and 1991 (Frank and McCoy, 1993), while Hawai'i and Florida are thought at present to acquire new species at a rate of c. 15 species per annum (CSIR, 2009). While a significant number of pests come from different parts of the world, Central and South America and the Caribbean region appear to be the major sources of pests intercepted in the USA (McCullough et al., 2006). Miami is by far the most important port of interceptions with c. 117,498 insects and 2800 mites intercepted between 1984 and 2000 (McCullough et al., 2006). While the orders Hemiptera, Lepidoptera and Diptera collectively account for >75% of insect interceptions into the USA, Coleoptera, Heteroptera and Lepidoptera had the highest species diversity of intercepted insects (McCullough et al., 2006).

In Europe, 57 species from North America and 52 from Asia have established populations in European forests (Mattson et al., 2007). According to Jacas (Josep A. Jacas, Castelló, Spain, pers. comm.) c. 84 new arthropod pests have been introduced into Spain during the period 1965–2010. Of these, 55% are Hemipterans, 10% are Lepidopterans and 14% are mites. The importance of invasive Hemipterans in South-East Asia and West Africa is documented by Muniappan et al. (2009), and includes the mealybugs *Paracoccus marginatus* Williams and Granara de Willink, *Phenacoccus madeirensis* Green, *P. manihotii* Matile-Ferrero, *P. solenopsis* Tinsley, *P. jackbeardsleyi* Gimpel and Miller and the whitefly *Aleurodicus dugesii* Cockerell. The most invasive arthropod species in China include two lepidopterans, the banana moth *Opogona sacchari* (Bojer) and the fall webworm *Hyphantria cunea* Drury, the cotton mealybug *P. solenopsis* (a threat to cotton), the Colorado potato beetle *Leptinotarsa decemlineata* Say, and a predicted threat to apple of >US$600 million by *Cydia pomonella* (L.) (Yan et al., 2001; Runzhi et al. (2010a, b, c). In contrast, Richardson et al. (2009) consider that alien insects have caused little (if any) disruption to ecosystems in South Africa. However, it is accepted that for most countries, invasions or introductions of arthropod pests are currently affecting not only ecosystems but also agricultural crops.

While there is still controversy over allocating neutral terminology to define 'invasive', 'exotic', 'introduced', or 'naturalized' species, which leaves the terms open to subjective interpretation (Colauti and MacIsaac, 2004), in this book we use the term 'invasive' to designate those non-indigenous species that have the potential to have adverse effects on the invaded habitat (Davis and Thompson, 2000). Invasive species can be grouped into different categories. For example, Kiritari (2003) considered that invasive species with potential danger to agriculture can be grouped in two categories. The first category includes a relatively small group of species with known high potential for damage (e.g., tephritid fruit flies), and the second category includes species identified as potential quarantine pests, known to be destructive to crop plants only within a localized region or country.

Because of the current situation worldwide, we are proposing three categories for invasive pest arthropods:

1. Pests that might not be serious threats in their area of origin, but which in certain environments have gained pest status within months of their discovery. One example is the armored scale *Andaspis punicae* (Laing) (Heteroptera: Diaspididae), which invaded Florida, USA in the 1990s, and became a major pest of litchi (*Litchi chinensis* Sonn.). Because of the lack of knowledge and uncertainty of its area of origin, absence of host–pest relationship studies, and undetected damage in areas originally infested with it, studies of its biology, dynamics and biological control were only initiated when economic damage to litchi trees was already irreversible (Peña *et al.*, 2005). A more recent example is the redbay ambrosia beetle *Xyleborus glabratus* Eichhoff (Coleoptera: Curculionidae), which was first detected in 2002 in Georgia, USA. This wood-boring beetle is the vector of laurel wilt, a lethal disease causing major damage, not only to natural habitats in the south-eastern USA, but also threatening the complete destruction of the avocado industry of the USA. There is, however, no information regarding this insect as an important pest in Asia, where it was originally reported. Local researchers have pioneered studies on the biology and host associations of *X. glabratus* (Hanula *et al.*, 2008; Mayfield *et al.*, 2008), but studies on pest detection and management are in their early stages (Hanula and Sullivan, 2008; Kendra *et al.*, 2011). A common pattern is insufficient data on topics such as the biology, ecology, host range and area of origin of these pests.

2. Species that cause significant economic damage in their area of origin, and consequently have been studied extensively in their home country. Examples include the South American pinworm *Tuta absoluta* (Meyrick); the carambola fruit fly *Bactrocera carambolae* Drew and Hancock; the fruit fly *B. invadens* Drew, Tsuruta and White; the pea leaf miner *Liriomyza huidrobensis* (Blanchard); the avocado seed weevil *Heilipus laurii* Boheman; and the four-spotted coconut weevil *Diocalandra frumenti* (F.).

3. Species that have recently invaded neighbouring countries, e.g., the hibiscus mite, *Aceria hibisci* Nalepa (Welbourn *et al.*, 2010), the mango gall midges *Erosomyia mangiferae* Felt, *Gephyraulus mangiferae* (Felt), the red palm weevil *Rhynchophorus ferrugineus* (Olivier) and the avocado lace bug *Pseudacysta perseae* (Heidemann) invading islands of the Caribbean Region or northern South America, but not yet present in other areas (Uechi *et al.*, 2002). It is important to assemble knowledge of the biology of these pests, the damage they can cause, and control tactics before they reach the continental USA.

When facing alien invasions, one of the major problems is the lack of information available for exotic pests (Kiritani and Morimoto, 2004; Hoddle, 2010). For instance, Hoddle (2010) demonstrated that exported fresh fruit that originated from countries where the fruit are native or endemic typically have associations with numerous insect and mite species that have co-evolved with those fruit. Some of these could be well-known pests; others might be undescribed species and previously unknown as pests; while additional species may be named and known, but their host-plant associations are poorly understood. In other cases, there have been significant advances in describing pest identity, host range, geographical distribution and biology either in their area(s) of origin (Culik, 2010; Diaz, 2010; Avila and Santos, 2010; Carrillo and Torrado, 2010; Birke *et al.*, 2010; Castañeda *et al.*, 2010; Faleiro *et al.*, 2010), or in areas of invasion or possible invasion (Malumphy, 2009; Hoddle, 2010; Mannion and Glenn, 2010; Kishimoto *et al.*, 2010; Peña *et al.*, 2010; Roda *et al.*, 2010). This knowledge must be made available for researchers and regulatory personnel in potentially threatened regions in order to prevent pest establishment.

This book highlights studies conducted by experts on different pest species. This is the result of a workshop on potential invasive pests, where we focused on pests from South America, Central America and the islands of the Caribbean basin. The book focuses on some of the most recent invasive pests and on potential invaders in order to provide researchers, regulatory personnel, extension specialists and growers with the tools necessary to address these threats.

These include the Coleopterans such as red palm weevil *Rhynchophorus ferrugineus*, weevils affecting Lauraceae, *Heilipus* spp., exotic bark beetles, Scolytinae; and the Chrysomelid *Diabrotica speciosa* (Germar). This section is followed by Lepidopterans such as *Stenoma catenifer* Walsingham; the South American tomato pinworm *Tuta absoluta*; the solanaceous fruit borer *Neolucinodes elegantalis* (Guenee); *Tecia solanivora* Povolny (Lepidoptera: Gelechiidae), a key pest of potato tubers which recently invaded cultivation areas in South America and the Canary Islands, causing havoc with the potato industry; and the noctuid *Copitarsia* spp., pests of vegetables and cut flowers.

The importance of several dipterans is also treated. Tephritid fruit flies, including *Anastrepha obliqua* (Macquart), *A. ludens* (Loew), *A. grandis* (Macquart), *Bactrocera carambolae* and *B. invadens* are addressed, as well as a novel method for improved detection of *Anastrepha* larval infestation in citrus fruit. This section also presents preliminary information on gall midge pests, *Prodiplosis longifila* Gagné and *Contarinia* complex and the recent invasions of *Acarina* in South America.

The Hemiptera section covers basic knowledge of the passionvine mealybug *Planococcus minor* (Maskell), the citrus orthezia, *Praelongorthezia praelonga* (Douglas) and recent invaders such as whiteflies, together with a description of potential invasive armored scales into the USA and the Caribbean region, and techniques to determine their likelihood for establishment.

The last section includes analysis of the effect of alien species on native species and communities through herbivory, predation or parasitism. This is followed by addressing the likelihood of establishment following introductions of diaspidid scales into the USA. Inductive and deductive approaches in the modelling of insect pests and the risk of establishment and expansion are also discussed.

In summary, this book presents basic and preliminary information on potential arthropod pests with emphasis on those of Neotropical origin. Additional gathering of this type of information is needed, not only from the New World, but also from the Old World or from those areas with active trade agreements around the world. As increasing international trade is predicted to increase biological invasions, transnational cooperation among researchers working on these potential threats is crucial to protect our agroecosystems.

Jorge E. Peña
University of Florida Center for Tropical Agriculture
and Tropical Research and Education Center
18905 SW 280 Street
Homestead, Florida 33031-3314, USA

References

Avila, C.J. and Santos, V. (2010) *Diabrotica speciosa*: important soil pest in South America. Potential Invasive Pest Workshop, University of Florida – Center for Tropical Agriculture, October 10–14, Miami, Florida (Abstract), p. 6.

Birke, A., Guillen, L. and Aluja, M. (2010) *Anastrepha ludens* and *Anastrepha obliqua* (Diptera: Tephritidae): two pestiferous tropical fruit flies that could potentially invade temperate areas under a global warming scenario. Potential Invasive Pest Workshop, University of Florida – Center for Tropical Agriculture, October 10–14, Miami, Florida (Abstract), p. 9.

Carrillo, D. and Torrado, E. (2010) Biology and Management of *Tecia solanivora* (Lepidoptera: Gelechiidae), an important pest of potatoes *Solanum tuberosum* in the Colombian Andes. Potential Invasive Pest

Workshop, University of Florida – Center for Tropical Agriculture, October 10–14, Miami, Florida (Abstract), p. 12.

Castaneda, A., Peña, J.E. and Equihua-Martinez, A. (2010) Avocado weevils of the genus *Heilipus*. Potential Invasive Pest Workshop, University of Florida – Center for Tropical Agriculture, October 10–14, Miami, Florida (Abstract), p. 21.

CISR (Center for Invasive Species Research) (2009) Frequently asked questions about invasive species. University of California, Riverside, http://cisr.ucr.edu/invasive_species_faqs.html, accessed 20 October 2011.

Colautti, R.I. and MacIsaac, H. (2004) A neutral terminology to define 'invasive' species. *Diversity and Distribution* 10, 135–141.

Culik, M.P. (2010) Experiences with *Planococcus minor* (Hemiptera: Pseudococcidae) in Espirito Santo, Brazil, with respect to potential invasive species. Potential Invasive Pest Workshop, University of Florida – Center for Tropical Agriculture, October 10–14, Miami, Florida (Abstract), p. 18.

Davis, M.A. and Thompson, K. (2000) Eight ways to be a colonizer; two ways to be an invader: a proposed nomenclature scheme for invasion ecology. *Bulletin of the Ecological Society of America* 81, 226–230.

Diaz, A. (2010) The fruit borer, *Neoleucionodes elegantalis* (Guenee) (Lepidoptera: Crambidae), an insect pest of neotropical solanaceous fruits. Potential Invasive Pest Workshop, University of Florida – Center for Tropical Agriculture, October 10–14, Miami, Florida (Abstract), p. 19.

Faleiro, J.R., Giblin-Davis, F., Jacas, J., Peña, J. and Vidyasagar, P. (2010) Potential Invasive Pest Workshop, University of Florida – Center for Tropical Agriculture, October 10–14, Miami, Florida (Abstract), p. 25.

Frank, J.H. and McCoy, E. (1993) The introduction of insects into Florida. *Florida Entomologist* 76, 1–53.

Hanula, J., Mayfield III, A.E., Fraedrich, S.W. and Rabaglia, R.J. (2008) Biology and host associations of redbay ambrosia beetle (Coleoptera: Curculionidae: Scolytinae), exotic vector of laurel wilt killing redbay trees in the Southeastern United States. *Journal of Economic Entomology* 101, 1276–1286.

Hanula, J.L. and Sullivan, B. (2008) Manuka oil and phoebe oil are attractive baits for *Xyleborus glabratus* (Coleoptera: Scolytinae), the vector of laurel wilt. *Environmental Entomology* 37, 1403–1409.

Hoddle, M. (2010) Surveys for potentially invasive Lepidoptera associated with avocado fruit. Potential Invasive Pest Workshop, University of Florida – Center for Tropical Agriculture, October 10–14, Miami, Florida (Abstract), p. 33.

Hulme, P.E. (2009) Trade, transport and trouble: managing invasive species pathways in an era of globalization. *Journal of Applied Ecology* 46, 10–18.

Kendra, P.E., Montgomery, W.S., Niogret, J., Peña, J.E., Capinera, J.L., Brar, G., Epsky, N.D. *et al.* (2011) Attraction of the redbay ambrosia beetle, *Xyleborus glabratus*, to avocado, lychee, and essential oil lures. *Journal of Chemical Ecology* 37, 932–942.

Kiritani, K. (2001) Invasive insect pests and plant quarantine in Japan. *Extension Bulletin. Food and Fertilizer Technology Center* 498, 1–12.

Kiritani, K. (2003) Invasive alien species issues. Proceedings Forest Insect Population Dynamics and Its Influences, Kanazawa, 14–19 September, http://dspace.lib.kanazawa-u.ac.jp/dspace/handle/2297/6135, accessed 3 July 2012.

Kiritani, K. and Morimoto, N. (2004) Invasive insect and nematode pests from North America. *Global Environmental Research* 8, 75–88.

Kishimoto, C., Hara, A. and Nino-Duponte, R. (2010) Host range of the nettle caterpillar, *Darna pallivitta* (Moore) (Lepidoptera: Limacodidae) in Hawai'i. Potential Invasive Pest Workshop, University of Florida – Center for Tropical Agriculture, October 10–14, Miami, Florida (Abstract), p. 30.

Malumphy, C. (2009) First interception in Europe of *Philephedra tuberculosa* Nakahara and Gill (Hemiptera: Coccidae), a neotropical pest of ornamental plants and fruit crops. *Entomologisk Tidskrift* 130, 109–112.

Mannion, C. and Glenn, H. (2010) New whiteflies in the landscape of south Florida. Potential Invasive Pest Workshop, University of Florida – Center for Tropical Agriculture, October 10–14, Miami, Florida (Abstract), p. 43.

Mattson, W., Vanhamen, H., Veteli, S., Sivonen, S. and Niemela, P. (2007) Few immigrant phytophagous insects on woody plants in Europe: legacy of the European crucible? *Biological Invasions* 9, 957–974.

Mayfield III, A.E., Peña, J.E., Crane, J., Smith, J., Branch, C., Ottoson, E. and Hughes, M. (2008) Ability of the redbay ambrosia beetle (Coleoptera: Curculionidae: Scolytinae) to bore into young avocado (Lauraceae) plants and transmit the laurel wilt pathogen (*Raffaelea* sp.). *Florida Entomologist* 91, 485–487.

McCullough, D., Work, T., Cavey, J., Liebhold, A. and Marshall, D. (2006) Interceptions of nonindigenous plant pests at ports of entry and border crossings over a 17-year period. *Biological Invasions* 8, 611–630.

Muniappan, R., Shepard, M., Watson, G.W., Carner, G.R., Rauf, A., Sartiami, D., Hidayat, P. et al. (2009) New records of invasive insects (Hemiptera: Sternorrhyncha) in Southeast Asia and West Africa. *Journal of Agricultural and Urban Entomology* 26, 167–174.

Peña, J.E., Goenaga, R., Castillo, J. and Hodges, G. (2005) Steps toward managing the armored scale *Andaspis punicae* in Florida and Puerto Rico. *Caribbean Food Crops Society* 41, 138–148.

Peña, J.E., Duncan, R. and Crane, J. (2010) The African fig fly: surveys to ascertain the status of an invasive species in the US Potential Invasive Pest Workshop, University of Florida – Center for Tropical Agriculture, October 10–14, Miami, Florida (Abstract), p. 49.

Pimentel, D., Zuniga, R. and Morrison, D. (2005) Update on the environmental costs associated with alien-invasive species in the United States. *Ecological Economics* 52, 273–288.

Richardson, D., Bond, W., Dean, W.R., Higgins, S., Midgley, G., Milton, S., Powrie, L. et al. (2009) Invasive alien species and global change; a South African perspective. In: Mooney, H. and Hobbs, R. (eds) *Invasive Species in a Changing World*. Island Press, Washington, DC, pp. 457.

Roda, A., Kairo, M., Francis, A., Millar, J. and Rascoe, J. (2010) Developing survey and mitigation strategies for the passionvine mealybug (*Planococcus minor*). Potential Invasive Pest Workshop, University of Florida – Center for Tropical Agriculture, October 10–14, Miami, Florida (Abstract), p. 54.

Runzhi, Z., Yanpling, W. and Yalan, L. (2010a) Discovery of a new invasive mealybug, *Phenacoccus solenopsis* Tinsley (Hemiptera: Pseudococcidae) in China. Potential Invasive Pest Workshop, University of Florida – Center for Tropical Agriculture, October 10–14, Miami, Florida (Abstract), p. 71.

Runzhi, Z., Yingchao, L. and Ning, L. (2010b) Introduction, dispersal and potential impacts of Colorado potato beetle, *Leptinotarsa decemlineata*, in China. Potential Invasive Pest Workshop, University of Florida – Center for Tropical Agriculture, October 10–14, Miami, Florida (Abstract), p. 72.

Runzhi, Z., Jing, L., Lei, D. and Hongyu, Z. (2010c) The distribution and threat of invasive codling moth, *Cydia pomonella* (L.) in China. Potential Invasive Pest Workshop, University of Florida – Center for Tropical Agriculture, October 10–14, Miami, Florida (Abstract), p. 73.

Uechi, N., Kawamura, F., Tokuda, M. and Yukawa, J. (2002) A mango pest, *Procontarinia mangicola* (Shi) comb. nov. (Diptera: Cecidomyiidae), recently found in Okinawa. *Japanese Applied Entomology and Zoology* 37, 589–593.

Welbourn, C., Rodrigues, J.C. and Peña, J.E. (2010) *The Hibiscus Erineum Mite, Aceria hibisci (Acari:Eriophyidae) A New Introduction in the Caribbean and a Potential Threat to Florida Hibiscus*. ENY-852, University of Florida, Institute of Food and Agricultural Sciences, Gainesville, Florida, USA, pp.5.

Yan, X., Zhenyu, L., Gregg, W. and Dianmo, L. (2001) Invasive species in China – an overview. *Biodiversity and Conservation* 10, 1317–1341.

1 Biology and Management of the Red Palm Weevil, *Rhynchophorus ferrugineus*

Robin M. Giblin-Davis,[1] Jose Romeno Faleiro,[2] Josep A. Jacas,[3] Jorge E. Peña[4] and P.S.P.V. Vidyasagar[5]

[1]*Fort Lauderdale Research and Education Center, University of Florida/IFAS, Davie, Florida, USA;* [2]*Mariella, Arlem-Raia, Salcette, Goa 403 720, India;* [3]*Universitat Jaume I, Campus del Riu Sec, Castelló de la Plana, Spain;* [4]*Tropical Research and Education Center, University of Florida/IFAS, Homestead, Florida 33031, USA;* [5]*King Saud University, Riyadh, Kingdom of Saudi Arabia*

The red palm weevil (RPW) *Rhynchophorus ferrugineus* (Olivier) (Coleoptera: Curculionidae) is a palm borer native to South Asia, which has spread mainly due to the movement of cryptically infested planting material to the Middle East, Africa and the Mediterranean during the last two decades. Globally, the pest has a wide geographical distribution in diverse agroclimates and an extensive host range in Oceania, Asia, Africa and Europe. The RPW is reported to attack over 40 palm species belonging to 23 different genera worldwide. Although it was first reported as a pest of coconut (*Cocos nucifera*) in South Asia, it has become the major pest of date palm (*Phoenix dactylifera*), and the Canary Island date palm (CIDP) (*P. canariensis*) in the Middle East and Mediterranean basin, respectively. Recent invasions suggest that it is a potential threat to *P. dactylifera* plantations in the Maghreb region of North Africa and a variety of palm species in the Caribbean, continental USA and southern China. Strict pre- and post-entry quarantine regulations have been put in place by some countries to prevent further spread of this highly destructive pest. Early detection of RPW-infested palms is crucial to avoid death of palms and is the key to the success of any Integrated Pest Management (IPM) strategy adopted to combat this pest.

Because signs and symptoms of RPW infestation are only clearly visible during the later stages of attack, efforts to develop early-detection devices are being undertaken. Once infested by RPW, palms are difficult to manage and often die because of the cryptic habits of this pest. However, in the early stages of attack palms can recover after treatment with insecticides. IPM strategies, including field sanitation, agronomic practices, chemical and biological controls and the use of semiochemicals both for adult monitoring and mass trapping, have been developed and implemented in several countries. This chapter summarizes the research developed during the last century on different aspects of the RPW, including latest findings on its biology, taxonomy, geographic distribution, economic impact and management, and prevention options.

1.1 Introduction

In his seminal revision of *Rhynchophorus* and *Dynamis*, Wattanpongsiri (1966) laid out a comprehensive overview for the distribution, biology, morphology and taxonomy of these impressive palm-associated weevils. If you compare the distribution maps of *Rhynchophorus* species in Wattanpongsiri (1966) with what is known today, the RPW or Asian Palm Weevil (APW) *Rhynchophorus ferrugineus* (Olivier) (Coleoptera: Curculionidae/Rhynchophoridae/Dryophthoridae) is the only species that has significantly expanded its range. Although not explicitly stated, a quick review of the biology and the cryptic boring behavior of these weevils in Wattanapongsiri's tome adumbrates their invasive potential, especially if whole palms or offshoots are collected from areas where these weevils occur naturally, and moved long distances to areas where they do not occur. It turns out that the temptation to move date palms from infested areas in South and Central Asia to the Middle East and Mediterranean in the 1980s, Spain in the 1990s, France in the mid-2000s and Curaçao and Aruba in 2008 was too much for date growers and landscape developers. The weevil began to appear in new territories, aggressive treatments were attempted to fend off the invasions, epiphytotics often ensued and especially susceptible palms such as the ornamental CIDP (*P. canariensis*) were often available to fan the spread. Although RPW is native to Central, South and South-East Asia and is reported chiefly from *C. nucifera* (Wattanapongsiri, 1966), only 15% of the coconut-growing countries have reported this pest. On *P. dactylifera*, the spread has been rapid during the last two decades and it is now reported from 50% of the date palm-growing countries and the entire Mediterranean basin on CIDP, *P. canariensis*. Because *R. ferrugineus* has expanded its range into areas where other palm weevil species occur, such as the Americas and Africa, this has potentially exacerbated the problem of accurate identification of the *Rhynchophorus* species in some situations. The purpose of this chapter is to revisit what we know about *Rhynchophorus ferrugineus* and closely related species, with a panel of experts with differing vantage points, to gain deeper insight.

1.2 Basic Taxonomy of RPW and Relatives

Weevil borers of palms are members of seven natural lineages within the 'Curculionidae' sensu lato, with the Dryophthoridae (or Dryophthorinae, depending upon the taxonomic authority used) being the most damaging to palms worldwide (Giblin-Davis, 2001). Four tribes within the Dryophthoridae are well-known from palms: the Rhynchophorini which includes the genera *Rhynchophorus* (mostly tropical/subtropical worldwide distribution) and *Dynamis* (Neotropical distribution); the Sphenophorini which includes *Metamasius* (Neotropical distribution), *Rhabdoscelus* (Asian distribution) and *Temnoschoita* (African distribution); the Diocalandrini which includes *Diocalandra* (South-East Asian distribution); and the Orthognathini which includes *Rhinostomus* (worldwide tropical distribution) and *Mesocordylus* (Neotropical distribution). *Rhynchophorus* and *Dynamis* species are most often referred to as 'palm weevils' and are relatively large insects, with adults being up to 5 cm long and 2 cm wide; larvae are up to 6.4 cm long and 2.5 cm wide (Giblin-Davis, 2001). Adults of *Dynamis* species are usually glossy black, in contrast to *Rhynchophorus* species which can be highly variable in coloration, ranging from all black to almost all reddish brown; with a glossy to matte, textured finish. There are nine named species of *Rhynchophorus*, including: *R. cruentatus* from Florida and the coastal south-eastern USA and the Bahamas; *R. palmarum* from Mexico, Central and South America and the southernmost Antilles; *R. ferrugineus* (=*R. vulneratus*; see Hallet *et al.*, 2004) originally from South-East Asia but with a recently expanded range (see above); *R. phoenicis* from central and southern Africa; *R. quadrangulus* from west-central Africa; *R. bilineatus* from New Guinea; *R. distinctus* from Borneo; *R. lobatus* from Indonesia; and *R. ritcheri* from Peru (Wattanapongsiri, 1966; Hallett *et al.*, 2004; Thomas, 2010). *R. distinctus*, *R. lobatus* and *R. ritcheri* are considered rare and localized species and will not be dealt with here.

A recent pest alert was generated to help distinguish the three species occurring in the New World following the recent introduction of *R. ferrugineus* to Curaçao in the Caribbean (Thomas, 2010). This highlights the need to distinguish *R. ferrugineus* from other *Rhynchophorus* species where they may overlap because of expansion of the RPW range. Distinguishing *R. cruentatus*, *R. palmarum* and *R. ferrugineus* adults from each other is relatively easy and can be accomplished with dorsal characters of the pronotum (Thomas, 2010), but other morphological characters are necessary when trying to separate R. ferrugineus from the other common species in South-East Asia and Africa. The most reliable characters discussed by Wattanapongsiri (1966) include a combination of traits, including the pronotum, dorsal, lateral and ventral aspects of the head including the basal and distal submentum shape, sungenal suture, scutellum and mandibles (Figs. 1.1–1.8). In the following key we consider the six most common *Rhynchophorus* species that occur in continents or areas where RPW co-occurs or has the potential to co-occur. In essence, the first couplet used by Thomas (2010) works well to remove *R. palmarum* from all of the rest of the species.

Key for adults of the six most common *Rhynchophorus* species:

1. 'Pronotum lobed posteriorly'; mandibles usually bidentate in lateral view; body color black (Figs. 1.1–1.3)   *R. palmarum***
   – Pronotum flatly curved posteriorly; mandibles not bidentate in lateral view (broadly rounded or sharply tridentate); color black, red and black, or red (Fig. 1.2)   2
2. 'Pronotum abruptly narrowed anteriorly'; giving the appearance of broad shoulders (Fig. 1.4, arrow); mandibles unidentate or broadly rounded in lateral view; males without anterio-dorsal rostral setae*** (Fig. 1.5) 3
   – 'Pronotum gradually narrowed anteriorly' (Fig. 1.4, arrow); mandibles sharply tridentate in lateral view; males with anterio-dorsal rostral setae*** (Fig. 1.3)   4
3. a. Nasal plate present, subgenal sutures parallel-sided, wide (>25% of the head width at ventral base) (Figs. 1.5 and 1.6)   *R. quadrangulus*
   – b. Nasal plate absent, subgenal sutures tapering anteriorly, narrow (<15% of the head width at ventral base) (Figs. 1.5 and 1.6)   *R. cruentatus*
4. a. Scutellum tapers acutely to a fine point posteriorly (Fig. 1.7)   *R. phoenicis*
   – Scutellum tapers broadly to a blunt point posteriorly (Fig. 1.7)   5
5. a. Submentum with straight subgenal sutures (Fig. 1.8)   *R. bilineatus*
   b. Submentum with concave subgenal sutures (Fig. 1.8)   *R. ferrugineus*

**Both *R. palmarum* and members of the Neotropical genus *Dynamis* are black and have the posterior margin of the pronotum lobed posteriorly, but in *Dynamis* the posterior pronotal extension is about twice as deep and the scutellum is less than one-third the size (in volume) of that feature in *R. palmarum* (Fig. 1.1).

***Character requires presence of males. Rostral hairs can be absent in nutritionally deprived and very small males of *Rhynchophorus*.

1.3 General Biology, Detection and Distribution

RPW or APW, *Rhynchophorus ferrugineus*, is reported globally on at least 40 species of palms (i.e., *Areca catechu, Arecastrum romanzoffianum, Arenga pinnata, Borassus flabellifer, Calamus merrillii, Caryota cumingii, Caryota maxima, Chamaerops humilis, Cocos nucifera, Corypha ulan, Elaeis guineensis, Livistonia decipiens, L. chinensis, Metroxylon sagu, Oncosperma horrida, O. tigillarium, Roystonia regia, P. canariensis, P. dactylifera, P. sylvestris, Sabal blackburniana, Trachycarpus fortunei* and *Washingtonia robusta*) (Esteban-Duran et al., 1998b, Murphy and Briscoe, 1999; Malumphy and Moran, 2007; OJEU, 2008; EPPO, 2008, 2009; Dembilio et al., 2009). RPW is now known from all the continents of the world and is a key pest of coconut (*Cocos nucifera*) in South and

Dynamis borassi   *Rhynchophorus palmarum*

Fig. 1.1 Dorsal views of diagnostic traits (i.e., the posterior pronotum edge (=vertical arrows) and relative scutellum size (=horizontal arrows) between the genera *Dynamis* and *Rhynchophorus*. (Photos: R.M. Giblin-Davis.)

Fig. 1.2 Dorsal views of the posterior pronotum edge (=vertical arrows) showing the differences between *Rhynchophorus palmarum* and five other species of *Rhynchophorus*. (Photos: R.M. Giblin-Davis.)

South-East Asia, date palm (*Phoenix dactylifera*) in the Middle East and *P. canariensis* in Europe, and wherever they overlap. RPW, in common with all palm weevils in the genera *Rhynchophorus*, *Dynamis*, *Metamasius*, *Rhabdoscelus* and *Rhinostomus*, is an internal tissue borer of infested palms. If it is detected in the early stage of attack, the palm host can recover with an insecticide treatment. However, palms in the latter stages of attack exhibit extensive tissue damage in the region of the apical meristem, often harboring several overlapping generations of RPW. These palms are difficult to treat and usually die. The lethal nature of this pest, coupled with the high value of the attacked palm species, warrants early action against RPW.

Palm weevils in general and RPW in particular are attracted to wounded, damaged, or dying palms, and in cases such as the CIDP, to apparently healthy palms (Hunsberger et al., 2000). Males of these weevils produce aggregation pheromones that are synergistically attractive with the kairomones produced by suitable hosts, usually early fermentation products such as ethyl esters and ethanol (Giblin-Davis et al., 1996a). Once they arrive at a palm, males and females typically seek protection from water loss by burrowing down into the petiole bases in the crown region, into

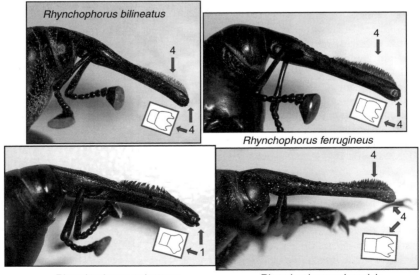

Fig. 1.3 Right lateral views of the heads of males of *Rhynchophorus bilineatus*, *R. ferrugineus*, *R. palmarum*, and *R. phoenicis* showing the dorsal rostral setae and distal mandibles. Insets depict a single mandible redrawn from Wattanapongsiri (1966). (Photos: R.M. Giblin-Davis.)

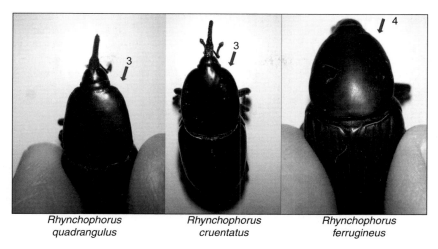

Fig. 1.4 Dorsal views of the anterior pronotal shoulders (=diagonal arrows) comparing *Rhynchophorus quadrangulus* and *R. cruentatus* with *R. ferrugineus*. (Photos: R.M. Giblin-Davis.)

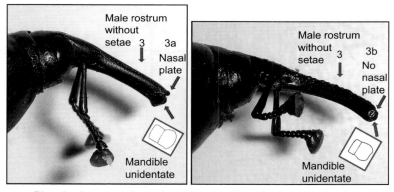

Fig. 1.5 Right lateral views of the heads of males of *Rhynchophorus quadrangulus* and *R. cruentatus* showing the lack of dorsal rostral setae, presence or absence of a nasal plate and the morphology of the distal mandibles. Insets depict a single mandible redrawn from Wattanapongsiri (1966). (Photos: R.M. Giblin-Davis.)

fleshy wounds, or into the junction between offshoots and the mother stem in palms such as the date palm. Like most weevils, RPW females use small mandibles at the distal tip of the distended rostrum to chew a hole into suitable host tissue before oviposition of a 2–3 mm long yellowish-colored egg. Eggs are often laid in close proximity to one another and take 2–4 days to eclose as small, first instar, legless larvae. The lower temperature threshold for the egg stage is 13.1°C and this stage has a thermal constant of 40.4 ± 2.0 DD (day degrees) (Dembilio and Jacas, 2011). In general, studies suggest that a gravid female will lay about 250 eggs (3–531) over her lifetime (which may last up to 120 days) and may require multiple inseminations to insure fertility. There are 13 larval instar stages of increasing head capsule and body size with increasing damage potential upon each molt (Dembilio and Jacas, 2011). The larvae have large chewing mandibles relative

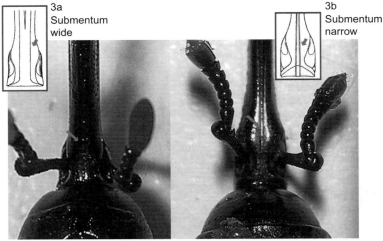

Fig. 1.6 Ventral view of the base of the head of males of *Rhynchophorus quadrangulus* and *R. cruentatus* showing the relative shapes of the submentum delineated by the gular sutures. Insets depict the same view redrawn from Wattanapongsiri (1966). (Photos: R.M. Giblin-Davis.)

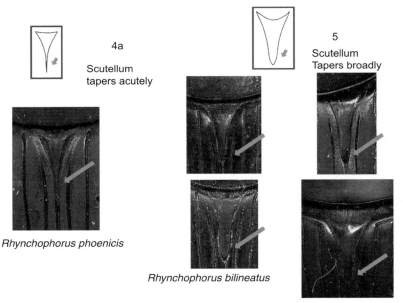

Fig. 1.7 Dorsal views of the scutellum of *Rhynchophorus phoenicis* where it tapers acutely versus *R. bilineatus* and *R. ferrugineus*, where it tapers broadly (see arrows). Insets depict the scutellum redrawn from Wattanapongsiri (1966). (Photos: R.M. Giblin-Davis.)

to the adult stage and move peristaltically through the randomly oriented galleries in the region of the crown (especially in the case of CIDP, but usually in the stem near or in offshoots in the lower 3 m of the stem in date palm and the Mediterranean fan palm *Chamaerops humilis*). The galleries are often filled with a frass that is composed of cross-oriented fibers and feces

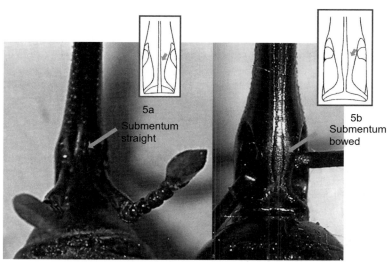

Fig. 1.8 Ventral view of the base of the head of males of *Rhynchophorus bilineatus* and *R. ferrugineus* showing the relative shapes of the submentum delineated by the gular sutures. Insets depict the same view redrawn from Wattanapongsiri (1966). (Photos: R.M. Giblin-Davis.)

resembling shredded wheat. Research is continuing on the acoustic detection of RPW larval stages using improved methods to discriminate 3–10 ms sound impulses of feeding and locomotory movements from background noise (Mankin *et al.*, 2008). Siriwardena *et al.* (2010) have developed and tested an acoustic detection system for field application in coconut that was 97% accurate for RPW larval infestations with a false-positive rate of about 8%. The odor of RPW-infested trees produces a scented signal that dogs can be trained to, but requires frequent retraining (Nakash *et al.*, 2000). Lethal infestations of RPW are in the range of 20–100 per palm but easily exceed 200. The duration of the larval phase is variable and depends upon available host nutrition, temperature and humidity, but is usually in the range of 25–105 days before the last-instar larva begins to create a large oblong cocoon (c. 75 × 35 mm) out of host fiber, most often in the petiole bases or stem. Dembilio and Jacas (2011) estimated that 666.5 DD were necessary for complete larval development in live *P. canariensis*. The last instar quickly transforms into the prepupal stage which lasts 2–11 days and retains much of the appearance of the larva but moves in the same characteristic twisting motion of the pupal stage. The next stage to occur in the cocoon is the pupa, which usually lasts for about 11–50 days before the molt to the adult. Dembilio and Jacas (2011) determined that the pupal stage required 282.5 DD, and therefore set the thermal constant of *R. ferrugineus* (egg to adult) feeding in *P. canariensis* at 989.3 DD. The adults can stay in the cocoon for several weeks before they emerge, according to abiotic conditions. Depending upon the condition of the host palm, the weevils may disperse for long distances (>900 m) or remain in the host to mate and recycle for another generation. It often takes 2–3 generations before a CIDP or a date palm will succumb to an RPW infestation. Depending on temperature, these generations can take place in one single year, but often require a minimum of 2 years (Dembilio and Jacas, 2011). Mark-release-recapture studies suggest that RPW can disperse about 7 km in less than a week (Abbas *et al.*, 2006).

A symbiotic relationship occurs between RPW and other *Rhynchophorus* species, *Dynamis*, *Metamasius hemipterus*, *Cosmopolites sordidus*, *Scyphophorus yuccae*, *Sphenophorus*, etc. (palm, agave, banana and bromeliad-associated weevils in the Dryophthoridae) and a clade of gamma-proteobacterial endosymbionts. *Nardonella dryophthoridicola* occurs intracellularly in bacteriocytes which comprise a bacteriome organ surrounding the larval intestine. The bacteria also occur in the oocytes of adult females for transovariole vertical transmission (Nardon et al., 2002; Lefèvre et al., 2004). The association is apparently quite ancient, being estimated to be about 125 million years old, and was first observed in RPW in 1965 (Buchner, 1965). The actual role that these symbionts play is unknown, but is presumed to involve supplementation of the weevil host's diet with essential nutrients, insect temperature resistance, host-plant detoxification, or parasite protection (Lefèvre et al., 2004).

1.4 *Rhynchophorus* Nematode Symbionts

RPW is a congener of R. palmarum, the chief vector of the red ring nematode (RRN) Bursaphelenchus cocophilus, causal agent of red ring disease (RRD) of palm trees in the neotropics (Giblin-Davis, 1993). RRN is a stylet-bearing, obligate plant-parasitic nematode (Superfamily: Aphelenchoidea) that is part of a clade of mostly mycophagous nematodes, i.e. Bursaphelenchus (Ye et al., 2007). Plant-parasitic nematodes have not been reported from Rhynchophorus species from North America, Asia or Africa. However, as RPW spreads to areas where R. palmarum and RRD occur, there is the potential for an association to develop between it and RRN which could change the dynamics of both symbionts. This concern led to a Federal Import Order (effective 10 February 2010) restricting the movement of 17 species of palms that might allow importation of R. ferrugineus, R. palmarum or B. cocophilus into the USA until pest risk analysis is completed to determine if risk avoidance measures are available for these pests.

RRD is one of the most important wilt diseases of coconut and African oil palms in the neotropics, causing annual losses of 10–15% (Giblin-Davis, 1993). It is vectored chiefly by R. palmarum, but also potentially by Dynamis borassi and Metamasius hemipterus, the latter two sharing 4-methyl-5-nonanol (ferrugineol) as their main pheromone with RPW. This suggests that RPW would be co-attracted to sites where it might obtain access to RRN and would likely be a suitable vector for the nematode, as has been suggested for R. cruentatus, if it co-occurred with R. palmarum and RRD (Giblin-Davis, 1991). Recent introduction of RPW into the New World has challenged our knowledge of how interspecific interactions between different Rhynchophorus species will play out, as well as how the exchange of associated organisms such as RRN will be manifested. RRN is transmitted to susceptible palms during oviposition or other activities, and only a few are necessary for successful transmission. In coconut palms, symptoms include a typical wilt with premature coconut drop (except mature nuts), and premature senescence of progressively younger leaves, which often break at the base and hang. Stem, petiole and root cross-sections often show a red ring of anthocyanin-rich pigments and these tissues usually yield large numbers of dispersal third-stage juveniles which occur intercellularly in ground parenchymal tissue. RRN infestation causes an irreversible wilt that kills the tree in a couple of months because of tyloses formation and vascular occlusion of water-conducting xylem in the vascular bundles. Palm weevils colonize RRD trees, and during feeding and tunneling become associated with the dispersal stage of the nematodes which appear synchronized for transmission (Giblin-Davis, 1993).

Acrostichus rhynchophori (Rhabditidae), previously referred to as '*Diplogasteritus* sp.' or '*Acrostichus* (*Diplogasteritus*) sp.' (Gerber and Giblin-Davis, 1990a, b; Giblin-Davis et al., 2006) was cultured from dauer juveniles (JIII) recovered from the genital capsule of R. cruentatus from southern Florida, and R. palmarum from Colombia, Costa Rica and Trinidad (Kanzaki et al., 2009). This association was shown to be phoretic in nature, and the nematodes feed on bacteria. These two weevil species also share another phoretic bacteriophagous nematode, Teratorhabditis palmarum (Gerber and Giblin-Davis, 1990b). In addition to these two species, four species of nematodes, Bursaphelenchus cocophilus (RRN, see above), B. gerberae (a mycophagous phoretic nematode), Caenorhabditis angaria (previously called 'PS1010' or Rhabditis sp.) (Sudhaus et al., 2011) and Mononchoides sp. (a nematode predator of nematodes) have been reported from R. palmarum (Gerber and Giblin-Davis, 1990a, b; Giblin-Davis et al., 2006; Kanzaki et al., 2008, 2009; Sudhaus et al., 2011). In the case of R. palmarum, the dauer juveniles of T. palmarum were isolated from the genital capsule (ovipositor or aedeagus) and body cavity of the weevils (Gerber and Giblin-Davis, 1990b); the dauer juveniles of Caenorhabditis angaria were recovered from the genital capsule (Gerber and

Giblin-Davis, 1990a); and the dauer juveniles of *B. cocophilus* appeared to infect the body cavity (Giblin-Davis, 1993; Griffith *et al.*, 2005). Interestingly, *C. angaria* is chiefly a bacterivorous phoretic associate of *Metamasius hemipterus* which was recently discovered as this weevil invaded south Florida, but is also known from *R. palmarum* and *M. hemipterus* from Trinidad (Sudhaus *et al.*, 2011). The lack of host fidelity in some of these phoretic nematode associations corroborates the notion from the pheromonal research above that these weevils, when occurring sympatrically, can be cross-attracted to the same host tree where they may develop together or in close proximity, allowing for the exchange of some symbionts.

Teratorhabditis synpapillata and *Praecocilenchus ferrruginophorus* are the only reported nematode associates of the RPW (Rao and Reddy, 1980; Kanzaki *et al.*, 2008). The host/vector association of *T. synpapillata* on RPW presumably involves phoresy and reproduction in dead or dying palms, because *T. synpapillata* was found under the elytra and in the frass of larval tunnels of RPW and is very similar to *T. palmarum*, which is carried as dauers by *R. palmarum* or *R. cruentatus* to dead or dying palms where it feeds on bacteria (Gerber and Giblin-Davis, 1990a, b). The sister-species relationship of *T. synpapillata* and *T. palmarum* in our molecular studies suggests that a specialized association with *Rhynchophorus* weevils was already present in the common ancestor of both species. Allopatry between American and Asian weevil host species could have allowed for discontinuity in nematode gene flow and eventual genetic drift and speciation. However, that *T. synpapillata* was twice independently isolated from dung or dung-enriched soil is suggestive of a much looser association where movement of nematodes between soil and arboreal biomes is accomplished by either unknown weevil activities, other insects that frequent and/or reproduce in both biomes, and/or as rotting palm trees are recycled into soil during decomposition (Kanzaki *et al.*, 2008).

Praecocilenchus ferrruginophorus is a stylet-bearing insect-parasitic nematode of the hemocoel of adult RPW (Reddy and Rao, 1980) which is very similar in morphology and apparent biology to *P. rhaphidophorus* which causes qualitative reductions of the ovaries of *R. bilineatus* in New Britain (Poinar, 1969). However, in *R. bilineatus* in New Britain, where *P. rhaphidophorus* occurs naturally in about 15% of populations, the weevils were still capable of reproduction but produced fewer eggs and had reduced fat bodies when compared with uninfested weevils. More work is needed to see if these two insect-parasitic Aphelenchoidids are conspecific, or not, and if the genus occurs in other *Rhynchophorus* species.

Entomopathogenic nematodes have been tested for biological control of RPW and *M. hemipterus* with positive results (Giblin-Davis *et al.*, 1996b; Murphy and Briscoe, 1999; Dembilio *et al.*, 2009). However, these evolutionarily divergent genera (i.e., *Heterorhabditis* and *Steinernema*) have never been recovered as natural nematode associates of *Rhynchophorus* species and will not be dealt with in detail here (see preventative treatments below). They are soil-inhabiting nematodes that have evolved in a mutualistic complex with symbiont bacteria, becoming a commercialized complex available for control of a wide range of insect hosts in moist and often cryptic habitats.

1.5 *Rhynchophorus* Aggregation Pheromones

The first reports suggesting that male weevils of the family Dryophthoridae produce aggregation pheromones attracting adults of both sexes, involved the sugarcane and palm stem-boring weevil, *Rhabdoscelus obscurus* (Chang *et al.*, 1971; Chang and Curtis, 1972). Since then, male-produced pheromones have been confirmed and identified for many weevils in this group including *R. obscurus* and all of the palm-associated *Rhynchophrous*, *Dynamis* and *Metamasius* species that have been examined so far (Giblin-Davis *et al.*, 1996a, 1997, 2000). Eight to 10-carbon, methyl-branched, saturated or unsaturated secondary alcohols comprise the major pheromones for these weevils (Table 1.1). In most cases, there is more than one chemical component to the natural blends that are produced and detected by the antennae of each weevil. It is the *S* enantiomer or *S,S* stereoisomer of the pheromone that is produced and detected in these weevils, and non-natural stereoisomers that result from synthetic production have been shown to be benign (non-interruptive in field applications) (citations in Giblin-Davis *et al.*, 1996a). Thus, relatively

Table 1.1. Male-produced aggregation pheromones and minor components identified from palm-associated or related weevils.

	Common chemical name	Rhynchophorus bilineatus	R. cruentatus	R. ferrugineus	R. palmarum	R. phoenicis	Dynamis borassi	Metamasius hemipterus	Rhabdoselus obscurus	Scyphophorus acupunctatus
Distribution		New Guinea	North America	Asia (expanded)	Neotropics	Central Africa	Neotropics	Neotropics	Asia (expanded)	Neotropics
Common name		New Guinea palm weevil	Palmetto weevil	Red or Asian palm weevil	American palm weevil	African palm weevil	Palm weevil	West Indian sugarcane weevil	New Guinea sugarcane weevil	Agave weevil
3-pentanol		–	–	–	–	–	–	minor		
3-pentanone		–	–	–	–	–	–	minor		
(4S, 2E)-6-methyl-2-hepten-4-ol	rhynchophorol	–	–	–	XX	–	–		XX[a]	
2-methyl-4-heptanol		–	–	–	–	–	–	minor	minor[a]	minor
2-methyl-4-heptanone		–	–	–	–	–	–	minor		XX
2,3-epoxy-6-methyl-4-heptanol		–	–	–	minor	–	–			
2-methyl-4-octanol		–	–	–	–	–	–	minor	XX[a]	minor
2-methyl-4-octanone		–	–	–	–	–	–	minor		XX
(3S, 4S)-3-methyl-4-octanol	Phoenicol	–	–	minor	–	XX	–			
(5S, 4S)-5-methyl-4-octanol	Cruentol	–	XX	–	–	–	–			
5-nonanol		–	–	–	–	–	–	minor		
(4S, 5S)-4-methyl-5-nonanol	Ferrugineol	XX	–	XX	minor	–	XX	XX		
4-methyl-5-nonanone		–	–	XX	–	–	–	minor		

[a] Compound may be a major (XX) or minor aggregation pheromone depending upon the geographic isolate of the weevil.

inexpensive racemic mixtures can be used for field trapping of weevils. In some species such as RPW, minor components such as 4-methyl-5-nonanone have been identified (Table 1.1) and some of these have improved trapping efficiencies (see below).

In studies where lethal traps baited with only pheromones or fermenting host tissue were compared with a combination of both, the combination lures synergized trap efficacy by about 8–20-fold. The major pheromone for RPW, ferrugineol, is also the major pheromone for several other weevils, including the regionally sympatric New Guinea palm weevil, *R. bilineatus* (Table 1.1). Phoenicol, the pheromone for the African palm weevil, *R. phoenicis*, was identified as a potential minor component for RPW (=*R. vulneratus*) (Hallet *et al*., 1993). In addition, synomonal pheromone cross-attraction has been reported for several species of palm-associated Dryophthoridae and may be adaptive in overcoming a palm's defense and time-efficient use of a temporarily suitable host (Giblin-Davis *et al*., 1996a). Interestingly, other non-palm-associating members of the Dryophthoridae such as the agave weevil use similar compounds as their pheromones (Ruiz-Montiel, 2008) (Table 1.1). Empirical studies will be needed to see how different *Rhynchophorus* species are recruited to lethal traps of invasive congeners.

conspecificity, which was subsequently confirmed using several levels of evidence, i.e., *R. vulneratus* is a junior synonym of *R. ferrugineus* (Hallet *et al*., 2004). Empirical studies from Saudi Arabia showed that the addition of the minor ketone (4-methyl-5-nonanone) in small amounts increased capture rates of food-baited RPW pheromone traps by nearly 65% (Abozuhairah *et al*., 1996). In both sexes, the response to ferrugineol appeared to increase with mating (Poorjavad *et al*., 2009). Controlled olfactometry studies showed that female weevils were more attracted to the pheromone than males, when tested separately. However, after mating, adults were less attracted to the pheromone (Faleiro, 2009). Soroker *et al*. (2005) suggested that preferential RPW female attraction to the pheromone may be due to pressure on females to disperse in search of mates, food resources and/or oviposition sites, but also suggested that further studies were needed. Trapping is most efficient for RPW if the aggregation pheromone is combined with the food bait and ethyl acetate (Oehlschlager, 2005) which is very similar to other palm and sugarcane weevils (Giblin-Davis *et al*., 1994, 1996a,c). An overview of the RPW pheromone trap design and major trapping protocols reported world-wide follows. Generating good bait-lure synergy is essential to sustained trapping efficiency of RPW pheromone traps.

1.6 RPW Trapping Overview

Trapping adult *Rhynchophorus ferrugineus* with food-baited traps to monitor activity of the pest, or mass trapping of adults in the field has been recommended since about 1975 as a component of the RPW-IPM program in coconut plantations of India. Abraham and Kurian (1975) reported that split coconut logs smeared with fresh toddy were effective for trapping RPW in the south Indian state of Kerala. Later, coconut logs treated with coconut toddy + yeast + acetic acid were reported to attract the highest number of RPW adults (Kurian *et al*., 1984). Subsequently, Hallett *et al*. (1993) identified the male-produced aggregation pheromone 'ferrugineol' (4-methyl-5-nonanol) and a minor component (4-methyl-5-nonanone) from RPW and *R. vulneratus*. They also observed interspecific matings between *R. ferrugineus* and *R. vulneratus*, suggesting

1.6.1 Trapping protocols

The following variables are important for successful RPW trapping and retention; trap design, lure efficiency and longevity, type of food bait, trap density, placement of traps, replacement of food baits and efficacy and repellency of insecticides used in RPW pheromone traps.

Trap design

RPW pheromone traps have been designed to facilitate easy entry of adult weevils into the trap while ensuring operational ease for handling and servicing (renewing food bait and insecticide solution) in the trap. Based on the experience of trapping *R. palmarum* in Latin America (Oehlschlager *et al*., 1993), *R. cruentatus* in Florida (Giblin-Davis *et al*., 1994; Giblin-Davis *et al*., 1996a) and initial RPW pheromone trap designs in Saudi Arabia (Anonymous, 1998) the upright bucket trap (5L) with a rough outer surface/jute sack wrapping was found to be the most suitable because it captured more RPW both in date and coconut plantations in Saudi Arabia and India (Abozuhairah et al., 1996; Faleiro et al., 1998; Hallett et al., 1999, Ajlan and Abdulsalam, 2000). In addition, the upright bucket traps are relatively easy to service (for renewal of food bait, water and insecticide). In the United Arab Emirates (UAE) and India, specially fabricated plastic traps with a rough exterior surface have been designed.

The upright bucket trap with four windows (1.5×5 cm^2) cut equidistantly below the upper rim of the bucket is baited with a new pheromone lure hung from inside the lid of the bucket with a piece of wire. About 200 g of kairomone-releasing food bait (dates, green coconut petiole, sugarcane, etc.) is also added to the trap and is vital to ensure entry of the adults into the trap. Moisture is another critical component in trap design for palm weevils (Weissling and Giblin-Davis, 1993; Giblin-Davis et al., 1994). The food bait is mixed in one liter of water laced with insecticide (0.05% carbofuran 3G or 0.1% carbaryl 50WP) solution (Anonymous, 1998; Oehlschlager, 1998) to immobilize and kill the captured weevils. Response of RPW to trap colors has been varied in different reports. In the UAE, black traps recorded higher captures when compared with white traps (Hallett et al., 1999). In India, trap color did not significantly influence weevil counts (Faleiro, 2005) while controlled wind-tunnel studies in Spain suggested that RPW adults prefer colored traps (Martinez Tenedor et al., 2008).

Pheromone lure efficiency, release rate and longevity

Attractiveness of RPW pheromone lures is important to sustain the efficiency of a pheromone-based trapping program. Studies carried out on this aspect in coconut and date plantations are summarized in Table 1.2. During 1994, Chem Tica International, Costa Rica, first commercially synthesized and formulated pheromone lures (Ferrolure) for RPW, and these are widely used in RPW pheromone-based control programs in several countries. Ferrolure is composed entirely of 4-methyl-5-nonanol, while Ferrolure+ is 10% 4-methyl-5-nonanone and 90% 4-methyl-5-nonanol. The Central Plantation

Table 1.2. Summary of efficacy trials of ferruginol-based pheromone lures for *Rhynchophorus ferrugineus*.

Series no.	Formulations tested	Country; Crop; Duration of trial	Superior lure	References
1	– Chem Tica International (high release/slow release)	Saudi Arabia; Date palm; 90 days	High release	Faleiro et al., 2000
2	– Agrisense (fast release/ slow release) - Chem Tica International (Ferrolure, Ferrolure+) - Calliope	Saudi Arabia Date palm 30 days	Ferrolure+	Faleiro et al., 2000
3	– Agrisense lures - Chem Tica International (Ferrolure improved and Ferrolure+)	India Coconut Two trials – 45 days each	Ferrolure improved	Faleiro and Chellappan, 1999
4	– Chem Tica International (Ferrolure+) - ISCA Technologies lure - CPCRI lure - Pherobank lure	India Coconut Two trials – 30 days each	Pherobank 400 mg lure	Faleiro, 2005
5	– CPCRI lure - Chem Tica International (Ferrolure+)	India Coconut 150 days	Ferrolure+	Faleiro et al., 2004
6	– Chem Tica International (Ferrolure+) - ISCA Technologies lure	India Coconut Trial discontinued after lure was exhausted	ISCA Technologies	Kalleshwaraswami et al., 2004
7	– Agrisense lures - Chem Tica International (Ferrolure+)	India Coconut 30 days	Agrisense lures and Ferrolure+	Abraham et al., 1999

Crop Research Institute (CPCRI), Kerala, India, synthesized RPW pheromone based on Sri Lankan technology during 2000. At present, Pest Control India (PCI), Bangalore, is the leading manufacturer of RPW pheromone in India. RPW pheromone lures have also been commercially produced by Agrisense, UK; Plant Research International (Pherobank), The Netherlands; and ISCA Technologies, USA (Faleiro, 2006). Thermoplastic spatulas, sachets, vials, glass ampules and plastic cans have been used to dispense RPW pheromone in the field (Faleiro, 2005). It is desirable to have a dispenser in which the lure can be seen so that exhausted lures can be identified easily and replaced with fresh dispensers to sustain the overall trapping efficiency.

It is essential to maintain a uniform release rate of RPW pheromone and also to have a lure that persists in the field. A release rate of ferrugineol at 3 mg per day was recommended by Hallett et al. (1999). Trials conducted in Goa, India, showed that ferrugineol sustained trapping efficiency even at a low dose of 0.48 mg per day (Faleiro, 2005). Field studies conducted with the PCI, Bangalore lure in Kerala, India, had a uniform release rate of 1.76 mg/day which sustained the trapping efficiency over a period of 14 weeks (Jayanth et al., 2007). An easy scoring technique to identify exhausted pheromone lures (Ferrolure) was devised by Faleiro et al. (1999) where it was recommended that lures with less than 5% of the chemical be replaced to maintain the trapping efficiency in area-wide RPW-IPM programs. Traps set in the shade retain the chemical lure for longer periods compared with traps exposed to sunlight. Hence, it is recommended that traps be set under shade to maintain a uniform and sustained release of pheromone into the environment.

Reports from Saudi Arabia suggest that both Ferrolure and Ferrolure+ (700 mg) were exhausted in about 12 weeks during the summer versus 24 weeks during winter, when traps were set under shade (Faleiro et al., 1999). Ferrolure+ and Ferrolure persisted in the coconut plantations of India for 150 and 84 days, respectively, as compared with Tripheron* (Trifolio-M GmbH, Lahnau, Germany) which had field longevity of 100 days (Krishnakumar and Maheswari, 2003). However, Tripheron+* had field longevity of 170 days and was superior to Ferrolure but on a par with Ferrolure+ with regard to weevil captures (Krishnakumar et al., 2004). The CPCRI lure (78.5 mg) from India was reported to have a field longevity varying from 90 to 150 days (Mayilvaganan et al., 2003; Faleiro et al., 2004). Further, studies from India have reported longevity in the field of 105 and 245 days for 250 mg and 800 mg Chem Tica lures, respectively, and 126–357 days for 400 mg and 1100 mg ISCA Technologies lures, respectively (Kalleshwaraswamy et al., 2004). Plant Research International (Pherobank), The Netherlands, dispensed RPW pheromone using a biodegradable polymer, which allows the pheromone to be emitted gradually at a high temperature (Toussaint, 2006). Under the agro-climatic conditions prevailing in Kerala, India, the PCI, Bangalore lure (800 mg) has field longevity of c. 450 days (Jayanth et al., 2007). Unfortunately, a standardized evaluation of all available release devices with set starting volumes and formulations in the field with homogeneous RPW distribution/abundance has not been conducted. Thus, it is difficult to compare them accurately. However, pheromone is a critical component to the synergy of food-baited traps and the cost-effective, long-term release of RPW aggregation pheromones needs to be factored into the trapping equation.

Food baits

Food baits added to RPW and other *Rhynchophorus* species pheromone traps play an important role in orienting the attracted weevils into the trap (Giblin-Davis et al., 1996a; Hallett et al., 1999) similar to other Coleopteran systems (Borden, 1985). The synergy that occurs between the pheromonal lure and the food bait is vital in enhancing trapping efficiency of food-baited RPW pheromone traps. Weak bait–lure synergy potentially results in the attracted weevils orienting themselves towards nearby palm trees instead of towards the trap. The importance of adding kairomone-releasing food baits in RPW pheromone traps has been emphasized by many, and a summary of recommendations is presented in Table 1.3. Moisture is also an important ingredient of a palm weevil pheromone trap, because RH (relative humidity) enhances bait–lure synergy (Weissling and Giblin-Davis, 1993; Giblin-Davis et al., 1996a). Ethyl acetate is one of several important esters that are released during fermentation of host tissues that appears to be

Table 1.3. Food baits recommended for use in *Rhynchophorus ferrugineus* pheromone traps.

Sr. no	Food bait recommended	Reference
1	Sugarcane	Oehlschlager, 1994; Faleiro and Chellappan, 1999; Hallett et al., 1999
2	Dates (*khajur*)	Faleiro, 2005
3	Date palm tissue	Anonymous, 1998
4	Coconut petiole	Faleiro, 2005
5	Plantains	Nair et al., 2000
6	Sugarcane – molasses	Muthiah et al., 2005
7	Coconut shavings	Jayanth et al., 2007
8	Palmyrah fruit juice	Muthiah et al., 2007

kairomonal to *Rhynchophorus* species (Gries et al., 1994; Giblin-Davis et al., 1994, 1996a). Increasing levels of ethyl acetate to about 400 mg/day increased attraction of the aggregation pheromone and bait tissue of the palm-associated weevil, *Metamasius hemipterus* (Giblin-Davis et al., 1996c), and the same phenomenon has been confirmed in RPW pheromone traps. Experiments conducted in the UAE, Egypt, Oman and Saudi Arabia showed that RPW captures were increased by 2–3 times when ethyl acetate was added to food-baited pheromone traps (Oehlschlager, 1998; El-Sebay, 2003; Abdallah and Al-Khatri, 2005; Shagag et al., 2008). Chem Tica International has incorporated both the pheromone and ethyl acetate into a single lure called Ferrolure + HP which can be used in tissue-baited traps (Oehlschlager, 2005). However, ethyl acetate is expensive and substantially adds to the operational costs.

In general, the food bait used for RPW trapping should be easily obtained and cost-effective. Hence, as low-grade dates might be recommended for RPW pheromone traps in the Middle East, green coconut petiole shavings might be ideal in India. RPW were most attracted to bucket traps baited with sugarcane followed by traps baited with coconut exocarp, whereas date fronds were the least preferred bait in date palm orchards in Gujarat state in India (Muralidharan et al., 1999). Ideally, the food bait will have a relatively high sugar content (Giblin-Davis et al., 1996a; Oehlschlager, 2005). Faleiro (2005) suggested that 200 g of coconut petiole was sufficient for one trap, and found that weevil captures declined progressively when traps were serviced (change of food bait + water) at 10, 20 and 30 days. However, addition of water in traps sustained the trapping efficiency when traps were not serviced beyond 15 days. This emphasizes the need to keep food baits hydrated, because *Rhynchophorus* species are moisture-loving (Wattanpongsiri, 1966; Weissling and Giblin-Davis, 1993), and because moisture is required for the host tissue to continue to ferment and release host kairomones (Giblin-Davis et al., 1996c). Oehlschlager (2005) extended the effective life of the food bait in RPW traps from 2 to 7 weeks by adding propylene glycol to the bait to slow down evaporation of water from the trap. Thus, a suitable food bait mixed with water will help improve the bait–lure synergy and enhance RPW captures (Faleiro, 2006). Rochat (2006) suggested that more work is needed to understand the mechanism of additive or synergistic responses to the combined odor of pheromone + palm.

Insecticide in traps

Once the adult weevils enter the RPW pheromone trap it is essential to prevent their escape. This can be achieved by immobilizing/killing the captured weevil with an insecticide mixed with the bait, or by mechanically preventing escape of adults (e.g. by using funnel traps; see Hallett et al., 1999). However, as bucket traps have been reported to be the most suitable RPW-pheromone traps, insecticides are commonly used to kill and retain captured adult weevils. Trials from Al-Qateef in Saudi Arabia have shown that among the insecticides evaluated in RPW pheromone traps, deltamethrin had the least repellency as compared with quinalphos, which exhibited the highest repellency to adult weevils (Abozuhairah et al., 1996). Trials conducted in India showed that carbofuran and carbaryl were suitable for use in RPW pheromone traps to kill and retain trapped weevils (Faleiro, 2005; Abraham and

Nair, 2001). Soap/detergent has been used to kill and retain captured weevils in bucket traps (Giblin-Davis et al., 1994; Rochat, 2006; Jayanth et al., 2007). Care should be taken to see that the insecticide/detergent used does not counter the odors produced by the lure/bait; this would lower the trapping efficiency.

Trap placement

In order to maximize the lure longevity in the field it is essential to set RPW pheromone traps in the shade. Care should also be taken to avoid setting traps on palms in the age group susceptible to RPW attack of <20 years, or near, or on very susceptible palms, such as *P. canariensis* (Hunsberger et al., 2000). Higher RPW captures were recorded when traps were placed at a height of 1 m from the ground in coconut plantations in Goa, India (Faleiro, 2005). In Israel, UAE and Spain, RPW pheromone traps are currently set at ground level. However, traps placed at waist height on tree trunks are convenient to service when compared with traps set at ground level (Faleiro, 2005). RPW was effectively mass trapped in area-wide pheromone-based IPM programs in Kerala, India, by setting traps about 1.5 m from the ground. In this trial, traps set in the middle of the coconut plantation caught fewer weevils when compared with traps on the periphery. Due to fragmentation of coconut farms, large-scale community-based mass trapping of RPW is likely to be more effective when compared with single-farm trapping (Jayanth et al., 2007).

Trapping density

Initially, Oehlschlager (1994) recommended a trapping density of one trap/ha and one trap/100 ha in mass trapping and monitoring programs, respectively. However, mass trapping of RPW adults in pheromone-based RPW-IPM programs has been implemented at different densities ranging from high density trapping of 10 traps/ha of date plantations in Israel to 1.5 traps/ha in date plantations of the Al-Hassa region in Saudi Arabia. These reports also indicate that monitoring of RPW was done at one trap/ha and one trap/100 ha in Israel and Saudi Arabia, respectively (Abraham et al., 2000; Soroker et al., 2005). In Egyptian date plantations, RPW was effectively mass trapped at a trap density of 0.5 traps/ha (Anonymous, 2004). In coconut plantations in Goa, India RPW pheromone trap density trials for mass trapping programs revealed that a trap density of one trap/ha was sufficient. However, weevil catches doubled by increasing the density to two traps/ha (Faleiro, 2005). Recently, Jayanth et al. (2007) reported successful management of RPW in area-wide programs by setting pheromone traps at 2.5 traps/ha. Rochat (2006) recommended that RPW pheromone traps should be set apart from areas bearing high infestation risks due to the characteristics of the palm, or only at low density. The agro-system involved, intensity of the pest, attractiveness of the lure and the resources available to service the traps may influence the trapping density. In a date plantation in Al Hassa, Saudi Arabia, with less than 1% infestation, one trap/ha was sufficient to mass trap the pest. However, in plantations with more than 1% infestation, 10 traps/ha recorded the best and significantly superior weevil captures as compared with other tested trapping densities (i.e., 1, 2, 4 and 7 traps/ha) (Faleiro, 2009). Based on the preceding discussion the best trapping protocols to be adopted are summarized in Table 1.4.

1.6.2 Monitoring, mass trapping of RPW and validating pheromone-based RPW-IPM programs

Monitoring the activity of RPW is essential for keeping a close watch on the establishment and subsequent build-up of the pest. After initial reports of infestation, it is imperative to monitor the activity of adult weevils. Food-baited RPW pheromone traps have been widely used in several countries in surveillance programs to assist and develop early-warning alerts. Adopting the best pheromone-trapping protocols is essential in such surveillance programs to minimize the risk of the pheromone traps becoming devices for spreading RPW (Rochat, 2006; Faleiro, 2006). Combining weevil activity as gauged through RPW pheromone traps along with geographic information systems (GIS) can serve as a valuable tool to improve and support decision-making capabilities, especially when trying to manage invasive pest populations in area-wide operations. GIS allows storage of vast amounts of data on the spatial and temporal spread of a pest, and also assists in predictive analysis.

Table 1.4. Suggested pheromone trapping protocols for *Rhynchophorus ferrugineus*.

Trap component/ protocol	Recommendation
Trap design	Capture adult weevils with bucket traps with four windows (1.5 × 5 cm^2) (5 L capacity) cut equidistantly below the upper rim of the bucket having rough outer surface
Pheromone lure	Use commercially available lures that are efficient and long lasting
Food bait	Should be easily available, cost effective and generate good bait–lure synergy. Dates (200 g) in one liter water are the best
Ethyl acetate	Significantly enhances weevil captures when incorporated in food-baited RPW pheromone traps
Insecticide in trap	Add non-repellant insecticide to the water and food bait in the trap. Soap/detergent may repel adult weevils
Trap servicing (renewing food bait and insecticide solution)	Once every 7–10 days (vital to generate strong bait–lure synergy)
Trap density for surveillance	1 trap/100 ha *or* 1 trap every 1–2 km along motorable roads.
Trap density for mass trapping	High density: 4–10 traps/ha; low density: 1 trap/ha (depending on the intensity of the population and resources available for servicing the traps)
Trap placement	Preferably set traps on the periphery of the plantation, 1 m above the ground, under shade. Hang traps on trunks of old palms with hardened trunk tissue (>25 years old)/non-host trees. Ground or surface trapping can also be practiced. Do not set traps near young palms with offshoots.
Lure replacement	Go by manufacturer's recommendation. Prefer long lasting and efficient lures.

Ferrugineol-based pheromone traps attract and capture twice as many females as adult male weevils (Oehlschlager, 1994; Hallett et al., 1999; Vidyasagar et al., 2000a; Abraham et al., 2001; Faleiro, 2005; Soroker et al., 2005, Jayanth et al., 2007). From the vantage of weevil management, this is desirable because female weevils initiate damage to the palms through oviposition and subsequent larval development. Examination of the reproductive status of female weevils captured in RPW pheromone traps has shown that most of these females are young, gravid and fertile (Abraham et al., 2001; Faleiro et al., 2003; Jayanth et al., 2007) and many had initiated egg development (Kalleshwaraswamy et al., 2005). Furthermore, Jayanth et al. (2007) reported that 74% of the pheromone traps captured female weevils that had not yet initiated oviposition. These studies highlight the potential benefits of using food-baited pheromone traps in mass-trapping programs to suppress population build-up in the field.

Successful management of RPW in large areas of date (El-Ezaby et al., 1998; Abraham et al., 2000; Vidyasagar et al., 2000a, b; Al-Khatri, 2004; Soroker et al., 2005; Oehlschlager, 2006) and coconut plantations (Rajapakse et al., 1998; Faleiro, 2005; Sujatha et al., 2006; Jayanth et al., 2007) have been reported from several countries. These programs are long term in nature and need substantial investment over a period of time. The report of enhanced infestations around RPW pheromone traps by Rochat (2006) from date plantations in Iran requires further investigation. It would be advisable to protect young palms in a radius of 50–100 m from the traps by periodic insecticide cover sprays (Faleiro, 2006). It is also important to further understand the mechanism of semiochemical attraction in RPW, to minimize the risk involved in attracting RPW female weevils to young palms near the traps, and also to examine potential repellants that could be used to reduce host apparency in critical locations. Based on experiments with repellants for *R. palmarum* in oil palm, Oehlschlager (2005) suggested that they could be used in 'push–pull' strategies for the management of palm weevils.

Faleiro (2006) recommended that pheromone-based area-wide IPM programs need to be implemented at a low infestation level of just 1% infested coconut palms. Implementing area-wide management of RPW on the basis of trap captures or infestation reports can be inaccurate, as they may either under- or over-estimate the pest intensity in the field. The aggregated nature of this pest (Faleiro et al., 2002) may also result in inaccurate estimation of infestation levels. Faleiro and Kumar (2008) proposed sampling plans based on the concept of sequential sampling which was developed using the formulae given in Southwood and Henderson (2000). The sampling plans allow for rapid and accurate classification of RPW infestation in coconut plantations of India, by inspecting palms to detect infestation levels in a sequence that allows a decision to be made to either implement or not to implement an area-wide management program of RPW control. A similar decision-making sampling plan to initiate area-wide management of RPW in date plantations in Al-Hassa, Saudi Arabia, was also developed (Faleiro, 2008). These plans classify the intensity of the pest in the field, allowing for a surveillance program to be upgraded to a mass trapping-based IPM operation. Conversely, sequential sampling plans could be used to validate RPW-IPM programs to arrive at an accurate decision of scaling-down mass trapping of the pest to surveillance mode or withdrawing the traps from the field. These plans are based on the hypothesis that 'area-wide management of RPW is not required' and are developed at risk factors of 'a' and 'b' set at 0.05, where 'a' is the probability of error that a low infestation level is wrongly categorized as high, and 'b' is the reverse situation where a high level of infestation is categorized as low. The plan also takes into account the assumed action threshold of infested palms for initiating control and the aggregation index (K).

The pheromone-based area-wide RPW-IPM program in date palms in Al-Hassa, Saudi Arabia, was validated during 2008 on a 100-ha scale (10,000 palms), where fewer than 42 RPW-infested palms/100 ha suggested that infestation levels were below the assumed action threshold of 1% and that the strategy was having an impact. Situations where there were more than 54 RPW-infested palms/100 ha implied that the damage levels were above the action threshold and that the IPM program needed major improvement. At intermediate infestations of 42–54 infested palms/100 ha, damage levels were approaching the action threshold (1%) and the strategy in these areas needed to be strengthened.

1.7 RPW-IPM Programs in Date Palms

Due to prevailing economic and social conditions in Saudi Arabia and other parts of the Middle East, date groves are often neglected or abandoned over time, making them reservoirs for RPW and leading to the capture of several times more adult weevils in pheromone traps than in farms that are well-tended (Abozuhairah et al., 1996). Systematic clearing of abandoned date gardens helps to temper area-wide epizootics of RPW by reducing their breeding sites.

Irrigation practices can play a key role in the spread of RPW infestations in date palms in the Middle East. Some farmers use flood irrigation, creating a situation where water is always in contact with the base of the date palm and its offshoots. This creates an environment very favorable for the weevils to lay their eggs (Aldryhim and Al-Bukiri, 2003a, b). Studies of the distribution of RPW in drip and flood irrigation within a palm grove showed that 89% of the total infested trees were detected in plots with flood irrigation. This suggested that irrigation management and soil moisture are key factors in the dispersion and colonization success of RPW in date palms (Aldryhim and Al-Bukiri, 2003a, b).

Cultural management practices can also involve the removal and destruction of infested offshoots and the application of soil mixed with a pesticide or gypsum which can be effective for preventing the entry of RPW into the date palm. Pesticides are used either for preventing the entry of RPW or for curing an already infested date palm. The chemicals used in the IPM program in Saudi Arabia include Supracide® (Gowan Co., Yuma, Arizona, USA), Metasystox® (Bayer, Isando, Germany), Cypermethrin, Dimethoate (27.8%), Chlorpyrifos (22.2%) and Trichlorphon (Vidyasagar et al., 2000a). Date palm stems with RPW infestations can be treated by wound cleaning and inundation with a pesticide. The identified infestation is first cleaned by removing the softer, damaged tissues. Slanting holes 20–25 cm deep are then created around the stem wound, where a pesticide is introduced following the labeled recommendations, and sealed with wet mud.

The treated palms are then monitored to confirm the cessation of oozing sap from these sites and the subsequent drying of the wound, which can suggest successful control of RPW (Abozuhairah et al., 1996). In regions where the humidity is very high, deeply damaged stems are treated by stem cleaning followed by filling these cavities with wet sand mixed with pesticide dust. These sand-filled stems are covered with a polyethylene sheet to retain the humidity. This method was successfully applied for several palms in the Eastern Province of Saudi Arabia (Vidyasagar, pers. obs., 2010).

1.8 Recent Invasions of RPW

1.8.1 Northern Mediterranean Basin and the Canary Islands

In Europe, palm trees are mainly grown for ornamental purposes in urban areas and resorts, and therefore cannot be considered as a conventional crop. Only palms grown in groves or nurseries should be considered as a traditional crop. This makes the European case completely different from that in most RPW-infested countries up until now, where palms are mostly grown in regular orchards or around oases for their products (oil, dates, fronds, etc.). In Europe, palms have been widely used in gardens, parks and avenues since the 18th century (Morici, 1998), especially the highly prized but very RPW-susceptible CIDP, *Phoenix canariensis*. Other exotic palm species, such as *Trachycarpus fortunei* and *Washingtonia* spp., or the indigenous *Chamaerops humilis* palm in Europe, are also highly valued as ornamentals (Fig. 1.9). In addition, palms in Europe have other environmental or historical value. For example, the Theophrastus palm tree forest at Vai in Crete (Greece) has the largest subpopulation of the threatened Cretan date palm, *P. theophrasti*, which can be found on other Aegean Islands. The date palm grove at Elx (south-eastern Spain, with about 240,000 palms) is catalogued as a World Heritage Site by UNESCO, and nearby groves at

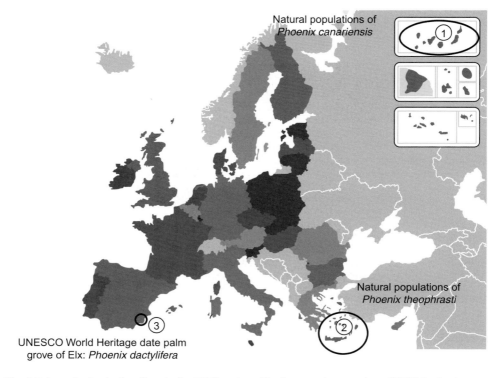

Fig. 1.9 Important palm locations in the EU threatened by the recent expansion of RPW (red palm weevil, *Rhynchophorus ferrugineus*): (1) Canary Islands off the north-western coast of Africa; (2) Crete and other Aegean Islands; and (3) South-eastern Spain. (Prepared by J.A. Jacas.)

Alacant and Orihuela have been declared Historic Sites by the Spanish Government. Last but not least, the wild forests of *P. canariensis*, in their native Canary Islands, constitute the most important source of genetic variation for this species. For all the above reasons, the detection in Europe of *R. ferrugineus* raised new questions in a scenario that was completely different to what had gone before.

Rhynchophorus ferrugineus was first detected in the Mediterranean basin in 1992, when it was found in date palms in Egypt (EPPO, 2008). Three years later, the occurrence of *R. ferrugineus* in south-east Spain (Almuñécar, Autonomous Community of Andalusia) (Fig. 1.10) was officially acknowledged and control measures against it were immediately put into place. These measures were partially successful and until 2000 the pest remained localized in municipalities around the initial focus of Almuñécar (Fig. 1.10). As a consequence, that year, Spanish restrictions to palm movement were partially lifted. Unfortunately, the relaxation of the containment measures led to massive imports of infested palms from Egypt (Fig. 1.11). These palms were extensively used in new urban areas and resorts set along the Mediterranean coastal districts of Spain in conjunction with the housing market bubble occurring at that time. Luckily, in 2003 fear of new *R. ferrugineus* outbreaks prompted Spanish authorities to prohibit the import of any palm species into the World Heritage palm grove of Elx, as an urgent preventive measure against the weevil. In 2006, this protection extended to the historical date palm groves of Alacant and Orihuela, and palm movement within a radius of 5 km around them (protecting an area of 23,562 ha for the three historical sites) was restricted. However, in 2004, an outbreak of *R. ferrugineus* occurred in Olocau (province of Valencia) (Fig. 1.10), more than 600 km north-east of the initial focus of Almuñécar and about 200 km north of Elx. Additional foci appeared during 2005 (Murcia and Catalonia) and 2006 (Balearic Islands), and by the end of that year all the Mediterranean Autonomous Communities of Spain were officially infested (Fig. 1.10).

In 2005, *R. ferrugineus* was detected in two resorts in the Canary Islands of Fuerteventura and Gran Canaria (Fig. 1.10). In 2006, strict measures to protect native forests of *P. canariensis*, including an eradication program and

Fig. 1.10. Spanish Autonomous Communities infested by RPW (red palm weevil, *Rhynchophorus ferrugineus*) (year of detection in parentheses) with important locations related to the spread of the pest in Spain (dots). (Prepared by J.A. Jacas.)

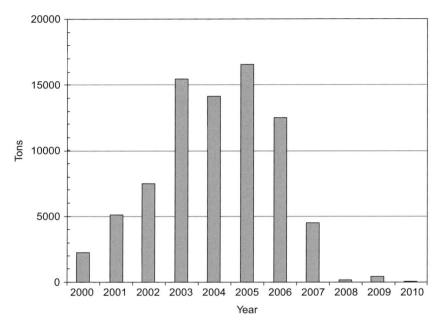

Fig. 1.11. Adult *Phoenix dactylifera* palms (tonnes) imported from Egypt to Spain from 2000 to May 2010 (source: EU Conference on the Red Palm Weevil, Valencia, May 2010). (Prepared by J.A. Jacas.)

the prohibition of importation of any palms from outside the Islands, were established for the whole Autonomous Community of the Canary Islands (BOC, 2007), which have their own phytosanitary regulations apart from those in force in the rest of Spain. New foci appeared up until 2008, including an outbreak in the hitherto uninfested island of Tenerife in 2007. However, no additional foci have been detected since 2008, and the islands of El Hierro, La Gomera, Lanzarote and La Palma have remained pest-free.

In parallel to the situation in Spain, *R. ferrugineus* spread through southern Europe. It was officially declared in Italy in 2004, in Greece in 2005, in France and Cyprus in 2006, in Portugal, Malta and Turkey in 2007 and in Georgia and Slovenia in 2009 (Fig. 1.12). As in all previous cases in Europe, infestations are presumed to have occurred earlier, from 1–2 years prior, depending on local temperatures. Dead palms or those close to death are easily detected by the untrained eye, and 1–2 years is the time necessary for *R. ferrugineus* to complete two to three generations in a single palm, the time that a new infestation takes to result in palm death (Dembilio and Jacas, 2011).

1.8.2 *R. ferrugineus* legal issues in the EU: local, regional, national and European regulations

First detection of *R. ferrugineus* within the EU took place in the Spanish Autonomous Community of Andalusia (Fig. 1.10). As a consequence the Department of Agriculture and Fisheries of the Andalusian Government (Junta de Andalucía) was the first European legal body to take action against this pest. Immediately after detection, palm movement from the nurseries in the provinces of Almería, Málaga and Granada (those around the initial focus of Almuñécar, Fig. 1.10) was prohibited for 2 months. During this time all nurseries had to protect their palms with a soil treatment of Aldicarb. In the meantime, the Spanish Ministry of Agriculture, Fisheries and Food (MAPA) published an Order (BOE, 1996) establishing a series of provisional

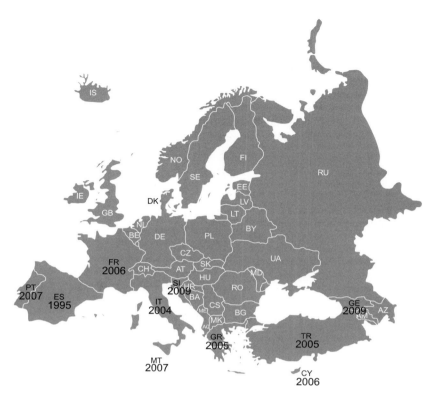

Fig. 1.12. Year of detection of RPW (red palm weevil, *Rhynchophorus ferrugineus*) in different European countries. (Prepared by J.A. Jacas.)

measures against *R. ferrugineus*. These measures were relatively strict and included:

- prohibition of the import of Palmaceae from non-EU countries (third countries);
- compulsory use of the EU Plant Passport for any movement of Palmaceae originating within the EU; and
- eradication measures, including chemical treatments, pheromone trapping and destruction of infested specimens both in public and private gardens in the infested area.

In June 1997, the Andalusian Government established further measures against *R. ferrugineus* (BOJA, 1997), including the obligation of all palm producers, dealers and importers in Andalusia to enroll with the Spanish Official Register of Plant Producers, Dealers and Importers regulated by the Real Decreto 2071/93 (BOE, 1993). As a consequence, all these agents were subjected to Official Phytosanitary Inspections, which are compulsory to qualify for the EU Plant Passport. Andalusian Government measures also included serious limitations for the movement of palms within infested areas, which was always subjected to the non-detection of symptomatic palms within the nursery, as described in the same Real Decreto (Article 7, point 6).

The presence of *R. ferrugineus* in Spain obviously affected EU regulations, and the Council Directive 77/93/EC on protective measures against the introduction into the Community of organisms harmful to plants or plant products and against their spread within the Community in force at that time, had to be adapted.

Research activities carried out during these years by the Spanish National Institute for Agronomic Research (INIA) (Esteban-Durán *et al.*, 1998a, b) and the Universidad de Almería (Cabello *et al.*, 1997; Barranco *et al.*, 1998, 2000) improved current knowledge on the weevil bioecology and control. This information, together with that gathered from technicians working in the infested area, allowed the Spanish MAPA to derogate the former Order (Orden de 18 de Noviembre de 1996; BOE, 1996) and to approve a new one (Orden de 28 de Febrero de 2000; BOE, 2000), where restrictions to palm movement both within EU and from third countries were partially lifted. These restrictions, though, were extended to other *Rhynchophorus* species (*R. bilineatus*, *R. cruentatus*, *R. palmarum*, *R. phoenicis*, *R. quadrangulus* and *R. vulneratus* [=variant of *R. ferrugineus* see Hallet *et al.*, 2004]), and referred to all specimens belonging to the Palmaceae family with an upper diameter of >5 cm, with the exception of seeds and fruits.

No further legal changes occurred until 2004, in coincidence with the explosive spread of *R. ferrugineus* in Europe starting that year (Fig. 1.12). Both local and national regulations changed with the detection of the weevil in their territories. Finally, the alarm created within the EU by this situation prompted the EU to publish the Commission Decision 2007/365/EC on emergency measures against the introduction and spread within the EU of *R. ferrugineus* (OJEU, 2007). This decision was modified in October 2008 (OJEU, 2008) and August 2010 (OJEU, 2010), and was incorporated into national, regional and local laws. The main points are:

- *Specific import requirements*: palms should have been grown throughout their life either in a country where *R. ferrugineus* is not known to occur, or in a *R. ferrugineus*-free area. Otherwise, palms should have been grown during a period of at least 1 year prior to export in a place of production subjected to official inspections certifying that no signs of *R. ferrugineus* presence have been observed.
- *Conditions for movement*: plants should be accompanied by a plant passport and, if originating from an infested area, they should have been grown during a period of 2 years prior to the movement in a site with complete physical protection against *R. ferrugineus*, and no signs of its presence should have been observed during official inspections.
- *Establishment of demarcated areas within the EU infested countries*: these areas should include the infested zone plus a buffer zone of at least 10 km beyond the boundary of the infested zone. Extensive monitoring and appropriate measures against *R. ferrugineus* aimed at its eradication should be carried out within these areas.

Although the aforementioned measures have been applied by affected EU countries, new foci of *R. ferrugineus* have been continuously detected, and the situation within the EU has worsened. For example, in Spain (Fig. 1.10) 49,800 palms, mostly *P. canariensis*, were killed by *R. ferrugineus* from 1996 to 2009. In the case of the Autonomous Community of Valencia, control measures taken against the weevil during this period, mainly an eradication program, cost about €11 million (Dembilio and Jacas, 2011). Until now, eradication has been successful only in the Canary Islands (Figs. 1.13–1.14), with the first pest-free foci being declared in 2010 after 3 years of non-detection of infested palms and no collection of adults in pheromone-baited traps in the demarcated areas.

Reasons for the failure of the eradication program up until now include: (i) the difficulty of the early detection of infested palms; (ii) the lack of a sound quarantine treatment against the weevil; (iii) the difficulty of involving homeowners in the process; (iv) the risks associated with the use of mass trapping in uninfested areas; (v) the lack of highly effective, environmentally safe plant protection strategies (biological control, semiochemicals, soft pesticides, etc.) suitable for public areas such as gardens, parks and avenues; and (vi) the incomplete knowledge of the bioecology of *R. ferrugineus* developing in *P. canariensis* under the Mediterranean climate. Research is therefore urgently needed to fill these gaps.

1.8.3 Current strategies against *R. ferrugineus*: the Spanish case

The current strategy against *R. ferrugineus* in Spain includes different actions including chemical, biological, cultural, biotechnological and legal/regulatory controls as well as sanitation:

- *Preventive treatments*. A minimum of eight treatments per season (from March to November) are recommended. Only five active substances were authorized by the Spanish Ministry of Agriculture in winter 2011.

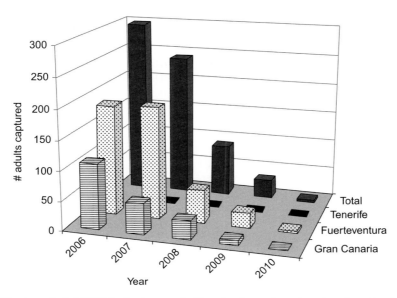

Fig. 1.13. Number of adult RPW (red palm weevil, *Rhynchophorus ferrugineus*) captured in pheromone- and tissue-baited traps in the Canary Islands from 2006 to May 2010 (source: EU Conference on the Red Palm Weevil, Valencia, May 2010). (Prepared by J.A. Jacas.)

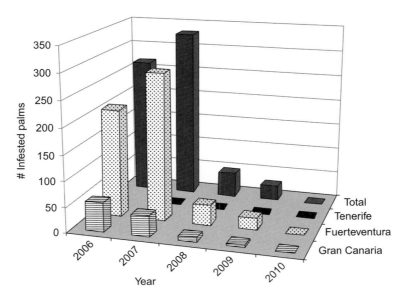

Fig. 1.14. RPW- (red palm weevil, *Rhynchophorus ferrugineus*) infested palms destroyed in the Canary Islands from 2006 to May 2010 (source: EU Conference on the Red Palm Weevil, Valencia, May 2010). (Prepared by J.A. Jacas.)

These were abamectin, chlorpyrifos, imidacloprid, phosmet and thiametoxam (MARM, 2011), as well as the entomopathogenic nematode *Steinernema carpocapsae* (Weiser) (Nematoda: Steinernematidae), which proved highly effective when applied in a chitosan formulation (Llácer *et al.*, 2009; Dembilio *et al.*, 2010a). The latter has also proved effective as a curative treatment (Llácer *et al.*, 2009). To solve the problem of reaching the top of the palm when treating tall palms with either nematodes or other pesticides, the use of a fixed 4-mm polyethylene pipeline holding 2–4 micro-sprinklers on the top of the stipe is becoming popular in many Spanish cities. In Valencia, for instance, most palms in public gardens have such a pipeline fixed on the top of the stipe down to a height of 2.5 m. When needed, this line is directly connected to a pump on a transport platform and the pesticide is applied from it with no need for a worker to get to the top of the palm stipe (Fig. 1.15).

- *Pruning.* Wounds, such as those from pruning, emit volatiles that attract adult *R. ferrugineus*, and thus pruning can increase the likelihood of a new infestation. This is the reason why pruning and any other activity producing wounds are best performed in winter, when adult *R. ferrugineus* activity is reduced and immature mortality is highest (Dembilio and Jacas, 2011). All wounds, regardless of the time of year (for example, after cutting an 'inspection window' into the palm canopy), should be immediately treated with an appropriate insecticide.
- *Sanitation: arboreal surgery.* Based on the traditional production of 'palm syrup' from *P. canariensis* in the Canary Island of La Gomera (Fig. 1.16), infested palms can be mechanically sanitized. This is a technique especially valuable for old monumental palms that must be saved from RPW. Provided that the inner meristematic tissues have not yet been affected by the weevil, the palm can recover in a few months.

Fig. 1.15. Treatment application pipeline (left-pointing arrow in left and top right photos) and inspection window in the crown region of a specimen tree of CIDP (Canary Islands date palm, *Phoenix canariensis*) (ellipses) in a public palm garden in the city of Valencia, Spain. See text for description of how this is used for frequent applications of pesticides or biological control agents into the vulnerable crown region of legacy palms. Bottom-right: a residential site where two CIDP have died, exhibiting typical crown symptoms due to RPW infestations (right-pointing arrows) One CIDP is a candidate for curative and prophylactic treatment using the inspection window and pipeline (left-pointing arrow). (Photos: J.A. Jacas.)

Fig. 1.16. Traditional palm syrup extraction in the Canary Islands, demonstrating the resiliency of mature CIDP (Canary Islands date palm, *Phoenix canariensis*) to occasional severe leaf and petiole removal. Upper images: palm management just above the apical meristem for syrup collection. Lower figures: palm recovery after syrup extraction. (Source: Mr Gerardo Mesa Noda.)

- *Monitoring for early detection*. Trained gardeners and technicians are necessary for early detection of *R. ferrugineus* infested palms. In general, the Departments of Agriculture of the different Spanish Autonomous Communities are responsible for the training courses addressed to the technicians who will later work either for private companies or for the municipalities involved.
- *New plantations*: Any palm used in a new plantation, landscape project or garden should have a valid EU Plant Passport. This should ensure that these plants have followed the strict rules regulating plant production and movement in agreement with the Commission Decision 2007/365/EC.
- *Removal and destruction of affected palms*. Infested palms should be removed and destroyed. Municipalities and homeowners are responsible for the cutting of the palms and their transportation to a designated destruction site where they are ground to powder. Burning is not recommended for destruction because palms do not burn easily, and complete destruction of *R. ferrugineus* cannot be guaranteed in this case.
- *Trapping*. Mass trapping is only allowed under direct supervision of the technicians of the Departments of Agriculture of the different Spanish Autonomous Communities. A trap set in an uninfested area can easily lead to its infestation by weevils responding to the attractive plumes coming from the trap. In addition, a trap can greatly increase the incidence of *R. ferrugineus* in an area if neighboring palms are not adequately protected. This is especially true in the case of the highly susceptible *P. canariensis*.

1.8.4 North Africa

Date farming is an important component of the agrarian economy in North Africa on which millions of farm families earn their livelihood. Egypt and the five Maghreb countries (Algeria, Morocco, Tunisia, Libya and Mauritania) in North Africa account for nearly 25% of the global date production. RPW apparently entered Saudi Arabia in 1987 through importation of ornamental palms and then spread to date palm plantations, causing serious loss to palms in the Eastern Province and later in most of the palm-growing regions of the country (Abraham and Vidyasagar, 1993) (Fig. 1.17). In date palm, propagation involves the use of offshoots produced from the base (bole region) of the stem of palms. RPW is attracted to cracks where these offshoots connect and wounds made during their harvest for transplantation. Unfortunately, hidden stages of RPW are easily transported for long distances and can establish in new locations (Abraham and Vidyasagar, 1993; Abraham et al., 1998). RPW was reported in North Africa from Egypt in the early 1990s (Cox, 1993) and has since spread to all of the major date-palm oases of that country, mainly through transportation of infested palm trees.

The Maghreb region of North Africa is critically situated between several RPW-infested countries including Egypt in the east, the Canary Islands in the west and countries of the Mediterranean basin including Spain, Portugal and Italy in the north. The Maghreb countries did well to keep this dreaded pest of palms at bay until 2008, when RPW was recorded on *P. canariensis* from the Tangier region in Northern Morocco bordering Spain during December 2008, and on *P. dactylifera* from Tabrouk in the northeast of Libya bordering Egypt during January 2009. In order to eradicate/control RPW in Morocco and Libya and prevent its spread to the other countries of the Maghreb region (Algeria, Tunisia and Mauritania), the Food and Agriculture Organization (FAO) of the United Nations initiated an expert consultation in early 2010 aimed at strengthening the national capacities for the management of RPW in these countries.

In the infested countries of Morocco and Libya, strict quarantine regulations banning imports of palms from abroad and movement of

Fig. 1.17. Typical red palm weevil (RPW) symptoms in date palm (*Phoenix dactylifera*) orchards in Saudi Arabia. Top left: Damage hole seen before excavation; Bottom left: same palm after excavation and cleaning showing a large hollowed-out cavity and toppled palm. Top right: an adult RPW near a damage hole at the base of the stem, Bottom right: brown viscous ooze coming from stem, indicating RPW infestation (Photos: P.S.P.V. Vidyasagar.)

palms from the infested regions of these countries are in place. Besides quarantine, a RPW pheromone-based strategy comprising mass trapping adult weevils, preventive insecticidal treatments and eradication of infested palms has been implemented to combat the pest in Morocco. In Libya, pheromone traps are used to monitor the situation throughout the country, including the infested area of Tabrouk where all ornamental (c. 4000) and date palms (c. 1500) are being eradicated, to ensure elimination of the pest from this area to prevent its spread to the nearest date palm oasis of Jagboub (Faleiro, 2010). The uninfested Maghreb countries of Algeria, Tunisia and Mauritania have also banned imports of palms from abroad, regulate the palm nursery industry and are in the process of building capacities to monitor the situation.

1.9 Future Research

In recent years, many scientific papers covering different aspects of *R. ferrugineus* bioecology and control have been published, and research has provided important tools to improve its management. Some of the areas explored include:

- *Early detection.* As already mentioned, an important problem associated with *R. ferrugineus* is the difficulty of early detection. Because recently infested palms can be easily mistaken as pest-free, inadvertent movement of infested plants has been common and has greatly contributed to the current distribution of this weevil. Different groups have focused on the development of acoustic sensors for early detection (Levsky et al., 2007; Mankin et al., 2008; Potamitis et al., 2009, Gutiérrez et al., 2010). However, the degree to which sensors developed so far have been used in practice remains unclear. Other detection techniques, including molecular tools, should be explored.
- *Plant quarantine.* In addition to current pre-departure and post-entry quarantine protocols (OJEU, 2007), the development of a quarantine treatment to disinfest palms, either chemical (Llácer and Jacas, 2010) or physical, would greatly reduce the enormous risks that palm movement imposes worldwide at present.
- *Chemical control.* Although highly effective pesticides exist (Barranco et al., 1998; Llácer et al., 2010, Dembilio et al., 2010a), there are many problems related both to the delivery of the product to the target (tunnelling larvae within the palm) and to the ecotoxicological profile of these biocides. New environmentally friendly products are urgently needed, and alternative application methods (such as trunk injection) or uptake mechanisms (as systemic products) should be investigated further.
- *Biological control.* There is an incomplete knowledge of the natural enemies of *R. ferrugineus* in its native habitat (Murphy and Briscoe, 1999; Faleiro, 2006), and this precludes any classical biological control program against it. On the other hand, different entomopathogenic nematodes (Abbas et al., 2001a, b; Llácer et al., 2009) and fungi (Gindin et al., 2006; El-Sufty, 2009; Sewify et al., 2009; Dembilio et al., 2010b) have been identified. This opens new possibilities of inundative or augmentative releases of these biocontrol agents, and of their combined use with semiochemicals in attract and infect strategies.
- *Semiochemicals and trapping.* Semiochemicals are key to the management of *R. ferrugineus* in palm commercial groves (see above), and will probably have a central role for its management in non-agricultural contexts. However, there are many open questions related to their use for monitoring and mass trapping (trap design, density, maintenance and servicing, location) as well as new possibilities for use (push-and-pull strategies, attract-and-kill, attract-and-infect, attract-and-sterilize).
- *Resistance and induced plant defenses.* Both antibiotic and antixenotic mechanisms of defense have been identified in some palm species (Barranco et al., 2000; Dembilio et al., 2009). Further research is needed to clarify the basis for such mechanisms, and studies on induced defenses could result in novel approaches for the management of the weevil.

There is a clear need for research to help us improve the management of *R. ferrugineus*. This should allow us to continue enjoying palms in our parks, gardens and natural landscapes.

References

Abbas, M.S.T., Hanounik, S.B., Mousa, S.A. and Mansour, M.I. (2001a) On the pathogenicity of *Steinernema abbasi* and *Heterorhabditis indicus* isolated from adult *Rhynchophorus ferrugineus* (Coleoptera). *International Journal of Nematology* 11, 69–72.

Abbas, M.S.T., Saleh, M.M.E. and Akil, A.M. (2001b) Laboratory and field evaluation of the pathogenicity of entomopathogenic nematodes to the red palm weevil, *Rhynchophorus ferrugineus* (Oliv.) (Col.: Curculionidae). *Journal of Pest Science* 74, 167–168.

Abbas, M.S.T., Hanounik, S.B., Shahdad, A.S. and Al-Bagham, S.A. (2006) Aggregation pheromone traps, a major component of IPM strategy for the red palm weevil, *Rhynchophorus ferrugineus* in date palms (Coleoptera: Curculionidae). *Journal of Pest Science* 79, 69–73.

Abdallah, F.F. and Al-Khatri, S.A. (2005) The effect of pheromone, kairomone and food bait on attracting males and females of red palm weevil. *Egyptian Journal of Agricultural Research* 83, 169–177.

Abozuhairah, R.A., Vidyasagar, P.S.P.V. and Abraham, V.A. (1996) Integrated management of red palm weevil *Rhynchophorus ferrugineus* in date palm plantations of the Kingdom of Saudi Arabia. In: *Proceedings of the XX International Congress of Entomology*, Florence, Italy, pp. 25–36.

Abraham, V.A. and Kurian, C. (1975) An integrated approach to the control *Rhynchophorus ferrugineus* F. the red weevil of coconut palm. In: *Proceedings of the 4th Session of the FAO Technical Working Party on Coconut Production, Protection and Processing*, 14–25 September, Kingston, Jamaica.

Abraham, V.A. and Nair, S.S. (2001) Evaluation of five insecticides for use in the red palm weevil pheromone traps. *Pestology* 25, 31–33.

Abraham, V.A. and Vidyasagar, P.S.P.V. (1993) Strategy for the control of red palm weevil of date palm in the Kingdom of Saudi Arabia. Part II. Consultancy report submitted to the Ministry of Agriculture and Water, Riyadh, Kingdom of Saudi Arabia, pp. 32.

Abraham, V.A., Al Shuaibi, M.A., Faleiro, J.R., Abozuhairah, R.A. and Vidyasagar, P.S.P.V. (1998) An integrated management approach for red palm weevil, *Rhynchophorus ferrugineus* Oliv., a key pest of date palm in the Middle East. *Journal of Agricultural and Marine Sciences* 3, 77–84.

Abraham, V.A., Nair, S.S. and Nair, C.P.R. (1999) A comparative study on the efficacy of pheromone lures in trapping red palm weevil, *Rhynchophorus ferrugineus* Oliv. (Coleoptera: Curculionidae) in coconut gardens. *Indian Coconut Journal* 30, 1–2.

Abraham, V.A., Faleiro, J.R., Al-Shuaibi, M.A. and Prem Kumar, T. (2000) A strategy to manage red palm weevil *Rhynchophorus ferrugineus* Oliv. on date palm *Phoenix dactylifera* L. – Its successful implementation in Al-Hassa, Kingdom of Saudi Arabia. *Pestology* 24, 23–30.

Abraham, V.A., Faleiro, J.R., Al Shuaibi, M.A. and Saad Al Abdan (2001) Status of pheromone trap captured female red palm weevils from date gardens in Saudi Arabia. *Journal of Tropical Agriculture* 39, 197–199.

Ajlan, A.M. and Abdulsalam, K.S. (2000) Efficiency of pheromone traps for controlling the red palm weevil *Rhynchophorus ferrugineus* (Olivier) (Coleoptera: Curculionidae), under Saudi Arabia conditions. *Bulletin of the Entomological Society of Egypt (Economics Series)* 27, 109–120.

Aldryhim, Y. and Al-Bukiri, S. (2003a) Effect of irrigation on within-grove distribution of red palm weevil *Rhynchophorus ferrugineus*. *Agricultural and Marine Sciences* 8, 47–49.

Aldryhim, Y. and Khalil, A. (2003b) Effect of humidity and soil type on survival and behavior of red palm weevil *Rhynchophorus ferrugineus* (Oliv.) adults. *Agricultural and Marine Sciences* 8, 87–90.

Al-Khatri, S.A. (2004) Date palm pests and their control. In: *Proceedings, Date Palm Regional Workshop on Ecosystem-Based IPM for Date Palm in Gulf Countries*, 28–30 March, Al-Ain, UAE, pp. 84–88.

Anonymous (1998) *Final Report of the Indian Technical Team (Part A), Red Palm Weevil Control Project*, Ministry of Agriculture and Water, Riyadh, Kingdom of Saudi Arabia, pp. 1–65.

Anonymous (2004) *The Middle East Red Palm Weevil Programme, July 1998 to June, 2004 (Final Report)*. The Peres Center for Peace, Tel Aviv-Jaffa, Israel, pp. 62.

Barranco, P., de la Peña, J., Martín, M.M. and Cabello, T. (1998) Efficiency of chemical control of the new palm pest *Rhynchophorus ferrugineus*. *Boletín de Sanidad Vegetal, Plagas* 24, 301–306. (In Spanish.)

Barranco, P., de la Peña, J.A., Martín, M.M. and Cabello, T. (2000) Host rank for *Rhynchophorus ferrugineus* (Olivier, 1790) (Coleoptera: Curculionidae) and host diameter. *Boletín de Sanidad Vegetal, Plagas* 26, 73–78. (In Spanish.)

BOC (2007) Orden de 29 de octubre de 2007 por la que se declara la existencia de las plagas producidas por los agentes nocivos *Rhynchophorus ferrugineus* (Olivier) y *Diocalandra frumenti* (Fabricius) y se establecen las medidas fitosanitarias para su erradicación y control. *Boletín Oficial de Canarias* (BOC) 2007/222, 6 November 2007, www.gobiernodecanarias.org/boc/2007/222/002.html, accessed 23 July 2012.

BOE (1993) Real Decreto número 2071/93 de 26/11/1993, relativo a las medidas de protección contra la introducción y difusión en el territorio nacional y de la Comunidad Económica Europea de organismos nocivos para los vegetales o productos vegetales, así como para la exportación y tránsito hacia países terceros. *Boletín Oficial del Estado* (BOE) 300, 35603 (Marginal 29872), www.croem.es/Web/CroemWebAmbiente.nsf/c3ac6fdb4e288069c1256bf3005412a5/b8544624d1b3407c, accessed 23 July 2012.

BOE (1996) Orden de 18 de noviembre de 1996 por la que se establecen medidas provisionales de protección contra el curculiónido ferruginoso de las palmeras [*Rhynchophorus ferrugineus* (Olivier)]. *Boletín Oficial del Estado* (BOE) 285, 35614–35615, www.boe.es/diario_boe/txt.php?id=BOE-A-1996-26385, accessed 23 July 2012.

BOE (2000) ORDEN de 28 de febrero de 2000 por la que se establecen medidas provisionales de protección contra el curculiónido ferruginoso de las palmeras [*Rhynchophorus ferrugineus* (Olivier)]. *Boletín Oficial del Estado* (BOE) 59, 9694–9695, www.boe.es/boe/dias/2000/03/09/pdfs/A09694-09695.pdf; www.boe.es/buscar/doc.php?id=BOE-A-2000-4552, accessed 23 July 2012.

BOJA (1997) Orden de 9 de junio de 1997 por la que se dictan medidas de protección fitosanitaria contra el curculiónido ferruginoso de las palmeras *Rhynchophorus ferrugineus* Olivier en el ámbito de la Comunidad Autónoma de Andalucía. *Boletín Oficial de la Junta de Andalucía* 72, 24/06/1997, www.juntadeandalucia.es/boja/boletines/1997/72/d/updf/d27.pdf, accessed 23 July 2012.

Borden, J.H. (1985) Aggregation pheromones. In: Kerkut, G.A. and Gilbert, L.I. (eds) *Comprehensive Insect Physiology, Biochemistry and Pharmacology, Vol. 9*. Pergamon Press, Oxford, UK, pp. 257–285.

Buchner, P. (1965) *Endosymbionts of Animals with Plant Microorganisms*. Interscience, New York, New York.

Cabello, T., de la Peña, J.A. and Barranco, P. (1997) Laboratory evaluation of imidacloprid and oxamyl against *R. ferrugineus*. *Annals of Applied Biology (Tests of Agrochemicals and Cultivars)* 130 (Suppl.) 18, 6–7.

Chang, V.C.S. and Curtis, G.A. (1972) Pheromone production by the New Guinea sugarcane weevil. *Environmental Entomology* 1, 476–481.

Chang, V.C.S., Curtis, G.A. and Ota, A.K. (1971) Insects. *Hawaiian Sugar Planters' Association Annual Report*, 43–44.

Cox, M.L. (1993) Red palm weevil, *Rhynchophorus ferrugineus* in Egypt. *FAO Plant Protection Bulletin* 41, 30–31.

Dembilio, Ó. and Jacas, J.A. (2011) Basic bio-ecological parameters of the invasive red palm weevil, *Rhynchophorus ferrugineus* (Coleoptera: Curculionidae), in *Phoenix canariensis* under Mediterranean climate. *Bulletin of Entomological Research* 101, 153–163, doi: 10.1017/S0007485310000283.

Dembilio, Ó., Jacas, J.A. and Llácer, E. (2009) Are the palms *Washingtonia filifera* and *Chamaerops humilis* suitable hosts for the red palm weevil, *Rhynchophorus ferrugineus* (Coleoptera: Curculionidae)? *Journal of Applied Entomology* 133, 565–567.

Dembilio, Ó., Llácer, E., Martínez de Altube, M.M. and Jacas, J.A. (2010a) Field efficacy of imidacloprid and *Steinernema carpocapsae* in a chitosan formulation against the red palm weevil *Rhynchophorus ferrugineus* (Coleoptera: Curculionidae) in *Phoenix canariensis*. *Pest Management Science* 66, 365–370.

Dembilio, Ó., Quesada-Moraga, E., Santiago-Álvarez, C. and Jacas, J.A. (2010b) Potential of an indigenous strain of the entomopathogenic fungus *Beauveria bassiana* as a biological control agent against the red palm weevil, *Rhynchophorus ferrugineus*. *Journal of Invertebrate Pathology* 104, 214–221.

El-Ezaby, F., Khalifa, O. and El-Assal, A. (1998) Integrated pest management for the control of red palm weevil in the UAE Eastern region, Al-Ain. *Proceedings of the First International Conference on Date Palms*, March 1998, Al-Ain, UAE, pp. 269–281.

El-Sebay, Y. (2003) Ecological studies on the red palm weevils *Rhynchophorus ferrugineus* Oliv. (Coleoptera: Curculionidae) in Egypt. *Egyptian Journal of Agricultural Research* 81, 523–529.

El-Sufty, R., Al-Awash, S.A., Al Bgham, S, Shahdad, A.S. and Al Bathra, A.H. (2009) Pathogenicity of the fungus *Beauveria bassiana* (Bals.) Vuill to the red palm weevil, *Rhynchophorus ferrugineus* (Oliv.) (Col.: Curculionidae) under laboratory and field conditions. *Egyptian Journal of Biological Pest Control* 19, 81.

EPPO (2008) Data sheets on quarantine pests. *Rhynchophorus ferrugineus*. *EPPO Bulletin* 38, 55–59.

EPPO (2009) *Rhynchophorus ferrugineus* found on *Howea forsteriana* in Sicilia, Italy. No. 3 2009/051. European and Mediterranean Plant Protection Organization, Paris, http://archives.eppo.int/EPPOReporting/2009/Rse-0912.pdf, accessed 23 July 2012.

Esteban-Durán, J., Yela, J.L., Beitia Crespo, F. and Jiménez Álvarez, A. (1998a) Exotic curculionids liable to be introduced into Spain and other EU countries through imported vegetables. *Boletín de Sanidad Vegetal, Plagas* 24, 23–40. (In Spanish.)

Esteban-Durán, J., Yela, J.L., Beitia Crespo, F. and Jiménez Álvarez, A. (1998b) Biology of red palm weevil, *Rhynchophorus ferrugineus* (Olivier) in the laboratory and field, life cycle, biological characteristics in its zone of introduction in Spain, biological method of detection and possible control. (Coleoptera: Curculionidae: Rhynchophorinae). *Boletin de Sanidad Vegetal Plagas* 24, 737–748.

Faleiro, J.R. (2005) Pheromone technology for the management of red palm weevil *Rhynchophorus ferrugineus* (Olivier) (Coleoptera: Rhynchophoridae) – A key pest of coconut. *Technical Bulletin* 4, ICAR Research Complex for Goa, Ela, Old Goa, India, pp. 40.

Faleiro, J.R. (2006) A review on the issues and management of red palm weevil *Rhynchophorus ferrugineus* (Coleoptera: Rhynchophoridae) in coconut and date palm during the last one hundred years. *International Journal of Tropical Insect Science* 26, 135–154.

Faleiro, J.R. (2008) Consultancy report on red palm weevil (IPM mission 8 January 2008–7 February 2008). Submitted to the UN FAO, National Date Palm Research Centre, Al Hassa, Saudi Arabia, pp. 31.

Faleiro, J.R. (2009) Testing and refining protocols for area-wide management of red palm weevil (RPW), *Rhynchophorus ferrugineus* (Olivier) in date agro-ecosystems of Al-Hassa, Saudi Arabia. Final report, Date Palm Research Centre, King Faisal University, Al Hassa, Saudi Arabia, pp. 36.

Faleiro, J.R. (2010) Consultation on strengthening of national capacities for the management of the red palm weevil in North Africa (Morocco, Libya and Tunisia) (IPM mission, 8–28 February 2010). FAO, Rome, pp. 43.

Faleiro, J.R. and Ashok Kumar, J. (2008) A rapid decision sampling plan for implementing area-wide management of red palm weevil, *Rhynchophorus ferrugineus*, in coconut plantations of India. *Journal of Insect Science* 8, available online at insectscience.org/8.15.

Faleiro, J.R. and Mani Chellappan (1999) Attraction of red palm weevil *Rhynchophorus ferrugineus* to different ferrugineol based pheromone lures in coconut gardens. *Journal of Tropical Agriculture* 37, 60–63.

Faleiro, J.R., Abraham, V.A. and Al Shuaibi, M.A. (1998) Role of pheromone trapping in the management of red palm weevil. *Indian Coconut Journal* 29, 1–3.

Faleiro, J.R., Mahmood Al Shuaibi, Abraham, V.A. and Premkumar, T. (1999) A technique to assess the longevity of the palm weevil pheromone (Ferrolure) under different conditions in Saudi Arabia. *Sultan Qaboos University Journal for Scientific Research, Agricultural Science* 4, 5–9.

Faleiro, J.R., Abraham, V.A., Nabil Boudi, Al Shuaibi, M.A. and Premkumar, T. (2000) Field evaluation of different types of red palm weevil *Rhynchophorus ferrugineus* pheromone lures. *Indian Journal of Entomology* 62, 427–433.

Faleiro, J.R., Ashok Kumar, J. and Rangnekar, P.A. (2002) Spatial distribution of red palm weevil *Rhynchophorus ferrugineus* Oliv. (Coleoptera: Curculionidae) in coconut plantations. *Crop Protection* 21, 171–176.

Faleiro, J.R., Rangnekar, P.A. and Satarkar, V.R. (2003) Age and fecundity of female red palm weevils *Rhynchophorus ferrugineus* (Olivier) (Coleoptera: Rhynchophoridae) captured by pheromone traps in coconut plantations of India. *Crop Protection* 22, 999–1002.

Faleiro, J.R., Mayilvaganan, M., Nair, C.P.R. and Satarkar, V.R. (2004) Efficacy of indigenous pheromone lure for red palm weevil, *Rhynchophorus ferrugineus* (Olivier) (Coleoptera: Rhynchophoridae). *Insect Environment* 10, 164–166.

Gerber, K. and Giblin-Davis, R.M. (1990a) Association of the red ring nematode, *Rhadinaphelenchus cocophilus*, and other nematode species with *Rhynchophorus palmarum* (Coleoptera: Curculionidae). *Journal of Nematology* 22, 143–149.

Gerber, K. and Giblin-Davis, R.M. (1990b) *Teratorhabditis palmarum* n. sp. (Nemata: Rhabditidae), an associate of *Rhynchophorus palmarum* and *R. cruentatus* (Coleoptera: Curculionidae). *Journal of Nematology* 22, 337–347.

Giblin-Davis, R.M. (1991) The potential for introduction and establishment of the red ring nematode in Florida. *Principes* 35, 147–153.

Giblin-Davis, R.M. (1993) Interactions of nematodes with insects. In: Khan, W. (ed.) *Nematode Interactions*. Chapman and Hall, London, pp. 302–344.

Giblin-Davis, R.M. (2001) Borers. In: Howard, F.W., Moore, D., Giblin-Davis, R.M. and Abad, R. *Insects on Palms*. CABI Publishing, London, pp. 267–304.

Giblin-Davis, R.M., Weissling, T.J., Oehlschlager, A.C. and Gonzalez, L.M.(1994) Field response of *Rhynchophorus cruentatus* (F.) (Coleoptera: Curculionidae) to its aggregation pheromone and fermenting plant volatiles. *Florida Entomologist* 77, 164–177.

Giblin-Davis, R.M., Oehlschlager, A.C., Perez, A.L., Gries, G., Gries, R., Weissling, T.J., Chinchilla, C.M. et al. (1996a) Chemical and behavioral ecology of palm weevils. *Florida Entomologist* 79, 153–167.

Giblin-Davis, R.M., Peña, J.E. and Duncan, R.E. (1996b) Evaluation of an entomopathogenic nematode and chemical insecticides for control of *Metamasius hemipterus sericeus* (Coleoptera: Curculionidae). *Journal of Entomological Science* 31, 240–251.

Giblin-Davis, R.M., Peña, J.E., Oehlschlager, A.C. and Perez, A.L. (1996c) Optimization of semiochemical-based trapping of *Metamasius hemipterus sericeus* (Coleoptera: Curculionidae). *Journal of Chemical Ecology* 22, 1389–1410.

Giblin-Davis, R.M., Gries, R., Gries, G., Peña-Rojas, E., Pinzon, I., Peña, J.E., Perez, A.L. et al. (1997) Aggregation pheromone of the palm weevil, *Dynamis borassi* (F.) (Coleoptera: Curculionidae). *Journal of Chemical Ecology* 23, 2287–2297.

Giblin-Davis, R.M., Gries, R., Crespi, B., Robertson, L.N., Hara, A.H., Gries, G., O'Brien, C.W. et al. (2000) Aggregation pheromones of two geographical isolates of the New Guinea sugarcane weevil, *Rhabdoscelus obscurus*. *Journal of Chemical Ecology* 26, 2763–2780.

Giblin-Davis, R.M., Kanzaki, N., Ye, W., Center, B.J. and Thomas, W.K. (2006) Morphology and systematics of *Bursaphelenchus gerberae* n. sp. (Nematoda: Parasitaphelenchidae), a rare associate of the palm weevil, *Rhynchophorus palmarum* in Trinidad. *Zootaxa* 1189, 39–53.

Gindin, G., Levski, S., Glazer, I. and Soroker, V. (2006) Evaluation of the entomopathogenic fungi *Metarhizium anisopliae* and *Beauveria bassiana* against the red palm weevil *Rhynchophorus ferrugineus*. *Phytoparasitica* 34, 370–379.

Gries, G., Gries, R., Perez, A.L., Gonzalez, L.M., Pierce Jr, H.D., Oehlschlager, A.C., Rhainds, M. et al. (1994) Ethyl propionate: synergistic kairomone for African palm weevil, *Rhynchophorus phoenicis* L. (Coleoptera: Curculionidae). *Journal of Chemical Ecology* 20, 889–897.

Griffith, R., Giblin-Davis, R.M., Koshy, P.K. and Sosamma, V.K. (2005) Nematode parasites of coconut and other palms. In: Luc, M., Sikora, R. and Bridge, J. (eds) *Plant Parasitic Nematodes in Subtropical and Tropical Agriculture*. 2nd ed., CABI Publishing, Wallingford, UK, pp. 493–527.

Gutiérrez, A., Ruiz, V., Moltó, E., Tapia, G. and Téllez, M.M. (2010) Development of a bioacoustic sensor for the early detection of red palm weevil (*Rhynchophorus ferrugineus* Olivier). *Crop Protection* 29, 671–676.

Hallett, R.H., Gries, G., Borden, J.H., Czyzewska, E., Oehlschlager, A.C., Pierce Jr, H.D., Angerilli, N.P.D. et al. (1993) Aggregation pheromones of two Asian palm weevils, *Rhynchophorus ferrugineus* and *R. vulneratus*. *Naturwissenschaften* 80, 328–331.

Hallett, R.H., Oehlschlager, A.C. and Borden, J.H. (1999) Pheromone trapping protocols for the Asian palm weevil, *Rhynchophorus ferrugineus* (Coleoptera: Curculionidae). *International Journal of Pest Management* 45, 231–237.

Hallett, R.H., Crespi, B.J. and Borden, J.H. (2004) Synonymy of *Rhynchophorus ferrugineus* (Olivier) 1790 and *R. vulneratus* (Panzer) 1798 (Coleoptera, Curculionidae, Rhynchophorinae). *Journal of Natural History* 38, 2863–2882.

Hunsberger, A., Giblin-Davis, R.M. and Weissling, T.J. (2000) Symptoms and within-tree population dynamics of *Rhynchophorus cruentatus* (Coleoptera: Curculionidae) infestation in Canary Island date palms *Florida Entomologist* 83, 290–303.

Jayanth, K.P., Mathew, M.T., Narabenchi, G.B. and Bhanu, K.R.M. (2007) Impact of large scale mass trapping of red palm weevil, *Rhynchophorus ferrugineus* Olivier in coconut plantations in Kerala using indigenously synthesized aggregation pheromone lures. *Indian Coconut Journal* 38, 2–9.

Kalleshwaraswamy, C.M., Jagadish, P.S. and Puttaswamy, S. (2004) Longevity and comparative efficacy of aggregation pheromone lures against red palm weevil, *Rhynchophorus ferrugineus* (Olivier) (Coleoptera, Curculionidae). *Pest Management in Horticultural Ecosystems* 10, 169–172.

Kalleshwaraswamy, C.M., Jagadish, P.S. and Puttaswamy, S. (2005) Age and reproductive status of pheromone trapped females of red palm weevil, *Rhynchophorus ferrugineus* (Olivier) (Coleoptera, Curculionidae). *Pest Management in Horticultural Ecosystems* 11, 7–13.

Kanzaki, N., Fukiko, A., Giblin-Davis, R.M., Hata, K., Soné, K., Kiontke, K. and Fitch, D. (2008)

Teratorhabditis synpapillata (Sudhaus 1985) (Rhabditida: Rhabditidae) is an associate of the red palm weevil, *Rhynchophorus ferrugineus* (Coleoptera: Curculionidae). *Nematology* 10, 207–218.

Kanzaki, N., Giblin-Davis, R.M., Zeng, Y., Ye, W. and Center, B.J. (2009) *Acrostichus rhynchophori* n. sp. (Rhabditida: Diplogastridae): a phoretic associate of *Rhynchophorus cruentatus* Fabricius and *R. palmarum* L. (Coleoptera: Curculionidae) in the Americas. *Nematology* 11, 669–688.

Krishnakumar, R. and Maheshwari, P. (2003) Efficacy of different pheromones in trapping the red palm weevil *Rhynchophorus ferrugineus* (Oliv.). *Insect Environment* 9, 28.

Krishnakumar, R., Maheshwari, P. and Dongre, T.K. (2004) Study on comparative efficacy of different types of pheromones in trapping the red palm weevil, *Rhynchophorus ferrugineus* Oliv. of coconut. *Indian Coconut Journal* 34, 3–4.

Kurian, C., Abraham, V.A. and Ponnamma, K.N. (1984) Attractants, an aid in red palm weevil management. *Placrosym* 6, 581–585.

Lefèvre, C., Charles, H., Vallier, A., Delobel, B., Farrell, B. and Heddi, A. (2004) Endosymbiont phylogenesis in the Dryophthoridae weevils: evidence for bacterial replacement. *Molecular Biology and Evolution* 21, 965–973.

Levsky, S., Kostyukovsky, M., Pinhas, J., Mizrach, A., Hetzroni, A., Nakache, Y., Rene, S. et al. (2007) Detection and control methods for the red palm weevil in date palm offshoots. *The XXVI Annual Meeting of the Entomology Society of Israel*, Haifa, Israel.

Llácer, E. and Jacas, J.A. (2010) Efficacy of phosphine as a fumigant against *Rhynchophorus ferrugineus* (Coleoptera: Curculionidae) in palms. *Spanish Journal of Agricultural Research* 8, 775–779.

Llácer, E., Martínez, J. and Jacas, J.A. (2009) Evaluation of the efficacy of *Steinernema carpocapsae* in a chitosan formulation against the red palm weevil, *Rhynchophorus ferrugineus*, in *Phoenix canariensis*. *BioControl* 54, 559–565.

Llácer, E., Dembilio, Ó. and Jacas, J.A. (2010) Evaluation of the efficacy of an insecticidal paint based on chlorpyrifos and pyriproxyfen in a microencapsulated formulation against *Rhynchophorus ferrugineus* (Coleoptera: Curculionidae). *Journal of Economic Entomology* 103, 402–408.

Malumphy, C. and Moran, H. (2007) Red palm weevil *Rhynchophorus ferrugineus*. Central Science Laboratory Plant Pest Notice 5. http://faculty.ksu.edu.sa/10439/Documents/fifty.pdf, accessed 23 July 2012.

Mankin, R.W., Mizrach, A., Hetzroni, A., Levsky, S., Nakache, Y. and Soroker, V. (2008) Temporal and spectral features of sounds of wood-boring beetle larvae: identifiable patterns of activity enable improved discrimination from background noise. *Florida Entomologist* 91, 241–248.

MARM (2011) Registro de Productos Fitosanitarios. Ministerio de Medio Ambiente y Medio Rural y Marino www.magrama.gob.es/es/agricultura/temas/medios-de-produccion/productos-fitosanitarios/registro/menu.asp, accessed 23 July 2012.

Martinez Tenedor, J., Gomez Vives, S., Ferry, M. and Diaz Espejo, G. (2008) Reversals in tunnel of wind for the improvement of the effectiveness of pheromone traps of the red palm weevil, *Rhynchophorus ferrugineus* (Coleoptera: Dryophthoridae). *Boletin de Sanidad Vegetal Plagas* 34, 151–161.

Mayilvaganan, M., Nair, C.P.R., Shanavas, M. and Nair, S.S. (2003) Field assay of locally synthesized ferrugineol for trapping *Rhynchophorus ferrugineus*. *Indian Coconut Journal* 33, 8–9.

Morici, C. (1998) *Phoenix canariensis* in the wild. *Principes* 42, 85–93.

Muralidharan, C.M., Vaghasia, U.R. and Sodagar, N.N. (1999) Population, food preference and trapping using aggregation pheromone ferrugineol on red palm weevil (*Rhynchophorus ferrugineus*). *Indian Journal of Agricultural Science* 69, 602–604.

Murphy, S.T. and Briscoe, B.R. (1999) The red palm weevil as an alien invasive: biology and the prospects for biological control as a component of IPM. *Biocontrol News and Information* 20, 35–46.

Muthiah, C. Natarajan, C. and Nair, C.P.R. (2005) Evaluation of pheromones in the management of red palm weevil in coconut. *Indian Coconut Journal* 35, 15–17.

Muthiah, C., Nair, C.P.R., Cannayane, I. and Rajavel, D.S. (2007) Evaluation of pheromone traps with food baits for monitoring coconut red palm weevil. *Hexapoda* 14, 15–19.

Nair, S.S., Abraham, V.A. and Radhakrishnan Nair, C.P. (2000) Efficiency of different food baits in combination with pheromone lures in trapping adults of red weevil, *Rhynchophorus ferrugineus* Oliv. (Coleoptera: Curculionidae). *Pestology* 24, 3–5.

Nakash, J., Osem, Y. and Kehat, M. (2000) A suggestion to use dogs for detecting red palm weevil (*Rhynchophorus ferrugineus*) infestation in date palms in Israel. *Phytoparasitica* 28, 153–155.

Nardon, P., Lefevre, C., Delobel, B., Charles, H. and Heddi, A. (2002) Occurrence of endosymbiosis in Dryopthoridae weevils: Cytological insights into bacterial symbiotic structures. *Symbiosis* 33, 227–241.

Oehlschlager, A.C. (1994) Use of pheromone baited traps in control of red palm weevil in the kingdom of Saudi Arabia. Consultancy report, Ministry of Agriculture, Riyadh, Saudi Arabia, pp. 17.

Oehlschlager, A.C. (1998) Trapping of date palm weevil. *Proceedings, FAO Workshop on Date Palm Weevil* (Rhynchophorus ferrugineus) *and Its Control*, 15–17 December 1998, Cairo, Egypt.

Oehlschlager, A.C. (2005) Current status of trapping palm weevils and beetles. *Proceedings, Date Palm Regional Workshop on Ecosystem-Based IPM for Date Palm in the Gulf Countries*, 28–30 March 2004, Al-Ain, UAE. *Planter* 81, 123–143.

Oehlschlager, A.C. (2006) Mass trapping as a strategy for management of *Rhynchophorus* Palm Weevils. *Proceedings, First International Workshop on Red Palm Weevil*, Valencia, Spain, 28–29 November, 2005. Fundacion Agroalimed, Valencia, Spain pp. 143–180.

Oehlschlager, A.C., Chinchilla, C.M., Gonzalez, L.M., Jiron, L.F., Mexon, R. and Morgan, B. (1993) Development of a pheromone-based trapping system for *Rhynchophorus palmarum* (Coleoptera: Curculionidae). *Journal of Economic Entomology* 86, 1381–1392.

OJEU (2007) Commission Decision 2007/365/EC on emergency measures against the introduction and spread within the EU of *R. ferrugineus* (Olivier) [notified under document number C (2007) 2161]. *Official Journal of the European Union* L 139, 24–27.

OJEU (2008) Commission Decision of 6 October 2008 amending Decision 2007/365/EC on emergency measures to prevent the introduction into and the spread within the Community of *R. ferrugineus* (Olivier) [notified under document number C (2008) 5550]. *Official Journal of the European Union* L 266, 51–54.

OJEU (2010) Commission decision of 17 August 2010 amending Decision 2007/365/EC on emergency measures to prevent the introduction into and the spread within the Community of *R. ferrugineus* (Olivier) [notified under document number C (2010) 5640]. *Official Journal of the European Union* L 226, 42–44.

Poinar Jr, G.O. (1969) *Pracocilenchus rhaphidophorus* n. gen., n. sp. (Nematoda: Aphelenchoidea) parasitizing *Rhynchophorus bilineatus* (Montrouzier) (Coleoptera: Curculionidae) in New Britain. *Journal of Nematology* 1, 227–231.

Poorjavad, N., Goldansaz, S.H. and Avand-Faghih, A. (2009) Response of red palm weevil *Rhynchophorus ferrugineus* to aggregation pheromone under laboratory conditions. *Bulletin of Insectology* 62, 257–260.

Potamitis, I., Ganchev, T. and Komtodimas, D. (2009) On automatic bioacoustic detection of pests: the cases of *Rhynchophorus ferrugineus* and *Sitophilus orzyae*. *Journal of Economic Entomology* 102, 1681–1690.

Rajapakse, C.N.K., Gunawardena, N.E. and Perera, K.F.G. (1998) Pheromone baited trap for the management of red palm weevil *Rhynchophorus ferrugineus* F. (Coleoptera: Curculionidae) population in coconut plantations. *Cocos* 13, 54–65.

Rao, P.N. and Reddy, N.Y. (1980) Description of a new nematode *Pracocilenchus ferruginophilus* n. sp. from weevils pests (Coleoptera) in South India. *Rivista di Parasitologia* 44, 93–98.

Rochat, D. (2006) Trapping: Drawbacks and prospects: need for more research. *Proceedings, First International Workshop on Red Palm Weevil*, Valencia, Spain, 28–29 November, 2005. Fundacion Agroalimed, Valencia, Spain, pp. 99–104.

Ruiz-Montiel, C., Garcia-Coapio, G., Rojas, J.C., Malo, E.A., Cruz-Lopez, L., del Real, I. and Gonzalez-Hernandez, H. (2008) Aggregation pheromone of the agave weevil, *Scyphophorus acupunctatus*. *Entomologia Experimentalis et Applicata* 127, 207–217.

Sewify, G.H., Belal, M.H. and Al-Awash, S.A. (2009) Use of the entomopathogenic fungus, *Beauveria bassiana* for the biological control of the red palm weevil, *Rhynchophorus ferrugineus* Olivier. *Egyptian Journal of Biological Pest Control* 19, 157–163.

Shagag, A., Al-Abbad, A.H., Dan Dan, A.M., Abdallah Ben Abdallah and Faleiro, J.R. (2008) Enhancing trapping efficiency of red palm weevil pheromone traps with ethyl acetate. *Indian Journal of Plant Protection* 36, 310–311.

Siriwardena, K.A.P., Fernando, L.C.P., Nanayakkara, N., Perera, K.F.G., Kumara, A.D.N.T. and Nanayakkara, T. (2010) Portable acoustic device for detection of coconut palms infested by *Rhynchophorus ferrugineus* (Coleoptera: Curculionidae). *Crop Protection* 29, 25–29.

Soroker, V., Blumberg, D., Haberman, A., Hamburger-Rishad, M., Reneh, S., Talebaev, S., Anshelevich, L. et al. (2005) Current status of red palm weevil infestation in date palm plantations in Israel. *Phytoparasitica* 33, 97–106.

Southwood, T.R.E. and Henderson, P.A. (2000) *Ecological Methods*. 3rd edn, Wiley-Blackwell, New York, New York.

Sudhaus, W., Kiontke, K. and Giblin-Davis, R.M. (2011) Description of *Caenorhabditis angaria* n. sp. (Nematoda: Rhabditidae), an associate of sugarcane and palm weevils (Coleoptera: Curculionidae). *Nematology* 13, 61–78.

Sujatha, A., Chalapathi Rao, N.B.V. and Rao, D.V.R. (2006) Field evaluation of two pheromone lures against red weevil, (*Rhynchophorus ferrugineus* Oliv.) in coconut gardens in Andhra Pradesh. *Journal of Plantation Crops* 34, 414–416.

Thomas, M.C. (2010) Giant palm weevils of the genus *Rhynchophorus* (Coleoptera: Curculionidae) and their threat to Florida palms. *Pest Alert: Florida Department of Agriculture and Consumer Services, Division of Plant Industry.* DACS-P-01682, pp. 2.

Toussaint, E. (2006) March of the red palm weevil to be halted by Wageningen innovation. *International Pest Control* 48, 268.

Vidyasagar, P.S.P.V., Al-Saihati, A.A., Al-Mohanna, O.E., Subbei, A.I. and Abdul Mohsin, A.M. (2000a) Management of red palm weevil *Rhynchophorus ferrugineus* Olivier. A serious pest of date palm in Al-Qatif, Kingdom of Saudi Arabia. *Journal of Plantation Crops* 28, 35–43.

Vidyasagar, P.S.P.V., Mohammed Hagi, Abozuhaiah, R.A., Al-Mohanna, O.E. and Al-Saihati, A.A. (2000b) Impact of mass pheromone trapping on red palm weevil adult population and infestation level in date palm gardens of Saudi Arabia. *Planter* 76, 347–355.

Wattanpongsiri, A. (1966) A revision of the genera *Rhynchophorus* and *Dynamis* (Coleoptera: Curculionidae). *Department of Agriculture Science Bulletin, Bangkok* 1, 1–328.

Weissling, T.J. and Giblin-Davis, R.M. (1993) Water loss dynamics and humidity preference of *Rhynchophorus cruentatus* (Coleoptera: Curculionidae) adults. *Environmental Entomology* 22, 94–98.

Ye, W., Giblin-Davis, R.M., Braasch, H., Morris, K. and Thomas, W.K. (2007) Phylogenetic relationships among *Bursaphelenchus* species (Nematoda: Parasitaphelenchidae) inferred from nuclear ribosomal and mitochondrial DNA sequence data. *Molecular Phylogenetics and Evolution* 43, 1185–1197.

2 Avocado Weevils of the Genus *Heilipus*

Alvaro Castañeda-Vildózola,[1] Armando Equihua-Martinez[2] and Jorge E. Peña[3]
[1]*Facultad de Ciencias Agricolas, Universidad Autonoma del Estado de México, Toluca, Estado de México, México;* [2]*Instituto de Fitosanidad, Colegio de Posgraduados, Montecillo, Texcoco, Estado de México, México;* [3]*Tropical Research and Education Center, University of Florida/IFAS, Homestead, Florida 33031, USA*

The genus *Heilipus* (Coleoptera: Curculionidae), originally described by Germar in 1854, comprises 91 species distributed in the Americas, 39 for North and Central America (O'Brien and Wibmer, 1982) and 52 for South America (Wibmer and O'Brien, 1986). Vanin and Gaiger (2005) reported *H. odoratus* as a new South American species, increasing the number of the original list to 92. The morphological characters that describe the genus *Heilipus* were proposed by Kuschel (1955) and are as follows: prementum glabrous, hind tibiae curved and formed in the inner corner unciniform strong mucro, premucron absent and mesosternal process tuberculiform. It has been suggested that most of these species show their close relationship with plants within the families Lauraceae and Annonaceae, and primitive angiosperms, however this association has been confirmed for only a few of them (Benchaya-Nunes, 2006; Castañeda-Vildózola *et al.*, 2007). *Persea americana* Mill, *P. scheideana* Nees., *Aniba rosedora* Ducke, *Cinnamomum zeylanicum* Ness., *Annona squamosa* L. and *A. muricata* L. are reported as hosts of *Heilipus* spp. Citrus, cocoa and cotton are included as hosts of *Heilipus* spp., but these reports are inconclusive.[i] Currently eight species of *Heilipus* associated with avocado have been reported on the American continent. These comprise three fruit-boring species, *H. lauri* Boheman, *H. pittieri* Barber and *H. trifasciatus* Fabricius; and five stem borers, *H. apiatus* Oliver, *H. albopictus* Champion, *H. elegans* Guerin, *H. catagraphus* Germar and *H. rufipes* Perty.

Larvae of these insects might also feed on branches and fruit. *Heilipus* spp. are considered the most important weevils on cultivated avocado in the Americas (Lourenção *et al.*, 1984, 2003; Castañeda-Vildózola *et al.*, 2007, Rubio *et al.*, 2009). *Conotrachelus*, represented by *C. perseae* and *C. aguacatae*, are the only species associated with avocado within that genus. *C. perseae* is distributed from Mexico to Costa Rica. *C. aguacatae* is endemic to Mexico, mainly in the states of Michoacán, Jalisco and Guanajuato where it is considered an important pest (Whitehead, 1979; González Herrera, 2003; Francia, 2008). Species of *Conotrachelus* and *Heilipus* associated with avocado are listed as pests with quarantine significance. In Mexico, the presence of *H. lauri*, *C. perseae*, *C. aguacatae* and other stem branch borers of avocado has fostered the creation of the Norma Oficial Mexicana FITO-2002, to eradicate and prevent their spread to new areas of cultivation. There is no full information on damages by *Conotrachelus* and *Heilipus* in American countries; however, in Mexico, *H. lauri* may cause up to 80% losses in fruit production in home gardens (Peña, 1998).

2.1 Origin and Distribution

Laurenção et al. (2003) mention that the genus *Heilipus* is Neotropical in origin with the greatest diversity in the area. Only three species, *H. apiatus*, *H. albopictus* and *H. lauri* have reached the Nearctic region; the other five species are found in subtropical and tropical areas of the continent. *Heilipus* spp. lay eggs in avocado stems and fruits. In Florida, USA, larvae of *H. apiatus* (Oliver) are stem borers and are able to kill young trees (Wolfenbarger, 1948; Woodruff, 1963). In Mexico, the stem borer *H. albopictus* Champion has been reported in Morelos and Hidalgo (Morrone, 2003), and the large seed borer *H. lauri* Boheman in Morelos, Puebla, State of Mexico and Veracruz (García, 1962; Medina, 2005). In Costa Rica, there are three species of avocado borers, the stem borer *H. elegans* Guerin and fruit borers *H. trifasciatus* Fabricius (= *H. perseae* Barber) and *H. pittieri* Barber (González-Herrera, 2003). *H. trifasciatus* has also been reported in Panama (Dietz and Barber, 1920). In Brazil, *H. elegans* Guerin, *H. rufipes* Perty and *H. catagraphus* Germar have been reported as avocado stem borers, although *H. catagraphus* has also been observed damaging avocado fruit and stems of *Annona squamosa* L. (Lourenção et al., 1984; Lourenção et al., 2003). Recently, *H. lauri*, *H. pittieri* and *H. elegans* were also reported as fruit and stem borers in Colombia (Rubio et al., 2009). Wibmer and O'Brien (1986) did not report the presence of *H. lauri* in South America at that time; it is possible that this species, endemic to Mexico, has been accidentally introduced through seed shipments to South America in recent years.

2.1.1 Large avocado seed borer *H. lauri* Boheman (Coleoptera: Curculionidae)

Description

Boheman's original description in 1854 was made from specimens collected in Mexico. Later, Champion (1907) redescribed *H. lauri* including distinctive characters of the species such as long rostrum; conical prothorax, narrow and transverse; wide elytra at its base and narrow at the apex with two ocher spots on each. Barber (1912, 1919) also provided a very general description for *H. lauri*: adults brown with two conspicuous spots on each elytron anteromedian and subapical position, the rostrum of the female is 1.5 times larger than the pronotum and the male is slightly smaller.

2.1.2 Description of developmental stages of *H. lauri* by Castañeda-Vildózola (2008)

Egg

Ovoid, 1.40 ± 0.06 mm long ($n = 50$, range 1.28–1.54 mm) and 0.87 ± 0.03 mm wide ($n = 50$, range 0.80–0.96 mm). At the time of oviposition, the chorion is bright white, and during the course of embryonic development takes on a brown to dark brown coloration. Its surface is finely reticulate and with pentagonal shapes (Fig. 2.1A).

Larva

The fourth instar larva reaches a length of 24.21 ± 1.49 mm ($n = 20$, range 21.00–26.29 mm). The body is robust, curved, white opaque (Fig. 2.1B). The head capsule is 1.87 ± 0.06 mm wide ($n = 70$, range 1.80–2.04 mm), is light brown and is not retracted into prothoracic segment. Rudimentary and conical antennas. Epicranial suture visible throughout its length, frontal suture U-shaped front with arms forming a lobe. It has five pairs of frontal setae (Fs); $Fs4$ and $Fs5$ are very long, the other three are smaller. In the epicranial region there are five pairs of dorsal setae (DS), $Ds1$, $Ds3$ and $Ds5$ present larger than the $Ds2$ and $Ds4$. It also has two pairs of lateral setae (Lt); $Lt1$ and $Lt2$ are of similar size, and two ventral pairs ($Vs1$ and $Vs2$) are short (Fig. 2.1C). The clypeus has three pairs of clypeal setae (Cl); setae $Cl1$ and $Cl2$ are located on the edge of clypeus and forehead (Fig. 2.1D). Labro (Lbr) is lobed, with three pairs of setae; $Lbr1$ and $Lbr2$ are longer than $Lbr3$ (Fig. 1D); epipharynx with two labral bars or conical tormae, with four pairs of anterolateral setae and six pairs anteromedial (Fig. 2.1E). The jaws have two distal teeth; one is well developed and the other more rudimentary. Each mandible has a pair of setae (Mb); $Mb1$ is located between the distal teeth, and the seta $Mb2$ is located in a marginal position (Fig. 2.1F). Maxillary palps are bisegmented, with the basal

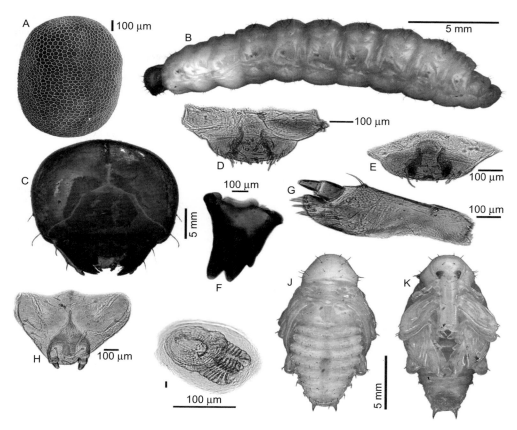

Fig. 2.1. Description of developmental stages of *H. lauri*. (A), Egg; (B), Larva; (C), Head capsule; (D), Labroclypeus; (E), Hepipharynx; (F), mandible; (G), Maxillary palp; (H), Labium; (I), Abdominal spiracle; (J), Pupa, dorsal view; (K), Pupa, ventral view.

segment square and the apical conical; cardo well defined; estipite (Est) has four setae, Est1 has a basal position and three setae are aligned transversely. The mala has three ventral setae and seven dorsal. In the labium, the premental sclerite is complete, and the membrane has a pair of long setae (Fig. 2.1G). The posmentum (Psm) has three pairs of long setae; Psm1 is located near the base of this structure, and Psm2 and Psm3 are placed near the labial palps (Fig. 2.1H). The prothorax has a prothoracic plate with three setae on each end and a pair of biphore spiracles, typical of the Curculionidae. The pleural region has two setae and pedal with six setae. The mesothorax and metathorax have two lobes on the surface; each thoracic segment has six setae of different lengths. The abdomen is composed of nine segments; segments I to VII have a pair of side bicameral spiracles, and the eighth is dorsal (Fig. 2.1I). The spiracles have setae in the lower edge; each abdominal segment has four lobes or folds on the cuticle and two pairs of postdorsal setae. Segment IX has two pairs of dorsal setae. Anus is surrounded by four lobes, the lateral with a pair of setae (Castañeda, 2008).

Pupa

The pupa is 16.97 ± 1.25 mm in length (n = 25, range 14.95–19.78 mm). The prothorax is subtriangular in shape, wider at its base with rounded edges. It has nine pairs of setae arranged as follows: two pairs anterodorsal, two discal pairs, four pairs dorsal and one posterodorsal. Metathorax with two pairs of setae, escutelum in transverse position and slightly elevated; metathorax with two pairs of setae, pteroteques surface with tubercles parallel striated on its edges (Fig. 2.1J). The head bears a pair of frontal setae, the eyes are oval and show a couple of setae on each, the female rostrum reaches the metathoracic coxae and on males just to the mesothorax. It has five pairs of setae in linear position, four pairs in basi-rostral position and the fifth is inserted at one-third of the apex of the rostrum, near the base of the antennae (Fig. 2.1K). Abdomen in dorsal view is composed of nine segments. Segments I to VII have spiracles, the first segment with pleural seta over the spiracle, segments II to VII with pleural seta on the right side of the spiracle. Segment VIII has no pleural seta. Segments I to VII have three pairs of dorsal setae; those on segment I are shorter and thinner than the rest. Segment VIII has a pair of dorsal setae. Segment IX has a pair of pseudocercis, caudally oriented, and a pair of ventral setae (Fig. 2.1J) (Castañeda, 2008).

Adult

The integument is opaque, reddish-black, and legs are red. Body length (excluding rostrum) is 14.77 ± 0.87 mm ($n = 17$, range 13.03–15.91 mm) in females and 13.78 ± 0.76 mm ($n = 17$, range 12.50–15.15 mm) in males. The rostrum is 7.29 ± 0.67 mm ($n = 17$, range 6.2–8.38 mm) in females and 5.32 ± 0.28 mm ($n = 17$, range 4.92–5.94 mm) in males (Figs. 2.2 A and 2.2B). The rostrum is curved in females and slightly thinner than in males, whose rostrum is short and straight. The eyes are oval. The prothorax reaches a length of 3.59 ± 0.18 mm ($n = 17$, range 3.25–3.92 mm) and 3.86 ± 0.20 mm wide ($n = 17$, range 3.51–4.11 mm) in females, and 3.28 ± 0.17 mm in length ($n = 17$, range 2.05–3.56 mm) and 3.65 ± 0.11 mm wide ($n = 17$, range 3.42–3.81 mm) in males. Its apex is constrained, bisinuated at the base, with a rough surface; the apical margin is curved and rounded at the base. The elytra in females reaches a length of 10.37 ± 0.59 mm ($n = 17$, range 9.28–11.16 mm) and 5.86 ± 0.32 mm ($n = 17$, range 5.33–6.20 mm) wide; in males it is 0.40 mm ± 9.78 ($n = 17$, range 9.13–10.46 mm) long and 5.49 ± 0.24 mm ($n = 17$, range 5.14–5.89 mm) wide. The sides are subparallel up to four-fifths of its length, gradually narrowed to the apex and together are rounded. Humeral angles of elytra form a humeral callus near to the apical third, and have a prominent periapical callus characteristic of the genus *Heilipus*. In ventral view the abdomen shows large scales, thin and rounded at the center; at the edges the scales are short, thick and white. In ventrites I and II a shallow depression is found in both males and females. The genitalia have been described by Castañeda et al. (2007) and are shown in Figs. 2.2C–2.2J (Castañeda, 2008).

2.1.3 Distribution

According to information provided by Champion (1907) and Barber (1912), the presence of *H. lauri* has only been documented in Mexico, its distribution covering the Mexican transition zone where the Nearctic and Neotropical entomofauna overlap (Halffter, 1987). In Mexico, *H. lauri* is limited to the Transmexican Volcanic Belt (Morrone et al., 2002), which is considered the center of origin of *P. americana* var. *drymifolia* (Storey et al., 1986).

The Transmexican Volcanic Belt includes the states of Mexico (Almoloya de Alquisiras, Coatepec Ixtapan Harinas, Ixtapan de la Sal, Villa Guerrero, Texcaltitlán and Zumpahuacán), Guerrero (Acapetlahuaya, Cacalotenango, Cuetzala progress, Chapa, Iguala, San Francisco de la Cuadra, San Felipe Chila, Pachivira, Taxco), Hidalgo (Acaxochitlan and Xochicotlan), Morelos (Cuernavaca, Tepoztlan and Puente de Ixtla) and Veracruz (Córdoba, Huatusco, Zapotitlan, Zongolica) where *H. lauri* is present (Garcia, 1962; Castañeda-Vildózola, 2008). As stated earlier, Rubio et al. (2009) report *H. lauri* in South America; however, it is suspected that the species was accidentally introduced through seed shipments from Mexico into that area (Castañeda and Equihua, pers. comm.).

2.1.4 Host plants

The large avocado seed borer is a monophagous species, and its only known host plant is *P. americana*. Both native genotypes and commercial cultivars are attacked. Salgado and Bautista (1993) reported that *H. lauri* causes damage in avocado Hass, Fuerte and Colin

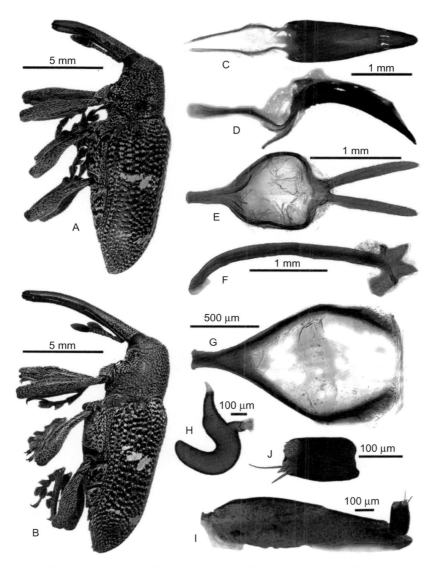

Fig. 2.2. Description of adult *H. lauri*. (A), Male side view; (B), Female side view; (C), Aedeagus dorsal view; (D), Aedeagus lateral view; (E), Tegmen; (F), Espicule gastrale; (G), Sternite VII; (H), Spermatheca; (I), Coxito; (J), Style.

V-33. Medina (2005) documented that *H. lauri* has a preference for native avocado fruits *P. americana* var. *drymifolia* and cultivars Choquette, Fuerte and Hass. Castañeda-Vildózola *et al*. (2009) reported *H. lauri* in fruits of *Persea schiedeana* Nees in Huatusco and Zongolica, Veracruz. The information from these authors represents the first report of *H. lauri* damaging fruit of a different *Persea* species from *P. americana*.

2.1.5 Biology

Egg

Perseatol exudates on the fruit are an indication of an *H. lauri* fruit infestation (Fig. 2.3A). Eggs are laid singly inside a hole made by the female in the developing avocado fruit. In the field 1–3 eggs are commonly found per fruit (Fig. 2.3B). Egg incubation lasts 10.87 ± 0.45 d (Table 2.1).

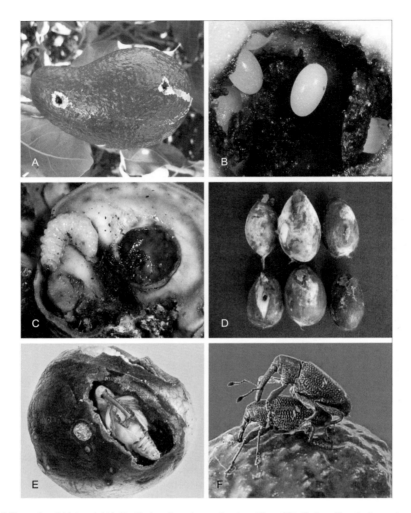

Fig. 2.3. Life cycle of *H. lauri*. (A), Fruit showing signs of oviposition; (B), Oviposition hole and egg; (C), Feeding larva; (D), Seeds infested with larvae close to pupation; (E), Pupa inside the seed; (F), Mating and female borer fruit for eggs oviposition.

Table 2.1. Average duration in days of development stages of *H. lauri* reared in the laboratory (26 ± 2°C, 60–70% RH, photoperiod 12:12).

Stage	Number of individuals observed	Duration (d)	
		Average ± DS	Range
Egg	64	10.87 ± 0.45	10–13
Larva	64	48.51 ± 2.30	44–55
Pupa	64	15.32 ± 1.58	11–18
Cycle egg–adult	64	74.68 ± 1.71	72–80
Longevity of adult	34	309.55 ± 86.72	181–464

In Mexico, ovipositing females were recorded from March to September and their frequency was linked with fruit phenology.

Larva

The larvae are cannibalistic, therefore, only one larva is found per cotyledon and a maximum of two per seed. (Fig. 2.3C). Under laboratory conditions larval development lasts 48.51 ± 2.30 d (Table 2.1). These results differ from those reported by Garcia (1962), who concluded that the *H. lauri* larval stage lasts 58.58 days with a range of 54–63. In contrast to *Conotrachelus perseae* larvae that are gregarious and completely destroy the seed, pupating in the soil, *H. lauri* do not fully destroy the seeds, and pupate inside them. In Ixtapan de la Sal, Mexico, larvae of *H. lauri* occurred from April to August 2004, while in 2005, larvae occurred in February and early March. According to this information it is believed that overlapping generations may occur during the year (Castañeda, 2008).

Pupa

Fruits containing larvae near to pupation fall to the ground (Fig. 2.3D). The larva forms a pupal chamber within the cotyledon (Fig. 2.3E). The pupal stage lasts 15 days (Table 2.1). The sex can be determined at the pupal stage; females have a longer rostrum reaching the metathoracic coxae, while in males it is shorter and only reaches the mesothoracic coxae. In the field, pupation occurred in late August and September (Castañeda, 2008).

Adult

The adult makes a circular hole in the pit to facilitate emergence. Adult longevity under laboratory conditions was 309.55 ± 86.72 d ($n = 34$, range 181–464 d) (Table 2.1) (Castañeda, 2008). Adults are diurnal and are observed more frequently mating on fruits, or depositing eggs from 09:00 to 19:00 h. The availability of food throughout the year (fruits and leaves), the presence of various avocado cultivars and environmental conditions favor the presence of this insect in the state of Mexico. Garcia (1962) concluded that in Tepoztlan and Cuernavaca, Morelos, Mexico, the large seed weevil has two generations per year, the first from June to August and the second from December to February.

2.1.6 Seed borer *Heilipus pittieri* Barber (Coleoptera: Curculionidae)

Barber (1919) and González-Herrera (2003) reported the presence and importance of this weevil in Costa Rica.

Description

Weevil length (excluding rostrum) ranges from 13 to 16.5 mm; rostrum ranges from 5 to 5.70 mm in males and from 8.10 to 9.80 mm in females. The coloration of the body is bright red and there are no spots. The elytra show interstria with white setae. The mesoesternal tubercle is strongly developed and the legs are dark and red (Barber, 1919). *H. lauri* can be distinguished from *H. pittieri* by its orange spots on the darker-colored elytra.

Distribution

Its presence is documented in Costa Rica (González-Herrera, 2003) and it has also been found in Nicaragua (Maes, 2004).

Host plants

Originally found in seeds of *P. scheideana* Nees (*Persea pittieri*), known locally as 'Chinene' (Guillermo Cruz-Castillo, pers. comm., 2008). Currently it has been found damaging Hass avocado fruit, constituting a major pest in Costa Rica (Allan González-Herrera, pers. comm., 2008).

Biology

There is no information on its biology.

Damage

The only known information is that adults lay eggs in the fruit, larvae feed on the seed and pupation occurs inside the seed (González-Herrera, 2003).

2.1.7 Seed borer *Heilipus trifasciatus* Fabricius (Coleoptera: Curculionidae)

Dietz and Barber (1920) reported the presence of this weevil in the Panama Canal region. The information we have on this insect is very brief, and represents the only evidence of its existence. Listed below are generalities about this species.

Description

This species measures 11.00–15.50 mm in length (excluding rostrum) and 4.80–5.70 mm wide. The rostrum length in males is 2.90–3.40 mm and is 3.20–4.10 mm in females. Morphologically it is very similar to *H. lauri* and *H. pittieri*, but *H. trifasciatus* is more robust, has three pairs of brownish spots, two pairs on the elytra and one on the prothorax, similar in size to those on the elytra. The rostrum is short in both sexes, the mesosternum not prominent and legs are short (Dietz and Barber, 1920).

Distribution

Panama (Dietz and Barber, 1920), Costa Rica (González-Herrera, 2003, Nicaragua (Maes, 2004) and Colombia (Rubio *et al.*, 2009).

Host plants

It feeds on native avocados in Panama (Dietz and Barber, 1920) and in Hass avocados in Costa Rica (González-Herrera, 2003).

Biology

There is no detailed information, but it is reported that adults have a longevity of 116 days. In the field it has only one generation per year (Dietz and Barber, 1920).

Damage

Adults feed on growing fruits, tender leaves and vegetative buds. Adults make holes for oviposition; emerging larvae bore through the pulp to reach the seed; one to four larvae are recorded per seed; and pupation occurs within the seed (Dietz and Barber, 1920).

2.1.8 Stem avocado borer *H. albopictus* Champion (Coleoptera: Curculionidae)

Description

This weevil is 16 mm long and 6.25 mm wide, color is dark gray with three white irregular spots on each elytra and a white stripe on the lateral edges of the prothorax. The elytra are striated and the prothorax almost conical (Champion, 1907) (Fig. 2.4A).

Distribution

The distribution on this species is well known in Mexico; it has been reported in Cuernavaca, Morelos (Champion, 1907) and Tlanchinol, Hidalgo (Morrone, 2003). Recently, has been reported from Coatepec Harinas, Ixtapan de la Sal and Ixtapan del Oro, Estado de Mexico (Castañeda-Vildózola *et al.*, 2011).

Host plants

Avocado is the only known host. Weevil larvae bore into stems of native avocados or into stems of cvs. Hass and Fuerte (Castañeda-Vildózola *et al.*, 2010).

Biology

There is no information on biology, but according to field observations in Ixtapan de la Sal, Estado de Mexico, adults are present on trees from late February through June, and from July to October, adults are commonly observed at the base of the stems. Eggs are found from June to October and larvae occur from late June through April. Pupae are only observed in April (Castañeda-Vildózola *et al.*, 2010).

Damage

The adults bore into the base of the stems for oviposition (Fig. 2.4B), then the larvae feed on the bark, causing serious injury (Fig. 2.4C) and weakening affected trees. The presence of reddish exudates in stems is an indication of the presence of larvae (Fig. 2.4D). Larval density fluctuates between 1 and 24 larvae per stem. Pupation occurs between the bark and wood, forming a cocoon (Fig. 2.4E). Adults only feed on growing fruits causing injury, but the loss of fruits is not significant (Fig. 2.4F) (Castañeda-Vildózola *et al.*, 2010).

2.1.9 Stem avocado borer *H. apiatus* Oliver (Coleoptera: Curculionidae)

Description

Newly deposited eggs are white and change to yellow later on. Emerging larvae are 1 mm in length, and well-developed larvae reach 12 mm (Woodruff,

Fig. 2.4. *H. albopictus* Champion. (A) Lateral view of adult; (B) Adult on avocado stem; (C) Boring larvae; (D) Symptoms caused by larvae of *H. albopictus*; (E) Pupa; (F) Adult feeding on fruit.

1963). The adult is 12 mm long, black with two white spots of irregular shape on each elytra; the prothorax is narrow and convex with two lateral white stripes. The white spots are composed of tiny setae on a black background with a variable distribution, giving different spot patterns on each specimen (Woodruff, 1963).The rostrum is curved and slightly larger than the prothorax.

Distribution

Apparently *H. apiatus* is the only species of the genus *Heilipus* established in the USA; it has been recorded in Georgia, Tennessee North Carolina, South Corolina and Florida (Wolfenbarger, 1953).

Host plants

Damaged avocado stems have only been recorded from Florida. It has also been collected from other plants such as mandarin satsuma, cotton and under pine bark (Wolfenbarger, 1948; Woodruff, 1963).

Biology

There is no information on the biology of this insect. Wolfenbarger (1953) reported that the presence of adults occurs in the months of April,

May and June, but damage has been recorded from February to March.

Damage

Adult weevils lay eggs at the base of the stem, from which reddish exudates can be observed; the larvae feed on the bark causing girdling and death of affected trees. In advanced instars, the damage is very noticeable, reddish colored exudates accumulation observed and excrement at the base of the stem. Loss of up to 10% of individual trees in affected groves has been recorded. Adults also feed on fruit, but the damage is not important (Woodruff, 1963).

2.1.10 Avocado stem borer *H. elegans* Guerin (Coleoptera: Curculionidae)

Description

Heilipus elegans is 9–17 mm long; the body is reddish and covered with small white scales. The elytra have creamy white spots. The apical spot is almost round and the periapical callus discovered. Basal spot is elongated, not straight and continue until the prothorax. The rostrum is curved and robust. The antennae are inserted near the middle of the rostrum (Champion, 1907).

Distribution

Their presence has been documented in Guatemala, Costa Rica, Panama, Colombia and Brazil (Champion, 1907; Lourenção et al., 2003; Rubio et al., 2009).

Host plants

Damage has been reported in Costa Rica in Hass avocado (González-Herrera, 2003) and in the cultivated varieties Lorena, Santana and Choquette in Colombia (Rubio et al., 2009).

Life cycle

Unknown.

Damage

The female lay eggs at the base of the stems of avocado and the larvae tunnel the bark, weakening trees and killing them by girdling. The presence of fecal pellets at the base of the stem is an indication of larval presence. In Colombia, 49.1% damage was reported in avocado trees affected by *H. elegans*. Lorena cultivar with an average infestation of 5 larvae per tree showed the greatest, while Choquette and Santana had less damage with three larvae per tree (Rubio et al., 2009).

2.1.11 Avocado stem borer *H. catagraphus* Germar (Coleoptera: Curculionidae)

Description

No information.

Distribution

Its presence has been recorded in the states of São Paulo, Minas Gerais and Rio de Janeiro, Brazil (Lourenção et al., 1984).

Host plants

The larvae of *H. catagraphus* bore into stems of avocado (*P. americana* Mill), canelilla (*Nectandra venulosa* Meiissn.) and sugar apple (*Annona squamosa* L.), soursop (*A. muricata* L.) and *Rollinia sieberi* D.C. (Lourenção et al., 1984).

Life cycle

Unknown.

Damage

The adults feed on young fruits of avocado and tender shoots of *A. squamosa*. This species does not lay eggs in fruits, but it does in the stems and large branches. The larvae feed on the bark, seriously damaging it and even killing trees (Laurenção et al., 1984).

2.1.12 Avocado stem borer *H. rufipes* Perty (Coleoptera: Curculionidae)

Description

No information.

Distribution

Reported in the state of Ceara, Brazil (Lourenção et al., 2003).

Host plants

The only known host is the avocado (Lourenção et al., 2003).

Life cycle

No information.

Damage

The larvae feed on the bark of the stems of avocado, seriously damaging them and killing the trees. For instance, 94 ha of 5–10-year-old avocados, numbering approximately 14,700 trees, suffered 90% damage, with each tree having a density of 5–12 adult weevils (Lourenção et al., 2003).

2.2 Sampling and Monitoring Techniques

In Mexico, ten fruits collected from each of ten trees/ha are selected. Fruits with symptoms of infestation are dissected and inspected for eggs or juveniles of *H. lauri*. The samples are taken every 15 d to detect the different stages of pest development during the period of avocado fruiting. Adults on foliage are sampled by selecting one branch located approximately 1.6 m above the ground. The branch is cut and then shaken over a 2 × 2 m plastic sheet to dislodge adults present. In Mexico (under the official Mexican standard NOM-066-FITO-2002, which recommends strategies for managing avocado pests), fruit in packing houses is inspected by collecting 10% of fruits in 1% of the total boxes containing avocados in the packing houses.

2.3 Assessment of Damage and Economic Thresholds

In Mexico, those orchards without phytosanitary measures could experience economic damage amounting to losses of 30–60% caused by *H. lauri* (Medina, 2005). In Colombia, damage of 49.1% was reported in avocado trees affected by *H. elegans* (Rubio et al., 2009). In Mexico, avocado-producing areas are categorized into areas free of fruit borers, areas with fruit borers and controlled areas. Areas free of fruit borers are those where the environmental conditions do not allow the weevils to establish or where they have been eradicated after application of control measures, and where for consecutive years the presence of avocado weevils has not been recorded. Controlled areas are those where low weevil densities (adults or immature stages) are present. The threshold for an area is one captured insect, or a fruit or stem with borers. Those growers who participate in an export program must demonstrate zero insect infestation (under NOM-066-FITO-2002, see Section 2.2 above) before they are allowed to export fruit. In backyard orchards, infestations can result in 80% damaged fruit, and captured adults can fluctuate from one to five per tree (Salgado et al., 1993; Castañeda-Vildózola, 2008).

2.4 Management Tactics

The implementation of management tactics to reduce damage caused by *Heilipus* spp. in countries where their presence is reported, is slow. Mexico, as the largest producer and exporter of avocado fruit and with a plant problem that makes it susceptible to quarantine restrictions, has generated knowledge aimed at reducing the presence of *H. lauri* and other borer species of avocado fruits, branches and stems.

The following are methods of control recommended for the eradication of *H. lauri* and control of fruit, branches and stem borers.

2.4.1 Cultural control

It is advisable to harvest and burn fruit with signs of borer damage in order to reduce infestation for the next harvest cycle. Fruits and seeds that have fallen on the soil should be collected and destroyed in order to reduce *H. lauri* infestation (under NOM-066-FITO-2002, see Section 2.2 above). In Colombia, Rubio et al. (2009) recommend manual removal of larvae of *H. elegans* on affected avocado trees.

Chemical control

In Mexico, there are approved insecticides for use in the management of avocado fruit borer. Permethrin is recommended for control of *H. lauri*, while methyl parathion and malathion are also recommended for control of avocado pests (Medina, 2005). Because of its diurnal activity, control of *H. lauri* is implemented in the early hours of the morning. In Brazil, oil palm with chlorpyrifos at 1% is recommended for the control of larvae of *H. catagraphus* in stems of soursop (*A. muricata*). Moura et al. (2006) reported efficiency of 80%.

Biological control

There are no reports of the presence of natural enemy populations significantly affecting *Heilipus* spp. (Castañeda, 2008). However, in Mexico, Garcia (1962) reported parasitoids of the genus *Bracon* emerging from 20% of the larvae of *H. lauri*. In the state of Mexico, *Beauveria bassiana* is recommended for control of *H. lauri* larvae (Díaz-Pérez, 2010. Personal communication).

2.5 Discussion

Weevils of the genus *Heilipus* are of American origin. This group of insects has since ancient times co-evolved with *Persea* spp. Currently, large areas are planted with avocado, creating a monoculture that provides food for many species of insects. Avocado production is commercially important and allows an exchange between several American countries.

Despite their importance, there is little information on the morphological features, biology, distribution, sampling and control strategies in several countries that have recorded the presence of *Heilipus* spp.

Mexico is the main producer of avocados worldwide. Avocados are also grown in the backyard and grow wild in several states of the Mexican republic, where several insects have become established, causing damage at a local level. The official Mexican standard NOM-066-FITO-2002 recommends strategies for managing avocado pests, and its application (chemical, cultural and monitoring) has allowed the successful control of *H. lauri*.

Chemical control is based on the use of malathion and permethrin to reduce insect damage. Cultural control involves the destruction of damaged fruits or removal of larvae inside the stems, while monitoring is performed, using pheromones as a key point in integrated pest management.

Note

[1] www.caripestnetwork.org, accessed 5 July 2012.

References

Barber, H.S. (1912) Note on the a vocado weevil (*Heilipus lauri* Boheman). *Proceedings of the Entomological Society of Washington* 14, 181–183.

Barber, H.S. (1919) Avocado weevils. *Proceedings of the Entomological Society of Washington* 21, 53–61.

Benchaya-Nunes, A. (2006) Ritmo diár io de emergência e alimentação, e deter minação sexual baseado na estr idulação de *Heilipus odoratus* Vanin & Gaiger, 2005 (Coleopter a: Curculionidae: Molytinae), broca da semente do pau rosa. MSc thesis, Universidade Federal do Amazonas, Manaus, Brazil.

Castañeda-Vildózola, A. (2008) Bioecología del barrenador grande de la semilla del aguacate *Heilipus lauri* Boheman (Coleopter a: Curculionidae) en la región Central de México. PhD thesis, Colegio de P osgraduados en Ciencias Agrícolas, Montecillo, México.

Castañeda Vildózola, A., Valdez-Carrasco, J., Equihua-Martínez, A., González-Hernández, H., Romero-Nápoles, J., Solís-Aguilar, J.F. and Ramírez-Alarcón, S. (2007) Genitalia de tres especies de *Heilipus* Germar (Coleoptera: Curculionidae) que dañan fr utos de aguacate (*Persea americana* Mill.) en Mé xico y Costa Rica. *Neotropical Entomology* 36, 914–918.

Castañeda Vildózola, A., Del Angel-Coronel, O.A., Cruz-Castillo, J.G. and Váldez-Carrasco, J. (2009) *Persea schiedeana* (Lauraceae), Nuevo Hospedero de *Heilipus lauri* Boheman (Coleoptera: Curculionidae) en Veracruz, México. *Neotropical Entomology* 38, 871–872.

Castañeda Vildózola, A., Franco-Mora, O., Equihua-Martínez, A., Valdez-Carrascco, J. and González-Herrera, A. (2010). New records of *Heilipus albopictus* Champion (Coleopter a: Curculionidae) infesting avocado trees in México. *Boletín del Museo de Entomología de le Universidad del Valle* 11, 11–14.

Champion, G.C. (1907) *Biología Centrali-Americana, Insecta. Coleoptera. Rhynchophora. Curculionidae. Curculioninae (Part)* 4, pp. 144.

Del Ángel-Coronel, O.A. (2006) Fisiología del desarrollo, plagas de campo y patología postcosecha de frutos de chinene (*Persea scheideana* Ness). MSc thesis, Universidad Autónoma Chapingo, Chapingo, México.

Dietz, H.F. and Barber, H.S. (1920) A new avocado weevil from the canal zone. *Journal of Agricultural Research* 20, 111–118.

García, A.P. (1962) *Heilipus lauri* Boheman un barrenador de la semilla del aguacate en México. BSc thesis, Escuela Nacional de Agricultura, Chapingo, México.

González-Herrera, A. (2003) Artrópodos asociados al cultivo del aguacate (*Persea americana* Mill.) en Costa Rica. *Proceedings of the World Avocado Congress V*. Malaga, Spain, pp. 449–454.

Halffter, G. (1987) Biogeography of the montane entomofauna of Mexico and Central America. *Annual Review of Entomology* 32, 95–114.

Kuschel, G. (1955) Nuevas sinonimias y anotaciones sobre Curculionoidea (Coleoptera). *Revista Chilena de Entomología* 4, 261–312.

Lourenção, A.L., Rossetto, C.J. and Soares, N.B. (1984) Ocorrência de adultos de *Heilipus catagraphus* Germar, 1824 (Coleoptera: Curculionidae) danificando frutos de abacateiro. *Bragantia* 43, 249–253.

Lourenção, A.L., Soares, N.B. and Rosado-Neto, G.H. (2003) Ocorrência e danos de larvas de *Heilipus rufipes* Perty (Coleoptera: Curculionidae) em abacateiro (*Persea americana* Mill.) no Estado de Ceará. *Neotropical Entomology* 32, 363–364.

Maes, J.M. (2004) Insectos asociados a algunos cultivos tropicales en el Atlántico de Nicaragua. *Revista Nicaraguense de Entomología* Sup.1, Parte IV, 181–184.

Medina, Q.F. (2005) Incidencia del barrenador grande del hueso del aguacate *Heilipus lauri* Boheman (Coleoptera: Curculionidae) en Tepoztlan, Morelos. BSc thesis, Universidad Autónoma de Morelos, Cuernavaca, México.

Morrone, J.J. (2003) *Heilipus albopictus* (Champion, 1902) Coleoptera: Curculinidae: Molytinae: Molytini. *Dugesiana* 10, 35–36.

Morrone, J.J., Espinosa, O.D. and Llorente, B.J. (2002) Mexican biogeographic provinces: preliminary scheme, general characterizations and synonymies. *Acta Zoologica Mexicana* 85, 83–108.

Moura, J.I.L., Sgrillo, R.B. and Cividanes, F.J. (2006) Chlorpiriphos in palm oil to the control of *Heilipus catagraphus* (Coleoptera:Curculionidae) in Soursop. *Agrotrópica* 18, 53–56.

O'Brien, C.W. and Wibmer, J.B. (1982) Annotated check list of the weevils (Curculionidae *sensu lato*) of North America, Central America, and the West Indies (Coleoptera: Curculionidae). *Memoirs of the American Entomological Institute* 34, 1–382.

Peña, J.E. (1998) Current and potential arthropod pests threatening tropical fruit crops in Florida. *Proceedings of Florida State of Horticultural Science* 111, 327–329.

Rubio, G.J.D., Posada, F.F.J., Osorio, L.O.I., Vallejo, E.L.F. and López, N.J.C. (2009) Primer registro de *Heilipus elegans* Guérin-Méneville (Coleoptera: Curculionidae) atacando el tallo de árboles de aguacate en Colombia. *Revista U.D.C.A Actualidad & Divulgación Científica* 12, 59–68.

Salgado, M.L. and Bautista, N. (1993) El barrenador grande del hueso del aguacate en Ixtapan de la sal, México. In: Memoria. Fundación Salvador Sánchez Colín. CICTAMEX, S.C., Coatepec Harinas, México, pp. 225–231.

Storey, W.B., Bergh, B. and Zentmyer, G.A. (1986) The origin, indigenous range and dissemination on the avocado. *California Avocado Society Yearbook* 70, 127–133.

Vanin, S.A. and Gaiger, F. (2005) A new spermophagus species of *Heilipus* Germar from the Amazonian Region (Coleoptera, Curculionidae, Molytinae). *Revista Brasileira de Entomologia* 49, 240–244.

Wibmer, G.J. and O'Brien, C.W. (1986) Annotated check list of the weevils (Curculionidae *sensu lato*) of South America (Coleoptera: Curculionidae). *Memoirs of the American Entomological Institute* 39, 1–563.

Wolfenbarger, D.O. (1948) *Heilipus squamosus* Lec. A new enemy of the avocado. *California Avocado Society Yearbook* 33, 98–102.

Wolfenbarger, D.O. (1953) On the distribution of *Heilipus squamosus* (Lec.) a pest of the avocado. *Florida Entomologist* 34, 139–141.

Woodruff, R.E. (1963) An avocado weevil (*Heilipus apiatus* Oliv.) (Coleoptera: Curculionidae). *Florida Department of Agriculture, Division of Plant Industry (Entomology Circular 11)*, pp. 1.

Wysoki, M., van der Berg, M.A., Ish-Am, G., Gazit, S., Peña, J.E. and Waite, G. (2002) Pests and pollinators of avocado. In: Peña, J.E., Sharp, J.L. and Wysoki, M. (eds) *Tropical Fruit Pests and Pollinators: Biology, Economic Importance, Natural Enemies and Control*. CABI Publishing, Wallingford, UK, pp. 223–294.

3 Exotic Bark and Ambrosia Beetles in the USA: Potential and Current Invaders

Robert A. Haack[1] and Robert J. Rabaglia[2]

[1]*USDA Forest Service, Northern Research Station, 1407 S Harrison Road, East Lansing, Michigan 48823, USA;* [2]*USDA Forest Service, Forest Health Protection, 1601 N Kent Street, RPC–7, Arlington, Virginia 22209, USA*

3.1 Introduction

Bark and ambrosia beetles (Coleoptera: Curculionidae: Scolytinae) are among the most important insects affecting trees and forests worldwide. There are approximately 6000 scolytine species worldwide, with species found on all continents except Antarctica (Table 3.1) (Wood and Bright, 1992; Bright and Skidmore, 1997, 2002; Wood, 2007). The majority of species are found in the tropics, but many also occur in boreal forests. Undoubtedly, there are hundreds of additional species that have not yet been described. Many authorities now consider the bark and ambrosia beetles a subfamily (Scolytinae) of the weevil family (Curculionidae) (Alonso-Zarazaga and Lyal, 2009), while others continue to treat them as a distinct family (Wood, 2007). In this chapter, we will use the subfamily ranking Scolytinae, but recognize that most plant protection agencies worldwide continue to use Scolytidae.

Although adults of all scolytine species bore into their host to lay eggs, they exhibit many different habits and utilize many different host tissues. True bark beetles bore through the outer bark to the phloem–cambial area where they construct characteristic galleries and lay eggs. Larval mines radiate out from the gallery as the larvae feed on the phloem (phloeophagous). Ambrosia beetle adults bore through the bark and into the xylem (wood) where they lay eggs. Ambrosia beetle adults and larvae cultivate and feed on symbiotic ambrosia fungi that grow in the galleries (xylomycetophagous). Most scolytine species in tropical regions exhibit the ambrosial habit, while most scolytines in temperate forests are true bark beetles. There are also a number of species that breed in seeds, cones, roots of woody plants, and stems and roots of non-woody plants (Wood, 1982; Sauvard, 2004).

Most species of bark and ambrosia beetles live in injured, weakened or dying woody plants, and are often among the first insects to colonize such host material (Haack and Slansky, 1987; Sauvard, 2004). A few species aggressively attack healthy trees, and during outbreaks cause extensive mortality of their host trees. Efficient host location is important, and is often mediated by olfactory responses to host odors, tree degradation products, or conspecific pheromones (Byers, 2004). Pheromones are used to attract potential mates, and in some scolytines are also used to mass attack host trees to overcome host resistance (Byers, 2004; Raffa *et al.*, 2008).

Mating systems and social organization vary among scolytines. Reproductive systems range from simple monogamy (one male with one female), to heterosanguineous polygyny (multiple members of one sex with one, unrelated member of the opposite sex), to consanguineous polygyny (multiple members of one sex with one,

Table 3.1. Approximate number of scolytine species worldwide and the number of those classified as ambrosia beetles by geographic area.

Continent or geographic area[a]	Approximate number of scolytine species[b]	
	Total	Ambrosia beetles[c]
Africa	1140	260
Asia	1920	500
Australia	130	50
Europe	230	25
North America	1700	400
South America	1250	450
Pacific Islands	220	100
Worldwide	6000	1800

[a]Geographic regions follow Wood and Bright (1992), Europe and Asia are divided by Ural Mountains, Asia includes Indonesia and Philippines, Africa includes Madagascar, North America includes Antilles Islands and Central America to Panama, New Zealand is included with Australia.
[b]Totals for each geographic area include all species in the region, not just the endemic species, therefore the sum of all world regions is greater than the worldwide totals.
[c]Ambrosia beetles are all genera in the tribes Corthylini, Hyorrhynchini, Premnobini, Scolytoplatypodini, Xyleborini, Xyloterini, and the genus *Camptocerus*.

related member of the opposite sex) (Wood, 1982; Kirkendall, 1983). Social organization ranges from parental care, to colonial breeding, and eusociality with division of labor and reproduction (Crespi, 1994; Kirkendall *et al.*, 1997).

The cryptic nature of scolytines, along with their wide variety of mating systems, host finding behaviors, and abilities to utilize different host tissues, allow them to be very successful in their native habitats as well as efficient invaders of new habitats (Wood, 1977, 1982; Haack, 2001; Roques *et al.*, 2009; Sauvard *et al.*, 2010). Ecologically and economically, these beetles are a very important group. In North America, various members of *Dendroctonus* kill vast expanses of forests each year. The current outbreak of the mountain pine beetle, *Dendroctonus ponderosae* Hopkins, has killed c. 18 million acres (7.3 million ha) of pine (*Pinus*) forests across the western USA, and a similar acreage in western Canada (Raffa *et al.*, 2008; USDA Forest Service, 2010). In Europe, the spruce bark beetle, *Ips typographus* (Linnaeus), historically has been the major pest of spruce (*Picea*), resulting in extensive areas of tree mortality (Wermelinger, 2004). Other species vector pathogenic fungi that cause tree diseases, for example the bark beetle *Scolytus multistriatus* (Marsham), which vectors the causal agents of Dutch elm disease (Evans and Finkral, 2010). Ambrosia beetles, and their associated fungi, often cause stain and degrade of valuable wood products, and in some cases, the fungal associates may be highly pathogenic to new hosts (Fraedrich *et al.*, 2008). In this chapter, we will: (i) summarize recent US interceptions of scolytines with a focus on those that were intercepted in association with wood; (ii) discuss the 58 scolytine species that were known to be established in the continental USA as of 2010; (iii) briefly discuss the biology and impact of the redbay ambrosia beetle, *Xyleborus glabratus* Eichhoff; and (iv) discuss current international efforts aimed at reducing the international movement of exotic plant pests.

3.2 Scolytine Interceptions on Wood from 1984 to 2008

Since 1984, the United States Department of Agriculture (USDA), Animal and Plant Health Inspection Service (APHIS) has maintained an electronic database for plant pests intercepted at US ports of entry (Haack, 2001, 2006; McCullough *et al.*, 2006). At first, this database was known as the Port Information Network, or PIN, but is now called the Pest Interception Database, or PestID. The internal policies of every country influence how cargo is inspected and which interceptions are recorded. In the USA, not all interceptions are entered into PestID, but rather there is a bias towards pests of live plants that are considered of quarantine significance. This policy has affected the number and types of scolytine interceptions recorded over the years. For example, because true bark beetles more often infest live trees than do ambrosia beetles, it is more likely that interceptions of bark beetles will be entered into PestID than will interceptions of ambrosia beetles. In addition, given that Canada and the USA share many insect species, few interceptions from Canada are ever considered of quarantine importance and entered into PestID. For example, of the 195 scolytine species recorded in Canada, only five are unique to Canada, while 190 are also present in the USA (Wood and Bright, 1992; Bright and Skidmore, 1997, 2002). It is important to remember these policies when examining the interception data discussed below.

In our analysis, we used PestID interception records for scolytines associated with wood for the 25-year period from 1984 through 2008. The data set consisted of 8286 records. Note, that our data set did not include all scolytines, such as those found in food items like coffee and nuts (Haack, 2001), but rather focused on those associated with wood in the form of wood packaging material (e.g., crating, pallets and dunnage), lumber and logs. Of the 8286 interceptions, 3446 (42%) were identified to the species level; 2239 (27%) to only the genus level; and 2601 (31%) to only the family level (Table 3.2). In the discussion below, these 8286 interception records are presented as a means of suggesting which scolytines are most likely to invade the continental USA; however, it is important to realize that the pool of potential invaders will change in relation to changes in trading partners, types of imported products and the tree species utilized to manufacture wood packaging material.

3.3 Continent of Origin

Scolytines from eight continents or world regions were intercepted on wood in the USA from 1984 to 2008 (Table 3.2). For each interception record, we assigned the corresponding country of origin to a continent in a manner similar to that used in Haack (2001, 2006), which is slightly different to the world regions used in Table 3.1 in this chapter. For example, for the data that follow, we assigned all interceptions from Russia and Turkey to Asia, Mexico to Central America, and Australia and New Zealand to the Pacific region. Most of the 8286 interceptions originated in Europe (48%), Central America (32%) and Asia (15%) (Table 3.2). Overall, 46 genera and 107 species were identified among the intercepted scolytines (Table 3.2). The diversity of intercepted scolytines was greatest for Asia and Europe (Table 3.2); however, keep in mind that USDA APHIS historically identified more bark beetles than ambrosia beetles. The five scolytine genera that represented the most interceptions from each continent are given in Table 3.2.

Over the 25-year period from 1984 to 2008, the annual number of wood-associated scolytine interceptions was highest in 1985 (647 interceptions), then decreased during the 1990s (low of 161 in 1991), and increased again in the 2000s (high of 517 in 2002; Fig. 3.1). Note that the number of interceptions shown for 1984 is artificially low because the electronic PestID database actually began midway through 1984. The strong reduction in overall interceptions during the late 1980s and 1990s was related to a dramatic decline in interceptions from Europe (Fig. 3.1). By contrast, the increase in interceptions during the 2000s was primarily the result of more interceptions from Central America, especially Mexico (Fig. 3.1).

There are many factors that influence changes in interception rates (Haack, 2001). For example, exporters can change the type or quality of packaging materials they use, or they can treat the wood prior to export. Importing countries can also influence interception rates by changing their inspection policies, import regulations and major trading partners. For example, during the 1990s, the USA implemented two regulations on wood packaging material that likely affected interception rates. First, beginning in 1996, the USA required that all unmanufactured solid wood packaging be 'totally free from bark and apparently free from live plant pests' or else be certified as treated for wood pests by the exporting country (USDA APHIS, 1995). In addition, beginning in 1999, the USA imposed stricter regulations on wood articles exported from China to the USA (USDA APHIS, 1998). As for major changes in US trading partners, the most dramatic change has been with China, given that the value of imports from China to the United States increased from US$3.1 billion in 1984 when China was the 20th leading US trading partner, to US$338 billion in 2008 when China was the 2nd leading USA trading partner (US Census Bureau, 2011).

3.4 Country of Origin

Scolytines were intercepted on wood at US ports of entry from at least 85 distinct countries that exist today, plus two countries that no longer exist (USSR and Yugoslavia) and the US state of Hawai'i, for a total of 88 'countries'. Although Hawai'i is a US state, goods from Hawai'i are inspected when exported to the continental USA. Of the 85 countries that exist today, nine were in

Table 3.2. Summary data by continent of origin for 8286 scolytine interceptions made in association with wood products or wood packaging materials at USA ports of entry from 1984 to 2008.

Continent[a]	Number of interceptions	Number identified to only			Number of identified		Five most common genera in decreasing order
		Family level	Genus level	Species level	Genera	Species	
Africa	70	17	34	19	9	3	*Hypothenemus, Orthotomicus, Pityophthorus, Hylastes, Polygraphus*
Asia	1218	243	619	356	32	43	*Phloeosinus, Orthotomicus, Hypocryphalus, Hypothenemus, Pityogenes*
Central America	2690	1094	1113	483	24	28	*Pityophthorus, Ips, Gnathotrichus, Pseudopityophthorus, Dendroctonus*
Caribbean	9	0	8	1	7	1	*Hypothenemus, Pityophthorus, Xyleborus, Pseudothysanoes, Cryptocarenus*
Europe	3977	1155	324	2498	31	72	*Pityogenes, Ips, Orthotomicus, Hylurgops, Hylurgus*
North America	19	8	3	8	5	5	*Dendroctonus, Phloeosinus, Polygraphus, Scolytus, Dryocoetes*
Pacific	11	3	3	5	5	4	*Hylurgus, Hylurgops, Phloeosinus, Xyleborus, Crypturgus*
South America	209	54	117	38	22	11	*Hypothenemus, Xyleborus, Hylurgus, Coccotrypes, Pityophthorus*
Unknown[b]	83	27	18	38	18	16	*Pityogenes, Ips, Orthotomicus, Hylurgops, Scolytus*
Total	8286	2601	2239	3446	46	107	*Ips, Pityophthorus, Pityogenes, Orthotomicus, Hylurgops*

[a]See text for details on how countries were assigned to continents as well as Haack (2001).
[b]For some interceptions the country or likely country of origin was not listed. This can happen, for example, when a live insect is found walking freely inside a container and not associated closely with specific cargo. This is especially common when there is mixed cargo from multiple countries within a single container.

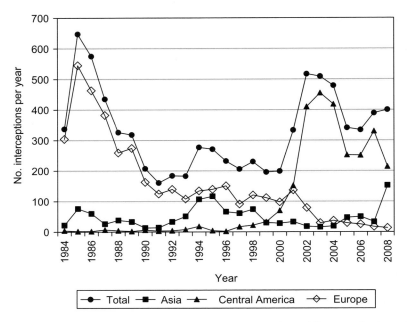

Fig. 3.1. Total number of annual wood-associated scolytine interception at US ports of entry and for selected continents of origin for the period 1984 through 2008.

Africa, 21 in Asia (which included Russia and Turkey), seven in Central America (including Mexico), four in the Caribbean, 29 in Europe, five in the Pacific region (including Australia, Hawai'i and New Zealand) and 11 in South America. There were 12 countries from which 100 or more interceptions were made from 1984 to 2008: Mexico (2581 interceptions), Italy (1322), Germany (715), Spain (533), China (428), Belgium (316), France (255), India (225), Turkey (180), Portugal (178), United Kingdom (137) and Russia (113). Additional details on scolytine interceptions by country can be viewed in Haack (2001) for the period 1985–2000.

3.5 Intercepted Scolytines by Receiving US State

Scolytines were intercepted on wood at ports of entry in 38 US states as well as in Puerto Rico and the US Virgin Islands. Puerto Rico and the US Virgin Islands serve as official US ports of entry for many foreign goods that are later shipped to the US mainland. Of the 8286 wood-associated scolytine interceptions, 5235 were made at maritime ports, 2572 at land borders with Canada and Mexico, 415 at airports, 55 at other US inspection stations, 6 at rail inspection centers, 2 at foreign sites (preclearance inspections) and 1 in Hawai'i during pre-departure inspection to the US mainland. Of the 2572 interceptions at land borders, only 19 (0.7%) were actually of Canadian origin, while 2534 (98.5%) were of Mexican origin, and the remainder (0.8%) were from other countries for products shipped to the USA through either Canada or Mexico.

There were 100 or more wood-associated scolytine interceptions made in 15 US states and Puerto Rico from 1984 to 2008, including Texas (3065 interceptions), Florida (890), Georgia (615), California (576), Louisiana (399), Ohio (340), South Carolina (339), Washington (272), New York (238), Kentucky (233), Maryland (208), North Carolina (191), Puerto Rico (160), New Jersey (131), Alabama (103) and Michigan (100). All of the states listed above have major maritime ports, international airports, or land border crossings with Canada or Mexico. Overall, most interceptions made along the US west coast were on goods shipped from Asia. By contrast, most interceptions made along the US east coast or at ports along the Great Lakes were of European origin, and most interceptions made along the US border with Mexico, including California, were of Central American origin, remembering that Mexico was classified as part of Central America in this analysis.

3.6 Genera of Intercepted Scolytines

Of the 8286 wood-associated scolytine interceptions, 5685 (69%) were identified to at least the genus level and represented 46 genera (Tables 3.2–3.3). Given that 31% of the interceptions were not identified beyond the family level it is likely that many more genera were intercepted. There were 14 genera with 100 or more interceptions from 1984 to 2008: *Ips* (917 interceptions), *Pityophthorus* (827), *Pityogenes* (618), *Orthotomicus* (591), *Hylurgops* (346), *Hylurgus* (291), *Gnathotrichus* (216), *Tomicus* (194), *Hypothenemus* (182), *Xyleborus* (181), *Dryocoetes* (166), *Phloeosinus* (160), *Hylastes* (135) and *Hypocryphalus* (115).

For each scolytine genus, the number of originating continents (world regions), countries and the top four originating countries are presented in Table 3.3. Individuals identified as *Xyleborus* were intercepted from seven continents, the most of any of the 46 identified genera, while two genera (*Phloeosinus* and *Pityophthorus*) were each intercepted from six continents. Eight genera were intercepted from 20 or more countries, including *Ips* (38 countries), *Pityogenes* (35), *Hypothenemus* (30), *Orthotomicus* (25), *Xyleborus* (25), *Hylurgops* (23), *Tomicus* (22) and *Dryocoetes* (21). A positive linear relationship was found between the number of interceptions for a given genus and the number of the originating countries ($F_{[1, 44]} = 57.7$, $P < 0.0001$, $R^2 = 0.57$).

The number of receiving US states and the number of years during the 25-year period from 1984 to 2008 that individuals of each scolytine genus were intercepted are provided in Table 3.3. *Ips* beetles were intercepted in the most US states (27) and specimens of five other genera were intercepted in 20 or more US states: *Orthotomicus* (25 US states), *Pityogenes* (25), *Dryocoetes* (22), *Hylurgops* (21) and *Hylurgus* (20) (Table 3.3). Five genera were intercepted during each year of the 25-year survey period (*Hylurgops*, *Hypothenemus*, *Ips*, *Pityogenes* and *Pityophthorus*), and individuals of three other genera were intercepted during 24 of the 25 years (*Hylurgus*, *Orthotomicus* and *Tomicus*) (Table 3.3).

3.7 Species of Intercepted Scolytines

Of the 8286 wood-associated scolytine interceptions, 3446 (42%) were identified to the species level and represented 107 species (Tables 3.2 and 3.4). As mentioned above for scolytine genera, many more species would likely have been identified as well, if more interceptions had been identified to the species level. Nevertheless, the species listed in Table 3.4 are probably representative of those commonly associated with wood packaging material used in international trade, but of course lack the scolytine species intercepted on US exports (Brockerhoff et al., 2006). The ten most commonly intercepted wood-associated scolytines were *Orthotomicus erosus* (Wollaston) (513 interceptions), *Pityogenes chalcographus* (Linnaeus) (512), *Ips typographus* (292), *Hylurgops palliatus* (Gyllenhal) (282), *Hylurgus ligniperda* (Fabricius) (239), *Ips sexdentatus* (Boerner) (189), *Tomicus piniperda* (Linnaeus) (159), *Ips integer* (Eichhoff) (82), *Hylastes ater* (Paykull) (67) and *Xyleborus eurygraphus* (Ratzeburg) (67) (Table 3.4). Of the 107 identified scolytine species, there were 15 species of *Ips*, 9 *Xyleborus*, 6 *Pityogenes*, 6 *Hylastes*, 5 *Scolytus*, and 4 each of *Crypturgus*, *Hylurgops*, *Orthotomicus* and *Polygraphus*. We recognize that several taxonomic changes have occurred in recent years for some of the species listed in Table 3.4; however, for the sake of consistency, we are presenting the species names as recorded by USDA APHIS.

As mentioned in Haack (2001), interception databases often include records that suggest range expansion of a species. The PestID database suggests several cases of range expansion if we consider the range data in Wood and Bright (1992) and Wood (2007) complete, and that there were no errors in identification or data entry. For example, there were interception records in the USA for several scolytine species originating from countries that were outside their published geographic ranges, including *Dendroctonus frontalis* Zimmermann from Turkey (1 interception), *Hylastes attenuatus* Erichson from South Africa (2), *Hylurgops palliatus* from Honduras (1) and Venezuela (1), *Hylurgus ligniperda* from Venezuela (1), *Phloeosinus rudis* Blandford from Belgium (1),

Table 3.3. Summary data by genus for the 8286 wood-associated interception records of scolytines intercepted at US ports of entry from 1984 to 2008.

Scolytine genus	Number of interceptions Total	Identified to species level	Number of continents	Number of countries	Top four countries of origin in decreasing order[a]	Number of receiving US states	Number of years intercepted from 1984 to 2008
Araptus	2	0	1	2	BR, PE	2	2
Carphoborus	38	18	3	8	TR, ES, IT, BE	13	16
Chaetophloeus	3	3	1	1	MX	3	3
Cnemonyx	2	0	2	2	PE, TT	2	2
Coccotrypes	17	0	5	6	BR, IT, VE, CR	6	11
Cryphalus	57	7	4	13	CN, IT, IN, PH	11	16
Cryptocarenus	1	0	1	1	TT	1	1
Crypturgus	66	36	5	15	PT, ES, DE, IT	14	18
Cyrtogenius	20	10	1	6	CN, KR, SG, HK	9	11
Dendroctonus	54	32	3	3	MX, CA, TR	5	10
Dryocoetes	166	47	4	21	CN, IT, DE, BE	22	23
Euwallacea	6	6	1	2	CN, IN	3	5
Gnathotrichus	216	80	2	8	MX, IT, FR, HN	7	15
Hylastes	135	122	5	15	ES, IT, FR, DE	18	22
Hylesinus	16	11	3	6	UK, BE, UA, IT	7	13
Hylocurus	1	1	1	1	MX	1	1
Hylurgopinus	4	2	2	2	MX, IT	2	3
Hylurgops	346	301	5	23	DE, IT, BE, MX	21	25
Hylurgus	291	240	5	16	IT, PT, ES, CL	20	24
Hypocryphalus	115	4	4	10	IN, BR, TW, MY	16	18
Hypothenemus	182	2	5	30	IN, BR, VE, CN	13	25
Ips	917	808	5	38	MX, IT, DE, ES	27	25
Monarthrum	13	0	2	3	MX, PE, BR	2	4
Orthotomicus	591	552	3	25	IT, ES, CN, TR	25	24

Pagiocerus	2	0	2	BR, IN	2	2
Phloeosinus	160	16	6	CN, JP, MX, CA	17	20
Phloeotribus	4	2	2	JO, ES, IL	4	4
Pityogenes	618	591	4	DE, IT, ES, BE	25	25
Pityokteines	18	15	1	IT, FR, GR, DE	7	10
Pityophthorus	827	20	6	MX, BR, ZA, IT	19	25
Polygraphus	66	49	4	IT, DE, CN, RU	17	20
Pseudohylesinus	3	3	1	MX	1	2
Pseudopityophthorus	99	2	1	MX	1	10
Pseudothysanoes	1	0	1	DO	0	1
Pteleobius	1	1	1	IT	1	1
Scolytodes	2	0	2	DE, BR	2	2
Scolytogenes	1	0	1	ID	1	1
Scolytoplatypus	3	0	2	CN, PT	2	2
Scolytus	98	20	5	BE, CN, IT, FR	19	22
Taphrorychus	64	35	4	BE, DE, FR, TR	17	17
Tomicus	194	166	3	IT, FR, ES, UK	19	24
Trypodendron	38	29	2	IT, TR, DE, FR	10	17
Xyleborinus	17	11	2	CN, IT, KR, GR	7	8
Xyleborus	181	120	7	IT, CN, TR, MX	19	20
Xylechinus	12	2	2	IN, IT	6	6
Xylosandrus	17	17	4	CN, BE, IT, NG	10	10
Scolytidae[b]	2601	0	7	BE, CN, IT, FR	33	25
Total	8286	3446	8	MX, IT, DE, ES	38	25

[a]Country codes: AT Austria, AU Australia, AZ Azerbaijan, BE Belgium, BR Brazil, BZ Belize, CA Canada, CL Chile, CN China, CR Costa Rica, DE Germany, DO Dominican Republic, EC Ecuador, EE Estonia, ES Spain, FI Finland, FR France, GT Guatemala, GR Greece, GY Guyana, HK Hong Kong, HN Honduras, HR Croatia, ID Indonesia, IL Israel, IN India, IT Italy, JP Japan, JO Jordan, KR South Korea, LT Lithuania, LV Latvia, MX Mexico, MY Malaysia, NG Nigeria, NL Netherlands, PE Peru, PH Philippines, PL Poland, PT Portugal, RO Romania, RU Russia, SE Sweden, SG Singapore, SK Slovakia, TR Turkey, TT Trinidad and Tobago, TW Taiwan, UA Ukraine, UK United Kingdom, VE Venezuela, VN Vietnam, ZA South Africa.

[b]The interception records labeled Scolytidae were only identified to the family level.

Table 3.4. Summary data by species for the 3446 wood-associated interception records of scolytines that were intercepted at USA ports of entry from 1984 to 2008 and identified to the species level.

Species	No. interceptions	No. continents	No. countries	Top five originating countries in decreasing order[a]	No. receiving US states
Carphoborus bifurcus Eichhoff	1	1	1	MX	1
Carphoborus minimus (Fabricius)	11	2	4	ES, TR, IT, BE	8
Carphoborus pini Eichhoff	5	1	2	ES, IT	3
Carphoborus rossicus (Semenov)	1	1	1	DE	1
Chaetophloeus mexicanus Blackman	3	1	1	MX	3
Cryphalus abietis (Ratzeburg)	2	1	2	DE, IT	1
Cryphalus piceae (Ratzeburg)	5	1	2	IT, FR	4
Crypturgus cinereus (Herbst)	11	3	5	DE, ES, RU, BE, AU	6
Crypturgus mediterraneus Eichhoff	19	1	5	PT, ES, IT, FR, NL	8
Crypturgus numidicus Ferrari	4	1	2	ES, GR	4
Crypturgus pusillus (Gyllenhal)	2	2	2	CN, LV	2
Cyrtogenius luteus (Blandford)	10	1	4	CN, KR, SG, VN	6
Dendroctonus frontalis Zimm.	2	2	2	MX, TR	2
Dendroctonus mexicanus Hopkins	28	1	1	MX	3
Dendroctonus pseudotsugae Hopkins	2	1	1	CA	1
Dryocoetes autographus (Ratzeburg)	29	3	11	IT, BE, ES, BR, HR	14
Dryocoetes hectographus Reitter	1	1	1	IT	1
Dryocoetes villosus (Fabricius)	17	1	5	BE, DE, UK, FR, IT	9
Euwallacea andamanensis (Blandford)	1	1	1	IN	1
Euwallacea validus (Eichhoff)	5	1	1	CN	2
Gnathotrichus denticulatus Blackman	3	1	1	MX	1
Gnathotrichus materiarius (Fitch)	25	2	5	FR, IT, ES, DE, GT	6
Gnathotrichus sulcatus Leconte	52	1	1	MX	3

Continued

Table 3.4. Continued.

Species	No. interceptions	No. continents	No. countries	Top five originating countries in decreasing order[a]	No. receiving US states
Hylastes angustatus (Herbst)	4	1	2	BE, FR	3
Hylastes ater (Paykull)	67	2	10	ES, IT, FR, DE, UK	14
Hylastes attenuatus Erichson	23	2	7	ES, IT, PT, FR, ZA	8
Hylastes cunicularius Erichson	7	1	4	IT, DE, UK, BE	5
Hylastes linearis Erichson	10	1	3	ES, PT, IT	3
Hylastes opacus Erichson	1	1	1	PL	1
Hylesinus crenatus Fabricius	1	1	1	UK	1
Hylesinus varius (Fabricius)	10	1	3	UK, BE, IT	5
Hylurgopinus rufipes (Eichhoff)	2	2	2	MX, IT	2
Hylurgops glabratus (Zetterstedt)	3	2	2	IT, RU	2
Hylurgops incomptus (Blandford)	9	1	1	MX	1
Hylurgops palliatus (Gyllenhal)	282	5	21	DE, IT, BE, UK, ES	21
Hylurgops planirostris (Chapuis)	7	1	1	MX	1
Hylurgus ligniperda (Fabricius)	239	4	15	IT, PT, ES, FR, CL	16
Hylurgus micklitzi Wachtl	1	1	1	IT	1
Hypocryphalus mangiferae (Stebbing)	4	2	2	IN, BR	3
Hypothenemus birmanus (Eichhoff)	1	1	1	SG	1
Hypothenemus obscurus (Fabricius)	1	1	1	PT	1
Ips acuminatus (Gyllenhal)	40	2	9	TR, IT, ES, FR, CN	11
Ips amitinus (Eichhoff)	2	1	2	IT, FI	2
Ips apache Lanier	9	1	2	HN, MX	4
Ips bonanseai (Hopkins)	27	1	1	MX	3
Ips calligraphus (Germar)	13	1	3	MX, HN, GT	4
Ips cembrae (Heer)	9	2	4	IT, DE, TW, BE	5
Ips cribricollis (Eichhoff)	39	1	3	MX, HN, GT	4
Ips grandicollis (Eichhoff)	14	1	2	MX, GT	3
Ips integer (Eichhoff)	82	1	1	MX	3
Ips lecontei Swaine	43	1	2	MX, GT	4

Continued

Table 3.4. Continued.

Species	No. interceptions	No. continents	No. countries	Top five originating countries in decreasing order[a]	No. receiving US states
Ips mannsfeldi (Wachtl)	7	2	2	ES, TR	4
Ips mexicanus (Hopkins)	5	1	1	MX	1
Ips pini (Say)	37	1	1	MX	2
Ips sexdentatus (Boerner)	189	2	12	IT, ES, FR, TR, PT	19
Ips typographus (Linnaeus)	292	2	25	IT, DE, BE, RO, RU	22
Orthotomicus erosus Wollaston	513	3	20	IT, ES, TR, CN, PT	24
Orthotomicus laricis (Fabricius)	29	2	10	IT, FR, DE, ES, PL	12
Orthotomicus proximus (Eichhoff)	2	1	2	IT, FI	2
Orthotomicus suturalis (Gyllenhal)	8	1	3	UK, DE, EE	4
Phloeosinus canadensis Swaine	2	1	1	CA	2
Phloeosinus rudis Blandford	14	2	2	JP, BE	7
Phloeotribus scarabaeoides (Bernard)	2	2	2	ES, JO	2
Pityogenes bidentatus (Herbst)	23	2	7	ES, FR, DE, IT, SE	9
Pityogenes bistridentatus (Eichhoff)	42	2	7	ES, TR, IT, FR, UK	9
Pityogenes calcaratus (Eichhoff)	4	1	3	ES, FR, IT	4
Pityogenes chalcographus (Linnaeus)	512	3	31	DE, IT, BE, RU, ES	25
Pityogenes quadridens (Hartig)	9	2	5	FI, TR, ES, LT, PT	7
Pityogenes trepanatus (Nordlinger)	1	1	1	LT	1
Pityokteines curvidens (Germar)	3	1	3	FR, IT, GR	3
Pityokteines spinidens (Reitter)	12	1	4	IT, FR, AT, DE	5
Pityophthorus mexicanus Blackman	6	1	1	MX	1
Pityophthorus pityographus (Ratzeburg)	14	2	5	IT, DE, MX, NL, FR	8
Polygraphus poligraphus (Linnaeus)	44	2	10	IT, DE, CN, UK, BE	14
Polygraphus proximus Blanchard	1	1	1	IT	1
Polygraphus rufipennis (Kirby)	2	1	1	CA	2

Continued

Table 3.4. Continued.

Species	No. interceptions	No. continents	No. countries	Top five originating countries in decreasing order[a]	No. receiving US states
Polygraphus subopacus Thomson	2	2	2	IT, AZ	1
Pseudohylesinus variegatus (Blandford)	3	1	1	MX	1
Pseudopityophthorus pruinosus (Eichhoff)	1	1	1	MX	1
Pseudopityophthorus yavapaii Blackman	1	1	1	MX	1
Pteleobius vittatus Fabricius	1	1	1	IT	1
Scolytus intricatus (Ratzeburg)	12	1	4	BE, FR, IT, DE	8
Scolytus multistriatus (Marsham)	2	2	2	IT, CA	2
Scolytus ratzeburgi Janson	1	1	1	FI	1
Scolytus schevyrewi Semenov	4	2	2	CN, CA	4
Scolytus scolytus (Fabricius)	1	1	1	UK	1
Taphrorychus bicolor (Herbst)	21	1	5	BE, DE, FR, FI, NL	10
Taphrorychus villifrons (Dufour)	14	2	5	BE, DE, FR, LV, TR	9
Tomicus minor (Hartig)	7	2	3	TR, BR, IT	3
Tomicus piniperda (Linnaeus)	159	2	21	IT, FR, ES, UK, DE	19
Trypodendron domesticum (Linnaeus)	10	2	4	IT, TR, BE, FR	4
Trypodendron lineatum (Olivier)	12	2	7	IT, TR, CH, SK, IL	4
Trypodendron signatum (Fabricius)	7	1	4	DE, FR, PO, BE	4
Xyleborinus saxesenii (Ratzeburg)	11	2	7	IT, CN, GR, TR, KR	6
Xyleborus affinis Eichhoff	22	6	5	GY, MX, ID, HN, BR	8
Xyleborus apicalis Blandford	1	1	1	MY	1
Xyleborus dispar (Fabricius)	2	1	1	IT	1
Xyleborus eurygraphus (Ratzeburg)	67	2	4	IT, TR, ES, GR	9
Xyleborus ferrugineus (Fabricius)	2	2	2	BR, BZ	2
Xyleborus glabratus Eichhoff	1	1	1	CN	1
Xyleborus intrusus Blandford	9	1	3	HN, NI, MX	3

Continued

Table 3.4. Continued.

Species	No. interceptions	No. continents	No. countries	Top five originating countries in decreasing order[a]	No. receiving US states
Xyleborus similis Ferrari	1	1	1	VN	1
Xyleborus volvulus (Fabricius)	15	4	11	PE, GY, HN, VE, EC	7
Xylechinus pilosus (Ratzeburg)	2	1	1	IT	1
Xylosandrus crassiusculus (Motschulsky)	10	3	4	CN, IT, NG, ID	8
Xylosandrus germanus (Blandford)	6	3	5	JP, PE, DE, BE, IT	5
Xylosandrus morigerus (Blandford)	1	1	1	BE	1

[a]Country codes for Tables: AT Austria, AU Australia, AZ Azerbaijan, BE Belgium, BR Brazil, BZ Belize, CA Canada, CL Chile, CN China, CR Costa Rica, DE Germany, DO Dominican Republic, EC Ecuador, EE Estonia, ES Spain, FI Finland, FR France, GT Guatemala, GR Greece, GY Guyana, HK Hong Kong, HN Honduras, HR Croatia, ID Indonesia, IL Israel, IN India, IT Italy, JP Japan, JO Jordan, KR South Korea, LT Lithuania, LV Latvia, MX Mexico, MY Malaysia, NG Nigeria, NL Netherlands, PE Peru, PH Philippines, PL Poland, PT Portugal, RO Romania, RU Russia, SE Sweden, SG Singapore, SK Slovakia, TR Turkey, TT Trinidad and Tobago, TW Taiwan, UA Ukraine, UK United Kingdom, VE Venezuela, VN Vietnam, ZA South Africa.

Pityogenes chalcographus from Brazil (1) and *Tomicus minor* (Hartig) from Brazil (1). Although the above interceptions should not be considered authoritative records that these species are established in these countries, it is of interest to note that Kirkendall and Faccoli (2010) recently reported that *Phloeosinus rudis* is established in France and The Netherlands.

The number of originating continents and countries for each of the 107 identified scolytine species intercepted in the USA, as well as the top five originating countries, are presented in Table 3.4. There were six countries from which 20 or more wood-associated scolytine species were intercepted and identified: Italy (50 species), France (29), Germany (28), Spain (28), Belgium (24) and Mexico (24).

The number of US states in which each of the 107 identified scolytines were intercepted is presented in Table 3.4. Overall, there was a significant and positive linear relationship between the number of interceptions for a given scolytine species and the number of the originating countries ($F_{[1,105]} = 300.7$, $P < 0.0001$, $R^2 = 0.74$), and similarly between the number of interceptions and the number of receiving US states ($F_{[1,105]} = 212.5$, $P < 0.0001$, $R^2 = 0.67$) (Table 3.5). There were seven US states where 25 or more species were identified among the intercepted scolytines: Texas (67 species), Florida (55), California (40), Louisiana (33), Georgia (30), South Carolina (30) and New York (25).

3.8 Currently Established Exotic Scolytines

As of 2010, we are aware of 58 scolytine species in 27 genera that are established in the continental USA (Table 3.5). Many of the species listed in Table 3.5, especially in the Xyleborini, had been transferred to new genera since publication of the world catalog by Wood and Bright (1992), including *Ambrosiophilus atratus* (Eichhoff), *Anisandrus dispar* (Fabricius), *A. maiche* Stark, *Cnestus mutilatus* (Blandford), *Cyclorhipidion californicus* (Wood), *C. pelliculosum* (Eichhoff), *Dryoxylon onoharaensum* (Murayama), *Wallacellus similis* (Ferrari), *Xyleborinus octiesdentatus* (Murayama) and *Xylosandrus amputatus* (Blandford). In Wood and Bright (1992), all of the above species were formerly members of the genus *Xyleborus* except *Cnestus mutilatus* (formerly in *Xylosandrus*), and *X. amputatus* (formerly in *Amasa*). The pine shoot beetle, *Tomicus piniperda*, is the only scolytine of

Table 3.5. Summary data for the 58 exotic scolytines recognized as being established in the continental United States as of 2010.

Tribe	Species	Feeding guild[a]	Record or report of first collection Year	Record or report of first collection State[b]	Native range	No. US states where found as of 2010[c]	Reference of first USA collection record[d]
Hylastini							
	Hylastes opacus Erichson	BB	1987	NY	Eurasia	17	37
	Hylurgops palliatus (Gyllenhal)[e]	BB	2001	PA	Eurasia	3	14
Hylesinini							
	Hylastinus obscurus (Marsham)	R	1878	NY	Europe	20	29
Tomicini							
	Hylurgus ligniperda (Fabricius)	BB	1994	NY	Eurasia	2	13
	Tomicus piniperda (Linnaeus)	BB	1991	MI	Eurasia	18	10
Phloeosinini							
	Phloeosinus armatus Reitter	BB	1992	CA	Asia	1	37
Scolytini							
	Scolytus mali (Bechstein)	BB	1868	NY	Europe	10	18
	Scolytus multistriatus (Marsham)	BB	1909	MA	Europe	48	6
	Scolytus rugulosus (Muller)	BB	1877	NY	Europe	48	7
	Scolytus schevyrewi Semenov[e]	BB	1994	CO	Asia	29	23
Ipini							
	Orthotomicus erosus (Wollaston)	BB	2004	CA	Eurasia	1	20
	Pityogenes bidentatus (Herbst)	BB	1988	NY	Europe	2	11
Dryocoetini							
	Coccotrypes advena Blandford	ST	1956	FL	Asia	1	36
	Coccotrypes carpophagus (Hornung)	ST	1926	FL	Africa	2	36
	Coccotrypes cyperi (Beeson)	ST	1934	LA	Asia	2	36
	Coccotrypes dactyliperda (Fabricius)	ST	1915	AZ, FL	Africa	4	16
	Coccotrypes distinctus (Motschulsky)	ST	1939	FL	Asia	2	36
	Coccotrypes rhizophorae (Hopkins)	ST	1910	FL	Asia	1	16
	Coccotrypes rutschuruensis Eggers	ST	1992	CA	Africa	1	38
	Coccotrypes vulgaris (Eggers)	ST	1985	FL	Asia	1	3
	Dactylotrypes longicollis (Wollaston)	ST	2009	CA	Europe	1	17
	Dryoxylon onoharaensum (Murayama)[f,g]	AB	1982	LA	Asia	18	5
							Continued

Table 3.5. Continued.

Tribe	Species	Feeding guild[a]	Record or report of first collection Year	Record or report of first collection State[b]	Native range	No. US states where found as of 2010[c]	Reference of first USA collection record[d]
Crypturgini							
	Crypturgus pusillus (Gyllenhal)	BB	1868	NY	Eurasia	9	18
Xyloterini							
	Trypodendron domesticum (Linnaeus)	AB	2008	WA	Europe	1	22
Premnobini							
	Premnobius cavipennis Eichhoff	AB	1939	FL	Africa	1	4
Xyleborini							
	Ambrosiodmus lewisi (Blandford)	AB	1990	PA	Asia	1	12
	Ambrosiodmus rubricollis (Eichhoff)	AB	1942	MD	Asia	21	4
	Ambrosiophilus atratus (Eichhoff)	AB	1988	TN	Asia	27	2
	Anisandrus dispar (Fabricius)	AB	1817	MA	Europe	24	25
	Anisandrus maiche Stark	AB	2005	PA	Asia	3	27
	Cnestus mutilatus (Blandford)[h]	AB	1999	MS	Asia	8	30
	Cyclorhipidion californicus (Wood)[h]	AB	1944	CA	Asia	28	34
	Cyclorhipidion pelliculosum (Eichhoff)[h]	AB	1987	PA	Asia	18	2
	Euwallacea fornicatus (Eichhoff)	AB	2002	FL	Asia	2	31
	Euwallacea validus (Eichhoff)	AB	1976	NY	Asia	23	36
	Wallacellus similis (Ferrari)[e,h]	AB	2002	TX	Asia	2	26
	Xyleborinus alni (Niisima)	AB	1996	WA	Eurasia	13	21
	Xyleborinus andrewesi (Blandford)	AB	2009	FL	Asia	1	25
	Xyleborinus octiesdentatus (Murayama)[e]	AB	2008	LA	Asia	2	28
	Xyleborinus saxesenii (Ratzeburg)	AB	1911	CA	Eurasia	43	16
	Xyleborus glabratus Eichhoff[e]	AB	2002	GA	Asia	4	26
	Xyleborus pfeilii (Ratzeburg)	AB	1992	MD	Eurasia	6	32
	Xyleborus seriatus Blandford[e]	AB	2005	MA	Asia	3	15
	Xylosandrus amputatus (Blandford)[e]	AB	2010	FL	Asia	1	8
	Xylosandrus compactus (Eichhoff)	AB	1941	FL	Asia	10	36
	Xylosandrus crassiusculus (Motschulsky)	AB	1974	SC	Asia	29	1
	Xylosandrus germanus (Blandford)	AB	1931	NY	Asia	32	9

Cryphalini

Species	Guild	Year	State	Origin	Refs	
Hypocryphalus mangiferae (Stebbing)	BB	1949	FL	Asia	1	33
Hypothenemus areccae (Hornung)	ST	1960	FL	Asia	1	36
Hypothenemus birmanus (Eichhoff)	ST	1951	FL	Asia	1	33
Hypothenemus brunneus (Hopkins)	ST	1904	TX	Africa	3	16
Hypothenemus californicus Hopkins	ST	1895	CA	Africa	14	16
Hypothenemus columbi Hopkins	ST	1881	SC	Africa	6	16
Hypothenemus crudiae (Panzer)	ST	1868	GA, LA	Asia	20	18
Hypothenemus erectus LeConte	ST	1876	TX	Africa	1	19
Hypothenemus javanus (Eggers)	ST	1977	FL	Africa	1	35
Hypothenemus obscurus (Fabricius)	ST	1977	FL	S America	1	35
Hypothenemus setosus (Eichhoff)	ST	1982	FL	Africa	1	36

[a] Feeding guilds: AB = ambrosia beetles, feed primarily on fungus (xylomycetophagous), BB = bark beetles, feed primarily on inner bark or phloem (phloeophagous), R = roots of herbaceous plants, e.g. Trifolium, (herbiphagous), ST = most are seed and twig beetles, feeding in seeds (spermophagous), or the pith (myelophagous) or phloem of twigs.
[b] US state abbreviations: AZ = Arizona, CA = California, CO = Colorado, FL = Florida, GA = Georgia, LA = Louisiana, MA = Massachusetts, MD = Maryland, MI = Michigan, MS = Mississippi, NY = New York, PA = Pennsylvania, SC = South Carolina, TN = Tennessee, TX = Texas, and WA = Washington.
[c] Number of US states where reported in literature or found in the USDA Forest Service Early Detection & Rapid Response surveys as of 2010, exclusive of Hawai'i.
[d] References: 1 = Anderson (1974), 2 = Atkinson et al. (1990), 3 = Atkinson et al. (1991), 4 = Bright (1968), 5 = Bright and Rabaglia (1999), 6 = Chapman (1910), 7 = Chittenden (1898), 8 = Cognato et al. (2011), 9 = Felt (1932), 10 = Haack and Poland (2001), 11 = Hoebeke (1989), 12 = Hoebeke (1991), 13 = Hoebeke (2001), 14 = Hoebeke and Acciavatti (2006), 15 = Hoebeke and Rabaglia (2008), 16 = Hopkins (1915), 17 = LaBonte (unpublished data), 18 = LeConte (1868), 19 = LeConte (1876), 20 = Lee et al. (2005), 21 = Mudge et al. (2001), 22 = NAPIS (2008), 23 = Negrón et al. (2005), 24 = Okins and Thomas (2010), 25 = Peck (1817), 26 = Rabaglia et al. (2006), 27 = Rabaglia et al. (2009), 28 = Rabaglia et al. (2010), 29 = Riley (1879), 30 = Schiefer and Bright (2004), 31 = Thomas (2004), 32 = Vandenberg et al. (2000), 33 = Wood (1954), 34 = Wood (1975), 35 = Wood (1977), 36 = Wood (1982), 37 = Wood (1992), 38 = Wood and Bright (1992).
[e] These species were first detected in the USA by the USDA Forest Service Early Detection & Rapid Response project.
[f] Bright and Rabaglia (1999) recorded this species from Delaware in 1977; the correct Delaware record is from 1997.
[g] These three species names have commonly been misspelled in the scientific literature. The current correct spellings are: *Xylosandrus mutilatus* (Dole and Cognato, 2010); *Xyleborus californicus* not *X. saxeseni*; and *Xyleborus pfeili* not *Xyleborus pfeili*.
[h] Recent taxonomic changes have affected the nomenclature of these species as follows: *Xylosandrus mutilatus* to *Cnestus mutilatus* (Dole and Cognato, 2010); *Xyleborus californicus* and *X. pelliculosus* to *Cyclorhipidion*; *Xyleborus similis* to *Wallacellus similis* (Hulcr and Cognato, 2010).

the 58 exotics for which there is a USDA federal quarantine that regulates movement of pine host material to areas outside the known current US range of this bark beetle (Haack and Poland, 2001). This US federal quarantine was enacted soon after discovery of *T. piniperda* in 1992 and it is still in effect as of 2010, even though this bark beetle has caused minimal damage in the currently infested portions of the eastern USA.

Overall, 19 of the 107 intercepted scolytines listed in Table 3.4, and an additional two intercepted species listed in Haack (2001), which included US interceptions for scolytines on all products, are among the 58 exotic scolytines now established in the continental USA (Table 3.5). The 21 species, listed in the order they appear in Table 3.5, include: *Hylastes opacus* Erichson, *Hylurgops palliatus*, *Hylurgus ligniperda*, *Tomicus piniperda*, *Scolytus multistriatus*, *S. schevyrewi* Semenov, *Orthotomicus erosus*, *Pityogenes bidentatus* (Herbst), *Coccotrypes carpophagus* (Hornung) (in Haack, 2001), *Dactylotrypes longicollis* (Wollaston) (in Haack, 2001), *Crypturgus pusillus* (Gyllenhal), *Trypodendron domesticum* (Linnaeus), *Euwallacea validus* (Eichhoff), *Xyleborinus saxesenii* (Ratzeburg), *Xyleborus glabratus*, *Wallacellus similis* (Ferrari), *Xylosandrus crassiusculus* (Motschulsky), *X. germanus* (Blandford), *Hypocryphalus mangiferae* (Stebbing), *Hypothenemus birmanus* (Eichhoff) and *H. obscurus* (Fabricius). Undoubtedly, if more of the intercepted scolytines had been identified to the species level (Tables 3.2, 3.4), then more of the scolytine species listed as established in the USA (Table 3.5) would have been found among the intercepted scolytines. This is especially true for species of *Coccotrypes* and *Hypothenemus*, which were seldom identified beyond genus in the PestID database. For example, only 3 of 502 *Coccotrypes* interceptions and 63 of 821 *Hypothenemus* interceptions were identified to species during 1985–2000 (Haack, 2001). For those scolytines considered true bark beetles, a significant positive relationship was found between interception frequency and likelihood of establishment (Brockerhoff *et al.*, 2006; Haack, 2006).

The species *Hypothenemus africanus* (Hopkins) was included by Haack (2001) in an earlier list of exotic scolytines that were considered to be established in the continental USA. However, even though the type specimen of *H. africanus* was first described from South Africa (Hopkins, 1915), it is now considered more likely that this species is native to the Americas given its current worldwide geographic distribution (Wood and Bright, 1992).

Kirkendall and Faccoli (2010) recently reported that there are 18 species of exotic scolytines established in Europe, of which seven are also established in the USA: *Ambrosiodmus rubricollis* (Eichhoff), *Ambrosiophilus atratus*, *Coccotrypes dactyliperda* (Fabricius), *Dactylotrypes longicollis*, *Xyleborus pfeilii* (Ratzeburg), *Xylosandrus crassiusculus* and *X. germanus*. In addition, Kirkendall and Faccoli (2010) list *Cyclorhipidion bodoanum* (Reitter) as established in Europe, and state that this species will be synonymized with *C. californicus*, which is established in the USA (Table 3.5). Once this occurs there will be eight exotic scolytines in common between Europe and the USA. Note that we listed Europe as the native range of *D. longicollis* in Table 3.5 given that this species is native to the Canary Islands, which are an autonomous territory of Spain. However, Kirkendall and Faccoli (2010) treat this species as an exotic to continental Europe.

Of the 58 exotic scolytines in the USA, one – *Hylastinus obscurus* (Marsham) – breeds in roots of herbaceous legumes, while the others breed primarily in woody plants and palms. We categorized 25 of the 58 scolytines as ambrosia beetles (43%), 13 as bark beetles (22%), 19 as seed and twig feeders (33%) and one as a root feeder of herbaceous plants (2%; Table 3.5). In general, considering the 58 species listed in Table 3.5, the ambrosia beetles in the tribes Premnobini and Xyleborini are polyphagous, inbreeding species and breed primarily in broadleaf woody plants; bark beetles are outbreeding with the species of Hylastini, Ipini, Phloeosinini, and Tomicini breeding in conifers while the Scolytini breed in broadleaf woody plants; *Coccotrypes* species are inbreeders and reproduce mostly in the seeds of palms and broadleaf woody plants; and *Hypothenemus* species are inbreeders, generally polyphagous, and reproduce in seeds and twigs of broadleaf woody plants (Wood, 1982; Wood and Bright, 1992; Jordal *et al.*, 2002). Considering the feeding guilds of the 58 exotic scolytines and their interception histories (Haack, 2001), it is likely that some of the ambrosia beetles and most seed and twig feeders arrived in association with seeds, fruit, cuttings and live plants, whereas the remaining ambrosia beetles and nearly all bark beetles arrived in association with wood packaging material, logs, lumber and perhaps fuel wood. In the case of fuel wood, for example, from

1996 to 2009 the USA imported fuel wood from 34 countries in Africa, Asia, the Caribbean, Central America, Europe, North America (Canada) and South America, with a total declared value on arrival of >US$98 million for the 14-year period (Haack et al., 2010b). Overall, the majority of exotic scolytines in both the USA and Europe are inbreeding, polyphagous species, which are traits that favor invasiveness (Kirkendall and Faccoli, 2010).

We attempted to locate in the scientific literature the earliest year of collection or the first year of reporting for each of the 58 exotic scolytines. In addition, we tried to identify the US state in which the first collection of each species was made. We recognize that nearly all exotic insects are present for many years before they are first collected, and similarly that the founding population of each exotic may enter the USA at a location that is very distant from where it is later first collected. For two species, *Coccotrypes dactyliperda* and *Hypothenemus crudiae* (Panzer), we were not able to determine the US state in which the first collections were made. In both cases, the first reports listed two US states in which each was collected but did not provide any dates of collection (LeConte, 1868; Hopkins, 1915). Nevertheless, we feel that such data can be used to approximate the historical arrival rate and the areas of the country where the founding population first became established.

Seven of the exotic scolytines first reported since 2000 were discovered as part of the United States Department of Agriculture, Forest Service 'Early Detection and Rapid Response' (EDRR) project (Table 3.5; Rabaglia et al., 2008). The EDRR project was initiated as a pilot project in 2001 and in 2007 the program was expanded to full implementation. Approximately 15 different US states now participate in the survey each year, which utilizes pheromone- and karimone-baited traps at 12 sites per state (Rabaglia et al., 2008). The relatively steep rate of new scolytine detections since 2000 is in part a reflection of the increased survey efforts such as EDRR. Given the success of the EDRR project in locating previously unknown exotic scolytines, as well as increasing public awareness and surveys by other groups, it is likely that more exotics will be found in the years ahead.

Anisandrus dispar, formerly in the genus *Xyleborus*, was the earliest recorded exotic scolytine in the continental USA (Peck, 1817; Table 3.5). Overall, about 16% ($n = 9$) of the 58 exotic scolytines were first collected or reported during the 1800s, 24% (14) during 1900–1949, 40% (23) during 1950–1999 and 21% (12) during 2000–2010. Taken as a whole, the species accumulation curve for exotic scolytines in the USA has shown accelerated growth over the past two centuries (Fig. 3.2), reflecting a similar steep increase in US imports (Haack, 2001; US Census

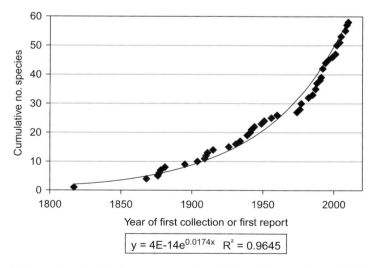

$y = 4E\text{-}14e^{0.0174x}$ $R^2 = 0.9645$

Fig. 3.2. Cumulative number of detections of new exotic scolytines in the continental USA by year of first collection or first published report. Data are from Table 3.5. Exponential equation fitted to the data.

Bureau, 2011). Similarly, accelerating species accumulation curves have been documented for exotic phloem and wood borers in the USA (Aukema et al., 2010) as well as for exotic scolytines in Europe (Kirkendall and Faccoli, 2010). If we consider the 57 exotic scolytines on palms and woody plants, and the three major feeding guilds they represent, a pattern emerged where most early invaders were seed and twig feeders while most recent invaders were ambrosia beetles (Table 3.5).

The 58 exotic scolytines were first reported from 16 US states (Fig. 3.3; Table 3.5). As expected, 14 of these 16 states have major maritime or land-based ports of entry given that they border oceans, the Great Lakes or neighboring countries. The five states where the most exotic scolytines were first found were Florida (17 species), New York (8), California (7), Louisiana (4) and Pennsylvania (4). There were only two interior US states where new exotics were first reported: *Ambrosiophilus atratus* in Tennessee and *Scolytus schevyrewi* in Colorado (Table 3.5); however, it is likely that Colorado and Tennessee were not the initial states where these two species first became established, given that each scolytine was found in several states when first reported (Atkinson et al., 1990; Negrón et al., 2005). When considering the US states in which the 58 exotic scolytines were first reported, along with their assigned feeding guild, there was a tendency for ambrosia beetles to establish in the south-eastern USA, bark beetles in the northeast, and seed and twig beetles along the southern tier of states from California to Florida (Fig. 3.3). Overall, Florida was the state in which most exotic ambrosia beetles (5 of 25), as well as seed and twig beetles (11 of 19) were first found,

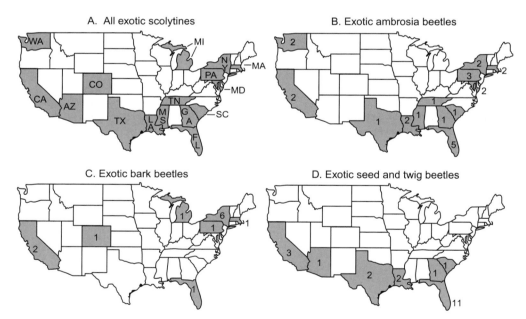

Fig. 3.3. Maps of the contiguous 48 US states indicating the states (shaded) where the established exotic scolytines were first reported in the continental USA by feeding guild. Map A indicates the 16 US states where one or more of the 58 exotic scolytines now established in the USA were first reported, and also provides the state abbreviations as given in Table 3.5. Similarly, maps are presented for the first US state of detection for the 25 exotic scolytines classified as ambrosia beetles (Map B), the 13 species of exotic bark beetles (Map C), and the 19 seed and twig feeders (Map D). Data are from Table 3.5. Numbers within each US state (Maps B–D) indicate the number of exotic scolytines that were first found in each state by feeding guild. For the two exotic seed and twig feeders (*Coccotrypes dactyliperda* and *Hypothenemus crudiae*) that were each initially reported from two US states (Table 3.5), we credited each state for each scolytine. No map is shown for *Hylastinus obscurus*, which is a root feeder of herbaceous legumes that was first found in New York.

whereas 6 of the 13 exotic bark beetles were first found in New York (Fig. 3.3). These geographic patterns may reflect historical trade routes for specific imports and their typical scolytine associates, or they may indicate strict biological (hosts) and environmental (temperature and humidity) conditions that some scolytines require for successful establishment. For example, it is logical that many of the exotic *Coccotrypes* species would be restricted to the relatively warm southern tier of states, given that they breed in seeds of palm trees.

Spread is the third step in the invasion process after arrival and establishment (Liebhold and Tobin, 2008). Using the scientific literature and trapping data from the US Forest Service EDRR project (Rabaglia *et al.*, 2008), we calculated the number of states in the continental USA in which each of the current 58 exotic scolytines was reported as of 2010 (Table 3.5). We recognize that these numbers underestimate the true range of these exotics, given that extensive sampling has not occurred in all US states. Moreover, because no exotic scolytines have been reported from Alaska as of 2010, the values given in Table 3.5 refer to only the 48 contiguous US states. Of the 58 exotic scolytines listed in Table 3.5, 19 species were still only reported from a single state, whereas 39 were reported from two or more states each (Table 3.5). *Scolytus multistriatus* and *S. rugulosus* (Muller) were the only two exotics found in all 48 contiguous US states as of 2010. As expected, there was a weak but significant positive linear relation between the number of years since initial discovery in the USA and the number of states currently occupied by the exotic scolytines as of 2010 ($F_{[1, 56]} = 8.3$, $P < 0.0055$, $R^2 = 0.13$).

The native range of the 58 exotic scolytines currently established in the USA include the continents of Africa (10 species), Asia (30), Europe (8), Eurasia (9) and South America (1) (Table 3.5). Note that we considered Eurasia separately from Europe and Asia for those species whose native range extends widely over both continents. Most species of ambrosia beetles were of Asian origin (76%), most bark beetles were native to Eurasia (46%), and most seed and twig feeders were of African origin (47%; Table 3.5). It is possible that some exotic scolytines now established in the USA arrived from locations outside their native ranges, given that several of the 58 species listed in Table 3.5 were established elsewhere in the world prior to their initial detection in the USA (Wood and Bright, 1992; Haack, 2001, Brockerhoff *et al.*, 2006; Kirkendall and Faccoli, 2010).

3.8.1 Redbay ambrosia beetle, *Xyleborus glabratus*

Very few of the 58 exotic scolytines currently established in the continental USA have been reported to cause widespread ecological and economic damage in recent decades. Three notable exceptions include one bark beetle (*Scolytus multistriatus*) and two ambrosia beetles (*Xylosandrus crassiusculus* and *Xyleborus glabratus*). *S. multistriatus* is associated with widespread mortality of elm (*Ulmus*) in North America, primarily as a result of vectoring fungi that are the causal agents of Dutch elm disease (Evans and Finkral, 2010). *X. crassiusculus* has caused extensive mortality of broadleaf trees, especially small-diameter trees, both in the USA and in other countries where it has been introduced (Kirkendall and Ødegaard, 2007). Details on the discovery, biology and impact of *X. glabratus* are presented below.

The redbay ambrosia beetle, *X. glabratus*, was first reported in the USA in 2002. The initial collection consisted of three adult specimens captured in traps near Port Wentworth, Georgia, that were deployed as part of the USDA Forest Service EDRR project (Rabaglia *et al.*, 2006, 2008). Additional delimiting trapping in 2002 and 2003 around Port Wentworth and Savannah, Georgia, found no additional beetles and therefore it was thought that *X. glabratus* was not actually established. In 2003 significant redbay (*Persea borbonia*) mortality was reported on Hilton Head Island, South Carolina. Although this mortality was first thought to be the result of drought or fluctuations in the local water table, *X. glabratus* and an unidentified fungus were recognized as the cause of the redbay mortality by 2004. Since 2004, *X. glabratus* was reported in Florida in 2005, Mississippi in 2009 and Alabama in 2010. Although natural spread has occurred through adult flight, human-assisted movement of infested host material is suspected in cases of long-distance spread. There is no USDA federal quarantine that regulates movement of potential host material of *X. glabratus* as of 2010.

It was soon realized that the symbiotic fungal associate of *X. glabratus* was extremely pathogenic to redbay (Fraedrich *et al.*, 2008). Harrington *et al.* (2008) described the fungus as a new species (*Raffaelea lauricola*) and also identified it as the causal agent of the disease now known as laurel wilt.

X. glabratus is native to Bangladesh, India, Japan, Myanmar and Taiwan, and in Asia it is reported to infest *Leucaena glauca*, *Lindera latifolia*, *Lithocarpus edulis*, *Litsea elongata*, *Phoebe lanceolata* and *Shorea robusta* (Rabaglia *et al.*, 2006). *X. glabratus* carries spores of *R. lauricola* in its mandibular mycangia and inoculates a tree as it bores through the bark and enters the xylem. The fungus causes disruption of water flow in the xylem, which results in vascular wilt and a characteristic discolouration or 'streaking' of the outer sapwood. In Asia, neither *R. lauricola* nor *X. glabratus* has been reported to cause laurel wilt. In North America, *R. lauricola* has been isolated only from members of the family Lauraceae, i.e., redbay, avocado (*Persea americana*), swampbay (*P. palustris*), camphor tree (*Cinnamomum camphora*), pondberry (*Lindera melissifolia*), pondspice (*Litsea aestivalis*) and sassafras (*Sassafras albidum*).

Of the 22 exotic Xyleborini established in the USA (Table 3.5), most species, with the exception of some *Xylosandrus* species, attack injured, weakened or dying trees. *Xyleborus glabratus* appears to be similar to these non-aggressive species with the exception of its highly pathogenic, symbiotic fungus. Dispersing *X. glabratus* adult females appear to be attracted to host odors, and after landing they initiate tunneling on the main trunk and branches. If the tree is healthy, colonization may be unsuccessful, but in the process of tunneling the beetle can inoculate the tree with the fungus. The fungus grows quickly in the sapwood, restricting water flow that leads to wilting foliage. In redbay, which seems to be the most susceptible North American host, the entire crown will wilt over a period of a few weeks to a few months, resulting in eventual tree death (Fraedrich *et al.*, 2008). As a tree declines, *X. glabratus* adults are attracted to the tree and soon colonize and produce brood in the sapwood. Adult females bore into the xylem where they lay eggs and cultivate the symbiotic fungus on which the larvae feed. As is typical for xyleborine ambrosia beetles, female progeny greatly outnumber the flightless males, and mate with their siblings in the parental gallery. In the south-eastern USA there are overlapping generations throughout the year with one generation requiring about 60 days (Hanula *et al.*, 2008). As the female brood adults feed in the gallery and become reproductively mature, *R. lauricola* fungal spores become packed in the mycangia near their mandibles. After mating, the female brood adults emerge from the dead or dying host trees and either reinfest the same tree if there is suitable wood moisture, or fly off to infest new host trees.

For many native and exotic xyleborine ambrosia beetles, ethanol is an effective attractant. The first specimens of *X. glabratus* in North America were collected in ethanol-baited funnel traps (Rabaglia *et al.*, 2006). However, additional trapping using ethanol showed that it was only weakly attractive to *X. glabratus*, whereas manuka oil and phoebe oil were shown to be highly attractive (Hanula and Sullivan, 2008). Hanula *et al.* (2008) also showed that *X. glabratus* was attracted to cut redbay bolts or injured redbay trees. Current survey efforts for laurel wilt and *X. glabratus* focus on visual detection of wilted host trees or traps baited with manuka oil.

Although *X. glabratus* has been found infesting sassafras, Hanula *et al.* (2008) noted that *X. glabratus* was not attracted to cut bolts of sassafras. Even though this may slow the spread of the disease in sassafras, observations in Georgia indicate that laurel wilt is able to persist in counties where sassafras is common but redbay is rare (Mayfield *et al.*, 2009).

Laurel wilt is well established from coastal South Carolina to Florida, as well as along the Gulf coast of Mississippi, and eradication is no longer feasible. Due to the extreme virulence of *R. lauricola*, the ability of *X. glabratus* to initiate populations from single unmated females and its efficiency in vectoring the fungus, continued high levels of redbay mortality are expected. Management of laurel wilt in forested areas where redbay is common will be difficult. Limiting long-distance, human-assisted spread of the beetle could delay the impacts of the disease in currently unaffected areas. Efforts such as germplasm conservation or screening for tree resistance may be the best long-term hope for redbay and other affected species. For high-value landscape trees, chemical treatments may be an option. Mayfield *et al.* (2008a) demonstrated the ability of macro-infusion of the fungicide propiconazole to protect redbay trees from laurel wilt. The movement of *X. glabratus* and laurel wilt into the commercial avocado-growing region of southern Florida threatens the avocado industry, as well as in other US states, Mexico and elsewhere in the Americas (Mayfield *et al.*, 2008b). Members of the Lauraceae are primarily woody plants and number about 3000 species worldwide, reaching their greatest diversity in the tropical and subtropical latitudes of Asia and the Americas (Chanderbali *et al.*, 2001), and therefore there are hundreds of plant species that are potentially at risk from this insect–fungus disease complex.

3.9 Future Prospects

The world community has taken positive steps to reduce the risk of pests being moved inadvertently in international trade. For example, in 2002, signatories to the International Plant Protection Convention (IPPC) adopted International Standards for Phytosanitary Measures No. 15 (ISPM No. 15), which was entitled 'Guidelines for Regulating Wood Packaging Material in International Trade' (IPPC, 2002). ISPM No. 15 was first revised in 2006 and most recently in 2009 (IPPC, 2009). The goal of ISPM No. 15 is to 'reduce significantly the risk of introduction and spread of most quarantine pests' (IPPC, 2009). The two currently approved phytosanitary treatments for wood packaging material include heat treatment to a minimum temperature of 56°C for 30 min throughout the profile of the wood and including the core, and fumigation with methyl bromide following prescribed schedules (IPPC, 2009). New methods are under development and when approved they will be added to ISPM No. 15.

After approval of ISPM No. 15 in 2002, individual countries or groups of countries began to require their trading partners to follow ISPM No. 15 protocols over the succeeding years. For example, Canada, Mexico and the USA began full implementation of ISPM No. 15 in 2006.

Very few surveys were ever conducted for insects of quarantine significance on wood packaging material prior to initiation of ISPM No. 15. In the paper by Bulman (1992), based on a random survey for pests in wood packaging material during 1989–1991 in New Zealand, 2.7% of the consignments with wood were infested with insects. By contrast, in three surveys conducted after implementation of ISPM No. 15, insect infestation rates of imported wood packaging material were reported as 0.1% in the USA (Haack and Petrice, 2009), 0.3% in the EU (IFQRG, 2006; Haack and Brockerhoff, 2011) and 0.5% in Australia (Zahid *et al.*, 2008). If these values are representative of actual pre- and post-ISPM No. 15 infestation rates, they suggest that implementation of ISPM No. 15 has lowered the occurrence of live wood-infesting insects moving in wood packaging material. Nevertheless, it is important to note that in all three surveys, live bark- and wood-infesting insects were still present in some wood packaging items that had been stamped with the ISPM No. 15 logo. The presence of live insects on wood marked with the ISPM No. 15 logo could indicate improper treatment, pest tolerance to the treatment, infestation after treatment or fraud. To reduce the possibilities that bark-infesting insects could survive or infest after treatment (Evans, 2007; Haack and Petrice, 2009), ISPM No. 15 was revised in 2009 to set maximum size limits for individual pieces of residual bark on wood packaging material (IPPC, 2009). The approved tolerance limits for residual bark allow for bark pieces of any length if they are less than 3 cm in width (as may occur along the edge of a board) or if a bark piece is wider than 3 cm, the total surface area must be less than 50 cm^2 (slightly larger than a typical credit card) (IPPC, 2009). Worldwide implementation of the recent revisions in ISPM No. 15 (IPPC, 2009) should further reduce the phytosanitary risk of wood packaging material.

It should be noted that most, if not all, of the exotic scolytines (Table 3.5) and other wood-infesting insects (Haack, 2006; Haack *et al.*, 2009, 2010a) discovered in the USA since approval of ISPM No. 15 in 2002 very likely became established prior to 2002. Therefore, future discoveries of exotic bark- and wood-infesting insects should not be considered as evidence that ISPM No. 15 has failed, unless the pest population can be dated accurately and linked to wood packaging material. Here again, it should be noted that the stated goal of ISPM No. 15 is not zero risk but rather to significantly reduce risk (IPPC, 2009).

Scolytines can move internationally in many other products besides wood. Many scolytines are intercepted in food items such as fruits and nuts, cut plants, nursery stock (plants for planting) and wooden handicrafts (Haack, 2001, 2006; Kirkendall and Faccoli, 2010; USDA APHIS, 2010). In recognition of the phytosanitary risk of international trade of live plants, a draft ISPM was released in 2010 for country consultation entitled 'Integrated Measures Approach for Plants for Planting in International Trade' (IPPC, 2010). This draft ISPM does not provide specific treatments, as does ISPM No. 15 for wood packaging material, but rather provides guidelines for exporting countries to follow in the areas of production practices, such as use of pest-free mother stock, isolation of introduced plant material, best management practices and treatments, and quality control (standard operating procedures, record keeping, training, internal audits and traceability). If approved and implemented, an ISPM that deals with plants for planting would certainly help reduce the phytosanitary risk now associated with international trade in live plants.

Acknowledgements

We thank Thomas Atkinson, Lawrence Kirkendall and Milos Knížek for providing technical data on scolytines; James LaBonte for providing unpublished data; Joseph Cavey for providing PestID data from USDA APHIS; and Toby Petrice and Therese Poland for providing critical comments on an earlier version of this chapter.

References

Alonso-Zarazaga, M.A. and Lyal, C.H.C. (2009) A catalogue of family and genus group names in Scolytinae and Platypodinae with nomenclatural remarks (Coleoptera: Curculionidae). *Zootaxa* 2258, 1–134.

Anderson, D.M. (1974) First record of *Xyleborus semiopacus* in the continental United States (Coleoptera: Scolytidae). *US Department of Agriculture, Cooperative Economic Insect Report* 24, 863–864.

Atkinson, T.H., Rabaglia, R.J. and Bright, D.E. (1990) Newly detected exotic species of *Xyleborus* (Coleoptera: Scolytidae) with a revised key to species in eastern North America. *Canadian Entomologist* 122, 93–104.

Atkinson, T.H., Rabaglia, R.J., Peck, S.B. and Foltz, J.L. (1991) New records of Scolytidae and Platypodidae from the U.S. and Bahamas. *Coleopterists Bulletin* 45, 152–164.

Aukema, J.E., McCullough, D.G., Von Holle, B., Liebhold, A.M., Britton, K. and Frankel, S.J. (2010) Historical accumulation of nonindigenous forest pests in the continental US. *BioScience* 60, 886–897.

Bright, D.E. (1968) Review of the tribe Xyleborini in America north of Mexico (Coleoptera: Scolytidae). *The Canadian Entomologist* 100, 1288–1323.

Bright, D.E. and Rabaglia, R.J. (1999) *Dryoxylon*, a new genus for *Xyleborus onoharaensis* Murayama, recently established in the southeastern United States (Coleoptea: Scolytidae). *Coleopterists Bulletin* 53, 333–337.

Bright, D.E. and Skidmore, R.E. (1997) *A Catalog of Scolytidae and Platypodidae (Coleoptera) Supplement 1 (1990–1994)*. National Research Council Press, Ottawa.

Bright, D.E. and Skidmore, R.E. (2002) *A Catalog of Scolytidae and Platypodidae (Coleoptera) Supplement 2 (1995–1999)*. National Research Council Press, Ottawa.

Brockerhoff, E.G., Bain, J., Kimberley, M.O. and Knížek, M. (2006) Interception frequency of exotic bark and ambrosia beetles (Coleoptera: Scolytinae) and relationship with establishment in New Zealand and world-wide. *Canadian Journal of Forest Research* 36, 289–298.

Bulman, L.S. (1992) Forestry quarantine risk of cargo imported into New Zealand. *New Zealand Journal of Forest Science* 22, 32–38.

Byers, J.A. (2004) Chemical ecology of bark beetles in a complex olfactory landscape. In: Lieutier, F., Day, K., Battisti, A., Grégoire, J.C. and Evans, H. (eds) *Bark and Wood Boring Insects in Living Trees in Europe, a Synthesis*. Kluwer Academic Publishers, Dordrecht, The Netherlands, pp. 89–134.

Chanderbali, A.S., van der Werff, H. and Renner, S.S. (2001) Phylogeny and historical biogeography of Lauraceae: evidence from the chloroplast and nuclear genomes. *Annals of the Missouri Botanical Garden* 88, 104–134.

Chapman, J.W. (1910) The introduction of a European Scolytid (the smaller elm bark beetle, *Scolytus multistriatus* Marsh.) into Massachusetts. *Psyche* 17, 63–68.

Chittenden, F.H. (1898) Fruit-tree bark beetle (*Scolytus rugulosus* Ratz.). US Department of Agriculture, Division of Entomology, Series 2, Circular 29, Washington, DC.

Cognato, A.I., O'Donnell, R. and Rabaglia, R.J. (2011) An Asian ambrosia beetle, *Xylosandrus amputatus* (Blandford) (Curculionidae:

Scolytinae: Xyleborini), discovered in Florida, USA. *Coleopterists Bulletin* (in press).

Crespi, B.J. (1994) Three conditions for the evolution of eusociality: are they sufficient? *Insectes Sociaux* 41, 395–400.

Dole, S.A. and Cognato, A.I. (2010) Phylogenetic revision of *Xylosandrus* Reitter (Coleoptera: Curculionidae: Scolytinae: Xyleborina). *Proceedings of the California Academy of Sciences Series 4* 61, 451–545.

Evans, A.M. and Finkral, A.J. (2010) A new look at spread rates of exotic diseases in North American forests. *Forest Science* 56, 453–459.

Evans, H.F. (2007) ISPM 15 treatments and residual bark: How much bark matters in relation to founder populations of bark and wood boring beetles. In: Evans, H. and Oszako, T. (eds), *Alien Invasive Species and International Trade*. Forest Research Institute, Sêkocin Stary, Poland, pp. 149–155.

Felt, E.P. (1932) A new pest in greenhouse grown grape stems. *Journal of Economic Entomology* 25, 418.

Fraedrich, S.W., Harrington, T.C., Rabaglia, R.J., Ulyshen, M.D., Mayfield, A.E., Hanula, J.L., Eickwort, J.M. *et al.* (2008) A fungal symbiont of the redbay ambrosia beetle causes a lethal wilt in redbay and other Lauraceae in the southeastern United States. *Plant Disease* 92, 215–224.

Haack, R.A. (2001) Intercepted Scolytidae (Coleoptera) at US ports of entry: 1985–2000. *Integrated Pest Management Reviews* 6, 253–282.

Haack, R.A. (2006) Exotic bark and wood-boring Coleoptera in the United States: recent establishments and interceptions. *Canadian Journal of Forest Research* 36, 269–288.

Haack, R.A. and Brockerhoff, E.G. (2011) ISPM No. 15 and the incidence of wood pests: recent findings, policy changes, and current knowledge gaps. International Research Group on Wood Protection 42nd Annual Meeting, Queenstown, New Zealand, 5–8 May 2011. IRG Secretariat, Stockholm, IRG/WP 11-30568.

Haack, R.A. and Petrice, T.R. (2009) Bark- and wood-borer colonization of logs and lumber after heat treatment to ISPM 15 specifications: the role of residual bark. *Journal of Economic Entomology* 102, 1075–1084.

Haack, R.A. and Poland, T.M. (2001) Evolving management strategies for a recently discovered exotic forest pest: the pine shoot beetle, *Tomicus piniperda* (Coleoptera). *Biological Invasions* 3, 307–322.

Haack, R.A. and Slansky, F. (1987) Nutritional ecology of wood-feeding Coleoptera, Lepidoptera, and Hymenoptera. In: Slansky, F. and Rodriguez, J.G. (eds) *Nutritional Ecology of Insects, Mites, Spiders, and Related Invertebrates*. John Wiley, New York, New York, pp. 449–486.

Haack, R.A., Petrice, T.R. and Zablotny, J.E. (2009) First report of the European oak borer, *Agrilus sulcicollis* (Coleoptera: Buprestidae), in the United States. *Great Lakes Entomologist* 42, 1–7.

Haack, R.A., Hérard, F., Sun, J. and Turgeon, J.J. (2010a) Managing invasive populations of Asian longhorned beetle and citrus longhorned beetle: a worldwide perspective. *Annual Review of Entomology* 55, 521–546.

Haack, R.A., Petrice, T.R. and Wiedenhoeft, A.C. (2010b) Incidence of bark- and wood-boring insects in firewood: a survey at Michigan's Mackinac Bridge. *Journal of Economic Entomology* 103, 1682–1692.

Hanula, J.L. and Sullivan, B. (2008) Manuka oil and phoebe oil are attractive baits for *Xyleborus glabratus* (Coleoptera: Scolytinae), the vector of laurel wilt. *Environmental Entomology* 37, 1403–1409.

Hanula, J.L., Mayfield, A.E., Fraedrich, S.W. and Rabaglia, R.J. (2008) Biology and host associations of redbay ambrosia beetle (Coleoptera: Curculionidae: Scolytinae), exotic vector of laurel wilt killing redbay trees in the southeastern United States. *Journal of Economic Entomology* 101, 1276–1286.

Harrington, T.C., Fraedrich, S.W. and Aghayeva, D.N. (2008) *Raffaelea lauricola*, a new ambrosia beetle symbiont and pathogen of the Lauraceae. *Mycotaxon* 104, 399–404.

Hoebeke, E.R. (1989) *Pityogenes bidentatus* (Herbst), a European bark beetle new to North America (Coleoptera: Scolytidae). *Journal of the New York Entomological Society* 97, 305–308.

Hoebeke, E.R. (1991) An Asian ambrosia beetle, *Ambrosiodmus lewisi*, new to North America (Coleoptera: Scolytidae). *Proceedings of the Entomological Society of Washington* 93, 420–424.

Hoebeke, E.R. (2001) *Hylurgus ligniperda*: a new exotic pine bark beetle in the United States. *Newsletter of the Michigan Entomological Society* 46, 1–2.

Hoebeke, E.R. and Acciavatti, R.E. (2006) *Hylurgops palliatus* (Gyllenhal), an Eurasian bark beetle new to North America (Coleoptera: Curculionidae: Scolytinae). *Proceedings of the Entomological Society of Washington* 108, 267–273.

Hoebeke, E.R. and Rabaglia, R.J. (2008) *Xyleborus seriatus* Blandford (Coleoptera: Curculionidae:

Scolytinae), an Asian ambrosia beetle new to North America. *Proceedings of the Entomological Society of Washington* 110, 470–476.

Hopkins, A.D. (1915) Classification of the Cryphalinae, with descriptions of new genera and species. US Department of Agriculture Report 99, Washing, DC.

Hulcr, J. and Cognato, A.I. (2010) New genera of Paleotropical Xyleborini (Coleoptera: Curculionidae: Scolytinae) based on congruence between morphological and molecular characters. *Zootaxa* 2717, 1–33.

IFQRG (International Forestry Quarantine Research Group) (2006) 3rd Meeting of the International Forestry Quarantine Research Group, FAO, Rome, 29 November–1 December 2005. Available at: https://www.ippc.int/file_uploaded/1290446486_IFQRG_December_2005_Meeting_Repo.pdf, accessed 14 July 2011.

IPPC (International Plant Protection Convention) (2002) *International Standards for Phytosanitary Measures: Guidelines for Regulating Wood Packaging Material in International Trade.* Publication No. 15. FAO, Rome.

IPPC (2009) *International Standards for Phytosanitary Measures: Revision of ISPM No. 15, Regulation of Wood Packaging Material in International Trade.* FAO, Rome.

IPPC (2010) *International Standards for Phytosanitary Measures; Draft Standard: Integrated Measures Approach for Plants for Planting in International Trade.* FAO, Rome.

Jordal, B.H., Normark, B.B., Farrell, B.D. and Kirkendall, L.R. (2002) Extraordinary haplotype diversity in haplodiploid inbreeders: phylogenetics and evolution of the bark beetle genus *Coccotrypes*. *Molecular Phylogenetics and Evolution* 23, 171–188.

Kirkendall, L.R. (1983) The evolution of mating systems in bark and ambrosia beetles (Coleoptera: Scolytidae and Platypodidae). *Zoological Journal of the Linnean Society* 77, 293–352.

Kirkendall, L.R. and Faccoli, M. (2010) Bark beetles and pinhole borers (Curculionidae, Scolytinae, Platypodinae) alien to Europe. *ZooKeys* 56, 227–251.

Kirkendall, L.R. and Ødegaard, F. (2007) Ongoing invasions of old-growth tropical forests: establishment of three incestuous beetle species in southern Central America. *Zootaxa* 1588, 53–62.

Kirkendall, L.R., Kent, D.S. and Raffa, K.F. (1997) Interactions among males, females and offspring in bark and ambrosia beetles: the significance of living in tunnels for the evolution of social behavior. In: Choe, J.C. and Crespi, B.J. (eds) *The Evolution of Social Behavior in Insects and Arachnids.* Cambridge University Press, Cambridge, UK, pp. 181–215.

LeConte, J.L. (1868) Synopsis of the Scolytidae of America north of Mexico. *American Entomological Society Transactions* 2, 150–178.

LeConte, J.L. (1876) Family IX: Scolytidae. *American Philosophical Society Proceedings* 15, 341–390.

Lee, J.C., Smith, S.L. and Seybold, S.J. (2005) Mediterranean pine engraver. US Department of Agriculture, Forest Service, State and Private Forestry, Pacific Southwest Region, Pest Alert R5-PR-016, Susanville, California.

Liebhold, A.M. and Tobin, P.C. (2008) Population ecology of insect invasions and their management. *Annual Review of Entomology* 53, 387–408.

Mayfield, A.E., Barnard, E.L., Smith, JA., Bernick, S.C., Eickwort, J.M. and Dreaden, T.J. (2008a) Effect of propiconazole on laurel wilt disease development on redbay trees and on the pathogen in vitro. *Arboriculture and Urban Forestry* 35, 317–324.

Mayfield, A.E., Peña, J.E., Crane, J.H., Smith, J.A., Branch, C.L., Ottoson, E.D. and Hughes, M. (2008b) Ability of the redbay ambrosia beetle (Coleoptera: Curculionidae: Scolytinae) to bore into young avocado (Lauraceae) plants and transmit the laurel wilt pathogen *Raffaelea* sp.). *Florida Entomologist* 91, 485–487.

Mayfield, A., Barnard, E., Harrington, T., Fraedrich, S., Hanula, J., Vankus, V., Rabaglia, B. et al. (2009) Recovery plan for laurel wilt on redbay and other forest species caused by *Raffaelea lauricola*, vector *Xyleborus glabratus*. Available at: www.ars.usda.gov/SP2UserFiles/Place/00000000/opmp/ForestLaurelWilt100107.pdf, accessed 14 July 2011.

McCullough, D.G., Work, T.T., Cavey, J.F., Liebhold, A.M. and Marshall, D. (2006) Interceptions of nonindigenous plant pests at US ports of entry and border crossings over a 17-year period. *Biological Invasions* 8, 611–630.

Mudge, A.D., LaBonte, J.R., Johnson, K.J.R. and LaGasa, E.H. (2001) Exotic woodboring Coleoptera (Micromalthidae, Scolytidae) and Hymenoptera (Xiphydriidae) new to Oregon and Washington. *Proceedings of the Entomological Society of Washington* 103, 1011–1019.

NAPIS (National Agricultural Pest Information System) (2008) Pest Tracker. Available at: http://pest.ceris.purdue.edu/searchpest.php?selectName=INBQRCA, accessed 14 July 2011.

Negrón, J.F., Witcosky, J.J., Cain, R.J., LaBonte, J.R., Duerr, D.A., McElwey, S.J., Lee, J.C. et al. (2005) The banded elm bark beetle: a new threat to elms in North America. *American Entomologist* 51, 84–94.

Okins, K.E. and Thomas, M.C. (2010) A new North American record for *Xyleborinus andrewesi* (Coleoptera: Curculionidae: Scolytinae). *Florida Entomologist* 93, 133–134.

Peck, W.D. (1817) On the insects which destroy young branches of the pear tree, and the leading shoot of the Weymouth pine. *Massachusetts Agriculture Report* 4, 205–211.

Rabaglia, R.J., Dole, S.A. and Cognato, A.I. (2006) Review of American Xyleborina (Coleoptera: Curculionidae: Scolytinae) occurring North of Mexico, with an illustrated key. *Annals of the Entomological Society of America* 99, 1034–1056.

Rabaglia, R.J., Duerr, D., Acciavatti, R. and Ragenovich, I. (2008) *Early Detection and Rapid Response for Non-Native Bark and Ambrosia Beetles*. US Department of Agriculture, Forest Service, Forest Health Protection, Washington, DC.

Rabaglia, R.J., Vandenberg, N.J. and Acciavatti, R.E. (2009) First records of *Anisandrus maiche* Stark (Coleoptera: Curculionidae: Scolytinae) in North America. *Zootaxa* 2137, 23–28.

Rabaglia, R.J., Knížek, M. and Johnson, W. (2010) First records of *Xyleborinus octiesdentatus* (Murayama) (Coleoptera, Curculionidae, Scolytinae) from North America. *ZooKeys* 56, 219–226.

Raffa, K.F., Aukema, B.H., Bentz, B.J., Carroll, A.L., Hicke, J.A., Turner, M.G. and Romme, W.H. (2008) Cross-scale drivers of natural disturbances prone to anthropogenic eruptions. *BioScience* 58, 501–517.

Riley, C.V. (1879) The clover-root borer (*Hylesinus trifolii*, Mueller) (Ord. Coleoptera; Fam. Scolytidae). In: *US Department of Agriculture, Annual Report of the Commissioner of Agriculture for the year 1878*, Washington, DC, pp. 248–250.

Roques, A., Rabitsch, W., Rasplus, J.Y., Lopez-Vaamonde, C., Nentwig, W. and Kenis, M. (2009) Alien terrestrial invertebrates of Europe. In: DAISIE (ed.) *Handbook of Alien Species in Europe*. Springer, Dordrecht, The Netherlands, pp. 63–79.

Sauvard, D. (2004) General biology of bark beetles. In: Lieutier, F., Day, K.R., Battisti, A., Grégoire, J.-C. and Evans, H.F. (eds) *Bark and Wood Boring Insects in Living Trees in Europe, a Synthesis.* Kluwer Academic Publisher, Dordrecht, The Netherlands, pp. 67–86.

Sauvard, D., Branco, M., Lakatos, F., Faccoli, M. and Kirkendall, L.R. (2010) Weevils and bark beetles (Coleoptera: Curculionoidea). *BioRisk* 4, 219–266.

Schiefer, T.L. and Bright, D.E. (2004) *Xylosandrus mutilatus* (Blandford), an exotic ambrosia beetle (Coleoptera: Curculionidae: Scolytinae: Xyleborini) new to North America. *Coleopterists Bulletin* 58, 431–438.

Thomas, M.C. (2004) Two Asian ambrosia beetles recently established in Florida (Curculionidae: Scolytinae). Available at: www.freshfromflorida.com/pi/enpp/ento/twonewxyleborines.html, accessed 14 July 2011.

US Census Bureau (2011) Statistical abstract of the United States 2011 – earlier editions. Available at: http://www.census.gov/compendia/statab/past_years.html, accessed 14 July 2011.

USDA APHIS (United States Department of Agriculture, Animal and Plant Health Inspection Service) (1995) 7 CFR Parts 300 and 319 - Importation of logs, lumber, and other unmanufactured wood articles. *Federal Register* 60, 25 May 1995, 27,665–27,682.

USDA APHIS (1998) 7 CFR Parts 319 and 354 – Solid wood packing material from China. *Federal Register* 63, 18 September 1998, 50,100–50,111.

USDA APHIS (2010) 7 CFR Part 319 – Proposed rule. Importation of wooden handicrafts from China. *Federal Register* 75, 23 September 2010, 57,864–57,866.

USDA Forest Service (2010) *Major Forest Insect and Disease Conditions in the United States: 2009 Update*. FS-952, US Department of Agriculture, Forest Service, Washington, DC.

Vandenberg, N.J., Rabaglia, R.J. and Bright, D.E. (2000) New records of two *Xyleborus* (Coleoptera: Scolytidae) in North America. *Proceedings of the Entomological Society of Washington* 102, 62–68.

Wermelinger, B. (2004) Ecology and management of the spruce bark beetle *Ips typographus* - a review of recent research. *Forest Ecology and Management* 202, 67–82.

Wood, S.L. (1954) A revision of North American Cryphalini (Scolytidae: Coleoptera). *University of Kansas Science Bulletin* 36, 959–1089.

Wood, S.L. (1975) New synonymy and new species of American bark beetles (Coleoptera: Scolytidae), Part II. *Great Basin Naturalist* 35, 391–401.

Wood, S.L. (1977) Introduced and exported American Scolytidae (Coleoptera). *Great Basin Naturalist* 37, 67–74.

Wood, S.L. (1982) The bark and ambrosia beetles of North and Central America (Coleoptera: Scolytidae), a taxonomic monograph. *Great Basin Naturalist Memoirs* 6, 1–1359.

Wood, S.L. (1992) Nomenclatural changes and new species in Platypodidae and Scolytidae (Coleoptera), Part II. *Great Basin Naturalist* 52, 78–88.

Wood, S.L. (2007) *Bark and Ambrosia Beetles of South America (Coleoptera: Scolytidae)*. Brigham Young University, Provo, Utah.

Wood, S.L. and Bright, D.E. (1992) A catalog of Scolytidae and Platypodidae (Coleoptera), Part 2: Taxonomic index. *Great Basin Naturalist Memoirs* 13, 1–1553.

Zahid, M.I., Grgurinovic, C.A. and Walsh, D.J. (2008) Quarantine risks associated with solid wood packaging materials receiving ISPM 15 treatments. *Australian Forestry* 71, 287–293.

4 *Diabrotica speciosa*: An Important Soil Pest in South America

Crébio José Ávila and Alexa Gabriela Santana

Embrapa Western Agriculture, Caixa Postal 661, Dourados 79824-100, Brazil

4.1 Introduction

The genus *Diabrotica*, typically of Neotropical origin, represents a large number of polyphagous beetles (Coleoptera: Chrysomelidae: Galerucinae: Luperini) and consists of about 338 described species (Wilcox, 1972). This genus is usually divided into three subgroups: signifera, fucata and virgifera. The fucata and virgifera groups are the most studied in the world (Krysan, 1986), containing 305 and 21 species, respectively, while the signifera group is represented by only 11 species (Wilcox, 1972; Krysan and Smith, 1987).

In Brazil, *D. speciosa* (Germar, 1824) is the predominant species of the genus *Diabrotica*. The adults attack the aerial part of various crops and the larvae attack roots and tubers (Gassen, 1989; Ávila and Milanez, 2004). In Brazil, the adults of *D. speciosa* have the common names 'vaquinha', 'brasileirinho' and 'patriota', while the larvae are commonly called corn rootworm.

4.2 Biological Aspects and Life Cycle

D. speciosa is a multivoltine species (several generations per annum). Eggs are yellowish and measure c. 0.36 mm wide and 0.65 mm in length. The incubation period varies according to the temperature. Milanez and Parra (2000a) found that the duration of the incubation period ranged from 19.6 d at 18°C and 5.7 d at 32°C. The base temperature (Tb) and thermal constant (K) were 11.1°C and 119.1 degree-days, respectively. Under natural conditions, females of *D. speciosa* lay their eggs on the soil, and its chemical, physical and biological properties influence the fecundity of the insect (Milanez and Parra, 2000b).

The larval stage has three instars, lasting c. 18 d, when the insect is reared at 25°C and fed with maize seedlings (Milanez, 1997). The pupal stage occurs naturally in the soil in pupal chambers, built by the larva at the end of the third larval instar. It has a duration of c. 12 d (pre-pupae + pupae), after which the adults emerge (Milanez, 1995; Silva-Werneck *et al.*, 1995). In the pupal stage it is possible to see a strong sexual dimorphism. The female has a ventral papilla near the end of the abdomen, whereas in the male pupae, this structure is absent (Krysan, 1986).

The longevity of *D. speciosa* adults, the pace of egg laying and fecundity depend on the substrate used during the larval stage and type of food (host) given to the insects. Milanez (1995) observed that larvae of *D. speciosa* reared on natural diet (corn seedlings), had shorter male (41.8 d) and female (51.6 d) longevity, than that of males (55.5 d) and females (58.5 d) reared on an artificial diet. Ávila *et al.* (2002) found that couples of *D. speciosa* fed with leaves of common bean and potatoes had significantly higher fecundity than those fed on corn or soybean foliage. These same authors also found that leaflets of young bean plants were preferred by adults of *D. speciosa* for feeding, and resulted in greater egg-laying capacity, when compared to leaflets from older plants (Ávila and Parra, 2001). This egg-laying capacity of insects is determined by ovogenesis (egg production), a physiological process that is regulated by the availability of nutrients present in the female body (Wheleer, 1996).

D. speciosa has a broad geographical distribution, occurring throughout South America, although it is present in greater populations in the Southern Cone (Christensen, 1943; Krysan, 1986). In Argentina, Christensen (1943) listed 60 host plant species of *D. speciosa*, belonging to 22 families. In Brazil, the adult *D. speciosa* has been found attacking the aerial parts of plants grown in all states of the Federation (Haji, 1981). This wide geographical distribution is probably due to the multivoltine and polyphagous nature of the insect or to their climatic adaptation in different regions of the country.

4.3 Host Plants and Damage

Adults of *D. speciosa* feed on leaves, new shoots, pods and fruits of various plant species (Haji, 1981; Gassen, 1989; Hickel *et al.*, 1997; Roberto *et al.*, 2001; Ávila and Milanez, 2004). In Brazil, the adult *D. speciosa* is reported to feed on the shoots of vegetables (Boff *et al.*, 1992; Folcia *et al.*, 1998; Picanço *et al.*, 1999; Grützmacher and Link, 2000), bean, soybean, sunflower, cotton, corn, tobacco, wheat and canola (Haji, 1981; Santos *et al.*, 1988; Gassen, 1989) and fruit trees (Hickel *et al.*, 1997; Roberto *et al.*, 2001). In addition to the direct damage caused by injury to plants, adults can act as vectors of virus (Fulton and Scott, 1977; Costa and Batista, 1979; Lin *et al.*, 1984; Boff *et al.*, 1992; Oliveira *et al.*, 1994; Ribeiro *et al.*, 1996).

Moreover, larvae of *D. speciosa* are considered more specialized (oligophagous), because they occur in a restricted number of host plants. The larvae can damage roots or tubers of plants (Gassen, 1989; Ávila and Milanez, 2004) and are reported to cause yield losses in corn (Fogaça and Calafiori, 1992; Gassen, 1994). The consumption of corn roots reduces the plant's capacity to absorb water and nutrients, making it less productive and also more susceptible to root diseases (Kahler *et al.*, 1985). As a result of the attack to the root, the shoot acquires a curved stem, known as 'goose-neck' (Fig. 4.1), which eventually affects plant architecture and photosynthesis. These losses may be intensified when the harvesting is done mechanically.

Marques *et al.* (1999) found that corn plants kept in greenhouse conditions, and infested with densities of 40, 80, 160 and 320 larvae per plant, had a significant reduction of the root system (decreased dry weight of roots). The root damage causes reduced plant height, and consequently dry weight, when compared to plants without infestation (Table 4.1).

According to Silva (1999), the severity of the damage caused by larvae of *D. speciosa* may vary depending on environmental conditions. Thus, Silva (1999) reported that in warmer regions of Paraná State the loss in productivity of corn was about 200 kg/ha, while in municipalities with lower temperatures (Ponta Grossa and Castro), the mean reduction was 600 kg/ha.

Fig. 4.1. Damage caused by larvae of *Diabrotica speciosa* to the corn roots and reflection in the aerial part of the plant.

Table 4.1. Mean of root dry weight, and height and dry weight of corn shoots, with different densities of infestations of larvae of *Diabrotica speciosa*. Piracicaba, Brazil (1997).

No. of larvae/pot or vase	Dry weight of root (g)	Plant height (cm)	Dry weight of aerial part (g)
0	2.64 a	108.6 a	6.95 a
40	0.94 b	88.5 b	3.46 b
80	0.54 c	56.1 c	1.28 c
160	0.35 d	40.5 d	0.61 d
320	0.16 e	23.6 e	0.27 e

Means followed by same letter do not differ from each other (Tukey, $\alpha = 0.05$)
Source: Marques *et al.* (1999)

A factor that predisposes the intensity of damage to the corn root system is the availability of host plants that serve as food for adults. Corn plants are adequate for the larva, but are inappropriate for adults, providing lower longevity as well as lower fertility (Ávila and Parra, 2002); however, broadleaf species such as bean are nutritionally adequate for the adult. The damage caused by larvae to the roots of corn is enhanced by the greater longevity and egg-laying capacity of the insect (Ávila, 1999). Polycultures of corn and beans show less damage than corn grown as a monoculture crop.

Adults can also cause damage to corn by piercing the leaves of young plants, or impairing fertilization when feeding from the style-stigma (new 'hair') of the ear during the pollination period. In potato, larvae of *D. speciosa* bore into the tubers, reducing their commercial value (Haji, 1981; Gassen, 1989; Hohmann 1989; Salles, 2000). Adults may also feed on the foliage of potato, and high infestations can cause severe defoliation and reduce crop productivity (Lara *et al.*, 2000a). The potato is an ideal host for the multiplication of the pest under field conditions, since both leaves and tubers are used by adults and larvae, respectively (Ávila and Parra, 2002).

Adults may cause defoliation throughout the bean crop cycle (Magalhães *et al.*, 1988); however, the critical phase is during the first 2 weeks of crop development, when they can cause severe defoliation, affecting both the development of the shoots and roots (Ávila, 1990; Adde *et al.*, 1994). Intense defoliation during flowering can also cause economic loss and impede the maturation of the pods (Nakano and Fornazier, 1983). The leaves of the host provide a greater longevity and high reproductive capacity for *D. speciosa* adults (Ávila and Parra, 2002). The roots of bean plants are not suitable for larval development, although *D. speciosa* can complete its life cycle in this host in the absence of other preferential hosts (Ávila, 1999).

The adults of *D. speciosa* can occasionally cause significant damage to other crops such as soybean, especially in the early stages of the crop (Ramiro *et al.*, 1987) or even during flowering (Izaguirre and Ramos, 1987). Cucurbitaceae species, including cucumber, pumpkin, cantaloupe and watermelon, and Cruciferae species are frequently visited by adults of *D. speciosa*, which may cause defoliation or destruction of flowers if not controlled (Baldin and Lara, 2001). In fruit trees, the adults may attack leaves, flowers or fruits of nectarine (Hickel *et al.*, 1997), passion fruit and grapes during flowering (Roberto *et al.*, 2001), which can reduce both yield and fruit quality.

In the USA, where some species of the genus *Diabrotica* are considered important pests (Kahler *et al.*, 1985), it is estimated that about US$1 billion is spent annually to control chrysomelids in corn alone (Metcalf, 1986). In Brazil, the economic impact caused by larvae and adults of *D. speciosa*, as well as the amount of resources spent for its control have not yet been estimated, although a significant amount and variety of insecticides are applied annually to control this pest in potatoes and corn, especially in the southeast and south.

4.3.1 Control strategies

The control of adults and larvae of *D. speciosa* is conducted almost exclusively through the use of chemical insecticides. The economic threshold level for this pest has not been studied in detail, although some attempts to determine this parameter have been made with respect to defoliation of bean by adults (Pereira et al., 1997) and damage caused by larvae to the corn root system (Marques et al., 1999).

Control of adults requires various applications of insecticides, because of their ability to migrate and move easily between crops, thus causing frequent reinfestations, especially when environmental conditions favor the development of the pest.

Insecticides that interfere with the development of immature insects (insect growth regulators) can also cause sterility of adult beetles, affecting their fecundity and egg viability (Elek and Longstaff, 1994). Adults of *D. speciosa* fed on bean leaves treated with lufenuron showed reduced fecundity and viability of eggs produced (Ávila et al., 1998; Ávila and Nakano, 1999). This deleterious effect on the progeny of *D. speciosa* can be of great significance in the field, reducing its biotic potential, since the larvae, due to their subterranean growth habit, are more difficult to control.

Chrysomelids of the Luperina tribe, which includes *D. speciosa*, show coevolution with plants containing the allelochemical cucurbitacin, often found in plants of the family Cucurbitaceae (Metcalf et al., 1982). Cucurbitacines deter a large number of arthropod defoliators from feeding; however, the Luperini beetles, also *Diabrotica* spp., are immune to their toxic effects and are able to use these chemicals to their benefit as defense substances (kairomones). The adults of *D. speciosa* can sequester these substances in their bodies and protect themselves against natural enemies, that is, predators (Nishida et al., 1986, 1992; Nishida and Fukami, 1990). *D. speciosa* uses cucurbitacines for recognition of its hosts, and these substances may also stimulate feeding (Ferguson and Metcalf, 1985). These substances, when mixed with chemical insecticides can also be used to control *Diabrotica* (Weissling and Meinke, 1991). The use of baits containing the insecticide plus cucurbitacin can also be used for monitoring the adult *D. speciosa* (Roel and Zatarin, 1989). Plant extracts sprayed on crops with an attractant, a repellent or with feeding inhibitor characteristics to adults of *D. speciosa* have been studied as tactics for monitoring and control of these pests (Potenza et al., 1988; Ventura et al., 1996; Ventura and Ito, 2000).

Chemical control

Chemical control of *D. speciosa* larvae in corn and potatoes should be preventive. For instance, seed treatment of corn has been shown to be inefficient because the larvae cause damage during the 1–2 months after sowing; the insecticides applied to protect the seed do not persist in the soil sufficiently long to ensure protection of the root system until the attack of the larva occurs (Gassen, 1994). Applications of granular or spray insecticide in planting furrows are effective for controlling *D. speciosa* larvae in corn (Ávila and Gomez, 2001) and potato fields (Link et al., 1989; Salles, 1998; Salles and Grützmacher, 1999; Nakano et al., 2001). However, the use of granular insecticides in soil has encountered technical limitations, including lack of machinery suitable for the application of the product (Ávila and Botton, 2000), as well as environmental and social restrictions, since most active ingredients in the granulated form are highly toxic for humans and present the risk of environmental contamination.

Behavioral control

The chemical ecology of *D. speciosa* has been studied very little compared to the species of Neotropical origin. Recent studies revealed the existence of a sexual pheromone produced by adults (Ventura et al., 2001). The identification and synthesis of these compounds may constitute an important strategy for the management and/or monitoring of *D. speciosa* in crop systems.

Studies to evaluate the effect of mineral or organic fertilizers on the development of larvae and adults of *D. speciosa* have been conducted on corn (Hohmann, 1989) and bean (Vardasca et al., 1989; Veronesi et al., 1990; Fagotti et al., 1994; Ajudarte et al., 1997); however, the results did not show clearly the benefits of this management tactic.

Plant resistance

The use of resistant varieties is considered to be an ideal tactic to control insects because it has no added cost to the producer, does not pollute the environment, does not cause biological imbalance to the ecosystem and permits a perfect integration with other control tactics in pest management (Lara, 1991). Studies to evaluate plant resistance to *D. speciosa* showed that genotypes of potato (Bonine, 1997; Salles, 2000; Lara et al., 2000b), soybean (Rezende and Rossetto, 1980; Rossetto et al., 1981; Lara et al., 1999) bean (Paron and Lara, 2001) and pumpkin (Baldin and Lara, 2001) responded differently to pest infestation.

Biological control

Several natural enemies attack adults and larvae of *D. speciosa* (Stock, 1993; Shaw, 1995; Picanço et al., 1998; Heineck-Leonel and Salles, 1997), although biological control of this pest has rarely been attempted (Pianoski et al., 1990; Tigano Milani et al., 1995; Silva-Werneck et al., 1995). Microbial control of *D. speciosa* larvae, especially with fungi, has great potential if the soil is a stable environment with respect to temperature and relative humidity, especially in no-tillage systems.

4.4 Rearing Techniques

One of the first published techniques for rearing the *Diabrotica* group under controlled conditions was developed by George and Ortman (1965) with *D. virgifera virgifera* LeConte and corn seedlings. Subsequently, several studies adapted this procedure for other *Diabrotica* species (Jackson and Davis, 1978; Dominique and Yule, 1983; Jackson, 1986; Branson et al., 1988).

The first attempt to rear *D. speciosa* in the laboratory was reported in Brazil by Haji (1981), who used rooted potato plants as larval food, and bean and soy leaves to feed the adults. Later, other rearing techniques for *D. speciosa* were developed using filter paper, germination paper (germitest), sand or soil as substrates for larval and pupal stages, employing corn seedlings as a diet for the larval stage, and bean leaves to feed adults (Carvalho and Hohmann, 1982; Pecchioni, 1988; Silva-Werneck et al., 1995; Milanez, 1995). Ávila et al. (2000) improved the method of rearing *D. speciosa* in laboratory conditions, using sterile vermiculite as a substrate for larval and pupal development, corn seedlings as food for the larvae and bean leaves for adults. With the use of vermiculite there is no need to prepare another environment (site) for pupal development, as proposed by Milanez (1995). Using this system it is possible to get a large number of insects, with >75% viability during the larva–adult period, as well as a significant reduction in manpower and laboratory space required. The protocol for establishing the rearing system of *D. speciosa* is given below (Ávila et al., 2000).

4.4.1 Cages for the maintenance of adults

The use of an aluminium framed cage with a clear acrylic front, covered on the sides and back with nylon screen (mesh ± 1 mm openings) (Fig. 4.2A) is preferred. The bottom of the cage should be lined with a galvanized aluminum sheet, and the food and oviposition substrate laid on top of it, the centre covered with nylon mesh. The nylon mesh on the bottom of the cage is to prevent build-up of excess moisture and to allow passage of small impurities and debris which may be collected on a sheet of paper placed under the base of the cage.

(A) (B)

Fig. 4.2. (A) Cages for the maintenance of adult *Diabrotica speciosa*; (B) Extraction of eggs from oviposition substrate in running water (Avila et al., 2000).

Feeding of adults

Adults can be fed either bean or potato leaves (Ávila and Parra, 2002). The base sheet that contains the leaflets should be immersed in water to ensure longevity, and the leaves should be changed at 2-day intervals.

4.4.2   Obtaining eggs

Substrate for oviposition is a moist black gauze placed over a small container (Milanez, 1995). Eggs can be removed from the oviposition substrate by washing the gauze in water over a thin cloth where the eggs will be retained (Fig. 4.2B). To avoid contamination by fungi during the incubation period, eggs should be treated with a 1% (w/v) solution of copper sulphate ($CuSO_4$) for 2–3 min.

4.4.3   Larval food

Corn seedlings (Fig. 4.3A) should be treated with the fungicide thiabendazole + captan (1 g + 1 g/kg of seed) to prevent contamination and sown in sterile vermiculite (Fig. 4.3B) and moistened with sterile distilled water. In about 3–5 d depending on the temperature at which the seeds were maintained, the seedlings will be ready to be offered to the larvae.

4.4.4   Larval and pupal development

Two plastic containers (Figs. 4.3C and 4.3D) are used in the rearing process, one being c. 15 cm diameter × 7 cm where newly hatched larvae will be inoculated, and the other c. 20 cm diameter × 10 cm, to which the larvae are transferred and maintained until the end of the pupal stage. At inoculation, place 40 g of sterile vermiculite at the bottom of the smaller container, and wet this with 80 ml of sterile distilled water. Distribute approximately 100 seedlings on the moist vermiculite and add 100 newly hatched larvae of *D. speciosa*. Add 50 g of moist vermiculite and later 100 ml sterile distilled water onto the seedlings and larvae. About 10 d after inoculation, the larvae should be transferred by sieving from the smaller to the larger container (Figs. 4.3E to 4.3F). The transfer vial should contain twice as much vermiculite and food in the inoculation container; this is sufficient food for the insect to complete its

Fig. 4.3. Substrates and containers used for *Diabrotica speciosa* rearing. (A) corn seedlings; (B) sterile vermiculite; (C) vial of inoculation; (D) vial of transfer; (EFG) screening of larvae from of vial inoculation to vial transfer (Avila *et al.*, 2000).

immature stage (larvae + pupae). When the insects enter the pupal stage, the corn seedlings should be cut above the vermiculite to facilitate capture of adults that emerge from the vermiculite.

Because of bacterial contamination, artificial diets have rarely been used for rearing *Diabrotica* spp. (Sutter *et al.*, 1971; Rose and McCabe, 1973; Marrone *et al.*, 1985; Schalk and Peterson, 1990). Regularly, there is low larval survival on artificial media, when compared to natural media. Milanez (1995) tested five artificial diets for *D. speciosa* rearing. Only the wheatgerm diet (Berger, 1963), allowed the development of the insect, but required more time than rearing on corn seedlings. Adults produced on an artificial diet had lower weight and lower egg-laying capacity compared to those reared on a natural diet (Milanez, 1995; Ávila and Parra, 2000).

The low levels of viability of the immature stages and the quality of adult *D. speciosa* obtained on artificial media should not serve to discourage further studies.

References

Adde, M.F., Cardoso, A.M. and Calafiori, M.H. (1994) Influência da *Diabrotica speciosa* (Germar, 1824) sobre a nodulação e raízes de feijoeiro, *Phaseolus vulgaris* L. *Ecossistema* 19, 84–87.

Ajudarte, J.C., Luz, E.B. and Calafiori, M.H. (1997) Influência de adubação potássica no dano da vaquinha *Diabrotica speciosa* (Germar, 1824) na cultura do feijoeiro, *Phaseolus vulgaris* L. *Ecossistema* 22, 13–16.

Ávila, C.J. (1990) Principais pragas e seu controle. In: *A Cultura do Feijoeiro em Mato Grosso do Sul, Dourados*. EMBRAPA, Unidade de Execução de Pesquisa de Âmbito Estadual de Dourados.

Ávila, C.J. (1999) Técnica de criação e influência do hospedeiro e da temperatura no desenvolvimento de *Diabrotica speciosa* (Germar, 1824) (Coleoptera: Chrysomelidae). PhD thesis, Escola Superior de Agricultura "Luiz de Queiroz", Piracicaba, Brazil.

Ávila, C.J. and Botton, M. (2000) *Aplicação de Inseticidas No Solo*. FEALQ, Piracicaba, Brazil, pp. 64.

Ávila, C.J. and Gomez, S.A. (2001) Controle químico de larvas de *Diabrotica speciosa* (Coleoptera: Chrysomelidae) na cultura do milho. In: *Reunião Sul Brasileira de Pragas de Solo*, Embrapa soja, Londrina, Brazil, pp. 254–257.

Ávila, C.J. and Milanez, J.M. (2004) Larva alfinete. In: Salvadori, J.R, Ávila, C.J. and Silva, M.T.B. (eds) *Pragas de Solo No Brasil*. Embrapa Trigo, Embrapa Agropecuária Oeste, and Cruz Alta, Fundacep-Fecotrigo, Passo Fundo, Brazil, pp. 345–378.

Ávila, C.J. and Nakano, O. (1999) Efeito do regulador de crescimento de insetos lufenuron na reprodução de *Diabrotica speciosa* (Germar) (Coleoptera: Chrysomelidae). *Anais da Sociedade Entomológica do Brasil* 28, 293–299.

Ávila, C.J. and Parra, J.R.P (2001) Influência da idade do feijoeiro sobre o consumo foliar e fecundidade de *Diabrotica speciosa*. *Revista de Agricultura* 76, 299–306.

Ávila, C.J. and Parra, J.R.P. (2002) Desenvolvimento de *Diabrotica speciosa* (Germar) (Coleoptera: Chrysomelidae) em diferentes hospedeiros. *Ciência Rural* 32, 739–743.

Ávila, C.J., Nakano, O. and Chagas, M.C.M. (1998) Efeito do regulador de crescimento de insetos lufenuron na fecundidade e viabilidade dos ovos de *Diabrotica speciosa* (Germar), 1924 (Coleoptera: Chrysomelidae). *Revista de Agricultura* 73, 69–78.

Ávila, C.J., Tabai, A.C.P. and Parra, J.R.P. (2000) Comparação de técnicas para a criação de *Diabrotica speciosa* (Germar) (Coleoptera: Chrysomelidae) em dietas naturais e artificiais. *Anais da Sociedade Entomológica do Brasil* 29, 257–267.

Ávila, C.J., Milanez, M.J. and Parra, J.R.P. (2002) Previsão de ocorrência de *Diabrotica speciosa* utilizando o modelo de graus-dia de laboratório. *Pesquisa Agropecuária Brasileira* 37, 427–432.

Baldin, E.L.L. and Lara, F.M. (2001) Atratividade e consumo foliar por adultos *Diabrotica speciosa* (Germ.) (Coleoptera: Chrysomelidae) em diferentes genótipos de abóbora. *Neotropical Entomology* 30, 675–679.

Berger, R.S. (1963) *Laboratory Techniques for Rearing Heliothis Species on Artificial Medium*. USDA Agricultural Research Service, Washington, DC, pp. 4.

Boff, M.I.C., Gandin, C.L.G. and Carissimi Boff, M.I. (1992) Principais pragas na cultura da melancia e seu controle. *Agropecuária Catarinense* 5, 39–41.

Bonine, D.P. (1997) Suscetibilidade de cultivares de batata (*Solanum tuberosum* L.) a *Diabrotica speciosa* (Germar) (Coleoptera: Chrysomelidae) e ocorrência de outras pragas subterrâneas. MSc thesis, Federal University of Pelotas, Pelotas, Brazil.

Branson, T.F., Jackson J.J. and Sutter, G.R. (1988) Improved method for rearing *Diabrotica virgifera*

virgifera (Coleoptera: Chrysomelidae). *Journal of Economic Entomology* 81, 410–414.

Carvalho, S.M. and Hohmann, C.L. (1982) Biologia e consumo foliar de *Diabrotica speciosa* (Germar, 1824) em feijoeiro (*Phaseolus vulgaris* L.,1753), em condições de labor atório. In: *Reunião Nacional de Pesquisa de Feijão*, Embrapa-CNPAF, Goiânia, Brazil, pp. 244–245.

Christensen, J.R. (1943) Estudo sobre o gênero *Diabrotica* Chev. en Argentina. *Revista de la Facultad de Agronomia y Veterinaria* 10, 465–516.

Costa, C.L. and Batista, M.F. (1979) Viroses transmitidas por coleópteros no Basil. *Fitopatologia Brasileira* 4, 177–179.

Dominique, C.R. and Yule, W.N. (1983) Laboratory rearing technique for the nor thern corn rootworm, *Diabrotica longicornis* (Coleoptera: Chrysomelidae). *Canadian Entomologist* 115, 569–571.

Elek, J.A. and Longstaff, B.C. (1994) Effect of chitin-synthesis inibitors on stored-products beetles. *Pesticides Science* 40, 225–230.

Fagotti, M.A.P., Delgado, J.P. and Calafior i, M.H. (1994) Influência do nitrogênio no dano da vaquinha, *Diabrotica speciosa* (Germar, 1824) na cultura do f eijoeiro, *Phaseolus vulgaris* L. *Ecossistema* 19, 61–66.

Ferguson, J.E. and Metcalf , R.L. (1985) Plant-derived defense compounds f or diabroticites (Coleoptera: Chrysomelidae). *Journal of Chemical Ecology* 11, 311–318.

Fogaça, M.S. and Calafiori, M.H. (1992) Danos de *Diabrotica speciosa* (Germar, 1824) em milho. *Ecossistema* 17, 69–72.

Folcia, A.M., Rodriguez, S.M., Rizzo, H.F., Russo, S. and Rossa, F.R. (1998) Presencia y fluctuación poblacional de artrópodos perjudiciales al cultivo de tomate . *Revista de la Facultad de Agronomia, Universidad de Buenos Aires* 18, 105–109.

Fulton, J.P. and Scott, H.A. (1977) Bean r ugose mosaic and related vir uses. *Fitopatologia Brasileira* 2, 9–16.

Gassen, D. N. (1989) *Insetos Subterrâneos Prejudiciais às Culturas no Sul do Brasil*. Embrapa-CNPT, Passo Fundo, Brazil, pp. 49.

Gassen, D. N. (1994) *Pragas Associadas à Cultura do Milho*. Aldeia Nor te, Passo Fundo, Brazil, pp. 92.

George, B.W. and Ortman, E.E. (1965) Rearing the western corn rootworm in laboratory. *Journal of Economic Entomology* 55, 375–377.

Grützmacher, A.D. and Link, D (2000) Levantamento da entomofauna associada a cultiv ares de batata em duas épocas de cultiv o. *Pesquisa Agropecuária Brasileira* 35, 653–659.

Haji, N.F.P. (1981) Biologia, dano e controle do adulto de *Diabrotica speciosa* (Germar, 1824) (Coleoptera: Chrysomelidae) na cultur a da batatinha (*Solanum tuberosum* L.). PhD thesis, Escola Super ior de Ag ricultura "Luiz de Queiroz", City?.

Heineck-Leonel, M.A. and Salles, L.A.B.O. (1997) Incidência de par asitóides e patógenos em adultos de *Diabrotica speciosa* (Germ.) (Coleoptera: Chrysomelidae) na região de Pelotas, RS. *Anais da Sociedade Entomológica do Brasil* 26, 81–85.

Hickel, E.R., Ducroquet, J .P.H.J. and Matos, C.S. (1997) Controle de pragas na floração da nectarina. *Agropecuária Catarinese* 10, 19–23.

Hohmann, C.L. (1989) Levantamento dos ar tropodos associados a cultura da batata no município de Ir ati, Paraná. *Anais da Sociedade Entomológica do Brasil* 18, 53–60.

Izaguirre, J.A.R. and Ramos, T.C. (1987) Estudos preliminares do compor tamento de cr i-somelídeos em áreas de campos de soja. *Revista Centro Agrícola* 14, 71–79.

Jackson, J.J. (1986) Rear ing and handling of *Diabrotica* virgifera and *Diabrotica undecimpunctata howardi*. In: Krysan, J.L. and Miller, T.A. (eds). *Methods for Study of Pest Diabrotica*. Springer Verlag, New York, New York, pp. 25–47.

Jackson, J.J. and Davis, D. G. (1978) Rearing western corn rootworm larvae on seedling cor n (Coleoptera: Chrysomelidae). *Journal of the Kansas Entomological Society* 51, 353–355.

Kahler, A.L., Olness, A.E., Suttter, G.R., Dybing, C.D. and Devine, O.J. (1985) Root damage by corn rootworm and nutrient content in maiz e. *Agronomy Journal* 77, 769–774.

Krysan, J.L. (1986) Introduction: biology, distribution, and identification of pest Diabrotica. In: Krysan, J.L. and Miller, T.A. (eds) *Methods for Study of Pest Diabrotica*. Springer Verlag, New York, New York, pp. 1–23.

Krysan, J.L. and Smith, R.F. (1987) Systematics of *virgifera* species groups of *Diabrotica* (Coleoptera: Chrysomelidae: Luperini). *Entomography* 5, 375–484.

Lara, F.M. (1991) *Princípios de Resistência de Plantas a Insetos*. 2nd ed. Ícone, São Paulo, Brazil, 336p.

Lara, F.M., Elias, J.M., Baldin, E.L.L. and Barbosa, J.C. (1999) Preferência alimentar de *Diabrotica speciosa* (Germ.) e *Cerotoma* sp. por genótipos de soja. *Scientia Agrícola* 56, 947–951.

Lara, F.M., Poletti, M. and Barbosa, J .C. (2000) Resistência de genótipos de batata (Solanum spp.) a *Diabrotica speciosa* (Germar, 1824) (Coleoptera: Chrysomelidae). *Ciência Rural* 30, 927–931.

Lara, F.M., Sargo, H.L.B and Boiça-J únior, A.L. (2000) Preferência alimentar de adultos de *Diabrotica speciosa* (Germar, 1824) (Coleoptera: Chrysomelidae) por genótipos de batata (*Solanum* spp.). *Anais da Sociedade Entomológica do Brasil* 29, 131–137.

Lin, M.T., Hill, J.H., Kitajima, E.W. and Costa, C.L. (1984) Two new serotypes of co wpea severe mosaic virus. *Phytopathology* 74, 581–585.

Link, D., Costa, E.C. and Correa Costa, E. (1989) Avaliação preliminar de inseticidas granulados no controle de pr agas do tubérculo da batatinha. *Revista do Centro de Ciências Rurais* 19, 305–309.

Magalhães, B.P., Carvalho, S.M., Peixoto Magalhães, B. and Mar tinez, S.C. (1988) Insetos associados a cultura. In: Zimmermann, M.J., Rocha, M. and Yamada, T. *Cultura Do Feijoeiro: Fatores Que Afetam a Produtividade*. Associação Brasileira para Pesquisa da Potas, Piracicaba, Brazil.

Marques, G.B.C., Ávila, C.J. and Parra, J.P.P. (1999) Danos causados por lar vas e adultos de *Diabrotica speciosa* (Coleoptera: Chrysomelidae) em milho . *Pesquisa Agropecuária Brasileira* 34, 1983–1986.

Marrone, P.G., Ferri, F.D., Mosley, T.R. and Menke, L.J. (1985) Improvements in laboratory rearing of the souther n corn rootworm, *Diabrotica undecimpunctata howardi* Barber (Coleoptera: Chrysomelidae), on ar tificial diet. *Journal of Economic Entomology* 78, 290–293.

Metcalf, R.L. (1986) F oreword. In: Krysan, J.L. and Miller, T.A. (eds) *Methods for the Study of Diabrotica*. Springer Verlag, New York, New York.

Metcalf, R.L., Rhodes , A.M., Metcalf , R.A., Ferguson, J.E., Metcalf, E.R. and Lu, P. (1982) Cucurbitacins contents and Diabroticite (Coleoptera: Chrysomelidae) feeding upon *Cucurbita* spp. *Environmental. Entomology* 11, 931–937.

Milanez, J.M. (1995) Técnicas de criação e bioecologia de *Diabrotica speciosa* (Germar, 1824) (Coleoptera: Chrysomelidae). PhD thesis , Escola Superior de Ag ricultura "Luiz de Queiroz", Piracicaba, Brazil.

Milanez, J.M. (1997) Ciclo biológico da v aquinha, praga do milho na região sul do país . *Agropecuária Catarinense* 10, 9–11.

Milanez, J.M. and Parra, J.R.P. (2000a) Biologia e exigências térmicas de *Diabrotica speciosa* (Germar) (Coleóptera: Chrysomelidae) em laboratório. *Anais da Sociedade Entomológica do Brasil* 29, 23–29.

Milanez, J.M. and Parra, J.R.P. (2000b) Preferência de *Diabrotica speciosa* (Germar) (Coleoptera: Chrysomelidae) para Oviposição em dif erentes tipos e umidade de solo . *Anais da Sociedade Entomológica do Brasil* 29, 155–158.

Nakano, O. and F ornazier, M.J. (1983) Pr agas. *Agropecuária* 5, 18–46.

Nakano, O., Florim, C.A. and Zambom, S . (2001) Atividade residual de fipronil sobre a *Diabrotica speciosa* alimentada com f olhas de batatinha – (*Solanum tuberosum*) In: *Reunião Sul Brasileira de Pragas do Solo, 8*. Londrina, Brazil, pp. 249–254.

Nishida, R. and Fukami, H. (1990) Sequestration of distasteful compounds b y some phar macophagous insects. *Journal of Chemical Ecology* 16, 151–164.

Nishida, R., Fukami, H., Tanaka, Y., Magalhães, B.P., Yokoyama, M. and Blumenschein, A. (1986) Isolation of feeding stimulants of Brazilian leaf beetles (*Diabrotica speciosa* and *Cerotoma arcuata*) from the root of *Ceratosanthes hilariana*. *Agricultural and Biological Chemistry* 50, 2831–2836.

Nishida, R., Yokoyama, M. and Fukami, H. (1992) Sequestration of cucurbitacin analogs by New and Old World chrysomelid leaf beetles in the tribe Luperini. *Chemoecology* 3, 19–24.

Oliveira, C.R.B., Marinho, V.L.A., Astolf Filho , S., Azevedo, M., Chagas, C.M. and Kitajima, E.W. (1994) Purification, serology and some properties of the purple granadilla (*Passiflora edulis*) mosaic vir us. *Fitopatologia Brasileira* 19, 455–462.

Paron, M.J.F.O. and Lara, F.M. (2001) Preferência alimentar de adultos de *Diabrotica speciosa* (Ger.) (Coleoptera: Chrysomelidae) por genótipos de f eijoeiro. *Neotropical Entomology* 30, 669–674.

Pecchioni, M.T.D. (1988) Cr ianza de *Diabrotica speciosa* (Coleoptera: Chrysomelidae) bajo condiciones de labor atorio. *Revista Peruana de Entomologia* 31, 86–90.

Pereira, M.F.A., Delfini, L.G., Antoniacomi, M.R. and Calafiori, M.H. (1997) Danos causados por vaquinha, *Diabrotica speciosa* (Germar, 1824), em feijoeiro, *Phaseolus vulgaris* L., em manejo integrado. *Ecossistema* 22, 17–20.

Pianoski, J., Bertucci, E., Capassi, M.C., Cirelli, E.A., Calafiori, M.H. and Teixeira, N.T. (1990) Eficiência da *Beauveria bassiana* (Bals) Vuill no controle da *Diabrotica speciosa* (Germar, 1824), em feijoeiro, *Phaseolus vulgaris* L., em diferentes adubações. *Ecossistema* 15, 24–35.

Picanço, M., Casali, V.W.D., Oliveira, I.R. and Leite, G.L.D. (1998) Himenópteros associados a *Solanum gilo* Raddi (Solanaceae). *Revista Brasileira de Zoologia* 14, 821–829.

Picanço, M., Leite, G.L.D., Bastos, C.S., Suinagra, F.A. and Casali, V.W.D. (1999) Coleópteros associados ao jiloeiro (*Solanum gilo* Raddi). *Revista Brasileira de Entomologia* 43, 131–137.

Potenza, M.R., Rossi, C.E. and Calafiori, M.H. (1988) Emprego de extrato de planta de girassol (*Helianthus annus* L.) no controle da cigarrinha (*Empoasca kraemeri* Ross and More, 1957) e da patriota (*Diabrotica speciosa* Germ., 1824) em feijoeiro (*Phaseolus vulgaris* L.). *Ecossistema* 12, 114–118.

Ramiro, Z.A., Batista Filho, A. and Machado, L.A. (1987) Levantamento de pragas e inimigos naturais em seis cultivares de soja. *Biológico* 53, 7–29.

Rezende, J.A.M. and Rossetto, C.J. (1980) Comportamento de populações paternais e F1 de soja em relação a *Colaspis* sp. e *Diabrotica speciosa* (Germar, 1824). *Bragantia* 39, 191–198.

Ribeiro, S.G., Kitajima, E.W. and Oliveira, C.R.B. (1996) A strain of eggplant mosaic virus isolated from naturally infected tobacco plants in Brazil. *Plant Disease* 80, 446–449.

Roberto, S.R., Genta, W. and Ventura, M.U. (2001) *Diabrotica speciosa* (Ger.) (Coleoptera: Chrysomelidae): new pest in table grape orchards. *Neotropical Entomology* 30, 721–722.

Roel, R. and Zatarin, M. (1989) Eficiência de iscas a base de abóbora d'água *Lagenaria vulgaris* (Cucurbitaceae) tratadas com inseticidas, na atratividade a *Diabrotica speciosa* (Germar, 1824) (Coleoptera: Chrysomelidae). *Anais da Sociedade Entomológica do Brasil* 18, 213–219.

Rose, R.I. and McCabe, J.M. (1973) Laboratory rearing techniques for the southern corn rootworm. *Journal of Economic Entomology* 66, 398–400.

Rossetto, C.J., Nagai, V., Igue, T., Rossetto, D. and Miranda, M.A.C. (1981) Preferência de alimentação de adultos de *Diabrotica speciosa* (Germar) e *Cerotoma arcuata* (Oliv.) em variedades de soja. *Bragantia* 40, 179–183.

Salles, L.A.B. (1998) Controle químico de pragas de solo na lavoura de batata. *Horticultura Brasileira* 16, 85–87.

Salles, L.A.B. (2000) Incidência de danos de *Diabrotica speciosa* em cultivares e linhagens de batata. *Ciência Rural* 30, 205–209.

Salles, L.A.B. and Grützmacher, A.D. (1999) Eficiência do inseticida clorpirifós no controle de larvas de *Diabrotica speciosa* (Germ.). *Ciência Rural* 29, 195–197.

Santos, A.B., Ferreira, E., Aquino, A.R.L., Sant'ana, E.P. and Baldt, A.F. (1988) População de plantas e controle de pragas em arroz com complementação hídric. *Pesquisa Agropecuária Brasileira* 23, 397–404.

Schalk, J.M. and Peterson, J.K. (1990) A meridic diet for banded cucumber beetle larvae (*Diabrotica balteata* LeConte). *Journal of Agricultural Entomology* 7, 333–336.

Shaw, S.R. (1995) A new species of Centistes from Brazil (Hymenoptera: Braconidae: Euphorinae) parasitizing adults of *Diabrotica* (Coleoptera: Chrysomelidae), with a key to New World species. *Proceedings of the Entomological Society of Washington* 97, 153–160.

Silva, O.C. (1999) Larva alfinete em milho. *Batavo* 7, 34–36.

Silva-Werneck, J.O., Faria, M.R., Abreu Neto, B.P., Magalhães, B.P. and Schimidt, F.G.V. (1995) Técnica de criação de *Diabrotica speciosa* (Germ.) (Coleoptera: Chrysomelidae) para bioensaios com bacilos e fungos entomopatogênicos. *Anais da Sociedade Entomológica do Brasil* 24, 45–52.

Stock, S.P. (1993) Mycoletzkya vidalae n. sp (Nematoda: Diplogasteridae), a facultative parasite of *Diabrotica speciosa* (Germ.) (Coleoptera: Chrysomelidae) from Argentina. *Research and Reviews in Parasitology* 53, 109–112.

Sutter, G.R., Krysan, J.L. and Guss, P.L. (1971) Rearing the southern corn rootworm on artificial diet. *Journal of Economic Entomology* 64, 65–67.

Tigano Milani, M.S., Carneiro, R.G., Faria, M.R., Frazao, H.S. and McCoy, C.W. (1995) Isozyme characterization and pathogenicity of *Paecilimyces fumosoroseus* and *P. lilacinus* to *Diabrotica speciosa* (Coleoptera: Chrysomelidae) and *Meloidogyne javanica* (Nematoda: Tylenchidae). *Biological Control* 5, 378–382.

Vardasca, L.A., Schiavetto, D.C. and Calafiori, M.H. (1989) Influence of *Diabrotica speciosa* (Germar, 1824) on maize (*Zea mays* L.) with organic and chemical fertilizers. *Ecossistema* 14, 158–162.

Ventura, M.U. and Ito, M. (2000) Antifeedant activity of *Melia azedarach* (L.) extracts to *Diabrotica speciosa* (Germ.) (Coleoptera: Chrysomelidae) beetles. *Brazilian Archives of Biology and Technology* 43, 215–219.

Ventura, M.U., Ito, M. and Montalvan, R. (1996) An attractive trap to capture *Diabrotica speciosa* (Ger.) and *Cerotoma arcuata tingomariana* Bechyne. *Anais da Sociedade Entomológica do Brasil* 25, 529–535.

Ventura, M.U., Mello, E.P., Oliveira, A.R.M., Simonelli, F., Marques, F.A. and Zarbin, P.H.G. (2001) Males are attracted by female traps: a new perspective for management of

Diabrotica speciosa (Germar) (Coleoptera: Chrysomelidae) using sexual pheromone. *Neotropical Entomology* 30, 361–364.

Veronesi, D.J., Sicchieri, M.A., Bexiga, M.A.S., Calafiori, M.H. and Teixeira, N.T. (1990) Efeito da adubação sobre a eficiência de inseticidas para o controle de *Diabrotica speciosa* (Germar, 1824) e mosaico dourado em feijoeiro, *Phaseolus vulgaris* L. *Ecossistema* 15, 5–10.

Weissling, T.J. and Meinke, L.J. (1991) Semiochemical-insecticidal bait placement and vertical distribution of corn rootworm (Coleoptera: Chrysomelidae) adults: implications for management. *Environmental Entomology* 20, 945–952.

Wheeler, D. (1996) The role of nourishment in oogenesis. *Annual Review of Entomology* 41, 407–431.

Wilcox, J.A. (1972) *Coleopterorum Catalogus Supplementa Pars 78 (Galerucinae: Luperini: Diabroticina)*. 2nd ed. BV Publishers, The Hague, pp. 296–431.

5 Potential Lepidopteran Pests Associated with Avocado Fruit in Parts of the Home Range of *Persea americana*

Mark S. Hoddle[1] and J.R.P. Parra[2]

[1]*Department of Entomology, and Center for Invasive Species Research, University of California, Riverside, California 92521, USA;* [2]*Departamento de Entomologia, Fitopatologia e Zoologia Agrícola, Escola Superior de Agricultura 'Luiz de Queiroz', Universidade de São Paulo, 13418-900, Brazil*

5.1 Introduction

Legal imports of fresh avocado (*Persea americana* Miller [Lauraceae]) fruit entering the USA, including California (the largest domestic producer of 'Hass'), are increasing steadily because of cumulative imports from several countries such as México, Chile, Perú, New Zealand and the Dominican Republic. Increasing avocado imports into the USA are due to advertising and promotion under the Hass Avocado Promotion and Research Order, and by the California Avocado Commission, along with various import associations realizing new business opportunities (Hoddle *et al.*, 2010). Another reason the USA market is rapidly expanding is the growing Hispanic population in the USA that regularly consumes avocados, as well as increasing consumer awareness that avocados are highly nutritious, tasty, versatile and easy to prepare. The overall growth in demand for 'Hass' in the USA has long exceeded domestic production, which simply cannot meet consumer demand. Consequently, imports of avocados from other countries are likely to increase for the foreseeable future to meet this growing demand for fresh fruit (Anonymous, 2006).

Importations of fresh produce into any area carry with them an easily identifiable risk: the threat of accidental introduction of unwanted arthropod pests, phytopathogenic diseases or a combination of both, in other words the pest threat as a vector for a pathogen. An excellent example of this latter possibility is the vector–pathogen complex represented by the red ambrosia bay beetle (*Xyleborus glabratus* Eichhoff [Coleoptera: Scolytidae]) and the laurel wilt fungus (*Raffaelea lauricola* Harrington and Fraedrich [Sordariomycetes: Ophiostomatales]) that is specific to native Lauraceae and avocados in the south-eastern USA (see Chapter 3). To minimize threats to importing regions, risk management programs are run by regulatory authorities representing the trading partners. The goal is to identify potential invasion threats which may be documented officially in risk assessment evaluations, with subsequent reports and analyses quantifying threat levels. The investigatory and evaluation processes at work are the basis of biosecurity practices enacted under law by countries that want to enforce quarantine restrictions. The application of these laws is to impose phytosanitary regulations governing imports and exports of commodities to prevent unwanted pest introductions, while minimizing restriction of free trade between cooperating nations.

This is an international trend, and overseen in North America by NAPPO (North American Plant Protection Organization), and specifically in the USA by USDA-APHIS.

Preparation of risk assessment reports necessitates that potential invasive pests be assigned to one of two general categories: (i) well-known pests in their home ranges that may or may not have a documented history of global movement; and (ii) the 'wild cards', species either unknown to science because they are undescribed or very poorly studied, and unrecognized as threats until they first become established outside their home range and cause unprecedented problems. Consequently, depending on the commodity under evaluation, risk assessments that are developed under free trade agreements may not be as robust as would be expected, because little effort has been invested in determining the risk posed by the unknown 'wild cards'. Realistically, there is little incentive for exporting nations to rigorously document their pest and disease fauna, as this would likely make exporting produce more difficult. Additionally, importing nations and their regulatory agencies lack time and resources to send scientists to partner countries to construct full pest inventories for commodities of mutual interest. Therefore, pest lists and risk assessments are constructed from literature reviews of the scientific and gray literature, interception documents and possibly museum collection records, for the commodity under consideration. The threat posed by potential invaders associated with avocado fruit, especially lepidopteran pests that could be moved via international trade, will be examined here in detail.

Avocados have an evolutionary range that includes the eastern and central highlands of México, Guatemala and the Pacific Coast of Central America as far south as Panama (Knight, 2002). Humans moved avocados from this area of origin into northern South America by 4000 BC where it became naturalized (Knight, 2002). Global avocado production statistics for 2009 indicated that four of the top ten producing nations are located in the native and naturalized range of this plant: México (ranked number 1); Central America (Guatemala, 10) and South America (Colombia, 6; and Perú, 8) (FAOSTAT, 2010). California in the USA is the fourth largest producer of avocados in the world (FAOSTAT, 2010). Importation of avocado fruit from México and Central America into the USA had been banned since 1914 to protect producers from the unwanted introductions of fruit-feeding pests, especially specialist internally feeding moths (this chapter), weevils (see Chapter 2) and tephritid flies (see Chapter 11), which have evolved with avocados (Hoddle and Hoddle, 2008a). However, recent legislative changes have allowed imports of fresh avocado fruit into the USA from areas where they had originally been banned. In 1997, the USA began importing avocados from México into restricted areas, and by 2007 year-round importations into all 50 US states were permitted (Morse et al., 2009).

Despite a long history of domestication and prodigious production of fruit for domestic and export markets, the pest fauna associated with avocados in the indigenous range of this plant is poorly documented and little studied. For example, three avocado foliage-feeding pests in California, *Tetraleurodes perseae* Nakahara (Hemiptera: Aleyrodidae), *Oligonychus perseae* Tuttle, Baker and Abbatiello (Acari: Tetranychidae) and *Scirtothrips perseae* Nakahara (Thysanoptera: Thripidae) were all species new to science at the time of first discovery in the USA. These three pests likely originated in the home range of *P. americana* (Hoddle, 2004), and molecular evidence confirmed this for *S. perseae* (Rugman-Jones et al., 2007). Recently, eight species of armored scales were found feeding on the skin of imported avocado fruit entering California from México (Morse et al., 2009). Of these eight species, at least one was previously unknown and was subsequently described (Evans et al., 2009), and two more are probably new species needing description (Morse et al., 2009). Only one of the eight scale species, *Hemiberlesia lataniae* Signoret (Hemiptera: Diaspididae), intercepted on fresh avocados exported from México, is established in California (Morse et al., 2009). This clearly suggests that imports of fresh avocados into the USA from the area of origin of this plant are contaminated with living insects, some of which are unknown species, and may pose a substantial new pest risk to California and other countries that receive fruit from these export areas.

The potential pest threat posed by fresh avocado fruit imports to local avocado producers is not unique to California. Israel and Spain are countries with domestic avocado industries that also import fruit from countries where this crop is naturalized, such as Perú, and they have suffered invasions from specialized avocado pests (e.g., *O. perseae*) that are native to the home range of this plant. This trend of increasing fruit importation from overseas – especially countries where avocados are indigenous and likely to have a rich but little-studied insect fauna – into nations where avocados are not native but where domestic production exists, is unlikely to stop or reverse itself as long as market demand and good financial returns exist. Consequently, in addition to established import sources for the USA, other countries including Colombia, Guatemala, South Africa, Spain and Uganda are currently seeking entry into the US market for fresh avocado fruit (USDA-APHIS, 2010). To meet head-on the threat posed by invasive pest species associated with fresh avocados, importing countries with domestic markets need to develop forward-leaning policies. These must simultaneously manage increasing import loads and the pest threats that will likely accompany an expanding pool of exporting countries that are both within (e.g., México and Central America) and outside (e.g., Uganda and Africa in general) the evolutionary area of origin of the avocado. Each exporting country presents a unique set of pest risks. Invaders could be co-evolved specialists from the home range of the avocado that may have a very limited host range, and represent a risk only to avocados and close relatives; or they may be generalist pests from within and outside the area of origin of the avocado, that can feed and breed on avocados as well as pose a threat to other existing crops and native flora. The remainder of this chapter will focus on potential invasion threats posed by well-recognized and poorly known lepidopteran pests associated with avocado fruit in the home range of this plant.

5.2 Lepidopteran Pests Associated with Avocado Fruit

Worldwide, 111 species of Lepidoptera in 73 genera representing 22 families have a documented association with avocados (HOSTS, 2010). The pest status and invasion potential of the majority of these species are not well understood. The most notorious of these fruit-feeding lepidopteran pests is *Stenoma catenifer*, but other species have recently been discovered attacking small and large avocado fruit in part of the native range of this plant (Hoddle and Brown, 2010).

5.2.1 *Stenoma catenifer* Walsingham (Lepidoptera: Elachistidae)

Common names

Avocado seed moth, avocado moth, taladrador del fruto del aguacate, da broca-do-abacate, la oruga barrenadora del hueso del aguacate, el barrenador del fruto del palto, pasador del fruto del aguacate, palomilla barrenadora del hueso del aguacate, chenille de la graine de avocatier, broca-do-fruto do abacateiro.

Distribution, description and taxonomy

Over 350 species of moths are described in the genus *Stenoma* (Becker, 1984), but this number could exceed 730 species (zipcodeZoo.com, 2010). Three species of *Stenoma* have been recorded in association with avocados: *S. catenifer*, *S. invulgata* Meyrick (Trinidad and Tobago) and *S. vacans* Meyrick (Neotropical) (HOSTS, 2010). *S. catenifer* was originally described by Walsingham (1909–1915) and there are no species synonyms for *S. catenifer* (J.W. Brown, pers. comm., 2010).

S. catenifer is native to Neotropical areas, and is considered to be one of the most important pests of avocados in México, Central and South America (Wysoki *et al.*, 2002). In total, there are at least 16 countries with commercial avocado industries that have native *S. catenifer* populations, and collectively they produce ~25% of the world's avocados (FAOSTAT, 2010). Several of these countries with native *S. catenifer* populations, such as México and Perú, export avocados from certified areas to countries that have domestic avocado production, but are outside the native range of *S. catenifer*. This pest has been recorded attacking avocados in México (Muniz Velez, 1958; Mendez Villa, 1961; Wolfenbarger and Colburn, 1966, 1979; Arellano, 1975), Guatemala (Popenoe, 1919), Belize (CABI, 2001), Honduras (Sasscer, 1921), Nicaragua (CABI, 2001), El Salvador (Wolfenbarger and Colburn, 1966), Costa Rica and Panama (CABI, 2001), Venezuela (Boscán de Martínez and Godoy, 1984), Argentina, Colombia and Ecuador

(CABI, 2001), Guyana (Cervantes Peredo et al., 1999), and Perú (Wille, 1952). In Brazil, *S. catenifer* is the major pest limiting commercial avocado production (Hohmann et al., 2003).

All life stages of *S. catenifer* have been described and line illustrated (Cervantes Peredo et al., 1999), and coloured photographs are available (Hoddle, 2011a).

Biology and ecology

Stenoma catenifer is a specialist on plants in the family Lauraceae (Cervantes Peredo et al., 1999). In avocados, primary economic damage results from larvae tunneling in fruit as they bore towards the seed where most of the feeding typically occurs. Infested fruit are characterized by accumulations of frass on the surface, and perseitol, a seven-carbon sugar alcohol, appears as a chalky-white residue that often oozes from feeding-tunnel entrances. Feeding-tunnel entrances (53%) are most often situated in the bottom-third of the fruit distal to the pedicel (Hoddle and Hoddle, 2008b). On average, ~1–2 larvae infest a single avocado fruit, but up to seven or eight larvae can be found in a single fruit (Hoddle and Hoddle, 2008b; Nava et al., 2006). Fruit of *Chlorocardium rodiei* (Schomb.) Rohwer, Richter and van der Werff are infested with ~3 larvae per fruit on average, with a maximum of 21 larvae being recorded (Cervantes Peredo et al., 1999). In Brazil, within-tree distribution of fruit damaged by *S. catenifer* differs between study sites, with infestations either being most concentrated in the upper half of trees (Hohmann et al., 2003) or in the middle to lower parts (Nava et al., 2006).

Larvae of this pest can also mine developing buds, green and woody avocado twigs and stems, and fruit pedicels (Wille, 1952; Wolfenbarger and Colburn 1966, 1979). Larval attacks can kill small trees and can cause considerable crop loss by causing premature fruit drop (Wille, 1952; Ventura et al., 1999). Because fruit are not present on trees year-round, *S. catenifer* probably maintains constant populations in orchards by feeding inside twigs and stems, which are always present. In the field, evaluation of 20 different avocado cultivars showed varying levels of susceptibility to *S. catenifer* infestation, with 5–54% of fruit infested depending on the cultivar (Hohmann et al., 2000). Commercial 'Hass' avocado orchards in Guatemala, under regular pesticide treatments, can have ~45% of fruit damaged by *S. catenifer*, suggesting that this cultivar is vulnerable to *S. catenifer* (Hoddle and Hoddle, 2008b). Similarly high levels of damage have been observed in avocado orchards in Brazil, with up to 60% fruit infestation (Hohmann et al., 2003; Nava et al., 2005c, 2006), Venezuela (~80%; Boscán de Martínez and Godoy, 1982), Perú (~80–100% on non-managed farms; M.S. Hoddle, pers. obs., 2010) and México (Wysoki et al., 2002).

In addition to avocados and *C. rodiei*, other hosts include *Persea schiedeana* Nees and *Beilschmedia* sp. (Cervantes Peredo et al., 1999), *Nectandra megapotamica* Mez and *Cinnamomum camphora* (L.) (Link and Link, 2008). In a non-avocado host, *C. rodiei*, fruit infestation rates did not exceed 10% (Cervantes Peredo et al., 1999). Similarly for *N. megapotamica* and *C. camphora*, fruit infestation rates did not exceed 5% (Link and Link, 2008). Because of high infestation rates, avocado fruit – especially commercial cultivars – may be a highly preferred host for *S. catenifer*.

In the laboratory, female *S. catenifer* display strong oviposition preferences for different avocado substrates, with the majority of eggs (~68%) being on woody branches to which fruit are attached, and ~9%, ~10%, and ~12% of eggs being laid on the fruit pedicel, between the pedicel and the fruit, and directly on the fruit, respectively (Hoddle and Hoddle, 2008b). In addition, female moths appear to discriminate between different types of avocado fruit when ovipositing. In choice tests, 'Hass' fruit are strongly preferred (2.69 times more favored) for egg laying when smooth-skinned non-'Hass' fruit are available simultaneously in experimental cages (Hoddle and Hoddle, 2008b). Avocado fruit are needed to stimulate ovipositional activity, and when fruit is present, female moths will oviposit onto artificial substrates with textured surfaces, of which quilted paper towels are most preferred. In the laboratory with a 14:10 L:D phase (scotophase initiated at 2000 h), oviposition commences at 1800 h and ends at 0600 h the following morning. Peak ovipositional activity occurs from 2000 h to 2400 h when 80% of eggs were laid (Nava et al., 2005a). Egg-laying habits appear to be mainly crepuscular or nocturnal (Nava et al., 2005a).

Adult *S. catenifer* are nocturnal, and flight activity begins immediately at dusk. During the day adult moths hide on the ground, the beige coloration and black spots on the forewings helping to camouflage them, especially when they are resting in dried grasses in avocado orchards (Hoddle, 2011a; Hoddle and Hoddle, 2008c). Resting moths will fly short distances when disturbed during the day (Cervantes Peredo et al., 1999).

Developmental and reproductive biology

Life-history characteristics of *S. catenifer* have been well studied in the laboratory (Cervantes Peredo et al., 1999; Boscán de Martínez, 1984; Nava et al., 2005a, b). Eggs can hatch in as few as 4 d at ~28°C or higher (Nava et al., 2005b); the five larval instars (Cervantes Peredo et al., 1999) can take 19–40 d to reach the pupal stage depending on the temperature (Nava et al., 2005b). In the laboratory, mature larvae, easily identified by the turquoise-blue color of the ventral surface, abandon seeds and fruit and typically walk for ~24–36 h before settling in a protected place to pupate within the protection of a loosely spun bivouac of silk (Hoddle and Hoddle, 2008b; Hoddle, 2011a). Larvae apparently pupate in the upper ~2 cm of soil in orchards (Boscán de Martínez and Godoy, 1984). Pupae require 8–20 d to complete development at 18–30°C, and the entire life cycle (egg to adult) requires 31–70 days over the same temperature range (Nava et al., 2005b). Adult males can live for 11–18 days, and female longevity is similar at 20–30°C (Nava et al., 2005b). Female moths can lay 133–319 eggs when reared on avocados at 20–30°C; when fruit of *C. rodiei* is used as the developmental substrate, average female fecundity is 206 eggs (Cervantes Peredo et al., 1999). The sex ratio is slightly female-biased at 1.46 females (Cervantes Peredo et al., 1999), and longevity and fecundity of adult male and female moths is not enhanced through the provision of carbohydrates (10% honey water) in comparison to adults given access to water only (Milano et al., 2010).

Laboratory-derived developmental data have been used to develop a degree-day model for *S. catenifer*, and up to five generations per annum may be possible in some parts of Brazil (Nava et al., 2005b).

Control measures

SEX PHEROMONE. The sex pheromone of *S. catenifer* is an unsaturated aldehyde, (9Z)-9, 13-tetradecadien-11-ynal. The presence of the alkyne is a very unusual functional group in lepidopteran pheromones, as is the high degree of unsaturation (Millar et al., 2008). The pheromone is a new class of natural compound, as the terminal conjugated dienyne has no precedent among known natural products (Millar et al., 2008). Synthesis instructions for the pheromone are available (Hoddle et al., 2009; Zou and Millar, 2010). The components of the pheromone (i.e., different ratios, and blends of the corresponding alcohols and acetates) have been field-tested, and results indicate that only (9Z)-9, 13-tetradecadien-11-ynal is needed to attract male moths. Concentrations of pheromone ranging from 10 µg to 1 mg are efficacious, and lures retain field activity for several weeks (Hoddle et al., 2009). Males show temporal responses to the pheromone with first arrival to traps in the field occurring at 0230 h and ceasing abruptly at 0430 h (Hoddle et al., 2009).

Operational parameters for the sex pheromone have been optimized under field conditions. Gray rubber septa are superior to polyethylene vials and small plastic bags for releasing pheromone. Trap height placement in avocado trees in commercial orchards is not important, as traps capture statistically equivalent numbers of moths when hung at ~0.15 m, ~1.75 m and ~4.5 m above the ground. Hanging traps at 1.75 m is recommended, as this height is convenient for placement and inspection (Hoddle et al., 2011). Probabilistic modeling of *S. catenifer* capture data, from different-sized commercial avocado orchards (~2–~76 ha) with varying levels of infestation, suggest that 10–13 randomly placed traps will capture at least one male *S. catenifer* with 90% confidence over a 7-d sampling period if the pest is present (Hoddle et al., 2011). It is likely that small orchards of ≤ 2 ha could be adequately monitored with fewer traps (e.g., 2–5 randomly placed traps), whereas for large orchards, 10–13 traps could be sufficient.

The pheromone has attracted males in Guatemala, México, Brazil and Perú, indicating that geographically separated pheromone races of *S. catenifer* probably do not exist (Hoddle et al., 2011). Interestingly, in avocado orchards in Guatemala and México, the pheromone was also highly attractive to another closely related moth, *Antaeotricha nictitans* (Zeller) (Lepidoptera: Elachistidae: Stenomatinae) (see below). This result could suggest that the *S. catenifer* sex pheromone is conserved among New World stenomatines and differing temporal responses to the pheromone separate sympatric species (Hoddle et al., 2011).

APPLICATIONS. With the sex pheromone, opportunities exist that may enable the development of new control strategies for *S. catenifer*. Such approaches could include mating disruption via the distribution of pheromone dispensers in orchards, which collectively make it difficult for males to find females for mating. Alternatively, the sex pheromone could allow the implementation of 'attract and kill' management for this pest. This concept is simple. A wax-based matrix impregnated with pheromone and a small amount of pesticide is applied as a few small droplets to trunks of avocado trees. The pheromone attracts male moths which pick up a debilitating dose of insecticide from contact with the droplets. Pest populations collapse because males are rapidly removed, thereby reducing mating rates with females.

It is recommended that the *S. catenifer* pheromone should be used to monitor for this pest in orchards exporting fresh avocado fruit from countries that have native populations, such as México and Perú, and year-round pheromone monitoring should be required for exporters as part of a certification program. In addition, regions with domestic avocado production that also import fresh fruit from high-risk areas (e.g., California and Spain) could use the pheromone to detect *S. catenifer* incursions. Early detection could facilitate the deployment of a rapid response to contain or eradicate the pest while populations are small and highly localized. In addition to monitoring *S. catenifer* in avocado orchards, the sex pheromone could be used to monitor population phenologies in natural areas, to better understand the basic ecology of this insect. Geographical and altitudinal distributions could be identified in countries with native populations of *S. catenifer* which may delineate areas naturally free of this pest, for example at altitudes > 2000 m where the climate is unfavourable but avocados are able to grow.

NATURAL ENEMIES. The eggs, larvae, pupae and adults of *S. catenifer* are attacked by a variety of natural enemies in the home range of this insect. In Brazil, eggs are parasitized by species of *Trichogramma* and *Trichogrammatoidea* (Hymenoptera: Trichogrammatidae). Although naturally occurring parasitism may exceed 60% in avocado orchards, it is not sufficient to prevent severe yield losses (Hohmann et al., 2003). In an attempt to remedy this shortcoming, Nava et al. (2007) evaluated in the laboratory the life-history traits of four different egg parasitoids (all Trichogrammatidae) and up to eight strains of individual species for mass rearing, to determine which would be most efficacious against *S. catenifer*. The best-performing parasitoids determined by egg parasitism rates were *Trichogramma atopovirilia* Oatman and Platner, and *Trichogrammatoidea annulata* De Santis. However, release rates of 28–30 parasitoids per host egg were estimated to be necessary to obtain satisfactory levels (~70%) of parasitism.

Larvae are attacked by a diverse guild of hymenopteran, and to a much lesser extent, dipteran parasitoids. In Guatemala, a gregarious *Apanteles* sp. (Hymenoptera: Braconidae) dominates the parasitoid fauna, accounting for >95% of parasitism (Hoddle and Hoddle, 2008b). Solitary *Macrocentrus* sp. (Braconidae), *Pristomerus* sp. and *Brachycyrtus* sp. (both Hymenoptera: Ichnuemonidae) collectively account for <5% parasitism (Hoddle and Hoddle, 2008b, c; Hoddle et al., 2011). In South America, *S. catenifer* larvae are attacked by an *Apanteles* sp. in Venezuela which inflicts ~30% parasitism (Boscán de Martínez and Godoy, 1982). In Perú, parasitism by *Apanteles* sp. is considered inconsequential (Wille, 1952). In Guyana, *Eudeleboea* sp. (Hymenoptera: Ichneumonidae) and *Chelonus* sp. (Braconidae) attack *S. catenifer* larvae infesting *C. rodiei* at very low levels (Cervantes Peredo et al., 1999). Five species of braconids, *Dolichogenidea* sp., *Hypomicrogaster* sp., *Apanteles* sp., *Chelonus* sp. and *Hymenochaonia* sp., and two species of ichneumonids, *Eudeleboea* sp. and *Pristomerus* sp., have been recorded attacking *S. catenifer* larvae in Brazil. *Dolichogenidea* sp. and *Apanteles* sp. dominate this guild and inflict peak parasitism of 30–40% (Nava et al., 2005c).

In Perú, the dominant larval parasitoid is an *Apanteles* sp., and a tachinid fly, *Chrysodoria* sp. has been reared from *S. catenifer* pupae (Hoddle and Hoddle, 2012). Spiders, in particular *Hogna* sp. (Lycosidae), are voracious predators of *S. catenifer* larvae, pupae and adults in the laboratory (Hoddle and Hoddle, 2008c). These spiders are extremely common on orchard floors in Guatemala, and may be important natural enemies attacking wandering larvae that have abandoned avocado seeds or fruit in search of pupation sites, of pupae in the soil, or adults resting on the orchard floor (Hoddle and Hoddle, 2008c). These life stages do not appear to be chemically protected via the sequestration of toxic avocado furans (Hoddle and Hoddle, 2008c).

CULTURAL CONTROL, STERILE INSECT TECHNIQUE AND INSECTICIDES. Cultural control recommendations for *S. catenifer* include removal and destruction (burning) of infested fruit, branches and shoots (Wille, 1952), or placing dropped fruit (~60 kg) infested with larvae in clear plastic bags that are airtight. After c. 4 days at 20–23°C, larval mortality is almost 100%. Elevated temperatures or the build-up of noxious gases inside sealed bags may contribute to increased rates of larval mortality (Nava et al., 2006). Control of weeds on the orchard floor may reduce hiding places for moths during the day, thereby lowering adult densities and subsequent mating and oviposition events (M.S. Hoddle, pers. obs.). Planting of highly susceptible avocado cultivars in orchards could act as trap crops, protecting economically valuable cultivars (Ventura et al., 1999), or resistant cultivars should be grown instead of susceptible cultivars (Hohmann et al., 2000).

Gamma radiation from cobolt-60 can reduce egg viability by 81% at doses of 50 Gy and 100% when eggs are irradiated with ≥ 75 Gy. When eggs are irradiated with 25 Gy (a dose sufficient to kill 12% of individuals), the levels of sterility in males and females that develop from these eggs are 93% and 100%, respectively (Silva et al., 2006). Doses of 75 Gy have been recommended for treating avocado fruit that could be contaminated with *S. catenifer* eggs (Silva et al., 2006). A higher dose, 150 Gy, is recommended for controlling larvae and pupa within fruit, a dose that is also sufficient to sterilize ~70% of treated pupae (Silva et al., 2007). At 200 Gy, irradiated males or females that breed with irradiated or unirradiated partners produce eggs with 0–3% viability, a result statistically equivalent to adults treated with 300 or 400 Gy. Therefore, for sterile insect production, a treatment dose of 200 Gy is recommended for *S. catenifer* (Silva et al., 2007).

Insecticides provide the primary control strategy for *S. catenifer*, although there are few published data documenting efficacy, application rates and timings for different products. Foliar applications will only be effective against eggs, small larvae penetrating fruit and stems, and adults landing on treated foliage. Once inside fruit and twigs, larvae are likely protected from sprays, while pupae will be protected in the soil. It is possible that systemic insecticides will not control larvae in fruit because leaves are better sinks than fruit.

Applications of pyrethroids can significantly reduce fruit infestation rates with applications being made at intervals of either 15 or 60 d, having statistically equivalent efficacy (Hohmann et al., 2000). Monthly applications of endosulfan and malathion on a rotating calendar schedule maintained *S. catenifer* infestation levels at ~45% in a commercial 'Hass' avocado orchard in Guatemala (Hoddle and Hoddle, 2008b). Similar calendar spray regimens have been reported from México as being necessary to minimize economic losses (Wysoki et al., 2002). Spray applications are affected by increasing tree height and architectural complexity, which can significantly limit control because coverage is reduced (Hohmann et al., 2000). In addition, it is assumed that broad-spectrum insecticides have severe negative effects on *S. catenifer* natural enemies, especially egg parasitoids, and applications may induce outbreaks of other avocado pests such as mites (Hohmann et al., 2000). To maximize impacts, pesticides should be applied at times of the year when *S. catenifer* populations are naturally low in avocado orchards (December, January and February in Brazil), thereby prematurely retarding population growth in advance of favorable conditions that promote pest outbreaks (Nava et al., 2006).

Invasion potential

Stenoma catenifer has demonstrated an ability to invade and establish in new areas in which it has not been previously recorded, either through the movement of infested fruit, or via the movement of new planting material that has infested branches, twigs or shoots.

In 2000, *S. catenifer* was recorded for the first time from Isla Santa Cruz, in the Galápagos Islands, Ecuador (Landry and Roque-Albelo,

2003). By 2003, the moth had established on additional islands in the Galápagos where avocados are grown, i.e., San Cristóbal, (Landry and Roque-Albelo, 2003). The mode of entry into the Galápagos is uncertain, but could have been from the importation of infested avocados from mainland Ecuador, an item intercepted previously by quarantine officials (Landry and Roque-Albelo, 2003). As recently as 2009, infested avocado fruit imported from mainland Ecuador were found for sale in a supermarket on Santa Cruz, indicating that quarantines are not robust enough to prevent continued introductions (Hoddle, 2011a). Sasscer (1921) reported interceptions of *S. catenifer* entering the USA in fruit imported from Honduras. Importation of fruit from areas with native or invasive *S. catenifer* populations increases the risk that this pest will be inadvertently introduced into new areas.

The establishment of *S. catenifer* in the Galápagos has reduced the viability of a commodity that could be locally produced and sold. It has also necessitated increased imports of fresh avocados from mainland South America, which have elevated costs of a popular food for local people, and further increased the risk of additional pests entering the Galápagos Islands. Fortunately, there are no endemic Lauraceae in the Galápagos, and avocados are the only representative of this plant family in these islands.

Limited fruit surveys suggest that infestation levels are extremely high, possibly reaching 90% around Bella Vista on Santa Cruz. In October 2009, 40 dropped fruit and exposed seeds were collected below avocado trees around Bella Vista. From these 40 fruit, 21 *S. catenifer* larvae were reared. Twenty larvae developed into pupae. One larva died from parasitism caused by a gregarious *Apanteles*(?) sp. However, all nine parasitoid larvae died as pupae, suggesting that parasitism may have been the opportunistic exploitation of an unsuitable host. Seven of the 20 pupae died and 13 adults emerged (62% larval-to-adult survivorship rate) (Hoddle, 2011a). *S. catenifer* may be flourishing in the Galápagos because of natural-enemy-free space.

Other Lepidoptera associated with avocados in parts of the native range of this plant

Comprehensive rearing studies for Lepidoptera associated with avocado fruit have been conducted in Guatemala (Hoddle and Hoddle, 2008a; Hoddle and Brown, 2010). Over the period November 2006 to January 2009, 1078 small 'Hass' avocado fruit (5–25 mm in length) and 7742 'Hass' and non-'Hass' avocados ≥ 100 mm in length (for a total of 8820 fruit) were collected and held for the emergence of Lepidoptera. A total of 1098 specimens representing ten moth species in ten genera and four families were reared (Table 5.1) (Hoddle and Brown, 2010). *S. catenifer* accounted for 91% of the reared material. However, this result is possibly biased, because this pest was the primary focus of collecting and rearing efforts, and other species associated with avocado fruit could be under-represented. Tortricidae had the most representatives, with six species. Two new moth species were discovered and subsequently described, *Histura perseavora* Brown (Tortricidae) (Brown and Hoddle, 2010), and *Holcocera plagatola* Adamski (Coleophoridae) (Adamski and Hoddle, 2009).

Cryptaspasma sp. nr *lugubris* was reared in large numbers across multiple collection sites from fruit that was collected intact from the ground and fruit picked directly from the same trees and co-mingled but separated by locality. It was also reared from fruit purchased from commercial vendors (Hoddle and Hoddle, 2008a). It is uncertain if this moth is an unrecognized avocado pest that attacks fruit hanging in trees, causing them to drop prematurely, or is a specialist of hard seeds and one that only attacks fruit after it has dropped to the ground (Brown and Brown, 2004). Laboratory studies indicated that this moth would oviposit on intact fruit and exposed avocado seeds. Field trials with intact fruit placed on the ground in avocado orchards indicated that fruit had a high probability of being consumed by small animals within several days. The length of time from placement in the field prior to consumption was inadequate for egg hatch to occur, and for *C.* sp. nr *lugubris* larvae to bore into the seeds, where they would be safe from accidental predation (Hoddle and Hoddle, 2008a). This observation tentatively suggests that oviposition on hanging fruit would need to occur if larvae are to have sufficient time to reach the protection of the avocado seed prior to fruit prematurely dropping to the ground as a result of larval feeding (Hoddle and Hoddle, 2008a). The pest status of *C.* sp. nr *lugubris* urgently needs resolving, as this species has been reared from avocados in Michoacán, México, an area with major 'Hass' production and international

Table 5.1. Species of Lepidoptera reared from 1078 small (5–25 mm in length) 'Hass' avocado fruit and 7742 large (≥ 100 mm in length) 'Hass' and non-'Hass' avocados collected in Guatemala. Rearing data for this table are from Hoddle and Hoddle (2008a) and Hoddle and Brown (2010). Also included is *Antaeotricha nictitans*, a species caught in *S. catenifer* pheromone traps placed in avocado orchards in Guatemala and México.

Species	Family	Pest status
Argyrotaenia urbana (Busck)	Tortricidae	Small-fruit specialist? First record from Guatemala
Polyortha n. sp.	Tortricidae	Small-fruit specialist? First record from Guatemala
Netechma pyrrhodelta (Myerick)	Tortricidae	Opportunist? First record from Guatemala and from avocados
Euxoa sorella Schaus	Noctuidae	Opportunist?
Micrathetis triplex Walker	Noctuidae	Opportunist?
Holcocera plagatola Adamski	Coleophoridae	Opportunist? New species and first record of this genus from Guatemala
Amorbia santamaria Phillips and Powell	Tortricidae	Known avocado pest, originally described from Guatemala
Histura perseavora Brown	Tortricidae	New avocado pest? Only known from Guatemala. First host plant record for this genus
Cryptaspasma sp. nr *lugubris* (Myerick)	Tortricidae	Previously unrecognized avocado pest?
Antaeotricha nictitans (Zeller)	Elachistidae	New avocado pest? Captured in *S. catenifer* pheromone traps in Guatemala and México
Stenoma catenifer Walsingham	Elachistidae	Well-recognized avocado pest

exports. Photographs of *C.* sp. nr *lugubris* eggs, larvae, pupae, adults, feeding damage and its parasitoids are available (Hoddle, 2011b).

Field testing of the *S. catenifer* sex pheromone in México and Guatemala resulted in the identification of a potential new avocado pest. Pheromone traps baited with (9Z)-9, 13-tetradecadien-11-ynal attracted substantial numbers of a related moth, *Antaeotricha nictitans* (Zeller) (Lepidoptera: Elachistidae: Stenomatinae). This moth was very abundant in traps set in avocado orchards at 400–700 m in Escuintla, Guatemala and Tapachula, México. Captures of *A. nictitans* were 7–9 times higher than captures of *S. catenifer* at these sites. *A. nictitans* is easily separable from *S. catenifer* based on size (*A. nictitans* is larger) and numbers of spots on the forewings (*A. nictitans* has just one central spot; *S. catenifer* has many, especially at the distal margin of the forewing). Photographs comparing *A. nictitans* and *S. catenifer* are available (Hoddle, 2011a). The significance of this moth in avocado orchards is unknown because there is very little information about its host-plant preferences (it has not been reared from avocados) or its associated natural enemy fauna. The avocado orchard in Guatemala that yielded *A. nictitans* was surrounded by pineapple and rubber tree plantations. In México, 'Hass' avocados were planted as a cover crop for coffee, and surrounding this coffee farm were native forest remnants and coffee plantations. Trapping results suggest that *A. nictitans* has a close association with avocados because this is the only plant common to sampling sites in Guatemala and México. If this assumption is correct, *A. nictitans* may be a previously unknown pest of avocados grown at low elevations. This possibility needs investigation.

5.3 Conclusions

The primary lepidopteran pest attacking avocados in the native range of this plant is *S. catenifer*. This pest has the potential to be accidentally moved into new areas via the importation of infested fruit or plants that originate in the home range of this pest. This risk is highest from countries with native populations of *S. catenifer*, and commercial avocado industries that export fruit to countries with domestic avocado production that are outside of the natural range of this pest. In the absence of natural enemies, abundant hosts and a permissive climate, *S. catenifer* is likely to flourish should it invade new areas. This has been observed in the Galápagos Islands. The sex pheromone for *S. catenifer* has been identified and field efficacy has been demonstrated in several countries. Mandatory monitoring with the pheromone should be implemented for all producers in countries with native populations of *S. catenifer* that export fresh avocados and planting stock. This moth is not only a risk to commercial avocado producers and home owners with backyard fruit trees, but it could also pose a significant incursion threat to native species of Lauraceae in countries that have no evolutionary association with insects that cause similar feeding damage.

Rearing studies for Lepidoptera associated with avocados in the native range of this plant have been very instructive, especially with respect to Guatemala. Survey results have clearly illustrated how poorly known are the moth fauna and associated natural enemies in countries where avocados are native and the biodiversity associated with this plant is expected to be greatest. At a minimum, the potential pest status of *C.* sp. nr *lugubris*, *H. perseavora* and *A. nictitans* needs to be resolved. It is recommended that surveys for these insects in other countries where avocados are native and have export industries (such as México) should be undertaken. These should determine their presence or absence, geographic distribution and abundance, and allow the development of monitoring and management tools (e.g., pheromones) if they are present and pestiferous, and pose an invasion risk to other countries.

References

Adamksi, D. and Hoddle, M.S. (2009) A new *Holcocera* Clemens from Guatemala and redescription of *H. icyaeella* (Riley) from the United States (Lepidoptera: Coleophoridae: Blastobasinae: Holcoerini): two congeners with an incidental preference for avocado. *Proceedings of the Entomological Society of Washington* 111, 254–262.

Anonymous (2006) Avocado situation and outlook for selected countries. World Horticultural Trade and U.S. Export Opportunities, www.fas.usda.gov/htp/Hort_Circular/2006/05-06/Avocados%205-10-06.pdf, accessed 1 December 2010.

Arellano, P.G. (1975) Key for the identification of larvae of avocado stone borers in México. *Folia Entomologica Mexicana* 31–32, 127–131.

Becker, V.O. (1984) Oecophoridae. In: Heppner, J.B. (ed.) *Atlas of Neotropical Lepidoptera. Checklist: Part I. Micropterigidae - Immoidae.* W Junk Publishers, The Hague, pp. 27–40.

Boscán de Martínez, N. and Godoy, F.J. (1982) *Apanteles* sp. (Hymenoptera: Braconidae) parasito del taladrador del aguacate *Stenoma catenifer* Walsingham (Lepidoptera: Stenomatidae) en Venezuela. *Agronomia Tropical* 32, 205–208.

Boscán de Martínez, N. and Godoy, F.J. (1984) Observaciones preliminares sobre la biologia de *Stenoma catenifer* Walsingham (Lepidoptera: Stenomidae) taladrador del aguacate (*Persea americana* Mill.). *Agronomia Tropical* 34, 205–208.

Brown, J.W. and Brown, J.L. (2004) A new species of *Cryptaspasma* Walsingham (Lepidoptera: Tortricidae: Olethreutinae) from Cental America, the Caribbean, and southeastern United States, with a catalogue of the world fauna of Microcorsini. *Proceedings of the Entomological Society of Washington* 106, 288–297.

Brown, J.W. and Hoddle, M.S. (2010) A new species of *Histura* Razowski (Lepidoptera: Tortricidae: Polyorthini) from Guatemala attacking avocados (*Persea americana*) (Lauraceae). *Proceedings of the Entomological Society of Washington* 112, 10–21.

CABI (2001) Crop Protection Compendium. www.cabi.org/cpc, accessed 12 July 2012.

Cervantes Peredo, L., Lyal, C.H.C. and Brown, V.K. (1999) The stenomatine moth, *Stenoma catenifer* Walsingham: a pre-dispersal seed predator of greenheart (*Chlorocardium rodiei* (Schomb.) Rohwer, Richter, and van der Werff) in Guyana. *Journal of Natural History* 33, 531–542.

Evans, G.A., Watson, G.W. and Miller, D.R. (2009) A new species of armored scale (Hemiptera: Coccoidea: Diaspididae) found on a vocado fruit from México and a key to the species of armored scales found on avocados worldwide. *Zootaxa* 1991, 57–68.

FAOSTAT (2010) Food and agricultural commodities production. http://faostat.fao.org/site/339/default.aspx, accessed 29 December 2010.

Hoddle, M.S. (2004) Invasions of leaf feeding arthropods: why are so many new pests attacking

California-grown avocados? *California Avocado Society Yearbook* 87, 65–81.

Hoddle, M.S. (2011a) The avocado seed moth, *Stenoma catenifer* Walsingham (Lepidoptera: Elachistidae). www.biocontrol.ucr.edu/Stenoma/Stenoma.html, accessed 31 January 2011.

Hoddle, M.S. (2011b) *Cryptaspasma* sp. http://biocontrol.ucr.edu/cryptaspasma/cryptaspasma.html, accessed 24 July 2012.

Hoddle, M.S. and Brown, J.W. (2010) Lepidoptera associated with avocado fruit in Guatemala. *Florida Entomologist* 93, 649–650.

Hoddle, M.S. and Hoddle, C.D. (2008a) Lepidoptera and associated parasitoids attacking Hass and non-Hass avocado in Guatemala. *Journal of Economic Entomology* 101, 1310–1316.

Hoddle, M.S. and Hoddle, C.D. (2008b) Bioecology of *Stenoma catenifer* (Lepidoptera: Elachistidae) and associated larval parasitoids reared from Hass avocados in Guatemala. *Journal of Economic Entomology* 101, 692–698.

Hoddle, M.S. and Hoddle, C.D. (2008c) Aspects of the field ecology of *Stenoma catenifer* (Lepidoptera: Elachistidae) infesting Hass avocados in Guatemala. *Florida Entomologist* 91, 693–694.

Hoddle, M.S. and Hoddle, C.D. (2012) Surveys for Stenoma catenifer (Lepidoptera: Elachistidae) and associated parasitoids infesting avocados in Perú. *Journal of Economic Entomology* 105, 402–409.

Hoddle, M.S., Millar, J.G., Hoddle, C.D., Zu, Y. and McElfresh, J.S. (2009) Synthesis and field evaluation of the sex pheromone of *Stenoma catenifer* (Lepidoptera: Elachistidae). *Journal of Economic Entomology* 102, 1460–1467.

Hoddle, M.S., Arpaia, M.L. and Hofshi, R. (2010) Mitigating invasion threats to the California avocado industry through collaboration. *California Avocado Society Yearbook* 92, 43–64.

Hoddle, M.S., Millar, J.G., Hoddle, C.D., Zu, Y., McElfresh, J.S. and Lesch, S.M. (2011) Field optimization of the sex pheromone of *Stenoma catenifer* (Lepidoptera: Elachistidae): evaluation of lure types, trap height, male flight distances, and number of traps needed per avocado orchard for detection. *Bulletin of Entomological Research* 101, 145–152.

Hohmann, C.L., Dos Santos, W.J. and Menegium, A.M. (2000) Availiação de técnicas de manejo para controle da broca-do-abacate, *Stenoma catenifer* (Wals.) (Lepidoptera: Oecophoridae). *Revista Brasileira de Fruticultura* 22, 359–363.

Hohmann, C.L., Menegium, A.M., Andrade, E.A., Garcia de Novaes, T. and Zandoná, C. (2003) The avocado fruit borer, *Stenoma catenifer* (Wals.) (Lepidoptera: Elachistidae): egg and damage distribution and parasitism. *Revista Brasileira de Fruticultura* 25, 432–435.

HOSTS (2010) HOSTS – A database of the world's lepidopteran host plants. www.nhm.ac.uk/research-curation/research/projects/hostplants, accessed 1 December 2010.

Knight, R.J. (2002) History, distribution, and uses. In: Whiley, A.W., Schaffer, B. and Wolstenholme, B.N. (eds) *The Avocado, Botany, Production, and Uses*. CABI Publishing, Wallingford, UK, pp. 1–14.

Landry, B. and Roque-Albelo, L. (2003) Presence of *Stenoma catenifer* Walsingham (Lepidoptera: Elachistidae: Stenomatinae), the avocado seed moth, in the Galápagos. *Noticias de Galápagos* 62, 15–16.

Link, D. and Link, F.M. (2008) Identicação de plantas hospedeiras da broca do abacate, *Stenoma catenifer* Walsingham (Lepidoptera: Elachistidae) no Rio Gande do Sul. *Neotropical Entomology* 37, 342–344.

Mendez Villa, M. (1961) Medidas practices para combater los barrenadores del hueso del aguacate. *Fitofilo* 14, 4–7.

Milano, P., Filho, E.B., Parra, J.R.P., Oda, M.L. and Cônsoli, F.L. (2010) Efeito da alimentação da fase adults na reprodução e longevidade de species de Noctuidae, Crambidae, Tortricidae e Elachistidae. *Neotropical Entomology* 39, 172–180.

Millar, J.G., Hoddle, M., McElfresh, J.S., Zou, Y. and Hoddle, C. (2008) (9Z)-9-13-Tetradecadien-11-ynal, the sex pheromone of the avocado seed moth, *Stenoma catenifer*. *Tetrahedron Letters* 49, 4820–4823.

Morse, J.G., Rugman-Jones, P.F., Watson, G.W., Robinson, L.J., Bi, J.L. and Stouthamer, R. (2009) High levels of exotic armored scales in imported avocados raise concerns regarding USDA-APHIS' phytosanitary risk assessment. *Journal of Economic Entomology* 102, 855–867.

Muniz Velez, R. (1958) La oruga barrenadora del hueso del aguacate, *Stenoma catenifer* Walsingham (Lepidoptera: Stenomatidae). *Memoria del Primer Congreso Nacional Entomologia y Fitopatologia*, pp. 170–174.

Nava, D.E., Para, J.R.P., Diez-Rodrigues, G.I. and Bento, J.M.S. (2005a) Oviposition behavior of *Stenoma catenifer* (Lepidoptera: Elachistidae): chemical and physical stimuli and diel pattern of egg laying. *Annals of the Entomological Society of America* 98, 409–414.

Nava, D.E., Haddad, M.L. and Parra, J.R.P. (2005b) Exigências térmicas, estimative do nmero de gerações de *Stenoma catenifer* e comprovação do model em campo. *Pesquisa Agropecuária Brasileira* 40, 961–967.

Nava, D.E., Parra, J.R.P., Costa, V.A., Guerra, T.M. and Cônsoli, F.L. (2005c) Population dynamics of *Stenoma catenifer* (Lepidoptera: Elachistidae) and related larval parasitoids in Minas Gerais, Brazil. *Florida Entomologist* 88, 441–446.

Nava, D.E., Parra, J.R.P., Bento, J.M.S., Diez-Rodriguez, G.I. and Haddad, M.L. (2006) Distribuição vertical, danos e controle cultural de *Stenoma catenifer* Walsingham (Lepidoptera: Elachistidae) em pomar de abacate *Neotropical Entomology* 35, 516–522.

Nava, D.E., Takahasi, K.M. and Parra, J.R.P. (2007) Linhagens de Trichogramma e Trichogrammatoidea para controle de *Stenoma catenifer*. *Pesquisa Agropecuária Brasileira* 42, 9–16.

Popenoe, W. (1919) The avocado in Guatemala. *USDA Bulletin* 743, pp. 69.

Rugman-Jones, P.F., Hoddle, M.S. and Stouthamer, R. (2007) Population genetics of *Scirtothrips perseae*: tracing the origin of a recently introduced exotic pest of Californian avocado orchards, using mitochondrial and microsatellite DNA markers. *Entomologia Experimentalis et Applicata* 124, 101–115.

Sasscer, E.R. (1921) Important insects collected on imported nursery stock in 1920. *Journal of Economic Entomology* 14, 353–355.

Silva, L.K.F., Arthur, V., Nava, D.E. and Parra, J.R.P. (2006) Tratamento quarentenário em ovos de *Stenoma catenifer* Walsingham (Lepidoptera: Elachistidae), com radiação gama do Cobolto-60. *Boletín de Sanidad Vegetal Plagas* 32, 507–512.

Silva, L.K.F., Arthur, V., Nava, D.E. and Parra, J.R.P. (2007) Uso da radiação gama do Cobolto-60 visando ao tratamento quarentenário e á esterilização de *Stenoma catenifer* Walsingham (Lepidoptera: Elachistidae). *Boletín de Sanidad Vegetal Plagas* 33, 427–438.

USDA-APHIS (2010) Current In-Progress PPQ RISK Analyses as of April 30 2010. www.aphis.usda.gov/import_export/plants/plant_imports/downloads/PRAlist.pdf, accessed 1 December 2010.

Ventura, M.U., Destro, D., Lopes, E.C.A. and Montalván, R. (1999) Avocado moth (Lepidoptera: Stenomidae) damage in two avocado cultivars. *Florida Entomologist* 82, 625–631.

Walsingham, L. (1909–1915) Tieina, Pterophorina, Orneodina, Pyralidina, and Hepialina (part). *Biologia Centrali Americana, Zoology Lepidoptera-Heterocera* 6, 168–169.

Wille, J.E. (1952) *Entomologia Agricola del Perú*. La Junta de Sanidad Vegetal, Ministerio de Agricultura, Aramburu Raygada HNOS S.A., Lima.

Wolfenbarger, D.O. and Colburn, B.E. (1966) Recent observations on some avocado pests in México and El Salvador. *Proceedings of the Florida State Horticultural Society* 79, 335–337.

Wolfenbarger, D.O. and Colburn, B. (1979) The *Stenoma catenifer*, a serious avocado pest. *Proceedings of the Florida State Horticultural Society* 92, 275.

Wysoki, M., van den Berg, M.A., Ish-Am, G., Gazit, S, Peña, J.E. and Waite, G.K. (2002) Pests and pollinators of avocado. In: Peña, J.E., Sharp, J.L. and Wysoki, M. (eds) *Tropical Fruit Pests and Pollinators, Biology, Economic Importance, Natural Enemies and Control*. CABI Publishing, Wallingford, UK, pp. 223–293.

ZipcodeZoo.com (2010) Members of the genus *Stenoma*, http://zipcodezoo.com/Animals/S/Stenoma_catenifer, accessed 29 December 2010.

Zou, Y. and Millar, J.G. (2010) Improved synthesis of (9Z)-9, 13-tetradecadien-11-ynal, the sex pheromone of the avocado seed moth, *Stenoma catenifer*. *Tetrahedron Letters* 51, 1336–1337.

6 Biology, Ecology and Management of the South American Tomato Pinworm, *Tuta absoluta*

Alberto Urbaneja,[1] Nicolas Desneux,[2] Rosa Gabarra,[3] Judit Arnó,[3] Joel González-Cabrera,[1] Agenor Mafra-Neto,[4] Lyndsie Stoltman,[4] Alexandre de Sene Pinto[5] and José R.P. Parra[6]

[1]*Unidad de Entomología, Centro de Protección Vegetal y Biotecnología, Instituto Valenciano de Investigaciones Agrarias (IVIA), Carretera Moncada-Náquera km 4,5, 46113 Moncada, Valencia, Spain;* [2]*INRA (French National Institute for Agricultural Research), UR 880, 400 route des chappes, 06903 Sophia-Antipolis, France;* [3]*Entomology IRTA, Ctra. de Cabrils km 2, 08348 Cabrils (Barcelona), Spain;* [4]*ISCA Technologies, Inc., 1230 W. Spring Street, Riverside, California 92507, USA;* [5]*Centro Universitário Moura Lacerda, Av. Dr. Oscar de Moura Lacerda, 1520, 14076-510, Ribeirão Preto, Brazil;* [6]*Departamento de Entomologia, Fitopatologia e Acarologia, Escola Superior de Agricultura 'Luiz de Queiroz' Universidade de São Paulo, 13418-900, Brazil*

6.1 Introduction

The South American tomato pinworm, *Tuta absoluta* (Meyrick) (Lepidoptera: Gelechiidae) is one of the most devastating pests of tomato crops (Estay, 2000; Leite *et al.*, 2004; EPPO, 2006; Desneux *et al.*, 2010). In the absence of control strategies, damage to tomatoes (*Lycopersicon esculentum*) can reach 100% because this pest attacks leaves, flowers, stems and especially fruits (López, 1991; Apablaza, 1992). *Tuta absoluta* has had several scientific names since its original description in Huancayo (Perú) in 1917 by Meyrick, namely, *Phthorimaea absoluta*, *Gnorimoschema absoluta*, *Scrobipalpula absoluta* or *Scrobipalpuloides absoluta*, but was finally placed under the genus *Tuta* (Barrientos *et al.*, 1998). This species is multivoltine, having tomato as its primary host.

Although originally from South America (Miranda *et al.*, 1998; Barrientos *et al.*, 1998; Estay, 2000), *T. absoluta* was detected in eastern Spain at the end of 2006 (Urbaneja *et al.*, 2008). Since then, it has rapidly invaded other European countries and spread throughout Mediterranean Basin countries, including areas in North Africa and the Middle East (Potting, 2009; Desneux *et al.*, 2010; Seplyarsky *et al.*, 2010; Kýlýc, 2010). Currently, *T. absoluta* is considered a serious threat of tomato in all of these newly infested areas.

Given the importance of the pest to infield- and greenhouse-grown tomatoes throughout the growing season, numerous studies related to its biology, ecology and management have been conducted in South America (Vargas, 1970; Fernández and Montagne, 1990). Accordingly, in newly invaded areas in Europe, several collaborative research projects have provided data, not only about *T. absoluta*'s biology and ecology, but also about the most cost-effective control strategies. In this chapter, we have focused on (i) pest origin, spread and current distribution; (ii) basic but critical information regarding *T. absoluta* bioecology; (iii) rearing techniques; (iv) known host plants; (v) sampling and monitoring techniques; and (vi) control methods available from both South America and newly infested areas, as well as their potential use for integrated pest management (IPM). Finally (vii), we have provided some hints on present and future management strategies.

6.2 Origin, Spread and Current Distribution

6.2.1 Current distribution and history of spread

Tuta absoluta is thought to be native to Central America, and has spread to the south where, since the early 1980s, it has been recorded as a pest in Argentina, Bolivia, Brazil, Chile, Colombia, Ecuador, Paraguay, Peru, Uruguay and Venezuela (Barrientos *et al.*, 1998; Estay, 2000; EPPO, 2008). In most cases, dissemination throughout South America was attributed to fruit commercialization (Cáceres, 1992).

Despite quarantine efforts [the pest is included in list A1 (EPPO, 2006)], the pest was detected in the north of Castellón de la Plana (eastern Spain) at the end of 2006 (Urbaneja *et al.*, 2008). During 2007, it was detected in several locations throughout the Spanish Mediterranean region, causing very serious damage. In the following growing season, it was found in all of the main coastal areas of tomato cultivation. In these areas, populations of *T. absoluta* immediately reached damaging levels. This has led to a progressive extension of their geographic distribution. In 2008 and 2009, the species spread not only to other Mediterranean countries such as Italy, France, Greece, Portugal, Morocco, Algeria and Tunisia (EPPO, 2008), but also to some European countries with colder climates, (i.e., Switzerland, UK, Germany and the Netherlands (Potting, 2009). In 2010, the establishment of *T. absoluta* in central Europe (Albania, Bulgaria, Romania), Lithuania (Ostrauskas and Ivinskis, 2010), and the Middle East (Bahrain, Kuwait, Israel, Jordan, Syria, Iraq, Saudi Arabia) was reported (Fig. 6.1).

6.2.2 Potential for future spread

The wide distribution of tomato crops along coastal areas of the Mediterranean region correlates with the expansion and movement of *T. absoluta* populations. Moreover, the high concentration of greenhouse cultivation provides longer seasonal availability of their preferred host plant species. The suitability of leguminous plants as a host needs further attention (EPPO,

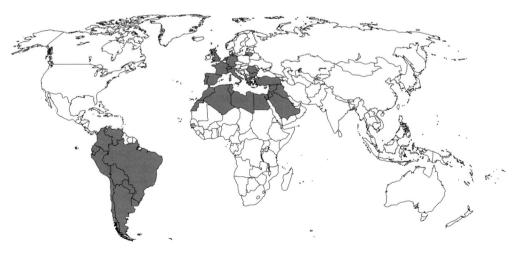

Fig. 6.1. Current distribution map of *Tuta absoluta* (updated from Desneux *et al.*, 2010).

2009). If *T. absoluta* expands its host range to several species of this family, it is likely that invasions into new areas of the world will occur in the future. Nevertheless, introductions into new regions appear to be linked to the import and export of tomato fruits. Several reports suggest that international trade is the main route for *T. absoluta* dissemination, for example in the UK, The Netherlands, Russia and Lithuania. Although data from sampling with pheromone traps seem to indicate that this species is characterized by an active dispersal capacity, more reliable information is needed to shed light on the contribution of *T. absoluta* adult dispersion to its spread in the palaeoarctic ecozone.

6.3 Bioecology

6.3.1 Description

The *T. absoluta* life cycle consists of four stages of development: egg, larva, pupa and adult (Fig. 6.2). The eggs measure 0.4 mm in length and 0.2 mm in diameter, and are cylindrical in shape, creamy-white when laid but becoming yellow-orange near hatching. The larval stage includes four instars which are well-defined in size and colour. The first instar is cream coloured and the head is dark; it measures c. 1.6 mm in length. After hatching, young larvae penetrate leaves, tomato fruits or stems to feed and develop, progressively changing to green. Larvae feed only on mesophyll tissues, leaving the epidermis intact. The second instar is c. 2.8 mm in length and the third instar is c. 4.7 mm. Upon reaching the fourth larval instar, a reddish spot appears at the dorsal level, extending longitudinally from the ocellus to the outer margin of the body. In this last instar, larvae can reach 8 mm in length.

The larva stops eating when it is ready to pupate, and usually drops to the ground using a silk thread to complete the pupal stage there; however, pupae can often be found on leaves as well. *T. absoluta* pupae are cylindrical and greenish in their early form, but become darker as they near adult emergence. The pupae measure c. 4.3 mm in length and 1.1 mm in diameter, and are usually covered by a white silky cocoon.

Adults measure 7 mm in length and present filiform antennae. The colour of scales varies

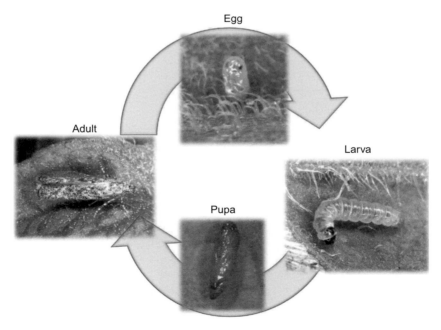

Fig. 6.2. Life cycle of *Tuta absoluta*.

from silver to grey, and black spots are present on anterior wings. Usually adults remain hidden during the day, showing greater morning–crepuscular activity. The abdomen of females is cream coloured, and they are typically wider and heavier than the males (Fig. 6.3). Adult dispersal is by flight.

6.3.2 Biology and life history parameters

T. absoluta has a high reproductive potential. Its life cycle is completed between 22 and 73 days depending on environmental conditions (Barrientos *et al.*, 1998). On tomato leaves, for example, the life cycle can be completed in as few as 22 days when temperatures are between 25 and 27°C. Temperature thresholds and thermal constants of the different stages of *T. absoluta* have been already established (Table 6.1) (Bentancourt *et al.*, 1996; Barrientos *et al.*, 1998; Mihsfeldt, 1998). Larvae do not enter diapause, and there can be 10–12 generations per annum in South America (Barrientos *et al.*, 1998). Mihsfeldt (1998) estimated that this species could have 6 to 9 generations per annum in São Paulo, Brazil.

Adult longevity is also influenced by environmental conditions. The mean lifespan is about 10–15 days for females and 6–7 days for males (Estay, 2000). Nevertheless, uncoupled males live significantly longer than coupled males and females, whatever their stage (Fernández and Montagne, 1990).

Females mate only once each day and they are able to mate six times during their life. Mating lasts an average of 4–5 h. The most prolific period is 7 days after the first mating, when females are able to lay 76% of their eggs (Uchoa-Fernandes *et al.*, 1995). At 25–27°C, the pre-oviposition period lasts for 2–3 days (Haji *et al.*, 1995; Mihsfeldt, 1998) with an egg-to-adult survival of 81.7%. Females oviposit preferentially on leaves (73%), and to a lesser extent on leaf veins and stem margins (21%), sepals (5%) or green fruits (1%) (Estay, 2000). Oviposition was found to be possible on unripened tomatoes only (Monserrat, 2009a). A single female can lay up to 260 eggs throughout its life (Uchoa-Fernandes *et al.*, 1995). However, the number of eggs laid by each female depends on the tomato variety, and under the

Fig. 6.3. Male and female of *Tuta absoluta*.

same conditions can vary from 130.1 ± 11.2 on leaves of the Santa Clara variety (Mihsfeldt, 1998) to 158.0 ± 10.9 on Santa Cruz Kada (Giustolin and Vendramim, 1996a). Rearing on artificial diet strongly reduces the number of eggs laid, reaching only 77.8 ± 14.1 (Mihsfeldt and Parra, 1999).

6.4 Rearing Techniques

Little research has been reported with artificial diets, and most of these studies have used plants of different tomato varieties (Vargas, 1970; Rázuri and Vargas, 1975; Quiroz, 1976; Muszinski *et al.*, 1982; Souza *et al.*, 1983; França *et al.*, 1984; Salas and Fernández, 1985; Paulo, 1986; Haji *et al.*, 1988; Imenes *et al.*, 1990; Pratissoli, 1995; Bentancourt *et al.*, 1996; Giustolin and Vendramim, 1996a; Ferreira and Anjos, 1997).

Among the various artificial diets tested, that developed by Greene *et al.* (1976) has been shown to be the most suitable (Mihsfeldt and Parra, 1999). Based on the work done by Greene *et al.* (1976), Mihsfeldt and Parra (1999) adapted a new diet for *T. absoluta*; they used dried beans with white husk and leaf extracts from the Santa Clara tomato variety to stimulate the larvae to feed. This diet appeared particularly suitable for *T. absoluta* larval development; however, in comparison with larvae fed on plant leaves, the development of both larval and pupal stages took longer and fewer eggs were laid. The same authors pointed out that further research should be conducted on reducing diet granulation, as

Table 6.1. Thermal requirements of *Tuta absoluta* according to various authors.*

Development stages	Tt (°C) Temperature threshold			K (DD) Thermal constant		
	On leaves[a]	On artificial diet[b]	On leaves[c]	On leaves[a]	On artificial diet[b]	On leaves[c]
Egg	6.9	7.9	9.7	103.8	76.6	72.3
Larva	7.6	11.2	6.0	238.5	366.6	267.2
Pupa	9.2	10.7	9.1	117.3	135.0	130.8
Egg–adult	8.1	10.8	8.0	453.6	574.5	463.0

* [a]Barrientos *et al.* (1998); [b]Mihsfeldt (1998); [c]Bentancourt *et al.* (1996).

large particles can make the ingestion process more difficult for recently hatched larvae. Accordingly, these authors considered that adapting to laboratory conditions may last from four to seven generations in this type of microlepidopteran (P. Singh, DSIR, New Zealand, pers. comm.). Therefore, a simple rearing system with tomato leaves may be the most effective (Pratissoli, 1995). It is worth mentioning that rearing on artificial diet leads to a significant reduction in the number of eggs laid by females, dropping from 130–158 to about 78 (Mihsfeldt and Parra, 1999).

In the laboratory, females lay eggs singly or in masses with up to ten eggs depending on attractive visual and olfactory stimuli (Mihsfeldt, 1998). Using white sulfite paper coated with smooth, green polyethylene and tomato leaf extract obtained in sulfuric ether as visual and olfactory stimuli (1 mL of ether per sulfite paper), Mihsfeldt (1998) reported 144.6 ± 15.6 eggs for each female, which suggested that tomato leaves could be replaced for the purposes of obtaining eggs in the laboratory.

According to Mihsfeldt (1998), the optimum temperature range for rearing is 18–25°C. Fertility life-table studies have demonstrated that a population reared on a natural diet can increase 72.28 times each generation (Ro = 72.28), with a finite rate of increase (λ) of 1220 (Mihsfeldt, 1998).

6.5 Plant Hosts and Damage

Tomato is the preferred host plant for *T. absoluta*. It is able to lay single eggs on almost every part of the tomato plant and can complete its life cycle feeding on leaves, stems, flowers and fruits. *T. absoluta* can also be found in other cultivated Solanaceae, such as aubergine (*Solanum melongena*), potato (*S. tuberosum*), pepper (*S. muricatum*) and tobacco (*Nicotiana tabacum*), as well as on other alternative plant hosts (Solanaceous weeds) such as *S. saponaceum*, *S. guitoense*, *S. nigrum*, *S. elaeagnifolium*, *S. bonariense*, *S. sisymbriifolium*, *Lepidium puberulum*, *Datura stramonium*, *D. ferox* and *Nicotiana glauca* (Vargas, 1970; Povolny, 1975; Campos, 1976; Garcia and Espul, 1982; Larraín, 1987; Desneux *et al.*, 2010). In Italy, *T. absoluta* has been recorded on *Physalis peruviana* (Tropea Garzia, 2009), beans (EPPO, 2009) and *Lycium* and *Malva* spp. (Caponero, 2009). On potato, *T. absoluta* only attacks aerial portions of the plant and does not develop on tubers. Damage is caused when larvae feed on the leaf mesophyll, expanding galleries, which affect the photosynthetic capacity of the plants and subsequently reduces yields (Fig. 6.4). Injuries made directly to the fruits may lead to severe yield losses. Tomato plants can be attacked at any developmental stage, from seedlings to mature plants. After hatching, young larvae penetrate the leaves, stems or tomato fruits creating conspicuous mines and galleries where they feed and develop. Larvae can also attack flowers and stop fecundation. At early stages, larval damage may go undetected due to the small diameter of the entrance hole made under the sepals, but when larval development is completed, a dark yellow halo can be observed surrounding the hole used by the adult to exit the host. Galleries in stems can alter general plant development if they become necrotic. Moreover, a single larva may create multiple galleries attacking several plant tissues, thereby increasing the damage level (Fig. 6.5). Fruits can also be attacked as soon as they are

Fig. 6.4. *Tuta absoluta* galleries.

Fig. 6.5. Damage caused by *Tuta absoluta* in the fruits and leaflets.

Fig. 6.6. Damage caused by *Tuta absoluta* in the fruits and leaflets.

formed. The galleries bored inside fruits are usually invaded by secondary pathogens, leading to fruit rot (Fig. 6.6) (Bahamondes and Mallea, 1969; Vargas, 1970).

Yield losses occur in fruits destined for the fresh market as well as in those used for industrial processing. The amount of these losses depends on the success of controlling the pest, ranging from 80% to 100% if there is no control whatsoever (Nakano and Paulo, 1983; Scardini et al., 1983; López, 1991; Apablaza, 1992).

6.6 Sampling and Monitoring Techniques

There are different sampling methods to determine the risk of damage in a tomato crop. One is monitoring for pest presence and density based on direct observations of the plant to estimate plant damage and how the damage develops. This can be complemented with the use of pheromone traps to estimate the level of male populations.

6.6.1 Pheromone traps

T. absoluta *sex pheromone*

Using capillary GC-MS, Attygalle et al. (1995) examined hexane extracts made from excised pheromone-producing glands of calling *T. absoluta* females, and found two significant chromatographic peaks in the region where Lepidopteran pheromones usually appear; each gland contained about 1-5 ng of the pheromone components. The major component was identified as (3E,8Z,11Z)-3,8,11-tetradecatrienyl acetate (Attygalle et al., 1995). Synthetic (3E,8Z,11Z)-3,8,11-tetradecatrienyl acetate was highly attractive to *T. absoluta* males and identical to the natural substance from female gland extracts. The complete female sex pheromone of *T. absoluta* was later found also to contain approximately 10% of (3E, 8Z)-3,8-tetradecadienyl acetate as a secondary component (Attygalle et al., 1996; Griepink et al., 1996; Svatos et al., 1996).

The market prior to the European invasion only required the manufacturing of small quantities of the synthetic pheromone for monitoring lures. As invasive *T. absoluta* populations became established, and then exploded across Europe and North Africa (Desneux et al., 2010), the demand for synthetic pheromone [(3E,8Z,11Z)-3,8,11-tetradecatrienyl acetate] increased significantly. Intense scale-up research and development resulted in increased availability, in tandem with a drastic reduction in the cost of synthetic sex pheromone for *T. absoluta*.

Monitoring

Monitoring for pest presence and density is a critical first step for its management. The primary value of sex pheromones, combined with traps, is to determine the presence and density of pest populations in the field. The sex pheromone of *T. absoluta* is highly species-specific. It attracts males to traps even when field population densities are extremely low. Pheromone-baited traps are, therefore, remarkable tools for detection of this insect, and should help in determining pest distribution at larger geographic scales.

Using pheromone traps as tools for monitoring the density of insect pest populations in the field requires an appropriate calibration, requiring correlation of certain combinations of trap and lure with the actual presence of the pest in the target area. This is usually done by correlating catches in pheromone traps with the population density of immature forms and the level of damage detected in the crop. Preliminary lure efficacy studies were conducted in Brazil to determine appropriate dosing and field longevity. Captures peaked with lures containing between 0.25 and 1.0 mg of pheromone per septum. No moths were captured in any of the non-baited traps. The analysis of these data revealed that captures in pheromone traps correlate with the density of leaf-mine and larvae in the field (Gomide *et al.*, 2001; Filho *et al.*, 2000, Ferrara *et al.*, 2001; Salas, 2004). In South America, pheromone traps are starting to be used for monitoring, with the economic injury level (EIL) established at 45 ± 19.5 adults/trap/day (Benvenga *et al.*, 2007). In Chile, the economic injury level (EIL) is established at 100 adults/trap/day (Bajonero *et al.*, 2008), or with two females or 26 larvae per plant. Despite previous experience, this correlation is not properly validated in newly infested areas in the Mediterranean region. The number of captures obtained per trap is, however, used to establish the risk level (Monserrat, 2009a, b). Normally, a delta trap baited with pheromone is used as standard (Fig. 6.7). The trap is placed at 0.5 m height at the beginning of the season and can later be raised to 1.5 m depending on the size of the plant. In fields or greenhouses where *T. absoluta* has not yet been established, it is common to find a time lapse of 3–4 weeks between the start of male captures and the detection of plant damage. When there are no male captures, the plant damage risk is very low or null.

Fig. 6.7. Delta trap lured with *Tuta absoluta* pheromone.

Plant sampling

Visually monitoring for the presence and density of *T. absoluta* in the field requires intensive training and specialized manual labour. Finding, counting and annotating the number of eggs on meristematic parts of the plant, as well as the estimation of gallery density and damaged fruit in the crop, are difficult and time-consuming tasks.

Direct plant sampling must be conducted to estimate the actual damage inflicted on the crop by *T. absoluta*. To this end, the percentage of infested plants should be counted. Only plants infested with healthy larvae should be considered in this sampling, ruling out those plants with dead larvae or old-damage symptoms. The periodicity of this sampling is correlated with the risk of damage, 5–7 days for periods of high risk and 14 days during those periods of no risk (Monserrat, 2009a). As with the pheromone trap index, depending on the percentage of infested plants, a particular control method should be adopted. Treatments with chemicals should only be applied when *T. absoluta* larvae can be found in the plants.

A level of 8% defoliation is used as the EIL in Colombia (Desneux *et al.*, 2010). In Brazil, sampling twice weekly on 60 plants/ha in staked tomatoes and 10 plants/ha in industrial tomatoes, the EIL used is 25% of the upper portion of the plant with larvae or eggs; or 25% of the leaves with larvae; or 5% of tomatoes (cluster) with eggs (Benvenga *et al.*, 2007).

Unfortunately, the injury level threshold (i.e., fruit damage in terms of percentage of plant or fruit infested) has not yet been scientifically established in the Mediterranean region and thus it is currently difficult to correlate the results of plant sampling and percentage of damage to the crop. However, a significant relationship between logarithmic estimation of the abundance of *T. absoluta* galleries in the seven upper leaves of the tomato plant with the percentage of damaged young fruits was observed when ten commercial tomato fields were surveyed periodically (Arnó, Matas and Gabarra, unpublished data). This relationship suggests that evaluation of the abundance of galleries may allow estimation of the level of damage on young fruits (Fig. 6.8).

6.7 Control Methods

The best methods for managing *T. absoluta* involve prevention of pest outbreaks, and long-term

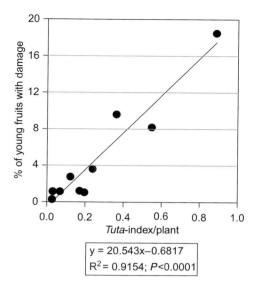

Fig. 6.8. Relationship between the mean percentage of young tomato fruits damaged and the logarithmic estimation of the abundance of *Tuta absoluta* galleries (mean *Tuta*-index). Data from ten open-field tomato crops surveyed periodically (Arnó, Matas and Gabarra, unpublished data).

$y = 20.543x - 0.6817$
$R^2 = 0.9154; P < 0.0001$

and economic control. These methods include biological, biotechnological, cultural and chemical control. The use of pesticides is recommended only when preventive measures are inadequate and careful field monitoring indicates that economic loss is likely.

6.7.1 Biological control

Biological control represents one of the main strategies for the integrated pest management of this insect pest (Haji *et al.*, 2002). Many natural enemies have been described for *T. absoluta* in the area of origin of this pest. Furthermore, since *T. absoluta* has been detected in the Mediterranean region, some indigenous natural enemies have been reported on this exotic pest as fortuitous biological control. In newly infested areas, evaluation is currently under way to determine if these natural enemies can be used in biological control programs that target *T. absoluta*.

Parasitoids

A list of c. 50 parasitoids of *T. absoluta* eggs, larvae and pupae in South America was reported by

Desneux et al. (2010). Parasitoids of eggs and larvae predominate, with only a few records for pupal and none for adult parasitoids. Before this study, and except for one citation (Polack, 2007), no publications had grouped all of the parasitoids together.

T. absoluta eggs are parasitized by Encyrtidae, Eupelmidae and Trichogrammatidae wasps (Trichogramma spp. being predominant). The species associated with Tuta absoluta in South America are Trichogramma exiguum Pinto and Platner, T. nerudai Pintureau and Gerding, and T. pretiosum Riley (Zucchi et al., 2010).

Successful control of Tuta absoluta with Trichogramma spp. releases has been reported in Colombia (Garcia Roa, 1989; Salas, 2001) and in Brazil (Haji et al., 2002; Parra and Zucchi, 2004), whereas unsuccessful control has been reported in Chile (Jimenez et al., 1998; Taco et al., 1998).

At the end of 1981, Tuta absoluta was recorded in Brazil's sub-medium São Francisco region where all the tomato extract industries were concentrated at that time, with a total area of 15,000 ha of industrial tomatoes. Due to T. absoluta resistance to Cartap (thicarbamate), the government developed an integrated pest management program which included cultural, legislative, microbiological and biological measures, the latter involving the import of Trichogramma pretiosum from Colombia (Haji et al., 2002). The adoption of these measures resulted in excellent control of Tuta absoluta with only 1–9% damaged tomatoes and with 30–49% parasitism observed in industrial tomato crops (Haji et al., 1995). Around 450,000 parasitoids were released per hectare, twice per week for 10 weeks, with the first release taking place when the first eggs of T. absoluta were detected on plants, 20–30 days after planting. Between releases, applications of Bacillus thuringiensis (Berliner) (Bt) were also made, totaling ten treatments per crop period, guaranteeing the control of those caterpillars which had emerged from eggs not parasitized by Trichogramma pretiosum.

It should be noted that for parasitism to be successful, the selection of the right parasitoid strains is fundamental (Pratissoli and Parra, 2000b, 2001; Pratissoli et al., 2005a). The best temperatures for successful parasitism by T. pretiosum range between 22°C and 25°C (Pratissoli, 1995) (Fig. 6.9). Developmental time ranges from 26.5 days at 18°C to 7.1 days at 32°C, and thermal requirements are 131.3 degree days for complete development (lowest temperature threshold of 13°C) (Table 6.2 and Fig. 6.10) (Pratissoli and Parra, 2000).

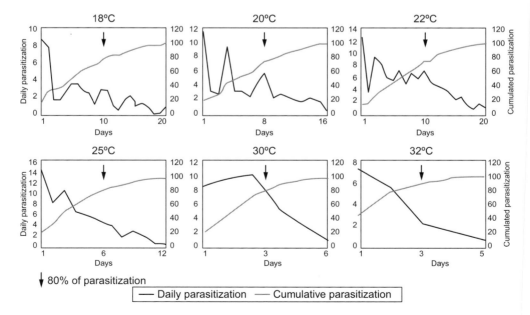

Fig. 6.9. Trichogramma pretiosum daily and total parasitization on Tuta absoluta at different temperatures (Pratissoli and Parra, 2000a).

Other studies that have focused on *T. pretiosum* have shown that nine inundative releases of the parasitoid at 800,000 individuals/ha once per week, on industrial tomatoes, and ten releases (same rate) on staked tomatoes, resulted in reductions of 82% and 85% in *Tuta absoluta* populations, respectively (Papa, 1998). Egg parasitism varied from 32% to 35%, and the level of control was similar to that obtained from conventional insecticides for *T. absoluta* management. In the case of tomatoes in greenhouse cultivation, the use of *Trichogramma pretiosum* resulted in 87% parasitism in Brazil (Parra and Zucchi, 2004). The dispersal capacity of *T. pretiosum* varies from 7 to 8 m, 24 h after release over an area of 120–140 m^2, indicating that 75 points/ha should be used for parasitoid release (Pratissoli et al., 2005b).

In the Mediterranean region, some egg parasitoids have already been detected (Desneux et al., 2010). *Trichogramma achaeae* Nagaraja and Nagarkatti has been identified as a potential biological control agent of *T. absoluta*. In greenhouse exclusion cages, seven releases of this parasitoid at a rate of 30 adults per plant every 3 or 4 days significantly reduced the number of *T. absoluta* larvae, galleries and damaged fruits, compared to control plots (Cabello et al., 2009b). Based on these findings, *T. achaeae* has become commercially available, with a recommended release dose, depending on the infestation level, of 250,000–500,000 adults/ha per week (Cabello et al., 2010). However, these high rates may not be economically sustainable, and thus parasitoid releases should be combined with other biological control methods.

Larval parasitoids can also be important, and species from the Bethylidae, Braconidae, Eulophidae, Ichneumonidae and Tachinidae have been recorded (Desneux et al., 2010). Several species of larval parasitoids have been recorded from Argentina (Berta and Colomo, 2000; Luna et al., 2007; Sanchez et al., 2009), Chile (Rojas, 1981; Larraín, 1986), Brazil (Uchoa-Fernandes et al., 1995; Miranda et al., 1998; Marchiori et al., 2004; Bacci et al., 2008), Colombia (Oatman and Platner, 1989), Perú and Venezuela (Desneux et al., 2010). *Apanteles gelechiidivoris* Marsh (Hymenoptera: Braconidae) has been reported to be an efficient biocontrol agent and is currently used in Colombia to control *T. absoluta* (Benavides et al., 2010). In laboratory experiments, Bajonero et al. (2008)

Table 6.2. Development of *Trichogramma pretiosum* on *Tuta absoluta* at different temperatures (Pratissoli and Parra, 2000b).

Temperature (°C)	Duration (days)	Viability (%)
18	26.47	79.80
20	17.78	86.12
22	16.32	94.34
25	10.30	96.24
30	7.50	88.97
32	7.11	83.37

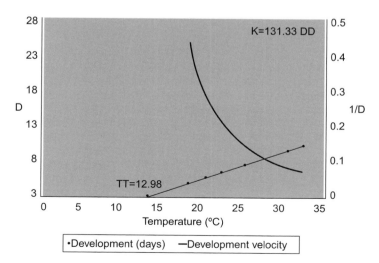

Fig. 6.10. *Trichogramma pretiosum* thermal requirements on *Tuta absoluta* (Pratissoli and Parra, 2000a).

demonstrated that *A. gelechiidivoris*, which lays its eggs singly on third instar *T. absoluta* larvae, can develop in a wide range of temperatures, with lower and higher developmental thresholds at 4°C and 33°C. *A. gelechiidivoris* was introduced into Chile from Colombia in the mid-1980s, and was found in abundance in 1997 in tomato crops in some areas of Chile (Rojas, 1997); however, a subsequent study indicated that *A. gelechiidivoris* resulted in low levels of parasitism and that *Dineulophus phthorimaeae* (De Santis) (Hymenoptera: Eulophidae) was the main natural enemy of *T. absoluta* with 70% parasitism (Larraín, 2001). In Argentina, the most important larval parasitoids are *Pseudoapanteles dignus* (Muesebeck) (Hymenoptera: Braconidae) and *D. phthorimaeae* (Polack, 2007). Data from surveys conducted over a 4-year span in tomato crops in Argentina showed that 53% of the parasitoids collected were *P. dignus* followed by *D. phthorimaeae* (Colomo *et al.*, 2002). Luna *et al.* (2007) demonstrated that *P. dignus* was able to synchronize its larval development time with that of the host, and to detect and parasitize the host even at low host densities. In addition, high levels of parasitism by *P. dignus* (up to 46%) were observed in commercial crops (Sanchez *et al.*, 2009), suggesting the importance of conservation biological control measures aimed at keeping populations of *P. dignus* in tomato crops. However, when studying field interactions of this parasitoid with *D. phthorimaeae*, Luna *et al.* (2010) determined that both species may coexist even at the leaf scale, but *D. phthorimaeae* is more efficient in situations of competition.

In the Mediterranean region, the parasitoid *Necremnus artynes* (Walker) (Hymenoptera: Eulophidae) (Fig. 6.11) has been sporadically observed in several locations along the coast (Mollá *et al.*, 2008; Arnó *et al.*, 2009a; Mollá *et al.*, 2010; Gabarra and Arnó, 2010). Another species of the same family has also been detected in northeastern Spain: *Stenomesius* sp. (Hymenoptera: Eulophidae), which was cited as *Hemiptarsenus zilahisebessi* Erdös (Hymenoptera: Eulophidae) in Gabarra and Arnó (2010) due to a misidentification. Both *N. artynes* and *Stenomesius* sp. are idiobiont ectoparasitoids, which parasitize second and third instars of *T. absoluta*. *Neochrysocharis formosa* (Westwood) (Hymenoptera: Eulophidae) has been identified as a potential larval parasitoid of *T. absoluta* in the same area (Lara *et al.*, 2010) and *Habrobracon hebetor* (Say) (Hymenoptera: Braconidae) has been observed parasitizing larvae on

Fig. 6.11. The larval parasitoid *Necremnus artynes*.

tomato field samples (Gabarra and Arnó, unpublished data). In addition, recent reports of various undetermined species, mainly Braconidae, attacking *T. absoluta* along the Mediterranean coast (Mollá *et al.*, 2008; Arnó *et al.*, 2009b; Gabarra and Arnó, 2010) may signal that the local parasitoid complex is progressively adapting to this new pest. Recently, two companies that produce natural enemies have identified *N. artynes*, *N. tidius* (Walker) and *Diadegma ledicola* Horstmann (Hymenoptera: Ichneumonidae) as good biocontrol candidates for *T. absoluta*, and plan to make *N. artynes* commercially available (Koppert, 2010; Bioplanet, 2010).

Although pupal parasitoids of *T. absoluta* have been neglected, Polack (2007) has recorded up to 30% pupal parasitism by *Conura* sp. (Hymenoptera: Chalcididae).

Predators

The biology of predators associated with *T. absoluta* in South America has been studied much less than that of parasitoids (Desneux *et al.*, 2010), although several studies have stressed the importance of predators in mortality of *T. absoluta*. Miranda *et al.* (1998) found that the highest natural mortality occurred during the larval stage, and the mortality associated with predation was much higher than that associated with parasitism. In contrast, predation on eggs was slightly lower than parasitism. According to these authors, this mortality was caused by the anthocorid *Xylocoris* sp. (Hemiptera: Anthocoridae), the coccinelid *Cycloneda sanguinea* (Linnaeus) (Coleoptera: Coccinellidae) and some members of the Phlaeothripidae family. Bacci *et al.* (2008) also found that the most important mortality factor for the larvae of *T. absoluta* was predation, particularly by the wasp *Protonectarina sylveirae* (Saussure) (Hymenoptera: Vespidae) and anthocorid and miridae bugs. Alternatively, several authors reported a significant decrease in populations of *T. absoluta* related to an increase in predator populations when conservation biological controls were applied, such as companion plants (Paula *et al.*, 2004; Miranda *et al.*, 2005; Medeiros *et al.*, 2009).

Several predators have been associated with crops infested by *T. absoluta* (Desneux *et al.*, 2010), but relatively few groups, genera or species have been demonstrated to prey on *T. absoluta* (Table 6.1), and even fewer are the number of species that have been studied in detail.

The predators that have been directly associated with *T. absoluta* are mainly predators of larvae, although some of them prey on eggs or pupae. As shown in Table 6.1, in the area of origin of this pest, most of the recorded predators belong to the order Hemiptera (mainly anthocorids) and wasps. No laboratory studies on the biology of the particular wasps have been conducted; but based on field observations, *P. sylveirae* may be a very efficient predator of second, third and fourth instar larvae, causing mortality of up to 30% in *T. absoluta* populations (Bacci *et al.*, 2008). Other wasps have also been observed to be predators of *T. absoluta* larvae (Medeiros *et al.*, 2009).

One of the most studied predators of *T. absoluta* in South America is the pentatomid, *Podisus nigrispinus* (Dallas) (Hemiptera: Pentatomidae). It is able to develop by preying solely on *T. absoluta* (Vivan *et al.*, 2002a, b, 2003). However, studies on life-table parameters showed that *T. absoluta* has more generations per annum than *P. nigrispinus*, and therefore the predator would most likely be unable to provide sufficient control. When comparing the effect of two types of prey [larvae of *T. absoluta* and larvae of the beetle, *Tenebrio molitor* (L.)] on the biological parameters of *P. nigrispinus*, these authors demonstrated that developmental time and reproduction were negatively affected when *P. nigrispinus* preyed on *T. absoluta*. Conversely, Torres *et al.* (2002) studied the dispersal potential of *P. nigrispinus* in tomatoes grown in a greenhouse environment, and found that second instar nymphs dispersed better if they were starved for 24 h prior to release, and if they were released either in the morning or in the evening.

Some studies of the biology of *Chrysoperla externa* (Hagen) (Neuroptera: Chrysopidae) report that eggs of *Sitotroga cerealella* (Oliv.) (Lepidoptera: Gelechiidae) were better prey for the development of this lacewing than *T. absoluta* eggs. Larval developmental time was longer and pupal weight was lower on a diet of *T. absoluta* eggs, despite the predator having consumed a higher number of eggs (Carneiro and Medeiros, 1997).

In the Mediterranean region, eight species, predominantly Hemiptera, have been identified as *T. absoluta* predators (Table 6.3). Adults of *Macrolophus pygmaeus* Rambur (Fig. 6.12) and *Nesidiocoris tenuis* Reuter (Fig. 6.13) (Hemiptera: Miridae) are predators of all pre-imaginal stages

Table 6.3. Known *Tuta absoluta* predators, and geographical area where they were recorded.

Taxa	Geographical area	First reference
Acari		
Phytoseiidae		
Amblyseius swirskii	Mediterranean	Mollá et al., 2010
Amblyseius cucumeris	Mediterranean	Mollá et al., 2010
Pyemotidae		
Pyemotes sp.*	South America	Oliveira et al., 2007
Araneae		
Araneidae		
Misumenops pallidus	South America	Medeiros, 2007
Coleoptera		
Coccinellidae		
Cycloneda sanguinea	South America	Miranda et al., 1998
Chilocorus sp.	South America	Vasicek, 1983
Hemiptera		
Anthocoridae		
Orius insidiosus	South America	Salas, 1995
O. majusculus	Mediterranean	Gabarra & Arnó, unpublished
O. laevigatus	Mediterranean	Gabarra & Arnó, unpublished
Lasiochilus sp.	South America	Bacci et al., 2008
Xylocoris sp.	South America	Miranda et al., 1998
Miridae		
Hyaliodocoris insignis	South America	Bacci et al., 2008
Annona bimaculata	South America	Bacci et al., 2008
Macrolophus pygmaeus	Mediterranean	Arnó et al., 2009; Urbaneja et al., 2009
Nesidiocoris tenuis	Mediterranean	Arnó et al., 2009; Urbaneja et al., 2009
Dicyphus maroccanus	Mediterranean	Mollá et al., 2010
Pentatomidae		
Podisus nigrispinus	South America	Vivan et al., 2002
Berytidae		
Metacanthus tenellus	South America	Oliver and Bringas, 2000
Nabidae		
Nabis sp.	South America	Vargas, 1970
N. pseudoferus	Mediterranean	Cabello et al., 2009
Hymenoptera		
Vespidae		
Polistes sp.	South America	Vargas, 1970
Protonectarina sylveirae	South America	Bacci et al., 2008
Brachygastra lecheguana	South America	Leite et al., 1998
Polybia sp.	South America	Medeiros, 2007
Neuroptera		
Chrysopidae		
Chrysoperla externa	South America	Carneiro and Medeiros, 1997
Thysanoptera		
Phlaeothripidae	South America	Miranda et al., 1998

*This genus is known to cause dermatitis in humans. Consequently, its use as biological control agent is not recommended (Cunha et al., 2006).

of *T. absoluta*, although they consume a higher number of eggs than larvae. The number of eggs fed upon each day by one *M. pygmaeus* or *N. tenuis* adult may exceed 100 eggs in laboratory conditions, whereas c. two first-instar larvae are preyed upon each day. Nymphs of *M. pygmaeus* consumed fewer eggs than nymphs of *N. tenuis* or adults of both Mirid species (Urbaneja et al., 2009; Arnó et al., 2009b). *N. tenuis* was able to complete its nymphal development with high survival by feeding on *T. absoluta* alone (Mollá et al., 2010) and was able to limit *T. absoluta* population growth at high pest densities (Nannini, 2009). In greenhouse exclusion cage experiments, when *T. absoluta* was released where predators were well-established, *N. tenuis* was able to reduce the pest populations better than *M. pygmaeus* (98% versus 66%, respectively) (Mollá et al., 2009, 2010). Similarly, Calvo et al. (2010) found that the inoculation of *N. tenuis*, either by release in seedling nurseries or in the crop, could maintain populations of *T. absoluta* under control.

M. pygmaeus and *N. tenuis* are among the most abundant natural enemies in tomato greenhouses and open fields in the Mediterranean region, and they are especially abundant in IPM tomato crops that are not heavily sprayed with chemical pesticides (Gabarra et al., 2008). Field data in commercial IPM cultivations showed that when Mirid bugs are well established in the crop, they can be an important control agent of *T. absoluta*. Abundance of galleries and damage in young tomato fruits were four times higher in crops managed with pesticides than in those where inoculation or conservation of *M. pygmaeus* and *N. tenuis* was used as a pest control strategy (Arnó et al., 2009b).

In the Mediterranean region, predation on *T. absoluta* has also been observed by *Nabis pseudoferus* Remane (Hemiptera: Nabidae) (Cabello et al., 2009a), *Dicyphus maroccanus* Wagner (Hemiptera: Miridae), *Amblyseius cucumeris* Oudemans and *A. swirskii* (Athias-Henriot) (Fig. 6.14) (Acari: Phytoseiidae) (Mollá et al., 2010) on tomato, and by *Orius majusculus* (Reuter) and *O. laevigatus* (Fieber) (Hemiptera: Anthocoridae) on potato (Gabarra & Arnó, unpublished data).

All *T. absoluta* predators that have been identified so far in Europe and North Africa prey only on eggs and very young larvae. This means that predators can only control the pest for a short period of its development, and therefore, high populations of these natural enemies are needed to reduce damage. To achieve these high densities while controlling the pest it will be necessary to release *T. absoluta* parasitoids and/or spray with *B. thuringiensis* early in the season before the full establishment of the predators (Mollá et al., 2010; Gabarra and Arnó, 2010; González-Cabrera et al., 2011).

Fig. 6.12. Adult of *Macrolophus pygmaeus* preying on *Tuta absoluta* eggs.

Fig. 6.13. Adult of *Nesidiocoris tenuis* preying on *Tuta absoluta* eggs.

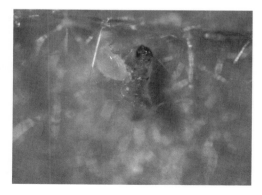

Fig. 6.14. *Amblyseius swirskii* preying on *Tuta absoluta* larva.

Entomopathogens

Several researchers have attempted to establish the basis for an effective control of *T. absoluta* with such biocontrol agents. Rodríguez et al. (2006) have shown that applications of *Beauveria bassiana* (Bals.) Vuill. as well as *Metarhizium anisopliae* (Metsch.) Sorok. sprayed on third instar larvae resulted in >90% mortality; however, this percentage decreased to 68% when larvae were fed leaves previously sprayed with Conidia (Rodríguez et al., 2006). In Brazil, the combination of moderately resistant tomato varieties with applications of *B. bassiana* or *B. thuringiensis* has resulted in additive or synergistic effects, causing significant mortality in larvae from all four instars (Giustolin, 1996; Giustolin et al., 2001a, b). Furthermore, screening programs in Chile showed that it is possible to improve effectiveness of Bt-based commercial formulations. Two newly isolated strains have demonstrated higher toxicity against neonate larvae than that of the Bt strain in the commercial product, Dipel® (Niedmann and Meza-Basso, 2006). Moreover, the expression of toxins in other *Bacillus* species naturally colonizing the phylloplane can result in longer persistence of the toxins (up to 45 days), which increases the chance for the larvae to ingest a lethal dose (Theoduloz et al., 2003). This approach should be further studied as a new delivery system, as it may help reduce the number of treatments necessary for control. In Brazil, a granulovirus isolated from the potato tuber moth, *Phthorimaea operculella* (Zeller) (Lepidoptera: Gelechiidae), was also highly virulent to *T. absoluta*, leading to significant mortality, delayed larval growth and inhibited pupation (Mascarin et al., 2010).

In the Mediterranean region, *T. absoluta* control has relied mostly on chemical pesticides, but safety issues as well as low residue levels authorized in fresh fruits have triggered the development of more environmentally sound control strategies. Despite the proven efficacy of Bt-based formulations against other Lepidopterans (González-Cabrera and Ferré, 2008), their use for controlling *T. absoluta* was initially recommended only when infestation levels were relatively low (Monserrat, 2009b, 2010); however, the results obtained in Spain from laboratory, greenhouse and open-staked tomato field assays have shown that Bt is effective against all instar larvae, reducing damage by up to 90% compared to controls when sprayed at 180.8 MIU/l (millions of international units per liter) (Fig. 6.15). Moreover, it was also shown that weekly sprays at 90.4 MUI/l can control *T. absoluta* throughout the growing season without any additional treatment, even at very high infestation levels (González-Cabrera et al., 2011). Recent greenhouse assays have shown that treatments performed every other week at 90.4 MUI/l (which lowered the cost) are enough to reach an optimum control of the pest. In these assays it was also shown that products based on *B. thuringiensis* var. *kurstaki* can be combined with those based on *B. thuringiensis* var. *aizawai* to obtain similar efficacy while reducing the risk of resistance development in insect populations. Furthermore, the integration of Bt treatments with releases of *N. tenuis* resulted in a very promising strategy. The assays performed in greenhouses showed that Bt applied in conjunction with *N. tenuis* can provide control of *T. absoluta* while the predator is established in the crop. Thereafter, the Bt treatments can be discontinued and *N. tenuis* alone is able to control the pest (Mollá et al., 2011).

The combination of Bt treatments and *B. bassiana* was also tested in open tomato fields in Spain, where it proved to be more effective than the fungi alone in terms of *T. absoluta* mortality observed per plant; however, there was variability in the protection of fruits during the experiment (Torres Gregorio et al., 2009). Further experiments should clarify the best way to integrate these two entomopathogens for controlling *T. absoluta*.

The suitability of three species of entomopathogenic nematodes for controlling *T. absoluta*

Fig. 6.15. Third instar larvae of *Tuta absoluta* killed by Bt.

was also tested under greenhouse conditions. These assays showed that nematodes were highly effective at killing larvae (up to 100% mortality), but they had almost no effect on pupae (<10% mortality). Additionally, the high level of parasitism reached (77.1–91.7%) was evidence that nematodes are able to kill the larvae inside the galleries, so that this approach could be integrated with other control strategies to reach better and more efficient control of the pest (Batalla-Carrera et al., 2010). Currently, pest control worldwide is moving toward an IPM strategy which includes biological, cultural or biotechnological methods. Thus, entomopathogens should play an important role in this global effort.

6.7.2 Semiochemical management

T. absoluta is an example of how, in the absence of an efficient conventional control, an immature pheromone technology for insect control can be firmly established (Witzgall et al., 2010). The use of pheromone-lured traps as a key tool to monitor (Attygalle et al., 1996; Michereff et al., 2000a,b; Benvenga et al., 2007), manage and control *T. absoluta* in Europe and North Africa has resulted in a global demand of >2,500,000 *T. absoluta* lures per annum. The availability of a more economical pheromone active ingredient is fostering the development of novel control techniques that use pheromone more intensively, such as attract-and-kill, and mating disruption (Stoltman et al., 2010).

Mass trapping

The Southern European and North African invasion of *T. absoluta* increased the demand for pheromone monitoring lures, which were then used in mass-trapping campaigns for greenhouse pest management. Mass trapping can be an effective management tool in isolated and controlled spaces such as greenhouses. For *T. absoluta* in a tomato greenhouse setting, at least one trap per 500 m^2 has been used to reduce moth populations significantly as part of an integrated pest management program (Stoltman et al., 2010). Mass-trapping programs must be deployed early in the plant growth cycle, when *T. absoluta* populations are present at low densities, otherwise the program is likely to fail.

Traditional paper and plastic delta traps can also be used in a mass-trapping program. These traps may be preferred in larger operations due to their relatively low cost, ease of deployment and disposal. Paper delta traps come with sticky interior walls and are designed for one-time use. They should be disposed of once the trap becomes saturated. Conversely, plastic delta traps come with removable sticky liners. Liners can be replaced after they are filled. One problem with the sticky traps is that the glue liner of the trap becomes rapidly saturated under high *T. absoluta* population densities, and requires frequent replacement.

In order to reduce the cost of mass trapping, and also avoid trap saturation, growers have started using water traps in their mass-trapping programs. Trap designs vary and can be as simple as deep plastic trays filled with soapy water and with the pheromone lure suspended over the centre of the tray, just above the water line, so that attracted moths are trapped when they touch the soapy water (Fig. 6.16).

Mating disruption

Mating disruption is the practice of continuously dispensing synthetic sex pheromones into an area to limit the male's ability to locate and subsequently fertilize the calling females. Under pheromone disruption by competitive attraction, female moths compete with sources of synthetic pheromone for the male's response, a density-dependent phenomenon. Non-competitive mechanisms are density-independent phenomena, where the presence of the synthetic pheromone in the field modulates the male's ability to perceive, or to respond to, natural blends due to desensitization by adaptation and habituation, sensory imbalance or camouflage of the plume.

The potential for using *T. absoluta* sex pheromone for control through mating disruption was first studied by Michereff et al. (2000b) in small plots (0.01 ha of fresh-market tomatoes using doses up to 80 g/ha of the major pheromone component alone. Some of the treatments achieved an impressive 90% trap shutdown in these exceedingly small plots, but there was no direct correlation between trap catch suppression and protection of the tomato crop. Failure in protecting these experimental plots was attributed to the use of the incomplete blend and/or to the strong edge effect resulting from high *T. absoluta*

Fig. 6.16. Water trap used in mass-trapping strategy attracts *Tuta absoluta* with a sex pheromone lure and a light source (photo courtesy of Rodrigo Oliveira Silva and Daniela Fernanda Klesener).

populations in neighboring fields. Since females are not affected by pheromone treatments, it is possible that migrating mated females invaded the pheromone-treated area (Michereff et al., 2000b). New efforts of mating disruption in greenhouses have demonstrated the feasibility of the technique for controlling the pest, despite cost constraints and availability of the pheromone (Navarro-Llopis et al., 2010). In both studies, lack of an economic scale-up synthesis of the *T. absoluta* sex pheromone precluded larger, more elaborated mating disruption trials.

When the economic scale-up synthesis of the major component of the *T. absoluta* sex pheromone was developed by ISCA Technologies, Inc. (Stoltman et al., 2010), it allowed researchers to create larger experimental field trials using a higher number of point sources and doses of pheromone active ingredient per treated area.

Attract and kill

Attract and kill is a density-dependent phenomenon that inherently requires substantially less pheromone than other non-competitive mating disruption formulations designed to achieve habituation/adaptation, sensory imbalance and camouflage, which prevent location of both natural and synthetic sources by overwhelming the male with pheromone. Attract and kill requires well balanced pheromone lures that not only elicit upwind flight behavior of the male, but also promote sustained contact with the point source.

Recently, ISCA Technologies began working on the development of an attract-and-kill formulation for *T. absoluta*. Preliminary field data indicate that this formulation has a longer field life than a similarly formulated mating disruption formulation, perhaps because of its sustained deleterious effect on attracted males (Stoltman et al., 2010) (Fig. 6.17).

6.7.3 Plant resistance

The highest levels of plant resistance to *T. absoluta* have been observed in non-commercial (wild-type) genotypes of tomato; however, there are studies reporting resistance in commercial genotypes as well as in hybrids resulting from crossings between wild-type and commercial genotypes. França et al. (1984) performed greenhouse assays to evaluate plant resistance in 22 tomato genotypes, including *Lycopersicon hirsutum* f. *typicum*, *L. hirsutum* f. *glabratum* and

Fig. 6.17. A dollop used for attract-and-kill of *Tuta absoluta* males in tomato (photo courtesy of Lyndsie Stoltman and Ya-Ting Yang).

L. pennellii. They observed a non-preference effect in *L. pennellii* leading to a high level of resistance to *T. absoluta*. In addition, *L. hirsutum* f. *typicum* and f. *glabratum* showed antibiosis and non-preference, respectively. Lourenção et al. (1984, 1985) found that plants of *L. peruvianum*, infested with *T. absoluta*, showed significantly lower losses (measured as damaged foliar area) than more susceptible genotypes. Barona et al. (1989) also found a high level of resistance in lines of *L. peruvianum* and *L. hirsutum* and also in *L. pimpinellifolium*.

Castelo Branco *et al.* (1987) evaluated varietal resistance in F2 plants (*L. pennellii* × *L. esculentum*) based on morphological (velvety leaves, very hairy) and physiological (strong exudation) characteristics. Comparisons between susceptible and selected populations indicate that it is possible to select insect-resistant plants on the basis of morphological and physiological leaf traits, but exposure of these plants to insect infestation is also recommended for plant selection.

Giustolin and Vendramim (1996b) observed that larvae reared on *L. hirsutum* f. *glabratum* showed longer development, lower viability, lighter pupae and lower fecundity compared to those reared on a susceptible cultivar of *L. esculentum* (Santa Cruz Kada). In addition, several authors reported similar results when assessing plant age as a resistance factor, and found that *T. absoluta* larval stages lasted longer and larvae suffered greater mortality on older plants of *L. hirsutum* f. *glabratum* and *L. esculentum* (Picanço et al., 1995; Leite et al., 1998; Suinaga et al., 1999).

Segeren et al. (1993) reported evidence of segregation for resistance to *T. absoluta* and the capacity to overcome the incompatibility barrier between tomato species by obtaining interspecific hybrids via embryo culture.

Giustolin and Vendramim (1996b) tested the effect of two allelochemicals on plant resistance. They used 2-tridecanone (2-T) and 2-undecanone (2-U), present in type VI trichomes of *L. hirsutum* f. *glabratum*. Larval mortality was 100% when fed with an artificial diet supplemented with mixtures of these two allelochemicals at concentrations ranging from 0.15% to 0.30% and 0.03% to 0.06%, respectively. Furthermore, there was concentration-dependent behavior when the diet was supplemented with 2-U. At 0.06% 2-U, there was high larval mortality (91.4%), while at 0.03% there was stimulation of insect development. Experiments performed with seven genotypes of two species (*L. esculentum*, *L. hirsutum* f. *glabratum*) and the F1 hybrid showed that *T. absoluta* oviposition and plant damage was greater in those genotypes with lower 2-T contents (Maluf et al., 1997). Supporting this finding, Labory et al. (1999) concluded that high levels of 2-T are associated with a non-preference for oviposition and with resistance mechanisms to feeding. According to Leite et al. (1999), 2-T production increases from the base to the top of the plant, and the authors associated *T. absoluta* oviposition with an antibiotic effect. Although genotypes of *L. hirsutum* f. *typicum* have shown lesser infestation levels by *T. absoluta* compared to *L. esculentum*, including less success in biological development, this material does not contain the 2-T and 2-U allelochemicals (Ecole et al., 2000). Leite et al. (2001) observed a higher mortality with increasing plant age. Their observation is most likely correlated with trichome density, which probably increases with age, resulting in a rise in the 2-T level and consequently a longer larval developmental period or high mortality.

Zingiberene is a sesquiterpene present in the glandular trichomes, especially in *L. hirsutum* f. *hirsutum*. Azevedo et al. (2003) found that crossing *L. esculentum* and *L. hirsutum* f. *hirsutum* results in an increase in zingiberene, which regulates resistance since it has a deterrent effect on *T. absoluta* oviposition and feeding.

Acyl sugars present in *L. pennellii* play an important role in the resistance of the tomato plant to *T. absoluta* (Resende et al., 2006; Pereira et al., 2008; Oliveira et al., 2009b). Oliveira et al. (2009a) discovered a positive correlation between tetracosane and hexacosane and the percentage of mined leaves. Similar results were obtained with acyl sugars by Gonçalves Neto et al. (2010) and Maluf et al. (2010).

6.7.4 Cultural practices

Prophylaxis is one of the most effective and cheapest ways of reducing pest infestation (Berlinger et al., 1999). Consequently, this is the aim of most of the cultural practices recommended for *T. absoluta* control. The adoption of prophylactic methods could be the key to success in controlling this pest, particularly in greenhouses, and as a result, there are several recommendations for cultural control (Arnó and Gabarra, (2010), and the online Endure Information Centre[j]).

One of the most accepted cultural methods for reducing *T. absoluta* populations is crop isolation. This can be achieved in greenhouses by screening vents and installing double doors. Monserrat (2009a) advised the use of mesh of at least 6×9 threads/cm^2 to exclude *T. absoluta* adults. One consideration, however, is that screening the greenhouse will reduce its ventilation, so measures to encourage air movement are essential to avoid adverse effects on the crop. In addition, nets will also hamper biological control, since they will prevent pest immigration and colonization by parasitoids and predators (Berlinger et al., 1999).

Reducing early pest infestations is very important. Seedlings, especially, must remain pest-free. Therefore, extreme control measures are advised for transplant producers. When the pest damage is low, particularly at the beginning of the growing season, it is important to remove leaves, stems and fruits affected by the presence of *T. absoluta* larvae or pupae, by placing the materials in sealed plastic bags exposed to direct sunlight. Also, before planting, and throughout the growing season, removal of weeds that may also host *T. absoluta* is also advised. Furthermore, it is recommended that infested crop residues should be removed either during the growing season, or immediately following harvest, by burying the residue or placing the material in closed containers covered with a transparent plastic film to allow fermentation. In Almería (south-eastern Spain), covering crop residues with plastic for no less than 3 weeks reportedly reduced the number of adult *T. absoluta* by 94% during the fall (Tapia et al., 2010). Crop residues can also be eliminated by burning or grinding combined with insecticide sprays (Robredo and Cardeñoso, 2008), although these methods may require a permit for burning, and involve the high cost of grinding.

Crop rotation with non-host crops is also imperative. In highly specialized farms where tomatoes are intensively produced, it is recommended that greenhouses should be emptied between crop cycles and sealed for 4–8 weeks, depending on the temperature (Monserrat, 2009b, 2010). Under these conditions, all adults emerging from the soil will die or will be captured by pheromone or light traps. In some situations, chemical treatments may be applied in order to reduce *T. absoluta* levels in soil. Soil solarization has been advised in warm climates, to kill pupae that remain in the soil after the harvest. To our knowledge, the impact of many of these measures on the infestation level of *T. absoluta* has not been quantified, but a broad consensus indicates that they may be useful as general prophylactic measures.

The use of genetic resistance may also be an alternative for controlling this pest, since some sources of resistance to *T. absoluta* have been reported in some wild tomato species. The two mechanisms of resistance detected so far have been antixenosis and antibiosis (Oliveira et al., 2009a, b).

In Brazil, sprinkler irrigation has been shown to have a significant impact on populations of *T. absoluta* eggs and larvae. This type of irrigation can lead to a 62% reduction in the number eggs on the plant, depending on plant age and intensity of irrigation (Costa et al., 1998). According to the same authors, water also negatively affects the number of larvae and galleries, although to a lesser extent (17–22%). To prevent *T. absoluta* fruit damage, Jordão and Nakano (2000) studied the use of paper bags covering the tomato, in combination with repellents to the pest. They concluded that the use of paper bags reduced damage caused by two Lepidopteran pests, *Helicoverpa zea* (Boddie) (Lepidoptera: Noctuidae) and *Neoleucinodes elegantalis* (Guenée) (Lepidoptera: Crambidae), but not by *T. absoluta*.

6.7.5 Chemical control

Many insecticides have traditionally been employed to control *T. absoluta* populations.

When the first pest outbreaks appeared in South America, organophosphate products (OP) and cartap were used, and were later substituted by pyrethroids in the 1970s (Desneux et al., 2010). In the 1980s, alternate applications of cartap and pyrethroids/thiocyclam were made (Lietti et al., 2005). New insecticides were introduced in the 1990s, such as acylurea, spinosad, abamectin, tebufenozide and chlorfenapyr. In addition, new pyrethroid molecules have been shown to be very efficient in Brazil (Silvério et al., 2009). The efficiency of OP products fell after 1980 with the development of resistance to such products by T. absoluta in Brazil and Chile, as well as resistance to cartap, abamectin and permethrin in Brazil (Siqueira et al., 2000, 2001), and to pyrethroids in Chile (Salazar and Araya, 1997) and Argentina (Lietti et al., 2005). This resistance raised doubts over insecticide use, but heavy chemical applications to control T. absoluta in these countries are still common. Recent research on plant extracts has demonstrated the efficiency of extracts of Trichilia pallens (Cunha et al., 2005, 2006, 2008) or neem (Gonçalves-Gervásio and Vendramim, 2007) on T. absoluta, but such products are rarely used in South America.

Since T. absoluta was detected in the Mediterranean region, the most common control practice has been based on the use of chemical insecticides (Bielza, 2010). Nevertheless, these treatments may disrupt existing IPM programs in tomato crops based on biological control (van der Blom et al., 2009) and may lead to resistance (Bielza, 2010), as it occurred in the area of origin of this pest. Therefore, there was an immediate need to choose pesticides which fulfilled two main objectives: (i) effectiveness against T. absoluta; and (ii) selectiveness, in order to preserve natural enemies in tomato crops. In addition to spinosad and indoxacarb, which were the two first available insecticides in the Mediterranean region, new effective and selective insecticides are currently available to control T. absoluta, such as flubendiamid, emamectin, rynoxapir, abamectin or etofenprox (Araujo-Gonçalves, 2010; Torné et al., 2010; Espinosa, 2010; Robles, 2010; López et al., 2010; Astor, 2010; Gutiérrez-Giulianotti, 2010). Moreover, azadiractine (neem) and sulfur treatments may also help to reduce T. absoluta incidence, although efficacies are much lower (Monserrat, 2009a).

However, repeated applications should be conducted each season in order to control T. absoluta exclusively by chemical means. Rotation of these active ingredients is compulsory to prevent resistance development (Ortega et al., 2008; Bielza, 2010), as well as the use of insecticides compatible with biological control, and integration with other control tactics.

6.8 Final Considerations

Given our knowledge of the behaviour and incidence of T. absoluta in tomato crops, and based on the information available on its control, it is impossible to control this pest with a single method. Therefore, to successfully manage this pest it is necessary to integrate several control strategies. The first step in T. absoluta management is to adopt the cultural practices described above, particularly in greenhouses. Second, methodical and periodical sampling methods should be adopted with the use of pheromone traps to monitor the incidence of adult males, and by direct damage assessments of the crop by means of plant sampling. Based on the data obtained from these samplings, appropriate control methods should be adopted. Preventive methods, such as the use of systematic B. thuringiensis treatments, alone or in conjunction with sex pheromone formulations for mating disruption or attract-and-kill, can prevent or significantly delay crop damage. Furthermore, the conservation and promotion of natural enemies in the system, especially combined with entomopathogens or selective insecticides, may contribute to suppression of T. absoluta populations. Further studies are needed to identify the most effective complement of environmentally sound management methods to control this pest. Finally, curative treatments with approved insecticides should be applied when necessary. In summary, the current impact of T. absoluta in tomato crops makes it necessary to adopt an IPM strategy to control and reduce its incidence.

Note

[i] www.endureinformationcentre.eu, accessed 14 July 2012.

References

Apablaza, J. (1992) La polilla del tomate y su manejo. *Tattersal* 79, 12–13.

Araujo-Gonçalves A. (2010) Alverde: la solución BASF para el control de *Tuta absoluta*. *Phytoma España* 217, 21–22.

Arnó, J., Mussoll, A., Gabarra, R., Sorribas, R., Prat, M., Garreta, A., Gómez, A. et al. (2009a) *Tuta absoluta* una nueva plaga en los cultivos de tomate. Estrategias de manejo. *Phytoma España* 211, 16–22.

Arnó, J., Sorribas, R., Prat, M., Matas, M., Pozo, C., Rodríguez, D., Garreta, A. et al. (2009b) *Tuta absoluta*, a new pest in IPM tomatoes in the northeast of Spain. *IOBC WPRS Bulletin* 49, 203–208.

Astor, E. (2010) Estrategias DuPont en el control de *Tuta absoluta*. *Phytoma España* 217, 107–111.

Attygalle, A.B., Jham, G.N., Svatos, A., Frighetto, R.T.S. and Meinwald, J. (1995) Microscale, random reduction to the characterization of (3E,8Z,11Z)-3,8,11-tetradecatrienyl acetate, a new lepidopteran sex pheromone. *Tetrahedron Letters* 36, 5474.

Attygalle, A.B., Jham, G.N., Svatos, A., Frighetto, R.T.S., Ferrara, F.A., Vilela, E.F., Uchoa Fernandes, M.A. et al. (1996) (3E,SZ,11Z)-3,8,11-tetradecatrienyl acetate, major sex pheromone component of the tomato pest *Scrobipalpuloides absoluta* (Lepidoptera: Gelechiidae). *Bioorganic and Medicinal Chemistry* 4, 305–314.

Azevedo, S.M., Faria, M.V., Maluf, W.R., de Oliveira, A.C.B. and de Freitas, J.A. (2003) Zingiberene-mediated resistance to the South American tomato pinworm derived from *Lycopersicon hirsutum* var. *hirsutum*. *Euphytica* 134, 347–351.

Bacci, L., Picanço, M.C., Sousa, F.F., Silva, E.M., Campos, M.R. and Tomé, H.V.T. (2008) Inimigos naturais da traça do tomateiro. *Horticultura Brasileira* 26, 2808–2812.

Bahamondes, L.A. and Mallea, A.R. (1969) Biología en Mendoza de *Scrobipalpula absoluta* (Meyrick) Povolny (Lepidoptera: Gelechiidae), especie nueva para la República Argentina. *Revista Facultad Ciencias Agrarias* 15, 96–104.

Bajonero, J., Cordoba, N., Cantor, F., Rodriguez, D. and Cure, J.R. (2008) Biology and life cycle of *Apanteles gelechiidivoris* (Hymenoptera: Braconidae) parasitoid of *Tuta absoluta* (Lepidoptera: Gelechiidae). *Agronomia Colombiana* 26, 417–426.

Barona, H.G., Parra, A.S. and Vallejo, F.A.C. (1989) Evaluacion de especies silvestres de *Lycopersicon* spp., como fuente de resistencia a *Scrobipalpula absoluta* (Meyrick) y su intento de transferencia a *Lycopersicon esculentum* Mill. *Acta Agronomica* 39, 34–45.

Barrientos, Z.R., Apablaza, H.J., Norero, S.A. and Estay, P.P. (1998) Temperatura base y constante térmica de desarrollo de la polilla del tomate, *Tuta absoluta* (Lepidoptera: Gelechiidae). *Ciencia e Investigación Agraria* 25, 133–137.

Batalla-Carrera, L., Morton, A. and García-del-Pino, F. (2010) Efficacy of entomopathogenic nematodes against the tomato leafminer *Tuta absoluta* in laboratory and greenhouse conditions. *BioControl* 55, 523–530.

Benavides, M.L.A., Rincon, F.C., Hakim, J.R.C., Rodriguez, D., Maldonado, D.E.P., Cuervo, J.B. NS Riano, D.A. (2010) *Integración de conocimientos y tecnologias de polinización y control biológico*, Bogotá.

Bentancourt, C.M., Scatoni, I.B. and Rodríguez, J.J. (1996) Influencia de la temperatura sobre la reproducción y el desarrollo de *Scrobipalpuloides absoluta* (Meyrick) (Lepidoptera, Gelechiidae). *Revista Brasileira Biologia* 56, 661–670.

Benvenga, S.R., Fernandes, O.A. and Gravena, S. (2007) Decision making for integrated pest management of the South American tomato pinworm based on sexual pheromone traps. *Horticultura Brasileira* 25, 164–169.

Berlinger, M.J., Jarvis, W.R., Jewett, T.J. and Lebiush-Mordechi, S. (1999) Managing the greenhouse, crop and crop environment. In: Albajes, R., Gullino, M.L., van Lenteren, J.C. and Elad, Y. (eds) *Integrated Pest and Disease Management in Greenhouse Crops*. Kluwer Academic Publishers, Dordrecht, The Netherlands, pp. 97–123.

Berta, D.C. and Colomo, M.V. (2000) Dos especies nuevas de *Bracon* F. y primera cita para la Argentina de *Bracon lucileae* (Hymenoptera, Braconidae), parasitoides de *Tuta absoluta* (Meyrick) (Lepidoptera, Gelechiidae). *Insecta Mundi* 14, 211–219.

Bielza, P. (2010) La resistencia a insecticidas en *Tuta absoluta*. *Phytoma España*, 103–106.

Bioplanet (2010) www.abim.ch/fileadmin/documents-abim/presentations2010/session5/4_Marco_Mosti_ABIM2010.pdf, accessed 16 February 2011.

Cabello, T., Gallego, J.R., Fernández-Maldonado, F.J., Soler, A., Beltrán, D., Parra, A. and Vila, E. (2009a) The damsel bug *Nabis pseudoferus* (Hem.: Nabidae) as a new biological control agent of the South American tomato pinworm, *Tuta absoluta* (Lep.: Gelechiidae), in tomato crops of Spain. *IOBC WPRS Bulletin* 49, 219–223.

Cabello, T., Gallego, J.R., Vila, E., Soler, A., del Pino, M., Carnero, A., Hernández-Suárez, E. et al.

(2009b) Biological control of the South American tomato pinworm, *Tuta absoluta* (Lep.: Gelechiidae), with releases of *Trichograma achaeae* (Hym.: Trichogrammatidae) in tomato greenhouses of Spain. *IOBC WPRS Bulletin* 49, 225–230.

Cabello, T., Gallego, J.R., Fernandez, F.J., Vila, E., Soler, A. and Parra, A. (2010) Aplicación de parasitoides de huevos en el control de *Tuta absoluta*. *Phytoma España*, 60–65.

Cáceres, S. (1992) La polilla del tomate en Corrientes. Biología y control. *Estación Experimental Agropecuaria Bella Vista, INTA*, pp. 19.

Calvo, J., Belda, J.E. and Giménez, A. (2010) Una nueva estrategia para el control biológico de mosca blanca y *Tuta absoluta* en tomate. *Phytoma España* 216, 46–52.

Campos, R.G. (1976) Control químico del "minador de hojas y tallos de la papa" (*Scrobipalpula absoluta* Meyrick) en el valle del Cañete. *Revista Peruana de Entomologia* 19, 102–106.

Caponero, A. (2009) Solanacee, rischio in serre. Resta alta l'attenzione alla tignola del pomodoro nelle colture protette. *Colture Protette* 10, 96–97.

Carneiro, J.R. and Medeiros, M.A. (1997) Potencial de consumo de *Chrysoperla externa* (Neuroptera: Chrysopidae) utilizando ovos de *Tuta absoluta* (Lepidoptera: Gelichiidae). In: *Congresso Brasileiro de Entomologia*, vol. 16. SEB, Salvador, Brazil, pp. 117–118.

Castelo Branco, M., França, R.H., Cordeiro, C.M.T., Maluf, W.R. and Resende, A.M. (1987) Seleção em F_2 (*Lycopersicon esculentum* x *L. pennellii*) visando a resistência à traça-do-tomateiro. *Horticultura Brasileira* 5, 30–32.

Colomo, M.V., Berta, D.C. and Chocobar, M.J. (2002) El complejo de himenópteros parasitoides que atacan a la polilla del tomate *Tuta absoluta* (Lepidoptera: Gelechiidae) en la Argentina. *I Acta zoológica lilloana* 46, 81–92.

Costa, J.S., Junqueira, A.M.R., Silva, W.L.C. and França, F.H. (1998) Impacto da irrigação via pivô-central no controle da traça-do-tomateiro. *Horticultura Brasileira* 16, 19–23.

Cunha, U.S.d., Vendramim, J.D., Rocha, W.C. and Vieira, P.C. (2005) Potential of *Trichilia pallida* Swartz (Meliaceae) as a source of substances with insecticidal activity against the tomato leafminer *Tuta absoluta* (Meyrick) (Lepidoptera: Gelechiidae). *Neotropical Entomology* 34, 667–673.

Cunha, U.S.d., Vendramim, J.D., Rocha, W.C. and Vieira, P.C. (2006) Fractions of *Trichilia pallens* with insecticidal activity against *Tuta absoluta*. *Pesquisa Agropecuária Brasileira* 41, 1579–1585.

Cunha, U.S.d., Vendramim, J.D., Rocha, W.C. and Vieira, P.C. (2008) Bioactivity of *Trichilia pallida* Swartz (Meliaceae) derived molecules on *Tuta absoluta* (Meyrick) (Lepidoptera: Gelechiidae). *Neotropical Entomology* 37, 709–715.

Desneux, N., Wajnberg, E., Wyckhuys, K., Burgio, G., Arpaia, S., Narváez-Vasquez, C., González-Cabrera, J. *et al.* (2010) Biological invasion of European tomato crops by *Tuta absoluta*: Ecology, geographic expansion and prospects for biological control. *Journal of Pest Science* 83, 197–215.

Ecole, C.C., Picanço, M., Moreira, M.D. and Magalhães, S.T.V. (2000) Componentes químicos associados à resistência de *Lycopersicon hirsutum* f. *typicum* a *Tuta absoluta* (Meyrick) (Lepidoptera: Gelechiidae). *Annals Sociedad Entomologica Brasileira* 29, 327–337.

EPPO (2006) European and Mediterranean Plant Protection Organization. Data sheets on quarentine pests. *Tuta absoluta*. www.eppo.org/QUARANTINE/insects/Tuta_absoluta/DS_Tuta_absoluta.pdf, accessed 21 February 2011.

EPPO (2008) European and Mediterranean Plant Protection Organization. Data sheets on quarentine pests. *Tuta absoluta*. www.eppo.org/publications/reporting/reporting_service.htm, accessed 21 July 2009.

EPPO (2009) *EPPO Reporting Service - Pest & Diseases. No 8.*

Espinosa, P.J. (2010) CAL-EX AVANCE EW: nueva alternativa en la estrategia de lucha control *Tuta absoluta*. *Phytoma España* 217, 76–80.

Estay, P. (2000) Polilla del Tomate *Tuta absoluta* (Meyrick). Impresos CGS Ltda. http://alerce.inia.cl/docs/Informativos/Informativo09.pdf, accessed 21 August 2007.

Fernández, S. and Montagne, A. (1990) Biología del minador del tomate, *Scrobipalpula absoluta* (Meyrick) (Lepidoptera: Gelechiidae). *Boletín de Entomología Venezolana* 5, 89–99.

Ferrara, F.A.A., Vilela, E.F., Jham, G.N., Eiras, A.E., Picanco, M.C., Attygalle, A.B., Svatos, A. *et al.* (2001) Evaluation of the synthetic major component of the sex pheromone of *Tuta absoluta* (Meyrick) (Lepidoptera: Gelechiidae). *Journal of Chemical Ecology* 27, 907–917.

Ferreira, J.A.M. and Anjos, N. (1997) Caracterização dos ínstares larvais de *Tuta absoluta* (Meyrick) (Lepidoptera: Gelechiidae). In: *Congresso Brasileiro de Entomologia, Vol. 16. Resumos*. SEB, Salvador, Brazil, pp. 64.

Filho, M.M., Vilela, E.F., Attygalle, A.B., Meinwald, J., Svatos, A. and Jham, G.N. (2000). Field trapping of tomato moth, *Tuta absoluta* with pheromone traps. *Journal of Chemical Ecology* 26, 875–881.

França, F.H., Maluf, W.R., Rossi, P.E.F., Miranda, J.E.C. and Coelho, M.C.F. (1984) Avaliação e seleção em tomate visando resistência a traça-do-tomateiro. In: *Congresso Brasileiro de Olericultura, 24*. Resumos. FCAV, Jaboticabal, Brazil, pp. v.1-p.143.

Gabarra, R. and Arnó, J. (2010) Resultados de las experiencias de control biológico de la polilla del tomate en cultivo de invernadero y aire libre en Cataluña. *Phytoma España* 217, 65–68.

Gabarra, R., Arnó, J. and Riudavets, J. (2008) Tomate. In: Jacas, J.A. and Urbaneja, A. (eds) *Control Biológico de Plagas Agrícolas*. Phytoma España, Valencia, Spain, pp. 410–422.

Garcia, M.F. and Espul, J.C. (1982) Bioecología de la polilla del tomate (*Scrobipalpula absoluta*) en Mendoza, República Argentina. *Rev Invest Agropecuarias INTA (Argentina)* 18, 135–146.

Garcia Roa, F. (1989) *Plagas del tomate y su manejo*. Palmira

Giustolin, T.A. (1996) Efeito de dois genótipos de Lycopersicon spp. associados aos entomopatógenos *Bacillus thuringiensis* var. *kurstaki* e *Beauveria bassiana* no desenvolvimento de *Tuta absoluta* (Meyrick, 1917) (Lep., Gelechiidae). PhD Thesis, Escola Superior de Agricultura "Luiz de Queiroz", Universidade de São Paulo, Sao Paulo, Brazil.

Giustolin, T.A. and Vendramim, J.D. (1996a) Biologia de *Scrobipalpuloides absoluta* (Meyrick), em folhas de tomateiro, em laboratório (Meyrick). *Revista Ecossistema* 21, 11–15.

Giustolin, T.A. and Vendramim, J.D. (1996b) Efeito de duas espécies de tomateiro na biologia de *Scrobipalpuloides absoluta* (Meyrick). *Anais Sociedad Entomologica Brasileira* 23, 511–517.

Giustolin, T.A., Vendramim, J.D., Alves, S.B. and Vieira, S.A. (2001a) Patogenicidade de *Beauveria bassiana* (Bals.) Vuill. sobre *Tuta absoluta* (Meyrick) (Lepidoptera: Gelechiidae) criada em dois genotipos de tomateiro. *Neotropical Entomology* 30, 417–421.

Giustolin, T.A., Vendramim, J.D., Alves, S.B., Vieira, S.A. and Pereira, R.M. (2001b) Susceptibility of *Tuta absoluta* (Meyrick) (Lep., Gelechiidae) reared on two species of *Lycopersicon* to *Bacillus thuringiensis* var. *kurstaki*. *Journal of Applied Entomology* 125, 551–556.

Gomide, E.V.A., Vilela, E.F. and Picanço, M. (2001) Comparison of sampling procedures for *Tuta absoluta* (Meyrick) (Lepidoptera: Gelechiidae) in tomato crop. *Neotropical Entomology* 30, 697–705.

Gonçalves-Gervásio, R.C.R. and Vendramim, J.D. (2007) Bioactivity of aqueous neem seeds extract on the *Tuta absoluta* (Meyrick, 1917) (Lepidoptera: Gelechiidae) in three ways of application. *Ciencia e Agrotecnologia* 31, 28–34.

Gonçalves Neto, A.C., Silva, V.F., Maluf, W.R., Maciel, G.M, Nizio, D.A.C., Gomes L.A.A. and Azevedo, S.M. (2010) Resistência à traça-do-tomateiro em plantas com altos teores de acilaçúcares nas folhas. *Horticultura Brasileira* 28, 203–208.

González-Cabrera, J. and Ferré, J. (2008) Bacterias Entomopatógenas. In: Jacas, J. and Urbaneja, A. (eds) *Control Biológico de Plagas Agrícolas*. Phytoma España 1, 85–97.

González-Cabrera, J., Mollá, O., Montón, H. and Urbaneja, A. (2011) Efficacy of *Bacillus thuringiensis* (Berliner) for controlling the tomato borer, *Tuta absoluta* (Meyrick) (Lepidoptera: Gelechiidae). *Biocontrol* 56, 71–80.

Greene, G.L., Leppla, N.C. and Dickerson, W.A. (1976) Velvetbean caterpillar: a rearing procedure and artificial diet. *Journal of Economic Entomology* 69, 487–488.

Griepink, F.C., Beek, T.A., Posthumus, M.A., Groot, A., Visser, J.H. and Voerman, S. (1996) Identification of the sex pheromone of *Scrobipalpula absoluta*, determination of double bond positions in triple unsaturated straight chain molecules by means of dimethyl disulphide derivatization. *Tetrahedron Letters* 37, 414.

Gutiérrez-Giulianotti, L. (2010) Programa de IPM Certis para el control de *Tuta absoluta*. *Phytoma España* 217, 60–65.

Haji, F.N.P., Parra, J.R.P., Silva, J.P. and Batista, J.G.S. (1988) Biologia da traça-do-tomateiro sob condições de laboratório. *Pesquisa Agropecuária Brasileira* 23, 107–110.

Haji, F.N.P., Freire, L.C.L., Roa, F.G., da Silva, C.N., Souza Júnior, M.M. and da Silva, M.I.V. (1995) Manejo integrado de *Scrobipalpuloides absoluta* (Povolny) (Lepidoptera: Gelechiidae) no Submédio São Francisco. *Anais da Sociedade Entomológica do Brasil* 24, 587–591.

Haji, F.N.P., Prezotti, L., Cameiro, J.D.S. and Alencar, J.A.D. (2002) *Trichogramma pretiosum* para controle de pragas no tomateiro industrial. In: Parra, J.R.P., Botelho, P.S.M., Ferreira, B.S.C. and Bento, J.M.S. (eds) *Controle Biológico no Brasil: Parasitóides e Predadores*. Manole Ltda, São Paulo, Brazil, pp. 477–494.

Imenes, S.D.L., Uchôa-Fernandes, M.A., Campos, T.B. and Takematsu, A.P. (1990) Aspectos biológicos e comportamentais da traça-do-tomateiro *Scrobipalpula absoluta* (Meyrick, 1917), (Lepidoptera-Gelechiidae). *Arquivos do Instituto Biológico* 57, 63–68.

Jimenez, M., Bobadilla, D., Vargas, H., Taco, E. and Mendoza, R. (1998) Nivel de daño de *Tuta absoluta* (Meyrick), (Lepidoptera: Gelechiidae), en cultivos experimentales de tomate sin aplicación de insecticidas convencionales. In: *XX Congreso Nacional de Entomología*, Sociedad Chilena de Entomología and Univ ersidad de Concepción, Concepción, Chile, pp. 43–51.

Jordão, A.L. and Nakano, O. (2000) Controle de lagartas dos frutos do tomateiro pelo ensacamento das pencas. *Anais Sociedade Entomologia Brasileira* 29, 773–782.

Koppert, B.S. (2010) Koppert complementa el control biológico de *Tuta absoluta* con un n uevo parasitoide. www.koppert.es/noticias/control-biologico/detalle/koppert-complementa-el-control-biologico-de-tuta-absoluta-con-un-nuevo-parasitoide, accessed 16 February 2011.

Kýlýc, T. (2010) First record of *Tuta absoluta* in Turkey. *Phytoparasitica* 38, 243–244.

Labory, C.R.G., Santa-Cecília, L.V.C., Maluf, W.R., Cardoso, M.G., Bearzotti, E. and Souza, J.C. (1999) Seleção indireta par a teores de 2-tridecanona em tomateiros segregantes e sua relação com a resistência à tr aça do tomateiro. *Pesquisa Agropecuaria Brasileira* 34, 723–739.

Lara, L., A.R., Salvador, E. and Téllez, M.M. (2010) Estudios de control biológico de la polilla del tomate *Tuta absoluta* Meyrick (Lepidoptera; Gelichiidae) en cultivos hortícolas de invernadero del Sureste Español. *Phytoma España* 221, 39.

Larraín, P. (1987) Plagas del tomate, primera parte: Descripción, fluctuación pob lacional, daño, plantas hospederas, enemigos natur ales de las plagas pr incipales. *IPA La Platina* 39, 30–35.

Larraín, P.S. (1986) Total mortality and parasitism of *Dineulophus phtorimaeae* (De Santis) in tomato moth larvae, *Scrobipalpula absoluta* (Meyrick). *Agricultura Tecnica (Chile)* 46, 227–228.

Larraín, P.S. (2001) Polilla del tomate y su manejo. *Informativo* n°1 pp. 4.

Leite, G.L.D., Picanço, M., Azevedo, A.A., Zurita, Y. and Marquini, F. (1998) Oviposicion y mortalidad de *Tuta absoluta* en *Lycopersicon hirsutum*. *Manejo Integrado de Plagas* 49, 26–34.

Leite, G.L.D., Picanco, M., la Lucia, T.M.C. and Moreira, M.D. (1999) Role of canop y height in the resistance of *Lycopersicon hirsutum* f. *glabratum* to *Tuta absoluta* (Lep., Gelechiidae). *Journal of Applied Entomology-Zeitschrift fur Angewandte Entomologie* 123, 459–463.

Leite, G.L.D., Picanco, M., Guedes , R.N.C. and Zanuncio, J.C. (2001) Role of plant age in the resistance of *Lycopersicon hirsutum* f. *glabratum* to the tomato leafminer *Tuta absoluta* (Lepidoptera: Gelechiidae). *Scientia Horticulturae* 89, 103–113.

Leite, G.L.D., Picanço, M., Jham, G.N. and Marquini, F. (2004) Intensity of *Tuta absoluta* (Meyrick, 1917) (Lepidopter a: Gelechiidae) and *Liriomyza* spp. (Diptera: Agromyzidae) attacks on *Lycopersicum sculentum* Mill. leaves. *Ciencia Agrotecnica*, 28, 42–48.

Lietti, M.M.M., Botto, E. and Alzogaray, R.A. (2005) Insecticide resistance in Argentine populations of *Tuta absoluta* (Meyrick) (Lepidoptera: Gelechiidae). *Neotropical Entomology* 34, 113–119.

López, E. (1991) Polilla del tomate: Problema crítico para la rentabilidad del cultiv o de v erano. *Empresa y Avance Agrícola* 1, 6–7.

López, J.M., Martín, L., López, A., Correia, R., González, F., Sanz, E., Gallardo , M. and Cantus, J.M. (2010) AFFIRM (Emamectina), una nueva arma contra la *Tuta absoluta* y otras orugas de lepidópteros. *Phytoma España* 217, 90–94.

Lourenção, A.L., Nagai, H. and Zullo, M.A.T. (1984) Fontes de resistência de *Scrobipalpula absoluta* (Meyrick, 1915) em tomateiro. *Bragantia* 43, 569–577.

Lourenção, A.L., Nagai, H., Siqueir a, W.J. and Fonseca, M.I.S. (1985) Seleção de linhagens de tomateiro resistentes a *Scrobipalpula absoluta* (Meyrick). *Horticultura Brasileira* 3, 77.

Luna, M.G., Sanchez, N.E. and Pereyra, P.C. (2007) Parasitism of *Tuta absoluta* (Lepidoptera, Gelechiidae) by *Pseudapanteles dignus* (Hymenoptera, Braconidae) under laboratory conditions. *Environmental Entomology* 36, 887–893.

Luna, M.G., Wada, V.I. and Sanchez, N.E. (2010) Biology of *Dineulophus phtorimaeae* (Hymenoptera: Eulophidae) and field inter action with *Pseudapanteles dignus* (Hymenoptera: Braconidae), larval parasitoids of *Tuta absoluta* (Lepidoptera: Gelechiidae) in tomato. *Annals of the Entomological Society of America* 103, 936–942.

Maluf, W.R., Barbosa, L.V. and Santa-Cecília, L.V.C. (1997) 2-tridecanone-mediated mechanisms of resistance to the South Amer ican tomato pinworm *Scrobipalpuloides absoluta* (Meyrick, 1917) (Lepidoptera-Gelechiidae) in *Lycopersicon* spp. *Euphytica* 93, 189–194.

Maluf, W.R., Silva, V.D., Cardoso, M.D., Gomes, L.A.A., Neto, A.C.G., Maciel, G.M. and Nizio, D.A.C. (2010) Resistance to the South Amer ican tomato pinworm *Tuta absoluta* in high acylsugar and/or high zingiberene tomato genotypes. *Euphytica* 176, 113–123.

Marchiori, C.H., Silva, C.G. and Lobo, A.P. (2004) Parasitoids of *Tuta absoluta* (Meyrick, 1917) (Lepidoptera: Gelechiidae) collected on tomato plants in Lavras, State of Minas Gerais, Brazil. *Brazilian Journal of Biology* 64, 551–552.

Mascarin, G.M., Alves, S.B., Rampelotti-Ferreira, F.T., Urbano, M.R., Demetrio, C.G.B. and Delalibera, I. (2010) Potential of a g ranulovirus isolate to control *Phthorimaea operculella* (Lepidoptera: Gelechiidae). *BioControl* 55, 657–671.

Medeiros, M.A. de, Sujii, E.R. and Mor ais, H.C. (2009a) Effect of plant diversification on abundance of South Amer ican tomato pinw orm and predators in tw o cropping systems . *Horticultura Brasileira* 27, 300–306.

Medeiros, M.A. de, Boas, G.L.V., Vilela, N.J. and Carrijo, O.A. (2009b) A preliminar sur vey on the biological control of South Amer ican tomato pinworm with the par asitoid *Trichogramma pretiosum* in greenhouse models. *Horticultura Brasileira* 27, 80–85.

Michereff, M., Vilela, E.F., Attygalle, A.B., Meinwald, J., Svatos, A. and Jham, G.N. (2000a) Field trapping of tomato moth, *Tuta absoluta* with pheromone traps. *Journal of Chemical Ecology* 26, 875–881.

Michereff, M., Vilela, E.F., Jham, G.N., Attygalle, A., Svatos, A. and Meinw ald, J. (2000b) Initial studies of mating disr uption of the tomato moth, *Tuta absoluta* (Lepidoptera: Gelechiidae) using synthetic sex pheromone. *Journal of the Brazilian Chemical Society* 11, 621–628.

Mihsfeldt, L.H. (1998) Biologia e e xigências térmicas de *Tuta absoluta* (Meyrick, 1917) em dieta artificial, Piracicaba, 87f. PhD thesis, Escola Superior de Ag ricultura "Luiz de Queiroz", Universidade de São P aulo, São P aulo, Brazil.

Mihsfeldt, L.H. and Parra, J.R.P. (1999) Biologia de *Tuta absoluta* (Meyrick, 1917) em dieta ar tificial. *Scientia Agrícola* 56, 769–776.

Miranda, M.M.M., Picanco, M., Zanuncio, J.C. and Guedes, R.N.C. (1998) Ecological lif e table of *Tuta absoluta* (Meyrick) (Lepidoptera: Gelechiidae). *Biological Science and Technology* 8, 597–606.

Miranda, M.M.M., Picanço, M.C., Zanuncio, J.C., Bacci, L. and Silva, da E.M. (2005) Impact of integrated pest management on the population of leafminers, fruit borers, and natural enemies in tomato. *Ciencia Rural* 35, 204–208.

Mollá, O., Montón, H., Beitia, F. and Urbaneja, A. (2008) La polilla del tomate *Tuta absoluta* (Meyrick), una nueva plaga invasora. *Terralia* 69, 3–42.

Mollá, O., Montón, H., Vanaclocha, P., Beitia, F. and Urbaneja, A. (2009) Predation b y the mir ids *Nesidiocoris tenuis* and *Macrolophus pygmaeus* on the tomato borer *Tuta absoluta*. *IOBC WPRS Bulletin* 49, 209–214.

Mollá, O., Alonso, M., Montón, H., Beitia, F., Verdú, M.J., González-Cabrera, J. and Urbaneja, A. (2010) Control biológico de *Tuta absoluta*. Catalogación de enemigos natur ales y potencial de los míridos depredadores como agentes de control. *Phytoma España* 217, 42–46.

Mollá, O., González-Cabrera, J. and Urbaneja, A. (2011) The combined use of *Bacillus thuringiensis* and *Nesidiocoris tenuis* against the tomato borer *Tuta absoluta*. *BioControl*, in press.

Monserrat, A. (2009a) *La Polilla del Tomate* Tuta absoluta *en la Región de Murcia: Bases para su Control*. Conserjería de Agricultura y Agua, Murcia, Spain.

Monserrat, A. (2009b) Medidas básicas par a el manejo de la polilla del tomate *Tuta absoluta*. *Agricola Vergel* 333, 481–491.

Monserrat, A. (2010) Estr ategias globales en el manejo de *Tuta absoluta* en Murcia. *Phytoma*, 81–86.

Muszinski, T., Lavendowski, I.M. and de Maschio, L.M. (1982) Constatação de *Scrobipalpula absoluta* (Meyrick, 1917) (= *G. norimoschema absoluta*) (Lepidoptera: Gelechiidae), como praga do tomateiro (*Lycopersicon esculentum* Mill.) no litoral do Paraná. *Anais da Sociedade Entomológica do Brasil* 11, 291–292.

Nakano, O. and Paulo, de A. (1983) As tr aças do tomateiro. *Agroquímica* 20, 8–12.

Nannini, M. (2009) Preliminar y evaluation of *Macrolophus pygmaeus* potential for control of *Tuta absoluta*. *IOBC WPRS Bulletin* 49, 215–218.

Navarro-Llopis, V., Alfaro, C., Vacas, S. and Primo, J. (2010) Aplicación de la confusión se xual al control de la polilla del tomate , *Tuta absoluta* Povolny (Lepidoptera: Gelechiidae). *Phytoma España*, 32–34.

Niedmann, L.L. and Meza-Basso, L. (2006) Evaluación de cepas nativas de *Bacillus thuringiensis* como una alter nativa de manejo integrado de la polilla del tomate (*Tuta absoluta* Meyrick; Lepidoptera: Gelechiidae) en Chile. *Agricultura Técnica* 66, 235–246.

Oatman, E.R. and Platner, G.R. (1989) Parasites of the potato tuberw orm, tomato pinw orm and other closely related Gelechiids. *Proceedings of the Hawaiian Entomological Society* 29, 23–30.

Oliveira, F.A., da Silva, D.J.H., Leite, G.L.D., Jham, G.N. and Picanco, M. (2009a) Resistance of 57 greenhouse-grown accessions of *Lycopersicon esculentum* and three cultiv ars to *Tuta absoluta* (Meyrick) (Lepidoptera: Gelechiidae). *Scientia Horticulturae* 119, 182–187.

Oliveira, F.A., da Silva, D.J.H., Picanço, M.C. and Jham, G.N. (2009b) Resistência tipo antixenose em acessos de tomateiro à *Tuta absoluta*. *Magistra* 21, 8–17.

Oliver, J.A.I. and Bringas, Y.M. (2000) Effects on the populations of the predator *Metacanthus tenellus* (Heteroptera: Berytidae) by the botanic insecticides rotenone and neem on tomato crop in Peru. *Revista Colombiana de Entomologia* 26, 89–97.

Ortega, F., Astor, E. and De Scals, D. (2008) El control de la polilla, *Tuta absoluta*. *Horticultura Internacional* 64, 30–31.

Ostrauskas, H. and Ivinskis, P. (2010) Records of the tomato pinworm (*Tuta absoluta* (Meyrick, 1917)) - Lepidoptera: Gelechiidae - in Lithuania. *Acta Zoologica Lituanica* 20, 151–155.

Papa, G. (1998) *Controle integrado de Tuta absoluta (Meyrick, 1917) com emprego de Trichogramma pretiosum Riley, 1879, inseticidas biológicos e fisiológicos*, Piracicaba, Brazil.

Parra, J.R.P. and Zucchi, R.A. (2004) Trichogramma in Brazil: Feasibility of use after twenty years of research. *Neotropical Entomology* 33, 271–281.

Paula, S.V.d., Picanço, de M.C., Oliveira, I.R. and Gusmao, M.R. (2004) Control of tomato fruit borers by surrounding crop strips. *Bioscience Journal* 20, 33–39.

Paulo, A.D. (1986) *Época de ocorrência de Scrobipalpula absoluta (Meyrick) (Lepidoptera-Gelechiidae) na cultura de tomate (Lycopersicon esculentumi Mill.) e seu controle*, Universidade de São Paulo, São Paulo, Brazil.

Pereira, G.V.N., Maluf, W.R., Goncalves, L.D., do Nascimento, D.R., Gomes, L.A.A. and Licursi, V. (2008) Selection towards high acylsugar levels in tomato genotypes and its relationship with resistance to spider mite (*Tetranychus evansi*) and to the South American pinworm (*Tuta absoluta*). *Ciencia e Agrotecnologia* 32, 996–1004.

Picanço, M.C., Silva, D.J.H., Leite, G.L.D., Mata, A.C.d. and Jham, G.N. (1995) Intensidade de ataque de *Scrobipalpuloides absoluta* (Meyrick, 1917) (Lepidoptera: Gelechiidae) ao dossel de três espécies de tomateiro. *Pesquisa Agropecuaria Brasileira* 30, 429–433.

Polack, A. (2007) Perspectivas para el control biológico de la polilla del tomate. *Horticultura Internacional* 60, 24–27.

Potting, R. (2009) *Pest Risk Analysis: Tuta absoluta, Tomato Leaf Miner Moth or South American Tomato Moth*. Plant Protection Service of the NethPlant Protection Service of The Netherlands, Ministry of Agriculture, Nature and Food Quality, Wageningen, The Netherlands.

Povolny, D. (1975) On three neotropical species of Gnorimoschemini (Lepidoptera, Gelechiidae) mining Solanaceae. *Acta Universal Agricola* 23, 379–393.

Pratissoli, D. (1995) Bioecologia de *Trichogramma pretiosum* Riley, 1879, nas tr aças *Scrobipalpuloides absoluta* (Meyrick, 1917) e *Phthorimaea operculella* (Zeller, 1873), em tomateiro. PhD thesis, Escola Superior de Agricultura "Luiz de Queiroz", Piracicaba, Brazil.

Pratissoli, D. and Parra, J.R.P. (2000a) Fertility life table of *Trichogramma pretiosum* (Hym., Trichogrammatidae) in eggs of *Tuta absoluta* and *Phthorimaea operculella* (Lep., Gelechiidae) at different temperatures. *Journal of Applied Entomology-Zeitschrift fur Angewandte Entomologie* 124, 339–342.

Pratissoli, D. and Parra, J.R.P. (2000b) Desenvolvimiento e exigencias térmicas de *Trichogramma pretiosum* Riley, criados em duas traças do tomateiro *Pesquisa Agropecuaria Brasileira* 35, 1281–1288.

Pratissoli, D. and Parra, J.R.P. (2001) Selection of strains of *Trichogramma pretiosum* Riley (Hymenoptera: Trichogrammatidae) to control the tomato leafminer moths *Tuta absoluta* (Meyrick) and *Phthorimaea operculella* (Zeller) (Lepidoptera: Gelechiidae). *Neotropical Entomology* 30, 277–282.

Pratissoli, D., Thuler, R.T., Andrade, G.S., Zanotti, L.C.M. and da Silva, A.F. (2005a) Estimate of *Trichogramma pretiosum* to control *Tuta absoluta* in stalked tomato. *Pesquisa Agropecuaria Brasileira* 40, 715–718.

Pratissoli, D., Vianna, U.R., Zago, H.B. and Pastori, P.L. (2005b) Dispersion capacity of *Trichogramma pretiosum* in propped up tomato. *Pesquisa Agropecuaria Brasileira* 40, 613–616.

Quiroz, C.E. (1976) Nuevos antecedentes sobre la biologia de la polilla del tomate, *Scrobipalpula absoluta* (Meyrick). *Agricultura Técnica* 36, 82–86.

Rázuri, V. and Vargas, E. (1975) Biología y comportamiento de *Scrobipalpula absoluta* Meyrick (Lep., Gelechiidae) en tomatera. *Revista Peruana de Entomologia* 18, 84–89.

Resende, J.T.V. de, Maluf, W.R., Faria, M.V., Pfann, A.Z. and do Nascimento, E.R. (2006) Acylsugars in tomato leaflets confer resistance to the South American tomato pinworm, *Tuta absoluta* Meyr. *Scientia Agricola* 63, 20–25.

Robles, J.L. (2010) FENOS: nuevo insecticida de Bayer Cropscience para el control de *Tuta absoluta*. *Phytoma España* 217, 87–89.

Robredo, F. and Cardeñoso, J.M. (2008) Estrategias contra la polilla del tomate , *Tuta absoluta*, Meyrick. *Agricultura* 903, 70–74.

Rodríguez, M., Gerding, M. and France, A. (2006) Effectivity of entomopathogenic fungus strains on tomato moth *Tuta absoluta* Meyrick (Lepidoptera: Gelechiidae) larvae. *Agricultura Técnica* 66, 159–165.

Rojas, S. (1981) Control de la polilla del tomate: enemigos naturales y patógenos. *IPA La Platina* 8, 18–20.

Rojas, S. (1997) Establecimiento de enemigos naturales. *Agricultura Técnica* 57, 297–298.

Salas, J. (1995) Presence of *Orius insidiosus* (Hemiptera: Anthocoridae) in central-western region of Venezuela. *Agronomia Tropical (Maracay)* 45, 637–645.

Salas, J. (2001) *Insectos Plagas del Tomate. Manejo Integrado*. Maracay, Venezuela.

Salas, J. (2004) Capture of *Tuta absoluta* (Lepidoptera: Gelechiidae) in traps baited with its sex pheromone. *Revista Colombiana de Entomologia* 30(1), 75–78.

Salas, J. and Fernández, S. (1985) Los minadores de la hoja del tomate . *Fonaip Divulga* 2, 21–22.

Salazar, E.R. and Araya, J.E. (1997) Detección de resistencia a insecticidas en la polilla del tomate. *Simiente* 67, 8–22.

Sanchez, N.E., Pereyra, P.C. and Luna, M.G.(2009) Spatial patterns of parasitism of the solitary parasitoid *Pseudapanteles dignus* (Hymenoptera: Braconidae) on *Tuta absoluta* (Lepidoptera: Gelechiidae). *Environmental Entomology* 38, 365–374.

Scardini, D.M.B., Ferreira, L.R. and Galveas, P.A.O. (1983) Ocorrência da traça-do-tomateiro *Scrobipalpuloides absoluta* (Meyr.) no Estado do Espírito Santo. In: *Congresso Brasileiro de Entomologia, 8*. DF: SEB. Brasília, Brazil, pp.72.

Segeren, M.I., Siqueir a, W.J., Sondahl, M.R., Lourenção, A.L., Medina Filho, H.P. and Nagai, H. (1993) Tomato breeding: 2.Characterization of F_1 and F_2 hybrid progenies of *Lycopersicon esculentum* x *L. peruvianum* and screening for virus and insect resistance. *Revista Brasileira Genet* 16, 773–783.

Seplyarsky, V., Weiss, M. and Haberman, A. (2010) *Tuta absoluta* Povolny (Lepidoptera: Gelechiidae), a new invasive species in Israel. *Phytoparasitica* 38, 445–446.

Silvério, F.O., Alvarenga, E.S.d., Moreno, S.C. and Picanço, M.C. (2009) Synthesis and insecticidal activity of new pyrethroids. *Pest Management Science* 65, 900–905.

Siqueira, H.A.A., Guedes, R.N.C. and Picanco, M.C. (2000) Cartap resistance and synergism in populations of *Tuta absoluta* (Lep., Gelechiidae). *Journal of Applied Entomology* 124, 233–238.

Siqueira, H.A.A., Guedes, R.N.C., Fragoso, D.B. and Magalhaes, L.C. (2001) Abamectin resistance and synergism in Brazilian populations of *Tuta absoluta* (Meyrick) (Lepidoptera: Gelechiidae). *International Journal of Pest Management* 47, 247–251.

Souza, de J.C., Reis, P.R., Gomes, J.M., Nacif, A.P. and Salgado, L.O. (1983) *Traça-do-tomateiro: Histórico, Reconhecimento, Biologia, Prejuízos e Controle*. Boletim 2, EPAMIG, Belo Horizonte, Brazil, pp. 14.

Stoltman, L., Mafra-Neto, A., Borges, R. and Zeni, D. (2010) Pheromone tools for early detection and control of the invasive tomato leafminer, *Tuta absoluta*. Entomological Society of America 58th Annual Meeting paper 49615, http://esa.confex.com/esa/2010/webprogram/Paper49615.html- A, accessed 2 December 2011.

Suinaga, F.A., Picanço, M., Jham, G.N. and Brommonschenkel, S.H. (1999) Causas químicas de resistência de *Lycopersicon peruvianum* (L.) a *Tuta absoluta* (Meyrick) (Lepidoptera: Gelechiidae). *Anais Sociedade Entomologia Brasileira* 28, 313–321.

Svatos, A., Attygalle, A.B., Jham, G.N., Frighetto, R.T.S., Vilela, E.F., Saman, D. and Meinwald, J. (1996) Sex pheromone of tomato pest *Scrobipalpuloides absoluta* (Lepidoptera: Gelechiidae). *Journal of Chemical Ecology* 22, 787–800.

Taco, E., Quispe, R., Bobadilla, D., Vargas, H., Jimenez, M. and Morales, A. (1998) Resultados preliminares de un ensayo de control biológico de la polilla del tomate *Tuta absoluta* (Meyrick), en el valle de Azapa. In: Sociedad (ed.) *IX Congreso Latino-Americano de Horticultura*. Santiago, Chile, pp. 54–56.

Tapia, G., Ruiz, M.A., Navarro, D., Lara, L. and Telléz, M.M. (2010) Estrategia de gestión de residuos vegetales en el control de *Tuta absoluta*. *Phytoma España* 217, 124–125.

Theoduloz, C., Vega, A., Salazar, M., González, E. and Meza-Basso, L. (2003) Expression of a *Bacillus thuringiensis* d-endotoxin cry1Ab gene in *Bacillus subtilis* and *Bacillus licheniformis* strains that naturally colonize the phylloplane of tomato plants (*Lycopersicon esculentum*, Mills). *Journal of Applied Microbiology* 94, 375–381.

Torné, M., Martín, A. and Fernández, J. (2010) Spintor 480SC: eficacia natural. *Phytoma España* 217, 27–31.

Torres, J.B., Evangelista, W.S., Barras, R. and Guedes, R.N.C. (2002) Dispersal of *Podisus* nigrispinus (Het., Pentatomidae) nymphs preying on tomato leafminer: effect of predator release time, density and satiation level. *Journal of Applied Entomology-Zeitschrift fur Angewandte Entomologie* 126, 326–332.

Torres Gregorio, J., Argente, J., Díaz, M.A. and Yuste, A. (2009) Aplicación de *Beauveria bassiana* en la lucha biológica contra *Tuta absoluta*. *Agrícola Vergel* 326, 129–132.

Tropea Garzia, G. (2009) *Physalis peruviana* L. (Solanaceae), a host plant of *Tuta absoluta* in Italy. *IOBC WPRS Bulletin* 49, 231–232.

Uchoa-Fernandes, M., Della Lucia, T. and Vilela, E. (1995) Mating, oviposition and pupation of *Scrobipalpula absoluta* (Meyrick) (Lepidoptera: Gelechiidae). *Annais da Socedade Entomologica do Brasil* 24, 159–164.

Urbaneja, A., Montón, H., Vanaclocha, P., Mollá, O. and Beitia, F. (2008) La polilla del tomate, *Tuta absoluta*, una nueva presa para los míridos *Nesidiocoris tenuis* y *Macrolophus pygmaeus*. *Agricola Vergel* 320, 361–367.

Urbaneja, A., Montón, H. and Mollá, O. (2009) Suitability of the tomato borer *Tuta absoluta* as prey for *Macrolophus caliginosus* and *Nesidiocoris tenuis*. *Journal of Applied Entomology* 133, 292–296.

van der Blom, J., Robledo, A., Torres, S. and Sánchez, J.A. (2009) Consequences of the wide scale implementation of biological control in greenhouse horticulture in Almeria, Spain. *IOBC WPRS Bulletin* 49, 9–13.

Vargas, H. (1970) Observaciones sobre la biología y enemigos naturales de la polilla del tomate, *Gnorismoschema absoluta* (Meyrick) (Lep. Gelechiidae). *IDESIA* 1, 75–110.

Vasicek, A.L. (1983) Natural enemies of *Scrobipalpula absoluta* Meyr. (Lep.: Gelechidae). *Revista de la Facultad de Agrónoma, Universidad Nacional de la Plata*, 59, 199–200.

Vivan, L.M., Torres, J.B., Barros, R. and Veiga, A.F.S.L. (2002a) Population growth rate of the predator bug *Podisus nigrispinus* (Heteroptera: Pentatomidae) and of the prey *Tuta absoluta* (Lepidoptera: Gelechiidae) under greenhouse conditions. *Revista de Biologia Tropical* 50, 145–153.

Vivan, L.M., Torres, J.B., Veiga, A.F.D.L. and Zanuncio, J.C. (2002b) Predator behavior and food conversion of *Podisus nigrispinus* preying on tomato leafminer. *Pesquisa Agropecuaria Brasileira* 37, 581–587.

Vivan, L.M., Torres, J.B. and Veiga, A.F.S.L. (2003) Development and reproduction of a predatory stinkbug, *Podisus nigrispinus*, in relation to two different prey types and environmental conditions. *BioControl* 48, 155–168.

Witzgall, P., Kirsch, P. and Cork, A. (2010) Sex pheromones and their impact on pest management. *Journal of Chemical Ecology* 36, 80–100.

Zucchi, R.A., Querino, R.B. and Monteiro, R.C. (2010) Diversity and hosts of *Trichogramma* in the New World, with emphasis in South America. In: Cônsoli, F.L., Parra, J.R.P. and Zucchi, R.A. (eds) *Egg Parasitoids in Agroecosystems with Emphasis on Trichogramma*. Springer, Dordrecht, The Netherlands, pp. 219–236.

7 *Tecia solanivora* Povolny (Lepidoptera: Gelechiidae), an Invasive Pest of Potatoes *Solanum tuberosum* L. in the Northern Andes

Daniel Carrillo[1]* and Edison Torrado-Leon[2]

[1]*Department of Entomology and Nematology, Tropical Research and Education Center, University of Florida, Homestead, Florida 33031, USA;*
[2]*Facultad de Agronomia, Universidad Nacional de Colombia, Bogota D.C., Colombia*

7.1 Introduction

Tecia solanivora Povolny (Lepidoptera: Gelechiidae) is a key pest of potato (*Solanum tuberosum* L.) tubers that recently invaded some cultivation areas in South America and the Canary Islands, with disastrous effects on the potato industry in these areas. The putative area of origin of *T. solanivora* is Guatemala, whence its common name (Guatemalan potato moth) is derived, probably ranging from the Isthmus of Tehuantepec in Mexico to Northern Honduras and El Salvador (Puillandre *et al.*, 2008). It was first described in 1973 and reported as a pest of potato in Guatemala, Costa Rica, Panama, Honduras and Salvador (Povolny, 1973; Niño, 2004). In 1983, it was introduced to Venezuela (Tachira State) through importation of infested potato tubers (variety Atzimba) from Costa Rica (Salazar and Torres, 1986); it rapidly became the most damaging insect pest of potato in the Venezuelan Andes. It was detected in Colombia in 1985, in the north-eastern state of Norte de Santander (Arias *et al.*, 1996), which borders Tachira. By 1990 it had reached the major cultivation areas in the center of the country, spreading to all potato cultivation areas of Colombia thereafter (Arias *et al.*, 1996). In 1996 it reached Ecuador (Pollet, 2001); and in 1999, the Canary Islands, Spain (EPPO, 2005), where it arrived through illegal importation of infested potato tubers, possibly from South America (Puillandre *et al.*, 2008). In all cases, the introduction of *T. solanivora* to new geographical areas was attributed to movement of infested tubers, and has resulted in population explosions that have significantly harmed potato production, often devastating potato crops in the invaded areas (Arias *et al.*, 1996; Pollet, 2001; Torres *et al.*, 1997).

7.2 Related Genera and their Distribution

In addition to *T. solanivora*, two other Gelechiid species, *Phthorimaea operculella* (Zeller) and *Symmetrischema tangolias* (Gyen), are considered potato pests in various areas of the world and are collectively referred as the potato tuber moth complex. Whereas there are similarities in their life history, the three species differ in their distribution, degree of polyphagy and the parts of the potato plant utilized. In contrast to *T. solanivora* which originated in Central America, both the potato tuberworm and the Andean potato moth (*P. operculella* and *S. tangolias*, respectively) likely originated in the mountainous region of South America (Sporleder *et al.*, 2004; Sporleder, 2008; Medina *et al.*, 2010). *P. operculella* has been a highly successful invasive species currently considered a cosmopolitan potato pest, whereas *S. tangolias* has a more restricted distribution in Bolivia, Peru, Ecuador, Colombia and Australia, where it was accidentally introduced (Osmelak, 1987; Griepink, 1995; Pollet *et al.*, 2003b; Sporleder, 2008). Co-occurrence of the three potato tuber moth species has been observed only in southern Colombia and Ecuador (Dangles *et al.*, 2008). Although little is known about the interaction of the three species, Dangles *et al.* (2008) studied their physiological responses to constant temperatures and found that *S. tangolias* was more cold tolerant, while *T. solanivora* had the highest growth rates at warmer temperatures. Surprisingly, the tuber moth species with the widest distribution, *P. operculella*, showed the poorest performance across the range of temperatures.

The most important difference among the species of the potato moth complex can be found in their feeding habits. *T. solanivora* is specialist on potato tubers, whereas *P. operculella* and *S. tangolias* also feed on potato leaves and stems (Rondon *et al.*, 2007; Sporleder, 2008). Moreover, *P. operculella* has at least 60 reported alternate hosts (Das and Raman, 1994) and *S. tangolias* is known to feed on other solanaceous plants (Osmelak, 1987; Sporleder, 2008).

7.3 Life History

T. solanivora has adapted to various mountainous ecosystems distributed throughout a wide altitudinal range. In the Northern Andes, *T. solanivora* occurs at medium and high altitudes between 1600 and 3400 m above sea level (ASL) (Torres *et al.*, 1997). Areas affected by *T. solanivora* in Costa Rica are located in medium altitudes between 1300 and 2300 m ASL (Povolny, 1973); in the Canary Islands, the highest damage occurs at low altitudes between 500 and 600 m ASL (EPPO, 2005). Studies conducted throughout an altitudinal range suggest a negative relationship between altitude and population densities of the pest (Gallegos, 2002; Barreto *et al.*, 2003). However, management practices and climatic variables are regarded as the key factors regulating population dynamics of *T. solanivora* in different areas of invasion (Notz, 1996; Torres, 1998; Nústez *et al.*, 1999; López-Ávila and Espitia, 2000; Barreto *et al.*, 2003).

7.4 Life Cycle

The life cycle duration of *T. solanivora* can vary greatly, depending on the temperature. Notz (1996) studied the development and reproduction of *T. solanivora* at a range of constant temperatures and concluded that the life cycle can last from approximately 42 days at 25°C to almost 200 days at 10°C. Thus, the number of generations per annum can vary from ten at 25°C to two at 10°C.

Eggs are laid individually or in small groups on the soil surface near the plant stem, or on uncovered tubers; seldom on the leaves or stems (Barreto, 2005). In stored potatoes, eggs are laid on or near the tuber eyes and on the sisal storage bags. Eggs are semi-spherical in shape and measure approximately 0.5 mm in diameter, which makes them almost imperceptible to the naked eye. Their coloration is white but they turn progressively yellow as they mature (Fig. 7.1.1). This stage can last from 5 to 25 days, depending on the environmental temperature (Notz, 1996).

Upon emergence, neonate larvae actively search for potato tubers, either burrowing into the soil or locating exposed tubers in storage conditions. Larvae enter tubers through the potato eyes or skin leaving imperceptible entrance holes, and start feeding and making galleries inside the tuber. The larval stage is composed of 4 instars that usually develop inside a single tuber; feeding starts in the outer layers (cortex and vascular ring), and advances into the inner structures (medulla) (Hilje, 1994). The first instar larvae are approximately 1.5 mm long and hyaline in appearance (Fig. 7.1.2); as they grow, larvae acquire a bluish-green coloration and reach approximately 16 mm by the last instar (Torres, 1998). The larval stage can last from approximately 18 to 80 days depending on the temperature (Notz, 1996). Mature larvae exhibit a pink coloration in the dorsum but retain the blue-green coloration ventrally (Fig. 7.1.3). At this point larvae cease

Fig. 7.1. Life cycle of *Tecia solanivora*: 1. Egg; 2. Neonate larva; 3. Larva; 4. Pupa; 5. Adult.

feeding, leave the tubers, and seek protected places to construct a silken cocoon covered with soil or dirt. Pupation occurs in the soil or dead plant tissue, and seldom inside the tubers; in potato storage, pupae can be found on tubers, packing sacks, walls, floors and crevices in the storage facility (López-Ávila and Espitia, 2000). During the pupal stage, differentiation between sexes starts to be conspicuous, as female pupae tend to be larger and heavier than male pupae. Moreover, females have the genital opening in the 8th abdominal segment and males in the 9th, which could be used as a diagnostic character to enable sex differentiation (Rincón and López-Ávila, 2004). The pupa has a brown-reddish coloration that becomes darker and shiny with time (Fig. 7.1.4). This stage can last from 10 to 90 days depending on the environmental temperature (Notz, 1996).

Adults are dull-colored moths with a narrow body and brownish-gray wings with small brown or black longitudinal markings (Figs 7.1.5 and 7.2). Females are larger and have a broader abdomen

Fig. 7.2. *Tecia solanivora* mating.

than males. Female moths can be differentiated by the presence of three dark stigmas in each wing, while males have only two (Povolny, 1973; López-Ávila and Espitia, 2000). Adults are active at dusk and remain hidden in dark and protected areas during daylight. *T. solanivora* is a sexually reproducing species in which females require a single mate to remain fertile for the majority of their adult life (Torres *et al.*, 1997; Rincón and Garcia,

2007) (Fig. 7.2). Adult longevity can vary from 10 to 25 days depending on the temperature. According to Notz (1996) the higher oviposition rates and longer fecundity periods occur at 15°C.

7.5 Sampling, Monitoring Techniques and Action Thresholds

The most widely used method for monitoring *T. solanivora* populations is trapping adult males using a synthetic sexual pheromone (Nesbitt *et al.*, 1995; Bosa *et al.*, 2005) as bait inside water or delta traps. This method has been used in population dynamics studies, for the establishment of action thresholds and in surveillance programs in places with risk of invasion.

Various studies using pheromone trapping identified key factors that regulate the population dynamics of *T. solanivora*. Factors such as the presence of residues from previous crops or other sources of infestation, (i.e., storage facilities, abandoned crops), together with pest-management practices, may influence the dynamics of this pest at a local level (Niño, 2004). A marked influence of abiotic factors on the dynamics of *T. solanivora* has been observed in the different areas. Most studies suggest that precipitation is a key factor regulating population growth of *T. solanivora*. High pest-population densities – population explosions – were often associated with dry periods, whereas populations were usually low in periods of high precipitation (Rodríguez *et al.*, 1988; Rodríguez and Lépiz, 1989; Alvarado *et al.*, 1993; Torres, 1998; Barragan *et al.*, 2002; Palacios, 2002). Moreover, these studies suggest that population dynamics of *T. solanivora* are related to the phenological stage of the crop. In general, adult activity (captures) is relatively low during the initial vegetative stages of the crop and increases when potato plants start the tuberization process. Adult activity and oviposition increase when the larval food substrate (tubers) becomes available, and remain high until tubers reach physiological maturity (Torres, 1998; Galindo and Española, 2004). This information has been key to design management tactics in the potato production systems affected by this pest. Pollet *et al.* (2003a) proposed a model to predict tuber damage in an early crop stage, based on pheromone trapping and climatic data measured during the first months of the plant cycle. Pheromone trapping has been used to establish nominal action thresholds that vary according to climatic variables and crop management. Based on experience, researchers have established thresholds from 50 to 100 adult males per week per trap (when using 4 traps per hectare) as an indicator for a pesticide application. In some cases, however, adult trapping does not correlate with tuber damage, finding low damage and high adult captures (Gallegos *et al.*, 2002) and vice versa.

In addition, pheromone trapping has been used as a key element of a preventive surveillance network established in Peru due to the high risk of introduction to this area (Oyola, 2002) and also to detect presence of the pest in storage facilities and other places.

7.6 Damage

The larval stage of *T. solanivora* is the only stage that causes direct damage to potato plants. Larvae feed and develop inside potato tubers, excavating galleries and leaving frass inside them, which allows the introduction of fungi and bacteria (Fig. 7.3). Larval feeding renders potato tubers unmarketable and useless for animal feeding purposes. Furthermore, additional resources are required to dispose of infested tubers, as they become a source of infestation for future crops or stored potatoes. Under favorable climatic conditions and poor management practices *T. solanivora* can cause complete yield loss in field and/or storage conditions (Vargas *et al.*, 2004; Arias, 1996; EPPO, 2005).

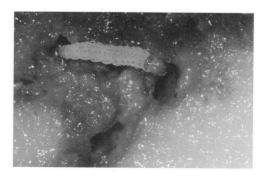

Fig. 7.3. Damage to potato tubers caused by *Tecia solanivora*.

7.7 Control Tactics

Management of *T. solanivora* requires the combination of multiple control methods both in field and storage conditions. In places where this pest is present, or threatening to invade, several research centers and institutions have proposed management alternatives including chemical, biological, autocidal, cultural and regulatory control tactics.

7.7.1 Chemical control

The effectiveness of chemical control practices in field conditions depends on the target and the timing of the applications. Most of the immature stages of *T. solanivora* develop inside the tuber and are out of reach of conventional (non-systemic) pesticide applications. Thus, to prevent the larvae from entering the tubers, applications should be directed to the base of the plant, targeting eggs and adults, the only stages exposed. Applications should be made only during the tuber formation phase. Early pesticide applications during the vegetative stages of the crop are unnecessary, as plants do not have tubers and therefore larval development is not possible. When tubers are formed and the larvae are already inside, applications do not reduce the damage. However, despite the recommendations made by research centers and extension agents, chemical control is often conducted on calendar-based applications (Niño, 2004). In Colombia, farmers rely on intensive pesticide use. This is unfortunate, since experiments conducted by Gallegos *et al.* (2002) suggested that cultural control measures are more efficient in controlling *T. solanivora* infestations than the application of several chemical pesticides.

Commonly used pesticides against *T. solanivora* are neurotoxins, including: organophosphates (Clorpirifos, Profenofos, Triclorphon) (Torres, 1998) pyrethroids (Permethrin) (Arevalo and Castro, 2003) and carbamates (López-Ávila and Barreto, 2004). Formulations for field application use water as a carrier, and dust is used as a carrier on pesticides to protect potato tubers that will be used as seed. In addition to chemically synthesized pesticides, plant-derived essential oils have also been shown to control *T. solanivora* in stored potatoes (Salazar and Betancourth, 2009; Ramirez *et al.*, 2010).

7.7.2 Autocidal

Studies have demonstrated that the reproduction of *T. solanivora* can be altered by disrupting the pheromone-mediated communication between sexes (Bosa *et al.*, 2005). The sexual pheromone produced by females to attract males, which can be synthesized and used for monitoring purposes, is a mixture of three chemical compounds in a specific ratio (Nesbitt *et al.*, 1985; Bosa *et al.*, 2005). When the ratio of the three components is modified, attraction of males by calling females is suppressed. Since *T. solanivora* only reproduces sexually, mating disruption could potentially reduce its populations. According to Bosa *et al.* (2006, 2008) this technique has potential for reducing populations of the pest both in storage and field conditions.

7.7.3 Biological control

The most developed and efficient biological control tactic against *T. solanivora* is the use of viral pathogens (Table 7.1). Initially, a granulovirus isolated from *P. operculella*, and formulated as a biopesticide by the International Potato Center (Lima, Peru), was considered promising to control *T. solanivora* under storage conditions (Sporleder, 2004). More recently, some Colombian granulovirus isolates found infecting populations of *T. solanivora* were selected for developing a biopesticide, as they were more pathogenic than Peruvian isolates (Gomez *et al.*, 2009; Espinel-Correal *et al.*, 2010).

Parasitoids of *P. operculella* were also evaluated for potential as biological control agents of *T. solanivora* (Navas *et al.*, 1986; Torres, 1998). The egg parasitoid *Copidosoma koehleri* (Hymenoptera: Encyrtidaeidae) was introduced to Venezuela in 1994 from Peru and Bolivia but showed poor development and low parasitism rates on *T. solanivora* (Torres, 1998) (Table 7.1). A Colombian native parasitoid, *Trichogramma lopezandinensis* (Hymenoptera: Trichogrammatidae), was tested for control of *T. solanivora* and showed potential to be used under storage and field conditions (Rincón and López-Ávila, 1999; Rubio *et al.*, 2004). In addition, the native entomopathogenic nematode *Steinernema feltiae* has shown some potential because of its ability to actively search

Table 7.1. Natural enemies reported in association with *T. solanivora* in the different countries where this pest is present.

Type of natural enemy	Scientific name	Location	Reference
Egg parasitoids	*Chelonus phthorimaea* Gahan	Guatemala	Navas *et al.*, 1986
	Copidosoma koehleri Blanchard	Venezuela	Torres, 1998
	Trichogramma lopezandinensis Sarmiento	Colombia	Rincón and López-Ávila, 1999
	Trichogramma aff. *pretiosum*	Colombia	Osorio *et al.*, 2001
Larval parasitoids	*Apanteles* sp.	Colombia	Osorio *et al.*, 2001
Predators (eggs and larvae)	*Buchananiella contigua* (Buchanan-White)	Colombia	Osorio *et al.*, 2001
Predators (eggs)	*Lyctocoris campestris* (Fabricius)	Colombia	Osorio *et al.*, 2001
Parasites (nematodes)	*Steinernema feltiae* Filipjev	Colombia, Venezuela	Saenz, 2003 Torres, 1998
Pathogens	Granuloviruses	Ecuador	Zeddam *et al.*, 2003
		Colombia	Espinel-Correal *et al.*, 2010
		Venezuela	Niño and Notz, 2000

for *T. solanivora* larvae inside potato tubers; however, methodologies for field applications still need to be developed (Saenz, 2003). Finally, several entomopathogenic fungi have been tested against *T. solanivora*, showing low control (Feris *et al.*, 2002).

Under natural conditions a number of natural enemy species have been reported in association with *T. solanivora* (Table 7.1). However, little is known about their influence in the population dynamics of this pest. For instance, Pollet *et al.* (2003a) suggested that natural enemies were almost non-existent in Ecuador. In contrast, studies conducted by Osorio *et al.* (2001) reported various parasitoids, predators and entomopathogens associated with *T. solanivora* in different areas of Colombia. The higher number of natural enemies reported in Colombia could be related to the longer time of establishment of the pest in this country. However, very few natural enemies have been reported in Venezuela where *T. solanivora* has been established for a longer period of time (Torres, 1998), and only one natural enemy has been reported from its native range in Guatemala (Navas *et al.*, 1986). The higher number of natural enemies found in Colombia could be explained by the sampling methodology that included placement of sentinel eggs in the field and collection of potato tubers from different locations, which resulted in significantly more findings than when only direct observation was used (Osorio *et al.*, 2001). The potential of most of the reported natural enemies of *T. solanivora* is still unknown, and requires further investigation.

7.7.4 Host plant resistance

Controlled infestation experiments have shown that commercial potato varieties in Venezuela (Niño, 2004) and Colombia (Bejarano *et al.*, 1999; Leiva and Echeverria, 1999) are susceptible to the attack of *T. solanivora*. An alternative that has been tested experimentally is the production of transgenic potato plants that express *Bacillus thuringiensis* (Bt) crystalline insecticidal proteins, achieving efficient control of *T. solanivora* (Valderrama *et al.*, 2007; Rivera and López-Ávila, 2010).

7.7.5 Cultural control

Cultural practices recommended for *T. solanivora* are often the same as those recommended for managing *P. operculella*. They are designed to destroy sources of infestation of the pest, create unfavorable conditions for pest development and interrupt the pest's life cycle.

Cultural practices proposed to reduce field infestations control *T. solanivora* include:

1. Good soil preparation to physically destroy immature stages present in the soil. Adequate soil preparation should also avoid formation of soil clumps that serve as refuges and oviposition sites for adult moths.
2. Using certified, pest-free seeds (tubers) previously treated with formulated granulovirus or chemical insecticides formulated as dusts.
3. Increasing planting depth by 15 cm to reduce the possibility of larvae reaching the tubers.
4. Increasing hilling height and frequency to protect tubers from ovipositing females and to reduce the possibility of larvae reaching the tubers.
5. Avoiding leftover potatoes that could serve as sources of infestation for the next planting season.

Cultural practices recommended to prevent infestations in potato storage include:

1. Storing tubers immediately after harvesting to avoid unnecessary exposure to ovipositing adult female moths in the field.
2. Washing tubers after harvesting; oviposition is significantly reduced on washed tubers (Vargas et al., 2004).
3. Disinfecting storage facility.
4. Avoiding reuse of packing material.
5. Allowing indirect light inside the storage facility, as adults are more active under dark conditions.
6. If possible, keeping store facility at temperatures below 8°C; *T. solanivora* cannot develop or reproduce below this temperature (Notz, 1996).

7.8 Regulatory Quarantine Methods

Studies suggest that *T. solanivora* could become established in various potato cultivation areas of South America and other parts of the world (Shaub et al., 2009). The introduction of this pest to Peru, the site of origin and where the highest diversity of potato varieties is found (Spooner et al., 2005), could have disastrous consequences. Restrictions to potato tuber movement and the implementation of surveillance networks in critical areas, together with extension programs, have helped to prevent the introduction of *T. solanivora* to this country (Naccha and Villar, 2005). The spread of *T. solanivora* to other parts of the world will depend largely on the preventive measures that the different potato-growing regions can implement to reduce the chances of an introduction. The European and Mediterranean Plant Protection Organization included *T. solanivora* in the list of species considered a threat to European agriculture, and recommended its regulation as a quarantine pest (EPPO, 2005). No major actions have been taken in other parts of the world.

7.9 Conclusion

The negative consequences experienced in areas affected by *T. solanivora* indicate the importance not only of preventing its introduction to new areas, but also of being prepared to respond to an eventual invasion. Scientists from various research centers and institutions have studied the biology and life history of this pest, and tested an array of management tactics for it. As a result, several alternatives to reduce the damage it causes are now available. However, the large-scale adoption of these management tactics in an integrated manner has not been achieved in the different areas. Unfortunately, due to the risk of high losses, lack of collaborative management, insufficient extension programs and the involvement of other important pest problems, potato growers tend to rely on the application of chemical pesticides as their only management option. Efforts are needed to increase the adoption of other tactics and to establish integrated pest management programs at regional levels.

References

Alvarado, J., Ortega, C. and Ace vedo, J. (1993) Evaluación de la densidad de trampas de feromona en la captura de la polilla Centroamericana de la papa. *Revista Latinoamericana de la Papa* 1, 77–88.

Arévalo, A. and Castro, R. (2003) Evaluación post-registro de los insecticidas con licencia de uso para controlar la polilla guatemalteca de la papa *Tecia solanivora* (Povolny 1973) (Lepidoptera: Gelechiidae) en Colombia. In: *Memorias II Taller Nacional, Tecia Solanivora*. Centro Virtual de la Cadena Alimentaria de la Papa, Bogotá, Colombia, pp. 86–89.

Arias, J.H., Jaramillo, J.A., Arevalo, E., Rocha, M.N.R. and Muñoz, G.L. (1996) *Evaluacion de la Incidencia y Severidad del Daño de la Polilla Gigante de la Papa Tecia Solanivora en el Departamento de Antioquia*. Boletin Tecnico, Corporacion Instituto Colombiano Agropecuario, Ministerio de Agricultura y Desarrollo Rural, Medellin, Colombia, pp. 24.

Barragán, A., Pollet, A., Prado, M., Onore, G., Aveiga, I. and Ruíz, C. (2002) Avances sobre la distribución y dinámica poblacional de *Tecia solanivora* en el Ecuador. In: *Memorias Del I Taller Internacional Sobre Prevención y Control de la Polilla Guatemalteca de la Papa*. Servicio Nacional de Sanidad Agraria, Centro Internacional de la Papa, Lima, Perú, 11–14 September 2001, pp. 43–50.

Barreto, N. (2005) Estudios bioecologicos de la polilla guatemalteca de la papa *Tecia solanivora* (Lepidoptera: Gelechiidae) en el altiplano Cundiboyacense Colombiano. In: López-Ávila, A. (ed.) *Memorias III Taller Internacional Sobre la Polilla Guatemalteca de la Papa*, Tecia solanivora. Cartagena de Indias, Colombia, 16–17 October 2003, pp. 95–105.

Barreto, N., Espitia, E., Galindo, R., Gordo, E., Cely, L., Sánchez, J. and López-Á vila, A. (2003) Fluctuación de la población de *Tecia solanivora* Povolny, (Lepidoptera: Gelechiidae) en tres intervalos de altitud en Cundinamarca y Boyacá, Colombia. In: *Memorias II Taller Nacional Tecia Solanivora 'Presente y Futuro de la Investigacion en Colombia sobre la Polilla Guatemalteca'*, Centro Virtual de la Cadena Alimentar ia de la Papa, Bogotá, Colombia, 24–25 Apr il 2003, pp. 119–121.

Bejarano, M., Ñustez, C. and Luque, J. (1999) Evaluación de la respuesta de trece genotipos de papa al daño de *Tecia solanivora* Povolny (Lepidoptera: Gelechiidae). In: *Conclusiones y Memorias del Taller Planeación Estratégica para el Manejo de Tecia solanivora en Colombia*. Universidad Nacional de Colombia, Federacion Colombiana de Productores de Papa, Instituto Inter americano de Ciencias Agricolas, Ministerio de Agricultura y Desarrollo Rural, Bogotá, Colombia, 22–24 J uly 1998, pp. 46.

Bosa, C.F., Cotes, A.M., Fukumoto, T., Bengtsson, M. and Witzgall, P. (2005) Pheromone-mediated communication disruption in Guatemalan potato moth, *Tecia solanivora*. *Entomologia Experimentalis et Applicata* 114, 137–142.

Bosa, C.F., Cotes, A.M., Osorio, P., Fukumoto, T., Bengtsson, M. and Witzgall, P. (2006) Disruption of pheromone communication in *Tecia solanivora* (Lepidoptera: Gelechiidae): flight tunnel and field studies. *Journal of Economic Entomology* 99, 1245–1250.

Bosa, C.F., Osorio, P., Cotes, A.M., Bengtsson, M., Witzgall, P. and Fukumoto, T. (2008) Control de *Tecia solanivora* (Lepidoptera: Gelechiidae) mediante su feromona para la interrupcion del apareamiento. *Revista Colombiana de Entomología* 34, 68–75.

Dangles, O., Carpio, C., Barragan, A.R., Zeddam, J.L. and Silvain, J.F. (2008) Temperature as a key driver of ecological sor ting among in vasive pest species in the Tropical Andes. *Ecological Applications* 18, 1795–1809.

Das, G.P. and Raman, K.V. (1994) Alternate hosts of the potato tuber moth, *Phthorimaea operculella* (Zeller). *Crop Protection* 13, 83–86.

EPPO (2005) Data sheets on quar antine pests: *Tecia solanivora*. *European Plant Protection Organization Bulletin* 35, 399–401.

Espinel-Correal, C., Léry, X., Villamizar, L., Gómez, J., Zeddam, J.L., Cotes, A.M. and López-Ferber, M. (2010) Genetic and biological analysis o f Colombian *Phthorimaea operculella* granulovirus isolated from *Tecia solanivora* (Lepidoptera: Gelechiidae). *Applied and Environmental Microbiology* 76, 7617–7625.

Feris, M.C., Gutierrez, C.G., Varela, A. and Espitia, E. (2002) Evaluacion de los hongos entomopatogenos *Beauveria bassiana* y *Metarhizium anisopliae* sobre *Tecia solanivora* (Lepidoptera: Gelechiidae). *Revista Colombiana Entomologia* 28, 179–182.

Galindo, J.R. and Española, J.A. (2004) Dinámica de la captura de *Premnotrypes vorax* (Coleoptera: Curculionidae) y la polilla guatemalteca *Tecia solanivora* (Lepidoptera: Gelechiidae) en tr ampas con dif erentes tipos de atr ayentes en un cultivo de papa cr iolla (*Solanum phureja*). *Revista Colombiana Entomologia* 30, 57–64.

Gallegos, P. (2002) Manejo de *Tecia solanivora* en el Carchi, Ecuador. In: *Memorias I Taller Internacional sobre Prevención y Control de la Polilla Guatemalteca de la Papa*. Servicio Nacional de Sanidad Ag raria, Centro Internacional de la P apa, 11–14 September 2001, Lima, Perú, pp. 101–103.

Gallegos, P., Suquillo, J., Chamorro, F., Oyarzún, P., Andrade, H., López, F., Sevillano, C. et al. (2002) Determinar la eficiencia del control químico para la polilla de la papa *Tecia solanivora*, en condiciones de campo. In: *Memorias del II Taller Internacional de Polilla Guatemalteca Tecia solanivora, Avances en Investigación y Manejo Integrado de la Plaga*, 4–5 June 2002, Quito, Ecuador pp. 7.

Gomez, J., Villamizar, L., Espinel, C. and Cotes, A.M. (2009) Comparacion de la eficacia y productividad de tres ganulovirus natives sobre larvas de *Tecia solanivora* Povolny (Lepidoptera: Gelechiidae). *Corpoica Ciencia y Tecnologia Agropecuaria* 10, 152–158.

Griepink, F.C., Van Beek, T.A., Visser, J.H., Voerman, S. and De Groot, A. (1995) Isolation and Identification of sex pheromone of *Symmetrischema tangolias* (Gyen) (Lepidoptera: Gelechiidae). *Journal of Chemical Ecology* 21, 2003–2013.

Hilje, L. (1994) Caracterización del daño de las polillas de la papa, *Tecia solanivora* y *Phthorimaea operculella* (Lepidoptera: Gelechiidae), en Cartajo, Costa Rica. *Revista Manejo Integrado de Plagas* 31, 43–46.

Leiva, E. and Echeverria, C. (1999) Apetencia de *Tecia solanivora* Povolny (Lepidoptera: Gelechiidae) a seis variedades de papa al Municipio de Pasto. In: *Conclusiones y Memorias del Taller "Planeación Estratégica para el Manejo de Tecia solanivora en Colombia"*. Universidad Nacional de Colombia, Federacion Colombiana de Papa–Instituto International de Ciencias Agricolas, Ministerio de Agricultura y Desarrollo Rural, Bogotá, Colombia, 22–24 July 1998, pp. 60.

López-Ávila, A. and Barreto, N. (2004) *Generación de Componentes Tecnológicos para el Manejo Integrado de la Polilla Guatemalteca de la papa Tecia solanivora con Base en el Conocimiento de la Biología, Comportamiento y Dinámica de Población de la Plaga*. Convenio Programa Nacional de Tecnologia Tecnica Agropecuaria-Corporacion Colombiana Instituto Agropecuario, Bogota, Colombia, pp. 159.

López-Ávila, A. and Espitia, M. (2000) *Plagas y Beneficios en el Cultivo de la Papa en Colombia*. Corporacion Colombiana de Investigacion Agropecuaria, Programa Nacional de Transferencia de Tecnologia, Programa Nacional de Manejo Integrado de Plagas, Produmedios, Bogotá, Colombia, pp. 37.

Medina, R.F., Rondon, S.I., Reyna, S.M. and Dickey, A.M. (2010) Population structure of *Phthorimaea operculella* (Lepidoptera Gelechiidae) in the United States. *Environmental Entomology* 39, 1037–1042.

Naccha, J.F. and Villar, A.C. (2005) Acciones preventivas del sistema de vigilancia fitosanitaria del Servicio Nacional de sanidad agraria del Peru contra la polilla guatemalteca de la papa *Tecia solanivora* Povolny (Lepidoptera Gelechiidae) In: López-Ávila, A. (ed.) *Memorias III Taller Internacional sobre la Polilla Guatemalteca de la Papa,* Tecia solanivora. Cartagena de Indias, Colombia, 16–17 October 2003, pp. 7–9.

Navas, L., Efraín, L. and Girón, L.F. (1986) Ciclo biológico del parásito (*Chelonus phthorimaea*) de la polilla de la papa (*Scrobipalpopsis solanivora* y *Phthorimaea operculella*). In: *Memorias IV Congreso Nacional de Manejo Integrado de Plagas*, Asociacion Guatemalteca de Manejo Integrado de Plagas, Guatemala, 1985, pp. 466–474.

Nesbitt, B., Beevor, P., Cork, A., Hall, D., Murillo, R. and Leal, H. (1985) Identification of components of the female sex pheromone of the potato tuber moth, *Scrobipalpopsis solanivora*. *Entomologia Experimentalis et Applicata* 38, 81–85.

Niño, L. (2004) Revision sobre la polilla de la papa *Tecia solanivora* en Centro y Sur america. Suplemento Revista Latinoamericana de la Papa, www.uach.cl/alap2004/Charlas%20Magistrales/08LNino%20In%20Extenso%20Revision%20Polilla%20de%20la%20Papa.pdf, accessed 23 July 2012.

Niño, L. and Notz, A. (2000) Patogenicidad de un virus granulosis de la polilla de la papa *Tecia solanivora* (Povolny 1973) (Lepidoptera: Gelechiidae) en el estado de Merida, Venezuela. *Boletín Entomología Venezolana* 15, 39–48.

Notz, A. (1996) Influencia de la temperatura sobre la biología de *Tecia solanivora* (Povolny) (Lepidoptera: Gelechiidae) criadas en tubérculos de papa *Solanum tuberosum* L. *Boletín Entomología Venezolana* 11, 49–54.

Núñez, C., Ariza, A., Becerra, J., Fuentes, L., García, G., González, D., Medina, X., *et al.* (1999) Resultados preliminares de la observación del comportamiento de *Tecia solanivora* en campos de Cultivo. In: *Conclusiones y Memorias del Taller "Planeación Estratégica para el Manejo de Tecia solanivora en Colombia"*. Universidad Nacional de Colombia, Federacion Nacional de Papa, Instituto Internacional Ciencias Agricolas, Ministerio de Agricultura y Desarrollo Rural, Bogotá, Colombia, 22–24 July 1998, pp. 48–49.

Osmelak, J.A. (1987) The tomato stemborer *Symmetrischema plaesiosema* (Turner), and the potato moth *Phthorimaea operculella* (Zeller), as stemborers of pepino: first Australian record. *Plant Protection Quarterly* 2, pp. 44.

Osorio, M., Espitia, M. and Luque, E. (2001) Reconocimiento de enemigos naturales de *Tecia solanivora* (Lepidoptera: Gelechiidae) en localidades productoras de papa en Colombia. *Revista Colombiana de Entomología* 27, 177–185.

Oyola, J. (2002) Riesgo de ingreso de Tecia solanivora en Perú. In: *Memorias I Taller Internacional sobre Prevención y Control de la Polilla Guatemalteca de la Papa*. Servicio Nacional de Sanidad Agraria, Centro

Internacional de la Papa, Lima, Perú, 11–14 September 2001, pp. 29–34.

Palacios, M. (2002) Uso de la feromona sexual el manejo de *Tecia solanivora*. En: *Memorias I Taller Internacional sobre Prevención y Control de la Polilla Guatemalteca de la Papa*. Servicio Nacional de Sanidad Agraria, Centro Internacional de la Papa, Lima, Perú, 11–14 September 2001, pp. 61–67.

Pollet, A. (2001) Guatemalan moth *Tecia solanivora* devastating potato crops in Ecuador. *International Pest Control* 43, 75–76.

Pollet, A., Barragan A., Lagnaoui, A., Prado, M., Onore, G., Aveiga, I., Lery, X. *et al.* (2003a) Prediccion de daños de la polilla guatemalteca *Tecia solanivora* (Povolny) 1973 (Lepidoptera: Gelechiidae) en el Ecuador. *Boletin de Sanidad Vegetal, Plagas* 29, 233–242.

Pollet, A., Barragan, A., Zeddam, J.L. and Lery, X. (2003b) *Tecia solanivora*, a serious biological invasion of potato cultures in South America. *International Pest Control* 45, 139–144.

Povolny, D. (1973) *Scrobipalpopsis solvanivora* sp.n. A new pest of potato (*Solanum tuberosum*) from Central America. *Acta Universitatis Agriculturae, Facultas Agronomica* 21, 133–146.

Puillandre, N., Dupas, S., Dangles, O., Zeddam, J.L., Capdevielle-Dulac, C., Barbin, K., Torres-Leguizamon, M. *et al.* (2008) Genetic bottleneck in invasive species: the potato tuber moth adds to the list. *Biological Invasions* 10, 319–333.

Ramirez, J.E., Gomez, J., Cotes, J.M. and Nústez, C. (2010) Efecto insecticidade los aceites escenciales de algunas Lamiaceas sobre *Tecia solanivora* Povolny en condiciones de laboratorio. *Agronomía Colombiana* 28, 255–263.

Rincón, D.F. and Garcia, J. (2007) Frecuencia de copula de la polilla guatemalteca de la papa *Tecia solanivora* (Lepidoptera: Gelechiidae). *Revista Colombiana de Entomología* 33, 133–140.

Rincón, C. and López-Ávila, A. (1999) Estudios biológicos del parasitoide *Trichogramma lopezandinensis* (Hymenoptera: Trichogrammatidae) orientados al control de la polilla guatemalteca de la papa *Tecia solanivora* (Lepidoptera: Gelechiidae). *Revista Colombiana de Entomología* 25, 67–71.

Rincón, R. and López-Ávila, A. (2004) Dimorfismo sexual en pupas de *Tecia solanivora* (Povolny) (Lepidoptera: Gelechiidae). *Revista Corpoica* 5, 41–42.

Rivera, F. and López-Ávila, A. (2010) Evaluación de ocho líneas de papa transgénica de las variedades Diacol Capiro y Parda Pastusa, transformadas con el gene cry 1Ac de *Bacillus thuringiensis*. In: *Memorias XXXVII Congreso de la Sociedad Colombiana de Entomologia*. Bogotá, Colombia, 1–2 June 2010, p. 100.

Rodríguez, V. and Lépiz, C. (1989) *Muestreo y Toma de Decisiones para Usar Insecticidas Contra las Polillas de la Papa*. Boletín divulgativo I. Ministerio de Agricultura y Ganadería, PRECODEPA, Costa Rica, pp. 17.

Rodríguez, V., Murillo, R. and Lépiz, C. (1988) Fluctuación de las capturas de las polillas de las papas *Scrobipalpopsis solanivora* Povolny y *Phthorimaea operculella* Zeller (Lepidoptera, Gelechiidae) en Cartago, Costa Rica. *Revista Manejo Integrado de Plagas (Costa Rica)* 9, 12–21.

Rondon, S.I., DeBano, G.H., Clough, P.B., Hamm, A., Jensen, A., Schreiber, J.M., Alvarez, J.M. *et al.* (2007) Biology and management of the potato tuberworm in the Pacific Northwest. *Pacific NorthWest* 594, 1–8.

Rubio, S.A., Vargas, B.I. and López-Ávila, A. (2004) Evaluation of the efficiency of *Trichogramma lopezandinensis* (Hymenoptera: Trichogrammatidae) to control *Tecia solanivora* (Lepidoptera: Gelechiidae) in storage potato. *Revista Colombiana de Entomologia* 30, 107–113.

Saenz, A. (2003) Eficacia de invasion de *Tecia solanivora* y *Clavipalpus ursinus* por el nematodo *Steinernema feltiae*. *Manejo Integrado de Plagas y Agroecologia (Costa Rica)*, 67, 35–43.

Salazar, C. and Betancourth, C. (2009) Evaluación de extractos de plantas para el manejo de polilla guatemalteca (*Tecia solanivora*) en cultivos de papa en Nariño, Colombia. *Agronomía Colombiana* 27, 219–226.

Salazar, J. and Torres, F. (1986) Adaptabilidad y distribución de la polilla guatemalteca de la papa *Scrobipalpopsis solanivora* (Povolný) en el Estado Táchira. *Agronomia Tropical* 36, 137–146.

Schaub, B., Chavez, D., Gonzales, J.C., Juarez, H., Simon, R., Sporleder, M. and Kroschel, J. (2009) Phenology modeling and regional risk assessment for *Tecia solanivora*. 15th Trienial Symposium of the International Society for Tropical Root Crops. Lima, Peru, 2–6 November 2009, www.cipotato.info/docs/abstracts/SessionVII/OP-49_B_Schaub.pdf, accessed 15 July 2012.

Spooner, D.M., McLean, K., Ramsay, G., Waugh, R. and Bryan, G.J. (2005) A single domestication for potato based on multilocus amplified fragment length polymorphism genotyping. *Proceedings of the National Academy of Sciences of the USA* 102, 14694–14699.

Sporleder, M. (2004) El granulovirus de *Phthorimaea operculella* (PoGV) caracteristicas biologicas, su interaccion con el medio ambiente y su possible uso en el manejo de *Tecia solanivora* (Lepidoptera: Gelechiidae). In: López-Ávila, A. (ed.) *Memorias III Taller Internacional sobre la Polilla Guatemalteca de la Papa,* Tecia solanivora. Cartagena de Indias, Colombia, 16–17 October 2003, pp. 37–49.

Sporleder, M. (2008) *Symmetrischema tangolias*, Andean/South American potato tuber moth. In: Wale, S., Platt, H.W. and Cattlin, N.D. (eds) *Diseases, Pests and Disorders of Potatoes: A Color Handbook.* Manson Publishing Ltd, London, pp. 132–136.

Sporleder, M., Kroschel, J., Gutierrez-Quispe, M.R. and Lagnaoui, A. (2004) A temperature-based simulation model for the potato tuberworm, *Phthorimaea operculella* Zeller (Lepidoptera: Gelechiidae). *Environmental Entomology* 33, 477–486.

Torres, F. (1998) *Biología y Manejo Integrado de la polilla Centroamericana de la Papa* Tecia solanivora *en Venezuela.* Serie A. N° 14, Fondo Nacional de Investigaciones Agropecuarias, Fundación para el Desarrollo de la Ciencia y la Tecnología del Estado Táchira, Maracay, Venezuela, pp. 60.

Torres, F., Notz, A. and Valencia, L. (1997) Ciclo de vida y otros aspectos de la biología de la polilla de la papa *Tecia solanivora* (Povolny) (Lepidoptera: Gelechiidae) en el Estado Táchira, Venezuela. *Boletín Entomología Venezolana* 12, 95–106.

Valderrama, A.M., Veásquez, N., Rodríguez, E., Zapata, A., Zaidi, M.A., Altosaar, I. and Arango, R. (2007) Resistance to *Tecia solanivora* (Lepidoptera: Gelechiidae) in three transgenic Andean varieties of potato expressing *Bacillus thuringiensis* CrylAc protein. *Journal of Economic Entomology* 100, 172–179.

Vargas, B.I., Rubio, S.A. and López-Ávila, A. (2004) Estudios de habitos y comportamiento de la polilla guatemalteca *Tecia solanivora* (Lepidoptera: Gelechiidae) en papa almacenada. *Revista Colombiana de Entomologia* 30, 211–217.

Zeddam, J.L., Vasquez, R.M., Vargas, Z., and Lagnaoui. A. (2003) Producción viral y tasas de aplicación del granulovirus usado para el control biológico de las polillas de la papa *Phthorimaea operculella* y *Tecia solanivora* (Lepidoptera: Gelechiidae). *Boletin de Sanidad Vegetal, Plagas* 29, 657–665.

8 The Tomato Fruit Borer, *Neoleucinodes elegantalis* (Guenée) (Lepidoptera: Crambidae), an Insect Pest of Neotropical Solanaceous Fruits

Ana Elizabeth Diaz Montilla,[1] Maria Alma Solis[2] and Takumasa Kondo[1]

[1]*Corporación Colombiana de Investigación Agropecuaria, Corpoica, Colombia;*
[2]*USDA/ARS, SEL, Room 133, Building 005, BARC-West, 10300 Baltimore Ave., Beltsville, Maryland 20705, USA*

8.1 Introduction

The tomato fruit borer, *Neoleucinodes elegantalis* (Guenée) (Lepidoptera: Crambidae) is one of the most important pests in the production of Solanaceae in South America. The larva of this insect develops inside the fruit, feeding on the mesocarp and the endosperm and caused damage that fluctuates between 13 and 77% (Costa Lima, 1949). This insect is considered a quarantine pest for several countries in the Americas (ICA and SOCOLEN, 1998; SAG, 2005; USDA *et al.*, 2005; SENASA, 2007). The objective of this chapter is to report several aspects of its biology, dynamics, damage, geographical range and integrated pest management and to provide information on species of the same genus.

8.2 Taxonomy

Order: Lepidoptera
Superfamily: Pyraloidea
Family: Crambidae
Subfamily: Spilomelinae
Genus: *Neoleucinodes* Capps

Species: *N. elegantalis* (Guenée)
Common name: tomato fruit borer

N. elegantalis was originally described as *Leucinodes elegantalis* Guenée, and recorded as a South American species attacking tomato in the states of Parana and Minas Gerais in Brazil (Capps, 1948). Capps indicated that there was confusion in the literature regarding the type species of the genus *Leucinodes* Guenée, and indicated that Hampson (1896) cited *L. elegantalis* Guenée as the type species of *Leucinodes* in his study of the moth fauna of British India, and again in his treatment of the subfamily Pyraustinae (Hampson, 1898). Furthermore, Klima (1939) also cited *L. elegantalis* as the type species of *Leucinodes* in his catalog of Pyraustinae (Capps, 1948). However, Capps (1948) noted that Walker (1859) had already designated the species *L. orbonalis* Guenée as the type species of the genus. Capps (1948) stated that under the Principle of Priority of the International Code of Zoological Nomenclature, the type species of the genus *Leucinodes* should be *L. orbonalis*, and not *L. elegantalis* as erroneously indicated by Hampson (1896). Capps (1948) regarded the New World *L. elegantalis* sufficiently different from other species in *Leucinodes* known from the Old

World, and described a new genus, *Neoleucinodes*, in order to accommodate the New World species, *N. elegantalis*.

8.3 Crops Attacked and their Economic Importance, Estimated Economic Impact and Pest Status

N. elegantalis is an oligophagous pest that attacks only fruits of plants belonging to the family Solanaceae. Some of these hosts are tropical fruits known for their exotic flavor, such as *Solanum quitoense* Lam., commonly known as lulo or naranjilla; the tree tomato (*S. betaceum* Cav.) and vegetable crops such as tomato (*S. lycopersicum* Lam.) (Fig. 8.1), eggplant (*S. melongena* L.) and green and red pepper (*Capsicum annuum* L.) (National Research Council, 1989). This pest also attacks a variety of wild solanaceous plants, mostly belonging to the *Leptostemonum* clade (D'Arcy, 1972; Nee 1979, 1991, 1999; Whalen, 1984).

Table 8.1 shows the complete list of solanaceous hosts of *N. elegantalis* in Latin America. There are five cultivated solanaceous plants that have been reported as hosts in countries like Brazil, Colombia, Ecuador, Honduras and Venezuela, and ten wild solanaceous hosts that have been recorded in Argentina, Brazil, Colombia, Panama, Peru and Venezuela.

Fig. 8.1 Damage to tomato caused by *N. elegantalis*.

Table 8.1. Solanaceous hosts of *N. elegantalis* in Latin America.

Solanaceae	Host	Ar	Br	Co	Ec	Ho	Pa	Pe	Ve	Reference
Cultivated	*Solanum lycopersicum*		×	×					×	Capps, 1948; ALAE, 1968; Posada *et al.*, 1981; Gallego and Vélez, 1992; Diaz, 2009
	S. melongena		×	×		×				Capps, 1948; Posada *et al.*, 1980; Gallego and Velez, 1992; Diaz, 2009; Espinosa, 2008
	S. annuum			×					×	Morales *et al.*, 2003; Díaz, 2009
	S. betaceum		×	×					×	Capps,1948; Gallego,1960; Posada *et al.*, 1981; Gallego and Vélez, 1992; Arnal *et al.*, 2005; Díaz, 2009
	S. quitoense			×	×					Sánchez, 1973; Posada *et al.*, 1981; Gallego and Vélez, 1992; Díaz, 2009; Revelo *et al.*, 2010
Wild hosts	*S. acerifolium*			×						Díaz, 2009
	S. atroporpureum			×						Díaz, 2009
	S. crinitum			×						Díaz, 2009
	S. torvum			×						Viáfara *et al.*, 1999; Díaz, 2009
	S. hirtum			×					×	Morales *et al.*, 2003; Díaz, 2009
	S. pseudolulo									Díaz, 2009
	S. viarum	×	×				×			Olckers *et al.*, 2002; Medal *et al.*, 1996
	S. sisymbrifolium	×								Capps, 1948
	S. gilo		×							Picanço *et al.*, 1997
	S. sessiliflorum							×		Diaz and Anteparra, 2008 (unpublished data)

Ar: Argentina, Br: Brazil, Co: Colombia, Ec: Ecuador, Ho: Honduras, Pa: Paraguay, Pe: Peru, Ve: Venezuela.

8.3.1 Economic losses

Fruit borer populations in Colombia can cause 70% loss in tomato (Restrepo, 2007), 21% in tree tomato and 13% in lulo or naranjilla (Fig. 8.2) (Colorado *et al.*, 2010). In Brazil, losses of 77% were reported by Picanço *et al.* (1998). In Venezuela, Marcano (1990) argued that the greatest loss of tomatoes (41% of damaged fruit) occurs during the rainy season, coinciding with the month of August, and the lowest losses (5.09% of damaged fruit) are recorded during the months of March and April. In Ecuador, losses of 90% in naranjilla (*S. quitoense*) have occasionally been recorded, with a maximum number of 18 larvae in a single fruit (Revelo *et al.*, 2010), and in Honduras, causing 1% loss in eggplant (SENASA, 2008).

8.4 Origin of the Pest and Current Distribution

N. elegantalis is widely distributed in the Caribbean (Cuba, Jamaica, Puerto Rico); North and Central America (Mexico, Guatemala, Honduras, Costa Rica, Panama) and South America (Argentina, Brazil, Colombia, Ecuador, French Guiana, Grenada, Guyana, Peru, Paraguay, Trinidad, Venezuela) (Capps, 1948; ALAE, 1968; Sanchez, 1973; Posada *et al.*, 1981; Gallego and Velez, 1992; Medal *et al.* 1996; Picanço *et al.*, 1997; Viáfara *et al.*, 1999; Olckers *et al.*, 2002; Morales *et al.*, 2003; Arnal *et al.*, 2005; Espinosa, 2008; Diaz, 2009).

Diaz (2009) found that in Colombia *N. elegantalis* is adapted to altitudes between 0 and 2560 m above sea level in six life zones, namely tropical dry forest (bs-T), montane dry forest (bs-PM), montane rain forest (bh-PM), very humid forest (BMH-PM), lower montane wet forest (bh-MB) and lower montane very wet forest (BMH-MB) which includes hot, temperate and cold climates. Table 8.2 describes the general characteristics of these life zones following Holdridge (1967).

Diaz (2009) also found that the presence of *N. elegantalis* can be determined according to the altitude where *Solanum* is grown. In the case of tree tomato (*S. betaceum*), *N. elegantalis* occurs between 1400 and 2279 meters above sea level (masl). Plantations of tree tomato occurring at altitudes above 2466 m that are located in the premontane wet forest (bh-MB) and premontane dry forest (bs-MB) are free of this insect pest. In lulo or naranjilla (*S. quitoense*), *N. elegantalis* occurs between 978 and 2560 masl, indicating that in Colombia this insect is widespread in most climates where these two solanaceous fruits are grown (Fig. 8. 3). In tomato *S. lycopersicum* (Fig.8.3),

Fig. 8.2 Damage to lulo caused by *N. elegantalis*.

Table 8.2. Geographical locations where *N. elegantalis* was collected in Colombia on cultivated and wild Solanaceae.

Altitude (masl*)	Temperature (°C)	Precipitation (mm)	Life zone	Climate type
0–850	> 24	1000–2000	Tropical dry forest (bs-T)	Hot
		500–1000	Premontane dry forest (bs-PM)	
850–2000	17–24	1000–2000	Premontane wet forest (bh-PM)	Temperate
		2000–4000	Premontane very wet forest (bmh-PM)	
2000–2800	12–17	1000–2000	Lower premontane wet forest (bh-MB)	Cold
		2000–4000	Lower premontane very wet forest (bmh-MB)	

*masl, meters above sea level

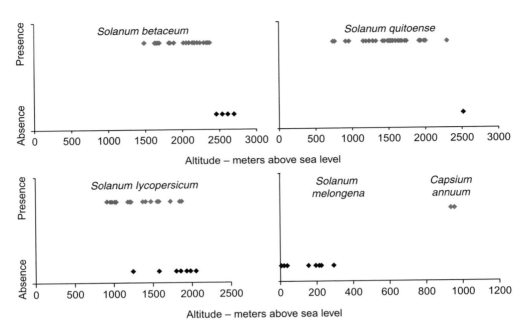

Fig. 8.3 Current distribution of *N. elegantalis* in Colombia (Diaz, 2009). The dots in the graph indicate the absence of *N. elegantalis* on solanaceous crops located within certain ranges of altitudes above sea level.

N. elegantalis appears widespread in temperate locations at altitudes between 936 and 1897 masl. Diaz (2009) also determined the occurrence of *N. elegantalis* on vegetable crops such as eggplant and red pepper in temperate areas located between 940 and 1875 masl. The insect was not present in warm climate zones of the tropical dry forest (bs-T) and tropical extremely dry forest (bms-T), characteristic of the Colombian Atlantic coast. Therefore, the author concluded that in Colombia there are areas where solanaceous plants are cultivated that have a potential for export, because *N. elegantalis* does not exist or its prevalence is low.

8.5 Other Species of *Neoleucinodes* in Latin America

Currently there are five other species of *Neoleucinodes* in the Western Hemisphere that are closely related to *N. elegantalis*. Most of these have been obtained from wild fruits of Solanaceae (Table 8.3).

8.5.1 *Neoleucinodes silvaniae* Diaz and Solis, 2007

Adult

Adults of *N. silvaniae* superficially look identical to *N. elegantalis*, but differ in that the 3rd labial palpus in the male and female of *N. silvaniae* is short, 0.4 mm in the female and 0.1 mm in the male. The scape of the female antenna is dorsally red, laterally red and white; the male scape is white with some red scales dorsally. Abdomen is grey and whitish. Forewing length is 2.2 cm in females and 1.7 cm in males (Diaz and Solis, 2007).

Male genitalia

Tegumen with an anterior margin completely sclerotized. Fibula is simple with the base not hollow, closer to the base of the valve than to the apex. Apex of the valva truncate; sclerotized costa extending ¾ of valva length. Cornutus of the phallus bladelike; apically curved slightly (Diaz and Solis, 2007).

Table 8.3. Other species of *Neoleucinodes* in Latin America (Capps, 1948; Diaz and Solis, 2007).

Species	Host plant	Country
N. silvaniae Diaz and Solis	*Solanum lanceifolium*	Colombia
N. prophetica (Dyar)	*S. umbellatum*	Colombia
	Unknown	Brazil, Costa Rica, Panama, Puerto Rico, Guatemala, Peru, Venezuela
N. imperialis (Guenée)	*S. subinerme*	Colombia
	Unknown	Costa Rica, Panama
N. dissolvens (Dyar)	Unknown	Ecuador, Brazil, French Guiana
N. torvis (Capps)	*S. rudepannum*	Colombia
	S. torvum	Cuba
	Unknown	Puerto Rico, Jamaica, Mexico, Guatemala, Costa Rica, Panama, Peru, Brazil, Guyana, Surinam, French Guiana

Female genitalia

Ostium bursae is membranous, bow-shaped, with a large aperture. Anterior and posterior apophyses short, with approximately the same length (0.7 and 0.6 mm, respectively). Bursa copulatrix (ductus + corpus bursae) three times the length of abdominal segment seven; signum is absent.

Larva

Last instar 6–8 mm long, body smooth, brownish-yellow. Body with conspicuously pigmented pinacula, particularly on mesothorax where the pigmentation is darker. Prothoracic shield is dark brown with dark markings and an extended reniform sclerotized point with blackish brown reticulations behind the XD2 seta. SD2 seta is present and easily visible through a stereoscope on a pigmented pinnaculum in front of the spiracle on A3 to A8. On the A9 segment, D1, D2, SD1 and L1 setae rise on a strongly sclerotized pinnaculum (Diaz and Solis, 2007).

Distribution

Colombia: Cundinamarca, Silvania, 1641 masl ex *S. lanceifolium* Jacq. (Diaz and Solis, 2007).

8.5.2 *Neoleucinodes prophetica* (Dyar, 1914: 278), Capps, 1948:76

Leucinodes elegantalis var. *prophetica* Dyar

Adult

Similar to small specimens of *N. elegantalis*. Coloration of wings is more opaque than in *N. elegantalis*. Male forewing is about 5.9 mm in length measured from the base to the wing apex, and 3.5 mm in width from the anterior margin to posterior margin of the median. Female forewing is about 7.6 mm in length measured from the base to the wing apex, and 3.0 mm in width from the anterior margin to posterior margin of the median line. Antennal scape is white. The 3rd labial segment short in both sexes, 1 mm in males and 1.2 mm in females (Diaz, 2009).

Male genitalia

Sclerotized, especially the vinculum. The fibula, a semicircular sclerotized ring, is located closer to the apex than to the base of the valva. Apex of the valva is rounded and constricted. Costa of the valva sclerotized is almost the entire length of the valva. Cornutus of the phallus is sword-shaped, without any curvature at the apex, and with an asymmetric sclerotized expansion at the base (Diaz, 2009).

Female genitalia

Ostium bursae is pouch-like, invaginated over a sclerotized ring located in the ventral medial area of the 8th tergite, which extends towards the back forming a shield with two compartments. The anterior and posterior apophyses are short, about the same length (0.78 and 0.63 mm, respectively). Length of the ductus and corpus bursae is three times the length of the 7th abdominal segment. Signum is absent. Corpus bursae is bag-shaped, asymmetrical oval (Diaz, 2009).

Distribution

Brazil, Costa Rica, Guatemala, Panama, Peru, Puerto Rico, Venezuela (Capps, 1948). The host plants in these countries are unknown. In Colombia, the species was collected in Calima, Darien, Valle del Cauca, at 1539 masl on *S. umbellatum* Mill. (Diaz, 2009).

8.5.3 *Neoleucinodes imperialis* (Guenée, 1854: 223), Capps, 1948: 78

Leucinodes imperialis Guenée 1854: 223.
Leucinodes discerptalis Walker, 1865: 1313.
Leucinodes discerptalis Walker, Hampson, 1898: 756.

Adult

Similar to young adults of *N. elegantalis*. Forewing hyaline white; approximately 7.2 mm in length measured from base to tip, and 2.5 mm wide from the anterior margin to posterior margin of the median line (Diaz, 2009). Capps (1948) indicated that the wings of specimens of *N. imperialis* are more stained than those of *N. elegantalis* and a little more brown in color, with the patch on the posterior margin of the forewing absent or imperceptible. The transversal anterior line of the forewing is blackish brown and rather dark. Abdomen, thorax, and brownish marks on the wings paler than in *N. elegantalis*, with some dark scales intermixed. Abdomen has an anterior whitish band.

Male genitalia

Sclerotized, especially the vinculum. Fibula is a long sclerotized plate with a triangular posterior tip that is located almost at the apex of the valva. Apex of valva is rounded, with a membranous extension, with pores, setae and hairs. Valva is short and with a slight sclerotization along the costa. Phallus has two opposed pairs of sclerotized processes.

Female genitalia

Resemble those of *N. torvis* but stouter. The genital opening sclerotized along lateral margins, interrupted ventrally. Bursa copulatrix has a sac-like extension (Capps, 1948).

Distribution

Brazil, Costa Rica and Panama. The host plants for these countries are unknown (Capps, 1948). In Colombia, the species was collected in Algeciras, Huila, 15 February 2006, at 2248 masl on fruits of *S. subinerme* Jacq. (Diaz, 2009).

8.5.4 *Neoleucinodes torvis* (Capps, 1948: 77)

The following description is adapted from Capps (1948) and Diaz (2009).

Adult

Antenna simple, slightly ringed with very short pubescence or cilia, segment length less than the width of the flagellum. Palps, head, thorax and abdomen dark yellow, whitish and pale, with some white and dark scales intermixed. The 3rd labial palp segment in the female measures 1.1 mm. White anterior band is not visible on the abdomen. Spots on the wings similar to those of *N. elegantalis* but a little darker; median patch on posterior margin of forewing narrow, inconspicuous, widened posteriorly. Transverse anterior line of forewing slightly oblique and thin, often not apparent in old specimens. Wings are approximately 6.8 mm in length measured from base to tip, and 1.8 mm wide from the anterior margin to posterior margin of the median line.

Male genitalia

Similar to those of *N. elegantalis* but valva broader distally and the fibula closer to apex than to the base of valva. Armature of the phallus small, narrow, with a concave sclerotization and a short, stout, hook-like process.

Female genitalia

Ostium bursae forms a central sclerotized shield, with an apparent broad genital opening and a median fold. Corpus bursae is narrow with bag-shaped extension. Anterior apophyses is approximately twice the length of anterior apophysis;

length of anterior apophyses 0.65 mm and posterior apophyses 0.35 mm; length of the ductus and corpus bursae is 3.8 times the length of the 7th abdominal segment. Signum absent.

Distribution

Brazil, Costa Rica, Dominican Republic, French Guiana, Grenada, Guatemala, Guyana Jamaica, Mexico, Panama, Puerto Rico, Peru and Surinarne Virgin Islands. In these countries the host plants are unknown. In Cuba it was collected from fruits of *S. torvum* Sw. (Capps, 1948). In Colombia, it was collected in Jardín, Antioquia, at 2282 masl on fruits of *S. rudepannum* Dunal (Diaz, 2009).

8.5.5 *Neoleucinodes dissolvens* (Dyar, 1914: 278) Capps, 1948: 76

Leucinodes dissolvens Dyar, 1914:278.
The following description follows mainly Capps (1948).

Adult

Closely resembles *N. elegantalis* but separable from it by the transverse anterior line of forewing. In *N. elegantalis* the line is strongly concave, while in *N. dissolvens* it is straight and outwardly oblique from costa to vein 2A, where it is angled inwardly to the hind margin of the wing. Male alar expanse 14–22 mm and female 20–25 mm.

Male genitalia

Similar to those of *N. elegantalis* but smaller in size, the valva relatively stouter, and the cornutus of the phallus expanded basally, somewhat ax-like.

Female genitalia

Similar to *N. elegantalis* but with expansion of the ductus bursae gradual anteriorly; its juncture with the bursa copulatrix not defined.

Distribution

N. dissolvens has been reported from Brazil, Ecuador and French Guiana, and the hosts are not known.

8.6 Other Genera Related to *Neoleucinodes* in Latin America

Other species in genera related to *Neoleucinodes* in the western hemisphere include *Proleucinodes melanoleuca* (Hampson) in Peru, hosts unknown; *P. xylopastalis* (Schaus) from Costa Rica, Guatemala and Mexico, hosts unknown; *P. lucealis* (Felder and Rogenhofer) from French Guiana, host unknown; *Euleucinodes conifrons* (Capps) from Peru and Colombia. In Colombia, specimens of *E. conifrons* were captured in a sex pheromone trap using Neoelegantol® (Agroecológica Platom C.A., Caracas, Venezuela) (Capps, 1948; Colorado et al., 2010).

8.7 Related Species Around the World: *Leucinodes orbonalis* Guenée

Leucinodes is the only other closely related genus to *Neoleucinodes* known to occur outside the Western Hemisphere. *Leucinodes orbonalis* is a destructive pest of eggplant and is known mainly from South and South-East Asia, and countries along the Indian Ocean in Africa (Waterhouse, 1998; FAO, 1982; Veenakumari et al., 1995; Talekar et al., 2002).

Eggs are laid on leaves and tender shoots of eggplants. Shortly after hatching, the neonate larvae migrate to the nearest shoot or fruit and bore inside. In fruit, the larvae typically enter just below the calyx (Tamaki and Miyara, 1982; Navasero, 1983; Talekar et al., 2002). *L. orbonalis* complete their larval stage in 15–20 days, during which time they pass through four larval instars. Fully grown larvae bore back to the fruit surface and emerge from the fruit or shoot leaving obvious exit holes. Larvae migrate to the soil surface to pupate in plant debris. Pupal cocoons are covered with a layer of thick material (Tamaki and Miyara, 1982; Navasero, 1983; Talekar et al., 2002). The pupal period lasts 7–10 days.

Young adults are generally found on the lower leaf surface following emergence from their cocoons. The weak-flying adults are active at night. Females are slightly bigger than the males. The abdomen of the male moth tends to be pointed, whereas the abdomen of the female moth is blunt (Talekar et al., 2002). The adults of *L. orbonalis* are morphologically similar to those of *N. elegantalis*. They can be differentiated by the type of frons, venation and characters of the genitalia (Capps, 1948).

L. orbonalis attacks mainly eggplant, where this occurs, and is the main insect pest of this crop. It has also been reported on eight other solanaceous cultivated plants, among them potato, tomatoes, green pepper and uchuva (*Physalis peruviana*). It has been recorded on wild solanaceous plants such as *S. torvum* and *S. nigrum* and also reported on other crops such as sugar beet (*Beta vulgaris*), sweet potatoes (*Ipomoea batatas*), mango (*Mangifera indica*), green peas (*Pisum sativum*) and cucumber (*Cucumis sativus*) (Fletcher, 1916; Pillai, 1922; Hussain, 1925; Hargreaves, 1937; Menon, 1962; Das and Patnaik, 1970; Mehto *et al.*, 1980; Isahaque and Chaudhuri, 1983; Whittle and Ferguson, 1987).

8.8 Biology of *N. elegantalis*

8.8.1 Life cycle

In tomato, larvae have an average duration of 16 ± 1.88 days and the pupa lasted 11.14 ± 1.23 days. The fecundity of a female was 23.24 ± 17.60 eggs, when reared on tomato at 25° C (Clavijo, 1984). At 27°C and 67% RH (relative humidity), the total life cycle lasts 33.91 days, with eggs lasting 5.54 ± 0.57 days, larva 16.41± 1.48 days, pupa: 8.12 ± 0.53 days and adults 4.30 ± 1.69 days (Fernandez and Salas, 1985). The average number of eggs per female was 34.26, the eggs had a 74.96% fertility rate and the male:female ratio was 1:1.16 (Fernandez and Salas, 1985). In eggplant, the species has four larval instars at 24°–25°C and five larval instars at 15°–30C° (Marcano, 1991b). The total developmental time was 39.16 days at 25°C (5.3 days for the eggs, 18.3 days for larvae, 9.5 for the pupa and 6.1 for the adult) (Marcano, 1991b). At temperatures of 25°C and 30.2°C, the preoviposition period was 2.8 and 2.5 days, oviposition period was 3.0 and 2.8 days and the average number of eggs per female was 75.5 and 60.0, respectively (Marcano, 1991b).The number of eggs per female ranged from 3 to 133 and from 4 to 159 at 25°C and 30.2°C, respectively (Marcano,1991b).

In lulo, Serrano *et al.* (1992) reported 6 days for the egg, 22 days for larvae, 12 days for pupa under 24°C and 74% RH conditions. Female and male lifespan 7 days and 4 days, respectively; the female preoviposition period lasts 3 days and she lays an average of 93 eggs (Serrano *et al.*, 1992).

8.8.2 Morphological description of the eggs of *N. elegantalis*

The eggs are flat, slightly sculptured and placed in groups or individually. They are 0.5 mm long and 0.3 mm wide (Fernandez and Salas, 1985). In a more in-depth study, Serrano *et al.* (1992) noted that the eggs are elliptical with a sculptured chorion, giving the appearance of a hammered surface, and with an average diameter of 0.47 mm (0.27–0.54 mm). Just-oviposited eggs are white, turning to a light yellow color and at the end of the incubation period they become brown. After hatching, the chorion is transparent and is not consumed by the newly hatched larvae (Serrano *et al.*, 1992).

8.8.3 Morphological description of the larva of *N. elegantalis*

This larval description is adapted from Capps (1948), incorporating the terminology of Stehr *et al.* (1987). Prothorax with two setae (L1 and L2) on the pre-spiracular shield and two subventral setae (SV1 and SV2). The meso- and meta-thorax with a subventral seta (SV1) (Fig. 8.4). Abdominal segments III-VI, with L1 seta close to the L2 seta on the same pinnaculum and below the spiracle. Segment IX with a pair of D2 setae on the same pinnacle and D1 seta near the SD1 on the same plate; L3 seta present and L1 and L2 setae absent. Triordinal hooks are arranged in a complete circle or often with weak or apparent break in at least some of the prolegs.

The mature larva is between 15 and 20 mm in length, tapering posteriorly, segments IX and anal segment small. Body color from white to pink. Body pinnacula without sclerotization and pigmentation. The color of the pinacula is similar to that of the body; they are present as a slightly raised blister particularly on the meso- and meta-thorax. Prothoracic shield pale yellow with light-brown markings, with no visible black reniform spot after the anterodorsal seta XD2 (Fig. 8.4). Head slightly wider than long, pale yellow, with indefinite reticulations (Fig. 8.5). In side view slightly rounded and

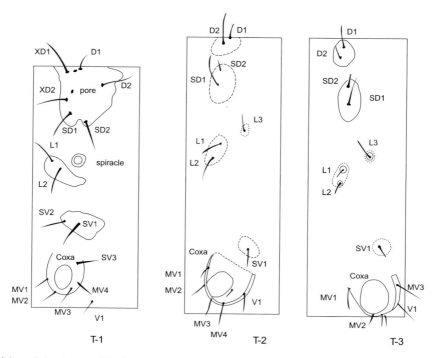

Fig. 8.4 Larval chaetotaxy of *N. elegantalis* (after Capps, 1948). Abreviations: D = dorsal seta; L = lateral seta; SD = subdorsal seta; SV = subventral seta; MV= proprioceptor seta; XD=seta located at the anterior margin of the prothoracic shield;T = thoracic segment; V = ventral seta. Drawing by Juán Manuel Vargas-Rojas.

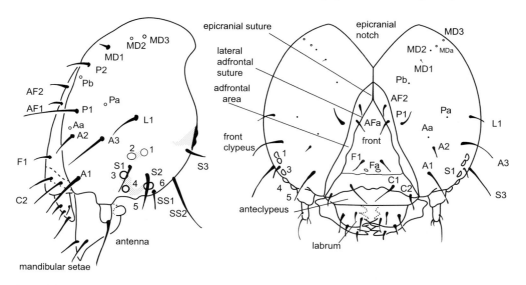

Fig. 8.5 Setal map of the cephalic capsule of *N. elegantalis* (after Capps, 1948). A = anterior seta; Aa = anterodorsal seta; AF = adfrontal seta; AFa = adfrontal puncture; C = clypeal seta; F = frontal seta; Fa = frontal puncture; L1 = lateral seta; MD = propioceptor seta dorsal; MDa = anterodorsal pore; MFs = Numbers 1, 2, 3, 4, and 5 = stemmata 1, 2, 3, 4, and 5; P = posteriodorsal seta; Pa, Pb = pore; S = stemmatal seta; SS = substemmatal seta. Drawing by Juán Manuel Vargas-Rojas.

not so flattened. Head with a darkened pigmentation a little wider at the posterior margin of head capsule. Seta (S) 2 is closer to S1 than to S3. There is an evident blackening of the estemmatal arc (area where the setae are located). However, there is a less intense pigmentation in the area between S2 and S3. S1 is located between S2 and S3. S1 can be located above the imaginary line or in the center of the line. Mandible with a smooth flange located ventrally, except for a denticle-like projection near the bottom of the tooth base.

8.8.4 Morphological description of the pupa of *N. elegantalis*

The pupa is obtect. The color varies from light to dark brown, measuring 12–15 mm, with a cremaster. Dorsum of the abdominal segments smooth. The 2nd and 3rd abdominal segment with a protruding cover above each spiracle (Capps, 1948).

8.8.5 Morphological description of adult *N. elegantalis*

The following description is taken from Capps (1948) and Maes (1995).

Adult male

Antennae filiform; lateral and ventral side covered with cilia, length approximately equal to the thickness of the flagellum near the base (Fig. 8.6A).

Labial palpi projecting upwards, brown, with a few scattered black scales. The 2nd and 3rd segments have scales facing downwards. The 1st segment quite strong, with feather-shaped scales projecting distally. The 2nd labial segment is the longest, the scales also projected distally and moderately scaled. The 3rd labial palpomeres extend horizontally and have no scales, its length is half the length of the 2nd segment. Frons eventually rounded, with dark-brown scales interspersed with black scales, occasionally with a few white scales. Posterodorsal area of the head and between the antennae covered with white scales.

The thorax in dorsal view is a mixture of brown, black and white scales. Prothorax with mostly white scales, giving the appearance of a white patch. Ventrally the thorax is white.

Dorsally the abdomen has a striking white band covering the entire 1st abdominal segment and part of the 2nd and 3rd segments, the rest of the segments covered by a mixture of dark-brown and black scales. The abdomen in ventral view, with the entire 1st abdominal segment and a large portion of the 2nd and 3rd segment white in color, the other segments paler than the dorsum. Laterally, the abdomen has small tufts of scales of the same color, often difficult to see in descaled specimens. White wings somewhat hyaline with scaly dark-brown or black areas. The anterior median line of the forewing is characteristically concave. The area near the edge of the wings can be distinguished faintly by the presence of dark scales. The wing expansion is 15–33 mm.

Genitalia

Valvae (= harpé of Capps, 1948) slender, elongate and apex much narrower than base; fibula (=clasper of Capps, 1948) slender, scalpel-like,

Fig. 8.6 *N. elegantalis*. (A): male; (B): female.

and in relation to lower margin of valvae, the fibula from near middle or distinctly near to base than to apex. Phallus slender, cornuti a simple spine, not conspicuously expanded at base.

Adult female

Antenna filiform, length of cilia slightly less than the width of the flagellum near the base (Fig. 8.6B). Labial palpi same as in the male except that the 3rd segment is longer, its length is equal to the 2nd segment. Color and spots, similar to the male. Wing expansion: 15–30 mm.

GENITALIA. Ostium bursae (= genital aperture of Capps, 1948) wide; ductus bursae elongate and cylindrical with a narrow sclerotized structure in the form of collar slightly anterior from where the ductus seminalis emerges; area between the ostium bursae and the collar membranous or slightly sclerotized, and if sclerotization occurs, more so anteriorly; bursa copulatrix simple, clearly showing the expansion at the point of connection to the ductus bursae.

8.8.6 Habits

Oviposition sites

Studies summarized below have shown that the oviposition sites of *N. elegantalis* are different in each host plant. On tomato, Rodrigues Filho *et al.* (2003) found that the moth oviposits on the stem, sepals and on the surface of fruits of different sizes (Blackmer *et al.*, 2001), but never on flowers. An average of 2.9 eggs is laid per every egg mass (Blackmer *et al.*, 2001). Salas *et al.* (1990) determined that the greater proportion of eggs (48%) was oviposited between the calyx (lower face) and fruit, 28% on the surface of the fruit, 20% on the calyx, 3% on the floral stalk and 1% on the flower buds. However, Marcano (1991a) argued that females of *N. elegantalis* oviposit on the fruit and eventually below the sepals. Viáfara *et al.* (1999) reported oviposition on leaves and stem when insect populations are high.

On lulo, Serrano *et al.* (1992) found that oviposition occurs on the flower stalks, and most commonly on 19-mm fruit. Individual or groups of 2–3 eggs are regularly found on the base of the calix area of the fruit. In tree tomato, oviposition occurs where the stalk connects to the fruit (Viáfara *et al.*, 1999). In eggplant, oviposition habits follow a pattern similar to those observed on other Solanaceae, according to Serrano *et al.* (1992).

8.8.7 Pupation sites

Pupation site varies according to the host plant and geographical area. Salas *et al.* (1990) reported that, when attacking tomato, *N. elegantalis* in Venezuela preferred the ground as a site for pupation. In Colombia, Viáfara *et al.* (1999) determined that the larvae of *N. elegantalis* pupate in the green leaves or dry fruits near the exit holes. In tree tomato, in Valle del Cauca, Colombia, the pupae are found on dry leaves on the soil surface (Viáfara *et al.*, 1999). In lulo, the fruit borer pupates on leaves and dry flower buds in the aerial part of the plant, but can also pupate in the spaces between the fruits of a cluster (Viáfara *et al.*, 1999) and in plant debris accumulated in the axils of the plants (Serrano *et al.*, 1992).

8.8.8 Rearing techniques for related species: *Leucinodes orbonalis*

A diet for mass rearing *N. elegantalis* has not been successfully developed. However, the Asian Vegetable Research and Development Center (AVRDC) has developed a diet for mass rearing *L. orbonalis* which could be adjusted for *N. elegantalis*. This diet consists of adding one part of eggplant flour to ten parts of a diet for *S. exigua* (Talekar *et al.*, 2002). For successful oviposition, 2–3 pairs of newly emerged adults are placed at 26–30°C in a nylon-netted cage containing paper as oviposition substrate. The nylon netting and paper should be checked for eggs 4–6 days later (Talekar *et al.*, 2002).

Slices of the diet are then placed in 9-cm-diameter plastic cups. Nylon and paper substrates containing eggs (bearing a total of about 50 eggs) are placed inside. Cups are then covered with snap-on lids lined with rough tissue paper (Talekar *et al.*, 2002).

When larvae reach the 3rd instar, two larvae are transferred to a 30-ml cup containing fresh diet. After 1 week, the larvae are again transferred to containers with fresh diet (Talekar *et al.*, 2002). *L. orbonalis* larvae will crawl onto the tissue paper lining the lid, and pupate there. Pupae can be collected from the lid (Talekar *et al.*, 2002). Adults usually emerge from 8–10-day-old pupae (Talekar *et al.*, 2002).

8.8.9 Monitoring and sampling techniques

Monitoring populations of *N. elegantalis* in tomato is performed using traps baited with sex pheromone as an attractant (Salas *et al.*, 1992). Salas *et al.* (1992) reported that trapping using the sex pheromone was more efficient in attracting adults than black light traps. Miras *et al.* (1997) evaluated the effectiveness of five traps to catch males of *N. elegantalis*, using virgin females in tomato plots. These authors concluded that a BM-94 trap with water (designed in 1994 by Professor Beatriz Miras, of the Behavior Laboratory, Universidad Simón Bolívar (USB), Venezuela) captured significantly more males than other traps tested.

The BM-94 trap (Fig. 8.7) consists of a plastic container 15 cm in diameter and 8 cm high, connected by four wires or cords to a roof comprising a 20 × 20 cm plastic sheet 1 mm thick, devoid of the corners of the vertices and a central hole 3 mm diameter. The container and the roof of the trap are 4 cm apart. The roof of the trap is tied by four nylon strings that come together in a hub, which is used as a support to hang the trap. Inside the plastic container 1 L of water is added with unscented soap or detergent to ensure retention of the captured insects.

Cabrera *et al.* (2001), Badji *et al.* (2003) and Jaffe *et al.* (2007) identified five components of the sex pheromone of *N. elegantalis* as: (E)-11-hexadecenol (E11–16:OH), (Z)-11-hexadecenol (Z11–16:OH), (E-11-hexadecenal (E11-16:Al), (E)-11-hexadecenyl acetate (E11-16:OAc) and (Z)-3,(Z)-6,(Z)-9-tricosatriene (Z3,Z6,Z9–23:Hy), of which Hexadecenol, E-11-16:OH exerts a strong EAG response to low doses. In field tests this compound was effective in attracting males. These authors reported that mixing (E)-11-Hexadecenol (E-11-16:0H) with (Z) -3, (Z) -6, (Z) -9-Tricosatrien (Z3, Z6, Z9-23: Hy) resulted in attraction 60 times greater than that exerted by the natural pheromone emitted by *N. elegantalis* females. Badji *et al.* (2003) determined that the most effective concentration for capturing *N. elegantalis* males was 100 and 1000 μg (micrograms). Both concentrations were effective for a period of 50 days. Jaffe *et al.*

Fig. 8.7 Traps used for *N. elegantalis*.

(2007) reported that (E)-11-hexadecenol has the ability to attract when used alone, and the isomer (Z)-11-hexadecenol can act as a repellent, because it attracts no males alone and reduces the catches when mixed with (E)-11-hexadecenol. Small amounts of (Z3, Z6, Z9) - Tricosatriene increased the attractiveness of (E)-11-hexadecenol, but large amounts inhibited the attraction. Tricosatriene failed as an attractant. Badji et al. (2003) reported a pheromone consisting of E11-16OH + 5% Tricosatriene that increased the number of catches, reducing damage to tomato. Jaffe et al. (2007) recognized a monogamous behavior in both males and females of *N. elegantalis* and selection mechanisms related to copulation, because females compete to produce a pheromone more attractive to males. According to these authors, these selection mechanisms may generate differences in the composition of the sex pheromone of populations of *N. elegantalis* along its geographic range and among different hosts.

Kuratomi (2001) evaluated the effect of mass trapping using pheromone traps in tomato in the Valle del Cauca, Colombia, using the commercial pheromone Neoelegantol®, which contains a mixture of three main components of the sex pheromone of *N. elegantalis*. An average of 2 and 6.5 adults/trap/day was caught using the BM-94 traps. Production increased 3–5 times in the experimental plots where trapping with sex pheromone was used. Arnal et al. (2005) evaluated mass trapping of *N. elegantalis* on tree tomato throughout a year, using handmade traps with a EUGO-TCC-2000® (Universidad Central de Venezuela, Maracay, Venezuela) design and baited with synthetic sex pheromone (Neoelegantol®). They found that this species is captured throughout the year, both in the dry season and in the rainy season, with fewer catches at high elevation. Salas (2008) found no differences in catches between the adhesive delta trap and the BM-94 trap with water, using the principal component of the sex pheromone (E11-16: OH) on tomato and paprika crops in the state of Lara, Venezuela.

Benvenga et al. (2010) conducted a study on the population dynamics of *N. elegantalis* on tomato in São Paulo, Brazil, using two delta traps/ha, each located on opposite corners of the field. These were fitted with synthetic sex pheromone Bio Neo® (Bio Controle, São Paulo-SP, Brazil) which is formulated with the inert material polypropylene and the sex pheromone components E 11 hexadecenol and Z3, Z6, Z9-Tricosatriene. These authors found that in the summer, the average catch/trap/day was 0.44, and for the winter season was 0.16. These results indicate that the population density of adult males in summer is larger than that found in winter. Insect captures were observed from the beginning of the flowering season until the harvest period. A study in Colombia by Colorado et al. (2010) showed that the sex pheromone Neoelegantol®, which contains a mixture of the three most important components of the females sex pheromone extracted from populations collected on tomato, has the potential to capture *N. elegantalis* populations not only on tomato, but also on lulo and tree tomato. The highest catches were recorded in lulo and tomato and lowest in tree tomato. However, the number of catches on lulo and tomato decreased when this crop was planted at higher elevations.

In the case of tree tomato, the pheromone trap catches were not influenced by altitude. It was also determined that the sex pheromone component of Neoelegantol®, besides attracting *N. elegantalis*, also attracted other moths of the families Noctuidae, Pyralidae and Geometridae.

The distributor of the sex pheromone Neoelegantol®, Agroecological Platom CA (2004), recommends a monitoring plan from the time of the transplanting of tomato, pepper or eggplant, placing 4 traps/ha, each placed opposite to the direction of night air currents, so that the traps are placed leeward. The traps should be placed at ground level when it is windy or hung at 0.5–1.0 m above the ground when wind is not a problem. Recording the date of placement of the traps is recommended in order to know when to change the septum containing the pheromone. Once a week, the soapy water of the traps should be changed, and the number of captured males of *N. elegantalis* recorded. If the average number of male *N. elegantalis* caught in the four traps is five or fewer, the surveillance program should continue; but if the average catch is >5 but <10, the density should be increased to 10 traps/ha. If after 1 week of setting the new traps the average number of catches of male *N. elegantalis* is maintained or decreased, then the program should continue using the same trap density. On the other hand, if the average number of male *N. elegantalis* >10, then 10 more traps should be placed per hectare. For the control of *N. elegantalis* a density of 20 traps/ha is ideal if the percentage of infested fruits is less than 20%.

8.8.10 Insect behavior and damage

Marcano (1991a) observed that adult *N. elegantalis* remains motionless throughout the day, with wings outstretched to the sides and the abdomen raised. The manifestation of the onset of activity by the moth is the extension of the abdomen, which is observed between 18:00 and 19:00 h, when the adult begins to move, whether walking or taking short flights. Mating occurs from 20:00 to 06:00, with a higher mating activity between 23:00 and 24:00. Oviposition occurs from 19:00 until dawn (Marcano, 1991a). According to Eiras (2000) mating is preceded by male wing vibration, and occurs between the 4th and 10th hour of the scotophase, with a peak of activity in the 7th hour. Only 2.8% of newly emerged adults mate (Eiras, 2000). Adult emergence occurs between 17:00 and 02:00, with peak emergence between 20:00 and 22:00 (Marcano, 1991a).

According to Eiras (2000), adult *N. elegantalis* emergence occurs during the scotophase. The females begin to emerge from the 1st to the 8th hour of the scotophase, with a maximum emergence in the 4th hour. The males show a similar trend, but emerge from the 2nd to the 11th hour, with a higher peak emergence in the 4th hour (Eiras, 2000).

Eiras and Blackmer (2003) reported that eggs hatch shortly after dawn. Between 7 and 123 min passed before newly emerged larvae began boring into the fruit, and they bored for 21–202 min. In Colombia, Serrano et al. (1992) determined that on lulo, newly hatched larvae secrete a silk thread and bore assuming a perpendicular position to the fruit exocarp, using their thoracic legs to scrape the epicarp in order to reach the mesocarp, until they make a completely circular entrance hole. Upon reaching the 3rd larval instar, the larva moves to the endocarp, completing the other instars inside.

8.9 Control Strategies

8.9.1 Chemical control

Studies conducted by Eiras (2000) and Eiras and Blackmer (2003) recommended the application of pesticides when fruits are 2.5 cm in diameter. Insecticides that work by ingestion or contact, or both, will be more effective at this stage. The most highly recommended insecticides are the chitin synthesis inhibitors Diflubenzuron (Dimilin® 25% WP; Uniroyal Chemical B.V. Holland, Chemtura Colombia Ltda, Bogotá, Colombia), Triflumuron (Alsystin® 480 SC; Bayer CropScience AG, Leverkusen, Germany), Lufenuron (Match®; Syngenta SA, Cartagena, Colombia), Methoxyfenozide (Intrepid® 2F 25% WP, Dow Agrosciences de Colombia S.A., Atlántico, Colombia) and the bacteria *Bacillus thuringiensis*. In Colombia, the neonicotinoids Thiametoxam + Lambdacyhalotrin (Engeo 247 SC®; Syngenta Protecao de Cultivos Ltda, Paulinia-SP, Brazil) and Emamectin Benzoate (Proclaim®; Gowan Milling, Yuma, Arizona) are recommended. Lima et al. (2001) evaluated different insecticides in tomato and suggested Alsystin® 4480 SC (triflumuron) at 30 ml p.c./100 L, Astro 450EW® (chorpyrifos; Dow Agrosciences Ltd., Kings Lynn, UK) at 120 and 150 ml commercial product/100 L and Match® at 80 ml p.c./100 L, resulting in the reduction of fruit damage 45 days after transplant. Martinelli et al. (2003) evaluated indoxacarb using 2.4–6.0 g active ingredient (AI)/100 L, esfenvalerate at 1.75AI/100 L, methomyl at 21.5 g AI/100 L and triflumuron at 15 g AI/100 L, and determined that *N. elegantalis* can be efficiently controlled after nine applications of the above-mentioned products. Motta et al. (2005) evaluated abamectin 18 EC (Vertimec 18 CE®; Syngenta Crop Protection AG, Birsfelden, Switzerland) in a dose of 1 L/ha for the control of *N. elegantalis*, using 5% of damaged tomato fruits as the economic threshold.

The authors concluded that abamectin failed to the control this insect, resulting in 70% fruit infestation after its application. On the other hand, Miranda et al. (2005) indicated that an integrated pest management strategy using an economic threshold of 20% of mined leaves and 5% of perforated fruits, could reduce pesticide applications by 65.5%, when compared to traditional control methods.

8.9.2 Biological control

Known natural enemies of *N. elegantalis* on different solanaceous hosts are recorded in Table 8.4. There is a great diversity of natural enemies of *N. elegantalis*, with many hymenopteran parasitoids

Table 8.4. List of natural enemies recorded for *Neoleucinodes elegantalis* on different solanaceous hosts.

Developmental stage	Natural enemy	Solanaceous host	Reference
Egg	*Chrysopa* sp. (Neuroptera: Chrysopidae)	*Solanum quitoense* Lam.	Serrano et al. (1992)
	Trichogramma sp. (Hymenoptera: Trichogrammatidae)	*S. lycopersicum* L. (cited as *Lycopersicum esculentum*)	Serrano et al. (1992); Diaz, (2009)
		Not determined	Noyes (2004)
	Trichogramma exiguum Pinto and Platner	*S. lycopersicum* Lam.	Viáfara et al. (1999)
		Not determined	Noyes (2004)
	Trichogramma pretiosum Riley	*S. lycopersicum* Lam.	Viáfara et al. (1999); Blackmer et al. (2001); Parra and Zucchi (2004)
		Not determined	Noyes (2004)
	Trichogramma minutum Riley	*S. lycopersicum* Lam.	Leiderman and Sauer (1953); Noyes (2004)
Egg–Larva	*Copidosoma* sp. (Hymenoptera: Encyrtidae)	*S. quitoense* Lam.	Serrano et al. (1992)
		S. betaceum Cav. (cited as *S. betaceum* Cav. Sendt.); *S. quitoense* Lam.	Viáfara et al. (1999)
		Not determined	Noyes (2004)
	Copidosoma sp. (Hymenoptera: Encyrtidae)	*S. betaceum* Cav. (cited as *S. betaceum* Cav. Sendt.)	Santamaria et al. (2007); Diaz and Brochero (unpublished data)
Larva	*Lixophaga* sp. (Diptera: Tachinidae)	*S. quitoense* Lam.	Serrano et al. (1992); Viáfara et al. (1999); Diaz and Brochero (unpublished data)
	Eupelmus littoralis (Hymenoptera: Eupelmidae)	Not determined	Noyes (2004)
	Apanteles sp.(Hymenoptera: Braconidae)	*S. quitoense* Lam.	Diaz and Brochero (unpublished data)
	Bracon sp. 1 (Hymenoptera: Braconidae)	*S. quitoense* Lam.	
	Bracon sp.2 (Hymenoptera: Braconidae)	*S. quitoense* Lam.	
	Chelonus spp. (Hymenoptera: Braconidae)	*S. quitoense* Lam.	

Continued

Table 8.4. Continued.

Developmental stage	Natural enemy	Solanaceous host	Reference
	Calliephialtes sp. (Hymenoptera: Ichneumonidae)	*S. quitoense* Lam.	Serrano *et al.* (1992)
	Brachymeria sp.; *Conura* sp. (Hymenoptera: Chalcididae)	*S. lycopersicum* Lam.	Viáfara *et al.* (1999)
	Brachymeria sp.	Unknown	Noyes (2004)
Pupa	*Brachymeria* nr *panamensis* Holmgren; *Brachymeria annulata* F.	*S. lycopersicum* Lam.	Romero (2000)
	Aprostocetus sp. (Hymenoptera: Eulophidae)	*S. betaceum* Cav. (cited as *S. betaceum* Cav. Sendt.); *S. quitoense* Lam.	Viáfara *et al.* (1999)
	Trichospilus diatraea (Hymenoptera: Eulophidae)	Not determined	Noyes (2004)
	Pimpla sanguinipes (Hymenoptera: Ichneumonidae)	*S. quitoense* Lam.	Diaz and Brochero (unpublished data)
	Lymeon sp. (Hymenoptera: Ichneumonidae)	*S. quitoense* Lam.	Diaz and Brochero (unpublished data)
	Neotheronia sp. (Hymenoptera: Ichneumonidae)	*S. quitoense* Lam.	Diaz and Brochero (unpublished data)
	Pos. *Beauveria* sp. (Hyphomycetes) entomopathogenic fungi	*S. quitoense* Lam.	Diaz and Brochero (unpublished data) Serrano *et al.* (1992)

found attacking the larvae and pupae in plantations of lulo. As parasitoids of larvae, the Braconidae were the most important, including *Apanteles* sp., *Bracon* sp. 1 and sp. 2 and *Chelonus* sp. (Diaz, 2009). Pupal parasitoids were Ichneumonidae, Chalcididae and Eulophidae. Species belonging to the family Ichneumonidae were the most abundant, including *Pimpla sanguinipes* Cresson, *Lymeon* Förster and *Neotheronia* Krieger (Diaz, 2009). Two species in the family Chalcididae, *Brachymeria* nr *panamensis* Holmgren and *B. annulata* F. (Romero, 2000) were collected. Within the Eulophidae *Trichospilus diatraea* Cherian and Margabandhu (Diaz, 2009) and *Aprostocetus* sp. (Viafara et al., 1999).

As an egg parasitoid, *Trichogramma* sp. is reported on tomato. This parasitoid has been widely studied as a biological control agent of *N. elegantalis* in tomato, not only in Colombia, but also in Brazil and Venezuela (Berti and Marcano, 1995; Cross, 1996; Corpoica Corporación Colombiana Agropecuaria, 2000). The *Trichogramma* species associated with *N. elegantalis* are *T. exiguum* Pinto and Platner (Viáfara et al., 1999; Noyes, 2004), *T. pretiosum* Riley (Viáfara et al., 1999; Blackmer et al., 2001; Parra and Zucchi, 2004; Noyes, 2004) and *T. minutum* Riley (Leiderman and Sauer, 1953; Noyes, 2004).

It has been observed that some parasitoids are more frequent in some crops than others, as in the case of the tachinid fly, *Lixophaga* sp., whereas braconids are associated almost exclusively with lulo. *Copidosoma* sp. is more abundant in tree tomato, while *Trichogramma* sp. is mostly found on tomato.

Regarding applied biological control, Berti and Marcano (1991) determined that *T. pretiosum* preferred to oviposit on eggs 2–3 days old over younger or older eggs. Studies carried out in Colombia indicate that inundative releases in tomato of the egg parasitoid, *T. pretiosum*, could result in 51–65% parasitism, resulting in 78% undamaged fruit (Cross, 1996). Tróchez et al. (1999) reported that *Copidosoma* sp. can reduce the control costs for *Neoleucinodes* by 17% in tree tomato when compared with chemical control costs, and reduced fruit damage to 8%.

Romero (2000) developed a mass-rearing protocol for *Brachymeria* nr *panamensis* using pupae of *Galleria mellonella* as a rearing host. However, cold storage of parasitized *Brachymeria* pupae affected negatively both the emergence and the sex ratio of the emerged offspring. Corpoica Corporación Colombiana Agropecuaria (2000) re-evaluated the efficacy of the parasitoids *T. exiguum* and *T. pretiosum* on tomato, and found that releasing the parasitoids at high rates 20 days after transplant resulted in an average egg parasitism of 40%. Martinez and Alvares Ramos (1996) reported 36% infected pupae of *N. elegantalis* using *Metarhizium anisopliae*.

8.9.3 Plant resistance

Research on plant resistance to *N. elegantalis* has been conducted in Brazil and Colombia. Salinas et al. (1993) described resistance to *N. elegantalis* in different *Solanum* accessions derived from interspecific crosses between *S. esculentum*, *S. pimpinellifolium* and *S. hirsutum*. The wild species were classified as highly resistant or resistant, whereas commercial varieties were classified as susceptible or moderately susceptible. Segregated varieties derived from interspecific crosses were classified as resistant or slightly susceptible, indicating the possibility of genetic introgression of resistance. This study showed that *N. elegantalis* prefers fruit phenotypes with greater weight and hard pericarp.

Parra et al. (1997) evaluated species of *Solanum* (as *Lycopersicon*) as possible sources of resistance to *N. elegantalis*. In their experiments, the wild genotype, *S. hirsutum* f. *typicum* was rated as resistant. The authors suggested that resistance in this case is related to small fruit size, high fruit pubescence preventing feeding, locomotion and oviposition of the insect.

De Souza et al. (2000) studied secondary metabolites in wild *Solanum* (as *Lycopersicon*) materials as responsible for mechanisms of resistance to *N. elegantalis*. They found that the type VI trichomes in *S. habrochaites* var. *glabratum* are associated with the metabolite 2 tridecanone, which is extremely toxic to eggs and neonate larvae. Restrepo (2007) evaluated 12 accessions of wild species *S. habrochaites* (varieties *typicum* and *glabratum*) and *S. peruvianum*, and one accession of cultivated tomato *S. lycopersicum* 'Wonder', as a susceptible control. The accessions PI 134417, PI 134418 and PI126449 of the variety *glabratum*, LA1624 and LA2092 of the variety *typicum* and the accession LA444-1 of *S. peruvianum* were classified as highly resistant to *N. elegantalis*. The control (tomato) was ranked as highly susceptible, with 60% of fruits attacked. However, interspecific crosses produced only hybrid seeds when tomato was crossed with six accessions of *S. habrochaites* and three accessions of the variety *typicum*. Crosses with two accessions of *S. peruvianum* did not produce hybrid seed, possibly due to the existence of a severe barrier of incompatibility between these two species. Casas (2008) developed a methodology for artificial infestation of tomato with *N. elegantalis* under controlled conditions, to identify genetic resistance in *Solanum*. This author confirmed that the glandular trichomes type I, IV and VI present in plants of wild genotypes PI134417, PI134418 of *S. habrochaites* var. *glabratum*, and the genotype LA1264 of *S. habrochaites* var. *typicum*, exerted an apparent antixenosis effect on *N. elegantalis*. Perez (2010) conducted four backcrosses using accessions PI134417 and PI134418 of *S. habrochaites* var. *glabratum* and the genotype LA1264 of *S. habrochaites* var. *typicum*. They also performed chain crosses between resistant plants in order to increase the frequency of resistance alleles. All crosses had a low efficiency, small quantity of seed per fruit and low germination percentage. Unfortunately, this type of work has shown that as the fruit regains size, shape and color in crosses, resistance to *N. elegantalis* becomes lost.

A study of molecular markers using the cDNA-AFLP technique to identify genes of interest involved in resistance to *N. elegantalis* has also begun, in order to promote the search for genetic resistance (Rodriguez, 2009).

In Brazil, plant materials classified as resistant to *N. elegantalis* are: PSX-76, IPA 7, Angela Hyper and Roma VF; susceptible plant materials are Olho Roxo (IPA selection), Campbell-28 and Europeel (Moreira *et al*., 1985; Lyra and Lima, 1998).

8.9.4 Cultural and physical control

Silva Júnior and Vizzotto (1986) determined that tomato pruning could reduce infestation of *N. elegantalis* by 50%. Diaz *et al*. (2003) compared the effectiveness of controlling *N. elegantalis* using stationary releases of *T. exiguum* and bagging tomato 'Chonto and Milano' flower clusters. This tactic resulted in 97% undamaged fruit compared with 62% of damaged fruit in the absolute control. Similar trials conducted in Brazil by Rodrigues Filho *et al*. (2001) found 1.5% infestation by *N. elegantalis* when bunches of tomatoes were bagged. Jordão and Nakano (2002) found that methamidophos residue levels in bagged fruit were three times less than the maximum allowed (0.3 mg/kg), compared to those without bagging, which contained six times more residue than the maximum level permitted. However, fruit bagging is not considered economic in commercial crops.

Paula *et al*. (2004) found that using sorghum as a crop barrier reduced the incidence of the fruit borers *Tuta absoluta* (Lepidoptera: Gelechiidae), *Helicoverpa zea* (Lepidoptera: Noctuidae) and *N. elegantalis* in tomato.

8.9.5 Quarantine methods

No quarantine methods have been developed for *N. elegantalis* in countries where it is considered a pest, because of the difficulty in finding an adequate mass-rearing technique. A post-harvest method is proposed which consists of separating the fruits with signs of infestation. This is a common post-harvest tactic used by Colombian farmers, and guarantees selection of uninfested fruit. This tactic could replace quarantine treatments for tropical fruits such as naranjilla and tree tomato, which are currently required in programs of pest risk analysis (ARP) by international markets.

8.9.6 Discussion

N. elegantalis is the main barrier to the export of tomato, eggplant and red pepper, and Andean fruits such as lulo and tree tomato, from Latin American countries to US and European markets. Farmers growing these fruits are small producers who lack the technical assistance to conduct more appropriate management for this insect. They invest between 5.2 and 56.7% of production costs in the purchase of insecticides and fungicides, which are often highly toxic. In many cases this investment is not economic; chemical control is inefficient, as it is applied without taking into account the biology and behavior of the insect. Although basic management information has been generated, there is an urgent need to develop programs that integrate all control technologies for *N. elegantalis*.

Colombia and Ecuador are carrying out a research project funded by the Regional Fund for Agricultural Technology, FONTAGRO, aimed at clarifying the occurrence of biotypes of *N. elegantalis* associated with different cultivated and wild Solanaceae. The preliminary results of a molecular analysis using CO1 barcoding show the presence of four haplotypes of *N. elegantalis*. These results are not related to the host plants where the insects were reared, but draws attention to geographical isolation, giving the location of the four haplotypes along the branches of the Andes in the territory of Colombia. The haplotype which showed the highest genetic divergence is located in the central cordillera of Colombia in the largest cultivated area of Solanaceae in the country. The initial conclusion of this research is that this haplotype could be related to incipient species, a result of geographical isolation and selection pressure by the use of insecticides.

We propose the development of an international research group that promotes grower education. More research is needed to study the biology of *N. elegantalis*, fruit selection, regulatory quarantine tactics for countries where the pest is present, mass-rearing techniques and insect-sterile techniques, determination of economic injury levels, economic thresholds, insect resistance to pesticides, mass parasitoid rearing, cultural control tactics to help natural enemies and the role of wild hosts as a source of parasitoids and predators.

Acknowledgments

Many thanks to Dr Jorge Peña for editing the chapter, and for his valuable contribution to improving the writing. The first author thanks FONTAGRO (The Latin America and the Caribbean Regional Fund for Agricultural Technology) for a research grant, FTG/RF-0719-RG 'Foundations for development strategies for biological control of the fruit borer *Neoleucinodes elegantalis* in exotic solanaceous Andean fruits' ('Fundamentos para el desarrollo de estrategias de control biológico del perforador del fruto *Neoleucinodes elegantalis* en frutas solanáceas andinas exóticas').

References

Agroecológica Platom, C.A. (2004) *Neoelegantol. Instrucciones de uso. Producto no tóxico para el control de perforador del fruto del tomate y pimentón. (*Neoleucinodes elegantalis*), específico para esta especie*. Agroecológica Platom, C.A., Caracas, Venezuela, pp. 4.

ALAE (Asociación Latinoamericana de Entomología) (1968) Catálogo de insectos de importancia económica en Colombia. *Agricultura Tropical* 1, p. 155.

Arnal, E., Ramos, F., Aponte, A., Suárez, H.Z., Cermeli, M. and Rojas, T. (2005) Reconocimiento de insectos y enemigos naturales asociados al tomate de árbol en Aragua y Miranda, Venezuela. In: Revista Digital CENIAP HOY, http://ceniap.gov, accessed 4 November 2007.

Badji, C.A., Eiras, A.E., Cabrera, A. and Klaus, J. (2003) Avaliação do feromônio sexual de *Neoleucinodes elegantalis* Guenée (Lepidoptera: Crambidae). *Neotropical Entomology* 32, 221–229.

Benvenga S.R., De Bortoli S.A., Gravena, S. and Barbosa, J.C. (2010) Monitoramento da broca-pequena-do-fruto para tomada de decisão de controle em tomateiro estaqueado. *Horticultura Brasileira* 28, 435–440.

Berti, J. and Marcano, R. (1995) Preferencia de *Trichogramma pretiosum* (Riley) (Hymenoptera: Trichogrammatidae) por huevos de diferentes hospederos. *Boletín de Entomología Venezolana* 10, 1–5.

Blackmer, J.L., Eiras, A.E. and de Sousa, C.L.M. (2001) Oviposition preference of *Neoleucinodes elegantalis* (Guenée) (Lepidoptera: Crambidae) and rates of parasitism by *Trichogramma pretiosum* Riley (Hymenoptera: Trichogrammatidae) on *Lycopersicon esculentum* in São José de Ubá, R.J. Brazil. *Neotropical Entomology* 30, 89–95.

Cabrera, A., Eiras, A., Gries, G., Gries, R., Urdaneta, N., Mirás, B., Badji, C. *et al.* (2001) Sex pheromone of tomato fruit borer, *Neoleucinodes elegantalis*. *Journal of Chemical Ecology* 27, 2097–2107.

Capps, H.W. (1948) Status of the Pyraustid moths of genus Leucinodes in the New World, with descriptions of new genera and species. *Smithsonian Institution Press, Proceedings* 98, 69–83.

Casas, N.E. (2008) Obtención de un método de infestación artificial con el pasador del fruto

Neoleucinodes elegantalis Guenée (Lep: Crambidae), para la deter minación de la resistencia genética en *Solanum* spp. MSc thesis, Universidad Nacional de Colombia, Palmira, Colombia.

Clavijo, J.A. (1984) *Algunos Aspectos de la Biología del Perforador del Fruto del Tomate Neoleucinodes elegantalis (Guenée), Lepidoptera: Pyralidae*. Trabajo de Ascenso, Universidad Central de Venezuela, Maracay, Venezuela, pp. 53.

Colorado, W., Diaz, A.E., Yepez, F. and Rueda, J. (2010) Evaluación de la f eromona sexual de *Neoleucinodes elegantalis* Neoeleganto® (Guenée) (Lepidoptera: Crambidae) en solanáceas cultivadas y silvestres. Abstracts XXXVII Congress of the Colombian Society of Entomology, SOCOLEN, Bogotá, Colombia, pp. 36.

Corpoica Corporación Colombiana Ag ropecuaria (2000) Generación de una tecnología par a la cría masiva de *Copidosoma* sp. para el control biológico de *Neoleucinodes elegantalis* plaga de solanáceas en Colombia.(Internal document.) Informe Final de Actividades de Pro yecto, ed. Diaz, A.E., Programa Nacional de Transferencia de Tecnología Agropecuaria, Pronatta. Corporación Colombiana de In vestigación Agropuecuaria, C.I., Palmira, Colombia, pp. 66.

Costa Lima, A.D. (1949) *Insetos do Brasil. 6° Tomo. Lepidópteros 2ª parte*. Escola Nacional de Agronomía, Rio de Janeiro, Brazil, pp. 420.

Cross, M.V. (1996) Ev aluación del par asitismo de *Trichogramma pretiosum* Riley sobre el perforador del fr uto del tomate *Neoleucinodes elegantalis* Guenée (Lepidopter a: Pyralidae) en el Valle. Agricultural engineer thesis, Universidad de Nariño, Pasto, Colombia.

D'Arcy, W.G. (1972) Solanaceae studies II: typification of subdivisions of Solan um. *Annals of Missouri Botanical Garden* 59, 262–278.

Das, M.S. and Patnaik, B.H. (1970) A ne w host of the brinjal shoot and fr uit borer *Leucinodes orbonalis* Guen., and its biology. *Journal of the Bombay Natural History Society*, 63, 601–603.

De Souza, C., Viana-Bailez, A. and Blac kmer, J. (2000a) Toxicity of tomato allelochemicals to eggs and neonato lar vae of *Neoleucinodes elegantalis* (Guenée) (Lep: Crambidae). Abstract Book I, XXI International Congress of Entomology, Brazil, 20–26 A ugust, Rio de Janeiro, Foz do Iguassu, Brazil, pp. 295.

Diaz, A.E. (2009) Car acterización morfométrica de poblaciones del perf orador del fruto *Neoleucinodes elegantalis* (Guenée) (Lepidoptera: Crambidae) asociadas a especies solanáceas cultivadas y silv estres en Colombia. MSc thesis, Universidad Nacional de Colombia, Bogotá.

Diaz, A.E. and Solis, M.A. (2007) A ne w species and species distr ibution records of *Neoleucinodes* (Lepidoptera: Crambidae: Spilomelinae) from Colombia feeding on *Solanum* sp. *Proceedings of the Entomological Society of Washington*, 109, 897–908.

Diaz, A.E., Peña, J. de J., Silva, J.G. and Trochez, A.L. (2003) Control biológico y mecánico del perforador del fr uto de tomate de mesa, *Neoleucinodes elegantalis*. *Revista Regional Novedades Técnicas*, 4, 22–26.

Eiras, A. (2000) Calling beha viour and e valuation of sex pheromone glands e xtract of *Neoleucinodes elegantalis* Guenée (Lepidopter a: Crambidae) in wind tunnel. *Anais da Sociedade Entomológica do Brasil*, 29, 453–460.

Eiras, A.E. and Blackmer, J.L. (2003) Eclosion time and larval behavior of the tomato fr uit borer, *Neoleucinodes* (Lepidoptera: Crambidae). *Scientia Agricola* 60, 195–197.

Espinoza, H.R. (2008) *Barrenador del Fruto de la Berenjena, Neoleucinodes elegantalis*. Hoja Técnica 2, Depar tamento de Protección Vegetal, Fundación Hondureña de In vestigación Agrícola, La Lima, Honduras, pp. 2.

FAO (Food and Ag ricultural Organization) (1982) Saudi Arabia — three insect pests on maiz e, okra and eggplant. *FAO Plant Protection Bulletin* 30, 24–25.

Fernandez, S. and Salas, J. (1985) Estudios sobre la biología del perforador del fruto del tomate *Neoleucinodes elegantalis* Guenée (Lepidoptera: Pyraustidae). *Agronomía Tropical*, 35, 77–82.

Fletcher, T.B. (1916) One hundred notes on Indian insects. *Agricultural Research Institute Pusa, Bulletin* 59, 27.

Gallego, F. and Vélez, R. (1992) *Lista de Insectos que Afectan los Principales Cultivos, Plantas Forestales, Animales Domésticos y al Hombre en Colombia*. Universidad Nacional de Colombia and Colombian Society of Entomology SOCOLEN Medellín, Antioquia, Colombia.

Hampson, G.F. (1896) *Fauna of British India (Moths)* 4. Taylor and Francis, London, xxviii and pp. 594.

Hampson, G.F. (1898) A revision of the moths of the subfamily Pyraustinae and f amily Pyralidae [part]. *Proceedings of the Zoological Society of London* 1898, 590–761.

Hargreaves, H. (1937) Some insects and their plants in Sierra Leone. *Bulletin of Entomological Research* 28, 513.

Holdridge, L. (1967) *Life Zone Ecology*. Tropical Science Center, San Jose, Costa Rica, pp. 206.

Hussain, M.L. (1925) Ann ual report of the Entomologist to Go vernment of Punjab,

Layallpur for the year ending 30 June 1924. Report of Department of Agriculture, Punjab 1923, 24, 59–90.

ICA (Instituto Colombiano Agropecuario) and SOCOLEN (Sociedad Colombiana de Entomología) (1998) *Documento Preliminar. Listado de Plagas Cuarentenarias por Productos para Colombia*, Bogotá, pp. 61.

Isahaque, N.M.M. and Chaudhuri, P. (1983) A new alternate host plant of brinjal shoot and fruit borer, *Leucinodes orbonalis* Guen. *Journal of Research, Assam Agricultural University*, 4, 83–85.

Jaffé, K., Mirás, B. and Cabrera, A. (2007) Mate selection in the moth *Neoleucinodes elegantalis*: Evidence for a supernormal chemical stimulus in sexual attraction. *Animal Behavior*, 73, 727–734.

Jordão, A.L. and Nakano, O. (2002) Ensacamento de frutos do tomateiro visando ao controle de pragas e à redução de defensivos. *Scientia Agrícola*, 59, 281–289.

Klima, A. (1939) Pyralidae: Subfam. Pyraustinae I. In: Bryk, F. (ed.) *Lepidopterorum Catalogus Pars. 89*. W. Junk, Berlin, Germany, pp. 224.

Kuratomi, N.H. (2001) Evaluación del uso de la feromona sexual "Neoelegantol" en la atracción de machos de *Neoleucinodes elegantalis* (Guenée) (Lep. Pyralidae) y su impacto en la reducción del daño de la plaga, en cultivos de tomate *Lycopersicon esculentum*. BSc thesis, Universidad Nacional de Colombia. Palmira, Valle, Colombia.

Leiderman, L. and Sauer, H.F.G. (1953) A broca pequena do fruto do tomateiro *Neoleucinodes elegantalis* (Guenée, 1854) *Biológico*, 19, 182–186.

Lima, M.F., de Boiça Júnior, A.L. and De Souza, R.S. (2001) Efeito de insecticidas no controle da roca pequena *Neoleucinodes elegantalis* na cultura do tomaterio. *Ecosistema* 26, 54–57.

Lyra Netto de A.M.C. and Lima, A.A.F. (1998) Infestação de cultivares de tomateiro por *Neoleucinodes elegantalis* (Lepidoptera: Pyralidae). *Pesquisa Agropecuaria Brasilera*, 33, 221–223.

Maes, K.V.N. (1995) A comparative morphological study of the adult Crambidae (Lepidoptera: Pyraloidea). *Annales de la Société Royale Entomologique de Belgique* 131, 159–168.

Marcano, R. (1990) *Estudio de la Biología, Ecología y Control del Taladrador del Fruto del Tomate Neoleucinodes elegantalis (Guenée, 1854) (Lepidóptera: Pyralidae) en la Zona Central del País*. Trabajo de Ascenso. Universidad Central de Venezuela, Maracay, Venezuela, pp. 138.

Marcano, R. (1991a) Estudio de la biología y algunos aspectos del comportamiento del perforador del fruto del tomate *Neoleucinodes elegantalis* (Lepidoptera: Pyralidae) en tomate. *Agronomía Tropical* 41, 257–263.

Marcano, R. (1991b) Ciclo biológico del perforador del fruto del tomate *Neoleucinodes elegantalis* (Guenée) (Lepidóptera: Pyralidae), usando berenjena (*Solanum melongena*) como alimento. *Boletín de Entomología Venezolana*, 6, 135–141.

Martinelli, S., Montagna, M.A. and Picinato, N.C. (2003) Efficacy of indoxacarb in the control of vegetable pests. *Horticultura Brasileira*, 21, 501–505.

Martínez, M. and Álvarez Ramos, J. (1996) *Metarhizium anisopliae: Control Alternativo Para el Perforador del Fruto del Tomate*. Publicación de Fondo Nacional de Investigaciones Agropecuarias, Estado de Lara, Venezuela, http://sian.inia.gob.ve/repositorio/revistas_tec/FonaiapDivulga/fd54/tomate.htm, accessed 16 November 2011.

Medal, J.C., Harudattan, R.C., Ullahey, J.J.M. and Itelli, R.A.P. (1996) An exploratory insect survey of tropical soda apple in Brazil and Paraguay. *Florida Entomologist* 79, 70–73.

Mehto, D.N., Singh, K.M. and Singh, R.N. (1980) Dispersion of *Leucinodes orbonalis* Guen during different seasons. *Indian Journal of Entomology* 42, 539–540.

Menon, P.P.V. (1962) *Leucinodes orbonalis* Guen. (Pyralidae: Lep.) new record on *Solanum indicum* Linn. in South India. *Madras Agricultural Journal* 49, 194.

Miranda, M.M.M., Picanço, M.C., Zanuncio, J.C. and da Silva, E.M. (2005) Impact of integrated pest management on the population of leafminers, fruit borers, and natural enemies in tomato. *Ciencia Rural* 35, 204–208.

Mirás, B., Issa, S. and Klaus, J. (1997) Diseño y Evaluación de trampas cebadas con hembras vírgenes para la captura del perforador del fruto del tomate. *Agronomía Tropical* 47, 315–330.

Morales Valles, P., Cermell, M., Godoy, F. and Salas, B. (2003) Lista de insectos relacionados a las solanáceas ubicados en el museo de insectos de interés agrícola del CENIAP INIA. *Entomotropica* 18, 193–209.

Moreira, J.O.T., Lara, F.M. and Churata-Masca, M.G.C. (1985) Resistência de cultivares de tomateiro (*Lycopersicon esculentum* Mill) à "broca-pequena-dos fruitos", *Neoleucinodes elegantalis* (Guéene, 1854) (Lepidoptera-Pyralidae). *Ciencia e Cultura* 37, 618–623.

Motta, M.M., Coutinho, P., Cola, J., Bacci, L. and Marques, É. (2005) Impact of integrated pest management on the population of leafminers, fruit borers, and natur al enemies in tomato . *Ciencia Rural* 35, 204–208.

National Research Council (US) (1989) *Lost Crops of the Incas: Little-Known Plants of the Andes with Promise for Worldwide Cultivation.* National Academies Press, Washington, DC, pp. 428.

Navasero, M.V. (1983) Biology and chemical control of the egg plant fr uit and shoot borer , *Leucinodes orbonalis* Guenée (Pyr austidae: Lepidoptera). BSc thesis , University of the Philippines at Los Baños, Laguna, Philippines.

Nee, M. (1979) Patterns in biogeography in *Solanum*, section Acanthophora. In: Hawkes, J.G., Lester, R.N. and Sk elding, A.D. (eds) *The Biology and Taxonomy of the Solanaceae.* Academic Press, London, pp. 569–580.

Nee, M. (1991) Synopsis of Solan um section Acanthophora: a group of interest for glycoalkaloids. In: Hawkes, J.G., Lester, R.N., Nee, M., and Estrada-R, N. (eds) *S olanaceae III: Taxonomy, Chemistry, Evolution.* Royal Botanic Gardens, Kew and Linnean Society of London, Richmond, UK, pp. 257–266.

Nee, M. (1999) Synopsis of *Solanum* in the Ne w World. In: Nee, M., Symon, D.E., Lester, R.N. and Jessop, J.P. (eds), *Solanaceae IV: Advances in Biology and Utilization.* Royal Botanic Gardens K ew, Richmond, UK, pp . 285–333.

Noyes, J.S. (2004) Univ ersal Chalcidoidea database. The Natural History Museum, London. http://flood.nhm.ac.uk/jdsml/perth/chalcidoidea/detail.dsml, accessed 1 May 2005.

Olckers, T., Medal, J.C. and Gandolfo, D.E. (2002) Insect herbivores associated with species of *Solanum* (Solanaceae) in Nor theastern Argentina and Southeaster n Paraguay, with reference to biological control of w eeds in South Africa and the United States of America. *Florida Entomologist* 85, 254–260.

Parra, A., López, C.M., García, M.A. and Baena, D. (1997) Evaluación de especies del género *Lycopersicon* como posibles fuentes de resistencia al pasador del fr uto *Neoleucinodes elegantalis* Guenée. *Acta Agronómica* 47, 45–47.

Parra, J.R.P. and Zucchi, R.A.(2004) *Trichogramma* in Brazil: feasibility of use after tw enty years of research. *Neotropical Entomology* 33, 271–281.

Paula, S.V. de, Picanço, M.C., Oliveria, I.R. and Gusmão, M.R. (2004) Control of tomato fr uit borers by surrounding crop str ips. *Bioscience Journal*, 20, 33–39.

Perez, M. (2010) Mejor amiento genético en *Solanum lycopersicum* para la resistencia al pasador del fr uto *Neoleucinodes elegantalis* Guenée (Lepidoptera: Crambidae). MSc thesis, Universidad Nacional de Colombia sede Palmira, Palmira, Colombia.

Picanço, M., Casali, V.W.D., Leite, G.L.D. and de Oliveira, I.R. (1997) Lepidopter a asociated with *Solanum gilo. Horticultura Brasileira* 15, 112–114.

Picanço, M., Leite , G.L.D., Guedes, R.N.C. and Silva, E.A. (1998) Yield loss in trellised tomato affected by insecticidal sprays and plant spacing. *Crop Protection* 17, 47–452.

Pillai, R.M. (1922) *Developmental activities, Entomology.* Department of Agriculture, Pretoria, South Africa.

Posada, L., P olania, I., López, A., Ruiz, N. and Rodriguez, D. (1981) *Nuevo huésped.* ICA Notas y Noticias Entomológicas, Bogotá, pp. 11.

Restrepo, E. (2007) Estudios básicos para iniciar la producción de cultivares de tomate *Solanum lycopersicum* L. con resistencia al pasador del fruto *Neoleucinodes elegantalis* (Guenée). PhD thesis, Universidad Nacional de Colombia, Palmira, Colombia.

Revelo, J., Viteri, P., Vásquez, W., León, J . and Gallegos, P. (2010) *Manual del Cultivo Ecológico de la Naranjilla.* INIAP, Quito, pp. 119.

Rodrigues Filho, I.L., Marchior, L.C. and Silva, L.V. da (2001) Estudo da viabilidade do ensacamento de pencas em tomateiro tutor ado para o controle de *Neoleucinodes elegantalis* (Guen., 1854) (Lepidopter a:Crambidae) in Paty do Alferes-RJ. *Agronomía, Rio de Janeiro* 35, 33–37.

Rodrigues Filho, I.L., Marchior, L.C. and Silva L.V. da (2003) Análise da o viposição de *Neoleucinodes elegantalis* (Gene, 1854) (Lepidoptera: Crambidae) para subsidiar estratégia de manejo . *Agronomía, Rio de Janeiro* 37, 23–26.

Rodriguez, D.M. (2009) cDNA – AFLP: Técnica molecular para analizar la e xpresión génica en tomate *Solanum lycopersicum* L. resistente al pasador de fruto *Neoleucinodes elegantalis* (Guenée). BSc thesis , Universidad Nacional de Colombia, Palmira, Colombia.

Romero, M. del P. (2000) Aspectos biológicos y cría masiva de Br achymeria próxima a *B. panamensis* (Hym: Chalcididae) par asitoide de pupas de *Neoleucinodes elegantalis* (Lep: Pyralidae) perforador del fr uto de las solanáceas. BSc thesis , Universidad Nacional de Colombia, Palmira, Colombia.

SAG (Servicio Agricola y Ganadero) División Protección Agrícola de Chile (2005) Cr iterios de regionalización actualizando las listas de plagas cuarentenarias. SAG, Santiago, pp. 19.

Salas, J. (2008) Capacidad de captur a de *Neoleucinodes elegantalis* (Lepidoptera: Pyralidae) en dos tipos de tr ampa provistas con su feromona sexual. *Bioagro* 20, 135–139.

Salas, J., Alvarez, C. and Parra, A. (1990) Contribución al conocimiento de la ecología del perf orador del fruto del tomate *Neoleucinodes elegantalis* Guenée (Lepidoptera: Pyraustidae). *Agronomía Tropical*, 41, 275–283.

Salas, J., Alvarez, C. and Parra, A. (1992) Estudios sobre la feronmona a sexual natural del perforador del fruto del tomate *Neoleucinodes elegantalis* (Guenée) (Lepidopter a: Pyralidae). *Agronomía Tropical* 42, 227–231.

Salinas, H., Vallejo, F.A. and Estr ada, E. (1993) Evaluación de la resistencia al pasador del fruto del tomate *Neoleucinodes elegantalis* (Guenée) en mater iales *L. hirsutum* Humb y Bonpl y *L. pimpinellifolium* (Just) Mill y su transferencia a mater iales cultivados de tomate, *L. esculentum* Mill. *Acta Agronómica*, 43, 44–56.

Sanchez, G. (1973) *Las Plagas del Lulo y su Control*. Boletín Técnico 25/26. Instituto Colombiano Agropecuario Programa Nacional de Entomología, Bogotá.

Santamaría, M.Y., Ebratt, E.E. and Benavides, M. (2007) Importancia económica de *Copidosoma* n. sp. (Hymenoptera: Encyrtidae) en Cundinamarca. Abstracts XXXIV Cong ress of the Colombian Society of Entomology SOCOLEN, Cartagena, Colombia, pp. 68.

SENASA (Servicio Nacional de Sanidad y Calidad Agroalimentaria, Chile) (2007) Cr iterios de regionalización, en relación a las plagas cuarentenarias para el terr itorio de Chile . www.ippc.int/file_uploaded/1180561484850_ RES5._792_07.pdf, accessed 11 No vember 2005.

SENASA (Servicio Nacional de Sanidad Agropecuaria de Honduras) (2008) Combaten insecto que ataca al cultivo de las berenjenas. www.senasa-sag.gob.hn/index.php?option=com_content&task=blogcategory&id=251&Itemid=389, accessed 9 January 2011.

Serrano, A., Muñoz, E., Pulido, J. and de la Cruz, J. (1992) Biología hábitos y enemigos natur ales del *Neoleucinodes elegantalis* (Guenée). *Revista Colombiana de Entomología* 18, 32–37.

Silva Júnior, A.A. and Vizzotto, V.J. (1986) *Poda Apical em Tomateiro*. Empresa Catar inense de pesquisa Ag ropecuária, EMPASC, Santa Catarina, Brazil, pp. 68.

Stehr, F.W., Martinat, P.J., Davis, D.R., Wagner, D.C., Heppner, J.B., Brown, R.L., Toliver, M.E. *et al.* (1987) Order Lepidoptera. In: Stehr, F.W. (ed.) *Immature Insects*. Kendall/Hunt, Dubuque, Iowa, pp. 288–596.

Talekar, N.S., Lin, M.Y. and Hw ang, C.C. (2002) Rearing of eggplant fr uit and shoot borer . T. Kalb Asian Vegetable Research & Development Center (A VRDC), www.avrdc.org/LC/eggplant/rear_efsb/01title.html, accessed 21 January 2011.

Tamaki, N. and Miyara, A. (1982) Studies on the ecology of the eggplant fruit-borer, *Leucinodes orbonalis* Guenée (Lepidopter a: Pyralidae). *Proceedings of the Association for Plant Protection of Kyushu (Japan)* 28, 158–162.

Tróchez, G.A., Diaz, A.E. and García, F . (1999) Recuperación de *Copidosoma* sp. (Hymenoptera: Encyrtidae), parasitoide de huevos de *Neoleucinodes elegantalis* (Lepidóptera: Pyralidae) en tomate de árbol (*Cyphomandra betacea*). *Revista Colombiana de Entomología*, 25, 179–183.

USDA, APHIS and PPQ (2005) *Importation of Peppers from Certain Central American Countries*. Federal Register Doc 05-20388 2005 70, 59283–59290, Washington, DC.

Veenakumari, K., Pr ashanth, M. and Ranganath, H.R. (1995) Additional records of insect pests of vegetables in the Andaman Islands (India). *Journal of Entomological Research* 19, 277–279.

Viáfara, H.F., Garcia, F . and Díaz, A.E. (1999) Parasitismo natural de *Neoleucinodes elegantalis* (Guénee) (Lepidópter a: Pyralidae) en algunas zonas productoras de Solanáceas del Cauca y Valle del Cauca Colombia. *Revista Colombiana de Entomología* 25,151–159.

Walker (1859) Pyralides. *List of the Specimens of Lepidopterous Insects in the Collection of the British Museum, London* 18, 509–798.

Waterhouse, D.F. (1998) *Biological Control of Insect Pests: Southeast Asian Prospects*. Australian Centre for International Agricultural Research, Canberra, pp. 548.

Whalen, M.D (1984) Conspectus of species groups in Solanum section Leptostemon um. *Gentes Herbarum* 12, 247.

Whittle, K. and F erguson, D.C. (1987) P ests not known to occur in the United States or of limited distribution. *U.S. Department of Agriculture, APHIS* 85, 1–9.

9 *Copitarsia* spp.: Biology and Risk Posed by Potentially Invasive Lepidoptera from South and Central America

Juli Gould,[1] Rebecca Simmons[2] and Robert Venette[3]

[1]*USDA Animal and Plant Health Inspection Service, Center for Plant Health Science and Technology, Buzzards Bay, Massachusetts, USA;* [2]*University of North Dakota, Grand Forks, North Dakota, USA;* [3]*USDA Forest Service, Northern Research, St. Paul, Minnesota, USA*

9.1 Introduction

Members of the genus *Copitarsia* (Lepidoptera: Noctuidae) represent a potential threat to US agriculture. Although they are not known to be established in the USA, they are frequently intercepted on vegetables and cut flowers at ports of entry. These species are not generally outbreak pests in their native ranges; however, it is possible that these moths could greatly impact domestic agriculture after invasion, due to the release from selection pressures posed by native predators and parasitoids. This threat is complicated by difficulties in identifying species and understanding their life history and host preferences. Here we summarize the state of knowledge for members of this genus in terms of their pest status and control. We have reviewed the taxonomic difficulties involved with members of *Copitarsia*, their geographic range, host plant preferences and associated economic impacts. We have summarized the life histories of these species and procedures for rearing them in colonies for future study. Finally, we have discussed interception, control and assessment of risk for large-scale damage to domestic agriculture.

9.2 Taxonomy

The genus *Copitarsia* is in the cutworm family, the Noctuidae. Over time, *Copitarsia* has included from six to 25 species, depending on which taxonomic authority is consulted (Table 9.1). Hampson (1906) included six species in *Copitarsia*: *C. humilis* (Blanchard), *C. consueta* (Walker), *C. turbata* (Herrich-Schäffer) (the type species), *C. naenoides* (Butler), *C. patagonica* Hampson and *C. purilinea* (Mabille) (Table 9.1). Subsequently, Köhler (1959) described *C. basilinea*. Poole transferred *C. editae* (Angulo et al., 1985) from *Euxoa*. Castillo and Angulo (1991) redescribed and revised *Copitarsia*, using both adult and immature stages as sources of characters. In this revision, Castillo and Angulo transferred *C. clavata* (Köhler) from *Cotarsina*. The authors also described two new species: *C. anguloi* Castillo and *C. paraturbata* Castillo and Angulo. Angulo and Olivares (2003) synonymized *Cotarsina* Köhler with *Copitarsia*, bringing the total number of species to 22; subsequently, these authors also placed two members of *Tarsicopia* Köhler within *Copitarsia* (Angulo and Olivares, 2009). In both cases, Angulo and

Table 9.1. List of species currently included in *Copitarsia*, as in Angulo and Olivares (2009).

Species, Author(s), Date
C. anatunca Angulo and Olivares (1999)
C. anguloi Castillo and Angulo (1991)
C. basilinea (Köhler, 1959)
C. belenensis (Köhler, 1973)
C. borisaniana (Köhler, 1958)
C. clavata (Köhler, 1951)
**C. corruda* Pogue and Simmons (2008)
**C. decolora* (Guenée, 1852)
C. editae Angulo and Jana-Saenz (1982)
C. fliessiana (Köhler, 1958)
C. fuscierena (Hampson, 1910)
C. gentiliana (Köhler, 1961)
C. gracilis (Köhler, 1951)
**C. humilis* (Blanchard, 1852)
**C. incommoda* (Walker, 1865)
C. maxima (Köhler, 1961)
C. mimica Angulo and Olivares (1999)
C. murina Angulo *et al.* (2001)
**C. naenoides* (Butler, 1882)
C. paraturbata Castillo and Angulo (1991)
C. patagonica Hampson (1906)
C. purilinea (Mabille, 1885)
C. roseofulva (Köhler, 1952)
C. tamsi (Giacomelli, 1922)
C. vivax Köhler (1951)

* = reported as pest species

Olivares (2003, 2009) cite male genital characters as the basis for these taxonomic decisions. Finally, Pogue and Simmons (2008) described *C. corruda*, which was originally part of *C. decolora* (see below), resulting in 25 species.

Of the 25 members of *Copitarsia*, only five species are listed in the literature as agricultural pests (Table 9.1): *C. corruda*, *C. decolora*, *C. humilis*, *C. incommoda* and *C. naenoides*. Typically, only three of these species are encountered at US ports of entry: *C. corruda*, *C. decolora* and *C. incommoda*. The taxonomy of *C. humilis* and *C. naeonoides* is not controversial, and because these species are not encountered at USA ports of entry, they are unlikely to have an impact on USA agriculture. The taxonomy of the remaining three pest species, *C. corruda*, *C. decolora* and *C. incommoda*, is problematic, complicated by use of non-valid names for species, historical misidentifications and morphologically cryptic lineages.

The use of invalid names complicates literature searches and communication for both *C. incommoda* and *C. decolora*. *C. incommoda* is often called by a name used earlier by taxonomists, i.e., *C. consueta* (Poole, 1989; Simmons and Pogue, 2004). Similarly, *C. decolora* is often referred to by the taxonomically invalid name *C. turbata* (Simmons and Pogue, 2004). Authors who are unaware of these taxonomic changes may use the incorrect names (Table 9.2).

The identification of pest *Copitarsia*, in addition to the taxonomic issues described above, is also complicated by the confusion of one species for another. Simmons and Pogue (2004) found that several revisions containing illustrations of male genitalia, which possessed defining characteristics for species of *Copitarsia*, were switched between *C. incommoda* and *C. decolora* (Artigas and Angulo, 1973; Angulo and Weigert, 1975; Parra *et al.*, 1986). Subsequent revisions of *Copitarsia* maintained these errors (Castillo and Angulo, 1991; Angulo and Olivares, 2003). Thus the ability to use geographic and host plant information for *C. decolora* and *C. incommoda* from early literature is severely limited, unless male genitalia are illustrated to confirm species identifications.

Species with large geographic distributions and wide host plant preferences, i.e., *C. decolora*, may be complexes of morphologically similar species. Table 9.2 provides geographic information for *C. decolora*, *C. incommoda* and *C. corruda*. These cryptic species may differ substantially in life history or habit, but may not be recognized until fine-scale studies are performed. Simmons and Scheffer (2004), using mitochondrial DNA, found that *C. decolora* was actually composed of two lineages. These two lineages were found to have differing host plant preferences, as well as differing genital morphology (Pogue and Simmons, 2008). Pogue and Simmons (2008) named the cryptic lineage *C. corruda*, based on the caterpillar's apparent preference for asparagus; however, Angulo and Olivares (2009) treat *C. corruda* as a junior synonym of *C. decolora* because the two species geographically overlap in portions of their ranges. Here, we treat *C. corruda* as independent of *C. decolora*, as these two species have differing morphologies, mitochondrial lineages and habits (Pogue and Simmons, 2008; Table 9.2). Based on these consistent differences, we believe that *C. decolora* and *C. corruda* are distinct, but sympatric species.

Table 9.2. Commonly encountered *Copitarsia*, with previously used names, likely country of origin and host plants (commonly found host plant for each species in boldface type; identity of caterpillars on hosts confirmed with molecular identification (Simmons and Pogue, 2008).

Species	Previous names	Confused with	Geographic range	Host plants
C. corruda Pogue and Simmons (2008)	*C. decolora* (in part) *C. turbata* (in part)	*C. incommoda* = *C. consueta*	Colombia, Ecuador, Peru	**Ammi**, *Asparagus*, *Aster*, *Callostephus*, *Iris*, *Lysimachia*
C. decolora (Guenée)	*C. turbata*	*C. incommoda* = *C. consueta*	Central America, Colombia, Ecuador, Mexico	*Allium*, *Alstroemeria*, *Apium*, *Argemone*, *Aster*, **Brassica**, *Campanula*, *Chamaemelum*, *Coriandrum*, *Dianthus*, *Eryngium*, *Helianthus*, *Hypericum*, *Limonium*, *Mentha*, *Mollucella*, *Origanum*, *Pisum*, *Rosa*, *Solidaster*, *Suadea*
C. incommoda (Walker)	*C. consueta*	*C. decolora* = *C. turbata*	Argentina, Bolivia, Chile Colombia, Ecuador, Peru	Host plants are unclear because of previous confusion

9.3 Biology and Life History

The *Copitarsia* life cycle is initiated when eggs are deposited either singly (*C. incommoda*: Lopez-A., 1996a) or in egg masses (*C. decolora*: Velasquez-Z., 1988) (Fig. 9.1). Individual females (Fig. 9.2) have high rates of reproduction. *C. incommoda* females are reported to lay 1638 eggs per female on artificial diet (Rojas and Cibrian-Tovar, 1994). *C. decolora* also has a high fecundity, with females laying 1038 eggs on artificial diet (Larrain-S., 1996), 572 eggs per female on lettuce (Arce de Hamity and Neder de Roman, 1992), and 1579 eggs per female on onion (Velasquez-Z., 1988). Although only 572 eggs were laid on lettuce, the females contained an average of 1552 ovarioles, suggesting that lettuce is not an ideal host plant. Fecundity of *C. corruda* was highest (1714 + 140 eggs per female) when it was reared on asparagus at 20°C (Gould *et al.*, 2005). *C. corruda* eggs developed between 9.3°C and 30.1°C; time to hatch varied with temperature, ranging from 33.1 ± 1.4 days at 9.3°C to 3.0 ± 0.2 days at 30.1°C. From these data, Gould *et al.* (2005) estimated that *C. corruda* eggs require 64.4–72.6 degree days (DD) at 7.2°–8.3°C base temperature. Similar to *C. corruda*, *C. decolora* eggs on lettuce required 8 days at 24.5°C to hatch (Arce de Hamity and Neder de Roman, 1992).

Fig. 9.1 Eggs of *C. corruda* (photo D. Winograd).

The number of days required to complete development depends on many factors, including temperature, relative humidity and host plant. *C. decolora* completed development in 60.7 days on onion at 20.4°C (Velasquez-Z., 1988), yet it took 82.5 days on artificial diet at the same temperature. *C. decolora* completed development in 42.9 days on lettuce at 24.5°C (Arce de Hamity and Neder de Roman, 1992). In the field *C. incommoda* completed development in 62 days on potato (Lopez-Avilla, 1996a). Total days from neonate to pre-pupa ranged from 23.2 ± 0.7 days (29.5°C) to 215.3 ± 1.8 days (9.7°C) for *C. corruda* (Gould *et al.*, 2005). Gould *et al.* (2005) reported that *C. corruda* larvae require 341.4 DD above a

Fig. 9.2 Adult *C. corruda* (photo D. Winograd).

Fig. 9.3 Final instar larva of *C. corruda* (photo: D. Winograd).

base temperature of 7.3°C when reared on asparagus; on artificial diet, *C. corruda* required fewer DD (245.5DD, base temperature 7.7°C). Larval development in *C. decolora* is slightly shorter in duration than *C. corruda*, ranging from 18.8 ± 0.7 days (29.3°C) to 35.1 ± 0.76 days (18°C) (Urra and Apablaza, 2005).

It is likely that species of *Copitarsia* are multivoltine throughout much of their geographic ranges. *C. decolora* is described as having 1–3 generations per annum on quinoa in Bolivia (Liberman-Cruz, 1986). *C. incommoda* has three generations per annum on lettuce in Argentina (Arce de Hamity and Neder de Roman, 1992) and four generations on rapeseed in Chile (Artigas and Angulo, 1973). *C. humilis*, on the other hand, seems to have only one generation per annum in alfalfa in Chile (Hichins-O. and Mendoza-M., 1976).

In general, *Copitarsia* larvae have 5–6 instars (Arce de Hamity and Neder de Roman, 1993; Lopez-A. 1996a), and reach a length of c. 37.2–41.99 mm at the final instar (Fig. 9.3; Pogue and Simmons, 2008). The two final larval instars consume the most and therefore cause the most damage. The larvae tend to be green in color, but Lopez-Avila (Lopez-A, 1996a) reported green, black and gray phases that also varied with respect to habitat and hosts. For *C. corruda*, larvae have fewer instars when raised on artificial diet (see Rearing, below; Gould et al., 2005), but complete six larval instars when raised on asparagus.

Temperature has a significant effect on the number of instars of *C. corruda*, especially when combined with the effect of artificial diet (Gould et al., 2005). The combination of higher temperatures (25–30°C) and an asparagus diet leads to significantly fewer pre-pupal instars.

Copitarsia larvae move into the soil and create cells in which they pupate (Fig. 9.4). Temperature and diet also play significant roles in development during the pupal stage, and Gould et al. (2005) reported a significant effect on pupal stage length when diet and temperature are combined. *C. corruda* reared on artificial diet have a shorter pupal stage at 14.6–24.9°C (13.9 ± 0.1 to 41.6 ± 0.4 days) than those reared on asparagus (14.1 ± 0.2 to 45.0 ± 0.4 days) (Gould et al., 2005). At the lowest temperature tested (9.7°C), *C. corruda* reared on asparagus have a shorter pupal stage (105.6 ± 0.8 days) versus those reared on artificial diet (107.4 ± 1.4 days) (Gould et al., 2005). *C. corruda* pupae require between 215.5 and 255.0 DD (95% confidence interval) when reared on artificial diet; on asparagus, pupae require between 220.5 and 252.5 DD (95% confidence interval). *C. decolora* pupae have slightly longer development times. At 23.9°C, the pupal stage lasted between 18 and 20 days (mean = 18.8 ± 0.7 days), while at 18°C, the pupal stage was 34–36 days (mean = 35.1 ± 0.76 days) (Urra and Apablaza, 2005). The ability of *Copitarsia* to successfully complete development at these temperatures is

Fig. 9.4 Pupae of *C. corruda* (photo: D. Winograd).

perhaps one of the reasons that it is successful throughout most of Latin America.

9.3.1 Rearing

Copitarsia larvae can be reared quite successfully on artificial diet. Acatitla Trejo *et al.* (2004) tested *C. incommoda* (= *C. decolora*) larvae on five artificial diets and found that a diet based on dehydrated alfalfa, wheat germ, brewer's yeast and pinto bean powder (Cibrian and Sugimoto, 1992) produced the heaviest pupae and led to the highest adult survival and reproduction. *C. corruda* larvae were successfully reared by the senior author for many generations on a high-wheatgerm diet (Bell *et al.*, 1981). To test survival of *C. corruda* on different vegetables (asparagus, broccoli, spinach), artificial diets from ground fresh vegetables, agar (21 g), methyl paraben (3.5 g), sorbic acid (2.1 g), ascorbic acid (6.3 g), tetracycline (0.07 g) and water (332.5 ml) were prepared. Varying amounts of vegetables were used because they contained different percentages of water. The larvae burrow into the diets to pupate, but need to be removed prior to eclosion because the diet is too moist. Adults can be reared in 18 cm diameter × 17 cm high cardboard canisters with cardboard bottoms and plastic wrap covers. The adults fed readily on a liquid diet of orange Gatorade® (The Gatorade Company, Chicago, Illinois) and laid eggs on the Kraft paper lining the cage. After chilling the adults in a refrigerator, the Kraft paper was easily removed to collect the eggs.

9.3.2 Geographic distribution

Species in the genus *Copitarsia* can be found along the western edge of South and Central America from the southern tip of Argentina north through central Mexico (Fig. 9.5). There are no records in the literature of *Copitarsia* in Brazil, Paraguay, Uruguay, Suriname, French Guiana, Guyana, Belize, Nicaragua, Panama, El Salvador and Honduras; however, the USDA-APHIS interception database shows *Copitarsia* arriving at the US border on commodities from Belize, Brazil, Costa Rica, El Salvador, Guatemala, Honduras, Nicaragua and Panama. It is not known if the commodities were harvested in those countries or simply shipped from packing facilities there. *Copitarsia* has also been recovered in commodity shipments from several Caribbean islands: the Dominican Republic, Haiti, Jamaica, St Lucia, and Trinidad and Tobago. A lack of records in the literature from these countries does not necessarily mean that *Copitarsia* is not present. Some possible explanations are: (i) *Copitarsia* simply may not reach high densities or damage crops and has therefore been ignored; (ii) scientific expertise in some countries may be lacking; (iii) scientists may not publish results in readily accessible journals; and (iv) the climate is marginal for *Copitarsia*. In much of its range, *Copitarsia* species are thought to be kept in check by natural enemies (Cortes-P. *et al.*, 1972; Artigas and Angulo, 1973; Cortes, 1976; Hichins-O. and Mendoza-M., 1976; Apablaza and Stevenson, 1995; Vimos-N. *et al.*, 1998). Figure 9.5 presents the minimum range of *Copitarsia*, with the actual range almost certainly including the countries between Mexico and South America and possibly islands in the Caribbean region as well.

The three important pest species, *C. corruda*, *C. decolora* and *C. incommoda*, have partially overlapping geographic ranges (Fig. 9.6), and may even be found on the same farm on different host-plant species (R. Simmons, North Dakota, pers. comm. 2011). *C. corruda* seems to have the most restricted geographic distribution, but the full extent of its range is uncertain because it has only recently been described (Pogue and Simmons, 2008); *C. corruda* has been reported from Colombia, Ecuador and Peru (Fig. 9.4, Table 9.2), and has most often been encountered on asparagus at ports of entry (Pogue and Simmons, 2008). *C. decolora* is the most widespread of the three pest

Fig. 9.5 Current distribution of *Copitarsia* as described in the literature.

Fig. 9.6 Distribution of *C. corruda* (dark grey dots), *C. decolora* (light grey dots), and *C. incommoda* (black dots), based on specimen data in Simmons and Pogue (2004), Simmons and Scheffer (2004) and Pogue and Simmons (2008).

species, and has been found from Mexico to Ecuador. *C. decolora* also appears to be the most polyphagous of the genus, feeding on 21 crops in 13 plant families (Table 9.2; Pogue and Simmons, 2008).

C. incommoda may be restricted to South America, particularly Venezuela, Colombia, Ecuador, Peru, Chile, Bolivia, Argentina and Uruguay. While *C. incommoda* has been reported to feed on several different commodities, many of these records are suspect because of the previous misidentification with *C. decolora* (Table 9.2). The issues with taxonomy highlight the importance of retaining voucher specimens so that the identity of the species can be validated in the future.

9.4 Associations With Plants

9.4.1 Host plants

Moths in the genus *Copitarsia* are considered to be polyphagous, and there are records of them feeding on many host plants. We gathered host plant records from several sources: taxonomic literature, including Artigas and Angulo (1973) and Castillo and Angulo (1991), where host plant information was provided for different species of *Copitarsia*. Several other manuscripts present anecdotal evidence of host plant associations and there are references of *Copitarsia* at US border crossings on several hosts that are not mentioned in the literature. Because of the difficult taxonomy of this genus, we have presented host plant information for *C. humilis* and *C. naeonoides* (species for which the taxonomy is clear) and for the genus *Copitarsia*. More specific information for *C. decolora*, *C. corruda*, and *C. incommoda* based on recent studies can be found in Table 9.2.

C. humilis has been found at low frequency in Chile, usually feeding on alfalfa (Cortes *et al.*, 1972, Cortes, 1976, Hitchins and Mendoza, 1976). *C. naeonoides*, known as the brown flax armyworm, has been found in central and southern Chile, and northern Argentina; this species feeds on radishes, asparagus, beetroots, potatoes, tobacco and flax (Klein Koch and Waterhouse, 2000; Angulo and Olivares, 2003).

The published literature lists 39 crop plants as hosts for *Copitarsia*, and *C.* spp. have been found on >9 additional crop species at US ports of entry (Table 9.3). These crop species represent 19 different families. Most of the crop species listed as hosts of *Copitarsia* are grown in the USA. We analyzed data from the

Table 9.3. Plants that are hosts for *Copitarsia* species larvae.

English common name	Spanish common name	Scientific name	Family	Countries	References
Alfalfa	Alfalfa	*Medicago sativa*	Fabaceae	Chile, Mexico, Peru	(Hichins-O. and Rabinovich, 1974; Angulo and Weigert, 1975; Hichins-O. and Mendoza-M., 1976; Cortes, 1976; Porter, 1980; Castillo and Angulo, 1991; Arce de Hamity and Neder de Roman, 1992, 1993; Rojas *et al.*, 1993; Apablaza and Stevenson, 1995; Larrain-S., 1996)
Apple	Manzana	*Malus sp.*	Rosaceae	Chile	(Larrain-S., 1996)
Artichoke	Alcachofa	*Cynara scolymus*	Asteraceae	Chile	(Larrain-S., 1984; Machuca-L. *et al.*, 1988, 1989a, b, 1990; Castillo and Angulo, 1991; Larrain-S. and Araya-C., 1994; Larrain-S., 1996)
Asparagus	Espárragos	*Asparagus officinalis*	Liliaceae	Chile	(Castillo and Angulo, 1991; Larrain-S., 1996)
Aster[a]		*Aster* spp.	Asteraceae		PIN-309[b]
Baby's-breath[a]		*Gypsophila* spp.	Caryophyllaceae		PIN-309[b]
Beet	Remolacha, betabel	*Beta vulgaris*	Chenopodiaceae	Chile, Mexico, Argentina	(Angulo and Weigert, 1975; Castillo and Angulo, 1991; Neder de Roman and Arce de Hamity, 1991; Arce de Hamity and Neder de Roman, 1992; Rojas *et al.*, 1993; Larrain-S., 1996)
Bell pepper[a]	Pimiento verde	*Capsicum* sp.	Solanaceae	Mexico	(PIN-309[b]; Riley, 1998)
Broad or Lima beans	Habas	*Vicia faba*	Fabaceae	Chile, Peru, Argentina	(Gomez-T., 1972; Neder de Roman and Arce de Hamity, 1991; Arce de Hamity and Neder de Roman, 1992; Lamborot *et al.*, 1995)
Broccoli	Broccoli	*Brassica oleracea*	Brassicaceae	Mexico	(Rojas *et al.*, 1993; Castrejon-G. *et al.*, 1998)
Cabbage	Repollo	*Brassica oleracea*	Brassicaceae	Chile, Mexico	(Carrillo-S., 1971; Aruta-M *et al.*, 1974; Monge-V. *et al.*, 1984; Grez, 1992; Rojas *et al.* 1993; Larrain-S., 1996; Castrejon-G. *et al.*, 1998)
Carnation	Clavel	*Dianthus caryophyllus*	Caryophyllaceae		(Castillo and Angulo, 1991)
Carrot	Zanahoria	*Daucus carota* subsp. *sivus*	Apiaceae	Argentina	(Neder de Roman and Arce de Hamity, 1991; Arce de Hamity and Neder de Roman, 1992)
Cauliflower	Coliflor	*Brassica oleracea botrytis*	Brassicaceae	Mexico	(Rojas *et al.*, 1993; Castrejon-G. *et al.*, 1998)

Common name	Spanish name	Scientific name	Family	Country	References
Chard	Acelga	*Beta vulgaris* subsp. *cicla*	Chenopodiaceae	Mexico, Chile	(Rojas *et al.*, 1993; Lamborot *et al.*, 1995)
Chick pea	Garbanzo	*Cicer arietinum*	Fabaceae	Chile	(Larrain-S., 1996)
Clover	Trébol	*Trifolium pratense*	Fabaceae	Chile	(Castillo and Angulo, 1991)
Coriander[a]	Cilantro	*Coriandrum sativum*	Apiaceae	Mexico	(Riley, 1998)
Corn	Maíz	*Zea mays*	Poaceae	Chile	(Castillo and Angulo, 1991; Larrain-S., 1996; Olivares and Angulo, 1995)
Eggplant	Berenjena	*Solanum melongena*	Solanaceae	Chile	(Lamborot *et al.*, 1995)
Field smartweed	Malezas	*Polygonum segetum*	Polygonaceae	Colombia	(Zenner de Polenia, 1990; Castillo and Angulo, 1991)
Flax or linen	Lino	*Linum usitatissimum*	Linaceae	Chile, Peru	(Wille-T., 1943; Angulo and Weigert, 1975; Castillo and Angulo, 1991)
Garlic	Ajo	*Allium sativum*	Liliaceae	Colombia, Chile	(Larrain-S., 1996; Lopez-A., 1996b)
Gladiolus[a]	Gladiolo	*Gladiolus* spp.	Iridaceae	Mexico	(PIN-309[b]; Riley 1998)
Grape	Uva, vid	*Vitis* spp., *Vitis vinifera*	Vitaceae	Chile	(Castillo and Angulo, 1991; Larrain-S., 1996)
Groundcherry[a]		*Physalis* spp.	Solanaceae		PIN-309[b]
Husk tomato[a]		*Physalis pubescens*	Solanaceae	Mexico	(Riley, 1998)
Jojoba	Jojoba	*Simmondsia californica*	Buxaceae	Chile	(Quiroga *et al.*, 1989; Castillo and Angulo, 1991; Larrain-S., 1996)
Kiwi	Kiwi	*Actinidia chinensis*	Actinidiaceae	Chile	(Larrain-S., 1996)
Lettuce	Lechuga	*Lactuca* spp.	Asteraceae	Argentina, Mexico	(Neder de Roman and Arce de Hamity, 1991; Arce de Hamity and Neder de Roman, 1992)
Lily of the Incas[a]		*Alstroemeria* spp.	Liliaceae		PIN-309[b]
Marigold	Maravilla	*Calendula* spp.	Asteraceae	Chile	(Larrain-S., 1996)
Onion	Cebolla	*Allium cepa*	Liliaceae	Chile, Mexico, Colombia	(Quiroz-E., 1977; Castillo and Angulo, 1991; Lamborot *et al.*, 1995; Larrain-S., 1996; Lopez-A., 1996a; Castrejon-G. *et al.*, 1998)
Peas	Arverjas, chicaro	*Pisum* spp.	Fabaceae	Chile, Mexico	(Rojas *et al.*, 1993; Lamborot *et al.*, 1995)
Pistacio	Pistacho or pistacio	*Pistacia* spp.	Anacardiaceae	Chile	(Larrain-S., 1996)

Continued

Table 9.3. Continued.

English common name	Spanish common name	Scientific name	Family	Countries	References
Potato	Papa	*Solanum tuberosum*	Solanaceae	Mexico, Colombia, Bolivia, Chile, Argentina, Peru	(Munro, 1968; Angulo and Weigert, 1975; Arestegui-P., 1976; Loo and Aguilera, 1983; Sanchez-V. and Maita-Franco, 1987; Zenner de Polenia, 1990; Castillo and Angulo, 1991; Arce de Hamity and Neder de Roman, 1992; Leyva-O. and Sanchez-V., 1993; Rojas *et al.*, 1993; Olivares and Angulo, 1995; Rojas *et al.*, 1996; Larrain-S., 1996; Lopez-A., 1996a)
Quinoa or quinua	Quinoa or quinua	*Chenopodium quinoa*	Chenopodiaceae	Chile, Bolivia	(Angulo and Weigert, 1975; Castillo and Angulo, 1991; Liberman-Cruz, 1986; Lamborot *et al.*, 1999)
Rapeseed	Raps	*Brassica napus*	Brassicaceae	Chile	(Artigas and Angulo, 1973; Castillo and Angulo, 1991; Larrain-S., 1996)
Raspberry	Frambuesa	*Rubus idaeus*	Rosaceae	Chile	(Castillo and Angulo, 1991; Larrain-S., 1996)
Rose[a]		*Rosa* spp.	Rosaceae		PIN-309[b]
Rosemary	Romerito	*Rosmarinus officinalis*	Lamiaceae	Mexico	(Rojas *et al.*, 1993)
Ryegrass	Ballica	*Lolium multiflorium*	Lamiaceae	Chile	(Angulo and Weigert, 1975; Castillo and Angulo, 1991)
Sea lavender[a]		*Limonium* spp.	Plumbaginaceae		PIN-309[b]
Spinach	Espinaca	*Spinacia oleracea*	Chenopodiaceae	Mexico	(Castillo and Angulo, 1991; Rojas *et al.*, 1993)
Strawberry	Frutilla	*Fragaria chiloensis*	Rosaceae	Chile	(Castillo and Angulo, 1991; Larrain-S., 1996)
Tobacco	Tobacco	*Nicotiana tabacum*	Solanaceae	Chile, Peru	(Angulo and Weigert, 1975; Castillo and Angulo, 1991; Larrain-S., 1996)
Sunflowers	Girasol	*Helianthus annuus*	Asteraceae		(Angulo and Weigert, 1975)
Tomato	Tomate	*Lycopersicon esculentum*	Solanaceae	Chile	(Lamborot *et al.*, 1995; Larrain-S., 1996)
Ulluco	Melloco, olluco, ulluma, chuguas	*Ullucus tuberosus*	Basellaceae	Ecuador	(Vimos-N. *et al.*, 1998)
Wheat	Trigo	*Triticum aestivum*	Poaceae	Chile	(Larrain-S., 1996)

[a] Host plants listed in Riley (1998) or the APHIS PIN-309 database of interceptions at US border ports have not been reported in the published literature unless otherwise noted.
[b] The common names of plants reported in the PIN-309 database were found in the Plants Database published by the Natural Resources Conservation Service.

USDA-National Agricultural Statistics Service (USDA-NASS, 1998) and the Texas Agricultural Statistics Service (Texas Agricultural Statistics Service, 1999) to determine the extent of production of host crop species in the USA. Figure 9.7 shows that species supporting *Copitarsia* larvae grow in almost every county in the continental USA.

9.4.2 Type and extent of damage

Copitarsia larvae cause both direct and indirect damage that affects both the quality and the yield of various crop species (Table 9.4). In some species, e.g., rapeseed, up to 60% defoliation does not affect yield. In other species, i.e., artichoke and cabbage, however, a single larva can either destroy a plant or make it unmarketable. Other crop species, including asparagus, jojoba, lettuce, potato, quinoa and tomato also suffer direct damage to the part of the plant sold as a commodity.

Except for one reference about damage to quinoa (80–90%) and another to artichoke (24%), we did not discover references to the extent of crop damage caused by *Copitarsia* (Table 9.4). Given the variability in pest density from year to year due to changes in biotic and abiotic factors, one would expect high variability in the damage that could be caused by *Copitarsia*; however, given its high reproductive capacity, the fact that it is polyphagous, has several generations per annum, and can directly damage marketable commodities, the potential for inflicting damage is high.

Arce de Hamity and Neder de Roman (1993) report that *C. turbata* was found for the first time in Jujuy, Argentina, in 1983 and that it caused serious damage to horticultural crops in that location. Whether the outbreak was the result of release from natural enemies as *Copitarsia* expanded its range, a disruption of natural enemies by insecticides, introduction to new host plants, and/or more favorable climatic conditions, is unknown. It is indicative, however, that *Copitarsia* has the ability to cause considerable damage when introduced to a new area.

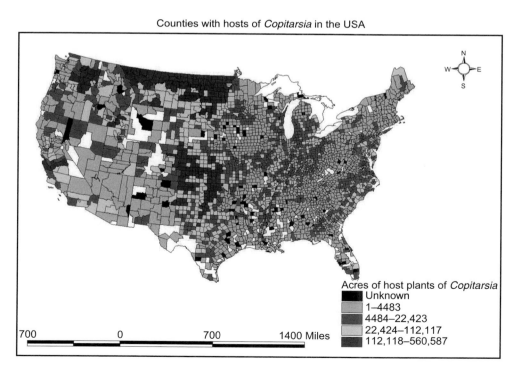

Fig. 9.7 Acreage of crops that could support populations of *Copitarsia* species.

Table 9.4. Damage caused by *Copitarsia* to various crop plants.

Crop	Species	Damage type	Comments	Reference
Alfalfa	*Copitarsia*	Direct	Young larvae eat the parenchyma of the leaves, while older larvae eat the entire buds and leaves	(Cortes-P., 1972)
Artichoke	*C. consueta*	Direct	Up to 24% of marketable heads with damage by *Copitarsia*	(Larrain-S., 1984)
Artichoke	*C. turbata*	Direct	Attacks inflorescence, reduces commercialization	(Larrain-S., 1996)
Artichoke	*C. consueta*	Direct	Up to 54% of the artichoke heads damaged by noctuid larvae during the harvest season. Newly formed flower buds also partially or totally destroyed by *C. consueta*	(Machuca et al., 1990)
Asparagus	*C. turbata*	Direct	Feeds on turiones	(Larrain-S., 1996)
Broccoli	*C. decolora*	Direct	A single larva can destroy the plant	(Acatitla Trejo et al., 2006)
Cabbage	*Copitarsia*	Direct	Damage is quite severe: larvae penetrate the head and feed on interior leaves; can lead to decay that kills the plant. Feeding on unformed heads leads to lateral buds that do not produce commercial cabbages	(Monge-V. et al., 1984)
Cabbage	*Copitarsia*	Direct	Larvae can destroy the plant	(Monge-V. et al., 1984)
Jojoba	*C. consueta*	Direct	Larvae feed on developing fruit and can destroy capsules and seeds	(Quiroga et al., 1989)
Lettuce	*C. turbata*	Direct	Attack begins on external leaves, will go between leaves; one larva will consume 5.32% of a lettuce plant	(Arce de Hamity and Neder de Roman, 1992)
Melloco	*C. turbata*	Indirect	Larvae break the small plants or cut the leaves	(Vimos-N. et al., 1998)
Onion	*C. turbata*	Indirect	Feeds on inside of onion stalks leaving only the epidermis. Hidden within the leaf; also cuts leaves, drills stems and perforates tubercles	(Velasquez-Z., 1988)
Onion	*Copitarsia*	Indirect	Holes for entrance into the leaf provide an entrance for plant pathogens, which can cause the plant to drop the leaf, even after *Copitarsia* has left. Usually only one larva/plant in one leaf, so not too important. Control justified only when >10% of the plants are infested	(Sanchez-V. and Maita-Franco, 1987)
Potato	*C. consueta*	Direct	Populations highest at the end of the season, which is detrimental because they infest the tubers	(Sanchez-V. and Aldana-M., 1985)
Potato	*C. consueta*	Indirect	Severe attacks occur in the dry periods when the potato plants are young	(Lopez-A., 1996a)
Potato	*Copitarsia*	Indirect	Damages potato shoots, cutting tender buds to the height of the soil line	(Arestegui-P., 1976)
Quinoa	*C. turbata*	Direct	Damages the floral buds and flowers, in addition to buds, stems and leaves	(Lamborot et al., 1999)
Quinoa	*C. turbata*	Indirect	80–90% damage from lepidoptera if no pesticides applied (where they are on the pesticide treadmill)	(Liberman-Cruz, 1986)
Rapeseed	*C. consueta*	Indirect	Defoliation up to 60% does not affect yield	(Artigas and Angulo, 1973)
Tomato	*C. turbata*	Direct	Damages mostly young fruit	(Larrain-S., 1996)

9.4.3 Pest status/estimated economic impact

A literature survey uncovered 21 references that discuss the economic status and frequency of outbreaks of *Copitarsia* in its native range (Table 9.5). *Copitarsia* was reported to cause high economic damage in Argentina, Bolivia, Colombia and Mexico, and little damage in Ecuador and Peru. In Chile, nearly an equal number of reports indicated high or low damage from the pest. In six of the cases where damage was described as high, the authors attributed high pest densities to the activities of man. The use of insecticides that released *Copitarsia* from control by natural enemies and other cultivation practices (monocultures, weed removal and planting at low density) were thought to favor the development of pest populations (Table 9.5). For example, in Chile, *Copitarsia* was not considered a pest of artichokes harvested early for local consumption, but it was a problem in exported artichokes that were harvested later in the year after *Copitarsia* populations had a chance to build (Table 9.5).

Although populations of *Copitarsia* can reach high densities, outbreaks are often considered infrequent (Table 9.5). Weather, particularly temperature and rainfall, seems to affect the survival of over-wintering pupae, with favorable conditions leading to high densities of adults during the spring, and thus to outbreaks (refer to Table 9.5). Research to more clearly define conditions that promote outbreaks of *Copitarsia* would be useful, to anticipate densities of insects that might be sent to the USA.

9.4.4 Interceptions at US ports of entry

Pest interception records provide an indication of the frequency of arrival of certain pests. These frequencies may be useful to assess the relative risks of *Copitarsia* arriving at a particular port on certain host plants from a given country. An analysis of *Copitarsia* interception records was conducted to answer the following questions: (i) Which ports most frequently report interceptions of *Copitarsia*? (ii) Which countries are most commonly reported as the origin of *Copitarsia*-infested goods/commodities? (iii) On which goods/commodities are interceptions of *Copitarsia* most frequently noted?

The analysis was based on information available in the APHIS AdHoc-309 database. The database was queried for all taxa of *Copitarsia* detected between September 2000 and August 2010. Four taxa were identified: *C.* sp., *C. consueta*, *C. incommoda* and *C. decolora*. Four informational fields were included in summary reports: port, interception date (reported as month and year), host and country of origin. The total number of interceptions with identical information in all fields was also provided. The four summary tables were combined into one master table for analysis.

From September 2000 to August 2010, 10,655 interceptions of *Copitarsia* were reported from 50 ports. Without adjusting for the volume or type of imported good/commodity, reported interceptions were most common in Miami (Table 9.6). Reports from Miami were five times greater than the second most commonly reporting port: Pharr, in Texas. Interceptions of *Copitarsia* from ports along the US/Mexico border were greatest from border ports in Texas. Notably, Nogales, Arizona, and San Diego, California, which receive a significant volume of agricultural commodities from Mexico, reported few interceptions of *Copitarsia* (Table 9.6).

Copitarsia-infested goods/commodities originated from at least 29 countries (Table 9.7). Over twice as many reports were based on commodities from Colombia compared to Mexico. Colombia, Mexico, Ecuador, Guatemala and Chile accounted for >98% of all interception records (Table 9.7); however, during the decade following 2000 all asparagus from Peru was fumigated without inspection, and therefore pest loads on asparagus from Peru are under-represented.

Approximately 150 genera of plants were indicated in reports of *Copitarsia* interceptions. Genera accounting for the top 90% of interceptions are reported in Table 9.8. The majority of reported interceptions (~60%) were from ornamentals and cut flowers.

9.5 Monitoring

Most growers scout for *Copitarsia* by using light traps, which are not specific and capture a wide variety of nocturnal insects. Rojas *et al.* (2006) identified the pheromone of *C. decolora* in Mexico as (Z)-9-tetradecenyl acetate (Z9D14:Ac) and

Table 9.5. Economic status and frequency of outbreaks of *Copitarsia* in its native range.

Country	Crop	*Copitarsia* species	Pest status	Damage low/high	Damage frequent/ infrequent	Suspected cause of outbreak	References
Argentina	lettuce	C. turbata	Can cause severe damage	High			(Arce de Hamity and Neder de Roman, 1993)
Bolivia	potato	C. consueta	Listed as the most troublesome tuber pest	High			(Munro, 1954)
Bolivia	quinoa	C. turbata	Can cause up to 80–90% loss if not controlled; conclude that changes in agricultural practices have led to pest outbreaks	High		Agricultural Practices	(Liberman-Cruz, 1986)
Chile	alfalfa	C. turbata	Did not cause detectable damage; diverse entomophagous insects contributing to control	Low	Infrequent		(Apablaza and Stevenson, 1995)
Chile	alfalfa	C. consueta + humilis	Up to 5 million noctuids/ha. Noctuids kept below economic threshold by parasites when no insecticide applied; when applying insecticides must apply 12–15 times/year	High		Agricultural Practices	(Cortes-P., et al., 1972; Cortes, 1976)
Chile	alfalfa	C. humilis	Irregular or occasional occurrence; always in a low density of population possibly regulated by natural enemies	Low	Infrequent		(Hichins-O. and Mendoza-M., 1976)
Chile	alfalfa	C. consueta	Low density during the year with the lowest density in the spring	Low	Infrequent		(Hichins-O. and Rabinovich, 1974)
Chile	artichoke	C. consueta	*Copitarsia* damaged up to 24% of marketable small heads. Only a problem for export because of rejection at the US ports; damage not severe during harvest season for local consumption	High		Agricultural Practices	(Machuca-L et al., 1989)
Chile	onion	Copitarsia spp.	Only occasional outbreaks		Infrequent		(Quiroz-E., 1977)
Chile	quinoa	C. turbata	Irregular but sometimes severe damage	High	Infrequent		(Lamborot et al., 1999)
Chile	rapeseed	C. consueta	Frequently explosive increases in populations: 90–100 larvae per plant. Lack of insecticides allows populations of parasites and disease to be maintained	High	Frequent	Agricultural Practices	(Artigas and Angulo, 1973)
Chile	various	C. naeniodes	Listed as a prevalent detrimental insect	High			(Duran-M., 1972)

Country	Crop	Species	Notes		Frequency	Factor	Reference
Chile	various	C. turbata	Outbreaks happen infrequently, but related to environmental conditions such as rainy winters free of frosts - leads to high survival of overwintering pupae		Infrequent	Weather	(Larrain, 1998)
Colombia	garlic and onion	C. consueta	Found at low incidence	Low	Infrequent		(Lopez-A., 1996b)
Colombia	potato	C. consueta	Severe attacks occur in dry periods when potato plants are young	High		Weather	(Lopez-A., 1996a)
Colombia	potato	C. consueta	Can cause economic damage, but mainly after severe dry seasons and only during the first month after germination	High		Weather	(Zenner de Polenia, 1990)
Ecuador	melloco	C. turbata	Except for rare cases are not of importance. In most cases controlled with natural enemies	Low	Infrequent		(Vimos-N. et al., 1998)
Mexico	cabbage	Copitarsia spp.	Percent damaged plants approximately 45%	High			(Carrillo-S., 1971)
Mexico	cabbage	Copitarsia spp.	Cultivation practices (monoculture, weed free areas, low planting densities, application of insecticides) favor the development of pest populations	High		Agricultural Practices	(Monge-V. et al., 1984)
Peru	potato	Copitarsia spp.	Losses from Copitarsia are relatively low and it is considered a secondary pest	Low	Infrequent		(Arestegui-P., 1976)
Peru	potato	C. turbata	Density was variable		Infrequent		(Leyva-O. and Sanchez-V., 1993)

Table 9.6. Reported interceptions of *Copitarsia* spp. at various ports in the USA.*

Commodity origin	Interceptions		Commodity origin	Interceptions	
Miami, Florida	7088	66.37%	Houston, Texas	17	0.16%
Pharr, Texas	1471	13.77%	Dallas, Texas	15	0.14%
Laredo, Texas	651	6.10%	Dulles, Virginia	13	0.12%
San Juan, Puerto Rico	411	3.85%	St. Louis, Missouri	10	0.09%
Hidalgo, Texas	219	2.05%	Newark, New York	9	0.08%
Rio Grande, Texas	146	1.37%	Ft Lauderdale, Florida	5	0.05%
Roma, Texas	114	1.07%	Aguadilla, Puerto Rico	5	0.05%
Los Angeles, California	104	0.97%	Calexico, California	4	0.04%
Atlanta, Georgia	93	0.87%	Port Everglades, Florida	4	0.04%
Otay Mesa, California	68	0.64%	Memphis, Tennessee	4	0.04%
San Diego, California	55	0.52%	Eagle Pass, Texas	4	0.04%
New York, New York	46	0.43%	Philadelphia, Pennsylvania	3	0.03%
Progreso, Texas	44	0.41%	Del Rio, Texas	3	0.03%
Brownsville, Texas	24	0.22%	Port Hueneme, California	2	0.02%
Chicago, Illinois	21	0.20%	San Diego, California	2	0.02%

*Two or fewer interceptions were also reported from San Ysidro, California; New Orleans, Lousiana; New Bedford, Massachusetts; Los Indios, Texas; Aruba; San Luis, Arizona; San Francisco, California; Orlando, Florida; Savannah, Georgia; Baton Rouge, Louisiana; Detroit, Michigan; Charlotte, North Carolina; Cincinnati, Ohio; Mayaguez, Puerto Rico; Ponce, Puerto Rico; Roosevelt Roads, Puerto Rico; Walterboro, South Carolina; St Croix, Virginia; Highgate Springs, Vermont; and Seattle, Washington.

Table 9.7. Country of origin on reported interceptions of *Copitarsia* spp. between 1 September 2000 and 31 August 2010.

Commodity origin	Interceptions		Commodity Origin	Interceptions	
Colombia	6577	60.92%	Israel	2	0.02%
Mexico	2811	26.04%	Jamaica	2	0.02%
Ecuador	935	8.66%	Panama	2	0.02%
Chile	292	2.70%	Aruba	1	0.01%
Costa Rica	44	0.41%	Bolivia	1	0.01%
Peru	44	0.41%	Cuba	1	0.01%
Guatemala	41	0.38%	El Salvador	1	0.01%
Netherlands	7	0.06%	Haiti	1	0.01%
Unknown	7	0.06%	Kenya	1	0.01%
Argentina	5	0.05%	Moldova	1	0.01%
Dominican Republic	4	0.04%	New Zealand	1	0.01%
Honduras	4	0.04%	Nicaragua	1	0.01%
Puerto Rico	4	0.04%	Trinidad and Tobago	1	0.01%
Brazil	2	0.02%	Venezuela	1	0.01%
Guyana	2	0.02%	–	–	–

(Z)-9-tetradecenol (Z9D14:OH). Field trials found that traps baited with pheromones with a higher ratio of acetate versus alcohol caught significantly more male moths (Muñiz-Reyes et al., 2007). Scientists studied several trap types and trap placement (height above the ground) (Acatitla Trejo et al., 2006) and recommended using a simple, inexpensive, low maintenance and effective 'economic trap' made of a re-usable plastic juice bottle placed at 150 cm above the soil surface.

9.6 Control

9.6.1 Regulatory

All green asparagus and white asparagus showing signs of green that originate in Peru must be treated with methyl bromide regardless of whether or not they are known to contain insects (Davis and Venette, 2004; APHIS, 2010a). This

Table 9.8. Commodities reported for port interceptions of *Copitarsia* spp.*

Host	Common name	Number intercepted	Percentage interceptions
Alstroemeria	Lily of the Incas	3358	29.42%
Brassica	Cole Crops	1046	9.16%
Chenopodium	Quinoa	750	6.57%
Chrysanthemum	Chrysanthemum	748	6.55%
Coriandrum	Coriander/Cilantro	683	5.98%
Aster	Aster	573	5.02%
Dianthus	Pinks/Carnation	441	3.86%
Limonium	Sea Lavender	359	3.14%
Mentha	Mint	272	2.38%
Helianthus	Sunflower	198	1.73%
Antirrhinum	Snapdragon	168	1.47%
Asparagus	Asparagus	167	1.46%
Callistephus	China Aster	153	1.34%
Molucella	Bells-of-Ireland	149	1.31%
Solidago	Goldenrod	138	1.21%
Ocimum	Basil	105	0.92%
Origanum	Oregano	90	0.79%
Apium	Celery	77	0.67%
Portulaca	Purslane	75	0.66%
Gypsophila	Baby's Breath	73	0.64%
Thymus	Thyme	73	0.64%
Physalis	Husk Tomato	67	0.59%
Rosa	Rose	67	0.59%
Suaeda	Seepweed	62	0.54%
Campanula	Bellflower	59	0.52%
Dendranthema	Arctic Daisy	59	0.52%
Amaranthus	Pigweed	57	0.50%
Rosmarinus	Rosemary	54	0.47%
Allium	Onion	51	0.45%
Chamaemelum	Chamomile	51	0.45%
Pisum	Pea	49	0.43%

*This table represents 90% of the interceptions. The remaining interceptions of *Copitarsia* spp. were reported from *Hypericum*, *Gladiolus*, *Lisianthus*, *Hydrangea*, *Anethum*, *Anemone*, *Iris*, *Prunus*, *Opuntia*, *Ranunculus*, *Matthiola*, *Lactuca*, *Anigozanthos*, *Matthiola*, *Phlox*, *Artemisia*, *Carthamus*, *Ammi*, *Ornithogalum*, *Bupleurum*, *Lilium*, *Gerbera*, *Zea*, *Anthurium*, *Capsicum*, *Delphinium*, *Eryngium*, *Raphanus*, *Rubus*, *Vaccinium*, *Achillea*, *Liatris*, *Spinacia*, *Zantedeschia*, *Agapanthus*, *Leucadendron*, *Celosia*, *Marjorama*, *Persea*, *Polianthes*, *Anthemis*, *Cichorium*, *Dahlia*, *Solidaster*, *Beta*, *Tagetes*, *Trachelium*, *Veronica*, *Cicer*, *Cynara*, *Eustoma*, *Balus*, *Amaryllis*, *Freesia*, *Petroselinum*, *Statice*, *Basilicum*, *Centaurea*, *Crocosmia*, *Godethia*, *Helichrysum*, *Lippia*, *Protea*, *Salvia*, *Solanum*, *Aconitum*, *Astilbe*, *Daucus*, *Draceana*, *Dysphania*, *Fragaria*, *Grevillea*, *Heliconia*, *Lathyrus*, *Lysimachia*, *Porophyllum*, *Strelitzia*, *Tulipa*, *Abelomoschus*, *Ajuga*, *Aloe*, *Annanas*, *Annona*, *Asclepias*, *Bougainvillea*, *Bouvardia*, *Calluna*, *Chamomilla*, *Codiaeum*, *Colocasia*, *Consolida*, *Digitalis*, *Eruca*, *Eucalyptus*, *Euphorbia*, *Glycine*, *Helianthemum*, *Hippeastrum*, *Laurus*, *Leucaena*, *Majorana*, *Matricaria*, *Paeonia*, *Phaseolus*, *Pinus*, *Pyrus*, *Ruscus*, *Salicornia*, *Satureja*, *Tillandsia*, *Vicia*, *Vigna*, *Vitis*.

rule was put in place because the majority of shipments were found to be infested with *Copitarsia*. Commercial consignments from Mexicali, Mexico, must be accompanied by a phytosanitary certificate stating that they originated in Mexicali and they must be inspected for pests. If no phytosanitary certificate is available, the shipment must be treated with methyl bromide. For commercial shipments from parts of Mexico other than Mexicali, the consignment must be accompanied by a phytosanitary certificate declaring that it is free of *Copitarsia* larvae and adults. *Copitarsia* eggs are not regulated from Mexico. For products coming from all other countries, the produce or flowers are inspected and treated if found to contain *Copitarsia* (APHIS, 2010a, b).

9.6.2 Chemical

Pesticides are frequently recommended for control in areas where *Copitarsia* is currently established (Munro, 1954, 1968; Cortes-P. *et al.*, 1972; Artigas and Angulo, 1973; Quiroz-E., 1977; Liberman-Cruz, 1986; Perlta-S., 1987; Sanchez-V. and Maita-Franco, 1987; Machuca *et al.*, 1990; Zenner de Polenia, 1990; Vimos-N. *et al.*, 1998; Acatitla Trejo *et al.*, 2006). For asparagus in Peru, organic insecticides such as chitin synthesis inhibitors or *Bacillus thuringiensis* are recommended; synthetic organic insecticides are used only when pest density and damage are high (Sanchez Valasquez and Apaza Tapia, 2000). Several authors have attributed outbreaks of *Copitarsia* to release from natural enemies following application of broad-spectrum insecticides (Artigas and Angulo, 1973; Cortes, 1976; Monge-V. *et al.*, 1984; Lamborot *et al.*, 1995).

9.6.3 Biological

Naturally occurring diseases and insect predators and parasitoids are widely reported in the literature. Two diseases caused by *B. thuringiensis* and *Beauvaria bassiana* are mentioned in passing (Lopez-A., 1996a); however, the fungus *Entomophthora sphaerosperma* can kill up to 80% of *Copitarsia* larvae on *Brassica* in Chile following summer rains (Aruta-M. *et al.*, 1974). Predatory Coleoptera (beetles) and Neuroptera (lacewings) reportedly attack eggs and larvae (Sanchez-V. and Maita-Franco, 1987; Lopez-A., 1996a); however, reports of parasitic flies and wasps are more numerous. Eight flies in the family Tachinidae parasitize *Copitarsia* larvae on tomatoes, broad beans, peas, potatoes, alfalfa and onions in Chile and Peru (Cortes, 1976; Alcala, 1978; Sanchez-V. and Maita-Franco, 1987; Leyva-O. and Sanchez-V., 1993; Lamborot *et al.*, 1995). The diversity of Tachinid species was more than three times greater in alfalfa in Chile in areas without intense and continuous use of residual insecticides (Cortes, 1976). In addition, low parasitism levels by Tachinids were attributed to intensive use of toxic agrochemicals (Lamborot *et al.*, 1995). At least nine parasitic Hymenoptera (wasps) attack eggs and larvae of *Copitarsia* on onion, eggplant, broad beans, potatoes, peas, alfalfa, swiss chard, quinoa and tomato. Only the egg parasitoids *Trichogramma brasiliensis* and *T. minutum*, *T. evanescens* and *T. pretiosum* have been mass reared and released for control of *Copitarsia* (Loo and Aguilera, 1983; SENASA, 2001).

9.6.4 Cultural

Controlling weeds that can serve as reservoirs for *Copitarsia* populations is an important tool for controlling infestations (Sanchez Valasquez and Apaza Tapia, 2000). Machuca *et al.* (1990) found that intensive, i.e., more frequent, weed control significantly reduced damage to artichokes. Post-harvest chilling of artichokes at 0.1°C for 21 days caused 72% and 81% mortality in 4th and 2nd instars, respectively (Urra and Apablaza, 2005).

9.6.5 Behavioral

Light traps are not only used to detect *Copitarsia* but are recommended as a measure to control populations in asparagus in Peru (Sanchez Valasquez and Apaza Tapia, 2000). Detailed recommendations have been made about the timing of deployment, how many traps per hectare, type of bulb, collection container and height of the trap in relation to the crop.

Perhaps the most unusual control method being studied is simulating the calls of echolocating bats in the packing shed for cut flowers in Columbia. In early trials, evasive reactions were observed in adult moths when the bat sounds were played (Paz *et al.*, 2007).

9.7 Pathway Analysis of Risk Posed by *C. corruda* Imported on Asparagus From Peru

Peruvian asparagus was so frequently infested that all shipments now require treatment with methyl bromide prior to importation. An initial risk assessment (Venette and Gould, 2006) concluded that the risks associated with this potential pest were probably high, but the quality and availability of data used to reach that conclusion were inadequate. A project was undertaken to collect the data necessary to produce an accurate science-based risk assessment to either validate or refute the need for high vigilance and expensive treatments (Gould et al., 2006). We developed a pathway analysis starting with the shipment of asparagus from Peru to the USA and determined the effects of environmental conditions and procedures followed at importer, wholesale and retail facilities on the survival and establishment of immature *C. corruda*.

Probabilistic models were constructed to estimate the risk of *C. corruda* escaping into the wild from importer, wholesale and retail facilities. For this insect to escape, larvae must hatch from eggs and be able to crawl into the environment. Because of the cold temperatures at which asparagus is shipped to preserve freshness, and the short time from harvest to consumption, a very small percentage of eggs has an opportunity to produce larvae that could escape. This opportunity comes when asparagus spears with *C. corruda* eggs are discarded into a dumpster. Given the large volume of asparagus from Peru, a few insects might be able to exploit this possibility. Several information sources were consulted, and new information was gathered to estimate the chances that one or more female larvae and at least one male (with whom she could mate) might escape. The number of potential mated females was also calculated. Information was obtained from the Foreign Agricultural Service database regarding the amount of green asparagus imported monthly from Peru. Separate models were constructed for three seasons, because 80% of fresh asparagus is imported from Peru during September–December and much less asparagus is imported from February to May. The number of eggs per spear was determined from data collected by Peruvian cooperators. The amount of asparagus discarded in dumpsters was determined by surveying importers, wholesalers and retailers. Known infested asparagus was shipped to the USA following strict biosecurity procedures to determine how many eggs were still viable and how soon eggs might hatch. A study in Peru measured the percentage of newly hatched larvae that could crawl out of a garbage receptacle filled with asparagus (Gould and Huamán Maldonado, 2006). Climate models were used to predict where the climate is potentially suitable for population establishment. Because data do not exist on availability of host plants, larval survival, or mate finding, these parameters were not included in the model. The model is therefore conservative in the estimate of risk.

The probability that at least one potential mated female could escape from the asparagus pathway at one or more importer facilities approached 100%, and the average time to the first potential mated female was 1 year for all three seasons. The number of potential mated females escaping at a single facility was greatest during September–December (average = 37); however, even during the least risky season (February–May) the model predicted an average of 5 females, and as many as 70, per facility. Clearly the risk was greatest during the September–December time frame, when most of the asparagus from Peru is imported, but the risk remained high during the other periods. The model estimated that a potential mated females could escape from a wholesale facility in 4 years during September–December, but the maximum number of potential mated females at a single facility averaged <1 and did not exceed three. During January/June–August and February–May, the numbers of years to the first potential mated female were 20 and 82, respectively, again with an average of less than one potential mated female escaping the produce pathway when females did escape. The model did not predict that any potential mated females would be produced at individual retail facilities.

The probabilistic model indicated that the risk of at least one female and male *C. decolora* larva escaping into the wild was greatest at importer facilities, and during the September–December time frame. While escape at importer facilities is most likely to occur on imports during the period from September to December, the differences between that and the other time periods did not seem to justify seasonal differences in handling and disposal of asparagus. Asparagus from Peru enters the USA through Miami and Los Angeles. The climate-modeling software CLIMEX™ (Hearne Scientific Software, South Yarra, Australia) strongly indicated that the Los Angeles area was climatically very suitable for establishment. Miami may be climatically suitable for establishment, but we are uncertain about that classification. Regardless, from both pathway and climate modeling, we conclude that the risk of establishment by *C. decolora* is greatest for asparagus discarded by importers in the Los Angeles area.

Establishment is less likely, but possible on discards by importers in the Miami area. Escape of potential *C. decolora* mated females is even less likely from asparagus discarded by wholesalers anywhere, especially during the transitional and February–May time periods, and even less probable at retailers' facilities.

Because of the uncertainty in the climate simulation as to whether *C. corruda* could or could not establish in Miami, Florida, and other locations in the south-east, we simulated the climate (day length, humidity, temperature) from locations where *Copitarsia* was very likely to survive (Long Beach, California) and areas where the climate model was uncertain (Sacramento, California; Houston, Texas; Mobile, Alabama; Miami, Florida; Jacksonville, Florida; and Charlottesville, South Carolina). As predicted, populations of *C. corruda* persisted through the summer in Long Beach, but none of the eggs hatched during the heat of the summer in all of the remaining locations, including Miami, Florida, where the majority of the asparagus is imported. Although it is possible that establishment could be initiated in Miami, Florida, populations would not be expected to persist because of high summer temperatures.

Acknowledgement

Thanks to Dario Corredor for providing information about control of *Copitarsia* in Colombia.

Note

[i] http://plants.usda.gov, accessed 19 July 2012.

References

Acatitla Trejo, C., Bautista Mar tinez, N., Vera Graziano, J., Romero Napoles, J. and Calyecac Cortero, H.G. (2004) Biological cycle and rates of survivorship and reproduction on *Copitarsia incommoda* (Walker) (Lepidoptera: Noctuidae) in five artificial diets. *Agrociencia* 38, 355–363.

Acatitla Trejo, C., Bautista Mar tínez, N., Carr illo Sánchez, J.L., Cibrián Tovar, J., del Río Galván, S.L, Osorio Córdoba, J ., Sánchez Arroyo, H. *et al*. (2006) *El Gusano del Corazón de la Col,* Copitarsia decolora. *Biología, Taxonomía y Ecología.* Colegio de Postgraduados, Texcoco, México.

Alcala, P. (1978) Tachinidos parasitos de *Copitarsia turbata* en el Valle del Mantaro . *Revista Peruana de Entomologia* 21, 126.

Angulo, A.O. and J ana-Saenz, C. (1982) A ne w species of *Euxoa* from Chile (Lepidopter a, Noctuidae). *Boletin de la Sociedad de Biologia de Concepción* 53, 13–17.

Angulo, A.O. and Olivares, T.S. (1999) A new genus and new species of hight [sic] Andean II (Lepidoptera: Noctuidae). *Guyana* 63, 51–61.

Angulo, A.O. and Oliv ares, T.S. (2003) Taxonomic update of the species of *Copitarsia* Hampson (Lepidoptera: Noctuidae: Cuculliinae). *Guyana* 67, 33–38.

Angulo, A.O. and Oliv ares, T.S. (2009) La polilla *Copitarsia decolora*: revisión del complejo de especies con base la morfología genital masculina y de los huevos (Lepidoptera: Noctuidae). *Revista de Biologia Tropical* 58, 769–776.

Angulo, A.O. and Weigert, G.T. (1975) Mimetismo y homocromismo larval en noctuidos Chilenos (Lepidoptera: Noctuidae). *Boletin de La Sociedad de Biologia de Concepción*, 49, 171–175.

Angulo, A.O., Jana-Sanchez, C. and P arra, L.E. (1985) *Copitarsia consueta* (Walker) y *Copitarsia naeonoides* (Butler): Espineretes larvales como car acteres diagnosticos (Lepidoptera: Noctuidae). *Agro Sur (Chile)* 13, 133–134.

Angulo, A.O., Olivares, T.S and Badilla, R. (2001) A new species of *Copitarsia* Hampson from Chile (Lepidopter a, Noctuidae, Cucullinae). *Gayana (Concepción)* 65, 1–4. doi: 10.4067/S0717/17-65382001000100001.

Apablaza, J.U. and Ste venson, T.R. (1995) Fluctuaciones poblacionales de áfidos y de otros ar tropódos en el f ollaje de alf alfa cultivada en la Region Metropolitana. *Ciencia de Investigacion (Chile)* 22, 115–121.

APHIS (2010a) Fresh Fruits and Vegetables Import Manual. www.aphis.usda.gov/import_export/plants/manuals/ports/downloads/fv.pdf, accessed 19 July 2012.

APHIS (2010b) Cut Flo wers and Greener y Import Manual. www.aphis.usda.gov/import_export/plants/manuals/ports/downloads/cut_flower_imports.pdf, accessed 19 July 2012.

Arce de Hamity, M.G. and Neder de Roman, L.E. (1992) Aspectos bioecologicos de *Copitarsia turbata* (Herrich-Schaffer) (Leipdoptera: Noctuidae) importantes en la deter minacion del dano economico en cultiv os de *Lactuca* sativa L. de la Quebrada de Humahuaca, Jujuy, Argentina. *Revista de la Sociedad Entomologica Argentina* 50, 73–87.

Arce de Hamity, M.G. and Neder de Roman, L.E. (1993) Morfologia de los estados inmaduros y aspectos etologicos de *Copitarsia turbata* (Herrich-Schaffer) (Lepidoptera: Noctuidae). *Neotropica* 39, 29–33.

Arestegui-P.A.(1976) Plagas de la papa en Andahuaylas, Apurimac. *Revista Peruana de Entomologia* 19, 97–98.

Artigas, J.N. and Angulo, A.O. (1973) *Copitarsia consueta* (Walker), biologia e importancia economica en el cultivo de raps (Lepidoptera, Noctuidae). *Boletin de la Sociedad de Biologia de Concepción* 46, 199–216.

Aruta-M. C., Carillo-L. R. and Gonzalez-M., S. (1974) Determinacion para Chile de hongos entomopatogenos del genero Entomophthora. *Agro Sur* 2, 62–70.

Bell, R.A., Owens, C.D., Shapiro, M. and Tardif, J.R. (1981) Development of mass-rearing technology. In: Doane, C.C. and McManus, M.L. (eds) *The Gypsy Moth: Research Toward Integrated Pest Management*. USDA Forest Service Technical Bulletin 1584, Washington, DC, pp. 599–655.

Blanchard, E. (1852) Noctuelianos. In: Gay, C. (ed.) *Historia Físca y Política de Chile* 7, pp. 71–86.

Butler, A.G. (1882) Heterocerous Lepidoptera collected in Chile by T. Edmonds. *Transactions of the Entomological Society* 2, 129–130.

Carrillo-S., J.L. (1971) Pruebas de thuricide (*Bacillus thuringiensis*) para combatir gusanos de la Colen Chapingo. *Agricultura Tecnica Mexicana* 3, 58–60.

Castillo, E.E. and Angulo, A.O. (1991) Contribution to the knowledge of the genus *Copitarsia* Hampson 1906 (Lepidoptera, Glossata, Cucullinae). *Guyana Zoologia* 55, 227–246.

Castrejon-G. J.R., Cibrian-T., L., Valdes, J. and Camino-L., M. (1998) Morfologia, distribucion y cuantificacion de los sensulos antenales en adultes de *Copitarsia consueta*. *Folia Entomologica Mexicana* 103, 63–73.

Cibrián, T.J. and Sugimoto, A. (1992) Elaboración de una dieta artificial para la cría de *Copitarsia consueta* Walker (Lepidoptera: Noctuidae). In: Martínez, C.G., A. Lastras R., J. Valle M., P. Medellín M., J. Hernández G., R. J. Loza, M. and R. García, L. (eds) Memorias del XXVII Congreso Nacional de Entomología, Sociedad Mexicana de Entomología, San Luis Potosí, Mexico, pp. 416.

Cortes, R. (1976) Multi-control of cutworms and armyworms (Noctuidae) in alfalfa in the desert valleys of Lluta and Camarones, Arica. In: *Ecological Animal Control by Habitat Management*. Proceedings of the Tall Timbers Conference Ecological Animal Control by Habitat Management, Tall Timbers Research Station, Tallahassee, Florida, pp. 79–85.

Cortes-P., R., Aguilera, A., Vargas, H., Hichins, N., Campos, L., Aguilera, A. and Pacheco, J. (1972) Las "Cuncunillas" (Noctuidae) de la alfalfa en Lluta y Camarones, Arica-Chile.- Un Problema bio-ecologico de control. *Sociedad Entomologia del Peru Anales* 15, 253–266.

Davis, E.E. and Venette, R.C. (2004) Methyl bromide provides phytosanitary security: a review and case study for Senegalese aspar agus. *Plant Health Progress* doi:10.1094/PHP-2004-1122-01-RV.

Duran-M., L. (1972) Problemas de la entomologia agricola en Chile austral. *Folia Entomologica Mexicana* 23/24, 45–46.

Giacomelli, E. (1922) Trois Lépidoptéres nouveaux de La Roja Rep. *Argentine Entomologist* 55, 225–227.

Gomez-T., J. (1972) Moscas minadoras en el cultivo de la haba (*Vicia faba* L.) en la Sierra Central del Peru. *Congreso Latino Americano de Entomologia* 15, 239–243.

Gould, J.R. and Huamán Maldonado, M. (2006) *Copitarsia decolora* (Lepidoptera: Noctuidae) larvae escaping from discarded aspar agus: data in support of a pathway risk analysis. *Journal of Economic Entomology* 99, 1605–1609.

Gould, J., Venette, R., and Winograd, D. (2005) Effect of temperature on development and population parameters of *Copitarsia decolora* (Lepidoptera: Noctuidae). *Environmental Entomology* 34, 548–556.

Gould, J.R., Caton, B.P. and Venette, R.C. (2006) A pathway assessment of risk of establishment in the contiguous United States by *Copitarsia decolora* (Guenée) on aspar agus from Peru. USDA-APHIS, Riverdale, Maryland, pp. 41.

Grez, A.A. (1992) Riqueza de especies de insectos herbivoros y tamano de parche de vegetacion huesped: una contrastacion experimental. *Revista Chilena de Historia Natural* 65, 115–120.

Guenée, M.A. (1852) Noctuélites. In: Boisduval, M. and Guenée, M.A. (eds) *Historie Naturelle des Insects, Lépidoptéres, Tome 5*. Librairie Encyclopédique de Roret, Paris, pp. 1–407.

Hampson, G.F. (1906) *Catalogue of the Lepidoptera Phalaenae in the British Museum, Volume 6*. Taylor and Francis, London.

Hampson, G.F. (1910) *Catalogue of the Lepidoptera Phalaenae in the British Museum. Volume 9.* Taylor and Francis, London, pp. 552, xv, plates 137–147.

Hichins-O., N. and Mendoza-M., R.(1976) Algunas observaciones sobre habitos y costumbres de estadios larvarios de noctuidos asociados a la alfalfa en Lluta y Camarones (Lepidoptera: Noctuidae). *Departamento Agricultura, Universidad Del Norte-Arica Idesia* 4, 163–169.

Hichins-O., N. and Rabino vich, J.E. (1974) Fluctuaciones de la pob lacion de lar vas de cinco especies de Noctuidos de impor tancia economica asociadas a la alfalfa en el Valle de Lluta. *Departamento Agricultura, Universidad Del Norte-Arica Idesia* 3, 35–79.

Klein Koch, C. and Waterhouse, D.F. (2000) The distribution and importance of arthropods associated with ag riculture and f orestry in Chile (Distribución e importancia de los ar trópodos asociados a la ag ricultura y silvicultur a en Chile). ACIAR Monograph 68, Canberra.

Köhler, P. (1952) Las Noctuidae Argentinas . Subfamilia "Cucullianae". *Acta Zoologica Lilloana* 12, 135–182.

Köhler, P. (1958) Noctuidarium Miscellanea I (Lep. Het.). *Revista de la Sociedad Entomologica Argentina* 21, 51–63.

Köhler, P. (1959) Noctuidarum miscellanea I. (Lep. Het.). *Revista de la Soceidad Entomologica Argentina* 20, 9–15.

Köhler, P. (1961) Noctuidar ium Miscellanea III. *Anales de la Sociedad Ceintífica Argentina* 172, 69–74.

Köhler, P. (1973) Noctuidarium Miscellanea V (Lep. Het.). *Acta Zoologica Lillioana* 30, 13–21.

Lamborot, L., Arretz, P ., Guerrero, M.A. and Araya, J.E. (1995) P arasitismo de hue vos y larvas de *Copitarsia turbata* (Herrich y Schaffer) (Lepidoptera: Noctuidae) en cultivos horticolas en la Region Metropolitana. *Acta Entomologica Chilena* 19, 129–133.

Lamborot, L., Guerrero, M.A. and Araya, J.E. (1999) Lepidopteros asociados al cultivo de la quinoa (*Chenopodium quinoa* Willdenow) en la z ona central de Chile . *Boletin Sanidad Vegetal de Plagas* 25, 203–207.

Larrain-S., P. (1984) Plagas de la alcachof a. Investigacion y prog reso agropecuario la platina (Chile). *Investigacion y Progreso Agropecuario La Platina (Chile)* 25, 19–22.

Larrain-S., P. (1996) Biologia de *Copitarsia turbata* (Lep. Noctuidae) bajo ambiente controlado . *Agricultura Tecnica* 56, 220–223.

Larrain-S., P. and Araya-C., J.E. (1994) Prospeccion y control quimico de plagas de la alcachofa en la region Metropolitana. *Investigacion Agricola (Chile)* 14, 35–41.

Leyva-O., C. and Sanchez-V., G. (1993) Ocurrencia estacional de las principales plagas del cultivo de papa en Cajamarca. In: Quevedo-I., F. and Arroyo-V, R. (eds) *Resumen de Investigaciones Apoyadas por FUNDEAGRO [Fundacion para el Desarrollo del Agro] 1988–1992*. Proyecto Transformacion de la Tecnologia Agropecuaria (TTA), Lima, pp. 91–92.

Liberman-Cruz, M. (1986) Impacto ambiental del uso actual de la tierra en el Altipiano sur de Bolivia. Con enfasis en el cultivo de *Chenopodium quinoa* Willd. *Rivista di Agricoltura Subtropicale e Tropicale* 80, 509–538.

Loo, P.E. and Aguiler a, P.A. (1983) Multiplicacion experimental de *Trichogramma brasiliensis* (Ashm.) (Hymenoptera: Trichogrammatidae) en la IV Region de Chile. *Idesia* 7, 45–52.

Lopez-A., A. (1996a) Insectos plagas del cultivo de la papa en colombia y su manejo . In: *Papas Colombianas con el Mejor Entorno Ambiental*. Comunicaciones y Asociados Ltda, Santaf e de Bogota, Colombia, pp . 146–148, 150–154.

Lopez-A., A. (1996b) Plagas del ajo y las cebollas . In: *El Cultivo del Ajo y las Cebollas en Colombia*. Corporacion Colombiana de Investigacion Agropecuaria, Instituto Colombiano Agropecuario, Santa f e de Bogota, Colombia, pp. 61–71.

Mabille, M.P. (1885) Diagnoses de Lépidoptéres nouveaux. *Bulletin de la Société Philomatique de Paris, series 7* 9, 55–70.

Machuca-L., J.R., Arretz-V., P. and Ar aya-C., J.E. (1988) Parasitismo de noctuidos en cultiv os de alcachofas en la Region Metropolitana: Identificacion y observaciones preliminares de los parasitos. *Revista Chilena de Entomologia* 16, 83–87.

Machuca, J.R., Arretz, P. and Ar aya, J. (1989a) A new Ichneumonid w asp for Chile. *Acta Entomologica Chilena* 15, 269–270.

Machuca, J.R., Arretz, P. and Ar aya, J. (1989b) Presencia de *Campoletis sonorensis* (Cameron, 1886) (Hymenopter a: Ichneumonidae) en Chile . *Acta Entomologica Chileana* 15, 269–270.

Machuca, J.R., Araya, J.E., Arretz-V., P. and Larrain, P.I. (1990) Evaluation of chemical and cultural control for noctuid larvae in Chilean artichokes produced for foreign markets. *Crop Protection* 9, 115–118.

Monge-V., L.A., Vera-G., J., Infante-G., S. and Carrillo-S., J.L. (1984) Efecto de las practicas culturales sobre las poblaciones de insectos y dano causado al cultivo del repollo (*Brassica oleraceae* var. *capitata*). *Centro De Entomologia y Acarologia* 57, 109–126.

Muñiz-Reyes, E, J. Cibrián-Tovar, J. Rojas-Leon, O. Diaz-Gómez, J., Valdés-Carrasco and N. Bautista-Martínez. (2007) Capturas de *Copitarsia decolora* (Lepidoptera: Noctuidae) en trampas cebadas con diferentes proporciones de feromona sexual. *Agrociencia* 41, 575–581.

Munro, J.A. (1954) Entomology problems in Bolivia. *FAO Plant Protection Bulletin* 2, 97–101.

Munro, J.A. (1968) Insects affecting potatoes in Bolivia. *Journal of Economic Entomology* 61, 882.

Neder de Roman, L.E. and Arce de Hamity, M.G. (1991) *Meteorus chilensis* Porter (Hymenoptera: Braconidae) enemigo natural de *Copitarsia turbata* (Herrich-Schaffer) (Lepidoptera: Noctuidae) en zonas de la Quebrada de Humahuaca, Jujuy. *Neotropica* 37, 137–144.

Olivares, T.S and Angulo, A.O. (1995) El organo timpanico en la clasificacion de Lepidoptera: Noctuidae. *Boletin Entomologia Venezuela* 11, 155–183.

Parra, L.E., Angulo, A.O. and Jana-Sáenz, C. (1986) Lepidoptera of agricultural importance: A practical key to its identification in Chile (Lepidoptera: Noctuidae). *Guyana Zoology* 50, 81–116.

Paz, H., Rodríguez, M., González, D., Galarza, C. and Torrado-León, E. (2007) Control de *Copitarsia decolora* en cultivos de flores mediante la emisión de frecuencias. #27 revista de ingeniería. Universidad de los Andes. Bogotá, Colombia, pp. 17–26.

Perlta-S., T. (1987) Plagas del maiz y su control en el Valle del Mantaro, Peru. *Revista Peruana de Entomologia* 28, 53–54.

Pogue, M.G. and Simmons, R.B. (2008) A new pest species of *Copitarsia* (Lepidoptera: Noctuidae) from the Neotropical Region feeding on asparagus and cut flowers. *Annals of the Entomological Society of America* 101, 743–762.

Poole, R.W. (1989) *Lepidopterum Catalogus, Fascilce 118: Noctuidae, Part 1*. E.J. Brill, Leiden, The Netherlands.

Porter, C.C. (1980) Joppini (Hymenoptera: Ichneumonidae) of Tarapaca. *Florida Entomologist* 6, 226–243.

Quiroga, P., Arretz, V. and Araya, J.E. (1989) Chewing insects on jojoba, *Simmondsia chinensis* (Link) Schneider, in the north-central and central regions of Chile, and characterization of damage. *FAO Plant Protection Bulletin* 37, 121–125.

Quiroz-E., C. (1977) Plagas de la cebolla. *Investigacion y Progreso Agricola* 9, 43–47.

Rasmussen, C., Lagnaoui A., and Esbjerg P., (2003) Advances in the knowledge of quinoa pests. *Food Reviews International* 19, 61–75.

Riley, D.R. (1998) Identification key for *Copitarsia*, *Spodoptera exigua*, and *Peridorma saucia* (unpublished).

Rojas, J.C. and Cibrian-Tovar, J. (1994) Reproductive behavior of *Copitarsia consueta* (Walker) (Lepidoptera: Noctuuidae): Mating frequency, effect of age of mating, and influence of delayed mating on fecundity and egg fertility. *Pan-Pacific Entomologist* 70, 276–283.

Rojas, J.C., Cibrian-Tovar, J., Valdez-Carrazco, J. and Nieto-Hernandez, R. (1993) Analisis de la conducta de cortejo de *Copitarsia consueta* y aislamiento de la feromona. *Agrociencia Proteccion Vegetal* 4, 23–39.

Rojas, J.C., Cruz-Lopez, L., Malo, E.A., Díaz-Gómez, O., Calyecac, G. and Cibrian-Tovar, J. (2006) Identification of the sex pheromone of *Copitarsia decolora* (Lepidoptera: Noctuidae). *Journal of Economic Entomology* 99, 797–802.

Sanchez-V., G.A. and Aldana-M., R. (1985) Algunas plagas de la papa en el Valle Mantaro, durante 1982–1983. *Revista Peruana de Entomologia* 28, 49–52.

Sanchez-V., G.A. and Maita-Franco, F. (1987) *Copitarsia turbata* (Lep.: Noctuidae) en papa del Valle Mantaro durante 1983–1984. *Revista Peruana de Entomologia* 30, 111–112.

Sanchez Valasquez and Apaza Tapia, W. (2000) *Plagas y Enfermedades del Esparrago en el Peru*. Graffiti Communicación Integral S.A.C. pp. 66–72.

SENASA (2001) Control biologico de plagas del esparrago. Hoja Divulgativa 8, Ministerio de Agricultura, Servicio Nacional de Sanidad Agraria, Programa Nacional de Control Biológico.

Simmons, R.B. and Pogue, M.G. (2004) Redescription of two often confused noctuid pests, *Copitarsia decolora* (Guenée) and *C. incommoda* (Walker) (Lepidoptera: Noctuidae). *Annals of the Entomological Society of America* 97, 1159–1164.

Simmons, R.B. and Scheffer, S.J. (2004) Evidence of cryptic species within the pest *Copitarsia decolora* (Guenée) (Lepidoptera: Noctuidae). *Annals of the Entomological Society of America* 97, 675–680.

Texas Agricultural Statistics Service (1999) County estimates: Texas Agricultural Statistics Service. www.nass.usda.gov/Statistics_by_State/Texas/index.asp, accessed 16 August 2012.

Urra, F. and Apablaza, J. (2005) Effects of cold treatment on mortality of *Dysaphis cynarae*

(Hemiptera: Aphididae) and *Copitarsia decolora* (Lepidoptera: Noctuidae) on fresh artichokes. *Ciencia e Investigacion Agraria* 32, 149–155.

USDA-NASS (1998) The 1997 Census of Agriculture. USDA, Washington, DC.

Velasquez-Z., L.D. (1988) Ciclo biologico de *Copitarsia turbata* (Lep.: Noctuidae) sobre cebolla, en Arequipa. *Revista Peruana de Entomologia* 30, 108–110.

Venette, R.C. and Gould, J.R. (2006) A pest risk assessment for *Copitarsia* spp., insects associated with importation of commodities into the United States. *Euphytica* 148, 165–183.

Vimos-N., C., Nieto-C., C., Rivera-M., M. (1998) El Melloco: Caracteristicas, tecnicas de cultivo y potencial en Ecuador. www.idrc.ca/library/document/096951/index_s.html, accessed June 2000.

Walker, F. (1865) *List of the Specimens of Lepidopterous Insects in the Collection of the British Museum, Vol. 11*. Edward Newman, London, pp. 493–764.

Wille-T., J.E. (1943) *Entomolgia Agricola del Peru: Manual para Entomologos, Ingenieros Agronomos, Agricultures y Estudiantes de Agricultura*. La Estacion Experimental Agricola de la Molina, Lima, pp. 334–348.

Zenner de Polania, I. (1990) Research and management strategies for potato insect pests in Colombia. In: Hahn, S.K. and Caveness, F.E. (eds) Integrated pest management for tropical root and tuber crops; workshop on the global status of and prospects for integrated pest management of root and tuber crops in the tropics, 25 October–30 October 1987, Ibadan, Nigeria. International Institute of Tropical Agriculture, Ibadan, Nigeria, pp. 139–148.

10 Host Range of the Nettle Caterpillar *Darna pallivitta* (Moore) (Lepidoptera: Limacodidae) in Hawai'i

Arnold H. Hara,[1] Christopher M. Kishimoto[2] and Ruth Y. Niino-DuPonte[1]
[1]*Department of Plant and Environmental Protection Sciences, Komohana Research and Extension Center, University of Hawai'i at Manoa, 875 Komohana Street, Hilo, Hawai'i 96720, USA;*
[2]*Honolulu, Hawai'i 96819, USA*

10.1 Introduction

The stinging nettle caterpillar, *Darna pallivitta* (Moore) (Lepidoptera: Limacodidae), was first discovered on the Island of Hawai'i in September 2001 (Pana'ewa 119° 39 min 13 s N / 155° 3 min 32 s W), and probably arrived from Taiwan on a shipment of rhapis palm seedlings (Conant *et al.*, 2002). The native range of *D. pallivitta* is China, Taiwan, Thailand, Java and Indonesia (Godfray *et al.*, 1987), where it is regarded as a minor pest mainly on palms and grasses, including maize. *D. pallivitta* quickly became established, and caused extensive feeding damage on numerous agricultural and nursery crops, and on landscape plants. Moreover, *D. pallivitta* was the first (and is currently the only) stinging caterpillar to appear in the state, a cause for major concern among Hawai'i's residents, visitors and agricultural workers. Larvae of *D. pallivitta* inflict a painful sting when their urticating spines release venom (a mixture of histamines) upon contact with the skin (Epstein, 1996) with reactions varying from burning and itching to blisters and persistent rash; serious complications may result from stings to eyes or allergic reactions affecting breathing. *D. pallivitta* is also an inter-island and interstate quarantine pest in Hawai'i. California commonly intercepts and rejects shipments of palms and other floral products from Hawai'i infested with *D. pallivitta* (Epstein and Kinee, 2003). As with other invasive insect species in Hawai'i, *D. pallivitta* will eventually spread to the other major Hawaiian Islands (Hawai'i Department of Agriculture, 2000).

Similar to other larvae in the family Limacodidae, *D. pallivitta* larvae are polyphagous (Holloway, 1986) and have been reported feeding on coffee (Holloway, 1986), bananas (Stephens, 1975), sugarcane (Pawar *et al.*, 1981), species of *Psidium*, *Eugenia*, *Imperata*, *Theobroma*, *Camellia* and *Citrus* (Holloway, 1986), coconut and oil palms, species of *Adenostemma*, *Areca*, *Breynia* and *Ficus*, and grasses, including maize (Godfray *et al.*, 1987). On preferred plant species, heavy infestations may totally defoliate leaves to the midrib (Igbinosa, 1985). Prior to this study, no data were available identifying plant species upon which *D. pallivitta* is able to complete its development. Confirming host plant species and preferred food sources of *D. pallivitta* will enable entomologists and plant quarantine regulators to monitor and regulate movement of these plants to prevent the spread of *D. pallivitta*.

10.2 Biology

The life history and biology of *D. pallivitta* in Hawai'i were well-documented by Nagamine and

Epstein (2007). Its life cycle duration from egg to egg-laying adult averages 80 days, ranging between 72 to 99 days. Eggs, approximately 1.0 × 1.6 mm, are laid singly, in a mass or in a single line, and are initially translucent to yellow (Fig. 10.1).

Larvae hatch in 7 days; the first instar does not feed and molts after two days. Total larval duration is 45–72 days with 8–11 instars (Fig. 10.2). Up to the 4th or 5th instar, larvae feed on one layer of leaf mesophyll in tracks parallel to the midrib, creating a 'window pane' effect. Later instars consume virtually all leaf tissue with only the midrib of the leaf remaining. Pupation occurs 5 days after formation of the cocoon, with adult moth emerging in approximately 19 days (Fig. 10.3). Female moths have filiform antennae and are larger, with average forewing length of 11.9 mm as compared to 10.5 mm for males who have bipectinate antennae. For both sexes, forewings are rust-colored and bisected with a light-colored diagonal line; hind wings and ventral surface are light brown. As observed with other limacodids, *D. pallivitta* moths often rest with their wings held below their bodies and may hang horizontally while grasping a vertical stem with their hind legs. Mass spectral analysis and subsequent synthesis identified the female-produced

Fig. 10.1 Life stages of *Darna pallivitta*. (A) newly deposited translucent eggs (1.1 mm wide x 1.6 mm long); (B) newly hatched larvae; (C) 6th instar with dorsal coloring (W 10.0 x L 15.0 mm) (photo credit: Hawai'i Department of Agriculture); (D) various instars feeding on ti (*Cordyline terminalis*); (E) prepupa; (F) pupa; G) adult moth (length 12.7 mm).

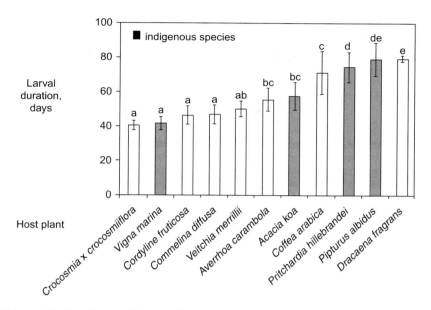

Fig. 10.2 Larval duration (days) of *Darna pallivitta* on certain host plants.

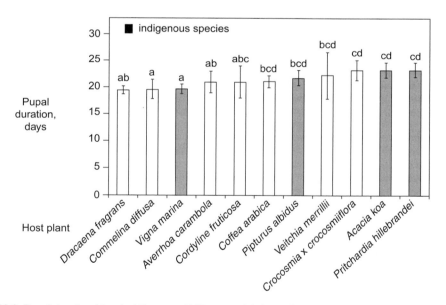

Fig. 10.3 Pupal duration (days) of *Darna pallivitta* on certain host plants. Mean length of pupation to adult emergence ranged from 19.46 ± 1.54 days to 23.36 ± 6.64 days.

sex pheromone as n-butyl (*E*)-7,9-decadienoate (major component) and ethyl (*E*)-7,9-decadienoate (minor component), both structurally similar to sex pheromone components previously reported from related *Darna* spp. (Siderhurst et al., 2007). Mating begins 1 day after emergence, with females ovipositing over approximately a 6-day period. A female can lay 479 eggs in its lifetime, with a hatching rate of 55%. Male and female longevity is 9.7 and 11.0 days, respectively.

10.3 Control Strategies

Short-term control strategy for *D. pallivitta* included chemical and biological insecticides. While *Bacillus thuringiensis* (Bt) was effective against *D. pallivitta* larvae, which stopped feeding after ingestion of Bt, they remained moribund for up to 14 days prior to death. During that time, however, the larvae remained capable of passive stinging upon contact. Another biological insecticide, spinosad, and the carbamate insecticide, carbaryl, were also not immediately lethal to *D. pallivitta* larvae, which remained moribund for 14 days. The pyrethroid, cyfluthrin, and the organophophate, chlopyrfos, killed *D. pallivitta* larvae within 3 days of treatment.

Classical biological control, a long-term control strategy, was initiated for *D. pallivitta* by the Hawai'i Department of Agriculture (HDOA) in 2004. An effective natural enemy, *Aroplectrus dimerus* Lin (Hymenoptera: Eulophidae) was discovered in Taiwan and imported to Hawai'i. Host-specificity studies conducted at the HDOA Insect Containment Facility determined *A. dimerus* to attack only *D. pallivitta* and none of the other 25 species of non-target Lepidoptera tested. HDOA conducted an environmental assessment and obtained the necessary Federal and State permits for field releases in 2010 (Hawai'i Department of Agriculture, 2007). Surveys in September 2010 on the Island of Hawai'i revealed high rates of parasitism of *D. pallivitta* larvae infesting *Dracaena* spp. nursery stock field plantings, as indicated by swarms of *A. dimerus* (K. Onuma and A. Hara, unpublished data).

Darna pallivitta is considered an intrastate, interstate and international quarantine pest because of its current limited distribution, wide host range and serious impact on human health. An effective quarantine treatment is needed preceding movement of preferred food and host plant species. Heat treatments have been documented as an effective quarantine disinfestation treatment for tropical flowers and foliage (Hara, 1994; Arcinas *et al.*, 2004; Tsang *et al.*, 1995). In vitro trials conducted in Hawai'i demonstrated that 97.8% mortality of late instar *D. pallivitta* larvae was achieved when infested host plants were subjected to hot water delivered as dip or shower at 47°C for 12 min; pupae required 49°C for 10 min to achieve 94% mortality (C. Kishimoto and A. Hara, unpublished data). Most palms that are hosts of *D. pallivitta* are able to tolerate heat treatments between 47 and 49°C (A. Hara, unpublished data).

10.4 Field Observations

Field observations of plants with *D. pallivitta* feeding activity were conducted from September 2001 through April 2006 by random sampling of vegetation throughout the Island of Hawai'i. *D. pallivitta* larvae were observed feeding on 57 species of plants representing 54 genera in 26 families, with monocots (57.9%) favored over dicots (42.1%) (Table 10.1). The most represented families were Arecaceae (9 spp., monocots), Fabaceae (7 spp., dicots) and Poaceae (6 spp., monocots). Five species were native to Hawai'i (*Pritchardia hillebrandei*, *Vigna marina*, *Acacia koa*, *Alyxia oliviformis*, and *Pipturus albidus*).

10.5 No-choice Host Range Tests

No-choice host range tests were conducted in the laboratory to identify plant species that could support the larval development of *D. pallivitta* to adulthood. Twenty-three plant species representing 15 families (Table 10.2) were selected based on their economic value as export potted plants to the mainland USA and their prevalence in Hawai'i's landscape (including weed species and ten native species). Leaf bouquets consisting of plant terminals were placed in glass vials filled with tap water. Only mature leaves were used, since field surveys indicated a preference for mature foliage, an observation supported by Godfray *et al.* (1987). Five first-instar larvae were transferred onto each leaf bouquet, placed in a plastic container and covered with a lid modified with a 9-cm diameter Lumite®-screened circular window (52 × 52 mesh, SI Corporation, Gainesville, Georgia). As larvae matured, the leaf bouquet size was tripled to accommodate their higher level of consumption. Leaf bouquets were replaced after being totally consumed, or when leaves began to desiccate (2–3 days). There were five replicates per plant species. Number of days until pupation and moth emergence were recorded. Emerging adult males and females on the same host plant were transferred to fresh leaf

Table 10.1. Recorded feeding list of *Darna pallivitta* (alphabetical by family) in Hawai'i.

Common name	Scientific name	Family	Monocot / Dicot
Ti	*Cordyline* Fruticosa (L.) A. Chev	Agavaceae	Monocot
Dracaena 'Lisa', 'Compacta'	*Dracaena deremensis* Engl.	Agavaceae	Monocot
Dracaena Massangeana	*Dracaena* Fragans (L.) Ker Grwl	Agavaceae	Monocot
Maile*	*Alyxia oliviformis* Gaudich.	Apocynaceae	Dicot
Chinese star jasmine	*Trachelospermum jasminoides* Lem.	Apocynaceae	Dicot
Monstera	*Monstera deliciosa* Liebm.	Araceae	Monocot
Fish tail palm	*Caryota mitis* Lour.	Arecaceae	Monocot
Cat palm	*Chamaedorea cataractarum* Mart.	Arecaceae	Monocot
Coconut palm	*Cocos nucifera* L.	Arecaceae	Monocot
Areca palm	*Dypsis lutescens* (H. Wendl.) Beentje & Dransf.	Arecaceae	Monocot
Pony tail palm	*Nolina recurvata* Hemsl.	Arecaceae	Monocot
Phoenix palm	*Phoenix roebelenii* O'Brien	Arecaceae	Monocot
Loulu palm*	*Pritchardia hillebrandii* (Kuntze) Becc.	Arecaceae	Monocot
Rhapis palm	*Rhapis excelsa* Henry ex Rehder	Arecaceae	Monocot
Manila palm	*Veitchia merrillii* (Becc.) H.E. Moore	Arecaceae	Monocot
Wedelia	*Wedelia trilobata* (L.) Hitchc.	Asteraceae	Dicot
Pineapple	*Ananas sativus* Schult. f.	Bromeliaceae	Monocot
Pink quill	*Tillandsia cyanea* E. Morren	Bromeliaceae	Monocot
Chick weed	*Drymaria cordata* Willd. ex Schult.	Caryophyllaceae	Dicot
Honohono grass	*Commelina diffusa* Burm. f.	Commelinaceae	Monocot
Shampoo ginger	*Zingiber zerumbet* (L.) Sm.	Costaceae	Monocot
Razor grass (Hawai'i nutrush)	*Scleria testacea* Nees	Cyperaceae	Monocot
Rhododendron	*Rhododendron* sp.	Ericaceae	Dicot
Poinsettia	*Euphorbia pulcherrima* Willd. ex Klotzsch	Euphorbiaceae	Dicot
Koa*	*Acacia koa* A. Gray	Fabaceae	Dicot
Perennial peanut	*Arachis pintoi* Krapov. and W.C. Greg.	Fabaceae	Dicot
Maunaloa vine	*Canavalia cathartica* Thouars	Fabaceae	Dicot
Silverleaf desmodium	*Desmodium uncinatum* E. Mey.	Fabaceae	Dicot
Wiliwili	*Erythrina sandwicensis* O. Deg.	Fabaceae	Dicot
Sleeping grass	*Mimosa pudica* L.	Fabaceae	Dicot
Vigna (beach pea)*	*Vigna marina* (Burm. f.) Merr.	Fabaceae	Dicot
Whale back	*Curculigo capitulata* (Lour.) Kuntze	Hypoxidaceae	Monocot
Iris	*Crocosmia (Tritonia)* × *crocosmiiflora* (V. Lemoine) G. Nicholson	Iridaceae	Monocot
Gladiolus	*Gladiolus* sp.	Iridaceae	Monocot
Walking iris	*Neomarica gracilis* Sprague	Iridaceae	Monocot
Lillyturf	*Liriope muscari* (Decne.) L.H. Bailey	Liliaceae	Monocot
Mondo, monkey grass	*Ophiopogon japonicus* (L. f.) Ker Gawl.	Liliaceae	Monocot
Koster's curse	*Clidemia hirta* (L.) D. Don	Melastomataceae	Dicot
Glory bush	*Tibouchina urvilleana* (DC) Cogn.	Melastomataceae	Dicot
Banana	*Musa* sp.	Musaceae	Monocot

Continued

Table 10.1. Continued.

Common name	Scientific name	Family	Monocot / Dicot
Strawberry guava (waiawi)	*Psidium cattleianum* Sabine	Myrtaceae	Dicot
Common guava	*Psidium guajava* L.	Myrtaceae	Dicot
Bamboo orchid	*Arundina graminifolia* (D. Don) Hochr.	Orchidaceae	Monocot
Starfruit	*Averrhoa carambola* L.	Oxalidaceae	Dicot
California grass	*Brachiaria mutica* (Forssk.) Stapf	Poaceae	Monocot
Torpedo (Wainaku) grass	*Panicum repens* L.	Poaceae	Monocot
Hilo grass	*Paspalum conjugatum* P.J. Bergius	Poaceae	Monocot
Vasey's grass	*Paspalum urvillei* Steud.	Poaceae	Monocot
Napier grass	*Pennisetum purpureum* Schumach.	Poaceae	Monocot
Sorghum	*Sorghum bicolor* (L.) Moench	Poaceae	Monocot
Rabbit foot fern	*Phlebodium aureum* (L.) J. Sm.	Polypodiaceae	Monocot
Strawberry	*Fragaria* sp.	Rosaceae	Dicot
Rose	*Rosa* sp.	Rosaceae	Dicot
Coffee	*Coffea arabica* L.	Rubiaceae	Dicot
Gardenia	*Gardenia jasminoides* Ellis	Rubiaceae	Dicot
Mamaki*	*Pipturus albidus* (Hook. Arn.) A. Gray	Urticaceae	Dicot
Red ginger	*Alpinia purpurata* (Vieill.) K. Schum.	Zingiberaceae	Monocot

*native to Hawai'i

bouquets to mate and oviposit, to determine emerging adult fertility.

Of the 23 plant species tested, *D. pallivitta* was able to successfully complete larval development and pupation on 11 species from 11 genera in 8 families. Seven of the 11 host plants (63.6%) were monocots. Four of the 11 confirmed host plants are native to Hawai'i, the monocots *Vigna marina* and *Pritchardia hillebrandei* and dicots *Acacia koa* and *Pipturus albidus*. Of the 12 plants upon which *D. pallivitta* did not complete its life cycle, the mean days of survival per plant species ranged from 5.6 days to 28.7 days. In the case of *Dianella sandwicensis*, *Psidium guajava*, *Clidemia hirta* and *Murraya paniculata*, two to five larvae were able to survive for up to 2–3 months, but none of them pupated successfully.

Darna pallivitta appears better adapted to mature on monocots than on dicots. Mean duration to pupation on monocots was shorter than on dicots. Mean time to pupation varied from 40.6 ± 5.4 days to 79.5 ± 3.5 days (Fig. 10.2) and was dependent on the plant species ($F_{10;41} = 42.84$, $P < 0.0001$).

Emergence times also depended on the plant species on which the larvae were reared ($F_{10;\ 41} = 5.54$; $P < 0.0001$). The mean rate of successful adult emergence for all host plants varied from 8% to 96%. Emerged *D. pallivitta* adults were able to mate and produce offspring.

10.6 Discussion

No-choice test results were not always indicative of preferred hosts in the field. During no-choice host range tests, starfruit (*Averrhoa carambola*) was found to be a good host plant (68% adult emergence) but observations of starfruit trees in the field never recorded a single sighting of *D. pallivitta*. Conversely, *D. massangeana* demonstrated to be a suboptimal host plant in the laboratory; however, numerous field observations recorded a number of middle- and late-instar *D. pallivitta* larvae feeding on *D. massangeana*, indicating that species to be a better host plant than laboratory results would suggest.

As observed in Hawai'i, *D. pallivitta* is reported to be highly polyphagous in its native range in East Asia (Holloway, 1986; Godfray *et al.*, 1987). In this study, *D. pallivitta* was recorded

Table 10.2. Host range test plant species for *Darna pallivitta*.

Common name	Scientific name	Family	Monocot/Dicot	Indigenous/Endemic
Ti	*Cordyline Fruticosa* (L.) Chev	Agavaceae	Monocot	
Dracaena	*Dracaena fragrans* (L.) Ker Gawl.	Agavaceae	Monocot	
Monster	*Monstera deliciosa* Liebm.	Araceae	Monocot	
Peace lily	*Spathiphyllum* sp. 'Clevelandii'	Araceae	Monocot	
Loulu palm	*Pritchardia hillebrandii* (Kuntze) Becc.	Arecaceae	Monocot	Endemic
Loulu hiwa palm	*Pritchardia martii* (Gaudich.) H. Wendl.	Arecaceae	Monocot	Endemic
Manila palm	*Veitchia merrilii* Becc.) H.E. Moore	Arecaceae	Monocot	
Pink quill	*Tillandsia cyanea* E. Morren	Bromeliaceae	Monocot	
Honohono grass	*Commelina diffusa* Burm. f.	Commelinaceae	Monocot	
Tree fern (hapu'u)	*Cibotium chamissoi* Kaulf.	Dicksoniaceae	fern	Endemic
Koa	*Acacia koa* A. Gray	Fabaceae	Dicot	Endemic
Vigna (beach pea)	*Vigna marina* (Burm. f.) Merr.	Fabaceae	Dicot	Indigenous
Iris	*Crocosmia* (*Tritonia*) × *crocosmiiflora* (V. Lemoine) G. Nicholson	Iridaceae	Monocot	
'Uki	*Dianella sandwicensis* Hook. and Arn.	Liliaceae	Monocot	Indigenous
Koster's curse	*Clidemia hirta* (L.) D. Don	Melastomataceae	Dicot	
'Ohi'a lehua	*Metrosideros polymorpha* Gaudich.	Myrtaceae	Dicot	Endemic
Common guava	*Psidium guajava* L.	Myrtaceae	Dicot	
Starfruit	*Averrhoa carambola* L.	Oxalidaceae	Dicot	
'Ie'ie	*Freycinetia arborea* Gaudich.	Pandanaceae	Monocot	Indigenous
Pandanus (hala)	*Pandanus tectorius* Parkinson ex Zucc.	Pandanaceae	Monocot	Indigenous
Coffee	*Coffea arabica* L.	Rubiaceae	Dicot	
Mock orange	*Murraya paniculata* L.	Rutaceae	Dicot	
Mamaki	*Pipturus albidus* (Hook. and Arn.) A. Gray	Urticaceae	Dicot	Endemic

feeding on 57 plant species in 26 families, preferring monocots, which laboratory studies confirmed. Many other species of limacodids are listed as pests of monocotyledonous crops (Stephens, 1975; Igbinosa, 1985; Godfray et al., 1987). *D. pallivitta* was observed feeding on five native species (*P. hillebrandei*, *V. marina*, *A. koa*, *A. oliviformis* and *P. albidus*) in the field, which (with the exception of *A. oliviformis*) were also confirmed hosts. *D. pallivitta*, therefore, will not only impact agricultural and nursery crops and exotic landscapes, but also affect Hawai'i's native forests.

Classical biological control in Hawai'i by a eulophid wasp, *A. dimerus*, will lower *D. pallivitta* populations and thereby reduce their negative impact on crops, landscapes and native forests. In Hawai'i, use of natural enemies has controlled more than 200 invasive pests over the past 100 years. No purposely introduced natural enemy species approved for release in the past 21 years has attacked any native or other desirable species (Funasaki et al., 1988).

Acknowledgements

We thank Stacey Chun, Patrick Conant, Christopher Jacobsen, Clyde Hirayama and Kyle Onuma for field assistance, and Walter Nagamine, Ron Heu and Brian Bushe for use of their photographs. This study was funded in part by the United States Department of Agriculture under CSREES Special Grant, managed by the Tropical/Subtropical Agricultural Research (T-STAR) Pacific Basin Administrative Group (PBAG) Project No. HAW00920-1017S. Finally, we thank Jorge E. Peña for inviting us to participate in the 'Potential Invasive Pest Workshop' in Coconut Grove, Florida, 10–14 October 2010.

References

Arcinas, A.C., Sipes, B.S, Hara, A.H. and Tsang, M.M.C. (2004) Hot water drench treatments for the control of *Radopholus similis* in rhapis and fishtail palms. *HortScience* 39, 578–579.

Conant, P., Hara, A.H., Nakahara, L.M. and Heu, R.A. (2002) Nettle Caterpillar *Darna pallivitta* Moore (Lepidoptera: Limacodidae). New Pest Advisory No. 01–03 (revised 2008). State of Hawai'i Department of Agriculture, Honolulu, Hawai'i.

Epstein, M.E. (1996) *Revision and Phylogeny of the Limacodid-Group Families, with Evolutionary Studies on Slug Caterpillars (Lepidoptera: Zygaenoidea)*. Smithsonian Contributions to Zoology 582, Smithsonian Institution Press, Washington, DC.

Epstein, M.E. and Kinnee, S.E. (2003) On the lookout for *Darna pallivitta* Moore: a recently established moth in Hawai'i. In: Kodira, U.C. (ed.) *2003 Plant Pest Diagnostics Laboratory Report*. Plant Pest Diagnostics Branch, California Department of Food and Agriculture, Sacramento, California, pp. 50–53.

Funasaki, G., Lai, P.Y., Nakahara, L.M., Beardsley, J.W. and Ota, A.K. (1988) A review of biological control introductions in Hawai'i: 1890 to 1985. *Proceedings of the Hawaiian Entomological Society* 28, 105–160.

Godfray, H.C.J., Cock, M.J.W. and Holloway, J.D. (1987) An Introduction to the Limacodidae and their biometrics. In: Cock, M.J.W., Godfray, H.C.J. and Holloway, J.D. (eds) *Slug and Nettle Caterpillars: The Biology, Taxonomy and Control of the Limacodidae of Economic Importance on Palms in South-East Asia*. CAB International, Wallingford, UK, pp. 1–8.

Hara, A.H. (1994) Ornamental and flowers. In: Paull, R.E and Armstrong, J.W. (eds) *Insect Pests and Fresh Horticultural Products: Treatments and Responses*. CAB International, Wallingford, UK, pp. 329–347.

Hawai'i Department of Agriculture (2000) *Distribution and Host Records of Agricultural Pests and Other Organisms in Hawai'i*. Division of Plant Industry, Plant Pest Control Branch, Survey Program, Honolulu, Hawai'i.

Hawai'i Department of Agriculture (2007) Field release of *Aroplectrus dimerus* Lin (Hymenoptera: Eulophidae) for biological control of the nettle caterpillar, *Darna pallivitta* (Moore) (Lepidoptera: Limacodidae), in Hawai'i. Draft Environmental Assessment. Honolulu, Hawai'i. Available at: http://gen.doh.hawaii.gov/Shared%20Documents/EA_and_EIS_Online_Library/Statewide/2000s/2008-04-23-DEA-Nettle-Caterpillar-Biocontrol-and-Agent-Host-Specificity-Report.pdf, accessed 24 January 2011.

Holloway, J.D. (1986) The moths of Borneo: Part 1 Key to families Cossidae, Metarbellidae, Ratardidae, Dudgeoneidae, Epipyropidae, and Limacodidae. *Malayan Nature Journal* 40, 1–165.

Igbinosa, I.B. (1985) On the ecology of the nettle caterpillar, *Latoia* (= *Parasa*) *viridissima*

Holland (Limacodidae: Lepidoptera). *Insect Science and its Application* 6, 605–608.

Nagamine, W.T. and Epstein, M.E.(2007) Chronicles of *Darna pallivitta* (Moore 1877) (Lepidoptera: Limacodidae): biology and larval morphology of a new pest in Hawai'i. *Pan-Pacific Entomologist* 83, 120–135.

Pawar, A.D., Mishra, M.P. and Tripathi, R.S. (1981) Records of parasitism by *Eupteromaus parnarae* Gahan (Hymenoptera: Peromalidae) on slug caterpillar, *Laotta bicolor* Walker (Lepidoptera:Limacodidae) from Karuchhapra, Lakshmiganj, Uttar Pradesh, India. *Journal of Advanced Zoology* 2, 65.

Siderhurst, M.S., Jang, E.B., Hara, A.H. and Conant, P. (2007) N-butyl (e)-7,9-decadienoate: sex pheromone component of the nettle moth, *Darna pallivitta*. *Entomologia Experimentalis et Applicata* 125, 63–69.

Stephens, C.S. (1975) Natural control of limacodids on bananas in Panama. *Tropical Agriculture (Trinidad)* 52, 167–172.

Tsang, M.C.C., Hara, A.H., Hata, T.Y., Hu, B.K.S., Kaneko, R.T. and Tenbrink, V.L. (1995) Hot-water immersion unit for disinfestation of tropical floral commodities. *Applied Engineering in Agriculture* 11, 397–402.

11 Fruit Flies *Anastrepha ludens* (Loew), *A. obliqua* (Macquart) and *A. grandis* (Macquart) (Diptera: Tephritidae): Three Pestiferous Tropical Fruit Flies That Could Potentially Expand Their Range to Temperate Areas

Andrea Birke,[1] Larissa Guillén,[1] David Midgarden[2] and Martin Aluja[1]

[1]*Red de Manejo Biorracional de Plagas y Vectores, Instituto de Ecología A.C., Xalapa, Veracruz, México;* [2]*USDA APHIS Medfly Program, Guatemala City, Guatemala*

11.1 Introduction

The family Tephritidae (Diptera) comprises over 4000 species of which c. 250 belong to the genus *Anastrepha*. Of these, fewer than ten species are considered to be economically important pests. In this review, we have concentrated on three pestiferous *Anastrepha* species considered highly polyphagous and identified as potential exotic invaders: *Anastrepha ludens* (Loew), *Anastrepha obliqua* (Macquart) and *Anastrepha grandis* (Macquart). *Anastrepha ludens*, known as the Mexican fruit fly, is an important pest of citrus that poses a considerable threat to production areas in the southern USA, particularly California, Florida and Texas. *Anastrepha obliqua*, the West Indian, mango or Antillean fruit fly, causes significant economic damage to mango, tropical plums (*Spondias* spp.) and regionally to guava (*Psidium guajava* L.), and could potentially invade subtropical USA, (i.e., Florida and Hawai'i). *Anastrepha grandis*, known as the South American cucurbit fruit fly, is a subtropical pest of cucurbits and has the potential to damage production of many *Cucurbita* spp. if it were to become established in the southern USA.

We provide a comprehensive overview of the published information on these three species to aid further research efforts and to facilitate decision-making processes in the areas of management and regulatory entomology. We place emphasis on recent studies of behavior, especially those related to oviposition and host selection (and use), life history (including variation on the duration of larval development according to type of host), ecology and population dynamics. We also discuss traditional control methods such as field sanitation, chemical and biological control, SIT, more recent efforts on biorational management and use of new tools. Finally, we consider short- and long-term research needs.

11.2 *Anastrepha ludens* (Loew), Mexican Fruit Fly

11.2.1 Distribution and hosts

The natural range of *Anastrepha ludens* spans between Mexico and Costa Rica (Hernández-Ortíz and Aluja, 1993). Recently, it was declared eradicated from the USA, where in the past it was occasionally reported in the south of Texas (NAPPO, 2012).

Anastrepha ludens is considered to be native to north-eastern Mexico where *Casimiroa greggii* (S. Watson) (yellow chapote), its ancestral host, is abundant (Baker et al., 1944). This species is highly polyphagous, with more than 38 tropical and subtropical fruits and vegetables reported as hosts (Norrbom, 2004; Aluja et al., 1987; Aluja and Mangan, 2008). It must be noted, however, that many of these records are of non-natural hosts: that is, fruits that have never been reported unequivocally to be infested under non-manipulated field conditions, but for which there is reliable experimental evidence that they could provide adequate properties to be infested and produce reproductive adults under manipulated (artificial) laboratory conditions (Aluja and Mangan, 2008), or conditional hosts (Aluja and Mangan, 2008). Considering only natural hosts, 'a fruit taxon that has been unequivocally reported to be infested under totally natural field conditions' (Aluja and Mangan, 2008), *A. ludens* mainly exploits species of Rutaceae, Anacardiaceae and Rosaceae (Aluja et al., 2000a; Norrbom, 2004). Its host plants of greatest economic importance are citrus (oranges and grapefruits) of the Rutaceae family, which comprise the largest number of *A. ludens* host plants (12 species), including yellow chapote. Mango (*Mangifera indica* L.), in the family Anacardiaceae, is another economically important host plant. Other economically important, albeit alternative hosts, are *Prunus persica* L. (peach) (McPhail and Bliss, 1933) and manzano peppers (*Capsicum pubescens* Ruis and Pavon cv. Rocoto) (Thomas, 2004).

11.2.2 Synonymy (from Norrbom, 2000)

Anastrepha fraterculus var. *mombinpraeoptans* Seín, 1933 (Seín, 1933)
Acrotoxa ludens Loew
Trypeta ludens (Loew 1873)
Anastrepha lathana Stone 1942

11.2.3 Morphology

Egg stage

Egg morphology is quite similar among different *Anastrepha* species: *A. ludens* eggs are white, spindle-shaped, broad anteriorly and tapering posteriorly; faint reticulation near the mycropyle consists primarily of irregular pentagons and hexagons. The length of eggs ranges between 1.37 and 1.60 mm and the width between 0.18 and 0.21 mm. More details can be obtained in Carroll and Wharton (1989).

Larvae

Larvae undergo three instars all of which are creamy white with a typical fruit fly larval shape (i.e., cylindrical maggot-shape, elongate, anterior end narrowed and somewhat recurved ventrally). They are composed of 11 body segments.

FIRST INSTAR. The length ranges from 0.68 to 2.42 mm, and the width 0.10–0.48 mm, the cuticle is transparent and the shape resembles a third instar although more blunt anteriorly. Anal lobes are usually simple. The posterior spiracle has two slit-like openings surrounded by oval rimae. Spiracular hairs are about twice as long as rima and are arranged in four groups with 2–4 short simple or bifid hairs per group.

SECOND INSTAR. Length 2.42–5.76 mm; width 0.48–1.24 mm; translucent white with a shape similar to that of a third instar.

THIRD INSTAR. Length 5.88–11.1 mm; width 1.2–2.5 mm, white, becoming ivory to yellow white with maturity; subcylindrical, tapering gradually to cephalic segment. The anterior spiracle is light golden-brown, usually with 16–18 tubules in a single row and reticulation of felt chamber. Anal lobes are usually bifid. The posterior spiracles are elongated (c. 1 × 5) and separated medially by c. three times the length of one spiracle. Spiracular hairs in four groups, axes of two dorsal groups at right angles, axes of two ventral groups nearly parallel; trunks 2–3 times as numerous in dorsal and ventral groups as in the lateral groups, each with 1–5 branches (Carroll and Wharton, 1989; Steck et al., 1990).

Puparium

The pupae are a cylindrical capsule with 11 segments. The length ranges from 6.5 to 7.5 mm. The anterior and posterior spiracles are as in larvae. Color can vary from reddish brown to brownish black.

Adult

The adult is slightly larger than a housefly, and is mostly yellowish-brown with a pale-yellow scutellum and a subscutellum with dark brown lateral spots occasionally extended to the mediotergite, but decreasing gradually in width. The wing band color is pale yellow for *A. ludens*; the S-band is complete and generally lightly connected with the C-band; there is a hyaline spot in the apex of R1; S and V bands are always unconnected and with a complete distal arm in the band V. Females have a notably long aculeus (3.35–5 mm) with a finely serrated apex (Hernández-Ortíz, 1992; White and Elson-Harris, 1992).

11.2.4 Life history

The life cycle of *A. ludens* is similar to that of other *Anastrepha* spp. Females lay their eggs in the epi- or mesocarp area of fruit. The developmental time for eggs is around 3 days (Celedonio et al., 1988), but other authors report a range from 6 to12 days (Christenson and Foote, 1960). After hatching, larvae feed on the fruit pulp, but they are also able to feed on seeds of fruits (e.g., *C. greggii*) (Aluja et al., 2000b). Larvae pass through three instars. Larval development time under laboratory conditions is variable (8–30 days) (Celedonio et al., 1988; Christenson and Foote, 1960) and can be influenced by environmental conditions and by the larval host (Darby and Kapp, 1933; Leyva, 1988; Leyva et al., 1991; Thomas, 1997). For example, development can last c. 27 days in preferred hosts, such as 'Ruby Red' and 'Marsh' grapefruit (Leyva et al., 1991) and >40 days in apples (M. Aluja, Veracruz, Mexico, pers. comm.). When larvae complete their development, they exit the fruit to pupate in the soil at different depths, depending on environmental conditions and soil characteristics (Hodgson et al., 1998). Depending upon temperature (18–26°C), pupal developmental time ranges from 12 to 32 days (Celedonio et al., 1988; McPhail and Bliss, 1933). Depending on environmental conditions, adults emerge in the largest numbers early in the morning (6–10 h) (McPhail and Bliss, 1933; Aluja et al., 2005). Under laboratory conditions, adults have a mean life expectancy of 17.3 days (Celedonio et al., 1988), but Darby and Kapp (1934) observed that *A. ludens* flies can live up to 11 months, and that males in general live longer than females. In one case they recorded a male that lived 14 months. The mean generation time in the laboratory is 46.9 days (Celedonio et al., 1988), allowing several pest generations per annum.

11.2.5 Behavior

Oviposition and host marking

Anastrepha ludens can lay several clutches, ranging from 6 to 30 eggs. Oviposition depth ranges from 3 to 3.5 mm (Birke et al., 2006). Females are not able to resorb oocytes and are forced to dump eggs when hosts are absent (Aluja et al., 2011a). Egg load can be influenced by adult nutrition (females fed with a mixture of sucrose and protein produce more eggs than those only fed sucrose), as well as by female density, availability of host, age and semiochemical context (females exposed to fruit odors and sex pheromone produce more eggs than those exposed to fruit alone, pheromone alone or no stimuli) (Aluja et al., 2001a). The number of eggs laid in each host depends on different factors: (i) host quality: females lay larger clutches in larger and non-infested fruit (Díaz-Fleischer and Aluja, 2003a, b); (ii) host firmness: females lay more eggs and larger clutches in firm (unripe) than in soft (ripe) host fruit (Díaz-Fleischer and Aluja, 2003c); and (iii) oviposition experience and presence of conspecifics: naive females deposit larger egg clutches than experienced ones (Díaz-Fleischer and Aluja, 2003b). During its life time, *A. ludens* may lay as many as 1500 eggs (Christenson and Foote, 1960). Due to its long ovipositor, it can generally deposit its eggs within the pulp even when fruit skin is relatively thick. When fruits are small and seeds are large, (e.g., *C. greggii* fruit), females can also oviposit into seeds. Oviposition may take place 8 days after emergence, but peaks at day 15 (Celedonio et al., 1988).

Adults are most active between 14 h and 16 h (Malo and Zapien,1994; Aluja et al. 2000b), with the preferred time for oviposition between 11 h and 14 h (Birke, 1995). After laying eggs, females deposit an oviposition-deterring pheromone (host-marking pheromone = HMP) by dragging the aculeus around the oviposition area on the fruit (Papaj and Aluja, 1993; Aluja and Díaz-Fleischer, 2006).

11.2.6 Mating

Anastrepha ludens and *A. obliqua* have a mating strategy based on lekking, similar to that exhibited by other polyphagous species, such as *Anastrepha fraterculus* and *Anastrepha suspensa* (Loew) (Aluja *et al.*, 1983). Males aggregate on foliage to perform courtship and attract females to mate. During courtship, males emit a sex pheromone from pleural glands and evaginated anal membranes. The pheromone is deposited on leaves. Males emit wing-fanning acoustic signals (songs) which are produced both prior to and during coupling (Aluja *et al.*, 1983, 2000b; Aluja, 1994). Male calling occurs during the afternoon (Aluja, 1994). Under laboratory conditions, copulation occurs at dusk when adults are sexually mature (6–10 days) and is influenced by adult nutrition (Celedonio *et al.*, 1988; Aluja *et al.*, 2001b, 2009a). The mean duration of matings is 165 min (Aluja *et al.*, 2000b).

Under laboratory conditions random mating was observed among females and males stemming from distinct hosts and from geographically distant locations. Siblings engaged in significantly longer copulations than non-siblings, suggesting that adults discriminated in favour of mates with similar genetic compositions (Aluja *et al.*, 2009b).

11.2.7 Population dynamics

Fruit fly population size fluctuates strongly from year to year, and is correlated with host plant availability and climatic factors, especially rainfall (Aluja, 1994; Aluja *et al.*, 1996). Within a single season, sampling with McPhail traps in a tropical area in Chiapas, Mexico, revealed that *A. ludens* and *A. obliqua* were more abundant between April and July than during the rest of the year (Aluja *et al.*, 1990). However, these seasonal peaks can vary and are dependent on host phenology and abundance. For example, during a study carried out in different types of orchards over 4.6 years, *A. ludens* captures in a citrus orchard peaked in February, following the period of maximum fruit availability, while the peak in a guava orchard surrounded by coffee and citrus trees was in May (Celedonio *et al.*, 1995). In addition, Aluja *et al.* (2012) reported considerable variance in the scale of population fluctuations every year, perhaps due to local rainfall and temperature, and global climatic patterns such as El Niño Oscillations.

11.3 *Anastrepha obliqua* (Macquart), West Indian, Antillean or Mango Fruit Fly

11.3.1 Distribution and hosts

Anastrepha obliqua is widespread in the neotropics, where it occurs in Mexico, Central and South America and a number of Caribbean islands. It was detected in the Florida Keys, but an intensive insecticide and host removal campaign during 1932–1936 achieved eradication (Clark *et al.*, 1996).

In the case of *A. obliqua*, no reports of any particular ancestral host are found but considering the number of natural hosts attacked and the distribution of the Anacardiaceae in the neotropics, it is reasonable to think that it could be a member of this family. The preferred host plants are mainly *Spondias* spp. (jobos or hog plums), followed by mango (*Mangifera indica*) (Aluja *et al.*, 1987, 2000a; Aluja and Birke, 1993), rose apple (*Syzygium jambos* L. (Alston)), starfruit (*Averrhoa carambola* L.) (Oxalidaceae) (Bressan and da Costa Teles, 1991) and guava probably used as alternative hosts in Veracruz, Mexico (Birke and Aluja, 2011), the Caribbean Islands (Jenkins and Goenaga, 2008), and Central and South America (Jirón and Hedström, 1988; Soto-Manatiu and Jirón, 1989; Zucchi, 2000). There are >100 host plants reported for *A. obliqua* (references in Norrbom, 2004); nevertheless, many of these records are erroneous because they were based on misidentification, particularly due to the confusion with *A. fraterculus* or other species in this complex, and on unreliable identification of the host.

11.3.2 Synonymy (from Norrbom, 2000)

Anastrepha fraterculus var. *mombinpraeoptans* Seín, 1933
Anastrepha mombinpraeoptans Seín
Anastrepha acidusa authors (not Walker)
Anastrepha trinidadensis Greene, 1934
Anastrepha ethalea Greene (not Walker)

Anastrepha fraterculus var. *ligata* Costa Lima 1934
Acrotoxa obliqua (Macquart)
Tephritis obliqua Macquart
Trypeta obliqua (Macquart)

11.3.3 Identification

Egg stage

Anastrepha obliqua eggs are creamy white, 1.15 and 1.58 mm long and 0.19–0.25 mm wide (combined data of Norrbom, 1985; Murillo and Jirón, 1994), with a stout and moderately sculptured shape (Norrbom, 1985).

Larvae

Larvae are white and have a typical fruit fly larval shape (see *A. ludens*), with anterior mouth hooks, ventral fusiform areas and flattened caudal end.

THIRD INSTAR. Larvae range between 7.5 and 9.0 mm in length and 1.4–1.8 mm in width. The ventral portion has fusiform areas on segments 2–10, with an anterior buccal carinae and usually 9–10 segments. The anterior spiracles are asymmetrical in lateral view with a depressed centre, and with tubules averaging 12–14 in number (Steck *et al.*, 1990; White and Elson-Harris, 1992).

Adults

The adult body is yellow-brown. The mesonotum is 2.6–3.3 mm long, with a pale-yellow scutellum and a mediotergite with two dark brown lateral stripes which do not extend to the subscutellum. The abdominal tergites are of a single color. The V and S bands of the wings are always connected; the costal and S bands touch on vein R4+5. Females have a 1.3–1.7-mm long acculeus finely serrated in the apex. The ovipositor sheath is 1.6–1.9 mm long and teeth of the eversible membrane are triangular in shape (Hernández-Ortíz, 1992; Steck *et al.*, 1990; White and Elson-Harris, 1992).

11.3.4 Life history

The life cycle of *A. obliqua* is similar to that of other *Anastrepha* spp. (see Section 11.2.4 of this chapter). Under laboratory conditions, developmental time for *A. obliqua* eggs, larvae and pupae is similar to that of *A. ludens*. The mean generation time is 48.5 days (Celedonio *et al.*, 1988), but this is variable and can depend on factors such as host plant and environmental conditions. For example, Toledo and Lara (1996), comparing two populations of *A. obliqua* adults stemming from larvae developing in mango and *Spondias mombin* L., found a generation time for mango of 50.28 days and in *S. mombin* of 56.09 days. Additionally, these authors report better demographic parameters for individuals from *S. mombin*, such as greater life expectancy (103.76 days) in mombin than in mango (92.8 days). Adults emerge in the largest numbers early in the morning, depending on environmental conditions. Both sexes feed on juices oozing from over-ripe fruit and the surface of ripe and unripe fruit, and females also feed on bird feces deposited on the top of leaves (Aluja and Birke, 1993).

11.3.5 Behavior

Oviposition

Anastrepha obliqua lays a single egg in every oviposition event (Aluja *et al.*, 2000b, 2001a). Under field conditions, oviposition activity starts shortly after sunrise at 7 h, with most of the ovipositions occurring early in the morning or late in the afternoon (Aluja and Birke, 1993). The number of eggs laid can depend on host quality. Females lay fewer eggs in infested than in clean fruit (Díaz-Fleischer and Aluja, 2003a) and lay more eggs in hosts with higher nutrient content (Fontellas-Brandalha and Zucoloto, 2004). When fruits are abundant females produce more eggs (Díaz-Fleischer and Aluja, 2003a). The nutritional state of females also influences egg production (Cresoni-Pereira and Zucoloto, 2006). The age of first reproduction is c. 10 days and may take place as early as 8 days after emergence, but oviposition generally peaks at day 15 (Celedonio *et al.*, 1988). As is the case with *A. ludens*, *A. obliqua* females also mark fruit after laying an egg (Aluja *et al.* 2000b; Aluja and Díaz-Fleischer, 2006).

Mating

The sexual behavior of *A. obliqua* is similar to that of *A. ludens* (see Section 11.2.6 in this chapter).

Anastrepha obliqua males call in leks or individually during the day, with the highest peak occurring in the morning and a smaller one in the afternoon (Aluja and Birke, 1993). Males can copulate up to three times each day in field cages with an over-supply of virgin females (Aluja *et al.*, 2001b). The mean duration of matings varied between c. 203 min in Mexico to c. 60 min in Brazil (Aluja *et al.*, 2000b). Copula duration was shown to be influenced by male diet and larval host (Pérez-Staples *et al.*, 2008). In a study performed under laboratory conditions, c. 20% of females re-mated at least once (Aluja *et al.*, 2009a). Sperm production and frequent mating in *A. obliqua* appears not to be as costly as for other species and does not reduce adult longevity (Pérez-Staples *et al.*, 2006).

11.3.6 Population dynamics

Factors known to influence *A. obliqua* population fluctuations were mentioned in Section 11.2.7 of this chapter. In the Canyon of Mazapa de Madero, Chiapas, Mexico, a tropical semi-perturbed area, the peak of abundance of *A. obliqua* adults was in June, although populations were high from April to July (Aluja *et al.*, 1990). However, these peaks can vary with host phenology and abundance (Celedonio *et al.*, 1995) and with the influence of climate (Aluja *et al.*, 1996, 2011a). Other factors that influence *A. obliqua* population dynamics are related with pupal mortality caused by desiccation, disease, predation and parasitism (Bressan-Nascimento, 2001); or in the case of adults, predation and rain (Aluja *et al.*, 1990, 2005, 2011a).

11.4 *Anastrepha grandis* (Macquart), South American Cucurbit Fruit Fly

This fly is remarkable due to its large size (hence its specific name) and is phylogenetically quite different from the other pest species of *Anastrepha* (Morgante *et al.*, 1980). First described by Macquart in 1846, it was not considered a significant pest for many years, but its importance increased during the last half of the 20th century and it is now a serious pest of native and introduced fruits/vegetables in the family Cucurbitaceae (melons, squash and cucumbers) (Weems, 2006). *A. grandis* differs somewhat in its pest status in each country where it is present, and its status has sometimes changed within a country (Weems, 2006).

11.4.1 Distribution and hosts

A. grandis was first described from Nueva Granada, which at the time (1846) was a Spanish colony including most of Colombia and parts of Ecuador, Venezuela and Panama. Today, *A. grandis* (Macquart) is mainly found along the Andean Cordillera from northern Argentina to Venezuela, as well as in Paraguay and Brazil (Norrbom, 2004). It also seems to be less evenly distributed than other *Anastrepha* species, being found in small localized populations near hosts (Malavasi *et al.*, 1990). Regions included in its range are northern Argentina, Paraguay, Bolivia, Brazil (as far north as Bahia), Colombia, Ecuador, Peru and Venezuela (Norrbom, 2000; APHIS PPQ, 2002; Weems, 2006). This species was recently detected in Panama, where intensive surveys are currently under way (APHIS-PPQ, 2009; EPPO, 2009).

A. grandis attacks most cucurbit fruits, including native and introduced species, though it seems to exhibit geographic variation in host preference, as the hosts attacked vary by country. In Brazil and Venezuela, it is a key pest of pumpkin (Martinez *et al.*, 1994), while in Paraguay and Argentina, it is considered mainly a pest of small squash. Larvae have been detected in shipments to the USA in fruit from Argentina and Brazil, and once on banana debris from Panama (Weems, 2006).

Native hosts are *Cucurbita moschata* Duchesne ex Poir and *C. maxima* Duchesne, which include a large variety of edible squashes and pumpkins (butternut, naples squash, buttercup squash, banana squash). Hosts introduced to South America include *C. pepo* L. (acorn squash, zucchini) from North America, and *Citrullus lanatus* (Thunb.) Matsum. and Nakai (watermelon), *Cucumis melo* L. (melon), *C. sativus* L. (cucumber) and *Lagenaria siceraria* (Molina) Standl. (gourd) from the Old World (mainly Asia and Africa). The sporadic reports of non-cucurbit hosts in the last 50 years (guava, citrus and passion fruit) are not well documented and are considered doubtful (Norrbom, 2000, 2004).

11.4.2 Synonymy (from Norrbom, 2000)

Tephritis grandis Macquart 1846
Trypeta (Acrotoxa) grandis Loew 1873
Anastrepha grandis Bezzi 1909
Anastrepha schineri Hendel 1914. Synonymy (Fischer 1932)
Anastrepha latifasciata Hering 1935. Synonymy (Lima 1937)

11.4.3 Identification

Egg and larvae

Egg and larval morphology is similar to that of other *Anastrepha* species, with notable size differences.

Adults

Adults are remarkable due to their large size, with a mesonotum ranging from 3.3 to 4.0 mm, yellow-brown with humerus, median stripe widening to include acrostichal bristles but not reaching the scutellum; lateral stripes from just before the transverse suture to the side of the scutellum, stripe below notopleuron, metapleuron and scutellum except extreme base, yellow; a sublateral stripe from the level of the humeral bristle to the scutellum, broken at transverse suture, a band along scutoscutellar suture, intensified medially, and a spot on pteropleuron, dark brown; metanotum blackened laterally. Macrochaetae are dark brown; pile is yellowish brown. No sternopleural bristle. Wing 9–10.5 mm long, with the bands yellow brown, rather diffuse; costal and S bands broadly connected, and no distinct hyaline spot anterior to veins R4 + 5; distal arm of V band absent, the proximal arm not joining S band. Female terminalia: the ovipositor sheath is 5.8–6.2 mm long, tapering posterior to apical third, which is distinctly depressed and broadened; in profile the sheath is distinctly concave dorsally on median half and concave ventrally on apical third. The rasper is well developed, of slender, curved hooks in five or six rows. Ovipositor slightly longer than the length of ovipositor sheath, being somewhat curved dorsoventrally to permit fitting sheath; tip long and slender, without serrations; extreme base slightly widened (Stone, 1942).

11.4.4 Life history

Nascimento *et al.* (1988) and Silva and Malavasi (1993a, 1996) studied various biological parameters of this species involving oviposition, mating and duration of life stages. There are three larval instars, all of which feed inside the fruit. When mature, the third instars tunnel out of the fruit to pupate in the soil. Development time varies with temperature; however, Silva and Malavasi (1996) report periods of 3–7 days for the egg, 13–28 days (mean 17.7 days) for larval development and 14–23 days (mean 19.7 days) for pupation.

11.4.5 Behavior

Oviposition

Oviposition under laboratory conditions peaks from 11h to 14 h (lights on at 6 h). Females lay eggs in clutches of up to 110, and never lay fewer than 10 eggs per clutch. As in many other species of *Anastrepha*, after laying eggs, the female marks their surface with a pheromone that deters oviposition (HMP) by other females. Oviposition takes an average of 45.8 ± 27.9 min (Silva and Malavasi, 1993a).

Mating

Mating behavior under laboratory conditions peaked at 17–20 h (using a dark room and red lights). As in many other *Anastrepha* species, leks of 4–10 males are formed, and males release pheromone by anal surface touching. Copulas continue to occur outside leks and mean duration is 4 h 22 min ± 2 h 17 min (Silva and Malavasi, 1993a, b).

11.4.6 Population dynamics

In general, *Anastrepha* are good fliers able to disperse many kilometers; and although there are no published studies, the large size and relatively narrow wing aspect ratio of *A. grandis* suggests that they are strong long-distance fliers (Sivinski and Dodson, 1992).

11.5 Sampling and Monitoring Methods

Monitoring of fruit flies is essential to detect outbreaks, estimate population size and define type and intensity of control measures. Recent studies on the response of *A. ludens* and *A. obliqua* to traps show that it is important to consider both climatological and biological data (landscape structure), because fruit fly population dynamics are directly influenced by temperature, relative humidity and changes in vegetation structure (Aluja *et al.*, 2011a). Understanding fruit fly population dynamics may aid in predicting when outbreaks are more likely to occur in order to intensify deployment of detection tools (traps and fruit sampling). Phenological data, wild host distribution, aerial photography and geographic coordinates also comprise important sources of relevant information such as deforestation rates and removal of preferred cultivars. A detailed record of such factors over the years is essential to correctly interpret population fluctuations (Aluja *et al.*, 2011a).

11.5.1 Fruit sampling and damage evaluation

Cutting fruit to search for larvae is a primary detection method for *Anastrepha* species. In the case of *A. ludens* and *A. obliqua*, larval damage can be detected in citrus fruits and mangoes as well as in other wild host species prior to the opening of fruit, by searching for soft brown areas and respiration and larval exit holes (A. Birke, Veracruz, Mexico, pers. comm.). Recognition of *A. grandis*-damaged fruit, particularly squash, is detected when oviposition punctures appear as slightly transparent circles, about 1 mm in diameter, with a black centre. According to the APHIS Global Pest and Disease Database (Pest ID 334 *Anastrepha grandis*) the species may be detected by: (i) locating early signs of infestation by examining stem or flower petal attachment sites on fruit for damage; (ii) checking fruit detached from or still on the vine for oviposition punctures; (iii) checking fruit on the ground for exit holes and larval tunnels in the pulp; and (iv) looking for larvae in the pulp in fruit still on the vines, because infested fruit can appear undamaged externally (Weems, 2006).

Larval identification is based primarily on characters of mature 3rd instar larvae, and is extremely difficult. When possible it is best to rear immature stages to adulthood. Sample size for fruit cutting is determined on the basis of the level of risk and the probability of presence of larvae (Aluja, 1994; Mangan *et al.*, 1997; Follett and Neven, 2006). But in most cases, the main purpose of fruit infestation monitoring is to establish presence/absence in the field, infestation levels and distribution of infestation within orchards (Aluja, 1994).

A. ludens and *A. obliqua* sampling can be targeted, and experience is required to select the preferred host trees which may be likely to be infested. In all cases, sampling should include wild species (Aluja, 1993, 1994; Aluja *et al.*, 2000a; Norrbom, 2004). Shipment of fruit such as citrus, mangoes, chili and guava require a Probit 9 level of risk (33,000 fruit with no larvae or one larva is permitted). If more larvae are found, then export permits are cancelled (Aluja and Mangan, 2008).

One of the main problems of sampling fruit is the rapid and accurate identification of larvae. In many cases, morphological features do not allow discrimination; therefore, genetic bar coding for rapid identification of pest species intercepted as immature stages in commercial shipments is being developed (Haymer *et al.*, 1994; Armstrong and Ball, 2005; Barr *et al.*, 2006).

11.5.2 Trapping techniques

Trapping of fruit fly adults is achieved primarily by baiting. These traps are continuously positioned throughout areas where fruit flies are likely to appear. If adults are identified, the number of baited traps throughout that area is greatly increased to capture adults and remove them from the environment, and to serve as a monitoring tool for the effectiveness of the control program (Aluja, 1994).

Development of traps and baits for *Anastrepha* species has been investigated for many years and still requires further development (Díaz-Fleischer *et al.*, 2009a). In the case of *A. ludens*, *A. obliqua* and *A. grandis*, the most widely used trap is the McPhail trap (in use for >100 years) (McPhail, 1937). Similar in design, the Multilure trap is of very low efficiency (low captures, female biased, etc.) (Aluja *et al.*, 1989; Díaz-Fleischer *et al.*, 2009a). Commercial McPhail traps are made of glass while Multilure traps are plastic and have several variants (yellow, green and transparent base) (Martínez *et al.*, 2007). Great efforts have been made to improve lures for *Anastrepha*, with emphasis on extending attractant life, reducing non-target organism captures and developing dry traps that are easier to handle. The most common attractants used for fruit flies are protein-based lures (McPhail, 1939; López and Spishako, 1963; Ríos *et al.*, 2005). Results have been summarized in several studies and reviews (Heath *et al.*, 1995; Epsky *et al.*, 1995; Epsky and Heath, 1998; Thomas *et al.*, 2001; Martinez *et al.*, 2007, Díaz-Fleischer *et al.*, 2009a).

New baits that have been effective for attracting several species of *Anastrepha* are BioLure® (Suterra LL Inc., Bend, Oregon Ceratrap Bioiberica, Palafolls, Barcelana), with two component lures (putrescine and ammonium acetate) compared to traps baited with torula yeast (Malavasi *et al.*, 1990; Martinez *et al.*, 2007; Díaz-Fleischer *et al.*, 2009a).

In Panama, urea is believed to enhance the attractiveness of Torula yeast tablets to *A. grandis* (Miguel de Souza Filho, Instituto Biologico, Campinas, Brazil, pers. comm.); however this has not been confirmed in field cages.

Sticky traps of various colors (unbaited and baited with two-component Biolure®) are no more attractive than protein-baited traps (Thomas *et al.*, 2001; Miguel de Souza Filho, Sao Paulo, Brazil, pers. comm.).

A special effort has been made in evaluating several lures based mainly on fruit volatiles for *A. ludens* and *A. obliqua*, including yellow chapote (Robacker *et al.*, 1990) white sapote (González *et al.*, 2006), *Citrus aurantium* L. (Robacker and Heath, 1996; Rasgado *et al.*, 2009), grape juice aroma (Massa *et al.*, 2008), guava (Malo *et al.*, 2005) and blends of synthetic volatile compounds identified from ripe *S. mombin* fruits (Cruz-López *et al.*, 2006; Toledo *et al.*, 2009). In the case of *A. grandis*, testing host plant volatiles is still pending.

For undercapitalized growers, alternative attractants such as fermented pineapple peel (Ortega-Zaleta and Cabrera-Mireles, 1996; Ríos *et al.*, 2005) and human urine (Piñero *et al.*, 2003) have been evaluated with good results.

In summary, there is no magic lure or detection method for adult *Anastrepha* spp. More studies are needed to improve the design of traps and lures considering specific hunger, fly physiological stage, age and fruit fly species to increase fly captures (IAEA, 2003; Díaz-Fleischer *et al.*, 2009a).

11.6 Bait Stations and Male Annihilation

Bait stations (containers with attractants and toxins targeting specific pests) are used in area-wide fruit-fly eradication programs to suppress *A. ludens* populations. *A. ludens* bait stations (sponges soaked in protein hydrolysate, sugar, adjuvant and a toxicant) have been shown to reduce populations by up to 70% (Mangan and Moreno, 2007).

11.7 Cultural and Physical Practices

Cultural control is simple and is widely used by low-income growers. It consists of: (i) fruit sampling and burial; (ii) weed control; and (iii) soil exposure, or exposure of pupae to the sun, and predators. If it is used over a whole region/area, *A. ludens* and *A. obliqua* infestation may be reduced by 60–80% (Aluja, 1993, 1996). In some countries such as Brazil and the Antillean islands, fruit bagging has emerged as a good alternative in backyard gardens (Coelho *et al.*, 2008).

11.8 Chemical Control

Anastrepha spp. are chemically controlled using bait insecticides (López *et al.*, 1969; Aluja, 1993). Products are typically dispersed as 'ultra-low volume' droplets applied via sprayers from airplanes, ground vehicles or directly with backpack sprayers. Spray patterns can be in alternate bands, patches, isolated hotspots or full coverage. If sprays are done in time and in conjunction with other fruit fly control measures, fruit fly populations can be reduced by up to 98% (Aluja, 1993; Aluja et al., 1990). Chemical control has traditionally relied on organophosphate insecticides such as malathion due to its low cost. Since the World Health Organization (WHO) considers organophosphates harmful to human health and non-target insects, other alternatives have emerged in recent years. These include Cyromazine (Cyromazine, Neporex®, Novartis AG, Basel, Switzerland), which inhibits development and formation of chitin (Moreno et al., 1994; Miranda et al., 1996); Phloxine B (D&C Red #28; Warner Jenkinson, St Louis, Missouri), a phototoxic dye used for fruit flies that does not harm natural enemies when applied with protein bait-spray (Mangan and Moreno, 1995, 2001; Moreno et al., 2001); and Spinosad (GF-120, NF Naturalyte™, Dow Agrosciences, Indianapolis, Indiana), a bacterial toxin which currently is the most widely used (Mangan et al., 2006). There are reports of negative effects of Spinosad on non-target insects (Cisneros et al., 2002). Thomas and Mangan (2005) found it to be effective and harmless when used in conjunction with protein-baits GF-120. Recent developments include the use of Lufenuron (Syngenta, Basel, Switzerland), a chemo-sterilant used in baited traps, and whose efficiency is high in the case of A. striata Schiner but not for A. ludens and A. obliqua (Moya et al., 2010), and HMP (Host Marking Pheromone; Anastrephamide, Syngenta, Basel, Switzerland) (Aluja and Díaz-Fleischer, 2006, Aluja et al., 2009a), which can be used in orchard structure manipulation schemes (see Section 11.11, Habitat Manipulation).

In Brazil, growers not located in fly-free or low-prevalence areas (see IPPC, 2008, for definitions) treat the plants with insecticides not necessarily targeted at fruit flies but which largely protect the crop from A. grandis. Insecticides registered for use on cucurbit crops include pyrethroids, organophosphates and nicitinoids: Imidacloprid, Fenthion, Acephate, acetamiprid, beta-cyfluthrin and bifenthrin.

11.9 Biological Control

No biological control scheme has been developed for A. grandis. In Brazil, no larvae or pupae collected in the field have yielded parasitoids or other natural enemies of the pest (Miguel de Souza Filho, São Paulo, Brazil, pers. comm.). This is not the case for A. ludens and A. obliqua, for which there are several alternative control methods based on the use of natural enemies.

The most efficient fruit fly predators are fire ants (Solenopsis geminata) and Pheidole spp. (Formicidae: Hymenoptera), some nitidulid, staphylinid and carabid species (Coleoptera) and other insects that attack larvae and pupae (Aluja, 1993; Aluja et al., 2005; Aluja and Rull, 2009). Adults are also attacked by deer mice (Peromyscus leucopus (Rafinesque) and P. boylii (Baird)), spiders, lizards and birds, but their impact on Anastrepha species such as A. ludens, A. obliqua and A. grandis has never been formally determined in the field (Thomas, 1995; Aluja et al., 2005; Aluja and Rull, 2009). However, parasitic Hymenoptera have been the focus of biological control of A. ludens and A. obliqua (Ovruski et al., 2000), both in terms of classical biological control (i.e., introduction of exotic parasitoids for long-term control of pests), and, more recently, augmentative or inundative biocontrol (i.e., short- to mid-term control of pests through massive and/or repeated releases of natural enemies). Numerous species of Eulophidae, Braconidae and Pteromalidae attack Anastrepha spp., specifically those in the genera Diachasmimorpha, Pachycrepoideus, Dirhinus, Tetrastichus, Aceratoneuromyia, Doryctobracon, Bracon, Amblymerus, Bracanastrepha and Aganaspis (Sivinski et al., 1997, 1998, 2000).

Several parasitoids of A. ludens and A. obliqua are reared under laboratory conditions in Mexico (Xalapa, Veracruz and Tapachula, Chiapas, Mexico) (Aluja et al., 2009d; Cancino et al., 2009). Domestication, colonization and rearing conditions depend on species (Aluja et al., 2009d). Techniques have been described for Doryctobracon areolatus (Szépligeti), D. crawfordi (Viereck), Opius hirtus (Fischer), Utetes anastrephae (Viereck) (all Braconidae, Opiinae), Aganaspis pelleranoi (Bréthes) and Odontosema anastrephae Borgmeier (both Figitidae, Eucoilinae) (all larval–pupal parasitoids), and the pupal parasitoid Coptera haywardi (Oglobin) (Diapriidae, Diapriinae). For large-scale releases, parasitoids can be obtained from artificially reared A. ludens larvae and pupae that can be irradiated at early stages of development to prevent adult fruit fly emergence in the field (irradiated larvae at 20–25 Gy, irradiated pupae at 15 Gy) (Aluja et al., 2009d; Cancino et al., 2009).

Suppression programs for *A. ludens* and *A. obliqua* populations have been undertaken in mango-producing areas of Chiapas, Mexico, through augmentative releases of introduced *Diachasmimorpha longicaudata* (Ashmead) (Montoya et al., 2000a) and in north and central Mexico (Montoya et al., 2000b). Released parasitoid densities fluctuate between 1500 and 2000 adults per ha (Montoya et al., 2007). The impact of parasitoids on fruit fly populations can be assessed by comparing areas with and without releases, or historical trap captures and fruit infestation data (before and after) for the same site; that is, flies per trap per day (FTD) indices and number of larvae per fruit or kilogram of fruit, as well as by comparing percentage parasitism (Montoya et al., 2007).

Trials in Mexico with *D. longicaudata* have shown that *A. obliqua* populations are suppressed more effectively by use of parasitoids than those of *A. ludens*. Such an effect may have been related to the particular properties of host plant species (Montoya et al., 2000b).

Currently, augmentative biological control in conjunction with sterile insect technique (SIT) is applied over extensive areas in the north of Mexico, but although there are suggestive trends in target population densities, unequivocal evaluation of impact is missing (Montoya et al., 2007). The main limitation for widespread area-wide releases of biocontrol agents is the high cost of rearing. Parasitoid mass releases may be most cost-effective when pesticides cannot be used in conjunction with SIT. Examples include natural vegetation adjacent to commercial orchards, urban/suburban backyard sources of infestation and protected areas such as national parks (Aluja, 1993; Aluja and Rull, 2009 and references therein).

Nematodes such as *Heterorhabditis bacteriophora* and *Steinernema feltiae* have been studied as a control method against *A. ludens* and *A. obliqua* larvae by Toledo et al. (2001; 2005a, b; 2006); however, information related to their effectiveness is still missing.

Other control alternatives for *A. ludens* and *A. obliqua* include bacteria (*Serratia marcesens*) and fungi (*Beauveria bassiana* (Bassiano) and *Metharizium anisopliae* (Metsch.) (Fajardo and Canal, 2009; Wilson et al., 2009), but their application methods and efficiency are still not developed and proven.

11.10 Sterile Insect Technique (SIT)

SIT has been successfully applied for >50 years in eradication and control programs for several pest species including fruit flies (Knipling, 1979). In Mexico, it is used in area-wide management programs to control *A. ludens* and *A. obliqua*. One of the biggest challenges for SIT-based programs for *A. ludens* and *A. obliqua* in Mexico is quality problems emerging from the packing and shipping of sterilized insects from the Guatemalan border to remote areas in the north (Hernández et al., 2010) and the maintenance of male competitiveness (Rull et al., 2005). Male mating performance can be enhanced when juvenile hormone is applied (Aluja et al., 2009e), irradiation doses are reduced (Rull and Barreda-Landa, 2007), strains are regularly refreshed with the incorporation of wild material (Rull and Barreda-Landa, 2007), and rearing conditions and release methods are improved (Díaz-Fleischer et al., 2009b; Ruíz-Guzmán, 2010). Another technique which has recently been evaluated is the application of host plant essential oils (*C. aurantium*, *C. paradisi* and *M. indica*) to enhance male performance, but no effect was detected for *A. ludens* and *A. obliqua* (Morató, 2010). Efforts are under way to develop transgenic *A. ludens* strains where males (pupae) can be separated through sexually dimorphic coloration (Meza et al., 2011). Artificial diets have not been developed for *A. grandis*, but the flies can be reared in the laboratory from hosts such as pumpkin or squash (Miguel de Souza Filho, São Paulo, Brazil, pers. comm.). FAO/IAEA recently established a laboratory colony for providing material for behavioral, ecological and genetic studies, which may in the future lead to the development of mass rearing and SIT.

11.11 Habitat Manipulation

Anastrepha habitat manipulation methods can be considered from an entire geographic region down to a single tree or a fruit, and include trap cropping, interception using traps, selective insecticide application, push-pull strategies and manipulation of fruit phenology (Aluja and Rull, 2009).

11.11.1 Trap cropping

This technique involves the use of preferred hosts (cultivars or plant species) as attractants to concentrate fruit flies where they can be selectively trapped or killed. In the case of *A. ludens*, preferred commercial hosts include citrus, particularly grapefruit and sweet orange, but feral non-commercial alternative hosts such as bitter orange can be more attractive and could potentially be used as trap crops (Aluja *et al.*, 2003b; A. Birke, Veracruz, Mexico, pers. comm.). In the case of *A. obliqua*, the main commercial hosts are mango and guavas, but non-commercial mango cultivars (mango de coche) and wild *Spondias* spp. are preferred over commercial fruit and can be used as trap crops (Aluja and Birke, 1993). Trap cropping can be achieved by surrounding blocks of commercial crops with preferred non-commercial cultivars or species whose fruit can be destroyed, or where control actions such as bait sprays or massive release of natural enemies can be concentrated (Aluja and Rull, 2009).

11.11.2 Interception using traps

Fruit fly interception using baited traps can be used as a pest manipulation tool for *Anastrepha* species by deploying traps in peripheral areas of orchards where highly attractive hosts are present. Fruiting phenology, fruit fly oviposition behavior and release of attractive volatiles are important aspects that must be considered when attempting to apply this technique (Aluja *et al.*, 1996; Aluja and Rull, 2009).

11.11.3 Push-pull strategy

Push-pull strategies consider behavior-modifying stimuli to manipulate the distribution and abundance of pests and/or beneficial insects for pest management (Cook *et al.*, 2006). This strategy has been tested experimentally for *Anastrepha* spp. by applying synthetic HMP (host-marking pheromones) to the tree canopy (Aluja *et al.*, 2003a; Aluja and Díaz-Fleischer, 2006). *Anastrepha* females encountering 1 or 100 mg/ml of feces extract were deterred and did not oviposit in treated HMP fruit (Aluja and Díaz-Fleischer, 2006). Use of *A. ludens* HMP has been shown to be successful when applied to tropical plums ('jobo') and reduced *A. obliqua* infestation by 90% (Aluja *et al.*, 2009c). Combined with attractive traps, this push-pull strategy has actually been proven under commercial conditions and has shown that while treated trees repelled fruit flies, untreated trees were preferred (M. Aluja, Veracruz, Mexico, pers. comm.).

11.12 Plant Resistance

Several fruit species are naturally resistant to immature stages of *A. ludens* and *A. obliqua* (Carvalho *et al.*, 1996; F. Díaz-Fleischer, Veracruz, Mexico, pers. comm.). Resistance of fruits to *A. ludens* and probably *A. obliqua* are caused by plant secondary metabolites, and by physical and structural properties of the fruit tissue. In the case of 'Hass' avocado, Aluja *et al.* (2004) showed that the fruits are naturally resistant to fruit flies of the genus *Anastrepha*. Resistance is achieved through egg encapsulation caused by avocado fruit tissue, and similar observations have been obtained for wild, non-commercial hosts such as white sapote (Aluja *et al.*, 2004; A. Birke, Veracruz, Mexico, pers. comm.). Apples and peaches are probably resistant due to their phenolic content (M. Aluja, Veracruz, Mexico, pers. comm.). In commercial citrus, resistance is due to high essential oil content in the flavedo, which can be enhanced through the use of gibberellic acid (Greany *et al.*, 1983; Birke *et al.*, 2006), and in mangoes, resistance is due to lignin channels (Joel, 1980). Both cause egg and larval mortality to *A. suspensa* and *Ceratitis capitata* (Wiedemann), but this has not been confirmed with *A. ludens* and *A. obliqua* (Greany *et al.*, 1983; Birke *et al.*, 2006; Aluja *et al.*, 2011b).

11.13 Postharvest Treatment

Postharvest treatments are mainly used for *Anastrepha* spp. and include temperature treatments and fruit irradiation (Mangan and Hallman, 1998; Mangan *et al.*, 1998).

11.13.1 Fruit temperature treatment

Temperature treatments such as immersion in hot water, and cold storage, need to be applied once fruit has been harvested and require fruit that is not fully mature. The main principle consists of exposing immature stages to lethal upper or lower temperatures (generally based on the least susceptible stage) for periods long enough to cause mortality without damaging fruit. Heat treatment has been very successful for certain mango cultivars which are attacked by several *Anastrepha* species (Sharp et al., 1989; Aluja et al., 2010).

11.13.2 Fruit irradiation

Effective generic irradiation doses for fruit range from 50 to 150 Gy for tephritid fruit flies (Hallman, 1999; Hallman and Worley, 1999; Follett et al., 2007), and depend on fruit fly species and type of fruit. Hallman and Martínez (2001) found that 69 Gy achieved 99.9968% security against the Mexican fruit fly, *A. ludens*, with a 95% level of confidence based on prevention of adult emergence from irradiated third instars in grapefruits. Bustos et al. (2004) found similar results for *A. ludens* in mangoes. For *A. obliqua* 70–80 Gy doses should be enough to eliminate eggs and first-instar larvae (Hallman and Loaharanu, 2002).

11.14 Use of Chitosan Polymers

Recently, polymers which cover fruit before it is shipped and exported have been developed. These polymers may prevent immature fruit fly stage respiration within the fruit and thereby cause mortality. A recent example is application of Chitosan® (from crabshells; Sigma-Aldrich, Steinheim, Germany) on mango which causes *A. ludens* egg and larvae mortality and reduces infestation significantly (Ventura-González et al., 2009).

11.15 Regulatory Control

An essential component of phytosanitary programs for fruit flies, including *A. ludens*, *A. obliqua* and *A. grandis*, is the application of legal measures to control the spread of pests and diseases. Legal control is achieved through quarantine application, restricted movement of fruit through use of permits, certificates of origin, orchard and fruit fly-free areas certification, survey of effective application of control measures, mandatory control measures, treatment certification (e.g., fumigation, chemical treatment), and others (Sharp and Hallman, 1994; Follett and Neven, 2006; Hennessey, 2007; Follett and Hennessey, 2007; Aluja and Mangan, 2008; USDA-APHIS, 2012 and references therein). For example, *A. grandis* is the cause of significant trade barriers between South American countries. The main strategy for producers intending to export is to grow cucurbit fruits in areas officially declared fly-free or low prevalence. Guidelines for the establishment and declaration of pest-free areas can be consulted in IPPC (1996).

11.16 Future Perspectives

Currently, international markets tend to require residue-free fruit with high aesthetic quality. Although both objectives are desirable, efforts to achieve one often conflict with the other. The clearest example of this is fruit damage when fruit flies are left uncontrolled versus the harmful accumulation of pesticides when certain controls are applied. In addition, there are global concerns about environmental degradation, which in the case of pest control, requires development and use of environmentally friendly methods.

To fulfill these expectations, alternative control methods are being developed to replace insecticides that are harmful to human health and the environment.

Many management strategies discussed in this chapter, such as push-pull strategies in habitat manipulation schemes, using HMP in large scale application, formal testing of predator and infective vectors effects, identification of repellent plant/compounds, enhancing parasitoid effectiveness as well as searching for fruit resistance to the attack by fruit flies through genetic engineering, are pending. Use of GPS-GIS information is already being used for follow-up of *C. capitata* captures in Chiapas, Mexico (Villaseñor et al., 2009) and would be a great advance if implemented at the US border in areas where yearly introductions of *A. ludens* occur. The solution to the fruit fly problem does not rest on a single technique; many of the above would be most efficacious if integrated with others.

Acknowledgements

We thank John Sivinski and Juan Rull for their valuable comments and suggestions for improvement. Parts of the research reported here were financed through grants to MA from the Mexican Campaña Nacional Contra Moscas de la Fruta (Convenios SAGARPA-IICA/IICA-INECOL), the Mexican Consejo Nacional de Ciencia y Tecnología (CONACyT Grants D111-903537, 0436P-N9506 and SEP-2004-CO1-46846-Q), CONACyT – Texas A & M Project ('The natural enemies of *Rhagoletis* spp. (Diptera: Tephritidae) in Mexico, with emphasis on the apple maggot, *Rhagoletis pomonella*'), the International Foundation for Science (Grant Number C/1741-1), the US Department of Agriculture (ARS funds managed by John Sivinski) and the Instituto de Ecología, A.C.

References

Aluja, M. (1993) *Manejo Integrado de la Mosca de la Fruta*. 1st ed. Trillas, Mexico City, Mexico.
Aluja, M. (1994) Bionomics and management of *Anastrepha*. *Annual Review of Entomology* 39, 155–178.
Aluja, M. (1996) Future trends in fruit fly management. In: McPheron, B.A. and Steck, G.J. (eds) *Economic Fruit Fly Pests: A World Assessment of their Biology and Management*. St. Lucie Press, Delray Beach, Florida, pp. 309–320.
Aluja, M. and Birke, A. (1993) Habitat use by *Anastrepha obliqua* flies (Diptera: Tephritidae) in a mixed mango and tropical plum orchard. *Annals of the Entomological Society of America* 18, 799–812.
Aluja, M. and Díaz-Fleischer, F. (2006) Foraging behavior of *Anastrepha ludens*, *A. obliqua* and *A. serpentina* in response to feces extracts containing host marking pheromone. *Journal of Chemical Ecology* 32, 367–389.
Aluja, M. and Mangan, R. (2008) Fruit Fly (Diptera: Tephritidae) host status determination: Critical conceptual, methodological, and regulatory considerations. *Annual Review of Entomology* 53, 473–502.
Aluja, M. and Rull, J. (2009) Managing pestiferous fruit flies (Diptera: Tephritidae) through environmental manipulation. In: Aluja, M., Leskey, T. and Vincent, C. (eds) *Biorational Tree Fruit Pest Management 7*. CAB International, Wallingford, UK, pp. 171–213.
Aluja, M., Hendrichs, J. and Cabrera, M. (1983) Behavior and interactions between *Anastrepha ludens* (L) and *A. obliqua* (M) on a field caged mango tree - I. Lekking behavior and male territoriality. In: Cavalloro, R. (ed.) *Fruit Flies of Economic Importance*. A.A. Balkema, Rotterdam, The Netherlands, pp. 122–133.
Aluja, M., Guillén, J., de la Rosa, G., Cabrera, M., Celedonio, H., Liedo, P. and Hendrichs, J. (1987) Natural host plant survey of the economically important fruit flies (Diptera: Tephritidae) of Chiapas, Mexico. *Florida Entomologist* 70, 329–338.
Aluja, M., Cabrera, M., Guillén, J., Celedonio, H. and Ayora, F. (1989) Behaviour of *Anastrepha ludens*, *A. obliqua* and *A. serpentina* (Diptera: Tephritidae) on a wild mango tree (*Mangifera indica*) harbouring three McPhail traps. *Insect Science and its Application* 10, 309–318.
Aluja, M., Guillén, J., Liedo, P., Cabrera, M., Rios, E., de la Rosa, G., Celedonio, H. et al. (1990) Fruit infesting tephritids (Dipt: Tephritidae) and associated pararasitoids in Chiapas, México. *Entomophaga* 35, 39–48.
Aluja, M., Celedonio-Hurtado, H., Liedo, P., Cabrera, M., Castillo, F., Guillén J. and Rios, E. (1996) Seasonal population fluctuations and ecological implications for management of *Anastrepha* fruit flies (Diptera: Tephritidae) in commercial mango orchards in Southern Mexico. *Journal of Economic Entomology* 89, 654–667.
Aluja, M., Piñero, J., López, M., Ruíz, C., Zúñiga, A., Piedra, E., Díaz-Fleischer, F. et al. (2000a) New host plant and distribution records in Mexico for *Anastrepha* spp., *Toxotrypana curvicauda* Gerstacker, *Rhagoletis zoqui* Bush, *Rhagoletis* sp., and *Hexachaeta* sp. (Diptera: Tephritidae). *Proceedings of the Entomological Society of Washington* 102, 802–815.
Aluja, M., Piñero, J., Jácome, I., Díaz-Fleischer, F. and Sivinski, J. (2000b) Behavior of flies in the genus *Anastrepha* (Trypetinae: Toxotrypanini). In: Aluja, M. and Norrbom, A. (eds) *Fruit Flies (Tephritidae): Phylogeny and Evolution of Behavior*, CRC Press, Boca Raton, Florida, pp. 375–406.
Aluja, M., Díaz-Fleischer, F., Papaj, D.R., Lagunes, G. and Sivinski, J. (2001a) Effects of age, diet, female density and the host resource on egg load in *Anastrepha ludens* and *Anastrepha obliqua* (Diptera: Tephritidae). *Journal of Insect Physiology* 47, 975–988.

Aluja, M., Jácome, I. and Macías-Ordóñez, R. (2001b) Effect of adult nutrition on male sexual performance in four tropical fruit fly species of the genus *Anastrepha* (Diptera: Tephritidae). *Journal of Insect Behavior* 14, 759–775.

Aluja, M., Díaz-Fleischer, F., Edmunds, A.J.F. and Hagmann, L. (2003a) Isolation, structural determination, synthesis, biological activity and application as control agent of the host marking pheromone (and derivatives thereof) of the fruit flies of the type *Anastrepha* (Diptera: Tephritidae). Instituto de Ecología, A.C., assignee. 29 April 2003. U.S. Patent 6,555,120 B1.

Aluja, M., Pérez-Staples, D., Macías-Ordóñez, R., Piñero, J., McPheron, B. and Hernández-Ortiz, V. (2003b) Nonhost status of *Citrus sinensis* cultivar Valencia and C. *paradisi* cultivar Ruby Red to Mexican *Anastrepha fraterculus* (Diptera: Tephritidae). *Journal of Economic Entomology* 96, 1693–1703.

Aluja, M., Díaz-Fleischer, F. and Arredondo, J. (2004) Nonhost status of commercial *Persea americana* 'Hass' to *Anastrepha ludens*, *Anastrepha obliqua*, *Anastrepha serpentina*, and *Anastrepha striata* (Diptera: Tephritidae) in Mexico. *Journal of Economic Entomology* 97, 293–309.

Aluja, M., Sivinski, J., Rull, J. and Hodgson, P.J. (2005) Behavior and predation of fruit fly larvae (*Anastrepha* spp.) (Diptera: Tephritidae) after exiting fruit in four types of habitats in tropical Veracruz, Mexico. *Environmental Entomology* 34, 1507–1516.

Aluja, M., Rull, J., Sivinski, J., Trujillo, G. and Pérez-Staples, D. (2009a) Male and female condition influence mating performance and sexual receptivity in two tropical fruit flies (Diptera: Tephritidae) with contrasting life histories. *Journal of Insect Physiology* 55, 1091–1098.

Aluja, M., Rull, J., Pérez-Staples, D., Díaz-Fleischer, F. and Sivinski, J. (2009b) Random mating among *Anastrepha ludens* (Diptera: Tephritidae) adults of geographically distant and ecologically distinct populations in Mexico. *Bulletin of Entomological Research* 99, 207–214.

Aluja, M., Díaz-Fleischer, F., Boller, E.F., Hurter, J., Edmunds, A.J.F., Hagmann, L., Patrian, B. et al. (2009c) Application of feces extracts and synthetic analogues of the host marking pheromone of *Anastrepha ludens* significantly reduces fruit infestation by *A. obliqua* in tropical plum and mango backyard orchards. *Journal of Economic Entomology* 102, 2268–2278.

Aluja, M., Sivinski, J., Ovruski, S., Guillén, L., López, M., Cancino, J., Torres-Anaya, A. et al. (2009d) Colonization and domestication of seven species of native New World Hymenopterous larval-prepupal and pupal fruit fly (Diptera: Tephritidae) parasitoids. *Biocontrol Science and Technology* 19, 49–79.

Aluja, M., Ordano, M., Teal, P.E.A., Sivinski, J., García-Medel, D. and Anzures-Dadda, A. (2009e) Larval feeding substrate and species significantly influence the effect of a juvenile hormone analog on sexual development/performance in four tropical tephritid flies. *Journal of Insect Physiology* 55, 231–242.

Aluja, M., Díaz-Fleischer, F., Arredondo, J., Valle-Mora, J. and Rull, J. (2010) Effect of cold storage on larval and adult *Anastrepha ludens* (Diptera: Tephritidae) viability in commercially ripe, artificially infested *Persea americana* 'Hass'. *Journal of Economic Entomology* 103, 2000–2008.

Aluja, M., Birke, B., Guillen, L., Díaz-Fleischer, F. and Nestel, D. (2011a) Coping with an unpredictable and stressful environment: The life history and metabolic response to variable food and host availability in a polyphagous tephritid fly. *Journal of Insect Physiology* 57, 1592–1601.

Aluja, M., Bigurra, E., Birke, A., Greany, P. and McDonald, R. (2011b) Delaying senescence of 'Ruby Red' grapefruit and 'Valencia' oranges by gibberellic acid applications. *Revista Mexicana de Ciencias Agrícolas* 2, 41–55.

Aluja, M., Ordano, M., Guillén, L. and Rull, J. (2012) Understanding long-term fruit fly (Diptera: Tephritidae) population dynamics: implications for area-wide management. *Journal of Economic Entomology* 105, 823–836.

APHIS PPQ (2002) Pathway-initiated risk assessment of fresh melon (*Cucumis melo* L. subsp. melo) and watermelon (*Citrullus lanatus* (Thunb.) Matsum. and Nakai var. *lanatus*) fruit from Peru into the continental United States. https://web01.aphis.usda.gov/oxygen_fod/fb_md_ppq.nsf/d259f66c6afbd45e852568a90027bcad/e3b4861408bc187c85256e2a005f75b5/$FILE/0060.pdf, accessed 31 July 2012.

APHIS PPQ (2009) Outbreak of *Anastrepha grandis* (South American cucurbit fruit fly) in Panama. Deputy Administrator to the State Plant Regulatory Officials. 6 July 2009. https://web01.aphis.usda.gov/oxygen_fod/fb_md_ppq.nsf/d259f66c6afbd45e852568a90027bcad/e3b4861408bc187c85256e2a005f75b5/$FILE/0060.pdf, accessed 31 July 2012.

Armstrong, K.F. and Ball, S.L. (2005) DNA barcodes for biosecurity: invasive species identification. *Philosophical Transactions of the Royal Society, Series B.* 360, 1813–1820.

Baker, A.C., Stone, W.E., Plummer, C.C. and McPhail, M.I. (1944) A review of studies on the

Mexican fruit fly and related Mexican species. *U.S. Department of Agriculture Miscellaneous Publication* 531, 1–155.

Barr, N.B., Copeland, R.S., De Meyer, M., Masiga, D., Kibogo, H.G., Billah, M.K., Osir, E. *et al.* (2006) Molecular diagnostics of economically important fruit fly species (Diptera: Tephritidae) in Africa using PCR and RFLP analyses. *Bulletin of Entomological Research* 96, 505–521.

Birke, A. (1995) Comportamiento de oviposición de la mosca Mexicana de la fruta *Anastrepha ludens* y uso de ácido giberélico para disminuir la susceptibilidad de toronja *Citrus paradisi* al ataque de esta plaga. BSc thesis, Universidad Veracruzana, Xalapa, Veracruz.

Birke, A. and Aluja, M. (2011) *Anastrepha ludens* and *Anastrepha serpentina* (Diptera: Tephritidae) do not infest *Psidium guajava* (Myrtaceae), but *Anastrepha obliqua* occasionally shares this resource with *Anastrepha striata* in nature. *Journal of Economic Entomology* 104, 1204–1211.

Birke, A., Aluja, M., Greany, P.D., Bigurra, E., Pérez-Staples, D. and McDonald, R. (2006) Long aculeus and behavior of *Anastrepha ludens* renders gibberellic acid ineffective as an agent to reduce 'Ruby Red' grapefruit susceptibility to the attack of this pestiferous fruit fly in commercial groves. *Journal of Economic Entomology* 99, 1184–1193.

Bressan, S., and da Costa Teles, M. (1991) Lista de hospedeiros e índices de infestação de algumas espécies do gênero *Anastrepha* Schiner, 1868 (Diptera: Tephritidae) na região de Ribeirão Preto - SP. *Anais da Sociedade Entomológica do Brasil* 20, 5–15.

Bressan-Nascimento, S. (2001) Emergence and pupal mortality factors of *Anastrepha obliqua* (Macq.) (Diptera: Tephritidae) along the fruiting season of the host *Spondias dulcis* L. *Neotropical Entomology* 30, 207–215.

Bustos, M.E., Enkerlin, W., Reyes, J. and Toledo, J. (2004) Irradiation of mangoes as a postharvest quarantine treatment for fruit flies (Diptera: Tephritidae). *Journal of Economic Entomology* 97, 286–292.

Cancino, J., Ruíz, L., Sivinski, J., Galvez, F.O. and Aluja, M. (2009) Rearing of five hymenopterous larval-prepupal (Braconidae, Figitidae) and three pupal (Diapriidae, Chalcidoidea, Eurytomidae) native parasitoids of the genus *Anastrepha* (Diptera: Tephritidae) on irradiated *A. ludens* larvae and pupae. *Biocontrol Science and Technology* 19, 193–209.

Carroll, L.E. and Wharton, R.A. (1989) Morphology of the immature stages of *Anastrepha ludens* (Diptera: Tephritidae). *Annals of the Entomological Society of America* 82, 201–214.

Carvalho, R., da Silva, R.A., Nascimento, A.S., Morgante, J.S. and Fonseca, N. (1996) Susceptibility of different mango varieties (*Mangifera indica*) to attack of fruit fly, *Anastrepha obliqua*. In: McPheron, B.A. and Steck, G.J. (eds) *Fruit Fly Pests: A World Assesment of Their Biology and Management*. St. Lucie Press, Delray Beach, Florida, pp. 325–331.

Celedonio-Hurtado, H., Liedo, P., Aluja, M., Guillén, J., Berrigan, D. and Carey, J. (1988) Demography of *Anastrepha ludens*, *A. obliqua* and *A. serpentina* (Diptera: Tephritidae) in Mexico. *Florida Entomologist* 71, 111–120.

Celedonio-Hurtado, H., Aluja, M. and Liedo, P. (1995) Adult population fluctuations of *Anastrepha* species (Diptera: Tephritidae) in tropical orchard habitats of Chiapas, Mexico. *Environmental Entomology* 24, 861–869.

Christenson, L.D. and Foote, R.H. (1960) Biology of fruit flies. *Annual Review of Entomology* 5, 171–192.

Cisneros, J., Goulson, D., Dervent, L.C., Penagos, D.I., Hernández, O. and Williams, T. (2002) Toxic effect of Spinosad on predatory insects. *Biological Control* 23, 156–163.

Clark, R.A., Steck, G.J. and Weems Jr, H.W. (1996) Detection, quarantine, and eradication of exotic fruit flies in Florida. In: Rosen, D.L. (ed.) *Pest Management in the Subtropics: Integrated Pest Management - a Florida Perspective*. Intercept Ltd, Andover, UK, pp. 29–54.

Coelho, L.R., Leonel, S., Crocomo, W.B. and Labina, A.M. (2008) Controle de pragas do pessegueiro através do ensacamento dos frutos. *Ciência e Agrotecnologia* 32, 1743–1747.

Cook, S.M., Khan, Z.R. and Pickett, J.A. (2006) The use of push-pull strategies in integrated pest management. *Annual Review of Entomology* 52, 375–400.

Cresoni-Pereira and Zucoloto, F.S. (2006) Influence of male nutritional conditions on the performance and alimentary selection of wild females of *Anastrepha obliqua* (Macquart) (Diptera, Tephritidae). *Revista Brasileira de Entomologia* 50, 287–292.

Cruz-López, L., Malo, E.A., Toledo, J., Virgen, A., Del Mazo, A. and Rojas, J.C. (2006) A new potential attractant for *Anastrepha obliqua* from *Spondias mombin* fruits. *Journal of Chemical Ecology* 32, 351–365.

Darby, H.H. and Kapp, E.M. (1933) Observations of the thermal death points of *Anastrepha ludens* (Loew). *U.S. Department of Agriculture, Technical Bulletin* 400, Washington, DC, pp. 22.

Darby, H.H. and Kapp, E.M. (1934) Studies on the Mexican fruit fly, *Anastrepha ludens* (Loew). *U.S. Department of Agriculture, Technical Bulletin* 444, Washington, DC, pp. 20.

Díaz-Fleischer, F. and Aluja, M. (2003a) Behavioural plasticity in relation to egg and time limitation: the case of two fly species in the genus *Anastrepha* (Diptera: Tephritidae). *Oikos* 100, 125–133.

Díaz-Fleischer, F. and Aluja, M. (2003b) Influence of conspecific presence, experience, and host quality on oviposition behavior and clutch size determination in *Anastrepha ludens* (Diptera: Tephritidae). *Journal of Insect Behavior* 16, 537–554.

Díaz-Fleischer, F. and Aluja, M. (2003c) Clutch size in frugivorous insects as a function of host firmness: the case of the tephritid fly *Anastrepha ludens*. *Ecological Entomology* 28, 268–277.

Díaz-Fleischer F., Arredondo, J., Flores, F., Montoya, P. and Aluja, M. (2009a) There is no magic fruit fly trap: multiple biological factors influence the response of adult *Anastrepha ludens* and *Anastrepha obliqua* (Diptera: Tephritidae) individuals to multiLure traps baited with BioLure or NuLure. *Journal of Economic Entomology* 102, 86–94.

Díaz-Fleischer, F., Arredondo, J. and Aluja, M. (2009b) Enriching early adult environment affects the copulation behaviour of a tephritid fly. *Journal of Experimental Biology* 212, 2120–2127.

EPPO (2009) First report of *Anastrepha grandis* in Panama (2009/176). European and Mediterranean Plant Protection Organization, Paris, p. 6.

Epsky, N.D. and Heath, R.R. (1998) Exploiting the interactions of chemical and visual cues in behavioral control measures for pest tephritid fruit flies. *Florida Entomologist* 81, 273–282.

Epsky, N.D., Heath, R.R., Guzman, A. and Meyer, W.L. (1995) Visual cue and chemical cue interactions in a dry trap with food-based synthetic attractant for *Ceratitis capitata* and *Anastrepha ludens* (Diptera: Tephritidae). *Environmental Entomology* 24, 1387–1395.

Fajardo, A.O. and Canal, N. (2009) Enthomopathogenic fungi used on *Anastrepha obliqua* young adult-fruit flies (Macquart) (Diptera: Tephritidae) In: Montoya P., Díaz-Fleischer, F. and Flores, F. (eds) *7th Meeting of the Working Group on Fruit Flies of the Western Hemisphere*, Mazatlán, México, pp. 157.

Follett, P.A. and Hennessey, M.K. (2007) Confidence limits and sample size for determining nonhost status of fruits and vegetables to tephritid fruit flies as a quarantine measure. *Journal of Economic Entomology* 100, 251–257.

Follett, P.A., and Neven, L.G. (2006) Current trends in quarantine entomology. *Annual Review of Entomology* 51, 359–85.

Follett, P.A., Yang, M., Lu, K.H. and Chen, T.S. (2007) Irradiation for postharvest control of quarantine insects. *Formosan Entomology* 27, 1–15.

Fontellas-Brandalha, T.M.L. and Zucoloto, F.S. (2004) Selection of oviposition sites by wild *Anastrepha obliqua* (Macquart) (Diptera: Tephritidae) based on the nutritional composition. *Neotropical Entomology* 33, 557–562.

González, R., Toledo, J., Cruz-López, L., Virgen, A., Santiesteban, A. and Malo, E.A. (2006) A new blend of white sapote fruit volatiles as potential attractant to *Anastrepha ludens* (Diptera: Tephritidae). *Journal of Economic Entomology* 99, 1994–2001.

Greany, P.D., Styer, S.C., Davis, P.L., Shaw, P.E. and Chambers, D.L. (1983) Biochemical resistance of citrus to fruit flies. Demonstration and elucidation of resistance to the Caribbean fruit fly, *Anastrepha suspensa*. *Entomologia Experimentalis et Applicata* 34, 40–50.

Hallman, G.J. (1999) Ionization radiation quarantine treatments against tephritid fruit flies. *Postharvest Biology and Technology* 16, 93–106.

Hallman, G.J. and Loaharanu, P. (2002) Generic ionizing radiation quarantine treatments against fruit flies (Diptera: Tephritidae) proposed. *Journal of Economic Entomology* 95, 893–901.

Hallman, G.J. and Martínez, L.R. (2001) Ionizing irradiation quarantine treatment against Mexican fruit fly (Diptera: Tephritidae) in citrus fruits. *Postharvest Biology and Technology* 23, 71–77.

Hallman, G.J. and Worley, J.W. (1999) Gamma radiation doses to prevent adult emergence from Mexican and West Indian fruit fly (Diptera: Tephritidae) immatures. *Journal of Economic Entomology* 92, 967–973.

Haymer, D.S., Tanaka, T. and Teramae, C. (1994) DNA probes can be used to discriminate between tephritid species at all stages of the life cycle (Diptera: Tephritidae). *Journal of Economic Entomology* 87, 741–746.

Heath, R.R., Epsky, N.D., Guzman, A., Dueben, B.D., Manukian, A. and Meyer, W.L. (1995) Development of a dry plastic insect trap with food-based synthetic attractant for the Mediterranean and the Mexican fruit flies (Diptera: Tephritidae). *Journal of Economic Entomology* 88, 1307–1315.

Hennessey, M.K. (2007). *Guidelines for the Determination and Designation of Host Status of a Commodity for Fruit Flies (Tephritidae)*. USDA-CPHST, Orlando, Florida.

Hernández, E., Rivera, P., Bravo, B., Salvador, M. and Chang, C. (2010) Improvement and development of alter native larval diets for mass rearing fruit flies species of the genus *Anastrepha* In: Sabater, B., Navarro,V.F. and Urbaneja, A.(eds) *8th International Symposium on Fruit Flies of Economic Importance*. Valencia, Spain, pp. 233.

Hernández-Ortíz, V. (1992) *El Género Anastrepha Schiner en México. (Diptera: Tephritidae). Taxonomía, Distribución y Sus Plantas Huéspedes*. Instituto de Ecología, Xalapa, Veracruz.

Hernández-Ortíz, V. and Aluja, M.(1993) Listado de especies del género neotropical *Anastrepha* (Diptera: Tephritidae) con notas sobre su distribución y plantas hosperderas. *Folia Entomológica Mexicana* 88, 89–105.

Hodgson, P.J., Sivinski, J., Quintero, G. and Aluja, M. (1998) Depth of pupation and sur vival of fruit fly (*Anastrepha* spp.: Tephritidae) pupae in a range of agricultural habitats. *Environmental Entomology* 27, 1310–1314.

IAEA (2003) *Trapping Guideline for Areawide Fruit Fly Programmes*. Joint FAO/IAEA Programme, Vienna, pp. 47.

IPPC (1996) *Requirements for the Establishment of Pest Free Areas 1996*. ISPM 4, FAO, Rome.

IPPC (2008) *International Standards for Phytosanitary Measures*, ISPM 5, FAO, Rome, pp. 22.

Jenkins, D. and Goenaga, R. (2008) Host breadth and parasitoids of fruit flies (*Anastrepha* spp.) (Diptera: Tephritidae) in Puerto Rico. *Environmental Entomology* 37, 110–120.

Jirón, L.F. and Hedström, J. (1988) Occurrence of fruit flies of the genera *Anastrepha* and *Ceratitis* (Diptera: Tephritidae), and their host plant availability in Costa Rica. *Florida Entomology* 71, 62–73.

Joel, D.M. (1980) Resin ducts in the mango fruit: a defence system. *Journal of Experimental Botany* 31, 1707–1718.

Knipling, E.F. (1979) The basic principles of insect population suppression and management. USDA Agricultural Handbook 512.

Leyva, J.L. (1988) Temperatura umbral y unidades calor requeridas por los estados inmaduros de *Anastrepha ludens* (Loew) (Diptera: Tephritidae). *Folia Entomológica* 74, 189–196.

Leyva, J.L., Browning, H.W. and Gilstrap, F.E. (1991) Development of *Anastrepha ludens* (Diptera: Tephritidae) in several host fruit. *Environmental Entomology* 20, 1160–1165.

López, D.F., Chambers, D.L., Sánchez, M. and Kamasaki, H. (1969) Control of the Mexican fruit fly by bait sprays concentrated at discrete locations. *Journal of Economic Entomology* 62, 1255–1257.

López, F. and Spishako, L.M.O. (1963) Reacción de la mosca de la fruta *Anastrepha ludens* (Loew) a atrayentes proteicos y fermentables. *Ciencia* 22, 103–114.

Malavasi, A., Duarte, A.L., Cabrini, G. and Engelstein, M. (1990) Field evaluation of three baits for South American cucurbit fruit fly (Diptera: Tephritidae) using McPhail traps. *Florida Entomologist* 73, 510–512.

Malo, E.A. and Zapien, G.I.(1994) McPhail trap captures of *Anastrepha obliqua* and *Anastrepha ludens* (Diptera: Tephritidae) in relation to time of day. *Florida Entomologist* 77, 290–294.

Malo, E.A., Cruz-Lopez, L., Toledo, J., Del Mazo, A., Virgen, A. and Rojas, J.C. (2005) Behavioral and electrophysiological responses of the Mexican fruit fly (Diptera: Tephritidae) to guava volatiles. *Florida Entomologist* 88, 364–371.

Mangan, R.L. and Hallman, G.J (1998) Temperature treatments for quarantine security: new approaches for fresh commodities. In: Hallman, G.J. and Denlinger, D.L. (eds) *Temperature Sensitivity in Insects and Application in Integrated Pest Management*, Westview Press, Boulder, Colorado, pp. 201–234.

Mangan, R.L. and Moreno, D.S. (1995) Development of phloxine B and urannine bait for control of Mexican fruit fly. In: Heitz, J.K. and Downum, R. (eds) *Light Activated Pest Control*. ACS Books, Washington, DC, pp. 115–126.

Mangan, R.L. and Moreno, D.S. (2001) Photoactive dye insecticide formulations: adjuvants increase toxicity to Mexican fruit fly. *Journal of Economic Entomology* 94, 150–156.

Mangan, R.L. and Moreno, D.S. (2007) Development of bait stations for fruit fly population suppression. *Journal of Economic Entomology* 100, 440–450.

Mangan, L., Frampton, E.L., Thomas, D. and Moreno, D.S. (1997) Application of the maximum pest limit concept to quarantine standards for the Mexican fruit fly (Diptera: Tephritidae). *Journal of Economic Entomology* 90, 1433–1440.

Mangan, R.L., Shellie, K.C., Ingle, S.J. and Firko, M.J. (1998) High temperature forced-air treatments with fixed time and temperature for 'Dancy' tangerines, 'Valencia' oranges and 'Rio Star' grapefruit. *Journal of Economic Entomology* 91, 933–939.

Mangan, R.L., Moreno, D.S. and Thompson, G.D. (2006) Bait dilution, spinosad concentration, and efficacy of GF-120 based fruit fly sprays. *Crop Protection* 25, 125–133.

Martínez, A.J., Salinas, E.J. and Rendon, P. (2007) Capture of *Anastrepha* species (Diptera: Tephritidae) with multilure traps and biolure attractants in Guatemala. *Florida Entomologist* 90, 258–263.

Martínez, N.B., Rincon, J., Perez, A., Linares, B. and Giraldo, H. (1994) Reconocimiento de *Anastrepha grandis* Diptera: Tephritidae en areas productoras de melón en Venezuela. *Agronomia Tropical* 44, 337–342.

Massa, M.J., Robacker, D.C. and Patt, J. (2008) Identification of grape juice aroma volatiles and attractiveness to the Mexican fruit fly (Diptera: Tephritidae). *Florida Entomologist* 91, 266–276.

McPhail, M. (1937) Relation of time of day, temperature, and evaporation to attractiveness of fermenting sugar solution to Mexican fruit fly. *Journal of Economic Entomology* 30, 793–798.

McPhail, M. (1939) Protein lures for fruit flies. *Journal of Economic Entomology* 32, 767–769.

McPhail, M. and Bliss, C.I. (1933) Observations on the Mexican fruit fly and some related species in Cuernavaca, Mexico in 1928 and 1929. *U.S. Department of Agriculture Circular* 255, 1–24.

Meza, J.S., Nirmala, X., Zimowska, G.J., Zepeda-Cisneros, C.S. and Handler, A.M. (2011) Development of transgenic strains for the biological control of the Mexican fruit fly, *Anastrepha ludens*. *Genetica* 139, 53–62.

Miranda, M., Leyva-Vázquez, J.L., Mota-Sánchez, D. and Collado, J. (1996) Developmental responses of *Anastrepha ludens* (Diptera: Tephritidae) to insect growth regulators. *Agrociencia* 30, 12–24.

Montoya, P., Liedo, P., Benrey, B., Barrera, J.F., Cancino, J. and Aluja, M. (2000a) Functional response and super parasitism of *Diachasmimorpha longicaudata* (Hymenoptera: Braconidae), a parasitoid of fruit flies (Diptera: Tephritidae). *Annals of the Entomological Society of America* 93, 47–54.

Montoya, P., Liedo, P., Benrey, B., Cancino, J., Barrera, J.F., Sivinski, J. and Aluja, M. (2000b) Biological control of *Anastrepha* spp. (Diptera: Tephritidae) in mango orchards through augmentative releases of *Diachasmimorpha longicaudata* (Ashmead) (Hymenoptera: Braconidae). *Biological Control* 18, 216–224.

Montoya, P., Cancino, J., Zenil, M., Santiago, G. and Gutiérrez, J.M. (2007) The augmentative biological control component in the Mexican National Campaign against *Anastrepha* spp. fruit flies. In: Vreysen, M.J.B., Robinson, A.S. and Hendrichs, J. (eds) *Area-Wide Control of Insect Pests: From Research to Field Implementation*. Springer, Dordrecht, The Netherlands, pp. 661–670.

Morató, S.R. (2010) Competitividad sexual de machos de *Anastrepha ludens* Loew (Diptera: Tephritidae) al exponerlos al aroma del aceite esencial de dos hospederos preferenciales. BSc thesis, Universidad Nacional Autónoma de México (UNAM), México City, México.

Moreno, D.S., Martínez, A. and Riviello, M.S. (1994) Cyromazine effects on the reproduction of *Anastrepha ludens* (Diptera: Tephritidae) in the laboratory and in the field. *Journal of Economic Entomology* 87, 202–211.

Moreno, D.S., Celedonio, H., Mangan, R.L., Zavala, J.L. and Montoya, P. (2001) Field evaluation of a phototoxic dye, Phloxine B, against three species of fruit flies (Diptera: Tephritidae). *Journal of Economic Entomology* 94, 1419–1427.

Morgante, J.S., Malavasi, A. and Bush, G.L. (1980) Biochemical systematic and evolutionary relationships of some Neotropical *Anastrepha*. *Annals of the Entomological Society of America* 73, 622–630.

Moya, P., Flores, S., Ayala, I., Anchis, J., Montoya, P. and Primo, J. (2010) Evaluation of lufenuron as a chemosterilant against fruit flies of the genus *Anastrepha* (Diptera: Tephritidae). *Pest Management Science* 66, 657–663.

Murillo, T. and Jiron, L.F. (1994) Egg morphology of *Anastrepha obliqua* and some comparative aspects with eggs of *Anastrepha fraterculus* (Diptera: Tephritidae). *Florida Entomologist* 77, 342–348.

NAPPO (2012) *Anastrepha ludens* (Mexican fruit fly) eradicated in the United States. Phytosanitary Alert System. www.pestalert.org/oprDetail.cfm?oprID=511, accessed 24 August 2012.

Nascimento, A.S. do, Malavasi, A. and Morgante, J.S. (1988) Programa de monitoramento de *Anastrepha grandis* (Macquart, 1845) (Diptera: Tephritidae) e aspectos da sua biologia. In: Souza, H.M.L. (ed.) *Moscas-das-frutas no Brasil*, ANAIS. Fundação Cargill, Campinas SP, Brazil.

Norrbom, A.L. (1985) Phylogenetic analysis and taxonomy of the *cryptostrepha, daciformis, robusta*, and *schausi* species groups of *Anastrepha* Schiner (Diptera: Tephritidae). PhD thesis, Pennsylvania State University, University Park, USA.

Norrbom, A.L. (2000) *Fruit Fly (Diptera: Tephritidae) Host Plant Database: Anastrepha grandis*. United States Department of Agriculture, Agricultural Research Service, Systematic Entomology Laboratory, www.sel.barc.usda.gov/diptera/tephriti/Anastrep/grandis.htm, accessed 31 July 2012.

Norrbom, A.L. (2004) Fruit fly (Tephritidae) host plant database. Version November 2004, www.sel.barc.usda.gov:8080/diptera/Tephritidae/TephHosts/search.html, accessed 31 July 2012.

Ortega-Zaleta, D.A. and Cabrera-Mireles, H. (1996) Productos naturales y comerciales par a la captura de *Anastrepha obliqua* M. en trampas McPhail en Veracruz. *Agricultura Técnica de México* 22, 63–75.

Ovruski, S., Aluja, M., Sivinski, J. and Wharton, R. (2000) Hymenopteran parasitoids on fr uit-infesting Tephritidae (Diptera) in Latin America and the southern United States: diversity, distribution, taxonomic status and their use in fruit fly biological control. *Integrated Pest Management Reviews* 5, 81–107.

Papaj, D.R. and Aluja, M. (1993) Temporal dynamics of host-marking in the tropical tephr itid fly, *Anastrepha ludens*. *Physiological Entomology* 18, 279–284.

Pérez-Staples, D. and Aluja, A. (2006) Sperm allocation and cost of mating in a tropical tephitid fruit fly. *Journal of Insect Physiology* 52, 839–845.

Pérez-Staples, D., Aluja, M., Macías-Ordóñez, R. and Sivinski, J. (2008) Reproductive trade-offs from mating with a successful male: the case of the tephritid fly *Anastrepha obliqua*. *Behavioral Ecology & Sociobiology* 62, 1333–1340.

Piñero, J., Aluja, M., Vázquez, A., Equihua, E. and Varón, J. (2003) Human ur ine and chic ken feces as fruit fly (Diptera: Tephritidae) attractants for resource-poor fruit growers. *Journal of Economic Entomology* 96, 334–340.

Rasgado, M.A., Malo, E.A., Cruz-López, L., Rojas, J. and Toledo, J. (2009) Olf actory response of the Mexican fruit fly (Dipter a: Tephritidae) to *Citrus aurantium* volatiles. *Journal of Economic Entomology* 102, 585–594.

Robacker, D.C. and Heath, R. (1996) Attraction of Mexican fruit flies (Dipter a: Tephritidae) to lures emitting host-fr uit volatiles in citr us orchard. *Florida Entomologist* 79, 600–602.

Robacker, D.C., Garcia, J.A. and Hart, W.G. (1990) Attraction of a laboratory strain of *Anastrepha ludens* (Diptera: Tephritidae) to the odor of fermented chapote fruit and to pheromones in laboratory experiments. *Environmental Entomology* 19, 403–408.

Ríos, E., Toledo, J. and Mota-Sánchez, D . (2005) Evaluación de atrayentes alimenticios para la captura de la mosca me xicana de la fr uta (Diptera: Tephritidae) en el Socon usco, Chiapas, México. *Manejo Integrado de Plagas y Agroecología* 76, 41–49.

Ruiz-Guzmán, I. (2010) Efecto del hacinamiento sobre la longe vidad, fecundidad y f ertilidad de la Mosca Me xicana de la F ruta *Anastrepha ludens* Loew (Diptera: Tephritidae). BSc thesis, Benemérita Universidad Autónoma de Puebla, Puebla, México.

Rull, J. and Barreda-Landa A. (2007) Colonization of a hybrid strain to restore male *Anastrepha ludens* (Diptera: Tephritidae) mating competitiveness for sterile insect technique prog rams. *Journal of Economic Entomology* 100, 752–758.

Rull, J., Brunel, O. and Méndez, M.E. (2005) Mass rearing history negatively affects mating success of male *Anastrepha ludens* (Diptera: Tephritidae) reared for sterile insect technique programs. *Journal of Economic Entomology* 98, 1510–1516.

Seín Jr, F. (1933) *Anastrepha* fruit flies in Puer to Rico. *Journal Department of Agriculture of Puerto Rico* 17, 183–196.

Sharp, J.L. and Hallman, G.J . (1994) *Quarantine Treatments for Pests of Food Plants*. Westview, Boulder, Colorado.

Sharp, J.L., Ovye, M.T., Ingle, S.J. and Har t, W.G. (1989) Hot-water quarantine treatment f or mangoes from Mexico infested with Mexican fruit fly and West Indian fr uit fly (Dipter a: Tephritidae). *Journal of Economic Entomology* 86, 1657–1662.

Silva, J.G. and Malavasi, A. (1993a) Mating and oviposition behavior of *Anastrepha grandis* in the laboratory. In: Aluja, M. and Liedo , P. (eds) *Fruit Fly: Biology and Management*. Springer-Verlag, New York, New York, pp. 181–185.

Silva, J.G. and Mala vasi, A. (1993b) The status of honeydew melon as a host of *Anastrepha grandis* (Diptera: Tephritidae). *Florida Entomologist* 76, 516–519.

Silva, J.G. and Mala vasi, A. (1996) Life cycle of *Anastrepha grandis*. In: McPheron, B .A. and Steck, G.J. (eds) *Fruit Fly Pests: A World Assessment of Their Biology and Management*. St. Lucie Press , Delray Beach, Florida, pp. 347–351.

Sivinski, J.M. and Dodson, G. (1992) Sexual dimorphism in *Anastrepha suspensa* (Loew) and other tephritid fruit-flies (Diptera, Tephritidae) - possible roles of de velopmental rate, fecundity, and dispersal. *Journal of Insect Behavior* 5, 491–506.

Sivinski, J., Aluja, M. and López, M. (1997) Spatial and temporal distributions of par asitoids of Mexican *Anastrepha* species (Diptera: Tephritidae) within canopies of fr uit trees. *Annals of the Entomological Society of America* 90, 604–618.

Sivinski, J., Vulinec, K., Menezes, E. and Aluja, M. (1998) The bionomics of *Coptera haywardi* (Ogloblin) (Hymenoptera: Diapriidae) and other pupal par asitoids of tephr itid fruit flies (Diptera). *Biological Control* 11, 193–202.

Sivinski, J., Piñero, J. and Aluja, M. (2000) The distributions of parasitoids (Hymenoptera) of *Anastrepha* fruit flies (Diptera: Tephritidae) along an altitudinal gradient in Veracruz, Mexico. *Biological Control* 18, 258–269.

Soto-Manatiu, J. and Jirón, L.F. (1989) Studies on the population dynamics of the fruit flies in *Anastrepha* (Diptera: Tephritidae) associated with mango (*Mangifera indica* L.) in Costa Rica. *Tropical Pest Management* 35, 425–427.

Steck, G.J., Carroll, L.E., Celedonio-Hurtado, H. and Guillen-Aguilar, J. (1990) Methods for identification of *Anastrepha* larvae (Diptera: Tephritidae), and key to 13 species. *Proceedings of the Entomological Society of Washington* 92, 333–346.

Stone, A. (1942) *The Fruit Flies of the Genus Anastrepha*. U.S. Department of Agriculture Miscellaneous Publication 439, Washington, DC, pp. 112.

Thomas, D. and Mangan, R. (2005) Nontarget impact of Spinosad GF-120 bait sprays for control of the Mexican fruit fly (Diptera: Tephritidae) in Texas citrus. *Journal of Economic Entomology* 96, 1950–1956.

Thomas, D.B. (1995) Predation on the soil inhabiting stages of the Mexican fruit fly. *Southwestern Entomologist* 20, 61–71.

Thomas, D.B. (1997) Degree-day accumulations and seasonal duration of the pre-imaginal stages of the Mexican fruit fly (Diptera: Tephritidae). *Florida Entomologist* 80, 71–79.

Thomas, D.B. (2004) Hot peppers as a host for the Mexican fruit fly *Anastrepha ludens* (Diptera: Tephritidae). *Florida Entomologist* 87, 603–608.

Thomas, D.B., Holler, T.C., Heath, R.R., Salinas, E.J. and Moses, A.L. (2001) Trap–lure combinations for surveillance of *Anastrepha ludens* fruit flies (Diptera: Tephritidae). *Florida Entomologist* 84, 344–351.

Toledo, J. and Lara, J.R. (1996) Comparison of the biology of *Anastrepha obliqua* reared in mango (*Mangifera indica* L.) and in mombin (*Spondias mombin*) infested under field conditions. In: McPheron, B.A. and Steck, G.J. (eds) *Fruit Fly Pests: A World Assesment of Their Biology and Management*. St. Lucie Press, Delray Beach, Florida, pp. 359–362.

Toledo, J., Gurgúa, J.L., Liedo, P., Ibarra, J.E. and Oropeza, A. (2001) Parasitismo de larvas y pupas de la mosca mexicana de la fruta, *Anastrepha ludens* (Loew) (Diptera: Tephritidae) por el nemátodo *Steinernema feltiae* (Filipjev) (Rhabditida: Steinernematidae). *Vedalia* 8, 27–36.

Toledo, J., Pérez, C., Liedo, P. and Ibarra, J.E. (2005a) Susceptibilidad de larvas de *Anastrepha obliqua* Macquart (Diptera: Tephritidae) a *Heterorhabditis bacteriophora* (Pinar) (Rabditida: Heterorhabditidae) en condiciones de laboratorio. *Vedalia* 2, 11–22.

Toledo, J., Ibarra, J.E., Liedo, P., Gómez, A., Rasgado, M.A. and Williams, T. (2005b) Infection of *Anastrepha ludens* (Diptera: Tephritidae) larvae by *Heterorhabditis bacteriophora* (Rhabditida: Heterorhabditidae) under laboratory and field conditions. *Biocontrol Science and Technology* 15, 627–634.

Toledo, J., Rasgado, M., Ibarra, J.E., Gomez, A., Liedo, P. and Williams, T. (2006) Infection of *Anastrepha ludens* following soil applications of *Heterorhabditis bacteriophora* in a mango orchard. *Entomologia Experimentalis et Applicata* 119, 155–162.

Toledo, J., Malo, E., Cruz-López, L. and Rojas, J. (2009) Field evaluation of potential fruit-derived lures for *Anastrepha obliqua* (Diptera: Tephritidae). *Journal of Economic Entomology* 102, 2072–2077.

USDA-APHIS (2012) Plant protection and quarantine manuals. www.aphis.usda.gov/import_export/plants/manuals/ports/downloads/fv.pdf, accessed 4 August 2012.

Ventura-González, C., Hernández, E., Anaya, L. and Figueroa, S.M. (2009) Efecto de biopelículas a base de quitosan sobre el desarrollo del huevo y larva de *Anastrepha ludens* en frutos de *Mangifera indica* Cv. Ataulfo) In: Montoya P., Díaz-Fleischer, F. and Flores, F. (eds) *7th Meeting of the Working Group on Fruit Flies of the Western Hemisphere*, Mazatlán, México, pp. 40.

Villaseñor, A., Lara, A.G., Cuellar, R.O., Gutiérrez-Ruelas, J.M., Zavala, J.L and Hernández-Baeza, J.L. (2009) Sistema de Geolocalización (GPS) y Administración de la Información (GIS) en línea Web ligados en un solo sistema llamado T-SIGA, para las operaciones de campo del Programa Regional Moscamed Chiapas-México y Guatemala In: Montoya, P., Díaz-Fleischer, F. and Flores, F. (eds) *7th Meeting of the Working Group on Fruit Flies of the Western Hemisphere*, Mazatlán, Mexico, pp. 18–19.

Weems Jr, H.V. (2006) *South American Cucurbit Fruit Fly, Anastrepha grandis (Macquart) (Insecta: Diptera: Tephritidae) (EENY-205)*. University of Florida, Institute of Food and Agricultural Sciences, Gainesville, Florida.

White, I.M. and Elson-Harris, M.M. (1992) *Fruit Flies of Economic Significance: Their Identification and Bionomics*. CAB International, Wallingford, UK, pp. 601.

Wilson, W., Oropeza, A., Montoya, P., Flores, S. and Toledo, J. (2009) Infección de *Anastrepha ludens* con el hongo *Beauveria bassiana* aplicado a suelo con diferentes rangos de humedad y temperatura. In: Montoya, P., Díaz-Fleischer, F. and Flores, F. (eds) *7th Meeting of the Working Group on Fruit Flies of the Western Hemisphere*, Mazatlán, México, pp. 158.

Zucchi, R.A. (2000) Especies de *Anastrepha*, sinonimias, plantas hospedeiras e parasitóides. In: Malavasi, A. and Zucchi, R.A. (eds) *Moscas-Das-Frutas Importância Econômica no Brasil-Conhecimiento Básico e Aplicado*. Editora Holos, Ribeirão Preto, Brazil, pp. 41–48.

12 *Bactrocera* Species that Pose a Threat to Florida: *B. carambolae* and *B. invadens*

Aldo Malavasi,[1] David Midgarden[2] and Marc De Meyer[3]
[1]*Medfly Rearing Facility – Moscamed Brasil, Juazeiro, Bahia, Brazil;*
[2]*USDA/APHIS, Guatemala City, Guatemala;* [3]*Royal Museum for Central Africa, Tervuren, Belgium*

12.1 Introduction

Tephritidae is one of the largest families of Diptera and contains more than 500 genera and 4000 species, divided into three subfamilies (White and Elson-Harris, 1992; Norrbom et al., 1999). Tephritidae pests are particularly important because of their ability to invade regions far from their native distribution. Introduced populations attack commercial fruit species, which causes countries importing fruit to impose quarantine regulations (McPheron and Steck, 1996). These restrictions can inhibit the sale of produce and the development or expansion of fruit production in the areas in which the pest species are established.

As their name implies, many members of the family are frugivorous (feed on fruit), and the most important pest species have a high capacity to disperse to and colonize new areas. There are three major characteristics that give Tephritidae a status of good potential invasive species:

1. A large and rapid rate of population growth. This allows many Tephritid species to increase their population size dramatically in a short period of time. In addition to the increase in density, one or few gravid females can rapidly infest a large number of hosts, expanding the geographic distribution of the population from a single point, (e.g., a backyard or garden tree) to adjacent areas and commercial groves.

2. High natural ability of dispersion. Some frugivorous fruit fly species are good flyers and can disperse quickly and in large number when suitable host trees are not available or are out of season. Well-fed adults – males and females – can fly large distances in search of reproductive and oviposition sites or just for shelter. Experiments using the mark-release-recapture methodology have shown that either males or females can travel many kilometers when the environment is inadequate. In addition, being physically strong, the adults can be carried large distances by wind, hurricanes and masses of warm air, a fairly common phenomenon in the atmosphere. Because of such events, Japan keeps a trapping network in the southernmost island of its archipelago, close to Taiwan. The distance between Taiwan and the Yonaguni Island is 180 km. Japan is a fruit fly-free country as it had conducted a large eradication program some decades ago and Taiwan remains infested by some species of *Bactrocera*. Although the distance is large the Japanese Ministry of Agriculture, Forestry and Fisheries trapping system occasionally captures *Bactrocera* adults in the islands close to the strait.

3. High anthropogenic dispersion. In fruit fly species, the egg and larval stages are necessarily inside a fresh fruit. It is not always possible to distinguish when a fruit is infested with eggs or larvae of fruit fly. Some fruits, such as guavas (*Psidium guajava* L.), carambolas (*Averrhoa caramboa* L.) and oranges (*Citrus* spp.) usually do not reveal any external evidence that they are infested unless in advanced ripe stage. Others, such as apple *Malus* × *domesticum*, peaches *Prunus persica* L. and papayas *Carica papaya* L. show in early stages that they are infested.

Due to these characteristics, many infested fruits can be carried by people traveling large distances. In some cultures (e.g., Latin America, South-East Asia and sub-Saharan Africa), it is quite common for people to carry fresh fruit as an easy source of food, since it is ready for consumption at any time. Therefore, fruit infested with fruit fly eggs or larvae can travel very large distances, because the cultural habit of people is associated with the modern transportation system, allowing a large number of immature insects to move distances that would be impossible by natural flight. This was the case for *B. carambolae*, the carambola fruit fly (CFF) (Fig. 12.1), introduced in Suriname in the north of South America in the early 1970s, and discussed in Section 12.3, as well as for the introduction of the invasive fruit fly *B. invadens* into Africa.

Many outbreaks of fruit flies follow a similar pattern. Every year, Mediterranean fruit fly or medfly (*Ceratitis capitata*) and *B. dorsalis*, the oriental fruit fly (OFF) are brought into California, USA, by travelers bringing fruits from infested countries. Either returning US tourists or foreigners bringing their preferred fruits to the USA are the most common sources of fruit fly detections in California. In 1993, it was suggested by a Scientific Advisory Panel of the Medfly Program in California that APHIS would carry out 100% of inspections at Los Angeles International Airport for all arriving international flights for at least 1 day. The result was that in 24 h of full inspection, 73 fruit fly larvae were intercepted. Considering this number for one full year, it is estimated that more than 25,000 larvae are brought into California every year.

12.2 Host Range and Colonization Ability

Tephritids can be categorized by the number of fruit species that they attack. Monophagy refers to attacks on a single fruit species; oligophagy refers to attacks on different host plants belonging to the same plant family; polyphagous flies infest many fruit species belonging to different families. Polyphagous flies have the potential to invade new territory when compared with mono- or oligophagous flies. For example, the medfly, *C. capitata*, is now a cosmopolitan species that infests more than 300 host plants (Liquido *et al.*, 1991). The origin of medfly is sub-Saharan Africa and it was probably brought to Europe by navigators in the 17th or 18th centuries.

B. dorsalis, the Oriental fruit fly, and *B. cucurbitae*, the melon fruit fly, occur in the Hawai'ian Archipelago and are categorized as being polyphagous and oligophagous, respectively. The record of outbreaks of these two species in California, USA, is around 10:1, giving a good measure of the relative aggressiveness of both species.

12.3 Carambola Fruit Fly, *Bactrocera carambolae*

The carambola fruit fly (CFF) (Fig. 12.1), is a native of Indonesia, Malaysia and Thailand, and was first collected in South America in 1975 in Paramaribo, Suriname. The flies collected were not identified and CFF was not found again until 1981 when specimens were sent to the US Department of Agriculture and identified as *Dacus dorsalis*, the OFF. Although the OFF is one of the world's most serious pests, no action was taken at that time. In 1986, the international community realized that the presence of OFF in Suriname would represent an important threat

Fig. 12.1 *Bactrocera carambolae* adults.

to the production and marketing of fruit throughout tropical and subtropical America and the Caribbean Basin Region. The fly was later found to be a separate but closely related species from the OFF, *B. carambolae*, the CFF.

Unsurprisingly, there are direct ties between Suriname and the native area of CFF, as both Suriname and Indonesia are former colonies of The Netherlands. Today the country has an estimated 450,000 inhabitants, 15% of whom are descending from Indonesian colonizers from Java in the early 20th century. Suriname maintains strong ties with The Netherlands and > 90% of trade and travel is with that country. During the colonial period and even after the independence in 1972, the Dutch KLM was the only airline to operate flights to and from Suriname.

When Drew and Hancock reviewed the tephritids in Asia, the OFF in Suriname was identified as a species that occurred in Thailand and Indonesia and which attacked carambola, and was classified as *B. carambolae*.

By 1998 it was clear (Malavasi et al., 2000) that the route of CFF introduction into South America was a traveler from Indonesia flying from Djakarta to Amsterdam and then from Amsterdam to Paramaribo in a 12,000-km journey. The distance between South-East Asia (where CFF occurs naturally), and the north of South America, does not allow an introduction by the regular maritime shipping lines.

Lack of funding and coordination among the international community allowed the fly to expand its geographic distribution. CFF was first found in French Guyana in the late 1980s and in Brazil in March, 1996.

12.3.1 Taxonomic status

The subfamily Dacinae (Christenson and Foote, 1960; Fletcher, 1987; White and Elson-Harris, 1992) are Old World in origin, principally tropical Asia and Africa, and are represented by over 800 species (including *Bactrocera* and *Ceratitis* spp.). Until the population in the Guyanas region of South America was found (Hancock, 1990; van Sauers-Muller, 1991), no member of the *B. dorsalis* complex of fruit fly was permanently established in the New World (Bateman, 1972; Drew et al., 1978; Vargas et al. 1990). Isolated specimens and small infestations of several *B. dorsalis* complex species have been found in Chile and the USA in California and Florida, but action by respective federal and state authorities prevented permanent establishment.

For much of the 20th century, the OFF has been held responsible for enormous losses to fruit and vegetable crops throughout Asia and South-East Asia (Drew and Hancock, 1994). Doubts about the singularity of this species began in the 1950s and 1960s (Hardy, 1969) and Drew and Hancock (1994) have now identified 52 morphologically and behaviorally similar species in what they term the *B. dorsalis* complex. Forty of these species are newly described. CFF is the member of this complex (Drew, 1989) that has established in South America.

12.3.2 Biology

Mating

There are two basic strategies that have been identified in tephritid mating (Prokopy, 1980; Shelly and Kaneshiro, 1991): (i) resource guarding, where the males stake out territory on a resource such as a host fruit, and (ii) lekking, a grouping of males of a species for the purpose of attracting females (Sivinski and Burk, 1989; Drew, 1987). Females are attracted to these leks when they are physiologically prepared for mating. Mating in *B. carambolae* occurs just before dark when the light intensity becomes less than 1000 lux (McInnis et al., 1999) and commonly occurs on host plants. In Suriname, the first mating by female *B. carambolae* usually takes place 18 days after emergence.

Oviposition

Females puncture the unripe, healthy fruit with their ovipositors, making cavities in which they lay their eggs. Females can lay as many as 3000 eggs over their lifetime in the laboratory, though some consider 1200–1500 to be the usual lifetime production under field conditions.

Adult feeding

B. dorsalis complex adults require a rich diet that is high in amino acids, vitamins, minerals, carbohydrates and water to survive and reproduce (Fletcher, 1987). Both males and females must feed daily, and they forage on non-host as well as host trees. Their diet has been observed to include honeydew, plant exudates, extrafloral nectaries, pollen, fruit juice, ripe fruits, microorganisms and bird droppings (Hendrichs and Hendrichs, 1990; Hendrichs and Prokopy, 1994). Both sexes appear to respond equally well to protein baits, though hypothetically, females should be more attracted because of their need for protein for egg production.

Life stages

The life stages of *B. carambolae* are egg, larvae (three instars), pupa and adult; they complete the cycle from egg to reproductive adult in 30–40 days. Some experiments have shown that the adult flies can remain alive as long as 125 days; generally, however, lifespan varies with temperature and availability of food. In mountainous regions of Hawai'i, for example, adult *B. dorsalis* can live up to 1 year (Vargas *et al.*, 1984). Once the larvae hatch, they begin to feed and burrow into the pulp of the fruit. During development, the larvae tunnel in the fruit and feed on the tissues and associated bacteria. When mature, larvae leave the fruit and burrow several centimeters into the soil, where they pupate.

Host range and geographic distribution

The host range and geographic distribution of *B. dorsalis* complex is very broad, with more than 150 varieties of fruit attacked throughout tropical and subtropical Asia. The host range of *B. dorsalis* complex species in Malaysia is more limited. van Sauers-Muller (1991) has reported that the host range of the *B. carambolae* in Suriname is very similar to that of CFF in Indonesia and Malaysia.

A regional CFF eradication program was established in Paramaribo in 1997 with aim of eradicating the species from South America. In 3 years the distribution of CFF was reduced in 80% in Suriname, it was eradicated from Guiana and kept the population only in Oiapoque in the French–Brazilian border (Fig. 12.2).

In 2002 when CFF was present in the urban areas of French Guiana as well as in isolated agricultural areas, the program was terminated by the government of France. As a result of the program interruption, the CFF returned to its original distribution and spread toward the north and west in the Republic of Guyana and toward the south and east in Amapa (Brazil), reaching Macapa, the state capital. The present distribution of CFF in South America includes all Suriname and French Guyana; west, central and south of Guiana; and the State of Amapa in Brazil. The host list for the northern part of South America is presented in Table 12.1.

12.3.3 Sampling and monitoring techniques

Most members of the Dacinae complex are strongly attracted to methyl eugenol, cue lure [4-(p-acetoxypheyl)-2-butanone] and/or Wilson's lure (which is the hydroxy derivative of cue-lure) (Cunningham, 1989). These chemicals are termed pseudo- or parapheromones because they attract only one sex and are found in the environment rather than being produced by the opposite sex. A trap baited with methyl eugenol is the most frequently used method of sampling for *B. dorsalis* complex species. Both males and females can be trapped using McPhail traps that are baited with a protein diet and water. This method of sampling has advantages over the male-only traps, but the results are variable and highly dependent on rainfall and humidity.

Two types of traps are used to detect CFF in South America:

1. Jackson traps, impregnated with a minimal amount of a methyl eugenol-malathion mixture and with a density of 10 traps per km^2, and checked biweekly to detect the presence of male CFF.

2. McPhail traps, baited with 300 ml of protein hydrolyzate food lure, placed with a density of two units per km^2, and checked every week to determine the presence of both male and female CFF, as well as other fruit flies.

Larvae surveys are performed upon fruits collected from the host plants. Naked eye, microscopic inspection and pupae recovery of the fruits are used to determine the presence of larvae, helping to qualify the state of infestation of a given area.

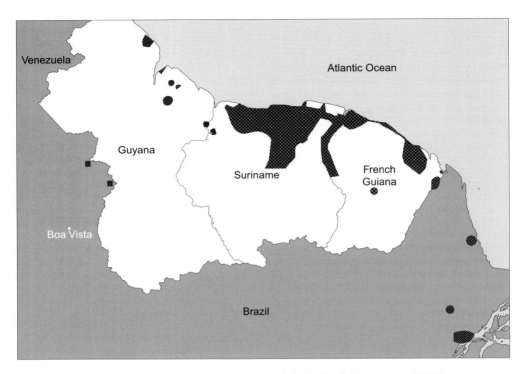

Fig. 12.2 Geographical distribution of *Bactrocera carambolae* in South America as of 2009.

12.3.4 Economic importance

CFF is listed along with other species in the *B. dorsalis* complex on the list of quarantine insects for the European Economic Community (EPPO, 2011). The presence of CFF in Suriname, French Guiana, Guyana and Brazil represents a threat to the production and marketing of fruit and vegetables throughout tropical and subtropical South and Central America and the Caribbean. Lack of action to control CFF when it was first detected allowed it to become established as a pest of significant economic importance (Vo and Miller, 1995; Sanchez, 1998). Discovery of this pest resulted in the imposition of quarantine restrictions by importing countries in 1986, reducing the exports of fruits and vegetables from Suriname and preventing the development and improvement of fruit exports in Suriname and French Guiana.

Spread of the CFF beyond its present territory would bring about dramatic changes in the fruit production levels and value, and in the current trade patterns of the newly infested countries. Economic losses would occur due to a decrease in value of the produce of the fruit crops that serve as host to the pest (caused by fruit damage and reduced yield); and export earnings would be drastically affected, as fruit exports from newly infected countries would rejected or subjected to restrictions in the international markets.

An economic analysis conducted by Sanchez (1998) showed that accumulated benefits on production over the twelve years projected a range from US$239 million to US$359 million, of which Brazil obtains the larger share, US$194–291 million. Venezuela, Cuba and Colombia, in that order, also profit substantially from the program. As expected, there is a close correlation between the proportion of the value of fruit produced and the value of the benefits obtained. Upon the completion and success of the eradication program, benefits will also be derived from the avoidance of export restrictions. Exports benefits are markedly larger than those for production. They range

Table 12.1. Host list for the northern part of South America.

Hosts	Scientific name	English	French
Primary			
Oxalidaceae	*Averrhoa carambola* L.	Carambola	Carambole
Myrtaceae	*Syzygium samarangense* (Blume) Merr. and Perry	Java apple	Pomme rosa
Secondary			
Anacardiaceae	*Mangifera indica* L.	Mango	Mangue
Malpighiaceae	*Malpighia punicifolia* L.	West indian cherry	Cerises pays
Myrtaceae	*Psidium guajava* L.	Guava	Goyave
Sapotaceae	*Manilkara achras* (Mill.) Fosberg	Sapodilla	Sapotille
Occasional			
Sapotaceae	*Chrysophyllum cainito* L.	Star apple	
Rhamnaceae	*Zizyphus mauritiana* (jujuba) Lam.	Indian jujube	Jujube
Myrtaceae	*Syzygium malaccense* (L.) Merr. and Perry	Malay apple	Pomme de madagascar Pomme d'amour
Myrtaceae	*Eugenia uniflora* L.	Surinam cherry	Cerise du Suriname
Rutaceae	*Citrus aurantium* L.	Sour orange	Orange amere
	Citrus reticulata Blanco	Mandarin	Mandarine
	Citrus sinensis (L.) Osbeck	Orange	Orange
	Citrus paradisi Macfad.	Grapefruit	Pamplemousse
Combretiaceae	*Terminalia catappa* L.	Tropical almond	
Anacardiaceae	*Anacardium occidentale* L.	Cashew	Noix de cajou
	Spondias mombin L.	Hogplum	
Clusiaceae	*Garcinia dulcis* (Roxb.) Kurz		
Guttiferae	*Mammea americana* L.	Mamey apple	

from a minimum of US$1604 million to a maximum of US$1848 million. The proportion between value of total fruit exports and benefits derived from the eradication of the pest is again high. Brazil receives the larger share of the benefits, which range from US$1192 million to US$1362 million in the twelve years encompassed, while Cuba, exporting a high proportion of its fruit production, ranks second with advantages ranging from US$262 million to US$315 million.

12.3.5 Control tactics

Two primary control tactics are used in the CFF eradication campaign: male annihilation and foliage baiting.

Male annihilation technique

CFF strongly responds to the male lure methyl eugenol. Male annihilation involves the widespread distribution of bait stations impregnated with methyl eugenol and the insecticide malathion. The lure attracts males to the stations where they feed, ingest the insecticide and die quickly. For the technique to be effective, bait stations must be evenly distributed throughout the whole infested area.

Over a period of time, progressive reduction of the male fly population results in fewer and fewer fertile females, until eventually the population reaches a very low level and may be completely eradicated (Koyama *et al.*, 1984). The technique is highly specific to a small number of fruit fly species, is environmentally benign and presents minimal health risks to human and animals. The male annihilation technique, MAT, is the basic method of eradication when an outbreak occurs in free areas.

In South America two kinds of bait stations are used for MAT. One is a wooden fiber block that is soaked in the malathion methyl-eugenol mixture; the other uses a 5–10 ml spot sprayed onto a tree or electricity pole. In the spray, the methyl eugenol and insecticide are mixed together with Min-U-Gel® (ITC Industrials-Floridin, Hunt Valley, Maryland). This is a powder that, when mixed in the right proportions, makes a type of jelly or grease. This gel lasts longer after being sprayed than the liquid by itself. More recently a combination of methyl-eugenol and spinosad, an organic insecticide, has been used to control the outbreaks in areas free from *Bactrocera*.

Protein baiting technique

Protein (hydrolyzed yeast protein), mixed with the insecticide malathion or the new generation of insecticides (spinosad) is used to spray the foliage, preferably of fruiting host plants. The protein acts as a food attractant to both males and females, particularly immature females which need protein to develop eggs. Flies feed on the protein, ingest the insecticide and die quickly. All fruit fly species, not just CFF, are attracted to protein baits. Where protein baiting and male annihilation can be carried out at the same time in infested areas, control of the CFF can be achieved more quickly and with greater efficiency. Where trap catches or larva surveys in fruit indicate a significant localized breeding population (i.e. a 'hot spot'), a protein-bait strategy is also employed.

12.3.6 Potential risk of introduction into Florida

To make an assessment on the risk of introduction, the following points should be considered:

- CFF has already been present in the north of South America for almost 40 years.
- There are no records of CFF presence in Venezuela and the Caribbean.
- The control program carried out in Brazil, if not achieving eradication, has at least successfully kept the species in a containment area.
- There is no evidence of CFF presence in the Amazon forest.
- Few new hosts have been reported after the first wave of invasion.
- There are no direct flights between Paramaribo and Miami.
- Agricultural products exported from Suriname to the USA are negligible.

Based on the above facts, the risk of introduction of CFF into Florida is considered low.

12.4 *Bactrocera invadens*

The first specimens of *B. invadens* (Bi) (Fig. 12.3) were found along the Kenyan coast during surveys conducted in 2003 within the framework of the African Fruit Fly Initiative (an international program co-ordinated by the International Centre of Insect Physiology and Ecology, Nairobi, Kenya) (Lux *et al.*, 2003). A few months later, the species was found in Tanzania (Mwatawala *et al.*, 2004), and was subsequently reported from countries throughout the African continent. Although the genus *Bactrocera* occurs in Africa, the indigenous species belong to different subgenera and Bi does not have any close relatives in this zoogeographical region (White, 2006). Taxonomists recognized the specimens as being an alien species, belonging to the *B. dorsalis* complex (Lux *et al.*, 2003). Drew *et al.* (2005) concluded that the African specimens belonged to a hitherto undescribed species and that identical specimens were collected in Sri Lanka. They described the new species as *B. invadens*.

Although previously unknown and apparently of no economic significance in its region of origin, Bi has made a tremendous impact in Africa

Fig. 12.3 *Bactrocera invadens* (copyright R.S. Copeland, reproduced by permission).

since it was first introduced. The species has demonstrated an unprecedented dispersal rate and is currently considered one of the major pest species, especially for the horticultural industry of tropical fruits such as mangoes. It is also a strong competitor displacing indigenous pest species of fleshy fruits, mainly representatives of the genus *Ceratitis*. It further aggravates the problems that local growers are facing because of both the indigenous pests and the exotic fruit flies such as *B. cucurbitae* and *B. zonata*, which were introduced to the continent earlier. Because of its economic significance and invasive behavior, it is considered a potential threat for many other regions with significant fruit-producing economies.

12.4.1 Taxonomic status

B. invadens belongs to the subgenus *Bactrocera* s.s. (White, 2006) and is very similar to *B. dorsalis* and related taxa. At first, it was identified as being an aberrant form of *B. dorsalis* (Lux et al., 2003), but was later recognized as a distinct species (Drew et al., 2005) belonging to the *B. dorsalis* complex as described by Drew and Hancock (1994). Specimens of Bi, however, demonstrate a high variability in thoracic and abdominal patterns (Drew et al., 2005), and morphological differences to unambiguously differentiate it from allied species are minor (White, 2006; Drew et al., 2008) and not easy to recognize by non-specialists. Recent studies regarding hybridization (Jessup et al., 2010), nucleotide sequences of mitochondrial COI (cytochrome oxidase region I), morphometrics (Schutze et al., 2012) and rectal volatiles (Tan et al., 2010) do not provide any evidence of distinct separation between particular species of the *Bactrocera dorsalis* complex, leading to assumptions that Bi and other species within the complex could be identical to *B. dorsalis* (Tan et al., 2010; Schutze et al., 2012). However, the extreme complexity of species boundaries within the *B. dorsalis* species complex and the actual taxonomic status of Bi should be investigated further.

12.4.2 Invasion history

Bi has an Asian origin, and a genetic study of Sri Lankan and African populations have confirmed that the former belongs to the native range (Khamis et al., 2009). The pathway by which Bi entered Africa is not clear, nor is there any definite evidence that eastern Africa was the original port of entry. The East African coast has historical ties with the Middle East and Central Asia, dating back several centuries (Gilbert, 2004) when there was a thriving trading business with sailing boats (dhows) transporting merchandise between the regions. However, no specimens of Bi could be found in any historical collection from the 20th century or before. Between 1999 and 2000, the African Fruit Fly Initiative occasionally deployed methyl eugenol traps in Kenya, Tanzania and Uganda (Lux et al., 2003) and an intensive fruit sampling and rearing program ran in Kenya between 1999 and 2003 (Copeland et al., 2006). No specimens were detected in that period. Similarly, fruit fly surveys were conducted in different countries in western Africa from 2000 onwards (Vayssières et al., 2005) but the species was not reported. All this seems to indicate that it was not present (or at very low abundance) prior to 2003. Since its first detection, however, numbers have risen sharply. In addition, within a timespan of 2 years (February 2003–January 2005), the species was recorded in very high numbers in no fewer than 11 countries ranging from Senegal in the west to Ethiopia and Tanzania in the east (Drew et al., 2005; De Meyer and White, 2008) (Fig. 12.4). Currently, it is known from more than 20 countries throughout the continent, as well as from Madagascar and the Comoro Islands in the Indian Ocean.

Most remarkable is the fact that, only 9 months after its first detection in Kenya, specimens were collected in Nigeria in western Africa (Umeh et al., 2008). It is unlikely that the species dispersed by natural means over this distance in such a short time. Khamis et al. (2009) showed that the population subdivision in Africa is not on a geographical basis but that three distinct African clusters are apparent, which can be interpreted as different invasion bursts or outbreaks. One of these originated from Nigeria. It is, therefore, not unlikely that the Nigerian detection represents an introduction event independent of those in East Africa. However, more studies are required to resolve the invasion history of the species in Africa further.

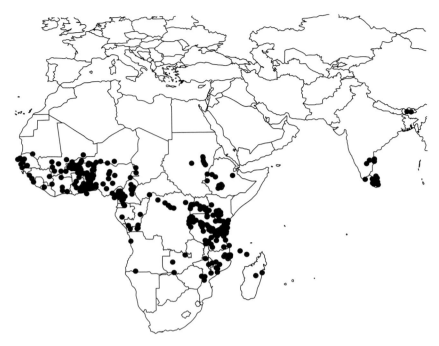

Fig. 12.4 Distribution of *Bactrocera invadens*.

12.4.3 Distribution

Knowledge of the distribution of Bi in its region of origin is fragmentary. The species seems to be widespread in Sri Lanka (De Meyer *et al.*, 2010) but records from other parts of the Indian subcontinent are rare. Some specimens from southeast India were recognized as Bi (Sithanantham *et al.*, 2006), while a few specimens were found among a large series of *B. dorsalis* from Bhutan (Drew *et al.*, 2007). Again, confusion with other species of the *B. dorsalis* complex might disturb a correct observation of the distribution pattern of Bi. Several closely related species, such as *B. kandiensis*, *B. dorsalis* s.s., or *B. rubigina* occur in the region, and a more in-depth study and survey should be conducted in order to establish the exact range of Bi in Central Asia.

Because of the absence of congeners (species within the genus *Bactrocera*) in Africa, the distribution of the species is well known there. As mentioned earlier, the species became widespread throughout the continent in a fairly short time. It is currently officially reported from most countries south of the Sahara, but excluding the southernmost part of Africa. The occurrence southwards is more restricted, with data from the northern parts of Namibia, Botswana and Mozambique. In the latter country, the spread of the fly has been surveyed since the first record in 2007 (Correia *et al.*, 2008), using an extensive network of methyl eugenol-baited traps along the major transportation routes and in horticultural areas (Cugala *et al.*, 2011). It is abundant and permanently established in the northern provinces, while still absent from the south. The area roughly between the Zambezi and Save Rivers has so far shown irregular occurrences of the fly, but the number of interceptions has been increasing in this region over the last years. It is not clear whether these records reflect true natural dispersal of the fly, or if they are the result of (increased) accidental introductions by human activities.

De Meyer *et al.* (2010) developed an ecological niche model, based on associations between known occurrence data and a set of environmental variables. The results of this modeling suggest that Bi prefers hot and humid environments with high – although not necessarily continuous – annual precipitation. This preference is indirectly supported by the relative abundance of Bi along altitudinal transects. Observations in eastern Africa have shown that its presence is lower at higher elevations that often also have high humidity but lower temperatures (Ekesi et al., 2006; Mwatawala et al., 2006a). The majority of known occurrence sites of Bi in Africa coincides with the predicted areas. Those sites outside the predicted areas might be due to anthropogenic microclimates, such as the observed high abundance in irrigated regions along the Nile River in Sudan.

12.4.4 Host list

Bi is a truly polyphagous species, attacking a wide variety of fleshy fruits. All known host data originate from rearing experiments in Africa. No host records are available from the area of origin. An overview of all host records reported so far can be found on the Invasive Fruit Fly Pests in Africa website[i] which is updated at regular intervals. So far, 65 species belonging to 26 families have been listed. This includes most tropical, subtropical and temperate climate fruits that are grown commercially in Africa, but also a number of ornamental exotics and wild indigenous fruits. In addition, some records are known from plants categorized as vegetables, such as solanaceous and cucurbitaceous crops.

With regard to incidence and infestation rates, the impact of Bi on the different hosts is very varied. Some hosts seem to be infested preferentially while others are only rarely attacked. Also, the number of emerging flies per unit of fruit (e.g. weight) can differ widely. Mwatawala et al. (2006b, 2009) present an in-depth study on the host range and utilization of Bi and co-occurring fruit flies in Tanzania, while Vayssières et al. (2009) present extensive data for Benin in western Africa. Both studies show a high preference of Bi for the following hosts (only those occurring in both areas): mango, guava and tropical almond. Other commercial crops such as avocado, banana or carambola (starfruit) are non-preferential hosts with limited infestation in both regions. On the other hand, while most citrus species show relatively low infestation rates in Tanzania and the drier Sudanian areas of Benin, they are considered more preferential hosts in the Guinean area of Benin (Vayssières et al., 2009). Similar differences in infestation indices of *Citrus* were also observed in Kenya (Rwomushana et al., 2008). Host preference, therefore, seems to be influenced by factors other than just presence or absence of the host.

12.4.5 Biology

Development

Ekesi et al. (2006) provide data on the development of Bi immature stages on a yeast–carrot-based artificial diet at 28°C. They showed that immature development takes on average 25 days (egg incubation 1.2 d; larval stage 11.1 d; pupa to adult 12.4 d) with 55% survival, and the mean generation time is 31 days. Life expectancy at pupal eclosion is lower for females (75.1 d) than for males (86.4 d). Females were observed to have a net fertility of 608 eggs. A further study by Rwomushana et al. (2008) showed that developmental times for all stages is affected by temperature, with the shortest at 30°C (17.76 d) gradually becoming longer under colder temperatures (75.74 d at 15°C). Lower developmental thresholds, based on linear regression equations, were between 8.7 and 9.4°C for the different immature life stages, while 35°C proved to be the higher threshold for adult emergence.

Temporal and spatial abundance

Bi shows distinct seasonal fluctuations in abundance. Mwatawala et al. (2006b) and Vayssières et al. (2005, 2009) demonstrated that a distinct peak period coincides with the mango fruiting season. In both regions, this also marks the onset of the rainy season (unimodal in Benin, bimodal in Central Tanzania, with the mango fruiting season falling in the short rainy season). In Tanzania, a second peak occurs with the onset of the guava fruiting season. However, the relative abundance strongly declines along an altitudinal transect, and there are indications that presence at higher elevations is the result of dispersal from lower regions during high abundance.

Interspecific competition

Duyck et al. (2004) state that life-history strategies play a role in determining the competitive interactions between invasive species and the pre-established species. Many invasive *Bactrocera* species, including Bi, demonstrate K traits, such as large adult size, that seem to favor both exploitation and interference competition. The competitive displacement of other fruit flies by *Bactrocera* species has been demonstrated in a number of cases (Duyck *et al.*, 2004, 2007). In Kenya, evidence for such displacement of the indigenous mango fruit fly (*Ceratitis cosyra*) by Bi has been shown by Ekesi *et al.* (2009) and Rwomushana *et al.* (2009). This plays an important role in evaluating the potential impact of Bi in new areas.

12.4.6 Control tactics

As with several other species of the *B. dorsalis* species complex, Bi is also strongly attracted to methyl eugenol. This parapheromone is the optimal tool for surveillance and monitoring activities, but only attracts males. Both sexes can be trapped using protein bait or other food baits such as three-components lure, but the attractiveness is much lower than methyl eugenol, as demonstrated by Mwatawala *et al.* (2006a). The use of methyl eugenol has shown its effectiveness in surveillance programs in South Africa, where it is intensively used for early detection (Venter *et al.*, 2010).

Control strategies are very similar to those outlined earlier in this article for the Carambola fruit fly. MAT, in combination with protein bait spraying and other possible eradication actions such as fruit stripping, are considered the appropriate plan of action. South Africa has developed an action plan along these lines to combat any possible detection (Manrakhan *et al.*, 2010).

12.4.7 Economic importance

The economic impact of Bi in Africa cannot be underestimated. Although the horticultural industry had already suffered considerable losses due to indigenous fruit fly pests such as medfly (*Ceratitis capitata*), mango fruit fly (*C. cosyra*) and Natal fruit fly (*C. rosa*), the problem has become dramatic since Bi spread throughout the continent. De Meyer *et al.* (2010) showed that many other parts of the world, mainly in tropical areas, are at least climatically suitable for the survival of Bi. Other regions, therefore, are anxious to avoid the introduction of this invasive pest in their regions. Several alerts have been issued, such as the quarantine order by the USA,[ii] or the A1 listing by EPPO. EPPO also developed a pest risk analysis to evaluate the different pathways along which Bi can enter the region. Because of the similar environmental conditions with areas where Bi occurs in Africa, the Ministry of Agriculture of Brazil has set up a detection system in those Brazilian cities with direct flights from infested areas in Africa.

Florida is potentially at risk. GARP, one of the ecological niche models developed by De Meyer *et al.* (2010), indicates southern Florida as suitable for establishment of the pest, with regard to prevailing climate conditions only. Considering that Bi is attracted to methyl eugenol, the trapping system in the state will allow early detection of any invasion. It is considered that the eradication plan developed for *B. dorsalis* will also be applicable for Bi.

Notes

[i] www.africamuseum.be/fruitfly/AfroAsia.htm, accessed 26 July 2012.
[ii] www.aphis.usda.gov/import_export/plants/plant_imports, accessed 28 July 2012.

References

Bateman, M.A. (1972) The ecology of fr uit flies. *Annual Review of Entomology* 12, 493–519.
Christenson, L.D. and Foote, R.H. (1960) Biology of fruit flies. *Annual Review of Entomology* 5, 171–192.
Copeland, R.C., Wharton, R.A., Luke, Q., De Meyer, M., Lux, S., Zenz, N., Macher a, P. *et al.* (2006) Geographic distribution, host fruit, and parasitoids of African fruit fly pests *Ceratitis anonae*, *Ceratitis cosyra*, *Ceratitis fasciventris*, and *Ceratitis rosa* (Diptera: Tephritidae) in Kenya. *Annals of the Entomological Society of America* 99, 262–278.
Correia, A.R.I., Rega, J.M. and Olmi, M. (2008) A pest of significant economic impor tance detected for the first time in Mozambique: *Bactrocera invadens* Drew, Tsuruta & White

(Diptera: Tephritidae: Dacinae). *Bollettino di Zoologia agraria e di Bachicoltura, Ser. II* 40, 9–13.

Cugala, D., Mansell, M. and De Meyer, M. (2011) *Bactrocera invadens* surveys in Mozambique. Fighting fruit flies regionally in Sub-Sahar an Africa. *CIRAD/Coleacp Information Letter* 1, 3.

Cunningham, R.T. (1989) P arapheromones. In Robinson, A.S. and Hooper, G. (eds) *Fruit Flies: Their Biology, Natural Enemies and Control.* Elsevier, Amsterdam, The Netherlands/New York, New York, pp. 221–230.

De Meyer, M. and White, I.M. (2008) True fruit flies of the Afrotropical Region. http://projects.bebif.be/fruitfly/index.html, accessed 25 July 2012.

De Meyer, M., Rober tson, M.P., Mansell, M.W., Ekesi, S., Tsuruta, K., Mwaiko, M., Vayssières, J.-F. *et al.* (2010) Ecological niche and potential geographic distribution of the invasive fruit fly *Bactrocera invadens* (Diptera, Tephritidae). *Bulletin of Entomological Research* 100, 35–48.

Drew, R.A.I. (1987) Behavioural strategies of fruit flies of the genus *Dacus* (Diptera: Tephritidae) significant in mating and host–plant relationships. *Bulletin of Entomological Research* 77, 73–81.

Drew, R.A.I. (1989) The taxonomy and distribution of tropical and subtropical Dacinae (Dipter a: Tephritidae). In: Robinson, A.S. and Hooper, G. (eds) *Fruit Flies, Their Biology, Natural Enemies and Control, Vol. 3A.* Elsevier, Amsterdam, The Netherlands/New York, New York, pp. 83–90.

Drew, R.A.I. and Hancock, D.L. (1994) The *Bactrocera dorsalis* complex of fruit flies (Diptera: Tephritidae: Dacinae) in Asia. *Bulletin of Entomological Research* Suppl. 2, pp. 1–68.

Drew, R.A.I., Hooper, G.H.S. and Bateman, M.A. (1978) *Economic Fruit Flies of the South Pacific.* Watson Ferguson, Brisbane, Australia, pp. 139.

Drew, R.A.I., Tsuruta, K. and White, I.M. (2005) A new species of pest fruit fly (Diptera:Tephritidae:Dacinae) from Sri Lanka and Africa. *African Entomology* 13, 149–154.

Drew, R.A.I., Romig, M.C. and Dorji, C. (2007) Records of Dacine fruit flies and new species of *Dacus* (Diptera: Tephritidae) in Bhutan. *Raffles Bulletin of Zoology* 55, 1–21.

Drew, R.A.I., Raghu, S. and Halcoop, P. (2008) Bridging the morphological and biological species concepts: studies on the *Bactrocera dorsalis* (Hendel) complex (Diptera: Tephritidae: Dacinae) in South-east Asia. *Biological Journal of the Linnean Society* 93, 217–226.

Duyck, P.F., David, P. and Quilici, S. (2004) A review of relationships between interspecific competition and invasions in fruit flies (Diptera: Tephritidae). *Ecological Entomology* 29, 511–520.

Duyck, P.F., David, P. and Quilici, S. (2007) Can more K-selected species be better invaders? A case study of fruit flies in La Réunion. *Diversity and Distributions* 13, 535–543.

Ekesi, S., Nderitu, P.W. and Rwomushana, I. (2006) Field infestation, life history and demographic parameters of the fruit fly *Bactrocera invadens* (Diptera: Tephritidae) in Africa. *Bulletin of Entomological Research* 96, 379–386.

Ekesi, S., Billah, M.K., Nderitu, P.W., Lux, S.A. and Rwomushana, I. (2009) Evidence for competitive displacement of the mango fruit fly, *Ceratitis cosyra* by the invasive fruit fly *Bactrocera invadens* (Diptera: Tephritidae) on mango and mechanisms contributing to the displacement. *Journal of Economic Entomology* 102, 981–991.

EPPO (2011) List of Quarantine pest. www.eppo.org/QUARANTINE/listA1.htm, accessed 25 July 2012.

Fletcher, B.S. (1987) The biology of Dacine fruit flies. *Annual Review of Entomology* 32, 115–144.

Gilbert, E. (2004) *Dhows and the Colonial Economy of Zanzibar, 1860–1970.* Ohio University Press, Athens, Ohio.

Hancock, D. (1990) Oriental Fruitfly Survey and Detection. FAO project report TCP/RLA/8858, FAO, Rome.

Hardy, D.E. (1969) Taxonomy and distribuion of the Oriental fruit fly and related species (Tephritidae-Diptera). *Proceedings of the Hawaiian Entomological Society* 20, 395–428.

Hendrichs, J. and Hendrichs, M.A. (1990) Mediterranean fruit fly (Diptera: Tephritidae) in nature: location and diel pattern of feeding behavior and other activities on fruiting and nonfruting hosts and nonhosts. *Annals of the Entomological Society of America* 83, 632–641.

Hendrichs, J. and Prokopy, R.J. (1994) Food foraging behavior in frugivorous fruit flies. In: *Fruit Flies and the Sterile Insect Technique.* CRC Press, Boca Raton, Florida.

Jessup, A.J., Schutze, M.K., Clarke, A.R., Islam, A. and Wornoayporn, V. (2010) Field cage studies on mating isolation between species within the *Bactrocera dorsalis* complex. Abstract Volume, 8th International Symposium on Fruit Flies of Economic Importance. Valencia, Spain, pp. 91.

Khamis, F.M., Karam, N., Ekesi, S., De Meyer, M., Bonomi, A., Gomulski, L.M., Scolari, F., *et al.* (2009) Uncovering the tracks of a recent and rapid invasion: the case of the fruit fly pest *Bactrocera invadens* (Diptera: Tephritidae) in Africa. *Molecular Ecology* 18, 4798–4810.

Koyama, J., Teruya, T. and Tanaka, K. (1984) Eradication of the oriental fruit fly (Diptera: Tephritidae) from the Okinawa Islands by male annihilation. *Journal of Economic Entomology* 77, 468–472.

Liquido, N.J., Shinoda, L.A. and Cunningham, R.T. (1991) Host plants of the Mediterranean fruit fly (Diptera, Tephritidae). An annotated world review. *Miscellaneous Publications of the Entomological Society of America* 77, pp. 1–52.

Lux, S.A., Copeland, R.S., White, I.M., Manrakhan, A. and Billah, M.K. (2003) A new invasive fruit fly species from the *Bactrocera dorsalis* (Hendel) group detected in East Africa. *Insect Science and its Application* 23, 355–360.

Malavasi, A., van-Sauers, A., Midgarden, D., Kellman, V., Didelot, D., Caplong, P. and Ribeiro, O. (2000) Regional programme for the eradication of the Carambola fruit fly in South America. In: Tan, K.H. (ed.) *Area-Wide Control of Fruit Flies and Other Insect Pests*. Penerbit Universiti Sains Malaysia, Penang.

Manrakhan, A., Venter, J.H. and Hattingh, V. (2010) *Bactrocera invadens* Drew, Tsuruta and White, the African invader fly Action plan. Unpublished report, Citrus Research International South Africa, Nelspruit, South Africa, pp. 16.

McInnis, D.O., Rendon, P., Jang, E., van Sauers-Muller, A., Sugayama, R. and Malavasi, A. (1999) Interspecific mating of introduced sterile *Bactrocera dorsalis* with wild *B. carambolae* (Diptera: Tephritidae) in Suriname: a case for cross-species SIT. *Annals of the Entomological Society of America* 92, 758–765.

McPheron, B.A. and Steck, G.J. (1996) Preface. In: *Fruit Fly Pests: A World Assessment of Their Biology and Management*. St. Lucie Press, Delray Beach, Florida, xxi–xxii.

Mwatawala, M.W., White, I.M., Maerere, A.P., Senkondo, F.J. and De Meyer, M. (2004) A new invasive *Bactrocera* species (Diptera: Tephritidae) in Tanzania. *African Entomology* 12, 154–156.

Mwatawala, M.W., De Meyer, M., Makundi, R.H. and Maerere, A.P. (2006a) Biodiversity of fruit flies (Diptera, Tephritidae) at orchards in different agro-ecological zones of the Morogoro region, Tanzania. *Fruits* 61, 321–332.

Mwatawala, M.W., De Meyer, M., Makundi, R.H. and Maerere, A.P. (2006b) Seasonality and host utilization of the invasive fruit fly, *Bactrocera invadens* (Dipt., Tephritidae) in central Tanzania. *Journal of Applied Entomology* 130, 530–537.

Mwatawala, M.W., De Meyer M., Makundi, R.H and Maerere, A.P. (2009) Host range and distribution of fruit-infesting pestiferous fruit flies (Diptera, Tephritidae) in selected areas of Central Tanzania. *Bulletin of Entomological Research* 99, 629–641.

Norrbom, A.L., Carroll, L.E. and Freidberg, A. (1999) Status of knowledge. In: Thompson, F.C. (ed.) *Fruit Fly Expert Identification System and Systematic Information Database*. Leiden, The Netherlands, pp. 9–47.

Prokopy, R.J. (1980) Mating behavior of frugivorous Tephritidae in nature. In: Koyama, J. (ed.) *Fruit Fly Problems*. Proceedings Symposium National Institute Agricultural Sciences, Kyoto, Japan, pp. 37–46.

Rwomushana, I., Ekesi, S., Gordon, I. and Ogol, C.K.P.O. (2008) Host plant and host plant preference studies for *Bactrocera invadens* (Diptera: Tephritidae) in Kenya, a new invasive fruit fly species in Africa. *Annals of the Entomological Society of America* 101, 331–340.

Rwomushana I., Ekesi, S., Ogol, C.K.P.O. and Gordon, I. (2009) Mechanisms contributing to the competitive success of the invasive fruit fly *Bactrocera invadens* over the indigenous mango fruit fly, *Ceratitis cosyra*: the role of temperature and resource pre-emption. *Entomologia Experimentalis et Applicata* 133, 27–37.

Sanchez, O. (1998) Economic feasibility of the Carambola fruit fly eradication program. Report to IF AD/IICA/CFF Program, Paramaribo, Suriname.

Schutze, M.K., Krosch, M.N., Armstrong, K.F., Chapman, T.A., Englezou, A., Chomic, A., Cameron, S.L. et al. (2011) Population structure of *Bactrocera dorsalis* s.s., *B. papayae* and *B. philippinensis* (Diptera: Tephritidae) in southeast Asia: evidence for a single species hypothesis using mitochondrial DNA and wingshape data. *BMC Evolutionary Biology* 12, 130 (doi: 10.1186/1471-2148-12-130).

Shelly, T.E. and Kaneshiro, K.Y. (1991) Lek behavior of the Oriental fruit fly, *Dacus dorsalis*, in Hawaii (Diptera: Tephritidae). *Journal of Insect Behavior* 4, 235–241.

Sithanantham, S., Selvaraj, P. and Boopathi, T. (2006) The fruit fly *Bactrocera invadens* (Tephritidae: Diptera) new to India. *Pestology* 30, 36–37.

Sivinski, J. and Burk, T. (1989) Reproductive and mating behavior. In Robinson, A.S. and Hooper, G. (eds) *Fruit Flies: Their Biology, Natural Enemies and Control*. Elsevier, Amsterdam, The Netherlands, pp. 343–351.

Tan, K.H., Tokushima, I., Ono, H. and Nishida, R. (2010) Comparison of phenylpropanoid volatiles in male rectal pheromone gland after methyl eugenol consumption, and molecular phylogenetic relationship of four global pest fruit fly species: *Bactrocera invadens*, B. dorsalis, B. correcta and B. zonata. *Chemoecology* 21, 25–33, DOI 10.1007/s00049-010-0063-1.

Umeh, V., Garcia, L.E. and De Meyer, M. (2008) Fruit flies of citrus in Nigeria: species diversity, relative abundance and spread in major producing areas. *Fruits* 63, 145–153.

van Sauers-Muller, A. (1991) An overview of the Carambola Fruit Fly *Bactrocera* species (Diptera:Tephritidae), found recently in Suriname. *Florida Entomologist*, 74, 432–440.

Vargas, R.I., Miyashita, D. and Nishida, T. (1984) Life history and demographic parameters of three laboratory-reared Tephritids (Diptera: Tephritidae). *Annals of the Entomological Society of America* 77, 651–656.

Vargas, R.I., Stark, J.D. and Nishida, T. (1990) Population dynamics, habitat preference, and seasonal distribution patterns of oriental fruit fly and melon fruit fly (Diptera: Tephritidae) in an agricultural area. *Environmental Entomology* 19, 1820–1828.

Vayssières, J.F., Goergen, G., Lokossou, O., Dossa, P. and Akponon, C. (2005) A new *Bactrocera* species in Benin among mango fruit fly (Diptera: Tephritidae) species. *Fruits* 60, 371–377.

Vayssières, J.-F., Sinzogan, A. and Adandonon, A. (2009) *Range of Cultivated and Wild Host Plants of the Main Mango Fruit Fly Species in Benin*. Regional Fruit Fly Control Project in West Africa, CIRAD-IITA, Leaflet 8, pp. 4.

Venter, J.H., Manrakhan, A., Matabe, Y., Brown, L. and Stones, W. (2010) The threat of *Bactrocera invadens* Drew, Tsuruta & White, to South Africa: surveillance, contingency and trade response. Abstract Volume, 8th International Symposium on Fruit Flies of Economic Importance, Valencia, Spain, pp. 213.

Vo, T.T. and Miller, C. (1995) Economic feasibility of eradicating carambola fruit fly (*Bactrocera carambolae*) from South America. Report. USDA-APHIS, Washington, DC, pp. 56.

White, I.M. (2006) Taxonomy of the Dacina (Diptera: Tephritidae) of Africa and the Middle East. *African Entomology Memoir* 2, 1–156.

White, I.M. and Elson-Harris, M.M. (1992) *Fruit Flies of Economic Significance: Their Identification And Bionomics*. CAB International, Wallingford, UK.

13 Signature Chemicals for Detection of *Citrus* Infestation by Fruit Fly Larvae (Diptera: Tephritidae)

Paul E. Kendra,[1*] Amy L. Roda,[2] Wayne S. Montgomery,[1] Elena Q. Schnell,[1] Jerome Niogret,[1] Nancy D. Epsky[1] and Robert R. Heath[1]

[1]*USDA-ARS, Subtropical Horticulture Research Station, Miami, Florida 33158, USA;* [2]*USDA-APHIS-PPQ, Center for Plant Health Science and Technology, Miami, Florida 33158, USA*

13.1 Introduction

Tropical tephritid fruit flies are invasive pests that impact fruit production and global export. Current US appropriations for exotic fruit fly risk management programs exceed US$57 million per annum (USDA-APHIS, 2006). Primary threats to US agriculture include the *Anastrepha* species, which occur throughout the American tropics and subtropics (Aluja, 1994), and the Mediterranean fruit fly *Ceratitis capitata* (Wiedemann), considered one of the most destructive agricultural pests worldwide, with several hundred recognized hosts (Liquido *et al.*, 1991). Outbreaks of *C. capitata* have occurred in Florida, USA, as recently as June 2010 in Palm Beach County (FDACS, 2010), and eradication campaigns for two major invasions in 1997 (Hillsborough and Polk Counties) and 1998 (Lake, Highland, and Manatee Counties) were estimated to cost the state of Florida US$26 million and US$11.5 million, respectively (Silva *et al.*, 2003). Costs due to the major California infestation in 1980 included US$59 million for chemical control, US$38 million for quarantine and fumigation, US$260 million in crop losses, plus an additional one-time cost of US$497 million for construction of fumigation facilities (Burk and Calkins, 1983).

The Caribbean fruit fly, *Anastrepha suspensa* (Loew) is the only economically important species currently established in Florida (Weems *et al.*, 2001a). It is found throughout the southern peninsula, where its preferred hosts include common guava (*Psidium guajava* L.), Cattley guava (*P. cattleianum* Sabine), loquat (*Eriobotrya japonica* Lindl.) and Surinam cherry (*Eugenia uniflora* L.). Low levels of infestation also occur in grapefruit (*Citrus* × *paradisi* Macfad.), a principal export, thereby making *A. suspensa* a quarantine pest of Florida *Citrus* (Nguyen *et al.*, 1992; Greany and Riherd, 1993). Other *Anastrepha* pests pose a threat to Florida agriculture due to close proximity of populations in Mexico and the Caribbean basin. In addition, the large volume of foreign produce shipments entering the state's ports creates a potential pathway for pest entry and spread (Kendra *et al.*, 2007 and references therein). Species of particular concern include *A. ludens* (Loew), a *Citrus* pest which has been detected in Florida in recent years (Thomas, 2004), *A. grandis* (Macquart), a specialist on fruits in the Cucurbitaceae (Norrbom, 2000), and *A. obliqua* (Macquart), a serious pest of mango (*Mangifera indica* L.) and the most abundant *Anastrepha* species in the West Indies (Weems *et al.*, 2001b).

Due to the high economic impact of tephritid pests, much attention has been focused on development of trapping systems for detection and monitoring of adult populations (Heath et al., 1995; Casaña-Giner et al., 2001; IAEA, 2003; Thomas et al., 2008). However, improved methods are critically needed for detection of the immature stages as well. Adult females have well-developed ovipositors that insert eggs beneath the skin of host fruits. Larvae feed and develop concealed within the pulp, making infestation difficult to detect. At US ports of entry, quarantine inspectors currently check incoming produce shipments by examining a small sample (typically 2% or less) of fruit for external signs of pest boring/feeding, and if suspicious, by slicing open the fruit to search for tephritid larvae (USDA-APHIS, 2010). Efficacy of visual inspections is questionable, especially for first instar larvae which are clear to pale white and only 2–3 mm in length. Gould (1995) estimated that only about 35% of grapefruits infested with *A. suspensa* were detected by trained agricultural inspectors. If not subjected to appropriate quarantine treatments, infested fruit may be distributed to consumers. Kendra et al. (2007) estimated a likelihood of ~10% that a mated female *A. suspensa* could result from infested grapefruit discarded directly into the south Florida environment, which might occur as part of residential composting practices. A single mated female can lay an average of 1.9 ($\pm$ 0.3) eggs per day over a lifespan of 73.6 ($\pm$ 4.9) days (Sivinski, 1993). Due to the high risk of pest introduction should infested fruit evade detection, there is great demand for more sensitive, high-throughput screening methods for detection of tephritid larval infestation.

This study evaluates gas chromatography (GC) as a potential technology for improved detection of hidden insect infestation. It is well documented in the literature that insect herbivory can elicit changes in host plant chemistry and volatile emissions (reviewed in Karban and Baldwin, 1997; Howe and Jander, 2008). It has also been shown that chemical changes can occur within host fruit as a result of insect infestation (Boevé et al., 1996; Hern and Dorn, 2001; Carrasco et al., 2005). In this study we examined *Citrus* infested with *A. suspensa* to determine if infested fruit emitted a detectable chemical profile distinct from that of non-infested *Citrus*. Samples of headspace volatiles were collected at various stages of infestation and chemical analysis was performed with several types of GC equipment. Since the primary goal was development of a rapid screening protocol for 'signature chemicals', the majority of analyses were performed with a rapid (9 min) GC separation method. To evaluate the efficacy of separation with this rapid method, and to identify the volatile chemical components, a slower (25 min) high-resolution GC separation was performed in combination with mass spectral analysis. In addition, we conducted a preliminary evaluation of a portable ultra-high-speed GC analyzer for detection of these same chemicals using a method requiring less than 80 s for sampling and chemical analysis.

13.2 Materials and Methods

13.2.1 Infestation and sample preparation

Anastrepha suspensa were obtained from a laboratory colony maintained at the USDA-ARS, Subtropical Horticulture Research Station in Miami, Florida, USA. All flies were of known age and reared under the following conditions: 25 $\pm$ 2°C, 75 $\pm$ 5% RH, and a 12:12-h (L:D) photoperiod (Kendra et al., 2006). Adults of mixed sex (~1:1 sex ratio) were housed in screen rearing cages (30 $\times$ 30 $\times$ 30 cm) and provisioned with water (released from agar blocks) and food (refined cane sugar and yeast hydrolysate, 4:1 mixture) *ad libitum*, prior to collection for fruit infestation. Approximately 3500 mature (10–12 days old, presumed mated) females were collected by aspiration from the rearing cages and placed in each of two infestation cages (94 $\times$ 51 $\times$ 51 cm) constructed from PVC frames covered with mesh pollination bags (Delstar Technologies, Middletown, Delaware) (Fig. 13.1A). Each cage contained 50 ripe Florida-grown grapefruit (*Citrus* $\times$ *paradisi* 'Marsh Red', obtained from a local natural foods market). Oviposition (Fig. 13.1B) was allowed for 24 h, and then the fruit was removed and rinsed with distilled water to remove the fly excreta (and any potential volatiles it may have liberated).

Half of the infested fruit was randomly divided into five groups for chemical sampling at different stages of infestation: egg, first instar,

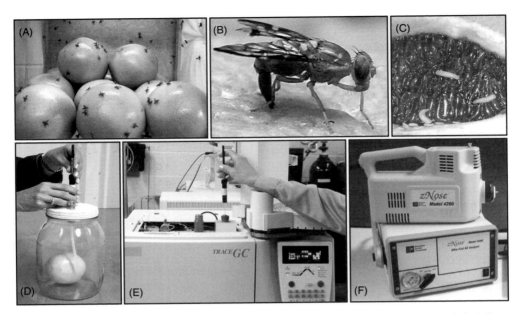

Fig. 13.1 Methods used for identification of signature chemicals from tephritid-infested citrus. Infestation cages (A) were set up with ripe grapefruit (*Citrus* × *paradisi* cv. Marsh Red) and mature females of the Caribbean fruit fly, *Anastrepha suspensa*, for 24 h oviposition (B). At 2–3-day intervals, several fruits were cut open to monitor the stage of larval development (C). Chemical sampling of infested fruit (and non-infested controls) was done by placing fruit in 3.85 L jars and collecting headspace volatiles by solid phase microextraction (SPME) (D). Chemicals were then separated and identified with standard gas chromatography-mass spectrometry (GC-MS) methods (E). A portable ultra-fast GC analyzer (F) was also evaluated for detection of signature chemicals using a method requiring less than 80 s for sampling and chemical analysis.

second instar, mid-third instar, and exiting third instar (final instar larvae exit the host fruit and enter a wandering stage prior to pupation in the soil). Two control treatments were also sampled; they consisted of non-infested fruit and mechanically injured fruit that were pierced with a tack five times to simulate oviposition wounds (tack length approximated length of *A. suspensa* ovipositor, 2.0 mm). The remaining half of the infested fruit was used to monitor progress of larval development. At 2–3 d intervals, several grapefruits were cut open to determine the larval instar and to estimate the level of infestation (Fig. 13.1C). Each segment was opened and the pulp separated and gently pressed to dislodge larvae. The albedo, pulp and juice were then examined under a microscope to detect larvae. Up until the time of chemical sampling, all fruit treatments were held in the laboratory at the same environmental conditions used for insect rearing. Following chemical collections, each sampled fruit was cut open and examined (as above) to document the developmental stage and the number of insects present.

13.2.2 Volatile collections and chemical analysis

Grapefruits were placed individually into 3.85 L jars with Teflon-lined lids fitted with short thru-hull ports (Swagelok; Solon, Ohio) and allowed to equilibrate for 30 min at 22°C. Headspace volatiles were collected using solid phase microextraction (SPME) with a 100-µm polydimethylsiloxane-coated (non-bonded) fiber (Supelco; Bellefonte, Pennsylvania). A sample was collected by inserting the SPME fiber through the port and exposing the fiber to headspace volatiles for 2 min adsorption (Fig. 13.1D).

Volatile profiles were obtained using a rapid separation method on a Trace™ GC (ThermoFisher; Waltham, Massachusetts) equipped with a DB-5 column (20 m × 0.18 mm × 0.18 μm) and a flame ionization detector (FID, 300°C). Chemicals were injected by thermal desorption (splitless injector, 250°C for 2 min) from the SPME fiber directly into the GC (Fig. 13.1E). Helium was used as the carrier gas at a constant flow rate of 0.05 mL s^{-1}. The temperature program consisted of an initial oven temperature of 50°C which was increased after injection at a rate of 0.583°C s^{-1} up to 220°C, and then held at 220°C for 4 min. Total run time was 9 min. Chemical analysis with the rapid GC method was performed on five replicate fruits per treatment (five infested and two control treatments).

To identify component peaks, additional SPME collections (as above) were analyzed by GC-mass spectrometry (GC-MS). Adsorbed chemicals were injected into a 5975B GC/MSD (Agilent; Santa Clara, California) equipped with an HP-5MS column (30 m × 0.25 mm × 0.25 μm) with helium as the carrier gas. The temperature program consisted of an initial oven temperature of 50°C which was increased at 0.167°C s^{-1} to 130°C, followed by a second ramp from 130°C to 210°C at 0.333°C s^{-1}. Total run time was 25 min. MSD source was set at 230°C, quadrapole at 150°C, and scans were recorded for mass range of 50–650 amu. Three replicate fruits per treatment were analyzed by GC-MS, and component peaks were identified using the NIST/EPA/NIH mass spectral library (NIST05) and confirmed by retention time and mass spectra of known standards.

A portable chemical profiling system incorporating an ultra-high-speed chromatography column (zNose® Model 4200; Electronic Sensor Technology, Newbury Park, California) was used for comparative analysis of selected samples (Fig. 13.1F). Headspace volatiles were collected by inserting the unit's intake needle into the sample port and allowing for 30 s adsorption (at a flow rate of 0.05 mL s^{-1}) onto an internal Tenax® trap. Chemicals were injected by thermal desorption and separated on a DB-5 column (1 m × 0.25 mm) using helium carrier gas, a temperature ramp of 40–175°C, and a surface acoustic wave (SAW) detector. Total run time from sample collection to GC separation was 79 s.

13.2.3 Statistical analysis

Prior to analysis, peak area of each chemical was normalized relative to the internal SPME standard for that GC run. For comparison of volatile profiles from different treatments, each chemical peak was evaluated separately by one-way analysis of variance (ANOVA) using Proc GLM (SAS Institute, 2001). Significant ANOVAs were followed by least significant difference test (LSD, $P < 0.05$) for mean separation. The Box-Cox procedure, which is a power transformation that regresses log-transformed standard deviations ($y + 1$) against log-transformed means ($x + 1$), was used to determine the type of transformation necessary to stabilize the variance before analysis (Box et al., 1978).

13.3 Results and Discussion

13.3.1 Level of infestation

Grapefruit dissections made immediately after chemical sampling indicated there were no significant differences in level of infestation among the five infested treatments ($F = 1.55$; df = 4, 24; $P = 0.226$). Mean (± SD) number of insects per fruit detected for each treatment was as follows: 17.0 (± 20.5) eggs, 25.8 (± 16.1) first instar larvae, 44.2 (± 24.9) second instar larvae, 43.4 (± 24.0) third instar larvae and 47.6 (± 22.1) exit holes from prepupal third instar larvae. Due to the difficulty with detection of the very early stages (Gould, 1995), the numbers reported for eggs and first instar larvae are likely to be under-representative of the actual infestation level in those two treatments. Dissections performed on the control fruits (non-infested and mechanically injured treatments) confirmed that they lacked immature stages of A. suspensa.

13.3.2 Identification of volatile constituents

There were 17 major peaks separated by the rapid GC analysis of grapefruit volatiles (Table 13.1). For all but three peaks, there were significant differences in quantities represented in the different treatments. One peak (RT 3.18 min) was greatly elevated in both the infested treatments and the injured fruit as compared to the non-infested controls. High resolution GC-MS analysis revealed that this broad peak represented two closely-eluting

Table 13.1. Normalized GC peak area (mean ± SD) of volatiles obtained by SPME collections and rapid (9 min) GC analysis from grapefruits (*Citrus x paradisi* cv. Marsh Red) infested with *Anastrepha suspensa* and non-infested controls. $n = 5$ per treatment.

RT (min)	Infested Treatments					Controls		F	P
	Egg	1st instar	2nd instar	3rd instar	Exit holes	Non-infest.	Injured		
3.18*	6136.4 ± 4661.1 a	2321.1 ± 2518.0 a	1914.5 ± 1587.0 a	715.2 ± 403.6 a	3507.2 ± 54.51.7 a	22.4 ± 50.2 b	7021 ± 3951.3 a	19.51	<0.0001
3.48*	66.4 ± 38.2 bc	329.8 ± 333.9 a	182.9 ± 129.8 ab	49.3 ± 52.3 c	63.6 ± 45.8 bc	45.1 ± 25.1 c	64.1 ± 39.6 bc	3.38	0.0124
3.61**	30.6 ± 14.2 a	2.2 ± 5.0 c	14.3 ± 20.4 bc	8.5 ± 10.7 bc	12.2 ± 13.2 ab	1.6 ± 3.6 c	6.7 ± 6.5 bc	3.71	0.0077
3.80**	19.1 ± 6.2 c	175.9 ± 101.9 a	76.9 ± 64.6 b	59.9 ± 55.0 bc	62.8 ± 42.1 bc	2.6 ± 5.8 d	50.4 ± 16.6 bc	9.03	<0.0001
4.00*	5.6 ± 3.4 a	27.5 ± 34.2 a	21.1 ± 21.3 a	9.5 ± 3.8 a	4.7 ± 3.7 a	0.0 ± 0.0 b	1.9 ± 4.2 b	10.58	<0.0001
4.22*	2.6 ± 5.8	3.3 ± 3.2	4.2 ± 3.6	12.2 ± 22.5	8.4 ± 8.9	0.0 ± 0.0	21.4 ± 36.4	1.95	NS
4.33*	8.4 ± 9.8 a	15.7 ± 6.5 a	6.3 ± 5.5 a	4.2 ± 4.9 a	2.5 ± 3.5 a	0.0 ± 0.0 b	13.6 ± 20.4 a	3.34	0.0131
4.53**	16.8 ± 19.4 abc	34.3 ± 28.9 a	25.8 ± 20.7 ab	9.7 ± 10.2 bcd	5.7 ± 4.3 cd	1.1 ± 2.6 d	15.9 ± 14.4 abc	3.69	0.0079
4.62*	31.5 ± 9.5 a	38.1 ± 20.6 a	46.6 ± 25.1 a	30.7 ± 23.2 a	26.1 ± 8.6 a	0.0 ± 0.0 b	20.1 ± 26.4 a	11.56	<0.0001
4.80**	42.6 ± 16.8 ab	63.4 ± 26.1 a	24.3 ± 10.9 bc	18.0 ± 12.2 c	23.4 ± 11.3 bc	20.7 ± 18.3 bc	57.5 ± 33.7 a	3.92	0.0057
4.89*	17.7 ± 7.7 a	19.4 ± 3.8 a	8.4 ± 3.7 ab	6.4 ± 4.7 b	8.7 ± 5.9 ab	0.6 ± 1.3 c	10.1 ± 11.8 b	5.94	0.0004
5.06*	62.4 ± 26.7	255.9 ± 123.2	146.7 ± 69.1	112.0 ± 80.1	113.7 ± 60.0	28.4 ± 24.6	101.9 ± 76.4	1.19	NS
5.18**	24.1 ± 8.3 ab	35.3 ± 17.0 a	19.0 ± 8.4 ab	16.7 ± 12.0 bc	18.7 ± 9.1 ab	6.7 ± 5.9 c	16 ± 7.2 bc	3.49	0.0106
5.28*	23.9 ± 11.5 a	8.8 ± 5.8 ab	8.4 ± 5.2 ab	4.5 ± 3.6 bc	7.1 ± 7.1 ab	0.0 ± 0.0 c	12.6 ± 17.8 ab	4.82	0.0017
5.45*	13.6 ± 5.8 a	6.1 ± 6.7 ab	1.9 ± 4.3 bc	1.9 ± 2.6 bc	1.0 ± 2.3 bc	0.0 ± 0.0 c	8.3 ± 12.0 ab	3.32	0.0135
5.69*	15.2 ± 23.9	3.1 ± 4.3	3.0 ± 2.8	0.9 ± 1.3	1.1 ± 2.5	0.0 ± 0.0	1.8 ± 4.0	1.6	NS
5.81*	34.4 ± 27.2 a	9.2 ± 4.6 a	6.5 ± 2.4 a	4.6 ± 2.8 a	6.2 ± 3.1 a	0.0 ± 0.0 c	2.1 ± 2.9 b	12.06	<0.0001

*Means followed by the same letter are not significantly different (LSD mean separation test on log $(x + 1)$ transformed data, $P < 0.05$; non-transformed means presented)
**Means followed by the same letter are not significantly different (LSD mean separation test on square-root $(x + 0.05)$ transformed data, $P < 0.05$; non-transformed means presented)

chemicals, D-limonene and β-ocimene (Fig. 13.2, peaks 5 and 6, respectively), and both chemicals were elevated with fruit infestation/injury. The highest levels were detected from fruit mechanically injured and fruit punctured by oviposition (Fig. 13.3A). Levels decreased with each progressive larval instar, apparently due to wound healing of the epidermis and epicarp (flavedo) of the grapefruit (Mulas et al., 1996). Levels again spiked when late third instar larvae began to exit the fruit and reinjure the peel. D-limonene and β-ocimene are known terpene constituents of *Citrus* peel, and they comprise up to 93% and 2.7% composition, respectively, of the peel oils in 'Marsh' grapefruits (Attaway et al., 1967). The large limonene/ocimene peak was therefore interpreted to be an indicator of *Citrus* peel damage, specifically a puncture wound (whether inflicted mechanically or by the female ovipositor).

Two broad peaks (RT 3.48 and 3.80 min, rapid method) were markedly elevated in the infested treatments relative to the non-infested controls and the mechanically injured fruit (Table 13.1). High resolution GC-MS showed that the 3.48 peak consisted of n-nonanal and an unidentified chemical (Fig. 13.2, peaks 7 and 8, respectively), but only the unknown compound was associated with larval infestation. The 3.80 peak consisted of hexyl butanoate (Fig. 13.2, peak 10), ethyl octanoate (Fig. 13.2, peak 11), and possibly another chemical (Fig. 13.2, unlabeled peak preceding peak 10), but hexyl butanoate was the primary chemical elevated with infestation. Hexyl butanoate (= hexyl butyrate) is a fruit ester, a major component of the aroma from apples *Malus domesticus* Borkh. (Matich et al., 1996), pears *Pyrus communis* L. (Argenta et al., 2003) and passion fruit *Passiflora edulis* Sims (Werkhoff et al., 1998), but in this study it was detected at very low levels in healthy intact grapefruit. Under natural conditions, hexyl butanoate emissions increase with the progression of fruit ripening, apparently an ethylene-mediated response (Lopez et al., 2007). With both hexyl butanoate and the unidentified chemical, the highest levels were observed in fruit infested with first-instar larvae (Fig. 13.3B, C), the stage of infestation most difficult to detect by visual inspection, and levels declined with subsequent instars. The ripening process would not account for the observed decrease in hexyl butanoate levels over time. This pattern of induced volatile emissions has been documented in another fruit commodity. Apples infested with larvae of the codling moth (*Cydia pomonella* L.) initially emitted high levels of esters and α-farnescene; the highest values were recorded from fruit infested with first-instar larvae, but with time the amounts decreased to levels equivalent to that from healthy fruits (Hern and Dorn, 2001). These elevated chemicals (in both commodities) may be interpreted as indicative of injury within the pulp or albedo layers as a result of larval feeding, but it is unclear why the levels would decrease during the later (larger) larval instars. Of the two peaks associated with tephritid-infested grapefruits, the unknown compound appeared to be the better candidate as a signature chemical for larval infestation. It was a much larger peak and there was better separation from neighboring peaks (Fig. 13.2A). Unfortunately, there was no match for this compound within the NIST mass spectral library. Additional work is needed to determine the chemical identity and source (host fruit, insect larvae, or microbial origin).

There were several other chemical peaks associated with fruit infestation and/or injury. These included isoamyl hexanoate (Table 13.1, RT 4.0; Fig. 13.2A, peak 13), co-eluting eugenol and hexyl hexanoate (Table 13.1, RT 4.53; Fig. 13.2A, peaks 14 and 15, respectively), and methyl eugenol (Table 13.1, RT 4.62; Fig. 13.2A, peak 17). Although significantly higher in the infested/injured fruit compared to non-infested fruit, the first three chemicals were detected at fairly low levels (i.e., small peaks), and with rapid GC separation the methyl eugenol co-eluted with β-elemene, a chemical found in significant quantities in non-infested fruit (Fig. 13.2B, peak 16). Therefore, it was concluded that none of these additional chemicals would serve well as reliable indicators of infestation. It was also noted that high-resolution GC-MS detected α-pinene, sabinene and β-myrcene (Fig. 13.2A, peaks 1–3, respectively) in infested fruit but not in the non-infested controls. Although potentially diagnostic of infestation, these small, fast-eluting chemicals were not resolved well by the rapid GC separation. The first clear peak eluting with the rapid GC method was the large limonene/ocimene peak. As with limonene, elevated levels of pinene and myrcene have been shown to be correlated with wounded *Citrus* fruit (Droby et al., 2008).

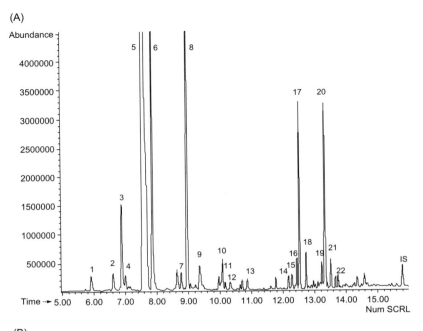

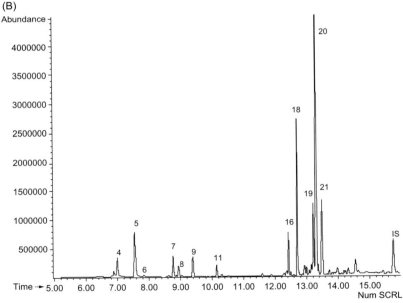

Fig. 13.2 High resolution (25 min) GC analysis of headspace volatiles obtained by SPME collections from grapefruit (*Citrus × paradisi* cv. Marsh Red). (A) Fruit infested with first instar larvae of the Caribbean fruit fly, *Anastrepha suspensa*. (B) Non-infested fruit. Peak identifications are as follows: 1. α-pinene, 2. sabinene, 3. β-myrcene, 4. ethyl hexanoate, 5. D-limonene, 6. β-ocimene, 7. n-nonanal, 8. unknown, 9. limonene oxide, 10. hexyl butanoate, 11. ethyl octanoate, 12. decanal, 13. isoamyl hexanoate, 14. eugenol, 15. hexyl hexanoate, 16. β-elemene, 17. methyl eugenol, 18. β-caryophyllene, 19. 2-isopropenyl-4a, 8-dimethyl-1,2,3,4,4a,5,6,7-octahydronaphthalene, 20. valencene, 21. α-panasinsen, 22. nerolidol, IS. internal standard.

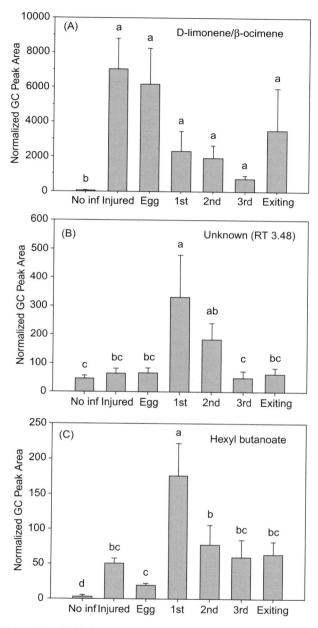

Fig. 13.3 Mean (± SE) quantity of (A) D-limonene/β-ocimene, (B) an unknown compound with retention time of 3.48 min, and (C) hexyl butanoate as determined by SPME collections and rapid (9 min) GC analysis of headspace volatiles from grapefruits (*Citrus* × *paradisi* cv. Marsh Red) that were non-infested, mechanically-injured, or infested with various immature stages of the Caribbean fruit fly, *Anastrepha suspensa*. Quantities are expressed as normalized GC peak areas relative to an internal SPME standard. Peaks for the unknown compound (B) and hexyl butanoate (C) contain small amounts of co-eluting n-nonanal and ethyl octanoate, respectively, but high resolution GC indicated neither was elevated with infestation. Bars topped with the same letter are not significantly different [LSD mean separation test on log $(x + 1)$ transformed data (A, B) or square-root $(x + 0.05)$ transformed data (C), $P < 0.05$; non-transformed means presented].

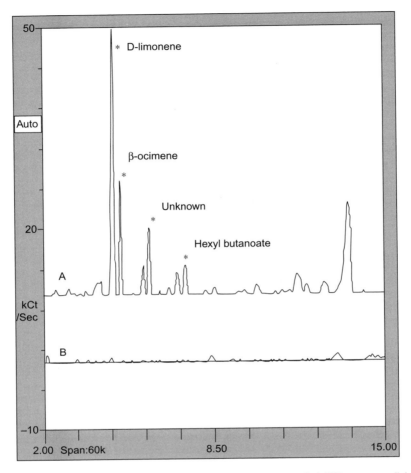

Fig. 13.4 Ultra-fast (79 s) GC analysis of headspace volatiles from grapefruit (*Citrus × paradisi* cv. Marsh Red). (A) Fruit infested with first-instar larvae of the Caribbean fruit fly, *Anastrepha suspensa*. (B) Non-infested fruit. Chemical peaks indicative of fruit injury and/or infestation are labeled. The peak for hexyl butanoate contains a small amount of co-eluting ethyl octanoate, but high resolution GC indicated it was not elevated with infestation.

Initial evaluation of the ultra-fast GC unit (Fig. 13.4) indicated that it was capable of detecting some of the same signature chemicals that were identified with high resolution GC. Compared to the 9 min separation method, the ultra-fast method actually gave better resolution of the lower molecular weight compounds. There was good separation between D-limonene and β-ocimene. Likewise, there was good resolution between n-nonanal and the unidentified chemical associated with infestation. Hexyl butanoate was also detectable with the ultra-fast method, but it co-eluted with ethyl octanoate as was observed with the 9 min GC method. The ultra-fast method gave poor resolution of the higher molecular weight compounds (RT > 9.0 s), but none of the peaks of interest eluted within that region.

Altered composition of host fruit odors (volatile profiles) as a result of tephritid larval infestation has been demonstrated previously, primarily through studies addressing host-seeking behavior in the fruit fly parasitoid *Diachasmimorpha longicaudata* (Ashmead) (Hymenoptera: Braconidae). Though chemical attractants were not identified, Eben et al. (2000)

used y-tube olfactometers to show that female *D. longicaudata* responded preferentially to odors from grapefruit and mango infested with *A. ludens* or *A. obliqua* versus non-infested control fruits. Carrasco et al. (2005) reported similar preferences in female *D. longicaudata* from bioassays that used hexanic and methanolic extracts from infested mangoes as compared to non-infested or mechanically damaged fruit. Comparative GC-MS analysis of the extracts indicated that there were both qualitative and quantitative differences in chemical content among the three mango treatments. Several compounds were elevated with infestation, including 3-carene, limonene, terpinolene and α-gurgenene, and one compound, 2-phenylethyl acetate, appeared to be unique to mangos infested with *A. ludens*.

13.4 Conclusions

Results of our study and that of Carrasco et al. (2005) indicate that there are GC-detectable volatile chemicals associated with tephritid infestation of fruit commodities. With infested grapefruits, the chemicals can be distinguished as those indicative of *Citrus* peel injury and those correlated with larval feeding (pulp/albedo injury). Of the chemicals identified, none appeared to be insect-produced, but rather natural fruit volatiles occurring at higher levels than normal. Elevated levels of D-limonene and β-ocimene are only indicative of puncture wounds or other external damage to the grapefruit peel. However, if these two chemicals are accompanied by elevated levels of hexyl butanoate and the (as of yet) unidentified compound, this volatile profile is potentially diagnostic of *Citrus* infestation. Preliminary tests indicate that these chemical signals, emitted from fruit with early stages of infestation, are detectable with the portable zNose® unit.

In recent years, sensitive chemical detection technology (zNose® and other forms of electronic nose) has been applied successfully in a variety of postharvest situations for early detection of insects and pathogens. These include detection of stink bugs (Hemiptera: Pentatomidae) and boll damage in cotton (Henderson et al., 2010), evaluation of fungal disease severity and stage of ripeness in mango (Li et al., 2009), and detection of lesser grain borers (Coleoptera: Bostrichidae) and extent of feeding damage in wheat (Zhang and Wang, 2007). If insect-infested commodities consistently release unique chemical profiles, this signature can be exploited to provide the basis for improved pest detection. Further evaluation of the tephritid/*Citrus* system is needed to (i) determine the sensitivity of larval detection by ultrafast GC methods; (ii) assess the applicability of these methods to other species of Tephritidae, other *Citrus* hosts and hosts of different ages (since background volatiles vary over time due to ripening and storage); and ultimately (iii) apply this technology toward development of rapid, more sensitive, non-destructive screening methods for detection of tephritid infestation at ports of entry.

Acknowledgments

The authors thank Aimé Vázquez, Luis Mazuera, Jorge Sanchez (USDA-ARS, Miami, Florida) and Scott Weihman (USDA-APHIS, Miami, Florida) for technical assistance; Rayko Halitschke (Cornell University, Ithaca, New York) and Jorge Peña (University of Florida, Homestead, Florida) for helpful suggestions with the manuscript; and USDA-APHIS-PPQ, Eastern Region, for purchase of the zNose® equipment. This article reports the results of research only. Mention of a proprietary product does not constitute an endorsement or recommendation by the USDA.

References

Aluja, M. (1994) Bionomics and management of *Anastrepha*. *Annual Review of Entomology* 39, 155–178.

Argenta, L.C., Fan, X. and Mattheis, J.P. (2003) Influence of 1-methylcyclopropene on ripening, storage life, and volatile production by d'Anjou cv. pear fruit. *Journal of Agriculture and Food Chemistry* 51, 3858–3864.

Attaway, J.A., Pieringer, A.P. and Barabas, L.J. (1967) The origin of citrus flavor components – III. A study of the percentage variations in peel and leaf oil terpenes during one season. *Phytochemistry* 6, 25–32.

Boevé, J.L., Lengwiler, U., Tollsten, L., Sorn, L. and Turlings, T.C.J. (1996) Volatiles emitted by apple fruitlets infested by larvae of the European apple sawfly. *Phytochemistry* 42, 373–381.

Box, G.E.P., Hunter, W.G. and Hunter, J.S. (1978) *Statistics for Experimenters. An Introduction to Design, Data Analysis, and Model Building.* J. Wiley & Sons, New York, New York.

Burk, T. and Calkins, C.O. (1983) Medfly mating behavior and control strategies. *Florida Entomologist* 66, 3–18.

Carrasco, M., Montoya, P., Cruz-Lopez, L. and Rojas, J.C. (2005) Response of the fruit fly parasitoid *Diachasmimorpha longicaudata* (Hymenoptera: Braconidae) to mango fruit volatiles. *Environmental Entomology* 34, 576–583.

Casaña-Giner, V., Gandía-Balanguer, A., Hernández-Alamós, M.M., Mengod-Puerta, C., Garrido-Vivas, A., Primo-Millo, J. and Primo-Yúfera, E. (2001) Attractiveness of 79 compounds and mixtures to wild *Ceratitis capitata* (Diptera: Tephritidae) in field trials. *Journal of Economic Entomology* 94, 898–904.

Droby, S., Eick, A., Macarisin, D., Cohen, L., Rafael, G., Stange, R., McColum, G. et al. (2008) Role of citrus volatiles in host recognition, germination and growth of *Penicillium digitatum* and *Penicillium italicum*. *Postharvest Biology and Technology* 49, 386–396.

Eben, A., Benrey, B., Sivinski, J. and Aluja, M. (2000) Host species and host plant effects on preference and performance of *Diachasmimorpha longicaudata* (Hymenoptera: Braconidae). *Environmental Entomology* 29, 87–94.

FDACS (2010) Mediterranean fruit flies found in Palm Beach County. Florida Department of Agriculture and Consumer Services. Press Release 15 June 2010. Available at: www.freshfromflorida.com/press/2010/06152010.html, accessed October 2010.

Gould, W.P. (1995) Probability of detecting Caribbean fruit fly (Diptera: Tephritidae) by fruit dissection. *Florida Entomologist* 78, 502–507.

Greany, P.D. and Riherd, C. (1993) Preface: Caribbean fruit fly status, economic importance, and control (Diptera: Tephritidae). *Florida Entomologist* 76, 209–211.

Heath, R.R., Epsky, N.D., Guzman, A., Deuben, B.D., Manukian, A. and Meyer, W.L. (1995) Development of a dry plastic insect trap with food-based synthetic attractant for the Mediterranean and the Mexican fruit fly (Diptera: Tephritidae). *Journal of Economic Entomology* 88, 1307–1315.

Henderson, W.G., Khalilian, A., Han, Y.J., Greene, J.K. and Degenhardt, D.C. (2010) Detecting stink bugs/damage in cotton utilizing a portable electronic nose. *Computers and Electronics in Agriculture* 70, 157–162.

Hern, A. and Dorn, S. (2001) Induced emissions of apple fruit volatiles by the codling moth: changing patterns with different time periods after infestation and different larval instars. *Phytochemistry* 57, 409–416.

Howe, G.A. and Jander, G. (2008) Plant immunity to insect herbivores. *Annual Review of Plant Biology* 59, 41–66.

IAEA (2003) *Trapping Guidelines for Area-Wide Fruit Fly Programmes.* Insect Pest Control Section, International Atomic Energy Agency, Vienna.

Karban, R. and Baldwin, I.T. (1997) *Induced Responses to Herbivory.* University of Chicago Press, Chicago, Illinois.

Kendra, P.E., Montgomery, W.S., Epsky, N.D. and Heath, R.R. (2006) Assessment of female reproductive status in *Anastrepha suspensa* (Diptera: Tephritidae). *Florida Entomologist* 89, 144–151.

Kendra, P.E., Hennessey, M.K., Montgomery, W.S., Jones, E.M. and Epsky, N.D. (2007) Residential composting of infested fruit: a potential pathway for spread of *Anastrepha* fruit flies (Diptera: Tephritidae). *Florida Entomologist* 90, 314–320.

Li, Z., Wang, N., Vijaya Raghavan, G.S. and Vigneault, C. (2009) Ripeness and rot evaluation of 'Tommy Atkins' mango fruit through volatiles detection. *Journal of Food Engineering* 91, 319–324.

Liquido, N.J., Shinoda, L.A. and Cunningham, R.T. (1991) *Host Plants of the Mediterranean Fruit Fly (Diptera: Tephritidae), an Annotated World Review.* Miscellaneous Publication 77. Entomological Society of America, Lanham, Maryland.

López, M.L., Villatoro, C., Fuentes, T., Graell, J., Lara, I. and Echeverría, G. (2007) Volatile compounds, quality parameters and consumer acceptance of 'Pink Lady 7' apples stored in different conditions. *Postharvest Biology and Technology* 43, 55–66.

Matich, A.J., Rowan, D.D. and Banks, N.H. (1996) Solid phase microextraction for quantitative headspace sampling of apple volatiles. *Analytical Chemistry* 68, 4114–4118.

Mulas, M., Lafuente, M.T. and Zacarias, L. (1996) Lignin and gum deposition in wounded 'Oroval' clementines as affected by chilling and peel water content. *Postharvest Biology and Technology* 7, 243–251.

Nguyen, R., Poucher, C. and Brazzel, J.R. (1992) Seasonal occurrence of *Anastrepha suspensa* (Diptera: Tephritidae) in Indian River

County, Florida, 1984–1987. *Journal of Economic Entomology* 85, 813–820.

Norrbom, A.L. (2000) *Anastrepha grandis* (Macquart). The Diptera Site. Systematic Entomology Laboratory, Agricultural Research Service, U.S. Department of Agriculture; and the Department of Entomology, National Museum of Natural History, Smithsonian Institution. Available at: www.sel.barc.usda.gov/diptera/tephriti/anastrep/grandis.htm, accessed October 2010.

SAS Institute (2001) *SAS/STAT Guide for Personal Computers, Version 8.2*. SAS Institute, Cary, North Carolina.

Silva, J.G., Meixner, M.D., McPheron, B.A., Steck, G.J. and Sheppard, W.S. (2003) Recent Mediterranean fruit fly (Diptera: Tephritidae) infestations in Florida – a genetic perspective. *Journal of Economic Entomology* 96, 1711–1718.

Sivinski, J.M. (1993) Longevity and fecundity in the Caribbean fruit fly (Diptera: Tephritidae): Effects of mating, strain and body size. *Florida Entomologist* 76, 635–644.

Thomas, D.B. (2004) Hot peppers as a host for the Mexican fruit fly, *Anastrepha ludens* (Diptera: Tephritidae). *Florida Entomologist* 84, 344–351.

Thomas, D.B., Epsky, N.D., Serra, C.A., Hall, D.G., Kendra, P.E. and Heath, R.R. (2008) Ammonia formulations and capture of *Anastrepha* fruit flies (Diptera: Tephritidae). *Journal of Entomological Science* 43, 76–85.

USDA-APHIS (2006) Exotic fruit fly strategic plan FY 2006–2010. U.S. Department of Agriculture Animal and Plant Health Inspection Service, Plant Protection and Quarantine. Available at: www.aphis.usda.gov/plant_health/plant_pest_info/fruit_flies/downloads/strategicplan06-19-06.pdf, accessed October 2010.

USDA-APHIS (2010) Fresh fruits and vegetables import manual. U.S. Department of Agriculture Animal and Plant Health Inspection Service, Plant Protection and Quarantine. Available at: www.aphis.usda.gov/import_export/plants/manuals/ports/downloads/fv.pdf, accessed October 2010.

Weems Jr, H.V., Heppner, J.B., Fasulo, T.R. and Nation, J.L. (2001a) Caribbean fruit fly, *Anastrepha suspensa* (Loew) (Insecta: Diptera: Tephritidae). University of Florida Featured Creatures. Available at: http://entnemdept.ufl.edu/creatures/fruit/tropical/caribbean_fruit_fly.htm, accessed October 2010.

Weems Jr, H.V., Heppner, J.B., Steck, G.J. and Fasulo, T.R. (2001b) West Indian fruit fly, *Anastrepha obliqua* (Macquart) (Insecta: Diptera: Tephritidae). University of Florida Featured Creatures. Available at: http://entnemdept.ufl.edu/creatures/fruit/tropical/west_indian_fruit_fly.htm, accessed October 2010.

Werkhoff, P., Güntert, M., Krammer, G., Sommer, H. and Kaulen, J. (1998) Vacuum headspace method in aroma research: Flavor chemistry of yellow passion fruits. *Journal of Agriculture and Food Chemistry* 46, 1076–1093.

Zhang, H. and Wang, J. (2007) Detection of age and insect damage incurred by wheat, with an electronic nose. *Journal of Stored Product Research* 43, 489–495.

14 Gall Midges (Cecidomyiidae) attacking Horticultural Crops in the Caribbean Region and South America

Juliet Goldsmith,[1] Jorge Castillo[2] and Dionne Clarke-Harris[3]

[1]*Plant Quarantine, Produce Inspection Branch, Ministry of Agriculture and Fisheries, Jamaica;* [2]*Universidad Nacional Agraria La Molina, Lima, Peru;* [3]*Caribbean Agricultural Research and Development Institute, Jamaica*

14.1 Introduction

Gall midges are a major problem of horticultural crops (tomatoes, potatoes, asparagus, pepper) in the Caribbean region and in South America. The hot pepper gall midge (HPGM) became a pest of significance in Jamaica when larvae were intercepted in shipments of hot peppers from Jamaica to the USA. The HPGM was initially identified as *Contarinia lycopersici* (Felt) by the United States Department of Agriculture – Animal and Plant Health Inspection Service (USDA-APHIS). After the pest interception, production and export of hot peppers from Jamaica declined by 25–70% (Ministry of Agriculture (undated)). Today, export of fresh hot pepper from Jamaica to the USA is still only 15% of 1997 exports. Another gall midge, *Prodiplosis longifila* Gagné was discovered and identified in Florida as early as 1936 attacking *Gossypium* spp., and was later found during the 1980s attacking lime, *Citrus aurantifolia* (Peña *et al.*, 1989). In Peru, Ecuador and Colombia, *P. longifila* attacks most horticultural crops, including asparagus, peppers, onions, beans, cucurbits, artichokes and marigold; and occasionally fruit trees such as *Citrus*; avocados; grapes; and weeds (e.g. *Chenopodium murale*, *Amaranthus* spp., *Ricinus comunis*) (Castillo, 2006; Valarezo *et al.*, 2003). In the coastal region of Peru, yields of asparagus and peppers have been reduced by 80% after *P. longifila* has attacked buds, blooms, flowers and fruits. In Peru, a series of solutions and control alternatives for *P. longifila* has been designed as part of an Integrated Pest Management (IPM) programme. In Ecuador the species is considered to be the most important problem for tomato cultivation (Valarezo *et al.*, 2003).

14.2 *Contarinia* spp., Undetermined Species

14.2.1 Taxonomy

In Jamaica, several attempts were made to confirm the identity of gall midges affecting pepper, with varying results. The HPGM was variously identified as *Contarinia lycopersici* (Felt), *Prodiplosis longifila* (Gagné), *Contarinia* sp. and *Prodiplosis* sp. This led the Jamaican government to designate the gall midge(s) affecting hot peppers as 'The Gall Midge Complex: *Contarinia lycopersici* and *Prodiplosis longifila*' (National Strategic Plan to Combat the Gall Midge Complex Affecting Hot Pepper, unpublished). Morphological studies carried out in Jamaica in 2001 suggested that the HPGM was neither *C. lycopersici* nor *P. longifila*. Preliminary investigations revealed morphological differences between the male genitalia of the HPGM and that of *P. longifila*. The differences from *C. lycopersici* were reportedly in relation to variations in the epidemiology of the two species. The study concluded that the HPGM was possibly an undescribed species (Goldsmith, 2001).

The uncertainties with respect to the identity of the pest in Jamaica prompted further consultations with Dr Raymond Gagné, taxonomist with the USDA Agricultural Research Service, Systematic Entomology Laboratory. Gagné agreed that the HPGM was not *C. lycopersici* nor *P. longifila*, but a new species of *Contarinia* (Raymond Gagné, Florida, pers. comm., 2010). Currently, the HPGM is classified as *Contarinia* spp., undetermined species.

14.3 Distribution

The presence of *Contarinia* spp., undetermined species of HPGM (characterized by host and epidemiology) has only been reported from Jamaica. It may be closely related to *C. lycopersici* (Felt), a pest of tomatoes (*Lycopersicon esculentum*), which was described in 1911 from specimens reared from tomato flowers collected in St. Vincent (Jensen, 1950). Known as the tomato midge, *C. lycopersici* was never reported from Jamaica. It is found mainly in Central America and the Caribbean islands of the Greater and Lesser Antilles. These include Barbados, Antigua and Barbuda, Belize, Dominica, Grenada, Montserrat, Saint Kitts and Nevis, Saint Lucia, Saint Vincent and the Grenadines, as well as Trinidad and Tobago. It may also be present in Guyana (CABI, 2006).

14.3.1 Hosts

In Jamaica the HPGM has only been found affecting hot peppers (*Capsicum chinense*). Extensive surveys of other cultivated and wild species of the genus *Capsicum*, as well as other members of the family Solanaceae, failed to detect HPGM infestation (Goldsmith, 2005). In Jamaica, the scotch bonnet and West Indies red peppers are the principal cultivated hot peppers, but all commercial cultivars/varieties of hot peppers grown in Jamaica serve as hosts to the gall midge.

14.4 Biology and Life History

The HPGM is multivoltine, with several generations per annum. Its life cycle is unknown; however, the females lay eggs within the pepper pedicel during the early evenings. The number of eggs laid by a single female has not been determined but they are usually found in masses of 5–10 eggs per mass. The eggs are small, clear or white and difficult to see with the naked eye. They hatch in 1–2 days, and the maggots feed on the stem tissue. Newly hatched HPGM larvae are transparent, becoming creamy white then yellow as they mature (Fig. 14.1). The duration of the larval period has not been determined; however, in the closely related *C. lycopersici*, the larval stage is estimated to last 8–10 days (Callan, 1941). Mature larvae range from 1.55 to 2.05 mm in length, and are able to flip several centimeters into the air to exit the pedicel and drop to the ground, where they burrow into the soil to pupate. Digging into the soil is facilitated by the presence of a clove-shaped sternal spatula.

The late-stage larva burrows beneath the soil surface and forms a cocoon using soil particles (Fig. 14.2). It is not known how long the larva remains in the soil before pupation, but adult emergence occurs 12–14 days after mature larvae enter the soil (Goldsmith, 2005).

14.5 Seasonality

The HPGM reproduces year-round in Jamaica but is most prevalent during the wet seasons and at higher elevations (Goldsmith, 2005; Williams, 2001). An island-wide survey of hot pepper farms in Jamaica in 2000 showed the highest incidence of gall midges occurring along the north coast of the island, which traditionally experiences the highest rainfall (Clarke-Harris *et al.*, 2000).

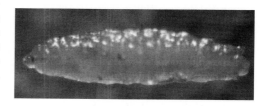

Fig. 14.1 *Contarinia* sp., undetermined species, 3rd instar larva.

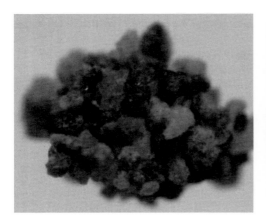

Fig. 14.2 *Contarinia* sp., cocoon.

During 2001, 149 Jamaican hot pepper farms were visited during a traditionally dry period, and then revisited during a wet period, and the incidence of gall midges was recorded. The dry period (January–April) had a mean rainfall of 24.0 mm, while the wet period (September–November) had a mean rainfall of 100.8 mm. There were significant differences in the number of farms that were infested with the midge during these two periods, as well as differences in the number of infested fruits per infested farm. During the dry period, 38% of farms sampled were infested with the gall midge, whereas during the wet period, the infestation level increased to 62%. The mean percentage of fruits infested per infested farm was 2.1% and 15.6% in the dry and wet season, respectively (Goldsmith, 2005).

14.6 Damage

HPGM larvae feed inside the pedicel or beneath the calyx of mature hot pepper fruits, causing a brown-to-black scar (Fig. 14.3). This scar may be secondarily invaded by pathogens. In Jamaica, the fungus *Alternaria* sp. has been associated with the scarred tissue of the fruit pedicel. In some cases, the scars can be extensive, extending past the calyx and into the fruit. The combination of larval feeding and the effects of the pathogens may weaken the connection between pedicel and fruit. Wind, rain or human activity can cause peppers with weakened stems to drop prematurely. These fruits tend to deteriorate, and are unmarketable.

Fig. 14.3 Damage caused by *Contarinia* spp. to peppers.

As a result, the fungi and the rotting of the fruit pedicel and fruits are more usual during very wet conditions, and severe rotting usually occurs only in abandoned fields.

14.7 Sampling and Monitoring Techniques

Hot pepper fields are monitored for the presence of adult gall midges using yellow sticky traps. Although this is not useful for determining infestation levels and does not reflect larval numbers, sticky traps give an indication of presence or absence of the pest (Goldsmith, 2005).

Goldsmith (2005) found a strong correlation ($R=0.880$) between larval infestation levels and green mature pepper fruits (Goldsmith, 2005). Thus, sampling only the green mature fruits will give a good indication of field infestation levels. Sampling for HPGM should commence when the pepper fruits begin to mature. Randomly select and tag 25–30 pepper plants/ha. Five mature pepper fruits should be randomly selected from each of the upper, middle and lower canopy levels of each plant. The sampling scheme is repeated on the same plants at 14-day intervals. Fruits are then inspected for the presence of a brown or black scar on or near the calyx of the fruit pedicel. Scars vary in size and degree, and are dependent on the number of larvae and the length of larval feeding time. Since not all infested fruits have a scar, and vice versa, it may be necessary to dissect the fruit pedicel to confirm the presence of larvae.

The calyxes of the pepper fruits should be removed and examined, and the pedicel dissected under a dissecting microscope in order to verify larval infestation. Where dissection is not immediately possible or if an accurate larval count is not necessary, the fruit may be placed in a resealable plastic bag and kept at room temperature. If the fruits are infested, the larvae will emerge from the pedicel and can be seen against the clear plastic bag (Goldsmith, 2005).

14.8 Rearing

A rearing method for adult gall midges was adapted from Callan (1941). Larvae were removed from flower buds and fruit pedicels. Third instars were placed in large petri dishes c. 14 cm diameter, containing c. 2 cm of fine damp sterilized sand. Glass covers were placed over the dishes, which were sealed with petroleum jelly to retain moisture. Every 2 days c. 1 ml of water was added to the cages as a fine spray. The cages were checked daily for signs of pupation and emergence of adult midges (Goldsmith, 2005).

14.9 Control Tactics

In Jamaica, cultural control involves field sanitation – prompt harvesting of mature fruits to reduce sources of re-infestation for the pepper gall midge (McDonald, 1999). Infested material may be destroyed by burning, or by placing it in plastic bags or sealed containers. The pepper gall midge thrives best under cool, shaded conditions. Fields are kept weed-free and the amount of shade available to the plants is minimized. Plastic mulch and drip irrigation may help in reducing midge populations by reducing the amount of moist soil surface available for pupation (McDonald, 1999). Insecticides (usually systemic rather than contact) are used to reduce HPGM populations. Systemic insecticides may be applied as a foliar spray against larvae as well as in the form of a soil drench (Goldsmith, 2005).

14.9.1 Chemical control

In studies conducted in Jamaica, a combination of cultural practices and the systemic insecticides fipronyl and imidacloprid were able to suppress gall midge larval populations below a 5% threshold (Clarke-Harris et al., 2000). In addition, systemic insecticides commonly used for other pepper pests have been reported to control the HPGM (Rural Agricultural Development Authority, Jamaica, pers. comm., 2006).

Treatment with a fungicide may be necessary during periods of rainfall to prevent the invasion of feeding scars by fungi (Goldsmith, 2005).

14.9.2 Biological control

No parasitoids have been found to be associated with the HPGM in Jamaica. Adults and larvae that left the pedicels are preyed on by generalist predators such as spiders and ants. Callan (1941) reported consistently finding the Platygasterid *Sactogaster* sp. to be associated with *Contarinia lycopersici* in Trinidad.

14.9.3 Quarantine control

The number of gall midge interceptions on hot peppers for export from Jamaica has been reduced significantly over the past few years. This is due mainly to the use of an integrated systems approach to the management of the pest. The main components of this system involve:

1. Grower registration and trace-back system. All farmers producing hot peppers for export must register with the Ministry of Agriculture Extension Authority. This rule facilitates inspection and monitoring of farms and inspection of produce.
2. Pest management. Farmers are trained in management of HPGM. This involves training in field inspection, pre-harvest treatments and effective control strategies.
3. Post harvest handling. Harvested peppers are examined in the field and in packing houses for HPGM scars and possible hitchhikers. From the field they are transported to certified packing houses where they are processed and packaged.

Hot peppers are inspected by first unloading onto a mesh surface which allows insects and smaller (unmarketable) sized peppers to fall through. They are then individually inspected for the presence of insects, insect damage or pesticide residue.

4. Port Inspection. At the export complex, hot peppers are inspected by Jamaican Plant Quarantine Inspection Officers. Inspections are done on inspection tables with slats or mesh bottoms that allow HPGH to be shaken through and captured on a solid secondary bottom. Ten percent of the pepper fruits were removed and dissected for larval detection.

Currently, the quarantine control measure in place in Jamaica to prevent the export of HPGM on hot pepper to the USA involves tarpaulin or chamber fumigation with methyl bromide. A 1999 study indicated that hot peppers fumigated with aluminium phosphide in the form of pellets that release phosphine gas kills >80% of HPGM larvae within fruits without significantly affecting fruit quality and shelf life (CARDI, 1999). This initial study suggests that phosphine gas could be a viable alternative to methyl bromide treatment.

14.10 *Prodiplosis longifila*

14.10.1 Taxonomy

Gagné (1986) conducted the revision of the genus *Prodiplosis*, considering *Geisenheyneira* to be its new synonym, and describing three new species, with variations in biology, anatomy, classification, taxonomy keys and hosts (Gagné, 1989). With the help of Gagné, Rodriguez (1992) identified the genus in Peru and reconfirmed that it was *Prodiplosis longifila*. In Peru, *Prodiplosis longifila* Gagné is commonly known in Spanish as 'Prodi', 'Caracha' (wounded rash, pellagra), 'Mosquilla de los brotes' (blossom fly) (Castillo, 2006), while in the USA it is known as the citrus gall midge (Peña *et al.*, 1989). To our knowledge no molecular identification of the species has taken place.

14.10.2 Distribution

In the USA, *P. longifila* is found in southern Florida (Peña *et al.*, 1989). In Peru, it is distributed in the coastal region which is characterized by average temperatures of 25–16°C, with lowest temperatures that reach 5°C. Insect developmental temperature thresholds are between 11°C and 33°C (Jose Castillo, Chavimochic, Lima, pers. obs., 2007). In Colombia is reported from the Cauca Valley, attacking tomatoes, whereas in Ecuador is found in the coastal region as well as in the Andean region, infesting field-grown tomatoes and those grown under cover (Valarezo *et al.*, 2003).

14.10.3 Hosts

In Peru, *P. longifila* attacks horticultural crops including asparagus (*Asparagus officinalis*), peppers (*Capsicum annuum*), onions (*Allium cepa*), beans (*Phaseolus vulgaris*), cucurbits (*Cucurbita* spp.), artichokes (*Cynara cardunculus*) and marigold (*Tagetes erecta*), as well as fruit trees including citrus and avocados (*Persea americana*), grapes (*Vitis vinifera*) and weeds (*Chenopodium murale, Amaranthus* spp., *Ricinus comunis*) (Castillo, 2006). In the coastal region of Peru, *P. longifila* reduces yields of asparagus and peppers by up to 80%, as it attacks the blooms, flowers and fruits. In Ecuador it affects tomatoes and potatoes, with minor damage observed on watermelon *Citrullus lanatus* (Thunb.), green pepper, beans and melon (Valarezo *et al.*, 2003). In the USA *P. longifila* has been found in citrus and cotton (*Gossypium*), there is one record from tomato flowers and none from other horticultural crops (Jorge E. Peña, Homestead, Florida, USA, pers. obs., 1990). In Colombia, it is often reported attacking tomatoes (Garcia, 2011).

14.10.4 Biology and life history

Biological studies were conducted on tomatoes (Valarezo *et al.*, 2003), asparagus (Rodriguez, 1992) and citrus (Peña *et al.*, 1989) under field and laboratory conditions. The egg stage lasts 1–2 days, the pupal stage 5–8 days, and the adult stage 1–2 days (Tables 14.1 and 14.2). The average life cycle fluctuates between 10 and 19 days, depending on temperature. In asparagus, the female deposits its eggs under the bracts, in the flower and in other structures. Upon hatching, the larvae start feeding on the epidermis, softening the epidermal tissue. The first larval stage is translucent,

Table 14.1. *Prodiplosis longifila* biology on *Lycopersicum esculentum*, Lambayeque, Peru, January–February 1992. Rodríguez.

Life-cycle stage	Days		
	Minimum	Maximum	Average
Egg	1	1	1
Larvae I	1	1	1
Larvae II	1	2	1.2
Larvae III	1	2	1.1
Pupae	4	5	4.8
Male adult	1	3	2.17
Female adult	4	6	5
Male cycle	10	12	11.7
Female cycle	13	16	14.25

Fig. 14.4 *P. longifila* larva.

Table 14.2. *Prodiplosis longifila* biology under asparagus field conditions (September 2001. APTCH) La Libertad, Peru.

Life stage	Time in days		
	Minimum	Maximum	Average
Egg	1.5	1.8	1.6
Larvae	4.3	5.2	4.7
Prepupae–Pupae	9.2	11.2	10.2
Developmental cycle			
Male	15.1	17.2	16.4
Female	15.4	17.7	16.7

changing to a progressive whitish color during its second instar, the most damaging, whereas the third larval instar is yellowish (Fig. 14.4). At the time of pupariation, the larvae leap or drop from the foliage to the ground and by successive leaps search for a shaded and humid area to penetrate the first few millimeters of soil and form a pupal cocoon with the small soil particles. On citrus, *P. longifila* pupariates at 0.30–0.90 m from the citrus trunk (Peña and Duncan, 1992). In Peru and Ecuador, the adult emerges between dusk and the early morning, initiating feeding on the floral or vegetative nectars (Rodriguez, 1992; Valarezo et al., 2003). In Florida, *P. longifila* emerges between 6 and 11 P.M. with negligible emergence during the day (Peña et al., 1989). An adult without feeding survives 1.2 d, when fed, a maximum of 16 d (Peña et al., 1989). In Peru, the female can oviposit 17–137 eggs with an average of 50 eggs/cluster. The female:male sex ratio in Peru is 1:1.53 (Rodriguez, 1992), while in Florida the ratio fluctuates between 2.3:1 and 1:1 (Peña et al., 1989).

14.10.5 Seasonality

In Peru, the seasonal appearance of *P. longifila* is always dependent upon temperature and relative humidity, and occurs mostly between spring and summer, decreasing during the winter. It is a multivoltine insect with 21–33 generations per annum (Jose Castillo, Lima, pers. obs., 2008). Linear correlation between infestation, temperature and rainfall showed that infestation levels are reduced at temperatures higher than 27°C and rainfall above 100 mm (Valarezo et al., 2003).

14.10.6 Damage

Damage to tomatoes and peppers is caused by the first and second instars of *P. longifila* larvae, which injure buds, leaves, flowers and fruits. These attacks deform the plants, particularly during the early stages of development. (Fig. 14.5). Soon after larval injury, a necrosis develops which later becomes corky. Larvae are protected by the vegetative parts of the plant they are attacking, and thus their control is difficult. Evaluation for damage is done directly on the parts they are attacking, such as sprouts, flowers, fruit and bracts. Due to the fluctuation of the produce market value, there are no economic damage levels, although a nominal threshold of 5% of damage to infested organs (flowers, buds, fruits) is frequently used (Cisneros, 1995). In asparagus the threshold is one to more than 20 adults per trap. The same values are extrapolated for other crops (Garcia, 2006).

In the USA, damage by *P. longifila* is evidenced first as larval feeding on the stamen and ovary of citrus flowers (Peña et al., 1987). Damage caused by *P. longifila* in limes results in floral abscission. However, distribution and prevalence of *P. longifila* infestation is considered independent of the severity of the fungus *Colletotrichum gloeosporioides* in the flowers (Peña and Duncan, 1989).

Fig. 14.5 *P. longifila* damage to tomato.

14.10.7 Sampling and monitoring techniques

At the onset of *P. longifila* infestations in horticultural crops, the sampling unit is the number of larvae in each infested organ (flower, leaf, bud); later on, however, because of the high densities and high infestation peaks, the sampling unit is changed to percentage of infested plants. Vilca (2000) reported that 1.71 larvae/growing terminal could be found at early potato and asparagus plant development and 2.74 larvae per terminal when the plants reach maturity.

Currently, 0.25 m^2 white sticky traps are placed in the middle of asparagus fields (Fig. 14.7). The infestation is based on the number of adults trapped each night. Interpretation of these results always depends on the crop phenology, and the crop phenology of neighboring fields. In the USA, Peña and Duncan (1992) demonstrated that orange or red colored traps were more efficient for trapping *P. longifila* than either yellow or whitish traps, which is the opposite of the practice in horticultural crops in Peru. Valarezo et al. (2003) reported no significant differences in adult trapping while using different colors (i.e., white, black, blue, red).

Fig. 14.6 *P. longifila* damage to asparagus.

Fig. 14.7 Sticky traps for monitoring adults of *P. longifila*.

14.10.8 Control tactics

The control strategies have been developed by the Sanitation Committee of APTCH (Association of Agro-eEporters in northern Peru), and adopted by the majority of the exporters in Peru. This integrated pest management programme (IPM) was initiated in 2000 after a crisis caused by the white fly (*Bemisia* spp.). Different tactics – ethological, physical and cultural control – were developed. Each crop has plant phenologies that enable the insect to find more structures suitable for attack. In asparagus, for instance, the insect can attack the bracts that surround the tips and can also attack its flowers. Consequently, each management tactic will be guided by the susceptibility of the crop stage to *P. longifila*.

Mass trapping of *P. longifolia* adults is used in horticultural crops in Peru, using a white-colored light placed behind plastic (Fig 14.8) that has been covered with vegetable oil. Use of one light trap for every 4 ha resulted in the capture of 81,000 adults, reducing damage and insecticide applications (Valarezo et al., 2003). Thousands of adults can be captured each night depending on the crop phenology. Toxic sugared baits are also applied to the base of plants to trap adults that aggregate there. Yellow or white sticky traps carried as flapping wings on an equine that is walked along infested rows are used to trap flying adults in *Capsicum*. However, care is recommended as this type of trap increases the presence of the mottled pepper virus (PMMoV), as it can be transmitted by contact.

Physical control is practiced by applying high-pressure water to living borderline fences where the adults hide during the day (Fig. 14.9).

Cultural control is practiced by harvesting the crops (asparagus, tomatoes, peppers) immediately and removing any crop residues from the field. The frequency of drip irrigation should be reduced to avoid waterlogged soils where the insect can pupariate; however, water use should be calibrated to avoid wilting and yield reduction. Several wild species such as *Chenopodium murale*, *Amaranthus* sp., *Ricinus comunis* and *Solanum nigrum* are hosts of *P. longifila* and serve as refuge for adults. Planting grasses is suggested as they are not hosts of *P. longifila*. With some crop species (e.g., asparagus), flowers are undesired, because they reduce shoot quality. Flower removal, particularly of male flowers, will reduce *Prodiplosis* infestations. The use of foliar fertilizers that cause premature flower drop reduces *Prodiplosis* densities; however, this strategy depends on application time, the type of foliar fertilizer, the dose and the asparagus variety (Prado, 2009).

High plant densities provide a favourable microclimate for insect development and reduce the effectiveness of insecticide applications or use of water pressure against adults and larvae. Rows that are parallel to the wind direction are detrimental for the establishment of the insect. When the rows are at right angles to the wind direction, insect infestation increases.

14.10.9 Chemical control

Use of neonicotonoids and ketoenols for larval control has proven to be advantageous in Peru, in addition to the application of plant extracts such as garlic and capsaicin as well as sulfur as repellents (Jose Castillo, Lima, pers. obs., 2010).

Research is needed in other fields such as the semiochemicals and plant resistance, as well as improving the search for the entomopathogens for the control of *P. longifolia*. The technique of using sterile males may be another possible way to control this insect.

14.10.10 Biological control

The presence of *Synopeas* sp. (Platygasteridae) at levels of 16–20% of parasitism (Rodriguez, 1992)

Fig. 14.8 Light sticky trap for *P. longifila* adult trapping.

Fig. 14.9 Washing borderline plants that harbor *P. longifila* adults.

is important in different crops, but is not enough to prevent economic damage. In Florida, *Synopeas* is considered to be the key mortality factor of *P. longifila* and the reason why this species is not an economic problem in citrus (Jorge E. Peña, personal communication). In Peru the fly *Coenosia* sp. (Muscidae) has been reported as an adult predator. Valarezo et al. (2003) recommend the introduction of *Omophalevaripes* (Hymenoptera: Eulophidae) and *Inostema opacum* (Hymenoptera: Platygastridae) as possible candidates for the biological control of *P. longifila* in Ecuador. Releases of *Chrysopa* spp. have been practiced without much success. Use of entomopathogenic fungi (*Beauveria bassiana*) has been practiced in Peru, also without success. Recently, *Heterohabditis* spp. have been isolated from soils in the Chavimochic region, and may be a promising option to improve the management of *Prodiplosis* (Buendia, 2007).

References

Buendia, O. (2007) Efectividad del nematodod *Heterorhabditis* spp. en la mortalidad de Galleria l bajo condiciones de laboratorio y aislados de la irrigación chavimochic. BSc thesis, Universidad Nacional Agraria La Molina, Lima.

CABI (CAB International) (2006) Crop Protection Compendium. CD-Rom. CAB International, Wallingford, UK.

Callan, E. (1941) The gall midges (Diptera: Cecidomyiidae) of economic importance in the West Indies. *Tropical Agriculture* 17, 117–127.

CARDI (1999) Hot Pepper Capsicum spp. Annual Technical Report, Caribbean Agricultural Research and Development Institute, Kingston.

Castillo, J. (2006) *Prodiplosis longifila* Gagné en la Irrigación Chavimochic, La Libertad, Trujillo, Perú. Arenagro. Cultivando el desierto. *Revista Institucional de la Asociación de Agricultores Agroexportadores Propietarios de Terrenos de Chavimochic* 2, 11–19.

Cisneros, F. (1995) *Control de Plagas Agrícolas*. 2nd ed., Lima.

Clarke-Harris, D., Chung, P., Lawrence, J., Goldsmith, J., Young, F., Thomas, C., Tolin, S. et al. (2000) An IPM strategy to combat the gall midge complex affecting hot pepper. IPM CRSP, Seventh Annual Report, 1999–2000. Virginia Polytechnic Institute and State University, Blacksburg, Virginia.

Gagné, R. (1986) Revision of *Prodiplosis* (Diptera: Cecidomyiidae) with descriptions of three new species. *Annals of the Entomological Society of America* 79, 235–245.

Gagné, R.J. (1989) *The Plant-Feeding Gall Midges of North America*. Cornell University Press, Ithaca, New York.

García, M. (2006) Manejo de poblaciones de *Prodiplosis longifila*, La Libertad, Trujillo, Perú. Arenagro. Cultivando el desierto. *Revista Institucional de la Asociación de Agricultores Agroexportadores Propietarios de Terrenos de Chavimochic* 2, 24–26.

Goldsmith, J. (2001) Preliminary investigations into the taxonomy of the gall midges (Diptera: Cecidomyiidae) affecting Capsicum sp. in Jamaica. *Proceedings of the Caribbean Food Crop Society* 37, 81–86.

Goldsmith, J. (2005) Identification, incidence and characterization of gall midges affecting hot peppers in Jamaica. MSc thesis, Northern Caribbean University, Mandeville, Jamaica.

Jensen, D. (1950) Notes on the life history of blossom midge, *Contarinia lysopersici* Felt (Diptera: Cecidomyiidae). *Proceedings, Hawaiian Entomological Society* 14, 91–100.

McDonald, F. (1999) Arthropod pests of hot pepper in the Caribbean with respect to their economic importance and management: A review. PROCICARIBE/CIPM Network, Kingston, Jamaica.

Mellonella, L. Bajo condiciones de laboratorio. thesis. Universidad Nacional Agraria La Molina, Lima, Peru.

Ministry of Agriculture (undated) Jamaica Agricultural Data. www.moa.gov.jm/data/data/agridata, accessed 18 March 2010.

Peña, J.E. and Duncan, R. (1989) Role of arthropods in the transmission of postbloom fruit drop. *Proceedings of the Florida State Horticultural Society* 102, 249–251.

Peña, J.E. and Duncan, R. (1992) Sampling methods for *Prodiplosis longifila* (Diptera: Cecidomyiidae) in limes. *Environmental Entomology* 21, 996–1001.

Peña, J.E., Baranowski, R.M. and McMillan, R. (1987) *Prodiplosis longifila* (Diptera: Cecidomyiidae): A new pest of citrus in Florida. *Florida Entomologist* 70, 527–529.

Peña J., Gagné, R. and Duncan, R. (1989) Biology and characterization of *Prodiplosis longifila* (Diptera:Ceciodmyiidae) on lime in Florida. *Florida Entomologist* 72, 444–450.

Prado, S. (2009) Efecto la aplicación de fertilizantes foliares en la caída de flores del espárrago (*Asparagus officinalis* L.) para el manejo de *Prodiplosis longifila* Gagné. BSc thesis, Universidad Nacional Agraria La Molina, Lima.

Rodríguez, S. (1992) Biología y morfotaxonomía de la 'caracha' (Diptera: Cecidomyiidae) en tomate (*Lycopersicon sculentum* Mill.) CV. Rio grande. BSc thesis, Universidad Nacional Pedro Ruiz Gallo, Lambayeque, Peru.

Valarezo, O., Cañarte, E., Navarrete, B. and Arias, M. (2003) *Prodiplosis longifila* (Diptera: Cecidomyiidae) principal plaga del tomate en el Ecuador. Instituto Nacional de Investigaciones Agropecuarias, Estacion Experimental Portoviejo, Quito, Ecuador, pp. 79.

Vilca, J. (2000) Fluctuación poblacional de *Prodiplosis longifila* Gagné (Diptera: Cecidomyiidae) en cultivos de papa y espárrago. Cañete-Perú. MSc thesis, Universidad Nacional Agraria La Molina, Lima.

Williams, R. (2001) Application of spatial analysis in the incidence of the gall midge in Jamaican hot pepper production. MSc thesis. Virginia Polytechnic Institute and State University, Blacksburg, Virginia.

15 Recent Mite Invasions in South America

Denise Navia,[1] Alberto Luiz Marsaro Júnior,[2]
Manoel Guedes Correa Gondim Jr,[3] Renata Santos de Mendonça[1]
and Paulo Roberto Valle da Silva Pereira[2]

[1]*Embrapa Recursos Genéticos e Biotecnologia, Parque Estação Biológica, W5 Norte Final, Cx. Postal 02372, 70770-917 Brasília, Distrito Federal, Brazil;* [2]*Embrapa Trigo, Rodovia BR 285, Km 294, Cx. Postal 451, 99001-970, Passo Fundo, Rio Grande do Sul, Brazil;* [3]*Universidade Federal Rural de Pernambuco – UFRPE, Avenida Dom Manoel de Medeiros s/n, Dois Irmãos, 52171-900, Recife, Pernambuco, Brazil*

15.1 Introduction

Phytophagous mites belong to a group of organisms which includes invasive species that can greatly impact agroecosystems and natural terrestrial ecosystems. These minute arachnids are not only extremely harmful to their host plants, but they also: (i) act as efficient vectors of major plant diseases; (ii) quickly develop resistance to pesticides; (iii) survive adverse environmental conditions; (iv) reproduce parthenogenetically (consequently establishing new colonies, since one or more individuals can disseminate easily through wind currents); and (v) adapt to new host plants in invaded areas. In addition, phytophagous mites can easily go unnoticed during a naked-eye inspection due to their small dimensions, and because they are often hidden on the host plant (Navia *et al.*, 2006a). Usually, mite infestation-related symptoms appear only when their numbers are high and eradication is no longer feasible.

There have been several cases where phytophagous mites were introduced inadvertently to new areas in the absence of efficient natural enemies, and found the appropriate conditions to develop, resulting in accentuated damage to the infested crops and causing serious socio-economic problems. Throughout acarological history, South America has been known as a continent of origin for important invasive phytophagous mites which have been the focus of international efforts – the cassava green mite, *Mononychellus tanajoa* (Bondar); probably the coconut mite, *Aceria guerreronis* Keifer; and the tomato red spider mite, *Tetranychus evansi* Baker and Pritchard. However, during the last few years South America has also become known as the stage for important mite invasions. Several phytophagous mites were recently detected in South America and have been a concern for national and regional plant protection organizations, researchers and growers. Most of these recently detected mites probably originated from tropical or subtropical Asia.

This chapter focuses on five phytophagous mites that were reported in South America during the last decade and that we considered relevant due to their current or potential socio-economic importance and/or their invasive history: the Hindustan Citrus Mite, *Schizotetranychus hindustanicus* Hirst; the Lychee Erinose Mite, *Aceria litchii* Keifer; the Red Palm Mite, *Raoiella indica* Hirst; the Wheat Curl Mite, *Aceria tosichella*

Keifer; and the Panicle Rice Mite, *Steneotarsonemus spinki* Smiley. This review includes information on taxonomy, distribution and invasion history, host plants, and economic impact on past and recently affected areas of these invasive mites. A brief review of morphological and biological aspects, symptoms, detection and control tactics in use or of potential for these invasive mites, is also presented.

15.2 The Hindustan Citrus Mite, Schizotetranychus hindustanicus (Hirst)

15.2.1 Introduction

The Tetranychidae mite *Schizotetranychus hindustanicus* (Hirst), commonly named the Hindustan Citrus Mite (HCM) or Citrus Nest-Webbing Mite (CNWM), was originally described from citrus from Coimbatore (southern India) by Hirst (1924). For almost 80 years after it was described, *S. hindustanicus* was found only in India, supposedly the area where the species originated. In 2002 it was reported in South America, infesting citrus in Zulia, northwestern Venezuela (Quirós and Geraud-Pouey, 2002). In 2008 the mite was identified in the State of Roraima, in the northern tip of Brazil, bordering Venezuela (Navia and Marsaro Jr, 2010). In 2010 the HCM was detected in Colombia on the country's northern border with Venezuela (N.C. Mesa-Cobo, Cali, Colombia, pers. comm., 2010). No further records of this species have been published in other countries or continents.

Currently HCM seems to have spread over areas where citrus is cultivated in Venezuela. The mite has been observed in the extreme west (Zulia), in the east (Sucre), in the north-central areas (Aragua), as well as in the south (Nienstaedt-Arreaza, 2007; M. Quirós, Maracaibo, Venezuela, pers. comm., 2011).

In Brazil the HCM still presents restricted distribution, being found only in three northeastern municipalities in the state of Roraima (Fantine *et al.*, 2010). Quarantine measures have been applied to avoid wider dissemination of the HCM in the country, especially to the main Brazilian citrus production areas.

15.2.2 Host plants

Citrus are the main host plants of the HCM. In addition to citrus, this mite has also been reported infesting coconut palms (*Cocos nucifera* L.) (Arecaceae); *Acacia* sp. (Mimosaceae); neem (*Azadirachta indica* A. Juss); Persian lilac (*Melia azedarach* L.) (Meliaceae); and sorghum (*Sorghum vulgare* Pers.) (Poaceae) in India (Cherian, 1931 in Migeon and Dorkeld, 2010; Gupta and Gupta, 1994; Bolland *et al.*, 1998).

Colonies of the HCM can develop on different citrus species and cultivars. In Venezuela, the mite has been reported infesting Tahiti lime (*Citrus latifolia* (Tanaka ex Yu. Tanaka)) (Quirós and Dorado, 2005), key lime (*Citrus aurantifolia* (Christm.) Swingle), mandarin (*Citrus reticulata* Blanco), lemon (*Citrus limon* (L.) and sweet orange (*Citrus sinensis* (L.) Osbeck) (Nienstaedt-Arreaza, 2007). When the HCM was reported in Brazil in 2008, colonies were found infesting the leaves and fruits of Tahiti lime and lemon plants (Navia and Marsaro Jr, 2010). Surveys were conducted in backyards and commercial orchards in Boa Vista, Roraima, from March 2009 to March 2010, in an effort to discover other citrus host plants for the HCM in Brazil. The presence of the HCM was confirmed on Rangpur lime (*Citrus limonia* Osbeck), Poncan mandarin, Valencia sweet orange and tangor murcott (*Citrus reticulata* Blanco × *Citrus sinensis* (L.) Osbeck) (Marsaro Jr *et al.*, 2010a).

15.2.3 Economic impact

Currently there is no information that citrus production is reduced due to HCM infestations throughout areas of occurrence in India or South America. However, severe damage on citrus leaves and fruits has been observed in Venezuela (Quirós and Geraud-Poney, 2002) and in Brazil (D. Navia, Brasilia, Brazil, pers. comm., 2010). Even if there are no confirmed citrus production reductions due to HCM infestations, there is no doubt that the mite can reduce the commercial value of fresh fruits due to depreciation in its aesthetic quality (Fig. 15.2).

Surely an important impact of the introduction and dissemination of the HCM in

Fig. 15.1 Citrus fruit damage caused by *S. hindustanicus*.

Fig. 15.2 Colony of *S. hindustanicus*.

South America can be the implementation of sanitary barriers in the international or domestic trade of fresh citrus fruits and plant material, or in the exchange of genetic material of host plants. Most Brazilian citrus exports consist of concentrated orange juice, a commodity that does not represent a gateway for the mite. However, fresh fruit exports have increased significantly, especially for lemon, which accounts for more than 60,000 t/year (Abanorte, 2008). Thus, the dispersal of the HCM to the main citrus production areas in Brazil could cause high economic impacts and/or commercial restrictions due to sanitary barriers.

15.2.4 Morphological and biological aspects

HCM adult and immature stages are yellowish or yellowish green, with dark internal spots on the sides of the idiosoma (Fig. 15.2). Legs I and II and gnathosoma can present amber coloration. The female body is oval, somewhat flattened, about 430 μm long, with stubby legs (Quirós and Geraud-Poney, 2002). The male is pear-shaped, paler and smaller than the female, with remarkably long legs (especially the first pair), typical red eyes, and measures less than 350 μm (Hirst, 1924; Quirós and Geraud-Poney, 2002; Nienstaedt-Arreaza, 2007). Eggs are round and slightly flattened, with a translucent to light-yellow coloration. Just before the larvae hatch, the eggs become yellowish green and show two small red spots which will become the red eyes of the larvae and other immature stages. Larvae present three pairs of legs, have a round shape, and are yellowish bright green after eclosion, gradually becoming oval and darker during feeding. During the deutonymph stage it is already possible to observe small differences between males and females, the deutonymph male being smaller with slightly sharper opisthosoma. Three quiescent stages occur between active stages: protochrysalis, deutochrysalis and teliochrysalis. These can be recognized by the opaque appearance and leg positioning, where the anterior ones are directed forward and the posterior ones backward (Nienstaedt-Arreaza, 2007).

Information on the biology of the HCM is scarce. In 2005–2006, Nienstaedt-Arreaza (2007) studied for the first time the biological and population aspects of the mite on key lime (*C. aurantifolia*), sweet orange (*C. sinensis*) and lemon (*C. limon*), in Maracay, Venezuela (Nienstaedt-Arreaza, 2007). Experiments were conducted under laboratory conditions 25 ± 2°C and $83 \pm 12\%$ RH. According to this author HCM is an arrhenotokous species; unfertilized females produce only male offspring and fertilized females produce both females and males. The average duration (in days) of the HCM life stages on different citrus plants is shown in Table 15.1. The period from egg stage to adult life takes between 14 and 15 days. Table 15.2

Table 15.1. Hindustan Citrus Mite period of development (in days) from egg to adult stage under laboratory conditions (25 ± 2°C and 83 ± 12% RH) in Maracay, Venezuela, on three citrus host species (Nienstaedt-Arreaza, 2007).

Citrus host	Egg	Larva	Protonymph	Deutonymph	Egg to adult
Orange	7.89 ± 0.86	2.46 ± 1.19	1.97 ± 0.80	1.59 ± 0.62	13.91 ± 3.47
Lime	7.48 ± 0.67	2.36 ± 0.85	1.88 ± 1.19	2.14 ± 0.78	13.86 ± 3.49
Lemon	7.69 ± 0.88	3.27 ± 1.22	2.10 ± 0.76	1.76 ± 1.01	14.82 ± 3.87

Table 15.2. Hindustan Citrus Mite preoviposition, oviposition and postoviposition periods (in days), and fecundity and fertility under laboratory conditions (25 ± 2°C and 83 ± 12% RH) in Maracay, Venezuela, on three citrus host species (Nienstaedt-Arreaza, 2007).

Citrus host	Preoviposition	Oviposition	Postoviposition	Fecundity	Fertility
Orange	1.75 ± 1.21	8.92 ± 3.71	1.85 ± 1.97	11.59 ± 5.35	99.09
Lime	2.42 ± 0.67	8.67 ± 3.89	4.08 ± 2.75	13.33 ± 8.79	100
Lemon	1.47 ± 1.03	9.28 ± 3.75	2.94 ± 2.27	10.92 ± 3.82	99.08

shows data on preoviposition, oviposition and postoviposition periods, and the fecundity and fertility of the HCM. The oviposition period for the studied plants lasted about 9 days, during which each female oviposited an average of 11 eggs on lemon and 13 eggs on lime. The different citrus plants did not affect the HCM life cycle and lifespan. The longest period of development was observed on the lemon plants for the larva, and for the deutonymph this occurred on the lime. Females presented a longer lifespan on lime plants. On all three citrus species the highest number of eggs was oviposited during the first 4 days of the oviposition period.

In 2005–2006, in Maracay, Venezuela, higher populations of the HCM were found in citrus orchards during the dry season. During rainy months, populations decreased abruptly and mites were not found. Temperature and relative humidity are not correlated with population fluctuation (Nienstaedt-Arreaza, 2007). This study suggested that precipitation is a key climatic parameter to the HCM populations.

HCM populations are more abundant on Persian lime and mandarin, leading Nienstaedt-Arreaza (2007) to suggest that the mite prefers these citrus species to lemon and sweet orange. However, biological studies under laboratory conditions did not reveal any differences between the life cycle period of mites on the different citrus species.

The common name given to *S. hindustanicus*, 'nest-webbing mite', is directly related to the female's behavior of spinning fine, circular 1–2 mm-diameter webs, under which they lay eggs (Quirós and Geraud-Poney, 2002; Navia and Marsaro Jr, 2010). The emerging larvae and nymphs feed and defecate on the cells and tissue protected by the web, whereas the adults can remain (or not) under the nest-web (Quirós and Geraud-Poney, 2002; Nienstaedt-Arreaza, 2007). Adults actively move on the leaves and fruit (Quirós and Geraud-Poney, 2002). The female HCM's web-spinning behavior was described by Nienstaedt-Arreaza (2007). It is possible that the HCM present social behavior similar to other tetranychid mites from the genus *Schizotetranychus* and *Stigmaeopsis*, which also build nests (see Mori and Saito, 2006; Saito, 1986).

15.2.5 Symptoms

HCM symptoms observed in citrus plants in Brazil (Navia and Marsaro Jr, 2010) are similar to those described in Venezuela (Quirós and Geraud-Poney, 2002). Chlorotic spots caused by HCM feeding first appear on the upper surface of the leaves, along the main ribs, and later extend to the entire leaf blade (Fig. 15.3). On fruits, females webbed over concavities or depressions

Fig. 15.3 *S. hindustanicus* damage on citrus leaves.

on the rind, and attacked fruits became silvered and hard (Quirós and Geraud-Poney, 2002).

According to studies conducted by Nienstaedt-Arreaza (2007) in citrus orchards with orange, lime, lemon and mandarin trees in Maracay, Venezuela, the HCM was uniformly distributed in the canopy and females began colonization on younger leaves.

15.2.6 Control tactics

Chemical

In Venezuela, the efficacy of azociclotin (1.5 g/l), liquid soap (30 ml/l) and mineral oil (12.5 ml/l) in controlling HCM on Tahiti lime (*C. latifolia*) was evaluated by Quirós and Dorado (2005). Results showed no significant differences between the efficiency of the miticide azociclotin (94%) and soap treatments (87%). The least efficient treatment was mineral oil, averaging at about 42% for mite mortality. In Brazil, dimetoathe and spirodiclofen were effective in controlling HCM in the field; however, these pesticides are not yet registered in the country for HCM control (A.L. Marsaro Jr, Boa Vista, Brazil, pers. comm., 2010).

Biological

In Brazil, predatory mites belonging to three families were observed in association with HCM: three species of Phytoseiidae, *Galendromus annectens* (De Leon), *Euseius concordis* (Chant)

and *Iphiseiodes zuluagai* Denmark and Muma; one species of Stigmaeidae of the genus *Agistemus* and one Bdellidae of the genus *Bdella* (Marsaro Jr *et al.*, 2009).

Regulatory

In Brazil, a post-harvest procedure for HCM-infested citrus fruits was evaluated. This procedure has become a requirement in domestic transportation of fresh fruits from infested to non-infested states (MAPA, 2009). The post-harvest procedure consists of: (i) immersing fruits in chlorine solution (200 ppm for 10 min); (ii) washing fruits with a ortho-phenylphenol solution (0.65%), a product generally used for cleaning citrus fruits; (iii) brushing; (iv) drying; and (v) waxing fruits with vegetable wax made of *Copernicia prunifera* Miller palm wax and oxidated colophony resin. After the post-harvest procedure, fruits are submitted to official inspection.

15.3 The Lychee Erinose Mite, *Aceria litchii* Keifer

15.3.1 Introduction

The Lychee Erinose Mite (LEM), *Aceria litchii* (Keifer), is a Eriophyidae mite described in 1943 from lychee, *Litchi chinensis* Sonn. (Sapindaceae) in Pensacola, Florida, and Honolulu, Hawai'i, USA (Keifer, 1943). This mite is also known in China as 'litchi hairy mite', 'hairy spider', or 'dog ear mite' (Waite and Hwang, 2002). According to Amrine (2003), the taxon *A. litchii* has a senior synonym *Eriophyes cordai* Oudemans and two junior synonyms, *E. chinensis* O'Gara and *Erineum sixtaliae* Corda.

Most of Eriophyoidea mites present high host-specificity (Oldfield, 1996). Although first described in Hawai'i, the LEM probably originated from the same place as the lychee or the longan. According to Menzel (2002), lychee originated from the region between South-east China, North-west Vietnam and Myanmar. Longan also originated from South-East Asia. Therefore, the LEM could be considered to be of South-East Asian origin.

The LEM is widely distributed in South and South-East Asia and occurs throughout India

(Puttarudriah and ChannaBasavanna, 1959; Sharma, 1985), Pakistan (Alam and Wadud, 1963), Bangladesh (Haque et al., 1998), Thailand (Keifer and Knorr, 1978), China and Taiwan (Huang, 1967; Huang et al., 1990). In Oceania the mite has also been reported in Hawai'i (Keifer, 1943; Nishida and Holdaway, 1945; Jeppson et al., 1975) and Australia (Pinese, 1981; Waite, 1986; Siddiqui, 2002).

The LEM is a major lychee pest in the main production areas of this fruit crop, which are concentrated in Asia (Alam and Wadud, 1963; Lall and Rahman, 1975; Prasad and Singh, 1981; Thakur and Sharma, 1990). In addition to lychee, the LEM has also been reported in Taiwan infesting longan, *Dimocarpus longan* Lour. (Huang, 2008).

Recently, the LEM has also been reported in South America. In 2007 severe symptoms of LEM infestations were observed in lychee orchards in the state of São Paulo, Brazil; however, at that time no eriophyid mites were found. In this same state, at the beginning of 2008, the LEM was found infesting a lychee orchard (Raga et al., 2010). Currently, the mite has disseminated in the states of São Paulo and Minas Gerais (Agrolink, 2010; O. Yamanishi, Brasilia, Brazil, pers. comm., 2010).

15.3.2 Economic impact

Although considered to be a major lychee pest, there are few quantitative data on economic losses due to LEM infestations in Asia or Oceania, or in the invaded area in South America. Prasad and Singh (1981) reported a yield reduction of 80%. In infested orchards in Brazil, the mite has caused losses of 70–80% (O. Yamanishi, Brasilia, Brazil, pers. comm., 2010). In Brazil, an increase of 20% has been estimated in lychee production costs as a result of methods to control LEM (Agrolink, 2010). LEM has been considered to be the first important lychee pest in the country (O Estado de São Paulo, 2010).

15.3.3 Morphological and biological aspects

LEM females are elongated, vermiform, 0.11–0.13 mm in length, and yellowish to reddish/pinkish in color (Keifer, 1943; Alam and Wadud, 1963). Eggs are round, spherical, translucent white at first and gradually becoming palish white, and are about 0.03–0.04 mm in diameter (Alam and Wadud, 1963; Butani, 1977). These mites present two immature stages – protonymph and deutonymph. In all stages two pairs of legs are present in the anterior body region.

LEM develops on young leaves and feeds on their inner surfaces (Alam and Wadud, 1963). The mites deposit eggs on leaf surfaces on the base of the erineal papillae. The incubation period lasts about 2.5 days, but can take as long as 4 days. The protonymph grows for 2–3 days before molting; the deutonymph lasts from 5 to 7 days. The preoviposition period takes 1.5 days. LEM complete a life cycle in a period between 13 and 19 days (Alam and Wadud, 1963; Jeppson et al., 1975). According to Prasad and Singh (1981) and Zhang (1997) in Waite and Hwang (2002), 13–15 overlapping generations of the LEM are produced each year in India and China.

Population peaks occur when host growth spurts are abundant; however, high temperatures, high relative humidity and heavy rainfall are unfavorable for mite development (Nishida and Holdaway, 1945; Alam and Wadud, 1963).

LEM seasonality was studied in Pakistan and India (Alam and Wadud, 1963; Sharma et al., 1986; Thakur and Sharma, 1990). In Pakistan, populations are very low during the winter (November to mid-February). Populations increase in late February. Mite populations increase gradually from March until the high relative humidity period in June. Population levels reach their peak during April–May. A gradual decrease begins at the end of May, falling to low levels in August. After the relative humidity decreases, population levels once again begin to increase, reaching a peak in September–October. Moderately hot and dry periods are more favorable for LEM multiplication. These authors observed that the mite is inactive during 4 months per annum. In Bihar, India, slightly different observations were made by Thakur and Sharma (1990) on seasonality and weather-influencing factors. These authors found that mites remain active throughout the year with populations ranging from 31 to 93 mites/2.5 cm^2, and the highest mite numbers were correlated with the highest temperatures, wind velocity and sunshine during May. The lowest population levels were observed in December. Relative humidity was considered an important factor, but not the main factor for regulating populations. Also, based on observations in Bihar, Sharma et al. (1986) considered that LEM populations are significantly influenced by temperature and relative humidity.

The development of the erineum on lychees as a result of feeding by the LEM has been questioned by Somchoudhury et al. (1989) and Sharma (1991), who proposed that the erineum does not arise from stimulated leaf cells, but is in fact formed by the thalli of the alga *Cephaleuros virescens* Kunze, with the alga and mite sharing a symbiotic relationship. Saha et al. (1996) studied the symptoms associated with the presence of *C. virescens* and the LEM at a fundamental leaf level, and agreed that the alga was involved in the production of erinose; however, Waite and Elder (1996) and Waite and Hwang (2002) have argued the possibility of *C. virescens* being responsible for the erinose associated with *A. litchii*. This is because chemical applications that kill the mites allow new foliage to develop without erinose, and erinose symptoms are not present on lychees in countries that have never recorded the mite, whereas the alga is present. Further studies should be conducted to clarify the relationship between LEM and *C. virescens*.

15.3.4 Dissemination

Trees may be infested with erinose mites at the time they are planted if they have been propagated as marcots from infested parent trees. Otherwise, for infestation to occur, mites must move directly between touching trees (Prasad and Singh, 1981), be either physically transported from tree to tree by human activity or other agents, or be carried by the wind (Wen et al., 1991 in Waite and Hwang, 2002).

For a long time it was considered that the main form of LEM dispersion was aerial, in common with most other eriophyid mites (Lindquist et al., 1996); however, Waite and McAlpine (1992) and Waite (1999) have presented evidence on the importance of LEM dispersion by honeybees. Waite and McAlpine (1992) found LEM on more than 23% of the honeybees foraging in heavily infested lychee trees, usually attached to their legs. Waite (1999) presented further evidence that this process resulted in mite infestation of flower panicles subsequently visited by bees. The development of erinose symptoms on flower panicles of non-infested trees from a single infested panicle showed that infestation was correlated with the sequence of flower opening and bee visitation. Because of this, pesticides need to be applied before and during inflorescence emergence and leaf expansion (Schulte et al., 2007).

Furthermore, LEM has surely been greatly disseminated through human activity. The use of quarantine procedures in the trade of fresh fruits, or exchange of any propagative material from infested to non-infested areas is extremely important.

15.3.5 Symptoms

The LEM infest the new growth of the lychee trees, causing a felt-like erineum to be produced on the undersurface of the leaves (Jeppson et al., 1975). Erinose is the first symptom of the mite infestation, and can begin to develop as leaflets are unfolding. Erineum, with respect to LEM infestation, consists of papillary strands that have some cell partitions. At first, the erineum is greener than what is deemed normal for plant coloration. Within 2 days, the erineum becomes thick, silvery white, but in a span of 3 or 4 days this hairy growth changes from light brown to a deep reddish brown. By that time the leaves are twisted (Jeppson et al., 1975). Very old erinea are almost black. When populations are at their highest levels erinea are light brown, verging on darkbrown erinose (Nishida and Holdaway, 1945); however, it has been observed that some mites do remain in the dark erinose until new abundant vegetation is produced (Wen et al., 1991 in Waite and Hwang, 2002; G.K. Waite, unpublished results, 1990).

At first, erineum may form as small blisters, but if the infestation is severe, it may eventually cover the entire leaflet, causing it to curl. Whole terminals may be deformed (Nishida and Holdaway, 1945). Fruits and flower buds can also be infested, in addition to shoots and young leaves. Erineum developing on leaf undersurfaces causes the leaves to curl and dry prematurely, ultimately resulting in stunted plants. Mite feeding also depresses flower bud development (Alam and Wadud, 1963; Jeppson et al., 1975). Damage to leaves can reach 45%, causing plants to fail in bearing fruits (Alam and Wadud, 1963). Many leaves may fall if infestations become very severe. While certain trees can tolerate substantial infestations without detriment to their growth, they could cause the growth of young trees to be restricted or could even kill them (Pinese, 1981). When flowering, if leaves immediately below a flower panicle are infested, that panicle will also be affected, eventually affecting the florets, preventing fruit set or producing malformed fruit. Even after fruit have set and developed to half their final size, the mites can colonize them, producing erinose on the skin that detracts from its appearance, which may make them unmarketable due to poor quality (Waite and Hwang, 2002).

15.3.6 Detection

Regular inspections of lychee foliage should be carried out to detect the presence of erinose mite on lychees around the time the trees are expected to flush. To determine if leaves carrying erinose are active, they should be picked and left to desiccate overnight. As the leaves dry, the mites move to the surface of the erinose where they are easily visible the next morning with the aid of a hand lens or stereoscope microscope. Populations exceeding 100,000 per leaflet have been assessed using this method in association with a washing and centrifuge technique (Waite, 1992). Waite (2005) recommends inspection of at least 20 lychee trees in an orchard.

15.3.7 Control tactics

Cultural

Pruning is the main cultural measure in integrated control of the LEM. Mites are easily spread in orchards by contact, wind or bees (Waite, 1999), and therefore it is very important to eliminate as many infested branches as possible from orchards. During the non-fruiting period, particularly during post-harvest pruning, infested branches should be cut off and burned (FAO, 2002; Wen et al., 1991 in Waite and Hwang, 2002; Waite, 2005).

Chemical

Satisfactory control of the LEM can be achieved through a program of successive spraying. The critical time for treatment with chemical sprays is when the trees are about to flush (Pinese, 1981; Waite, 1992; Waite and Elder, 1996). In Queensland, Australia, three sprays of dimethoate or wettable sulfur applied at 2–3 week intervals during bud emergence and leaf expansion protect the new flush from infestation by mites migrating upwards from infested leaves below (Waite and Hwang, 2002). The first spray should be applied to infested trees and their neighbors as a new flush begins to emerge. The second spray should be applied when the flush has fully emerged and just before the new leaves start to expand. The third spray should be applied after the new leaves have fully expanded but have not hardened (Waite, 2005). If this operation is carried out during the post-harvest flush, the mite population on the tree will be minimal when the flower panicles emerge. If infested leaves remain below the emerging flower panicle, a similar series of chemical applications should be made to prevent mites from moving up and damaging the flowers and fruit (Waite and Hwang, 2002).

Sprays of wettable sulfur have also been recommended for the control of LEM mite in China (Anonymous, 1978 in Waite and Hwang, 2002), Hawai'i (Nishida and Holdaway, 1945), and Taiwan (Wen et al., 1991 in Waite and Hwang, 2002). Evaluation of chemical products for the control of LEM has been performed in India. Prasad and Bagle (1981) and Sharma and Rahman (1982) found that dicofol had the best results for controlling LEM. The organophosphorous monocrotophos, phosphamidon, oxydemetonmethyl, dimethoate and diazinon, as well as chemicals from other groups – bromopropylate, propargite, azocyclotin, carbaryl, cyhexatin and chlordimeform – also performed satisfactorily (Mishra, 1980; Prasad and Bagle, 1981; Prasad and Singh, 1981; Sharma and Rahman, 1982). Sharma (1984) showed that combined applications of dicofol and phosphamidon or methyl demeton present effective control on LEM. According to Waite (2005), dichlorvos, chlorpyrifos, ometoate and isocarbophos are commonly used in China.

The efficacy of spiromesifen, a lipid biosynthesis-inhibiting agent, was evaluated in three concentrations (36 mg ai/l, 72 mg ai/l and 144 mg ai/l) for controlling LEM in northern Thailand. Experiments were conducted during the lychee flowering and fruiting period from March to May 2005, which corresponds to the end of the cool and dry seasons (Schulte et al., 2007). Single-dose applications reduced the number of mites per leaf by an average of 34% within 7 days. With fully concentrated, double-dosed applications, mite levels were reduced by >80% just 7 days after the second treatment, and complete elimination of the infestation was achieved 18 days after the second application at its highest concentration. Authors considered that the degree of erineum cover and algal density did not significantly affect the efficacy of the treatments.

Biological

Several predatory mites have been recorded as being associated with LEM; however, not all have been proven as predators. In India, these are Phytoseiidae *Euseius coccineae* (Gupta), *Euseius finlandicus* (Oudemans), *Amblyseius herbicolus* (Chant), *A. largoensis* (Muma), *A. paraaerialis* Muma, *Euseius pruni* (Gupta), *Typhlodromips syzygii* (Gupta), *Phytoseius intermedius* Evans and Macfarlane, *Typhlodromus fleschneri* Chant, *T. homalii* (Gupta), *T. sonprayagenis* Gupta; and the Cunaxidae *Cunaxa setirostris* (Hermann) (Lall and Rahman, 1975; Somchoudhury et al., 1987 in Waite and Hwang, 2002; Thakur and Sharma, 1989). In Australia, these are Phytoseiidae *Neoseiulus barkeri* (Hughes), *A. herbicolus*, *A. largoensis*, *A. nambourensis* Schicha, *Typhlodromips neomarkwelli* (Schicha), *Euseius neovictoriensis* (Schicha), *Phytoseius hawai'iensis* Prasad, *Phytoseius rubiginosae* Schicha, *Okiseius morenoi* Schicha and *Typhlodromus haramotoi* Prasad; an Anystidae of the genus *Anystis*; the Blattisocidae *Blattisocius dendriticus* (Berlese); an Ascidae of the genus *Lasioseius*; the Cheyletidae *Hemicheyletia wellsi* (Baker); a Cunaxidae of the genus *Cunaxa*; and the Stigmaeidae *Agistemus collyerae* Gonzalez (Schicha, 1987; Waite and Gerson, 1994). In China, these are *Amblyseius eharai* Amitai and Swirski, *A. herbicolus*, *A. largoensis*, *Euseius ovalis* (Evans), *Typhlodromips cantonensis* (Schicha), *T. okinawanus* (Ehara), *P. hawai'iensis*, *Phytoseius fujianensis* Wu and *Okiseius subtropicus* Ehara (Wu et al., 1991 in Waite and Hwang, 2002).

In addition to predator mites, a cecidomyiid fly larva, *Arthrocnodax* sp., was also a common predator of the LEM in Queensland, Australia (Waite and Gerson, 1994).

The Stigmaeidae predator mite *Agistemus exsertus* Gonzalez has been reported as being used to control LEM in Guangdong, Guangxi and Fujian Provinces in China (Ren and Tian, 2000 in Waite and Hwang, 2002). With the prospect of choosing an efficient natural enemy from China (where damage caused by LEM is not as severe as in other affected countries) and introducing it to Australia's habitats, Waite and Gerson (1994) compared predator mites associated with LEM from Australia and China. *Amblyseius eharai* was selected and introduced into Australia in 1993. Evaluations carried out under quarantine conditions in Queensland, Australia, confirmed that the predator feeds on LEM; however, *A. eharai* was never released in Australia due to the rise of other priorities in pest management in that country (G.K. Waite, unpublished results, 2001).

In the State of São Paulo, Brazil, the occurrence of predator mites was studied in lychee orchards with adult 12-year-old 'Bengal' trees (Picoli et al., 2010). Two branches 0.3 m long were collected each month from four trees between August, 2008 and August, 2009. Predators were mounted on microscope slides, identified and counted. During the project, 6543 mites of the family Phytoseiidae were recorded. The most abundant species was *Amblyseius compositus* (Denmark and Muma) (42.6%), followed by *P. intermedius* (31.2%), *Euseius concordis* (Chant) (14.1%), *A. herbicolus* (8.8%) and *Iphiseiodes zuluagai* Denmark and Muma (3.3%). *Amblyseius compositus*, *E. concordis* and *I. zuluagai* were positively correlated with LEM population, indicating a relationship of predation.

The functional and numerical response of *A. largoensis* preying on LEM was studied in China (Cheng et al., 2005). The Phytoseiidae mite showed a positive functional response. Adults of *A. largoensis* preyed on LEM in all life stages. With the growth of prey density within a certain range, the daily preying capacity of the predator increased gradually, reaching an average number of 37 adults per day.

In São Paulo, Brazil, Picoli (2010) observed that erinose was completely covered with whitish mycelia of the acaropathogenic fungi *Hirsutella thompsonii* (Fischer).

15.4 The Red Palm Mite, *Raoiella indica* Hirst

15.4.1 Introduction

The Red Palm Mite (RPM), *Raoiella indica* Hirst (Tenuipalpidae), was described in 1924 and found on coconut leaves from Coimbatore, India (Hirst, 1924). *Rarosiella cocosae* Rimando is used as a synonym, a species also described from coconut palms in the Philippines (Mesa *et al.*, 2009). The genus *Raoiella* Hirst has been restricted to the Eastern Hemisphere. According to Mesa *et al.* (2009), 12 valid species belong to this genus: one is of the western palearctic, two are afrotropical, eight are oriental, and one is from the Australian zoogeographic regions. Recently seven new *Raoiella* species have been discovered in Australia (Dowling *et al.*, 2010). The RPM is the only species in the genus that is currently reported to have been found in the western hemisphere. Some authors have suggested that the RPM comes from South-East Asia (Flechtmann and Etienne, 2005); however, recent molecular studies based on mitochondrial and nuclear genomic regions from different countries indicate that the most primitive RPM haplotypes were usually from the Middle East and later spread throughout the Old World and eventually into the neotropics (Dowling *et al.*, 2010).

After its original description, the RPM was reported in Pakistan and the USSR (Chaudhri, 1974; Mitrofanov and Strunkova, 1979, in Flechtmann and Etienne, 2004); the Near East (Israel, Oman, Iran and the United Arab Emirates) (Gerson *et al.*, 1983; Elwan, 2000; Arbabi *et al.*, 2002; EPPO, 2009); north-east Africa (Sudan and Egypt) (Sayed, 1942; Pritchard and Baker, 1958) and South Africa (Mauritius and La Réunion Islands) (Moutia, 1958; Quilici *et al.*, 1997). Plants that were reported to be infested include coconut (*Cocos nucifera* L.), date palm (*Phoenix dactylifera* L.) (Hirst, 1924; Moutia 1958; Pritchard and Baker 1958; Gerson *et al.*, 1983; Nageshachandra and ChannaBasavanna, 1984), princess or hurricane palm (*Dictyosperma album* (Bory) H. Wendl. and Drude) and areca palm (*Areca catechu* L. and *Areca* sp.) (Pritchard and Baker, 1958). Reports of the RPM infesting maple (*Acer* sp.), bean (*Phaseolus* sp.) and sweet basil (*Ocimum basilicum* L.) (Chaudhri, 1974; Mitrofanov and Strunkova, 1979, in Flechtmann and Etienne, 2004; Gupta, 1984) need to be confirmed.

In the Americas, RPM was detected in 2004 in Martinique (Flechtmann and Etienne, 2004, 2005), and afterwards in other islands of the Caribbean (Kane *et al.*, 2005; Etienne and Flechtmann, 2006; Rodrigues *et al.*, 2007; Myers, 2007; Welbourn, 2009; de la Torre, 2010). Not only did the RPM spread quickly, but it also greatly extended its range of hosts in this newly invaded area of the Americas. Cocco and Hoy (2009) listed 72 plant species as host plants of the RPM; however, within its areas of occurrence in the Eastern hemisphere, the mite was known to have fewer than ten host plants.

Field surveys were carried out in Kerala, India, between 2008 and 2010, investigating the range of hosts of the RPM. During those surveys, no RPM evidence was found on banana cultivars (Taylor, 2010). Other new hosts might be tropical ornamentals belonging to Heliconiaceae (*Heliconia rostrata* Ruiz and Pavón.; *H. bihai* (L.); *H. caribaea* Lam., *H. psittacorum* L. f.), Strelitziaceae (*Strelitzia reginae* Banks and *Ravenala madagascariensis* J.F. Gmel.) and Zingiberaeae (*Alpinia purpurata* (Vieill.) ex. K. Schum, *Etlingera elatior* (Jack.) R.M. Smith and *Zingiber* sp.) (Kane *et al.*, 2005; Etienne and Flechtmann, 2006; Cocco and Hoy, 2009). In Cuba, Ramos *et al.* (2010) also reported two cycad species – *Microcycas calocoma* (Miq.) A. DC. and *Cycas* sp. – as host plants for the RPM.

The RPM was detected in North America in 2007 in Palm Beach, Florida, USA. Since April 2009 the red palm mite has been spotted in five different counties of Florida (Broward, Martin, Monroe, Miami-Dade and Palm Beach) (Welbourn, 2009).

The RPM was also found in Cancun and Isla Mujeres, both in the state of Quintana Roo, Mexico (NAPPO, 2009; Estrada-Venegas *et al.*, 2010; Ramírez and Garcia, 2010; Juárez-Duran, 2010).

In early 2007 the RPM was found in South America, in the state of Sucre in north-eastern Venezuela (Vásquez *et al.*, 2008).

In July, 2009, the RPM was found in a sample of coconut leaves collected in the urban area of Boa Vista, Brazil (Navia *et al.*, 2010a, 2011; A.L. Marsaro Jr, unpublished results, 2010). Subsequent surveys have shown that the RPM is present in nine municipalities in the northeastern state of Roraima: Amajari, Alto Alegre,

Bonfim, Boa Vista, Cantá, Caracaraí, Iracema, Mucajaí and Pacaraima (A.L. Marsaro Jr, unpublished results, 2010). Brazilian host plants of the RPM were determined through surveys conducted from September, 2009 to March, 2010. The mite was found on 15 plant species: Arecaceae – *Bactris gasipaes* Kunth, *Caryota urens* L., *Dypsis lutescens* (H. Wendl.) Beentje and J. Dransf., *Elaeis guineensis* Jacq., *Euterpe oleracea* Mart., *E. precatoria* Mart., *Mauritia flexuosa* L. f., *Phoenix roebelenii* O'Brien, *P. pacifica* Seemann and H. Wendl., *Rhapis excelsa* (Thunb.) Henry ex. Rehder, *Veitchia merrillii* (Becc) H.E. Moore; Cannaceae – *Canna indica*; and Heliconiaceae – *H. bihai* (L.) cv. Napi, *H. psittacorum* (L.) cv. Golden Torch and *Heliconia* sp. Very low infestation levels were observed on *C. indica*, *B. gasipaes*, *D. lutescens* and *E. guineensis*, suggesting that these are probably less favorable RPM hosts than other plants, or are not multigenerational RPM hosts in Brazil. The presence of RPM on *M. flexuosa*, commonly known as 'buriti', is of particular interest. This species is a common and outstanding component of the Amazon forest. The detection of RPM on these trees suggests a natural dissemination of this pest in the region (Marsaro Jr *et al.*, 2010b).

The presence of the RPM in other northern South American countries, such as Colombia, was confirmed in 2010 (D. Carrillo *et al.*, unpublished results, 2010).

15.4.2 Damage and economic impact

High infestations of the RPM in coconut and banana cause severe yellowing of lower leaves followed by tissue necrosis (Flechtmann and Etienne, 2004; Peña *et al.*, 2006; Welbourn, 2009). Several basal leaves of banana plants can die due to high RPM infestation (Fig. 15.4) (D. Navia and F. Ferragut, Brasilia, Brazil, pers. comm., 2010). Young and old coconut palms under hydric and nutritional stress conditions are the most affected by RPM infestation (Moutia, 1958; Jeppson *et al.*, 1975). Damage on young coconut palms can be observed both in nurseries and in the field (Fig. 15.5) and plants can die due to severe attacks (Sarkar and Somchoudhury, 1988; Sathiamma, 1996).

Damage caused by high RPM infestations in coconut and banana plants in the invaded

Fig. 15.4 Colonies of *R. indica*.

Fig. 15.5 *R. indica* damage on banana.

areas of the Caribbean have been severe (Peña *et al.*, 2010). However, information on economic losses is scarce, and is based on estimates as opposed to results of experimental data. Reduced coconut production in Trinidad and Tobago due to RPM infestation has been estimated at around 70% (Philippe Agostini, President, Association of Coconut Growers, Cedros, Trinidad & Tobago, 2011). Similar losses were observed in coconut production areas in Venezuela (M. Quirós, Zulia, Venezuela, pers. comm. 2008). In Florida, USA, the estimated cost of regulatory sprays for ornamental palms growers has been estimated at US$500,000. There are no data on banana production loss due to RPM attacks; however, it is clear that plants can be seriously affected. Numerous basal leaves have died completely and mature leaves have become completely chlorotic (D. Navia and F. Ferragut, Brasilia, Brazil, pers. comm., 2010).

15.4.3 Morphological and biological aspects

The RPM can be easily recognized as a reddish tenuipalpid mite presenting long and whitish setae on dorsal opisthosoma (Fig. 15.6). RPM eggs are 90–100 μm long, reddish-pink, ovoid, smooth, and with one end slightly broadened and attached to the abaxial leaf surface by a slender stalk that is about twice as long as the egg itself (170–210 μm long). Eggs turn opaque 1 day before hatching (Nageshachandra and ChannaBasavanna, 1984; Kane and Ochoa, 2006). The newly emerged larva, 120–160 μm long, present three pairs of legs, and are bright orange-red. They move around the eggshell for a few minutes and then settle down for feeding. The protonymph differs from the larva in that it has four pairs of legs; the body length is 180–200 μm. The female protonymph has an ovoid body with a broad opisthosoma, while the male presents a narrow, pointed opisthosoma. The deutonymph is almost oval and is 240–250 μm long. Immature stages exhibit a smoother tegument that lacks the projecting setal bases that are apparent in the adults, and they also present shorter dorsal and lateral setae. The female is larger than the deutonymph, with a body length of 250–320 μm and a width of 190–220 μm. The male is pear-shaped and flattened in profile, and ends its development period with its reproductive structures. It is more active than females and is 220–230 μm in length and 140–150 μm in width. Adult females often exhibit dark patches on the opisthosoma (Nageshachandra and ChannaBasavanna, 1984; Kane and Ochoa, 2006). All dorsal body setae are slightly club-like and serrated (Peña et al., 2006).

The RPM presents an interesting sexual behavior. Males and females are sexually mature when they emerge, and males actively seek out females, suggesting that a sex pheromone is present (Hoy et al., 2006). When the male finds a female deutonymph in the quiescent stage it settles very close to it and waits for molting to begin. Courtship often starts in the deutonymphal stage. When female deutonymphs begin to molt, the waiting male becomes active and moves under her, bending his back up and forwards to mate. Mites remain in the mating position for about 16 min (Nageshachandra and ChannaBasavanna, 1984). Due to this typical behavior it is possible observe numerous male and quiescent female deutonymph couples on the infested leaves.

The biology of the RPM was studied in the Old World on coconut leaves under laboratory conditions (from 23.9 to 25.7°C and RU around 60%) in India (Nageshachandra and ChannaBasavanna, 1984), and under temperatures around 24°C during the months of February and March and 18°C in July and August in Mauritius. In Egypt, Zaher et al. (1969) studied biological characteristics of the RPM on date palms under environmental conditions during the summer. The RPM reproduces sexually and also through arhenotokous parthenogenesis, the production of males by non-fertilized eggs. The mite goes through the egg, larvae, protonymph, deutonymph and adult stage. Nageshachandra and ChannaBasavanna (1984) observed that the life cycle of females took 24.5 days and in males 20.6 days. Furthermore, the female's lifespan was of 50.9 days and that of the males was 21.6 days. In Mauritius, Moutia (1958) observed an average life cycle of 22 days in the summer and 33 days in the winter, and a lifespan of 27 days for females and up to 24 days for males. In India the preoviposition period of non-fertilized females lasted 2 days, and 5.9 days for the fertilized females (Nageshachandra and ChannaBasavanna, 1984); in Mauritius the period was 3 days during the summer and 7 days during the winter (Moutia, 1958); and in Egypt 3.3 days (Zaher et al., 1969). In India fertilized females produced an average of 22 eggs, and non-fertilized females produced 18.4 eggs (Nageshachandra and ChannaBasavanna, 1984); Moutia (1958) observed an average of 28 eggs and a maximum of 38 eggs per

Fig. 15.6 Coconut seedling pinna injured by *R. indica*.

female during an oviposition period of 27 days. According to Zaher et al. (1969) and Jeppson et al. (1975), RPM females lay an average of 2 eggs per day over an average period of 27 days for a total of c. 50 eggs per female. In Egypt, Zaher et al. (1969) reported that the mite can develop throughout the whole year, and that a complete generation is concluded within 3–4 weeks under temperatures varying from 23°C to 28°C.

In the New World, the biology of the RPM was studied in Cuba on areca palm (*A. catechu*) (Flores-Galano et al., 2010) and on coconut palms and banana plants (González-Reyes and Ramos, 2010). Studies on the areca palm were conducted in a laboratory at 25°C and 57% RH. The life cycle of the RPM on areca palms lasted on average 31 days, with females varying from 22 to 40 days, and males from 24 to 44 days. The life cycle observed in this study was longer and the oviposition period and fecundity significantly lower than in other biological studies on coconut (Moutia, 1958; Nageshachandra and ChannaBasavanna, 1984). Authors considered the need for further studies and discussed if areca palm is an appropriate host plant for the RPM. González-Reyes and Ramos (2010) compared the biological parameters of the RPM on coconut palms and banana plants (*M. acuminata* subgroup Cavendish 'Gran enano') under 26% and 75% UR. Authors did not observe significant differences in any stages of the development period of the RPM on these two host plants.

15.4.4 Population dynamics

In India, Nageshachandra and ChannaBasavanna (1984) observed that relative humidity and pluviosity are negatively correlated with the RPM populations, while temperature and photoperiod are positively correlated. Authors concluded that high levels of RPM populations are associated with low relative humidity, high temperature and long sunny days. Puttarudriah and ChannaBasavanna (1958) reported that RPM populations reached high levels during the summer months in Karnataka, and decreased abruptly at the beginning of the rainy season. A similar situation was observed in Mauritius, where RPM populations were more abundant in the field from September to March than during the period of heavy rain from November to January.

Population dynamics of the RPM was studied in Palm Beach, Broward and Miami-Dade counties in Florida, USA, from January, 2008 to March, 2010. RPMs were found throughout the year, and higher population levels were observed (i.e., average 4000 mites/pinna) in the area where the mite was first discovered, and lower population levels in newly invaded areas of the south. At all three sites the population levels were at their highest in the first 4 months after the initial infestation. A steady negative trend of population density has been observed since July 2008, which could be related to the subtropical conditions of the state as well as the build-up of predators (Duncan et al., 2010).

15.4.5 Dissemination

Transportation of infested plants or plant material appears to be an important way of spreading the RPM in the Caribbean basin region. RPMs have been found on coconut seeds destined for Florida, USA. In addition, handicrafts made from coconut leaves have been found to harbor live mites and viable eggs. Under natural conditions the RPM is spread by wind currents, along with most other plant-feeding mites (Welbourn, 2009).

Kane et al. (2005) considered that female adults most likely play a major role in the dispersal stage of the RPM. Insemination appears to occur during the female's final ecdysis, and prepares her for immediate dispersal by ensuring her ability to initiate a new colony. Authors consider that low numbers of females in high-density colonies, and the discovery of isolated females surrounded by clusters of 20–30 eggs, both support this hypothesis.

15.4.6 Rearing techniques

In order to develop an efficient method to rear the RPM in quarantine for a classical biological control project, Cocco and Hoy (2009) tested several banana and plantain varieties as hosts for the mite. Banana plants were more desirable than coconut palms, a favored host plant, because bananas are easier to rear in small cages, and produce new shoots quickly after pruning. The RPM females did not establish themselves on either the leaf discs of the banana and plantain cultivars ('Dwarf Cavendish', 'Dwarf Nino', 'Gran Nain', 'Dwarf Zan Moreno', 'Dwarf Green', 'Truly Tiny', *Musa sumatrana* × 'Gran Nain', 'Dwarf Puerto Rican', 'Rose', 'Nang Phaya', 'Misi Luki', 'Manzano', 'Lady Finger', 'Glui Kai' and 'Ebun Musak'), or on potted banana plants ('Glui Kai', 'Dwarf Green' and 'Nang Phaya'). However, they established multigenerational colonies on coconut leaf discs and trees. No RPM females survived on tested native palms (saw palmetto, *Serenoa repens* (W. Bartram) Small; cabbage palms, *Sabal palmetto* (Walter) Lodd. ex Schult. f. and dwarf palmetto *S. minor* (Jacq.) Pers.), but RPM completed a generation on the needle palm *Rhapidophyllum hystrix* (Frazer ex Thouin) H. Wendl. and Drude, with longer development time, higher mortality and lower fecundity than when reared on coconut discs. Results indicated that coconut leaf discs and trees are better hosts for rearing RPM banana plants, plantains and native palms. Authors also discussed possible reasons for the non-success of the multigenerational colonies of the RPM on banana plants. The need to confirm the suitability of plants listed as hosts for the establishment of multigenerational RPM colonies is also a point to consider, since the range of hosts for the mite might not be as broad as some reports indicate.

15.4.7 Detection

RPM establishes colonies on the undersides of leaves. This can be observed with the help of a hand lens magnifier (10 ×) due to the bright reddish color of all the development stages, which can be surrounded by numerous white exuviae. In coconut palms, infestation begins along the midrib (Welbourn, 2009); however in banana plants, infestation begins on the border and gradually reaches the midrib area (D. Navia and F. Ferragut, Brasilia, Brazil, pers. comm., 2010). Infested leaves initially become yellow and later develop more or less extensive areas of brown and dark necrotic tissue (Etienne and Flechtmann, 2006). Symptoms on coconut leaflets start as small yellow spots on the abaxial surface; these develop into larger, chlorotic spots (Rodrigues *et al.*, 2007). Symptoms on leaves of young coconut trees are first observed in the distal part of the leaflets, which become bronzed and then completely necrotic. The symptoms caused by a heavy infestation of the RPM in coconuts can be confused with nutritional deficiencies or lethal yellowing (Welbourn, 2009). Symptoms of RPM infestation observed on date palms in Egypt and Israel differ from those observed on coconuts, and consist of reddish and dark spots on the leaves (Zaher *et al.*, 1969; Gerson *et al.*, 1983).

Efficient sampling methods can be very helpful for early detection of pests in new areas, and allow the use of preventive measures or control methods to slow down spreading. Roda *et al.* (2010) conducted studies to provide and compare intra-tree dispersion of the RPM on coconut palms from three different geographical areas recently invaded by the mite, and to provide insight on the effects of intra-plant distribution, time of collection, number of individuals present in the samples and level of mite infestation. During the early stages of infestation, mite populations were highly clumped and concentrated in the lower strata of the palm. As populations increased, the distribution of mites remained clumped but the mites exploited other strata. If the sampling objective is to detect the presence of the RPM at very low densities, in order to verify the presence or absence of the pest in an area where it is not confirmed, the number of samples that needs to be collected is very large. Kane and Ochoa (2006) have also observed that in coconuts a greater number of mites is found on the lower leaves. There is no information on intra-tree distribution of the RPM in other host plants.

15.4.8 Control tactics

Establishing regulatory actions to prevent RPM dispersion in South America will determine their impact in the invaded or threatened areas of South America. Chemical control of the RPM can be useful for emergency control or for regulatory purposes, especially in nurseries; however, it would be difficult to apply this to most host palms because of their size, as well as the high costs for growers, and because it is not environmentally friendly. In the long run, it has been established that biological control deserves more attention. Therefore, efforts have been dedicated to determining the possibility of pest control either with local natural enemies (conservation and augmentation), or with natural enemies introduced from areas where the pest could be found at acceptable levels (classical biological control). Another approach almost unexplored by integrated RPM control is host plant resistance, and there have been few results obtained in the Old World.

Chemical

Most information on the effectiveness of chemical control of the RPM is based on evaluations conducted in India and the Near East. The efficacy of pesticides for controlling the RPM on coconut palms in India was evaluated and those with the best results were monocrotophos, phosphamidon, dicofol, dimethoate, quinalphos, ethion, phosalone and endosulfan (Sarkar and Somchoudhury, 1988; Jalaluddin and Mohanasundaram, 1990; Jayaraj *et al.*, 1991). Nadarajan *et al.* (1990) reported that several systemic insecticides exhibited toxicity to RPM larvae, nymphs and adults. In India, neem oil sprays mixed with sulfur after a thorough cleaning of the coconut crown also showed good results. The extract is sprayed from above, 5–6 times per annum, using a sprayer head attached to a long pole. Neem applications resulted in a yield increase of 25% (Peña *et al.*, 2006). The efficacy of ecofriendly compounds was evaluated for the control of the RPM on coconut seedlings in Oman by Pérez *et al.* (2010). The tested products were coconut oil + soap, and fenpyroximate, and only coconut oil + soap at 4.5 + 1% was effective and significantly different.

In the Americas, evaluations for the chemical control of the RPM have been conducted in Puerto Rico and in Florida, USA. Several products were tested for efficacy against the mite on potted coconut palms and banana plants in greenhouses in Rio Piedras, Puerto Rico, and on field-grown coconut palms in Broward County, Florida. In Puerto Rico the acaricides milbemectin, etoxazole, bifenazate, acequinocyl, dicofol and spiromesifin resulted in lower RPM populations than the control during a period of 21 days. In Florida, wettable sulfur, etoxazole, abamectin mixed with oil, and pyridaben were satisfactory during 42 days of evaluation (Peña and Rodrigues, 2010).

Biological

Several predators have been associated with the RPM in the Old World. Moutia (1958) reported that the Phytoseiidae mite *Amblyseius caudatus* Berlese was the main predator of RPM in coconut palms in Mauritius. This author observed that immature stages and adults of this phytoseiid could consume an average of 10.6 eggs/day and around 490 eggs during their life cycle. In India, Gupta (2001) presented a list of natural enemies of phytophagous mites associated with coconut palms, and cited the phytoseiid mite *Neoseiulus longispinosus* (Evans), *A. channabasavannai* Gupta and Daniel, and the lady beetles *Stethorus parcempunctatus* Puttarudriah and ChannaBasavanna, *S. tetranychi* Kapur and *Jauravia* sp. as predators of the RPM. Daniel (1981) considered *A. channabasavanni* as one of the most important predators of the RPM in India, along with the lady beetle *Stethorus keralicus* Kapur. There is no information on the predatory potential of these natural enemies in India.

Information on natural enemies of the RPM in the newly invaded areas of the Americas is still scarce. Carrillo *et al.* (2010) evaluated results of surveys of natural enemies on coconut palms in Florida, USA, pre- and post-RPM infestation. Some predators, including the phytoseiid mite *A. largoensis* and the bdellid mite *Bdella distincta* (Barker and Balock); the chrysopid *Ceraeochrysa claveri* Navas; the lady beetle *S. utilis* (Horn) and the thrips *Aleurodothrips fasciapennis* (Franklin) have been observed feeding on the invasive species. Among these, *A. largoensis* was the most abundant predator and its populations increased in number after the arrival of the RPM in south Florida (Peña *et al.*, 2009). Based on these observations, detailed studies were initiated for the biological control potential of *A. largoensis* against RPM. In order to verify whether *A. largoensis* can develop and reproduce when feeding exclusively on RPM, the biology of this predator was evaluated on various food sources, including the RPM. Five diets were tested – RPM, *Tetranychus gloveri* Banks, *Aonidiella orientalis* (Newstead), *Nipaecocus nipae* (Maskell) and oak (*Quercus virginiana* Mill.) pollen. *Amblyseius largoensis* was able to complete its life cycle and reproduce when feeding exclusively on RPM. The development of immature stages of this predator mite was faster, and fecundity and survivorship were higher when feeding on RPM or *T. gloveri* compared to the other food sources. The intrinsic rate of natural increase of *A. largoensis* was significantly higher when feeding on RPM rather than on other diets. Other native Florida predators either develop and reproduce poorly, or cannot complete development when feeding on the invasive species (Carrillo and Peña, 2010; Carrillo et al., 2010). Amblyseius largoensis has also been found associated with RPM in Trinidad and Tobago, Puerto Rico, Colombia and Cuba (Rodrigues et al., 2007; Peña et al., 2009; A. Roda, Miami, USA, pers. comm., 2010 in Carrillo et al., 2010; Hastie et al., 2010; D. Carrillo et al., unpublished results, 2010). These results suggest that A. largoensis can play a role in controlling RPM in invaded areas of the Americas.

Host plant resistance

Commercial coconut varieties have been evaluated in relation to RPM susceptibility in India, and all of them have shown to be highly susceptible (Sarkar and Somchoudhury, 1989). Preliminary studies on the resistance mechanisms of coconut palms to RPM infestations, based on the morphological and biochemical characteristics of the leaves, were conducted in India for the following tree varieties: 'Hooghly Tall', 'Hooghly Local', 'Hazari', 'Andaman Giant', 'Howrah Tall', 'Kerala Tall', 'Andaman Tall' and 'Chennangi'. The morphological characteristics tested were length, width, distance between midribs, and main midrib depth and width, while the biochemical characteristics tested were protein, nitrogen, calcium and phosphorus content. There was no correlation between the morphological characteristics and RPM populations; however, a positive correlation was observed amongst protein content, nitrogen and RPM populations (Sarkar and Somchoudhury, 1989).

15.5 The Wheat Curl Mite, *Aceria tosichella* Keifer, and Associated Viruses

15.5.1 Introduction

The Eriophyidae mite *Aceria tosichella* Keifer, commonly known as Wheat Curl Mite (WCM), was described on wheat (*Triticum aestivum* L.) leaves in Zemun-Beograd, Yugoslavia, in 1969 (Keifer, 1969). For many years, *A. tosichella* was mistaken for *Aceria tulipae* (Keifer), a species described in 1938 as a pest of tulip bulbs imported from The Netherlands into the USA (Keifer, 1969). Besides tulip bulbs, it was also thought to infest onions, garlic and grasses. Shevtchenko et al. (1970) revealed that the eriophyid species from Liliaceae was different from the one from wheat, and described *Aceria tritici* as the wheat species. Before Shevtchenko's description, Keifer (1969) had already described *A. tosichella* to accommodate the wheat species associated with the phytopathogenic virus. Since then, *A. tritici* has become a junior synonym of *A. tosichella*. Despite this, the name *A. tulipae* is still being used for the eriophyid mite associated with wheat, especially in North America (Kozlowski, 2000).

Biological, morphological and molecular studies have suggested that the phenotype recognized as *A. tosichella* contains different strains, or possibly even represents a complex of species. Harvey et al. (1995, 1999) showed that populations of the WCM from Kansas, Nebraska, Montana, Alberta, South Dakota and Texas vary in virulence, and can be identified by their response to genetic resistance in different grasses. Populations also vary in their ability to transmit Wheat streak mosaic virus (WSMV) and high plain virus (HPV) (Seifers et al., 2002). In Poland, Skoracka and Kuczynski (2006) found differences in infestation parameters and morphological variation in WCM populations from three host grasses – *Agropyron repens* (L.) P. Beauv., *Bromus inermis* Leyss. and *Arrhenatherum elatius* (L.) P. Beauv. ex J. Presl and C. Presl) – and suggested the occurrence of various WCM strains or host races. Carew et al. (2009) used data from three molecular markers – the mitochondrial 16S rRNA, the nuclear internal transcribed spacer 1 and adenine nucleotide translocase – to study the genetic variation of WCM in Australia. The results indicated the occurrence of at least two separate lineages that may represent putative species. A joint effort from researchers on the different continents where the WCM is present, and involving molecular, morphological and biological data, would be necessary to clarify the systematics of this important eriophyid vector.

At present the WCM is widespread in the main wheat production areas: Europe (Shevtchenko et al., 1970; Oldfield and Proeseler, 1996; Kozlowski, 2000; Golya et al., 2002); North America (Keifer, 1953; Oldfield and Proeseler, 1996; Sanchez-Sanchez et al., 2001); Asia (Oldfield, 1970); Middle East (Oldfield, 1970;

Denizhan et al., 2010); Oceania (Halliday and Knihinicki, 2004); and South America (Navia et al., 2006b; Pereira et al., 2009; Castiglioni and Navia, 2010).

The WCM occurs mainly on wheat, but populations can also develop on cultivated corn grass (*Zea mays* L.), oats (*Avena sativa* L.), barley (*Hordeum vulgare* L.), sorghum (*Sorghum* sp.), rye (*Secale cereale* L.) and pearl millet (*Pennisetum glaucum* (L.) R. Br.) (Jeppson et al., 1975), and infest a large quantity of grass of minor economic importance, and weeds, presenting about 120 host grasses (Amrine, 2003).

The WCM has been intercepted in wheat seeds in a quarantine station in Brazil (Navia and Flechtmann, 2008), which suggests that the mite was spread as a contaminant in seeds that were probably harvested in highly infested areas.

Quite often surveys for eriophyid mites in wheat and other host grasses are conducted just after viruses transmitted by the WCM are detected and damage appears, making it necessary to find the vector and alternate host plants in order to establish control tactics. An example of this was the report of the WCM in South America. The mite was first found in Argentina in 2004 (Navia et al., 2006b), 2 years after WSMV was detected in the country (Truol et al., 2004).

The occurrence of the WCM and associated viruses in Argentina is a warning of the threat the pathosystem of the WCM and WSMV/HPV present to cereal crops in other countries in South America, especially neighboring countries with close or contiguous cereal production areas. First, surveys were conducted in Brazil in October, 2006, and in August and October, 2007 on wheat, corn and oat crops, and on potential host grasses in 46 municipalities of the state of Rio Grande do Sul in the extreme south of Brazil that borders Argentina. The first report of WCM in the country occurred on wheat in four municipalities – Passo Fundo, Palmeira das Missões, São Luís Gonzaga and Santo Antônio das Missões (Pereira et al., 2009). Subsequent surveys were conducted in 2008–2010, The mite was found in 19 municipalities, all in Rio Grande do Sul, associated with 17 grasses: wheat, oat, barley, corn, *Agropyron* sp., *Andropogon bicornis* L., *Brachiaria decumbens* Stapf, *Brachiaria plantaginea* (Link) Hitchc., *Bromus unioloides* (Kunth), *Chascolytrum subaristatum* (Lam.) Desv., *Chloris polydactyla* (L.) Sw., *Digitaria insularis* (L.) Fedde, *D. horizontalis* Willd., *Lolium multiflorum* Lam., *Pennisetum americanum* (L.) Leeke, *Rhynchelytrum repens* (Willd.) C.E. Hubb. and *Sorghum halepense* (L.) Pers. (Pereira et al., 2010; Navia et al., 2010b). Symptoms caused by high WCM infestations were observed only in greenhouses (Fig. 15.7), while low infestations were detected in the field. Viruses associated with WCM were not detected in Brazilian cereal crops, although continuous field surveys were done.

In Uruguay, surveys were conducted from February, 2007 to November, 2008. The WCM was detected in the Departments of Colonia (four municipalities), Rio Negro and Soriano (one municipality in each), infesting wheat, *L. multiflorum* and *B. unioloides* (Castiglioni and Navia, 2010). Symptoms transmitted by the WCM virus were not observed in the field.

Surveys in Paraguay were conducted in August, 2007 on wheat and corn (D. Navia, Brasilia, Brazil, 2010, unpublished). The WCM was not found in Paraguay; however, complementary surveys should be conducted for more reliable results.

In Argentina, surveys were conducted in all wheat-producing areas from 2006 to 2010. In addition to Buenos Aires, the WCM and WSMV were detected in the Provinces of Córdoba, Entre Ríos, La Pampa, Santiago del Estero, Santa Fe, Tucumán and Salta (Navia et al., 2010b; A. Vanina, unpublished results, 2010). The presence of HPV was confirmed only in Córdoba and Buenos Aires provinces (Truol and Sagadin, 2008). WSMV is considered a regulated non-quarantine pest in Argentina (SENASA Resolución 248/2003), which implies permanent surveillance.

Fig. 15.7 Damage to wheat caused by *A. tosichella*.

15.5.2 Damage and economic impact

Yield losses in wheat due to high WCM population infestations can reach 30% (Harvey et al., 2002); however, there is no doubt that the main damage caused by the WCM comes from its ability to transmit and spread damaging virus diseases to cereal crops (Oldfield and Proeseler, 1996). The most important cereal viruses transmitted by the WCM are the wheat streak mosaic virus (WSMV) and the high plains virus (HPV) (Harvey et al., 1994; Seifers et al., 1997). WSMV causes major diseases in wheat and other cereals in important production areas (e.g., the USA). According to Velandia et al. (2010), WSMV losses are associated with the reduction in water-use efficiency, indicating that the disease reduced the plant's ability to uptake available soil moisture, resulting in grain and forage yield losses ranging from US$60.1/ha to US$339.9/ha in the Texas High Plains. In the Great Plains region, WSMV is responsible for average annual yield losses of approximately 5%, and complete yield loss in localized areas. Annual economic losses have amounted to US$80 million in the state of Kansas alone (University of Illinois, 1989; Christian and Willis, 1993; French and Stenger, 2003). Wheat yield reduction due to WSMV infections have also been estimated in other US states, varying from 50.2% to 91.4% in Colorado after the evaluation of 12 wheat cultivars (Shawan and Hill, 1984); from 31.9% to 98.7% in a 2-year field study in North Dakota (Edwards and McMullen, 1988); and a maximum reduction of 75% and 87% in fertile tillers and grain yield in a 2-year field study in Oklahoma (Hunger et al., 1992).

In 2006, a new virus transmitted by the WCM was detected in Kansas, USA, when one of the first wheat varieties to have high levels of genetic resistance to the WSMV developed severe symptoms of disease, which was discovered to have been caused by the *Triticum mosaic virus* (TriMV) (Seifers et al., 2008; 2009). Another virus recently confirmed as being transmitted by the WCM is the *Brome streak mosaic virus* (BrSMV) (Stephan et al., 2008). Mixed infections of WSMV and HPV, as well as of WSMV and TriMV, are commonly observed in wheat fields, and these viruses can have a synergistic effect on host plants causing more severe impact (Mahmood et al., 1998; Tatineni et al., 2010).

During the last decade, this invasive complex of WCM and WSMV/HPV, or at least one of its components, was introduced and then disseminated throughout the wheat-production area of Australia and South America, causing serious impacts and becoming a threat to production areas. In Australia the WSMV was first reported in 2003 (Ellis et al., 2003), probably due to an incursion of infected seeds from the Pacific northwest of the USA (Dwyer et al., 2007). Severe outbreaks in 2005 in the high rainfall cropping region of New South Wales affected at least 5000 ha of wheat, with many crops sustaining complete yield loss (Murray and Wratten, 2005; Murray, 2006). Detections of WSMV have been confirmed in all Australian states, except for the Northern Territory, and it is now evident that the entire Australian wheat belt is at risk of WSMV infections (Coutts et al., 2008). In 2002 the WSMV was detected in Argentina, in the central Province of Córdoba (Truol et al., 2004) and in a few years the virus has spread to the main wheat production areas. In 2007 severe WSMV and HPV outbreaks with 100% incidence caused total losses in several farms in the Mar y Sierras, Province of Buenos Aires (Truol and Sagadin, 2008).

15.5.3 Biological and morphological aspects

The WCM is a wormlike, whitish eriophyid mite (Fig. 15.8). Females are 200–230 μm in length and 44–48 μm in width (Keifer, 1969). The life cycle of the WCM comprises the stages of egg,

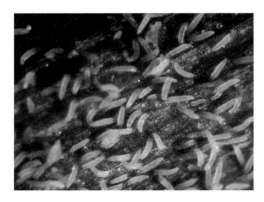

Fig. 15.8 Colonies of *A. tosichella*.

larva, nymph and adult. Deuterogyny was not reported for the WCM (Oldfield and Proeseler, 1996). Eggs of the WCM are deposited in straight lines along parallel leaf veins. Individual females lay from 3 to 25 eggs a day during their lives. Incubation lasts from 3 to 5 days at 9°C, but hatching stops at freezing temperatures. A complete life cycle takes 8–10 days under favorable conditions. The larval period takes about 2.25 days and the nymph requires 2.75 days. Adult females have a preoviposition period varying from 1 to 3 days (del Rosario and Sill, 1958, 1964; Jeppson et al., 1975). Boczeck and Chyczewski (1975) in Sabelis and Bruin (1996) reported that the WCM's period of development lasts 13 days at 20°C and 7 days at 27°C.

Warm, humid conditions appear to be ideal for optimal growth during WCM development (Somsen and Sill, 1970; Coutts et al., 2008; Schiffer et al., 2010). CLIMEX analysis of the WCM occurrence areas in Australia suggests that the species can persist in semi-arid and temperate areas, with distribution limited by heat and dry stress (Schiffer et al., 2010). During an experiment in Texas, USA, WCM colonies were maintained for several months at 5°C, although with low egg viability (Skare et al., 2002).

Adult WCM can be found on young plants just after germination. With plant maturation and subsequent tissue drying, adults prepare to migrate to other areas with the help of wind currents. They are known as the 'emigrating populations' or 'airborne mites'. First, mites tend to move to the highest parts of the host plants and can be observed on the apical leaves (Gillespie et al., 1997; Mahmood et al., 1998). In this way WCM spend the summer on volunteer wheat and green grass vegetation between rows of wheat, and move to early-planted wheat in the fall (Nault and Styer, 1969; Harvey et al., 2001)

Studies on WSMV transmission by the WCM showed that both immature and adult WCM can transmit the virus (Slykhuis, 1955). To become viruliferous the WCM must acquire WSMV during either one of the two immature stages, typically after at least 15–30 minutes of feeding on infected plant material. Once viruliferous, the WCM has the potential to transmit the virus for at least 7 days (Slykhuis, 1955; del Rosario and Sill, 1965).

15.5.4 Dissemination

Dissemination of WCM, or other eriophyoid mites, at small distances, can happen naturally through wind, pollinators and water (Lindquist et al., 1996). Windborne WCM is of key importance in the infestation and spread of WSMV. Thomas and Hein (2003) observed that the size of the source population is more important to the level of mite movement than host plant conditions. According to Gibson and Painter (1958) in CABI (2002), the WCM can also be disseminated through phoresy, mainly carried on aphids that infest cereal crops.

15.5.5 Symptoms

Symptoms due to WCM infestation include discoloration, curling or rolling of leaves, abnormal development of leaves and plant stunting. The stunting occurs because infested leaves do not expand normally, remaining inside older leaves causing the plant to be arched (Jeppson et al., 1975; CABI, 2002).

WSMV is a progressive disease that often starts at the edge of a field and moves inward. Infected plants exhibit symptoms ranging from chlorotic streaking and mosaic diseases to severe chlorosis, stunting, and, in extreme cases, plant death (Wiese, 1987). Severely infected plants produce fewer tillers that produce low-quality grains, or may remain in a vegetative state (Atkinson and Grant, 1967). In addition, infected plants have fewer roots and shoot biomass than healthy plants, resulting in drastic water-use efficiency reductions (Price et al., 2010). In the field it is extremely difficult to differentiate symptoms of WSMV-, HPV- and TriMV-infected plants. Plants that are infected with both WSMV and HPV or WSMV and TriMV diseases may die prematurely or produce little or no grain (Wolf and Seifers, 2008).

15.5.6 Control tactics

Controlling WSMV or other WCM-transmitted viruses depends basically on controlling mite vector populations. In general, chemical control of WCM is not efficient because colonies develop on protected parts of the host plants, making contact with the pesticide more difficult (Skare et al., 2003). In addition, populations are extremely numerous, with thousands of mites on a highly infested leaf (Fig. 15.8).

Chemical

In Texas, USA, carbofuran sprays were evaluated for controlling WCM on wheat inside greenhouses, without satisfactory results. As an alternative, aldicarb was applied to the soil and controlled the WCM. Carbofuran, when applied to the soil, was also efficient in reducing WCM and WSMV incidence (Harvey et al., 1979).

Host plant resistance

Currently, genes that confer resistance to mite infestations and disease infections that are present in wild grasses are introgressed into wheat (Harvey et al. 1994; Seifers et al., 1997, 1998; Malik et al., 2003a, b; Li et al., 2004). WCM-resistant cultivars presented lower rates of virus infections than susceptible cultivars (Conner et al., 1991; Harvey et al., 1995, 1997; Jezewska, 2000), showing the importance of the host resistance for controlling this problem (Malik et al., 2003a, b).

Some varieties resistant to WCM and its associated virus have been successfully developed, including Ron L and Mace (Wolf and Seifers, 2008; Graybosch et al., 2009). Problems have arisen during development, such as the resistance break of Ron L due to a new WCM-transmitted virus, TriMV (Seifers et al., 2008, 2009).

Cultural

Eliminating volunteer wheat, which is known to harbor large numbers of WCMs and viruses, can significantly reduce the risk of severe disease. Destroying these volunteers with herbicides or tillage at least 2 weeks before planting a new wheat crop will significantly reduce the risk of severe diseases (Wolf and Seifers, 2008). Grasses and grassy weeds around the borders of recently sown crops should also be controlled to further limit green host plants.

Regulatory

It is extremely important to adopt strict regulatory measures to avoid rapid dissemination of the WCM and associated virus, and to monitor the presence of this pest complex in cereal production areas that have not yet been affected. Considering that the WCM is present in fields within Uruguay and Brazil that have not yet been affected by the WCM-transmitted virus, an important prevention measure to avoid fast and broad dissemination of the complex in South America is the requirement of a phytosanitary certification for seeds produced in countries where WSMV, HPV or TriMV are present. Protocols to certify the seeds should consider that the WCM-associated virus must have a very low transmission rate through seeds, which means a large number of seeds should be tested to ensure they are not all infected. In Australia, Roger et al. (2005) showed that the WSMV transmission rate by wheat seeds is around 1.5%, which is similar to the highest rate obtained in Argentina by Sagadin et al. (2008). Transmission of the HPV has been reported only for sweet corn under greenhouse conditions in the USA, with a rate of 0.008% (Forster et al., 2001).

15.6 The Panicle Rice Mite (PRM), *Steneotarsonemus spinki* Smiley

15.6.1 Introduction

The tarsonemid mite *Steneotarsonemus spinki* Smiley, commonly known as Panicle Rice Mite (PRM), Rice Mite, Rice Sheath Mite or Rice Tarsonemidae, is a phytophagous species described from specimens collected on the delphacid planthopper *Tagasodes orizicolus* (Muir) (formerly *Sogata orizicola* Muir), in 1960 in Baton Rouge, Louisiana, USA (Smiley, 1967). A detailed review of the PRM and its status in the Americas is presented in Hummel et al. (2009) and Navia et al. (2009).

Rice is the most important host plant for the PRM and has long been considered its only host (Smiley et al., 1993). The survival rate of the mite on plants of 73 species – invasive plants and species that grow near rice fields, including 44 monocotyledons – was investigated in Taiwan (Ho and Lo, 1979) and results indicated that the mite was unable to develop on all of the host plants tested. In rice crops of Costa Rica, eggs, larvae and nymphs of the PRM were collected from a rice cogeneric weed plant, *Oryza latifolia* Desv., and the mite was able to complete its life cycle on this host (Sanabria, 2005). In addition, *Cynodon dactylon* (L.) Pers grass and two Cyperaceae – *Cyperus iria* L. and *Schoenoplectus articulates* (L.) Palla – have been reported as alternate hosts for the PRM in India (Rao and Prakash, 1996, 2002; CRRI, 2006). Another 14 plants belonging to the Poaceae (12), Caryophyllaceae (1) and Polygonaceae (1) families, listed in Hummel *et al.* (2009) have reportedly been infested with PRM in all development stages. Nevertheless, authors emphasized that these plants had not been proved to be 'true' PRM hosts and research was still needed to determine the range of PRM hosts.

The origin of the PRM is uncertain. The mite was originally described in the USA in 1967 from specimens collected in 1960 in Lousiania (Smiley, 1967); however, there is evidence that the PRM has occurred in Asian rice crops since the early 1900s. In 1930 it was reported that rice sterility in India was caused by a minute and agile arthropod, which was identified in 1978 as *S. spinki* (Teng, 1978 in Jagadiswari and Prakash, 2003). It is possible that the PRM began in Asia, where rice originated, and was later introduced to the Americas along with rice material as a contaminant, and imported for propagation some decades ago. Molecular studies could clarify the origin of the PRM as well as indicate if more than one introduction has occurred in the Americas.

The PRM is distributed throughout tropical Asia, where it is considered as a pest, and is present in the two major rice-producing countries of the world, China and India (Ou *et al.*, 1977; Rao and Das, 1977; Smiley *et al.*, 1993; Jagadiswari and Prakash, 2003). In India, new PRM-infested areas have been reported between 1999 and 2008, including Andhra Pradesh, Uttar Pradesh, Orissa, Jharkhand and West Bengal (Jagadiswari and Prakash, 2003; Karmakar, 2008). Other infested Asian countries include Taiwan, Philippines, Japan, Korea, Thailand and Sri Lanka (Lo and Hor, 1977; Sogawa, 1977; Shikata *et al.*, 1984; Smiley *et al.*, 1993; Cheng and Chiu, 1999; Cho *et al.*, 1999; Cabrera *et al.*, 2002b). The PRM was first reported in the Caribbean basin region in Cuba in 1997 (Ramos and Rodriguez, 1998). Later it was detected in the Dominican Republic and then Haiti (García *et al.*, 2002; Valès *et al.*, 2002; Ramos and Rodríguez, 2003). The report from Puerto Rico is from 2007 (NAPPO, 2007).

In continental Central America the PRM was first reported in Panama in October, 2003 (García, 2005); and then in Costa Rica in April, 2004; in Nicaragua in 2005 and also in Guatemala and Honduras (Sanabria, 2005; Rodriguez, 2005; Castro *et al.*, 2006). In Mexico, the RPM was first detected in Campeche in 2006 (Arriaga, 2007) and later in Tabasco and Veracruz (Ramirez and Garcia, 2010). In the USA, the PRM was rediscovered in 2007 in Texas (Texas Department of Agriculture, 2007); however, a re-examination of specimens mounted in 1993 from a rice greenhouse facility in Beaumont, Texas, revealed that the PRM had been collected but misidentified as *S. madecassus* Gutierrez (F. Beaulieu, pers. comm. in Hummel *et al.*, 2009). This became the first report of the mite in research rice fields in the USA (Texas Department of Agriculture, 2007). This was followed by other detections in Louisiana, Arkansas, New York and California (Hummel *et al.*, 2007; NAPPO, 2007; UCDavis, 2009).

In South America, the PRM was reported in Colombia in 2005, and in Venezuela in 2006 (Herrera, 2005; Almaguel *et al.*, 2007).

15.6.2 Damage and economic impact

Main PRM damage is caused indirectly due to its association with plant pathogens, especially with the pathogenic fungi *Sarocladium* (*Acrocylindrium*) *oryzae* (Sawada), and the bacteria *Burkholderia* (*Pseudomonas*) *glumae* (Kurita and Tabei) (Chen *et al.*, 1979; Rao *et al.*, 1993; Rao and Prakash, 2003).

PRM infestations have caused significant losses in rice crops in Asia, and in recently affected areas in the Caribbean basin region and Central America. Reductions in rice production vary each year according to country, suggesting that many factors are involved in the damage, such as rice varieties and population levels during previous years (Cheng and Chiu, 1999; Hummel *et al.*, 2009).

In Asia, important losses in rice crops due to the PRM infestations have been reported in Taiwan, China and India. In Tainan, Taiwan, 60% of crop damage was reported in 1974 (Cheng and Chiu, 1999). Another report from the 1976 outbreak in southern Taiwan reported that approximately 20–60% of harvested grains were empty, equivalent to US$9.2 million (Chen *et al.*, 1979). Economic losses in China range between 5% and

20% in both early- and late-season rice crops, but in some cases they reach up to 70% (Xu et al., 2001). In India, losses reached 50% in Godovari county (Rao et al., 2000).

In the Americas, PRM infestations have caused serious impacts since being reported in Cuba during the late 1990s. Damage is particularly severe during their first years in a country. In the first year when PRM infestations were confirmed in rice crops in Cuba (1997/8), only 40,000 of the expected 120,000 tonnes (t) of rice were harvested. A 70% reduction in yield was estimated. During subsequent years, yield losses caused by PRM infestations ranged from 30% to 60% (Ramos and Rodríguez, 2000). In the community of San Paul, Cuba, the presence of the PRM, and its association with fungi, was responsible for 85% of the loss of rice productivity, from 5.6 t/ha in 1997 to 0.8 t/ha in 1998 (Romero et al., 2003). In 2004, the first year of infestation in Costa Rica, rice yield losses reached 96,000 t in the Province of Guanacaste, which represents approximately 45% of the entire country's rice production at an estimated economic loss of US$11million (Barquero, 2004). In Haiti, 60% yield losses were attributed to the PRM (Almaguel and Botta, 2005), and 40–60% yield losses were reported in Panamá (Garcia, 2005).

There is no information on the economic impact of the PRM in South America. According to Almaguel (2010), the mite has not caused an economic impact in Venezuela, and morphological and behavioral anomalies were observed in populations. In Colombia, the PRM mite has not caused significant yield reductions (Correa-Victoria, 2007).

Damage caused by the PRM is correlated with mite density populations (Rao and Prakash, 2003; Karmakar, 2008). For example, in the Orissa Province, sterility of 4–90% was associated with 7–600 PRM individuals per tiller, in the Ghandari District of Andhra Pradesh Province; 15–50% sterility was associated with densities of 150–900 mites per tiller; and in Jharkhand Province, 19–28% sterility was associated with a density of only 3–7 mites per tiller (Rao and Prakash, 2003).

15.6.3 Morphological and biological aspects

PRM are very small, pale-brown mites. Both females and males are elongated and are wider in the hysterosoma region (Chow et al., 1980; Ramos and Rodríguez, 1998). According to the original description females are approximately 0.3 mm μm long and 0.1 μm wide, and males are shorter: 0.2 μm long and 0.1 μm wide (Smiley, 1967).

The life cycle of the PRM includes stages of egg, larva and pupa, or quiescent nymph (Lindquist, 1986). The quiescent nymph can be transported by males, as occurs with other tarsonemid mites (Ramos and Rodríguez, 2000; Xu et al., 2001).

The PRM reproduces both sexually and by arrenotokous parthenogenesis, implying that virgin females only produce males. The females can mate with their male descendants and subsequently produce males and females (Xu et al., 2001). Studies on the biological aspects of the PRC have been conducted mainly in China, Taiwan and Cuba (Lo and Ho, 1979; Ramos and Rodríguez, 2000; Xu et al., 2001; Almaguel et al., 2004). In Cuba, the PRM completed its life cycle from egg to adult in 11, 7 and 3 days, at 20°C, 24°C and 30°C, respectively (Ramos and Rodríguez, 2000; Almaguel et al., 2004). In contrast, Asian populations, under the same temperatures, developed more slowly, at 20, 13–17 and 3–8 days (Lo and Ho, 1979; Xu et al., 2001). In a laboratory study by Xu et al. (2001) in China, virgin females produced an average of 79 adult mites under temperatures varying from 24°C to 35°C in 17 days, while mated females produced 55 eggs under the same conditions (Xu et al., 2001). In Cuba, maximum oviposition observed was 78 eggs/female, with an average of 31 eggs/female (Ramos and Rodríguez, 2003). Female adults had longevity of 24–32 days and oviposition of 18–26 days at 25°C–30°C. Zhang (1984) reported the PRM female:male ratio was 22:1, 32:1 and 8:1 at 32°C, 28°C, and under field temperature conditions, respectively. PRM is able to produce 48–55 generations per annum under ideal climatic conditions (Almaguel et al., 2004). Thus, a large population of PRM can develop very quickly in a rice crop during a single growing season.

In Cuba, climatic conditions favoring development of the PRM were 25.5–27.5°C and 83.8–89.5% RH. They can complete development at temperatures ranging from 20°C to 34°C (Cabrera et al., 2003). Periods of less rain and more sunshine are more favorable to PRM proliferation (Ghosh et al., 1998). Santos et al. (1998) reported that 16°C is the minimum temperature for complete development; at lower temperatures, only embryonic development was observed.

The PRM is highly resistant to unfavorable abiotic conditions, tolerating high temperatures and long periods in a flooded environment. Xu et al. (2002) conducted a laboratory study to examine the ability of PRM to survive in a flooded rice field, with no plant material, and observed that the mites survived for 23–25 days and continued their development, with 94.3% of the eggs hatching successfully and larvae molting to either the stationary phase or the adult stage. PRMs are also able to survive exposure to high temperatures. Xu et al. (2002) observed a 50% survival rate at 41°C for 36 h.

Hummel et al. (2009) presented a detailed review of PRM population dynamics and abiotic factors affecting the development of the mite. Studies of PRM population dynamics conducted in Cuba by Ramos and Rodriguez (2001) showed that populations were low during the tillering stage, multiplied 24 times during the green ring stage, and then another three times during blooming. The population reached its maximum at grain filling and then decreased as the grain progressed from milk to soft dough stages of maturation. Leyva et al. (2003) found that in Cuba, rice planted from December to May escaped higher infestations, while rice planted from August through October experienced much greater levels of PRM infestation and damage.

15.6.4 Dissemination

Trade and transportation of rice seeds can disseminate PRM (Rao et al., 2000). Many mites, in various developmental stages, can be found in rice crop remnants (Lo and Hor, 1977, 1979). Hummel et al. (2009) published an electron scanning micrograph that showed numerous PRM individuals within an empty rice grain hull, and Kim et al. (2001) collected PRM from the surface of grain husks.

In flooded rice crops, PRM may disseminate from one bed to another through crop debris in the water that circulates through them (Xu et al., 2002; R.I. Cabrera, Havana, Cuba, pers. comm., 2006).

This mite, like several other tarsonemidae, is transported over short distances by wind, insects or birds that visit rice plantations (Almaguel et al., 2003). Air currents may transport the mite over short or possibly long distances. It is also worth emphasizing that mites may disperse from one plant to another alone, without the use of dispersal agents and regardless of the wind, especially when plant density is high and plants touch one another (Almaguel et al., 2003). In experiments aimed at understanding PRM wind dispersal between neighboring rice crops, adhesive traps were installed around rice crops for various time periods. A significant difference in time periods was observed, regardless of the force and direction of the wind, indicating that mite dissemination is not entirely stochastic and that there is dispersal behavior. For instance, the traps indicated that more mites disseminated between 12:00 and 3:00 PM (R.I. Cabrera, unpublished results, 2005). In flooded rice crops, PRM may disseminate from one bed to another through crop debris present in the water that circulates through them (Xu et al., 2002; R.I. Cabrera, Havana, Cuba, pers. comm., 2006).

15.6.5 Detection

Low PRM infestations in fields can be extremely difficult to detect. Intense surveys conducted in commercial rice fields in 2005 in Louisiana, USA, were negative (Castro et al., 2006). Because PRM had been collected in this state in 1960, which produced the original description of the species (Smiley, 1967), populations were most likely present, but below detectable levels. Low PRM populations were likely present in Texas before their report in 2007 and as early as 1993.

When neighboring areas have high PRM infestation, field inspections can be initiated when plants are very young and only 10 cm high. Nevertheless, inspection can also begin when plants are more developed, at 40 cm high, when there is no infestation around the field (R.I. Cabrera, Havana, Cuba, pers. comm. 2004).

Sheaths of leaves should be thoroughly inspected down to their basal portion, making it necessary to collect entire leaves close to the ground. In early infestations, numerous colonies can be found in this part of rice plants. The inner surface of leaf sheaths can become opaque when infested by the PRM, and small brown spots on the sheath are characteristics of the mite (Correa Victoria, 2007).

15.6.6 Control tactics

Adoption of control measures is needed in areas with high infestations of PRM. Chemical control is difficult, mainly due to the location of the colonies in the inner part of the sheath, making them almost inaccessible to chemical products used for control. Chemical control can be helpful, especially in emergencies (Ramos and Rodríguez, 1998, 2000; Cheng and Chiu, 1999; Almaguel et al., 2000), such as for quarantine purposes. The use of cultural practices in intensive production areas is essential; growers should work together to minimize losses caused by the PRM and associated pathogens. Plant resistance is an important tactic that can be employed with other control measures.

In Cuba, where cultural methods were adopted and less susceptible varieties were used, 3 t/ha of rice were harvested in the 1st year under high infestation pressure. In the 2nd year, with an average infestation of 3 mites/plant, yields of 8 t/ha with no pesticide application were obtained. Generally, more damage occurs in the second crop, but the adoption of control tactics resulted in a considerable increase in productivity (Barquero, 2004).

Biological

Although biological control is an important component of integrated pest management, rice has some characteristics that can make its implementation difficult, especially because the crop is grown in an annual cycle (Cheng and Chui, 1999; Ramos and Rodríguez, 2000). Several predatory mites have been reported in association with the PRM in Asia and Cuba (Lo and Ho, 1979, 1980; Ramos and Rodríguez, 1998; Cabrera et al., 2003; Ramos et al., 2005), but their efficiency was not evaluated. Acaropathogenic fungi, *Hirsutella nodulosa* Petch and *Entomophthora* sp., were reported to cause PRM mortality in Cuba and Sri Lanka (Cabrera et al., 2002b; Almaguel et al., 2003).

Chemical

Some authors have considered that pesticides do not provide adequate control of PRM infestations (Chow et al., 1980; Almaguel et al., 2000). Even systemic products that are used to control pests in hidden plant parts were ineffective for controlling populations of this mite (Chow et al., 1980; Almaguel et al., 2000).

Field trials conducted in India reported up to 90% mortality following treatments with Dicofol (Bhanu et al., 2006). An evaluation of seven pesticides reported that dimethoate caused 88% mortality in India (Ghosh et al., 1998). The chemical products recommended for use in integrated management of the rice mite in the Caribbean basin region and Central American countries were abamectin, biomite, dicofol, triazophos, endosulfan and ethoprophos (Cabrera et al., 1999, 2002a; Almaguel et al., 2005). Other products that have been tested under laboratory conditions and reported to cause >95% mortality of adults include bromopropylate, diafenthiuron and edifenphos (Cabrera et al., 2005). The treatment of seeds with benzimidazole plus tiram 200 ppm was also recommended, because it significantly decreases the percentage of sterile and spotted grains, and increases yields in infested areas (García et al., 2002).

Host plant resistance

Susceptibility of rice varieties to PRM damage has been evaluated in Asia and in the Americas (e.g., Cuba, Dominican Republic and Costa Rica). In Tainan, Taiwan, 60% of crop damage was reported in 1974 (Cheng and Chiu, 1999). Researchers concluded that variety Tainan #5 was extremely susceptible to PRM damage, and in response to extensive crop losses, the acreage of Tainan #5 was reduced from 400,000 ha in 1975 to 80,000 ha in 1981 (Cheng and Chiu, 1999). In India, japonica varieties were reported to be more susceptible to PRM infestations than indica varieties (Ou et al., 1977). In Cuba, RI-7, IACuba-19, Perla de Cuba, PP-2, RI-6, PP-1, RI-3, RI-13, IACuba-21, J-104, JMR IACuba-31 and Reforma provided the best results (Botta et al., 2002; Romero et al., 2003; Rosa, 2003). Evaluations for resistance to both PRM and the pathogenic fungi *S. oryzae* showed that Reforma and INCA LP-5 exhibited a lower incidence of mites, while the varieties Perla de Cuba and Reforma had less fungus infection (Hernández et al., 2002). In Costa Rica, Fedearroz 50, CFX 18 and CR 4477 are considered more tolerant to rice mite infestations and have been recommended in management programs (C. Sanabria, San José, Costa Rica, pers. comm. 2006). Ramos et al. (2001) surveyed 60 farms throughout the Dominican Republic. In every location, they sampled four varieties in different phonological stages. At the tillering and panicle initiation stages, variety ISA-40 was more susceptible to PRM infestation when compared to the varieties JUMA-57, Prosedoca and Prosequisa, with 100% infestation occurring in ISA-40.

Cultural

Cultural procedures have been established to reduce populations of the PRM, delay its arrival into the crop, and reduce yield losses and production costs (Ho and Lo, 1979; Cabrera et al., 1998; Ramos et al., 2001; Hernández et al., 2002, 2003; Romero et al., 2003; Sanabria, 2005). The main measures are elimination of crop debris and invasive plant species that can act as a source of infestation. These measures should be adopted in the production area and in neighboring areas:

- cleaning (disinfestation of) new crop areas;
- cleaning machinery and equipment if used by different farmers or in different areas;
- allowing prepared soil to remain plant-free for at least 2 weeks before planting;
- using seeds of controlled quality and origin, and disinfesting them before planting;
- preparing ahead of planting time;
- applying nitrogen fertilizers following different treatments;
- decreasing the height of the water layer;
- decreasing plant density in the crop;
- coordinating sowing time in neighboring production areas;
- avoiding rice planting in adjacent areas during or just after harvest time, and with the wind direction taken into account;
- using rice varieties with higher resistance to PRM damage;
- monitoring the crop 15 days after planting, especially in areas downwind of infested areas to obtain an early determination of its presence in the crop, and implementing control measures.

Acknowledgements

To Francisco Ferragut, Universidad Politécnica de Valencia, for his valuable suggestions. To National Council for Scientific and Technological Development (CNPq), Brazil, for the research and post-doc fellowships for authors and for the support to research projects involving invasive mites in South America.

References

Abanorte (2008) Cresce exportação de limão tahiti. Available at: www.abanorte.com.br/noticias/noticias-da-pagina-inicial/cresce-exportacao-de-limao-tahiti, accessed 6 February 2011.

Agrolink (2010) No va praga está atacando as plantações de lichia. Available at: www.agrolink.com.br/cereaisdeinverno/NoticiaDetalhe.aspx?CodNoticia=115177, accessed 6 February 2011.

Alam, M.Z. and Wadud, M.A. (1963) On the biology of litchi mite, *Aceria litchii* Keifer (Eriophyidae, Acarina) in East Pakistan. *Pakistan Journal of Science* 18, 232–240.

Almaguel, L. (2010) La impor tancia del ácaro del vaneo del arroz (*Steneotarsonemus spinki*) en Centroamérica y el Car ibe. In: Sánchez Gálvez, M.C., Sandoval-Islas, J.S. and Estrada-Venegas, E.G.E. (eds) Pr imer Simposio Internacional de Acarología en México – Memor ias. Universidad Autónoma Chapingo, Chapingo, México, pp. 39–47. Available at: armandocorrea.com/Acarologia/main.html, accessed 7 February 2011.

Almaguel, L. and Botta, E. (2005) Manejo integrado de *Steneotarsonemus spinki*, Smiley. Available at: www.inisav.cu/OtrasPub/Curso%20acarolog%C3%ADa.pdf, accessed May 2008.

Almaguel, L., Hernandéz, J., Torre, P.E., Santos, A., Cabrera, R.I., García, A., Riv ero, L.E., et al. (2000) Evaluación del compor tamiento del ácaro *Steneotarsonemus spinki* (Acari: Tarsonemidae) en los estúdios de regionalización desarrollados en Cuba. *Fitosanidad* 4, 15–19.

Almaguel, L., Santos , A., Torre, P.E., Botta, E., Hernández, J., Cáceres, I. and Ginar te, A. (2003) Dinámica de pob lación e indicadores ecológicos del ácaro *Steneotarsonemus spinki* Smiley 1968 (Acari: Tarsonemidae) en arroz de riego en Cuba. *Fitosanidad* 7, 23–30.

Almaguel, L., Torre, P.E. and Caceres , I. (2004) Suma de temper aturas efectivas y potencial de multiplicación del ácaro del v aneado del arroz (*Steneotarsonemus spinki*, Smiley) en Cuba. *Fitosanidad* 8, 37–40.

Almaguel, L., Botta, E.F ., Hernández, J. and Ginarte, A. (2005) Propuesta de manejo integrado del ácaro del v aneado del arroz par a los países de la región de latinoamér ica y el

Caribe. In: III Encuentro Inter national del Arroz: El ácaro del arroz *Steneotarsonemus spinki* (Tarsonemidae) retos y alter nativas para América Latina y el Car ibe, Libro de Resumen. INISAV, Havana, pp. 53–59.

Almaguel, L.R., Botta, E.F ., Hernández, J., González, T.A., Finalé, Y.D., Torre, P. and Ginarte, A. (2007) Asesor ía y capacitación sobre los ácaros de impor tancia económica en Centroamérica (2007). In: *Premio An ual. MINAG, 2007*. Instituto de In vestigaciones Sanidad Vegetal (INISAV) and Instituto de Investigaciones del Arroz (IIARR OZ), Havana.

Amrine Jr, J.W. (2003) Catalog of the Eriophyoidea. A working catalog of the Er iophyoidea of the world. Available at: http://insects.tamu.edu/research/collection/hallan/acari/eriophyidae, accessed 7 February 2011.

Arbabi, M., Khiaban, N.G.Z. and Askari, M. (2002) Plant mite f auna of Sistan-Baluchestan and Hormozgan Provinces. *Journal of the Entomological Society of Iran* 22, 87–88.

Arriaga, J.T. (2007) Detection of the rice tarsonemid mite (*Steneotarsonemus spinki* Smiley) in Palizada, Campeche, Mexico. Available at: www.pestalert.org/oprDetail. cfm%3FoprID¼268, accessed 18 July 2007.

Atkinson, T.G. and Grant, M.N. (1967) An evaluation of streak mosaic losses in winter wheat. *Phytopathology* 57, 188–192.

Barquero, M. (2004) Millonaria pérdida en arroz. *La Nación Lunes*. San José, Costa Rica. Available at: www.nacion.com, accessed 18 October 2004, p.2.

Bhanu, K.S., Reddy, P.S. and Zaher uddeen, S.M. (2006) Evaluation of some acaricides against leaf mite and sheath mite in r ice. *Indian Journal of Plant Protection* 34, 132–133.

Boczek, J. and Ch yczewski, J. (1975) *Beobachtungen zur Biologie einiger Gallmilbenarten (Eriophyioidea) der Gräser*. Tag Verlag, Akademie LandwirtschWiss 134, 83–90.

Bolland, H.R., Gutierrez, J . and Flechtmann, C.H.W. (1998) *World Catalogue of the Spider Mite Family (Acari: Tetranychidae)*. Brill Academic Publishers, Leiden, The Netherlands, pp. 392.

Botta, E., Almaguel, L., Her nández, J. and Torre, P.E. (2002) Ev aluación del compor tamiento de *Steneotarsonemus spinki* en dif erentes variedades de arroz. In: *Encuentro Internacional de Arroz, Memorias*. Instituto de Investigaciones del Arroz, Ha vana, p. 188. (Abstract).

Butani, D.K. (1977) Pest of litchi in India and their control. *Fruits* 32, 269–270.

CABI (2002) *Crop Protection Compendium*. (CD-ROM). CAB International, Wallingford, UK.

Cabrera, R.I., García, A., Almaguel, L. and Ginarte, A. (1998) Microorganismos patógenos del ácaro tarsonémido del arroz *Steneotarsonemus spinki*: (Acari: Tarsonemidae). In: *Encuentro Internacional de Arroz, Libro de Resúmenes*. Instituto de In vestigaciones del Arroz, Havana, p. 185. (Abstract).

Cabrera, R.I., Ginarte, A. and Hernandéz, J. (1999) Efecto de tr iazophos en el control del ácaro tarsonémido del arroz *Steneotarsonemus spinki* Smiley (Acari: Tarsonemidae). In: *Primer Congreso de Arroz de Riego y Secano del Area del Caribe, Libro de Resúmenes*. Universidad de Camagüe y, Camagüey, Cuba, p. 24.

Cabrera, R.I., Her nández, J.L. and García, A. (2002a) Resultado de las aplicaciones aéreas de triazophos y *Bacillus thuringiensis* para combatir el ácaro *Steneotarsonemus spinki* (Acari: Tarsonemidae) en el cultiv o del arroz. In: *Encuentro Internacional de Arroz, Memorias*. Instituto de In vestigaciones del Arroz, Havana, pp. 206–208.

Cabrera, R.I., Nugaliyadde, L. and Ramos, M. (2002b) Presencia de *Hirsutella nodulosa* sobre el ácaro tarsonémido del arroz *Steneotarsonemus spinki* (Acari: Tarsonemidae) en Sri Lanka. In: *Encuentro Internacional de Arroz, Memorias*. Instituto de Investigaciones del Arroz, Havana, pp. 186–188.

Cabrera, R.I., Ramos, M. and Fernandez, M.B. (2003) Factores que influy en en la ab undancia de *Steneotarsonemus spinki* em arroz, em Cuba. *Revista Manejo Integrado de Plagas y Agroecología* 69, 34–37.

Cabrera, R.I., Hernández, J., Ginarte, A., García, A., Páez, Y., Rivero, L.E. and Vega, M. (2005) Alternativas de los plaguicidas par a el control efectivo de *Steneotarsonemus spinki* Smiley en el cultiv o del arroz. In: *III Encuentro Internacional del Arroz: El ácaro del arroz Steneotarsonemus spinki (Tarsonemidae) retos y alternativas para América Latina y el Caribe, Libro de Resumen*. Havana, pp. 31–35.

Carew, M., Schiffer, M., Umina P., Weeks, A. and Hoffmann, A. (2009) Molecular mar kers indicate that the wheat curl mite *Aceria tosichella* Keifer may represent a species comple x in Australia. *Bulletin of Entomological Research* 99, 479–486.

Carrillo, D. and Peña, J. (2010) Studies on the biology of nativ e predators associated with *Raoiella indica* (Acari: Tenuipalpidae) in Florida, USA: implications on their potential as biological control agents of this exotic species.

In: Moraes, G.J., Castilho, R.C. and Flechtmann, C.H.W. (eds) Abstract Book of the XIII International Congress of Acarology (Abstract 79). Recife, Brazil, p. 45.

Carrillo, D., Peña, J., Hoy, M.A. and Frank, J.H. (2010) Development and reproduction of *Amblyseius largoensis* (Acari: Phytoseiidae) feeding on pollen, *Raoiella indica* (Acari: Tenuipalpidae), and other microarthropods inhabiting coconuts in Florida, USA. *Experimental and Applied Acarology* 52, 119–129.

Castiglioni, E. and Navia, D. (2010) Presence of the Wheat Curl Mite, *Aceria tosichella* Keifer (Prostigmata: Eriophyidae), in Uruguay. *Agrociencia* 14, 19–26.

Castro, B.A., Ochoa, R. and Cuevas, F.E. (2006) The threat of the panicle rice mite, *Steneotarsonemus spinki* Smiley, to rice production in the United States. In: *Proceedings of the Thirty-First Rice Technical Working Group*, The Woodlands, Texas, pp. 97–98.

Chaudhri, W.M. (1974) *Taxonomic Studies of the Mites Belonging to the Families Tenuipalpidae, Tetranychidae, Tuckerellidae, Caligonellidae, Stigmaeidae and Phytoseiidae*. University of Agriculture, Lyallpur, Pakistan, pp. 250.

Chen, C.N., Cheng, C.C. and Hsiao, K.C. (1979) Bionomics of *Steneotarsonemus spinki* attacking rice plants in Taiwan. In: Rodriguez, J.G. (ed.) *Recent Advances in Acarology, vol. 1*. Academic Press, New York, New York, pp. 111–117.

Cheng, C.H. and Chiu, Y.I. (1999) Review of changes involving rice pests and their control measures in Taiwan since 1945. *Plant Protection Bulletin Taipei* 41, 9–34.

Cheng, L., Zhang, X., Sha, L., Lu, A. and Chen, P. (2005) Functional and numerical response of *Amblyseius largoensis* to *Aceria litchii*. *Journal of Tropical Crops* 26, 53.

Cho, M.R., Kim, D.S. and Im, D.S. (1999) A new record of tarsonemid mite, *Steneotarsonemus spinki* (Acari: Tarsonemidae) and its damage on rice in Korea. *Korean Journal of Applied Entomology* 38, 157–164.

Chow, Y.S., Tzean, S.S., Chang, C.S. and Wang, C.H. (1980) A morphological approach of the tarsonemid mite *Steneotarsonemus spinki* Smiley (Tarsonemidae) as a rice plant pest. *Acta Arachnologica* 29, 25–41.

Christian, M.L. and Willis, W.G. (1993) Survival of wheat streak mosaic virus in grass hosts in Kansas from wheat harvest to fall wheat emergence. *Plant Disease* 77, 239–242.

Cocco, A. and Hoy, M.A. (2009) Feeding, reproduction, and development of the red palm mite (Acari: Tenuipalpidae) on selected palms and banana cultivars in quarantine. *Florida Entomologist* 92, 276–291.

Conner, R.L., Thomas, J.B. and Whelan, E.D.P. (1991) Comparison of mite resistance for control of wheat streak mosaic. *Crop Science* 31, 315–318.

Correa-Victoria, F. (2007) The rice tarsonemid mite *Steneotarsonemus spinki* Smiley. Available at: http://www.ricecap.uark.edu/Outreach/FactSheets/The%20rice%20tarsonemid%20mite.pdf, accessed 10 February 2011.

Coutts, B.A., Strickland, G.R., Kehoe, M.A., Severtson, D.L. and Jones, R.A.C. (2008) The epidemiology of wheat streak mosaic virus in Australia: case histories, gradients, mite vectors, and alternative hosts. *Australian Journal of Agricultural Research* 59, 844–853.

CRRI (Central Rice Research Institute) (2006) A new alternate host of rice panicle mite. *CRRI Newsletter* 27, 10.

Daniel, M. (1981) Bionomics of the predaceous mite *Amblyseius channabasavanni* (Acari: Phytoseiidae) predaceous on the palm mite. In: ChannaBasavanna, G.P. (ed.) *Contributions to Acarology in India*. Anubhava Printers, Bangalore, India, pp. 167–172.

de la Torre, P.E., González, A.S. and González, A.I. (2010) Presencia del ácaro *Raoiella indica* Hirst (Acari: Tenuipalpidae) en Cuba. *Revista de Protección Vegetal* 25, 1–4.

del Rosario, M.S. and Sill, W.H. (1958) A method of rearing large colonies of an Eriophyid mite, *Aceria tulipae* (Keifer), in pure culture from single eggs or adults. *Journal of Economic Entomology* 51, 303–306.

del Rosario, M.S. and Sill, W.H. (1964) Additional biological and ecological characteristics of *Aceria tulipae* (Acarina: Eriophyidae). *Journal of Economic Entomology* 57, 893–896.

del Rosario, M.S. and Sill, W.H. (1965) Physiological strains of *Aceria tulipae* (K.) and their relationships to the transmission of the wheat streak mosaic virus. *Phytopathology* 55, 1168–1175.

Denizhan, E., Szydlo, W., Diduszko, D. and Skoracka, A. (2010) Preliminary study on eriophyoid mites (Acari: Eriophyidae) infesting grasses in Turkey. In: Moraes, G.J., Castilho, R.C. and Flechtmann, C.H.W. (eds) Abstract Book of the XIII International Congress of Acarology (Abstract 125). Recife, Brazil, p. 70.

Dowling, A.P.G., Ochoa, R., Welbourn, W.C. and Beard, J.J. (2010) *Raoiella indica* (Acari: Tenuipalpidae): a rapidly expanding generalist among specialist congeners. In: Moraes, G.J., Castilho, R.C. and Flechtmann, C.H.W. (eds) Abstract Book of the XIII International Congress of Acarology (Abstract 131). Recife, Brazil, pp. 72–73.

Duncan, R.E., Carrillo, D. and Peña, J.E. (2010) Population dynamics of the red palm mite, *Raoiella indica* (Acari: Tenuipalpidae), in Florida, USA. In: Moraes, G.J., Castilho, R.C. and Flechtmann, C.H.W. (eds) Abstract Book of the XIII International Congress of Acarology (Abstract 134). Recife, Brazil, p. 74.

Dwyer, G.I., Gibbs, M.J., Gibbs, A.J. and Jones, R.A.C. (2007) Wheat streak mosaic virus in Australia: relationship to isolates from the Pacific Northwest of the USA and its dispersion via seed transmission. *Plant Disease* 91, 164–170.

Edwards, M.C. and McMullen, M.P. (1988) Variation and tolerance to Wheat streak mosaic virus among cultivars of hard red spring wheat. *Plant Disease* 72, 705–707.

Ellis, M.H., Rebetzke, G.J. and Chu, P. (2003) First report of wheat streak mosaic virus in Australia. *Plant Pathology* 52, 808.

Elwan, A.A. (2000) Survey of the insect and mite pests associated with date palm trees in Al-Dakhliya region, Sultanate of Oman. *Egyptian Journal of Agricultural Research* 78, 653–664.

EPPO (2009) *Raoiella indica* (Acari: Tenuipalpidae). Available at: www.eppo.org/QUARANTINE/Alert_List/insects/raoiella_indica.htm, accessed 29 September 2009.

Estrada-Venegas, E., Martínez-Morales, H.J., Villa Castillo, J. (2010) *Raoiella indica* Hirst (Acari: Tenuipalpidae): first record and threat in Mexico. In: Moraes, G.J., Castilho, R.C. and Flechtmann, C.H.W. (eds) Abstract Book of the XIII International Congress of Acarology (Abstract 140). Recife, Brazil, p. 77.

Etienne, J. and Flechtmann, C.H.W. (2006) First record of *Raoiella indica* (Hirst, 1924) (Acari: Tenuipalpidae) in Guadeloupe and Saint Martin, West Indies. *International Journal of Acarology* 32, 331–332.

Fantine, A.K., Sugayama, R., Zeidler, R., Trassato, L. and Vilela, E. (2010) Situação atual do ácaro-hindu dos citros (*Schizotetranychus hindustanicus*) (Acari: Tetranychidae), no estado de Roraima. In: Anais XXIII Congresso Brasileiro de Entomologia. Sociedade Entomológica do Brasil (Abstract), Natal, Brazil, p. 1856.

FAO (Food and Agriculture Organization) (2002) Expert consultation on lychee production in the Asia-Pacific Region. FAO Regional Office for Asia and the Pacific, Rome.

Flechtmann, C.H.W. and Etienne, J. (2004) The red palm mite, *Raoiella indica* Hirst, a threat to palms in the Americas (Acari: Prostigmata: Tenuipalpidae). *Systematic and Applied Acarology* 9, 109–110.

Flechtmann, C.H.W. and Etienne, J. (2005) Un nouvel acarien ravageur des palmiers en Martinique, premier signalement de *Raoiella indica* pour les Caraïbes. *Phytoma, La Défense des Vegetaux* 584, 10–11.

Flores-Galano, G., Montoya, A. and Rodríguez, H. (2010) Biología de *Raoiella indica* Hirst (Acari: Tenuipalpidae) sobre *Areca catechu* L. *Revista de Protección Vegetal* 25, 11–16.

Forster, R.L., Seifers, D.L., Strausbaugh, C.A., Jensen, S.G., Ball, E.M. and Harvey, T.L. (2001) Seed transmission of the High Plains virus in sweet corn. *Plant Disease* 85, 696–699.

French, R. and Stenger, D.C. (2003) Evolution of wheat streak mosaic virus: dynamics of population growth within plants may explain limited variation. *Annual Review of Phytopathology* 41, 199–214.

García, A., Hernández, J., Almaguel, L., Sandoval, I., Botta, E. and Arteaga, I. (2002) Influencia del ácaro *Steneotarsonemus spinki* Smiley (Acari: Tarsonemidae) y del hongo *Sarocladium oryzae* (Sawada) Gams & Hawks. sobre el vaneado y manchado de los granos de arroz. In: Encuentro Internacional de Arroz, Memorias. Instituto de Investigaciones del Arroz, Havana, pp. 189–193.

García, M.P.G. (2005) Vaneamento y manchado de grano en cultivos de arroz en Panamá. *Revista Arroz* 53, 455.

Gerson, U., Venezian, A. and Blumberg, D. (1983) Phytophagous mites on date palms in Israel. *Fruits* 38, 133–135.

Ghosh, S.K., Prakash, A. and Jagadishwari, R. (1998) Efficacy of some chemical pesticides against rice tarsonemid mite. *Environment and Ecology* 16, 913–915.

Gillespie, R.L., Roberts, D.E. and Bentley, E.M. (1997) Population dynamics and dispersal of wheat curl mites (Acari: Eriophyidae) in north central Washington. *Journal of the Kansas Entomological Society* 70, 361–364.

Golya, G., Kozma, E. and Szabo, M. (2002) New data to the knowledge on the eriophyoid fauna on grasses in Hungary (Acari: Eriophyoidea). *Acta Phytopathologica et Entomologica Hungarica* 37, 409–412.

González-Reyes, A.I. and Ramos, M. (2010) Desarrollo y reproducción de *Raoiella indica* Hirst (Acari: Tenuipalpidae) en laboratorio. *Revista de Protección Vegetal* 25, 7–10.

Graybosch, R.A., Peterson, C.J., Baenziger, D.D., Baltensperger, D.D., Nelson, L.A., Jin, Y., Kolmer, J., et al. (2009) Registration of

'Mace' Hard Red Winter Wheat. *Journal of Plant Registrations* 3, 51–56.

Gupta, S.K. and Gupta, Y.N. (1994) A taxonomic review of Indian Tetranychidae (Acari: Prostigmata) with description of new species, redescriptions of known species and k eys to genera and species. *Memoirs of the Zoological Survey of India* 18, 1–196.

Gupta, Y.N. (1984) On a collection of tetr anychoid mites from Tamil Nadu with description of new species of *Aponychus* (Acari: Tetranychidae). *Bulletin of the Zoological Survey of India* 6, 237–245.

Gupta, Y.N. (2001) A conspectus of natural enemies of phytophagous mites and mites as potential biocontrol agents of agricultural pests in India. In: Halliday, R.B., Walter, D.E., Proctor, H.C., Norton, R.A. and Colloff, E.M.J. (eds) *Acarology*. Proceedings of the 10th International Congress. CSIRO Publishing, Melbourne, pp. 484–497.

Halliday, R.B. and Knihinicki, D.K. (2004) The occurrence of *Aceria tulipae* (Keifer) and *Aceria tosichella* Keifer in Australia (Acari: Eriophyidae). *International Journal of Acarology* 30, 113–118.

Haque, M.M., Das, B.C., Khalequzzaman, M. and Chakrabarti, S. (1998) Er iophyid mites (Acarina: Eriophyoidea) from Bangladesh. *Oriental Insects* 32, 35–40.

Harvey, T.L., Martin, T.J. and Thompson, C.A. (1979) Controlling wheat curl mite and wheat streak mosaic virus with systemic insecticide. *Journal of Economic Entomology* 72, 854–855.

Harvey, T.L., Martin, T.J. and Seifers, D.L. (1994) Importance of plant resistance to insect and mite vectors in controlling vir us diseases of plants: resistance to the wheat curl mite (Acari: Eriophyidae). *Journal of Agricultural Entomology* 11, 271–277.

Harvey, T.L., Martin, T.J. and Seifers, D.L. (1995) Survival of five wheat curl mite, *Aceria tosichella* Keifer (Acari: Eriophyidae), strains on mite resistant wheat.*Experimental and Applied Acarology* 19, 459–463.

Harvey, T.L., Martin, T.J. and Seifers, D.L. (1997) Change in vir ulence of wheat cur l mite detect on TAM 107 wheat. *Crop Science* 37, 624–625.

Harvey, T.L., Seifers, D.L. and Mar tin, T.J. (1999) Survival of wheat cur l mites on diff erent sources of resistant wheat. *Crop Science* 39, 1887–1889.

Harvey, T.L., Seifers, D.L. and Mar tin, T.J. (2001) Host range differences between two strains of wheat curl mites (Acari: Eriophyidae). *Journal of Agricultural and Urban Entomology* 18, 35–41.

Harvey, T.L., Martin, T.J. and Seifers, D.L. (2002) Wheat yield reduction due to wheat cur l mite (Acari: Eriophyidae) infestations. *Journal of Agricultural and Urban Entomology* 19, 9–13.

Hastie, E., Benegas, A. and Rodríguez, H. (2010) Inventario de ácaros depredadores asociados a fitoácaros en plantas de las f amilias Arecaceae y Musaceae. *Revista de Protección Vegetal* 25, 17–25.

Hernández, J.L., Gómez, P., Galano, R., Botta, E., Duany, A., Ginarte, A. and Berbén, T. (2002) Influencia del marco y densidad de siembr a la fertilización nitrogenadada y el manejo de agua sobre la población y daños de *S. spinki* y los patógenos a él asociados. In: III Encuentro International del Arroz: El ácaro del arroz *Steneotarsonemus spinki* (Tarsonemidae) retos y alternativas para América Latina y el Car ibe, Libro de Resumen. Havana, pp. 129–131.

Hernández, J.L., Ginarte, A., Gómez, P.J., Cabrera, I., Galano, R., Vieira, R., Duany, A. *et al.* (2003) Recomendaciones agronómicas para el manejo del ácaro *Steneotarsonemus spinki* Smiley en el cultiv o de arroz. In: Agronomia, variedades y semillas: memorias. Instituto de Investigaciones del Arroz/Gr upo Agroindustrial Pecuário Arrocero, Havana, pp. 19–23.

Herrera, L.A.R. (2005) Ácaro del v aneamiento del arroz – *Steneotarsonemus spinki* Smiley (Prostigmata: Tarsonemidae). Available at: www.flar.org/pdf/foro-agosto-pdf05/acaro.pdf, accessed March 2006.

Hirst, S. (1924) On some new species of red spider. *Annals and Magazine of Natural History*, 14, 522–527.

Ho, C.C. and Lo, K.C. (1979) A sur vey of the host ranges of *Steneotarsonemus spinki* (Acari: Tarsonemidae). *Biological Abstracts* 71, 2452. (Paper published at National Science Council Member 7, 1022–1028, in Chinese).

Hoy, M.A., Peña, J. and Nguyen, R. (2006) Red palm mite, *Raoiella indica* Hirst (Arachnida: Acari: Tenuipalpidae). University of Flor ida IFAS Extension. Available at: http://edis.ifas.ufl.edu/IN711, accessed 7 February 2011.

Huang, K. (2008) Aceria (Acarina: Eriophyoidea) in Taiwan: five new species and plant abnormalities caused by sixteen species. *Zootaxa* 1829, 1–30.

Huang, T. (1967) A study on morphological features of erinose mite of litchi (*Eriophyes litchii* Keifer) and an obser vation on the conditions of its damage. *Plant Protection Bulletin* 9, 35–46.

Huang, T., Huang, K.W. and Horng, I.J. (1990) Two species of er iophyid mites injur ious to litchi trees in Taiwan. *Chinese Journal of Entomology* 3, 57–64.

Hummel, N.A., Castro, B.A., Stout, M.J. and Saichuk, J.K. (2007) *Rice Pest Notes, Pest Management and Insect Identification Series, the Panicle Rice Mite*. Pub. 3023, Louisiana State University Agricultural Center and Texas Cooperative Extension, Baton Rouge, Louisiana.

Hummel, N.A., Castro, B.A., McDonald, E.M., Pellerano, M.A. and Ochoa, R. (2009) The panicle rice mite, *Steneotarsonemus spinki* Smiley, a re-disco vered pest of r ice in the United States. *Crop Protection* 28, 547–560.

Hunger, R.M., Sherw ood, J.L., Evans, C.K. and Montana, J.R. (1992) Effects of planting date and inoculation date on se verity of wheat streak mosaic in hard red winter wheat cultivars. *Plant Disease* 76, 1056–1060.

Jagadiswari, R. and Prakash, A. (2003) Panicle mite causing sterility farmer's paddy fields in India. *Journal of Applied Zoological Research* 14, 212–217.

Jalaluddin, S.M. and Mohanasundaram, M. (1990) Control of the coconut red mite *Raoiella indica* Hirst (Tenuipalpidae: Acari) in the n ursery. *Indian Coconut Journal Cochin* 21, 7–8.

Jayaraj, J., Natarajan, K. and Ramasubr amanian, G.V. (1991) Control of *Raoiella indica* Hirst. (Tenuipalpidae: Acari) on cocon ut with pesticides. *Indian Coconut Journal Cochin* 22, 7–9.

Jeppson, L.R., Keifer, H.H. and Baker, E.W. (1975) *Mites Injurious to Economic Plants*. University of California Press, Berkeley, California.

Jezewska, M. (2000) Incidence of Wheat streak mosaic virus in P oland in the y ears 1998–1999. *Phytopathologia Polonica* 20, 77–83.

Juárez-Duran, M. (2010) Vigilancia epidemiológica del ácaro roj de las palmas (*Raoiella indica*) en México. In: Sánchez Gálv ez, M.C., Sandoval-Islas, J.S. and Estr ada-Venegas, E.G.E. (eds) Primer Simposio Internacional de Acarología en México – Memorias. Universidad Autónoma Chapingo, Chapingo, pp. 63–67. Available at: armandocorrea.com/Acarologia/main.html, accessed 7 February 2011.

Kane, E., Ochoa, R., Mathur in, G. and Erbe, E.F. (2005) *Raoiella indica* (Acari: Tenuipalpidae), an island hopping mite pest in the Car ibbean. In: Entomological Society of Amer ica Annual Meeting, Fort Lauderdale, Florida (Poster). Available at: www.sel.barc.usda.gov/acari/PDF/Raoiella indica-Kane et al.pdf, accessed 7 February 2011.

Kane, E.C. and Ochoa, R. (2006) Detection and identification of *Raoiella indica* Hirst (Acar i: Tenuipalpidae). Agency Repor t. USDA ARS Systematic Entomology Laboratory Beltsville, Maryland. Available at: http://entnemdept.ufl.edu/pestalert/Raoiella_indica_Guide.pdf, accessed 7 February 2011.

Karmakar, K. (2008) *Steneotarsonemus spinki* Smiley (Acari: Tarsonemidae) – a yield reducing mite of r ice crops in West Bengal, India. *International Journal of Acarology* 34, 95–99.

Keifer, H.H. (1943) Er iophyid Studies XIII. *The Bulletin, Department of Agriculture, State of California*, 32, 212–222 (pl. 171).

Keifer, H.H. (1953) Er iophyid studies XXI. *The Bulletin, Department of Agriculture, State of California*, 42, 65–79.

Keifer, H.H. (1969) *Eriophyid Studies C-3*. Agricultural Research Ser vice, C Ser ies. California Department of Agriculture, State of California, Sacramento, California, pp. 1–24.

Keifer, H.H. and Knorr, L.C. (1978) *Eriophyid Mites of Thailand*. Plant Protection Service Technical Bulletin 38, Bangkok, pp. 1–36.

Kim, D.S., Lee, M.H. and Im, D.J. (2001) Effect of dust mite incidence on grain filling and quality in rice. *Korean Journal of Crop Science* 46, 180–183.

Kozlowski, J. (2000) The occurrence of *Aceria tosichella* Keifer (Acari, Eriophyidae) as a vector of wheat streak mosaic vir us in P oland. *Journal of Applied Entomology* 124, 209–211.

Lall, B.S. and Rahman, M.F. (1975) Studies on the bionomics and control of the er inose mite *Eriophyes litchii* Keifer (Acarina: Eriophyidae). *Pesticides* 9, 49–54.

Leyva, Y., Zamora, N., Álvarez, E. and Jiménez, M. (2003) Resultados preliminares de la dinámica poblacional del ácaro *Steneotarsonemus spinki*. *Revista Electrónica, Granma Ciencia* 17, 1–6.

Li, H., Conner, R.L., Chen, Q., Gr af, R.J., Laroche, A., Ahmad, F. and K uzyk, A.D. (2004) Promising genetic resources for resistance to wheat streak mosaic vir us and the wheat curl mite in wheat-Thinop yrum partial amphiploids and their der ivatives. *Genetic Resources and Crop Evolution* 51, 827–835.

Lindquist, E.E. (1986) *The World Genera of Tarsonemidae (Acari: Prostigmata): A Morphological, Phylogenetic, and Systematic Revision, with a Reclassification of Family-Group Taxa in the Heterostigmata*. Memoirs of the Entomological Society of Canada 136, Entomological Society of Canada, Ottawa.

Lindquist, E.E., Sabelis, M.W. and Bruin, J. (1996) *Eriophyoid Mites, Their Biology, Natural Enemies and Control*. World Crop Pests vol 6. Elsevier, Amsterdam.

Lo, K.C. and Hor, C.C. (1977) Preliminar y studies on rice tarsonemid mite, *Steneotarsonemus spinki* (Acarina: Tarsonemidae). *Review of Applied Entomology. Series A, Agricultural*, 66, 274–284 (Paper published in National Science Council Monthly, in Chinese).

Lo, K.C. and Ho, C.C. (1979) Ecological observations on rice tarsonemid mite, *Teneotarsonemus spinki* (Acarina: Tarsonemidae). *Journal of Agriculture Research of China* 28, 181–192.

Lo, K.C. and Ho, C.C. (1980) Studies on the rice tarsonemid mite, *Steneotarsonemus spinki* Smiley. *Plant Protection Bulletin* 22, 1–9.

Mahmood, T., Hein, G.L. and Jensen, S.G. (1998) Mixed infection of hard red winter wheat with high plains virus and wheat streak mosaic virus from wheat curl mites in Nebraska. *Plant Disease* 82, 311–315.

Malik, R., Brown-Guedira, G.L., Smith, C.M., Harvey, T.L. and Gill, B.S. (2003a) Genetic mapping of wheat curl mite resistance genes Cmc3 and Cmc4 in common wheat. *Crop Science* 43, 644–650.

Malik, R., Smith, C.M., Brown-Guedira, G.L., Harvey, T.L. and Gill, B.S. (2003b) Assessment of *Aegilops tauschii* for resistance to biotypes of wheat curl mite (Acari: Eriophyidae). *Journal of Economic Entomology* 96, 1329–1333.

MAPA (Ministério da Agricultura, Pecuária e Abastecimento) (2009) Instrução Normativa No. 34, de 8 de setembro de 2009. Diário Oficial da União 06/09, 09/09/2009, Brasilia, Brazil.

Marsaro Jr, A.L., Sato, M.E., Mineiro, J.L.C., Navia, D., Aguiar, R.M. and Vieira, G.B. (2009) Ácaros predadores associados ao ácaro hindu dos citros, *Schizotetranychus hindustanicus* (Hirst, 1924) (Acari: Tetranychidae), no estado de Roraima, Brasil. In: Resumos do XI Simpósio de Controle Biológico (Abstract 09/093) (CD-ROM). Sociedade Entomológica do Brasil, Universidade do Vale do Rio dos Sinos, Bento Gonçalvez, Brazil.

Marsaro Jr, A.L., Navia, D., Sato, M.E., Silva Jr, R.J., Moreira, G.A.M. and Duarte, O.R. (2010a) Host plant range of the Citrus Hindu Mite, *Schizotetranychus hindustanicus* (Hirst) (Tetranychidae), in Brazil. In: Moraes, G.J., Castilho, R.C. and Flechtmann, C.H.W. (eds) Abstract Book of the XIII International Congress of Acarology (Abstract 266). Recife, Brazil, pp. 144–145.

Marsaro Jr, A.L., Navia, D., Gondim Jr, M.G.C., Duarte, O.R., Castro, T.M.M.G. and Moreira, G.A.M. (2010b) Host plants of the Red Palm Mite, *Raoiella indica* Hirst (Tenuipalpidae), in Brazil. In: Moraes, G.J., Castilho, R.C. and Flechtmann, C.H.W. (eds) Abstract Book of the XIII International Congress of Acarology (Abstract 267). Recife, Brazil, p. 145.

Menzel, C. (2002) *The Lychee Crop*. FAO, Rome.

Mesa, N.C., Ochoa, R., Welbourn, W.C., Evans, G.A. and Moraes, G.J. (2009) A Catalog of the Tenuipalpidae (Acari) of the World with a Key to Genera. Zootaxa 2098, Magnolia Press, Auckland.

Migeon, A. and Dorkeld, F. (2010) Spider Mites Web: a comprehensive database for the Tetranychidae. Available at: www.montpellier.inra.fr/CBGP/spmweb, accessed 6 February 2011.

Mishra, R.K. (1980) Gall formation and control of the erinose mite *Eriophyes litchii* Keifer. Acarina: Eriophyidae. In: Proceedings of the International Symposium IOBC/WPRS on Integrated Control in Agriculture and Forestry, International Organization for Biological and Integrated Control of Noxious Animals and Plants, Vienna, pp. 435–436.

Mori, K. and Saito, Y. (2006) Communal relationship in a social spider mite, *Stigmaeopsis longus* (Acari: Tetranychidae): an equal share of labor and reproduction between nest mates. *Ethology* 112, 134–142.

Moutia, L.A. (1958) Contribution to the study of some phytophagous acarina and their predators in Mauritius. *Bulletin of Entomological Research* 49, 59–75.

Murray, G. (2006) *Update on Wheat Streak Mosaic Virus*. Plant Disease Notes, New South Wales Department of Primary Industries, Sydney, Australia.

Murray, G. and Wratten, K. (2005) *Wheat Streak Mosaic Virus*. Plant Disease Notes, New South Wales Department of Primary Industries, Sydney, Australia.

Myers, J. (2007) Coconut shipments detained in Florida - Red palm mite contamination hurting exports. *Jamaica Gleaner News*. Available at: www.jamaica-gleaner.com/gleaner/20070803/business/business6.html, accessed 7 February 2011.

Nadarajan, L., ChannaBasavanna, G. and Chandra, B.K. (1990) Control of coconut pests through stem injection of systemic insecticides. *Mysore Journal of Agricultural Sciences* 14, 355–364.

Nageshachandra, B.K. and ChannaBasavanna, G.P. (1984) Plant mites. In: Griffths, D.A. and Bowman, C.E. (eds) *Acarology VI*. Ellis Horwood Publishers, Chichester, pp. 785–790.

NAPPO (North American Plant Protection Organization) (2007) Phytosanitary Alert System: Detections of Panicle Rice Mite, *Steneotarsonemus spinki*, in Stuttgart,

Arkansas and Ithaca, New York – United States. Available at: www.pestalert.org/oprDetail.cfm%3FoprID¼283, accessed 18 September 2007.

NAPPO (North American Plant Protection Organization) (2009) Phytosanitary Alert System: Detection of the red palm mite (*Raoiella indica*) in Cancun and Isla Mujeres, Quintana Roo, Mexico. Available at: www.pestalert.org/oprDetail.cfm?oprID=406, accessed 20 November 2009.

Nault, L.R. and Styer, W.E. (1969) The dispersal of Aceria tulipae and three other grass-infesting Eriophyid mites in Ohio. *Annals of the Entomological Society of America* 62, 1446–1455.

Navia, D. and Marsaro Jr, A.L. (2010) First report of the Citrus Hindu Mite, *Schizotetranychus hindustanicus* (Hirst) (Prostigmata: Tetranychidae), in Brazil. *Neotropical Entomology* 39, 140–143.

Navia, D., Moraes, G.J. and Flechtmann, C.H.W. (2006a) Phytophagous mites as invasive alien species: quarantine procedures. In: Morales-Malacara, J.B., Behan-Pelletier, V., Ueckermann, E., Pérez, T.M., Estrada, E., Gispert, C. and Badii M. (eds) Proceedings of the XI International Congress of Acarology, Acarology XI (Abstract). UNAM/Sociedad Latinoamericana de Acarología, Mexico DF, Mexico pp. 87–96.

Navia, D., Truol, G., Mendonça, R.S. and Sagadin, M. (2006b) *Aceria tosichella* Keifer (Acari: Eriophyidae) from wheat streak mosaic virus-infected wheat plants in Argentina. *International Journal of Acarology* 32, 189–193.

Navia, D., Mendonça, R.S. and Ochoa, R. (2009) The rice mite, *Steneotarsonemus spinki* Smiley, an invasive species in Americas. In: Sabelis, M.W. and Bruin, J. (eds) *Trends in Acarology*. Springer, Amsterdam, The Netherlands, pp. 379–384.

Navia, D., Moraes, G.J., Marsaro Jr, A.L., Gondim Jr, M.G.C., Silva, F.R. and Castro, T.M.M.G. (2010a) Current status and distribution of *Raoiella indica* (Acari: Tenuipalpidae) in Brazil. In: Moraes, G.J., Castilho, R.C. and Flechtmann, C.H.W. (eds) Abstract Book of the XIII International Congress of Acarology (Abstract 320). Recife, Brazil, p. 173.

Navia, D., Pereira, P.R.V.S., Lau, D., Castiglioni, E., Truol, G., Mendonça, R.S.M., Santos, A.S. et al. (2010b) The wheat curl mite *Aceria tosichella* in South America – occurrence areas and host plants. In: Moraes, G.J., Castilho, R.C. and Flechtmann, C.H.W.

(eds) Abstract Book of the XIII International Congress of Acarology (Abstract 322). Recife, Brazil, p. 174–175.

Navia, D., Marsaro Jr, A.L., Silva, F.R., Gondim Jr, M.G.C. and Moraes, G.J. (2011) First report of the Red Palm Mite, *Raoiella indica* Hirst (Acari: Tenuipalpidae), in Brazil. *Neotropical Entomology* 40, 409–411.

Nienstaedt-Arreaza, B.M. (2007) Estudio de algunos aspectos biológicos y ecológicos del ácaro hindú de los cítricos *Schizotetranychus hindustanicus* (Hirst, 1924) (Acari: Tetranychidae), en Maracay, Venezuela. BSc thesis, Universidad Central de Venezuela, Maracay, Venezuela.

Nishida, T. and Holdaway, F.G. (1945) *The Erinose Mite of Lychee*. Circular No. 48, Hawaii Agriculture Experiment Station, Honolulu, Hawaii, pp. 10.

O Estado de São Paulo (2010) Rústica lichia tem primeira praga no País. Available at: www.estadao.com.br/.../suplementos,rustica-lichia-tem-primeira-praga-no-pais.htm, accessed 10 February 2011.

Oldfield, G.N. (1970) Mite transmission of plant viruses. *Annual Review of Entomology* 15, 343–380.

Oldfield, G.N. (1996) Diversity and host plant specificity. In: Lindquist, E.E., Sabelis, M.W. and Bruin, J. (eds) *Eriophyoid Mites - Their Biology, Natural Enemies and Control*. World Crop Pests vol 6. Elsevier Science Publishing, Amsterdam, pp. 199–216.

Oldfield, G.N. and Proeseler, G. (1996) Eriophyoid mites as vectors of plant pathogens. In: Lindquist, E.E., Sabelis, M.W and Bruin, J. (eds) *Eriophyoid Mites - Their Biology, Natural Enemies and Control*. World Crop Pests vol 6. Elsevier Science Publishing, Amsterdam, pp. 34–45.

Ou, Y.T., Fang, H.C. and Tseng, Y.H. (1977) Studies on *Steneotarsonemus madecassus* Gutierrez of rice. *Plant Protection Bulletin* 19, 21–29.

Peña, J.E. and Rodrigues, J.C.V. (2010) Alternatives for the chemical control of the red palm mite, *Raoiella indica* (Acari: Tenuipalpidae) on palms and bananas. In: Moraes, G.J., Castilho, R.C. and Flechtmann, C.H.W. (eds) Abstract Book of the XIII International Congress of Acarology (Abstract 374). Recife, Brazil, pp. 202–203.

Peña, J.E., Mannion, C.M., Howard, F.W. and Hoy, M.A. (2006) *Raoiella indica* (Prostigmata: Tenuipalpidae): the red palm mite: a potential invasive pest of palms and bananas and other tropical crops in Florida. University of Florida IFAS Extension. Available at: edis.ifas.ufl.edu/IN681, accessed 7 February 2011.

Peña, J.E., Rodrigues, J.C.V., Roda, A., Carrillo, D. and Osborne, L.S. (2009) Predator–pre y dynamics and strategies for control of the red palm mite (*Raoiella indica*) (Acari: Tenuipalpidae) in areas of invasion in the Neotropics. In: Proceedings of the 2nd Meeting of IOBC/WPRS. Work Group Integrated Control of Plant F eeding Mites. International Organization for Biological and Integ rated Control of No xious Animals and Plants , Florence, Italy, pp. 69–79.

Peña, J.E., Carrillo, D., Rodrigues, J.C.V. and Roda, A. (2010) El ácaro rojo de las palmas , *Raoiella indica* (Acari: Tenuipalpidae), una plaga potencial par a América Latina. In: Sánchez Gálvez, M.C., Sandoval-Islas, J.S. and Estrada-Venegas, E.G.E. (eds) *Primer Simposio Internacional de Acarología en México – Memorias*. Universidad Autónoma Chapingo, Chapingo, Mexico, pp. 48–57. Available at: armandocorrea.com/Acarologia/main.html, accessed 7 February 2011.

Pereira, P.R.V.S., Navia, D., Salvadori, J.R. and Lau, D. (2009) Occurrence of *Aceria tosichella* in Brazil. *Pesquisa Agropecuária Brasileira* 44, 539–542.

Pereira, P.R.V.S., Lau, D ., Navia, D., Mendonça, R.S. and Bianchin, V. (2010) Monitoramento e distribuição do ácaro-do-enrolamento-do-trigo *Aceria tosichella* Keifer (Prostigmata: Eriophyidae) no Brasil no período 2009/2010. Online technical communication 283. Embrapa Trigo, Passo Fundo, Brazil. Available at: www.cnpt.embrapa.br/biblio/co/p_co283.htm, accessed 10 February 2011.

Pérez, C.A., Gabhoon, S .A., Raeedan, M.M.S ., Haddad, M.L. and Silv eira-Neto, S. (2010) Efficacy of ecofr iendly compounds against Red Palm Mite (*Raoiella indica*) on cocon ut seedlings in the Sultanate of Oman. In: Moraes, G.J., Castilho, R.C. and Flechtmann, C.H.W. (eds) Abstr act Book of the XIII International Congress of Acarology (Abstract 378). Recife, Brazil, pp. 204–205.

Picoli, P.R.F. (2010) *Aceria litchii* (Keifer) em lichia: ocorrência sazonal, danos provocados e identificação de possív eis agentes de controle biológico. MSc thesis, Universidade Estadual Paulista 'Julio de Mesquita filho', Ilha Solteira, Brazil.

Picoli, P.R.F., Vieira, M.R., Silv a, E.A. and Mota, M.S.O. (2010) Ácaros predadores associados ao ácaro-da-erinose da lichia. *Pesquisa Agropecuária Brasileira* 45, 1246–1252.

Pinese, B. (1981) Er inose mite - a ser ious litchi pest. *Queensland Agriculture Journal* 107, 79–81.

Prasad, V.G. and Bagle, B.G. (1981) Evaluation of acaricides for the control of litchi mite , *Aceria litchii* Keifer (Acarina: Eriophyidae). *Pesticides* 1, 22–23.

Prasad, V.G. and Singh, R.K. (1981) Pre valence and control of litchi mite , *Aceria litchii* Keifer in Bihar. *Indian Journal of Entomology* 43, 67–75.

Price, J., Workneh, F., Evett, S., Jones, D., Arthur, J. and Rush, C.M. (2010) Effects of wheat streak mosaic virus on root development and water-use efficiency of winter wheat. *Plant Disease* 94, 766–770.

Pritchard, A.E. and Baker, W. (1958) The false spider mite (Acar ina: Tenuipalpidae). *University of California Publications in Entomology* 14, 175–274.

Puttarudriah, M. and ChannaBasa vanna, G.P. (1958) Preliminary acaricidal test against the areca palm mite , *Raoiella indica*. *Arecanut Journal* 8, 87.

Puttaruddriah, M. and ChannaBasa vanna, G.P. (1959) A preliminar y account of polyphagous mites of Mysore. In: Proceedings of the First All India Cong ress on Zoology , Zoological Society of India, Bangalore , India, pp. 530–539.

Quilici, S., Kreiter, S., Ueckermann, E.A. and Vincenot, D. (1997) Predatory mites from various crops on Réunion Island. *International Journal of Acarology* 23, 283–291.

Quirós, M. and Dorado, I. (2005) Eficiencia de tres productos comerciales en el control del ácaro hindú de las cítr icas *Schizotetranychus hindustanicus* (Hirst), en el labor atorio. In: Resumen de Congreso de Entomología 2005 (Abstract). Universidad del Zulia, Mar acaibo, Venezuela.

Quirós, M. and Ger aud-Pouey, F. (2002) *Schizotetranychus hindustanicus* (Hirst) (Acari: Tetranychidae), new spider mite pest damaging citrus in Venezuela, South America. In: Morales-Malacara, J.B. and Rivas, G. (eds) Program and Abstr act Book of the XI International Congress of Acarology (Abstract). Universidad Nacional A utónoma de Mé xico, Mexico DF, Mexico, pp. 255–256.

Raga, A., Mineiro, J.L.C., Sato, M.E., Moraes, G.J. and Flechtmann, C .H.W. (2010) Pr imeiro relato de *Aceria litchii* (Keifer) (Prostigmata: Eriophyidae) em plantas de lichia no Br asil. *Revista Brasileira de Fruticultura* 32, 628–629.

Ramírez, M. and García, J . (2010) Campañas contra el ácaro del vaneo del arroz (*Steneotarsonemus spink*i Smiley) y contra el ácaro rojo de las palmas (*Raoiella indica* Hirst) en

México. In: Sánchez Gálvez, M.C., Sandoval-Islas, J.S. and Estrada-Venegas, E.G.E. (eds) Primer Simposio Internacional de Acarología en México – Memorias. Universidad Autónoma Chapingo, Chapingo, Mexico, pp. 57–62. Available at: armandocorrea.com/Acarologia/main.html, accessed 7 February 2011.

Ramos, M. and Rodríguez, H. (1998) *Steneotarsonemus spinki* Smiley (Acari: Tarsonemidae): nuevo informe para Cuba. *Revista de Protección Vegetal* 13, 25–28.

Ramos, M. and Rodríguez, H. (2000) Ciclo de desarollo de *Steneotarsonemus spinki* Smiley (Acari: Tarsonemidae) en laboratório. *Revista de Protección Vegetal* 15, 751–752.

Ramos, M. and Rodríguez, H. (2001) Aspectos biológicos y ecológicos de *Steneotarsonemus spinki* en arroz, en Cuba. *Manejo Integrado de Plagas* 61, 48–52.

Ramos, M. and Rodríguez, H. (2003) Análisis de riesgo de una especie exótica invasora: *Steneotarsonemus spinki* Smiley. Etudio de un caso. *Revista de Protección Vegetal* 18, 158–159.

Ramos, M., Gómez, C. and Cabrera, R.I. (2001) Presencia de *Steneotarsonemus spinki* (Acari: Tarsonemidae) en cuatro variedades de arroz en la República Dominicana. *Revista de Protección Vegetal* 16, 6–9.

Ramos, M., Rodriguéz, H. and Chico, R. (2005) Los ácaros depredadores y su potencial en la regulación de *Steneotarsonemus spinki*. III Encuentro Internacional del Arroz: El ácaro del arroz *Steneotarsonemus spinki* (Tarsonemidae) retos y alter nativas para América Latina y el Caribe, Libro de Resúmen. IIARROZ, Havana, pp. 10–13 (in Spanish).

Ramos, M., González, A.I. and González, M. (2010) Management strategy of *Raoiella indica* Hirst in Cuba, based on biology, host plants, seasonal occurrence and use of acaricide. In: Moraes, G.J., Castilho, R.C. and Flechtmann, C.H.W. (eds) Abstract Book of the XIII International Congress of Acarology (Abstract 405). Recife, Brazil, pp. 218–219.

Rao, J. and Prakash, A. (1996) *Cynodon dactylon* (Linn.) Pers. (Graminae): an alternate host of rice tarsonemid mite, *Steneotarsonemus spinki* Smiley. *Journal of Applied Zoological Research* 7, 50–51.

Rao, J. and Prakash, A. (2002) Paddy field weed, *Schoenoplectus articulatus* (Linn.) Palla (Cyperaceae): a new host of tarsonemid mite, *Steneotarsonemus spinki* Smiley and panicle thrips, *Haplothrips ganglbaureri* Schmutz. *Journal of Applied Zoological Research* 13, 174–175.

Rao, J. and Prakash, A. (2003) Panicle mites causing sterility in farmers' paddy fields in India. *Journal of Applied Zoological Research* 14, 212–217.

Rao, J., Prakash, A., Dhanasekhar an, S. and Ghosh, S.K. (1993) Observations on rice tarsonemid mite *Steneotarsonemus spinki*, whitetip nematode and sheath-rot fungus interactions deteriorating grain quality in paddy fields. *Journal of Applied Zoological Research* 4, 89–90.

Rao, P.R.M., Bhavani, T.R.M., Rao, T.R.M. and Reddy, P.R. (2000) Spikelet sterility/grain discoloration in Andhra Pradesh, India. *International Rice Research Notes. Notes from the fields* 25, 40.

Rao, Y.S. and Das, P.K. (1977) A new mite pest of rice in India. *International Rice Research Newsletter* 2, 8.

Roda, A.L., Peña, J.E., Hosein, F. and Rodrigues, J.C.V. (2010) Dispersion indices and sampling plans for the red palm mite (Acari: Tenuipalpidae) on coconut. In: Moraes, G.J., Castilho, R.C. and Flechtmann, C.H.W. (eds) Abstract Book of the XIII International Congress of Acarology (Abstract 421). Recife, Brazil, pp. 227–228.

Rodrigues, J.C.V., Ochoa, R. and Kane, E. (2007) First report of *Raoiella indica* Hirst (Acari: Tehuipalpidae) and its damage to coconut palms in Puerto Rico and Culebra Islands. *International Journal of Acarology* 33, 3–5.

Rodriguez, H. (2005) Evaluación del status fitosanitario del ácaro del vaneo del arroz, *Steneotarsonemus spinki* Smiley, en la Republica de Nicaragua. In: *III Encuentro International del Arroz: El ácaro del arroz Steneotarsonemus spinki (Tarsonemidae) retos y alternativas para América Latina y el Caribe, Libro de Resumen*. Instituto de Investigaciones del Arroz, Havana, pp. 39–42.

Roger, A., Jones, C., Coutts, B.A. and Mackie, A.E. (2005) Seed transmission of wheat streak mosaic virus shown unequivocally in wheat 1048. *Plant Disease* 89, 1048–1050.

Romero, L., Pérez, R., Arana, R., Cabrera, R.I., Hernández, J.L., Castillo, D. and Hernández, A.A. (2003) Experiencia adquirida de la interrelación entre la comunidad San Paul y el Instituto de Investigaciones del arroz. In: Forum Ramal del Cultivo del Arroz del Instituto de Investigaciones del Arroz, Memórias. Instituto de Investigaciones del Arroz, Camaguey, Cuba, pp. 40–45.

Rosa, B.F. (2003) Estudios de diferentes variedades de arroz en condiciones de prémontaña. In: Forum Ramal del Cultivo del Arroz del Instituto de Investigaciones del

Arroz, Memórias. Instituto de Investigaciones del Arroz, Camaguey, Cuba, pp. 23–25.

Sabelis, M.W. and Bruin, J. (1996) Evolutionary ecology: life history patterns, food plant choice and dispersal. In: Lindquist, E.E., Sabelis, M.W. and Bruin, J. (eds) *Eriophyoid Mites - Their Biology, Natural Enemies and Control.* Elsevier Science Publishing, Amsterdam, The Netherlands, pp. 329–366.

Sagadin, M.B., Rodríguez, S.M. and Truol, G.A.M. (2008) *Transmisión por Semillas de Wheat Streak Mosaic Virus (Wsmv) en Infecciones Naturales y Experimentales.* INTA/IFFIVE Informe 8, Córdoba, Argentina.

Saha, K., Somchoudhury, A.K. and Sarkar, P.K. (1996) Structure and ecology of *Cephaleuros virescens* Kunze and its relationship with *Aceria litchii* Keifer (Prostigmata: Acari) in forming litchi erineum. *Journal of Mycopathological Research* 34, 159–171.

Saito, Y. (1986) Biparental defence in a spider mite (Acari: Tetranychidae) infesting Sasa bamboo. *Behavioral Ecology and Sociobiology* 18, 377–386.

Sanabria, C. (2005) El ácaro del vaneo del arroz (*Steneotarsonemus spinki* L: Tarsonemidae). Actualidad Fitosanitaria Mai–Jun 2005. Available at: www.protecnet.go.cr/general/boletin/Bole22/Boletin22.htm, accessed 12 March 2006.

Sánchez-Sánchez, H.M., Henry, M., Cárdenas-Soriano, E. and Alvizo-Villasana, H.F. (2001) Identification of wheat streak mosaic virus and its vector *Aceria tosichella* in Mexico. *Plant Disease* 85, 13–17.

Santos, A., Almaguel, L. and Torre, P.E. (1998) Duración del ciclo de vida en condiciones controladas del ácaro *Steneotarsonemus spinki* (Acari: Tarsonemidae) en arroz (*Oryza sativa* L.) en Cuba. In: I Encuentro Internacional del Arroz, Resúmenes, IIARROZ, Havana.

Sarkar, P.K. and Somchoudhury, A.K. (1988) Evaluation of some pesticides against *Raoiella indica* Hirst on coconut palm in West Bengal. *Pesticides* 22, 21–22.

Sarkar, P.K. and Somchoudhury, A.K. (1989) Interrelationship between plant characters and incidence of *Raoiella indica* Hirst on coconut. *Indian Journal of Entomolology* 51, 45–50.

Sathiamma, B. (1996) Observations on the mite fauna associated with the coconut palm in Kerala, India. *Journal of Plantation Crops* 24, 92–96.

Sayed, T. (1942) Contribution to the knowledge of the Acarina of Egypt: The genus *Raoiella* Hirst (Pseudotetranychinae:Tetranychidae). *Bulletin de la Société Fouad, 1er Entomologie* 26, 81–91.

Schicha, E. (1987) *Phytoseiidae of Australia and Neighbouring Areas.* Indira Publishing House, West Bloomfield, Michigan.

Schiffer, M., Umina, P., Carew, M., Hoffmann, A., Rodoni, B. and Miller, A. (2010) The distribution of wheat curl mite (*Aceria tosichella*) lineages in Australia and their potential to transmit wheat streak mosaic virus. *Annals of Applied Biology* 155, 371–379.

Schulte, M.J., Martin, K. and Sauerborn, J. (2007) Efficacy of spiromesifen on *Aceria litchii* (Keifer) in relation to *Cephaleuros virescens* Kunze colonization on leaves of litchi (*Litchi chinensis* Sonn.). *Journal of Plant Diseases and Protection* 114, 133–137.

Seifers, D.L., Harvey, T.L., Martin, T.J., Jensen, S.G. (1997) Identification of the wheat curl mite as the vector of the high plains virus of corn and wheat. *Plant Disease* 81, 1161–1166.

Seifers, D.L., Harvey, T.L., Martin, T.J. and Jensen, S.G. (1998) A partial host range of the High Plains Virus of corn and wheat. *Plant Disease* 82, 875–879.

Seifers, D.L., Harvey, T.L., Louie, R., Gordon, D.T. and Martin, T.J. (2002) Differential transmission of isolates of the high plains virus by different sources of wheat curl mites. *Plant Disease* 86, 138–142.

Seifers, D.L., Martin, J.T., Harvey, T.L., Fellers, J.P., Stack, J.P., Ryba-White, M., Harber, S. et al. (2008) Triticum mosaic virus: a new virus isolated from wheat in Kansas. *Plant Disease* 92, 808–817.

Seifers, D.L., Martin, T.J., Harvey, T.L., Fellers, J.P. and Michaud, J.P (2009) Identification of the wheat curl mite as the vector of triticum mosaic virus. *Plant Disease* 93, 25–29.

Shahwan, I.M. and Hill, J.P. (1984) Identification and occurrence of Wheat streak mosaic virus in winter wheat in Colorado and its effects on several wheat cultivars. *Plant Disease* 68, 579–581.

Sharma, D.D. (1984) Control of litchi mite. *Indian Horticulture* 29, 27.

Sharma, D.D. (1985) Major pests of litchi in Bihar. *Indian Farming* 35, 25–26.

Sharma, D.D. (1991) Occurrence of *Cephaleuros virescens*, a new record of leaf-curl in litchi (*Litchi chinensis*). *Indian Journal of Agricultural Sciences* 61, 446–448.

Sharma, D.D. and Rahman, M.F. (1982) Control of litchi mite *Aceria litchii* (Keifer) with particular reference to evaluation of pre-bloom and post-bloom application with different insecticides. *Entomon* 7, 55–56.

Sharma, D.D., Singh, S.P. and Akhauri, R.K. (1986) Relationship between the population of *Aceria litchii* Keifer on litchi and weather factors. *Indian Journal of Agriculture Science* 56, 59–63.

Shevtchenko, V.G., Demillo, A.P., Razviaskina, G.M. and Kapkova, E.A. (1970) Taxonomic separation of similar species of eriophyid mites, *Aceria tulipae* Keif. and *A. tritici* sp. n. (Acarina, Eriophyoidea) - vectors of the viruses of onions and wheat. *Zoologicheskii Zhurnal* 49, 224–235. In: *International Journal of Acarology* (1996) 22, 149–160 (translated by Amrine Jr, J.W.).

Shikata, E., Kawano, S. and Senboku, T. (1984) Small virus-like particles isolated from the leaf sheath tissues of rice plants and from the rice tarsonemid mites *Steneotarsonemus spinki* Smiley (Acarina: Tarsonemidae) (in Japanese). *Annals of the Phytopathology Society of Japan* 50, 368–374; *Review of Plant Pathology* 64 (1985) (Abstract).

Siddiqui, S.B.M.A.B. (2002) Lychee production in Bangladesh. In: Papademetriou, M.K. and Dent, F.J. (eds) *Lychee Production in the Asia-Pacific Region.* FAO Regional Office for Asia and the Pacific, Rome, pp. 28–40.

Skare, J.M., Wijkamp, I., Rezende, J., Michels, G., Rush, C., Scholthof, K.B.G. and Scholthof, H.B. (2002) Colony establishment and maintenance of the eriophyid wheat curl mite *Aceria tosichella* for controlled transmission studies on a new virus-like pathogen. *Journal of Virological Methods* 108, 133–137.

Skare, J.M., Wijkam, I., Rezende, J., Michels, G., Rush, C., Scholthof, K.B.G. and Scholthof, H.B. (2003) Colony establishment and maintenance of the eriophyid wheat curl mite *Aceria tosichella* for controlled transmission studies on a new virus-like pathogen. *Journal of Virological Methods* 108, 133–137.

Skoracka, A. and Kuczynski, L. (2006) Infestation parameters and morphological variation of the wheat curl mite *Aceria tosichella* Keifer (Acari: Eriophyidae) In: Naukowa, G.G. and Ignatowicz, S. (eds) *Advances in Polish Acarology.* Wydawnictwo SGGW, Warsaw, pp. 330–339.

Slykhuis, J.T. (1955) *Aceria tulipae* in relation to the spread of wheat streak mosaic *Phytopathology* 45, 116–128.

Smiley, R.L. (1967) Further studies on the Tarsonemidae (Acarina). *Proceedings of the Entomological Society of Washington* 69, 127–146.

Smiley, R.L., Flechtmann, C.H.W. and Ochoa, R. (1993) A new species of *Steneotarsonemus* (Acari: Tarsonemidae) and an illustrated key to grass-infesting species in the western hemisphere. *International Journal of Acarology* 19, 87–93.

Sogawa, K. (1977) Occurrence of the rice tarsonemid mite at IRRI. *International Rice Research Newsletter* 2, 5.

Somchoudhury, A.K., Singh, P. and Mukherjee, A.B. (1989) Interrelationship between *Aceria litchii* (Acari: Eriophyidae) and *Cephaleuros virescens*, a parasitic alga in the formation of erineum-like structure on litchi leaf. In: Basavanna, G.P. and Viraktamath C.A. (eds) *Progress in Acarology 2*, OUP and IBH, Bangalore, India, pp. 147–152.

Somsen, H.W. and Sill Jr, W.H. (1970) *The Wheat Curl Mite, Aceria tulipae Keifer, in Relation to Epidemiology and Control of Wheat Streak Mosaic.* Research Publication 162. Kansas State University of Agriculture and Applied Science, Manhattan, Kansas.

Stephan, D., Moeller, I., Skoracka, A., Ehrig, F. and Maiss (2008) Eriophyid mite transmission and host range of a Brome streak mosaic virus isolate derived from a full-length cDNA clone. *Archives of Virology* 153, 181–185.

Tatineni, S., Graybosch, R.A., Hein, G., Wegulo, S.N. and French, R. (2010) Wheat cultivar-specific disease synergism and alteration of virus accumulation during co-infection with Wheat streak mosaic virus and Triticum mosaic virus. *Virology* 100, 230–238.

Taylor, B. (2010) Ecology of *Raoiella indica* and its natural enemies in Kerala, India. In: Moraes, G.J., Castilho, R.C. and Flechtmann, C.H.W. (eds) *Abstract Book of the XIII International Congress of Acarology* (Abstract 503). Recife, Brazil, pp. 272–273.

Texas Department of Agriculture (2007) Emergency action notification ordered to stop movement of rice products from Texas Research Facility TDA. Texas Department of Agriculture Press Release, Austin, Texas.

Thakur, A.P. and Sharma, D.D. (1989) New records of predatory mites on mite pests of economic crops in Bihar. *Biojournal* 1, 155–156.

Thakur, A.P. and Sharma, D.D. (1990) Influence of weather factors and predators on the populations of *Aceria litchii* Keifer. *Indian Journal of Plant Protection* 18, 109–112.

Thomas, J.A. and Hein, G.L. (2003) Influence of volunteer wheat plant condition on movement of the wheat curl mite, *Aceria tosichella*, in winter wheat. *Experimental and Applied Acarology* 31, 253–268.

Truol, G. and Sagadin, M. (2008) Monitoreo de *Aceria tosichella* Keifer vector de Wheat streak mosaic virus (WSMV) y de High plains virus (HPV) en trigos de la provincia de Buenos Aires y Córdoba. In: Resúmenes VII Congreso

Nacional de Trigo, Abstract. INTA/Universidad Nacional de la Pampa, Santa Rosa, Argentina, p. PV14.

Truol, G., French, R., Sagadin, M. and Arneodo, J. (2004) First report of Wheat Streak Mosaic Virus infecting wheat in Argentina. *Australian Plant Pathology* 33, 137–138.

UCDavis (2009) News & Information, 2009. UC Davis Cleanses Greenhouses to Eliminate Rice Pest. Available at: www.news.ucdavis.edu/search/news_detail.lasso%3Fid%9020, accessed June 2010.

University of Illinois (1989) Report on Plant Disease. Wheat streak mosaic. Department of Crop Sciences, University of Illinois, Urbana-Champaign, Illinois.

Valès, M., García, J. and Dossmann, J. (2002) Mejoramiento de los acervos genéticos. Desarrollo del arroz de secano para pequeños productores. Available at: www.ciat.cgiar.org/riceweb/esp/pdf/resultado_1.pdf, accessed 17 August 2004.

Vásquez, C., Quiros, M., Aponte, O. and Sandoval, D.M.F. (2008) First report of *Raoiella indica* Hirst (Acari: Tenuipalpidae) in South America. *Neotropical Entomology* 37, 739–740.

Velandia, M., Rejesus, R.M., Jones, D.C., Price, J.A., Workneh, F. and Rush, C. (2010) Economic impact of Wheat streak mosaic virus in the Texas High Plains. *Crop Protection* 29, 699–703.

Waite, G.K. (1986) Pests of lychees in Australia. In: Menzel, C.M. and Greer, G.N. (eds) The potential of lychee in Australia. Proceedings of the First National Seminar. Sunshine Coast Subtropical Fruits Association, Nambour, Australia, pp. 41–50.

Waite, G.K. (1992) Integrated pest management in lychee. In: Fullelove, G.D. (ed.) Proceedings of the Third National Lychee Seminar. Bundaberg and District Orchardists Association, Bundaberg, Australia, pp. 39–43.

Waite, G.K. (1999) New evidence further incriminates honey-bees as vectors of lychee erinose mite *Aceria litchii* (Acari: Eriophyidae). *Experimental and Applied Acarology* 23, 145–147.

Waite, G.K. (2005) Pests. In: Menzel, C.M. and Waite, G.K. (eds) *Litchi and Longan - Botany, Production and Uses*. CAB International, Wallingford, UK, pp. 237–259.

Waite, G.K. and Elder, R.J. (1996) Lychee/longan insect pests and their control. In: Welch, T. and Ferguson, J. (eds) Proceedings of the Fourth National Lychee Seminar Including Longans. Australian Lychee Growers' Association Inc., Yeppoon, Australia, pp. 102–109.

Waite, G.K. and Gerson, U. (1994) The predator guild associated with *Aceria litchii* (Acari: Eriophyidae) in Australia and China. *Enthomophaga* 39, 275–280.

Waite, G.K. and Hwang, J.S. (2002) Pests of litchi and longan. In: Peña, J.E., Sharp, J.L. and Wysoki, M. (eds) *Tropical Fruit Pests and Pollinators: Biology Economic Importance, Natural Enemies and Control*. CAB International, Wallingford, UK, pp. 331–359.

Waite, G.K. and McAlpine, J.D. (1992) Honey bees as carriers of lychee erinose mite *Eriophyes litchii* (Acari Eriophyiidae). *Experimental and Applied Acarology* 15, 299–302.

Welbourn, C. (2009) Pest Alert: Red palm mite, *Raoiella indica* Hirst (Acari: Tenuipalpidae). Available at: www.doacs.state.fl.us/pi/enpp/ento/r.indica.html, accessed 7 February 2011.

Wiese, M.V. (1987) *Compendium of Wheat Diseases*. 2nd ed., APS Press, St Paul, Minnesota.

Wolf, E. and Seifers, D.L. (2008) *Triticum* mosaic: a new wheat disease in Kansas. K State Research and Extension. EP-145. Available at: www.plantpath.ksu.edu/doc1189.ashx, accessed 17 August 2012.

Xu, G.L., Wu, H.J., Huan, Z.L., Mo, G. and Wan, M. (2001) Study on the reproductive characteristics of rice mite, *Steneotarsonemus spinki* Smiley (Acari: Tarsonemidae). *Systematic and Applied Acarology* 6, 45–49.

Xu, G.L., Wu, H.J. and Tong, X.L. (2002) Studies on stress resistance of *Steneotarsonemus spinki* Smiley. *Plant Protection* 28, 18–21.

Zaher, M.A., Wafa, A.K. and Yousef, A.A. (1969) Biological studies on *Raoiella indica* Hirst and *Phyllotetranychus aegyptiacus* Sayed infesting date palm trees in U.A.R. *Zeitschrift fuer Angewandte Entomologie* 63, 406–411.

Zhang, B.D. (1984) Preliminary observations on the biological characteristics of *Steneotarsonemus spinki* Smiley. *Insect Knowledge* 18, 55–56.

16 *Planococcus minor* (Hemiptera: Pseudococcidae): Bioecology, Survey and Mitigation Strategies

Amy Roda,[1] Antonio Francis,[2] Moses T.K. Kairo[2] and Mark Culik[3]

[1] *USDA-APHIS-PPQ, Center for Plant Health Science and Technology, Miami, Florida 33158, USA;* [2] *Center for Biological Control, College of Engineering Sciences, Technology and Agriculture, Florida Agricultural and Mechanical University, Tallahassee, Florida, USA;* [3] *Instituto Capixaba de Pesquisa, Assistência Técnica e Extensão Rural – INCAPER, Vitória, Espírito Santo, Brazil*

16.1 Introduction: Host Range, Economic Impact and Pest Status

Planococcus minor (Hemiptera: Pseudococcidae) is commonly referred to as the passionvine mealybug, pacific mealybug or guava mealybug. *P. minor* is one of 35 species belonging to a genus that is native to the Old World (Cox, 1989), which includes many well-known pests of economic importance (Williams and Watson, 1988; Cox, 1989). As a phloem feeder, *P. minor* can cause stunting and defoliation that eventually leads to reduced yield and fruit quality. The pest also causes indirect or secondary damage due to the sooty mold growth on honeydew produced by the mealybug. *P. minor* is also likely to transmit plant viruses such as swollen shoot virus of cacao, *Theobroma cacao* L. (Cox, 1989). In addition, multiple *Planococcus* species can transmit the same virus. For example, the Grapevine leafroll-associated virus is transmitted by both *P. citri* and *P. ficus* (Tsai *et al.*, 2008; Cid *et al.*, 2010).

Worldwide, the reported host plant range includes >250 species in nearly 80 families, some of which include important agricultural crops such as banana and plantain, *Musa* groups AAA, AAB and ABB, *Citrus*, cocoa, coffee (*Coffea arabica* L.), corn (*Zea mays* L.), grape (*Vitis vinifera* L.), mango (*Mangifera indica* L.), potato (*Solanum tuberosum* L.) and soybean (*Glycine max*) (Venette and Davis, 2004; Ben-Dov *et al.*, 2011). Although *P. minor* has a very broad host range, not all host records are necessarily reliable. Recent literature suggests that earlier records may be erroneous due to misidentification of closely related and difficult to distinguish mealybug, namely *P. citri* (Batra *et al.*, 1987; Cox, 1989; Williams and Granara de Willink, 1992; Santa Cecilia *et al.*, 2002). In addition, *P. minor* has a similar host range and geographical distribution as other *Planococcus* mealybugs, and multiple species may occur on the same plant (Cox, 1989). Infestation levels can also fluctuate spatially, even on plants in close proximity, and can vary from one year to the next (Miller and Kosztarab, 1979). Because of these issues, it is difficult to estimate the economic impact of *P. minor* alone. For instance, *P. minor* (formerly *P. pacificus*, reported as *P. citri* in 1966) reportedly made up approximately 90% of a scale complex on coffee in New Guinea, and caused an estimated yield reduction of 70–75%. In Taiwan, *P. minor* was considered as a major pest of important crops, including banana, *Citrus*, mango, celery (*Apium* spp.), melon (*Benincasa* spp.), pumpkin (*Cucurbita* spp.), soybean, betel nut (*Areca catechu*), star fruit (*Averrhoa carambola*), guava (*Psidium* spp.) and passionvine (*Passiflora* spp.) (Ho *et al.*, 2007). Although the host-plant ranges of *P. citri* and *P. minor* overlap, *P. minor* may prefer cacao more than *P. citri*, and many records of *P. citri* on this plant should refer to *P. minor* (Cox and Freeston, 1985). Similarly, although both species have been reported on citrus, this is a preferred host plant for *P. citri* and is rarely frequented by *P. minor*.

The mealybug can also exert an indirect economic impact due to trade restrictions. At US ports of entry, *P. minor* was intercepted >1160 times from 2005–2010, with 49% of the infested commodities arriving from Asian countries, 29% from the Caribbean basin region and the remainder from South America, North America and Europe (USDA, 2010). A US commodity-based pest risk assessment concluded that the likelihood of this pest becoming established in the USA was high, and the consequences of its establishment would be severe (Venette and Davis, 2004). As a result, the mealybug was considered a regulated pest and if found the commodity was either destroyed, re-exported or fumigated. When exporting products from infested countries the producer is often required to include phytosanitary measures that minimize the risk of movement of the mealybug to the USA. Similarly, US states may prohibit the movement of material or require compliance agreements that outline treatment and inspection requirements from infested states.

16.2 Origin and Distribution

P. minor is thought to be one of six species with origins in the Old World, and likely was introduced into the Neotropics through trade (Cox, 1989). It is now widely distributed throughout the Oriental, Austro-Oriental, Australian, Polynesian, Nearctic, Afrotropical, Malagasian, and Neotropical regions (Cox and Freeston, 1985; Williams, 1985; Williams and Watson, 1988; Cox, 1989; Williams and Granara de Willink, 1992; Ben-Dov *et al.*, 2011). *P. minor* was originally described in 1897 as *Dactylopius calceolariae* var. *minor* Maskell from a specimen collected in Mauritius, and was synonymized with *P. citri* by Morrison (1925). Cox (1981) redescribed the species as *P. pacificus* from material collected from Western Samoa, which was later recognized to be a synonym of *P. minor* (Cox, 1989).

The identification of many species in the genus *Planococcus* using morphological characters has been challenging (Cox and Wetton, 1988). *P. minor* is particularly difficult to separate from *P. citri* (Williams, 2004). A matrix system was developed based on six diagnostic characters, which were scored using a point system to identify adult females. The system was based on pioneer work by Cox (1981, 1983), who reared *P. citri* (Risso) and *P. minor* (Maskell) as well as *P. ficus* (Signoret) under different environmental conditions, to determine the limits of morphological variation within each species. Specimens having a total score of 35 or below were determined to be *P. minor*, and those having a total score of 35 or more to be *P. citri*. Cox and Freeston (1985) stated that when there are >13 ducts on the head and more than seven adjacent to the 8th pair of cerarii, then the species is undoubtedly *P. citri*. If there are 0–3 ducts on the head and 0–2 ducts adjacent to the 8th pair of cerarii, then the species is *P. minor*. This system is still relied upon by mealybug taxonomists to separate the two species.

P. minor has been routinely misidentified due to similarity in appearance, host plant range and geographic distribution (Williams, 1985; Cox, 1989; Williams and Granara de Willink, 1992; Ben-Dov *et al.*, 2011). Several authors highlighted inaccuracies in past literature, where the species of *Planococcus* commonly occurring in the Austro-oriental, Polynesian and the Neotropics regions was *P. minor* and not *P. citri*, despite most published records listing the latter (Williams, 1982; Cox and Freeston, 1985; Williams and Watson, 1988). The currently reported global distribution of *P. minor* suggests that the pest may be most closely associated with biomes characterized as desert and xeric shrubland; temperate grassland, savannahs, and scrubland; and tropical and subtropical moist broadleaf forest (Venette and Davis, 2004).

16.2.1 Molecular identification

Because it is difficult to distinguish *P. minor* from *P. citri* based on morphological characteristics, alternatives such as molecular identification of *P. minor* have been investigated (Rung et al., 2008, 2009; Malausa et al., 2010). Rung et al. (2008) found that sequences of the mitochondrial cytochrome oxidase-1 (COI) gene and the nuclear protein-coding gene elongation factor 1α (EF-1α) revealed three distinct clades within the *P. citri/P. minor* species complex. They found that '*P. citri*' and '*P. minor*' were clades, corresponding to morphologically identified species collected from various locations around the world and a 'Hawai'ian clade', which includes specimens morphologically indistinguishable from *P. citri* and occurring only in Hawai'i. In a few specimens, the results from COI conflicted in the placement, causing the authors to question if the gene would always give an accurate identification. If *P. minor* and *P. citri* hybridize under natural conditions, mitochondrion introgression could potentially occur, resulting in individuals that have the nuclear genome of *P. minor* and the mitochondrial genome of *P. citri* or vice versa (Rung et al., 2008). Recently, Malausa et al. (2010) found a set of markers that could reliably characterize complexes of cryptic taxa within the family Pseudococcidae. They used five markers, two regions of the mitochondrial COI gene, 28S-D2, the entire internal transcriber space 2 locus and the rpS15-16S region of the primary mealybug endosymbiont *Tremblaya princeps*. These markers distinguished between the species identified on morphological examination, including the most closely related species, *P. citri* and *P. minor*. The genus *Planococcus* appeared monophyletic. *P. citri* and *P. minor* clustered together for all genes, but were separated from *P. ficus*. As molecular analysis can be time-consuming and relatively expensive, the protocols used by Rung et al. (2009) and Malausa et al. (2010) were designed for use in routine work, as they require no gene cloning and make use of rapid, cost-efficient PCR procedures.

16.3 Biology, Life History and Rearing Techniques

The adult female mealybug is pinkish in color, wingless, and has a dark line running down the dorsal median of the insect (Fig. 16.1). The body is covered with white, cottony wax, and has a fringe of elongated waxy filaments that extend about the periphery of the body. An adult female mealybug is about 3 mm long and 1.5 mm in width. The mature female lays pinkish eggs in an egg sac of white wax, usually in clusters on the base of leaves, the twigs or bark of the host plant. The pest forms colonies on the host plant. If left undisturbed, the colonies can grow into large masses of white, waxy deposits on branches, fruiting structures and leaves. The mealybug and eggs sacs are also commonly found on flowers and fruits of a host plant. Eggs hatch into nymphs called crawlers which are very mobile. They may disperse over the host, especially toward tender growing parts, or be carried away by wind, people or animals. Ants may also play a role in mealybug dispersal. However, long-distance movement of the mealybug is most likely as a result of the movement of infested nursery stock and agricultural commerce. Nymphs of both sexes resemble female adults. Nymphs undergo three and four successive molts prior to emergence of adult females and males, respectively (Sahoo and Ghosh, 2001). The male third instar is referred to as the 'prepupa', while the fourth instar from which the adult emerges is termed 'pupa'. These are relatively inactive stages that develop in white cocoon-like structures (Sahoo and Ghosh, 2001). Adult males are c. 1 mm long with three distinct body divisions (Fig. 16.2), three pairs of legs and one pair of wings (Gill, 2004). Mouthparts are absent, therefore they only live for a few days (Sahoo and Ghosh, 2001).

The few studies undertaken on the life history of *P. minor* were conducted at either a single

Fig. 16.1 *Planococcus minor*, adult females.

Fig. 16.2 *P. minor* adult male on sticky trap.

temperature (Martinez and Suris, 1998; Sahoo and Ghosh, 2001) or fluctuating temperature regimes (Maity *et al.*, 1998; Biswas and Ghosh, 2000), and on different readily available host plants (Maity *et al.*, 1998; Biswas and Ghosh, 2000). Eggs required as few as 2–5 days to hatch at 26°C and 69% RH (Martinez and Suris, 1998). The development time for males was longer than for females (Maity *et al.*, 1998; Martinez and Suris, 1998), and the time to complete a single generation ranged from 31 to 50 days (Maity *et al.*, 1998; Martinez and Suris, 1998; Biswas and Ghosh, 2000). Most mealybugs are biparental (Gullan and Kosztarab, 1997). However, several types of parthenogenesis have been described in coccoids, including obligate and facultative parthenogenesis (Gullan and Kosztarab, 1997). Facultative parthenogenesis has been reported in *P. citri* (Myers, 1932; Panis, 1969), but other studies found no reproduction with unmated females of *P. citri* (Borges da Silva *et al.*, 2009). Studies have never been undertaken with *P. minor*, but both females and males occur in populations where males have been reported to be less numerous than females (Maity *et al.*, 1998; Martinez and Suris, 1998; Sahoo *et al.*, 1999; Sahoo and Ghosh, 2001). The preoviposition and oviposition periods of gravid females ranged from 6–11 and 8–14 days (Maity *et al.*, 1998), and 6–8 and 8–9 days (Biswas and Ghosh, 2000). Female fecundity varied depending on the host plants. Biswas and Ghosh (2000) reported 66–159 eggs on *Ixora signaporensis*, soybean and *Acalypha wilkesiana*. However, Maity *et al.* (1998) reported as many as 266–426 eggs on taro (*Colocasia esculenta*), sprouted potato and pumpkin. In warm climates, *P. minor* stays active and reproduces throughout the year (Ben-Dov, 1994). Sahoo *et al.* (1999) reported as many as ten generations occurring per year in India. The low-temperature tolerance and overwintering mechanisms for *P. minor* are unknown. *P. citri* overwinters primarily as eggs on the upper roots, trunk and lower branches of the host plant. Other mealybug species are known to overwinter in the soil or on the host plant, particularly under the bark as late-instar nymphs or adult females.

16.3.1 Rearing

P. minor can be reared on potted host plants; however, propagating and maintaining these host plants requires considerable greenhouse space, special lighting and a sizable workforce. Often, a fruit or vegetable can be substituted as the host plant substrate of choice for an insectary operation for mass-producing mealybugs (Meyerdirk *et al.*, 1998). *P. minor* has been successfully reared on squash and potatoes, using procedures adapted from those described by Meyerdirk *et al.* (1998). These plant materials have served as useful hosts for many different species of mealybugs, and are easy to maintain and manipulate. Mealybug cultures are typically maintained in closed, darkroom facilities. This reduces crawler movement and escape. Several alternative squash/pumpkin varieties can be used to rear the mealybug. The material should be purchased from an organic producer and should not be surface treated with wax or oil products. Potatoes should be grown in the dark to keep the sprouts from producing chlorophyll and turning green, which is undesirable for mealybug rearing. The mealybug crawlers and various instars will feed directly on the potato sprouts. Seed potatoes are preferred because they are not treated with sprouting inhibitors. Room humidity, and – most importantly – cage/cabinet humidity should be maintained above 50% RH. A crawler collection system consisting of a holding cabinet with a low-watt bulb modified with foil (so that a single beam of light projects downward onto a sheet of heavyweight paper) can be used to facilitate infestation of new plant material. Host material containing egg sacs that are about to hatch are placed around the periphery of the paper. Attracted by light, crawlers move from the old infested material unto the cardboard surface, and eventually to the paper surface under the beam of light. Crawlers can be collected daily by simply removing the paper and pouring them onto new host material.

16.4 Sampling and Monitoring Techniques

Surveys for live mealybugs require time-consuming and laborious examination of plant material (Millar et al., 2002). There are no simple and effective visual methods to detect most species (Geiger and Daane, 2001). *P. minor* has cryptic habits, therefore plants need to be examined closely in good light to find them. They are rarely found in direct sunlight and are more often present on leaf undersides, inside the calyx of sepals, in axils or under bark. Typical signs indicating the presence of *P. minor* include plant areas with dieback, leaf loss, localized discoloring/yellowing of leaves, wet patches and sooty mold on the bark, stems, leaves and fruit. Other important indicators of a *P. minor* infestation are ants attending mealybug colonies, and masses of mealybug waxy material. Live insect specimens cannot be identified to genus or species with confidence, because their taxonomy is based on microscopic characters that are only visible in specimens prepared on microscope slides (Watson and Chandler, 2000). Watson and Chandler (2000) recommend placing a small piece of infested plant material in a vial with 80% ethanol to kill and preserve the specimens, and not dislodge an individual insect, as they are often very soft and can be damaged by instruments.

Recent developments in the identification (Ho et al., 2007) and synthesis (Millar, 2008) of the female sex pheromone of *P. minor* may greatly aid in locating populations of the mealybug. Ho et al. (2008) isolated the sex pheromone by aeration of virgin females. The pheromone 2-isopropyl-5-methyl-2,4-hexadienyl acetate was identified, and the stereochemistry of the pheromone was assigned as (E) by comparison with synthetic standards of known geometry. The (E)-isomer was highly attractive to males in laboratory bioassays, whereas the (Z)-isomer appeared to antagonize attraction. In common with all of the scale and mealybug pheromones identified so far, this species produces unique pheromone chemicals, eliminating the possibility of competition for or interference with a particular pheromone channel (Millar, 2008). Because *P. minor* is strongly inhibited by the (Z)-stereoisomer form of its pheromone, the compound may be the pheromone of a related, sympatric species (Millar, 2008). A short and completely stereo-specific process to synthesize the pheromone was developed by Millar (2008). To produce the pheromone with high stereochemical purity is critically important, because the (Z)-isomer is a powerful behavioral antagonist. Solving the problem of synthesis provided a highly sensitive and effective method of detecting even small populations of *P. minor*. Although positive finds on a trap do not pinpoint the exact location of an infestation, they aid in defining the area where detailed field surveys need to be undertaken (Daane et al., 2006). Within the genus *Planococcus*, sex pheromones have been identified and synthesized for *P. citri* (Bierl-Leonhardt et al., 1981) and *P. ficus* (Hinkens et al., 2001), and successfully used in monitoring programs (Hinkens et al., 2001; Franco et al., 2004). Recently, the synthetic pheromone was used to locate populations of *P. minor* in south Florida (Stocks and Roda, 2011). The US state and national regulatory agencies required adult *P. minor* females to morphologically confirm the presence of this species in a new area, as there is no morphological way to identify male *Planococcus* species. Although not yet commercially available, the synthetic pheromone may provide a means to locate new infestations, as well as monitor changes in population levels.

16.5 Damage: Evaluation of Damage and Economic Thresholds

Planococcus spp. have piercing-sucking mouthparts which they insert into the plant vascular tissue, and which can remain in place through several molts, ingesting plant sap (Arnett, 1993). Feeding activity causes reduced yield, lower plant or fruit quality, stunted growth, discoloration and leaf loss (Venette and Davis, 2004). If left unchecked, *Planococcus* spp. often reach high densities, even killing perennial plants (Krishnamoorthy and Singh, 1987; Ben-Dov, 1994; Walton et al., 2006). Plant death may also be caused by viral diseases, because the mealybugs may also vector important viruses (Williams, 1985; Cox, 1989). In such cases, these mealybugs may be economic pests even at very low densities (Franco et al., 2009).

Up to 90% of the ingested plant sap may be excreted as honeydew (Mittler and Douglas, 2003). Sooty molds grow on the honeydew and can build up on the leaves, shoots, fruits and other plant parts (Mittler and Douglas, 2003). These molds can cover so much of the plant that they interfere with the plant's normal photosynthetic activity (Williams and Granara de Willink, 1992). Honeydew and sooty mold cause cosmetic defects to plants and/or their fruits, affecting the produce.

Franco et al. (2009) noted that most of the economically important mealybug species are associated with long lists of hosts, yet under low pressure of natural enemies they spread into new areas and are observed on relatively large numbers of host plants. With this potentially wide host-plant range, it is reasonable to anticipate that *P. minor* will find and utilize additional new hosts as it expands its distribution to new habitats (Venette and Davis, 2004). *P. minor* is reported to show distinct host preferences, commonly occurring on cocoa throughout its geographic range (Cox, 1989). In addition, plant host susceptibility to *P. minor* can vary widely, and infestation levels can fluctuate spatially, even on plants in close proximity (Venette and Davis, 2004).

Since multiple species from the genus *Planococcus* may occur on the same host plant, it is often difficult to estimate the impact of *P. minor* alone (Cox, 1989). Although widely distributed, this mealybug is not reported to be an economic pest in many countries. Some earlier host records in certain regions might be erroneous through misidentification of it as *P. citri* (Cox, 1989; Williams and Granara de Willink, 1992; Santa Cecilia et al., 2002). For example, *P. minor* as *P. citri* from Papua New Guinea where the mealybug comprised >90% of a mixed population with another pseudococcid and two different soft scales on coffee, and caused 70–75% reduction in crop yield (Szent-Ivany and Stevens, 1966). In India, this mealybug was reported as part of a *Planococcus* spp. complex or singly attacking custard apple (*Annona reticulata*) (Shukla and Tandon, 1984), grape (Batra et al., 1987; Tandon and Verghese, 1987), ber (*Ziziphus* sp.), guava, mango (Tandon and Verghese, 1987) and coffee (Reddy and Seetharama, 1997).

16.6 Control Tactics

16.6.1 Chemical

Chemical control is a common management strategy for mealybugs. Because of the generally cryptic habits and due to the protection of the mealy cover, effective chemical control relies on application of materials using high-vapor pressure, or timed when vulnerable stages such as crawlers are present (Franco et al., 2004). Major insecticides used against mealybugs include diazinon, dimethoate, azinfosmethyl, chlorpyrifos, parathion, pyrimifos-methyl and malathion, which are applied singly or in mixtures that include mineral oils (Franco et al., 2004; Buss and Turner, 2006; Daane et al., 2006). In India, *P. minor* has been shown to be resistant to several insecticides: organophosphates (Thirumurugan and Gautam, 2001), pyrethroids and organochlorines (Shukla and Tandon, 1984). Cultural practices such as pruning infested plant parts are used, to allow greater penetration of insecticides into the foliage (Franco et al., 2004). Soil drenches of systemic insecticides also work as they reach all parts of the plant, and control of mealybugs has improved with the introduction of many new systemic (Daane et al., 2006) neonicotinoids – acetamiprid, clothianidin, dinotefuran, imidacloprid, thiamethoxam – along with several insect growth regulators (IGR) (Buss and Turner, 2006).

16.6.2 Regulatory

A risk assessment by Venette and Davis (2004), developed under International Plant Protection Convention risk analysis standards, concluded that the economic consequences of *P. minor* introduction and establishment in the USA would be severe. Until April 2012 the mealybug was considered a high priority for exclusion by the US Department of Agriculture Animal and Plant Health Inspection Service (APHIS), Plant Protection and Quarantine (PPQ). In the USA *P. minor* was considered an 'actionable', quarantine-significant pest. If *P. minor* was found on imported products, the commodity was destroyed, re-exported or fumigated. Along with regulatory measures at ports of entry, the USA placed restrictions on the entry of plant products from countries known to have the pest. When exporting products from infested countries, the producer often is required to include pesticide or processing treatments that remove the mealybug from the commodity before exporting to the US. Irradiation treatments have been developed for *P. minor* as a potential phytosanitary measure that could be an alternative to current quarantine treatments (Ravuiwasa *et al.*, 2009). A dose of 150–250 Gy from a Cobalt 60 source decreased *P. minor* survival rate, percentage of adult reproduction, oviposition and fertility rate. The adult was the most tolerant life stage treated, and all treated life stages oviposited, but none of the F2 generation eggs hatched at the remanded dosage.

16.6.3 Biological

Mealybugs are amenable candidates for biological control, and this option has been deemed the best form of long-term control due to the reduction in costs associated with chemical control (Franco *et al.*, 2004; Buss and Turner, 2006). Very few natural enemies of *P. minor* were known (Ben-Dov *et al.*, 2011) until recent studies conducted in Trinidad (Francis, 2011). Despite the lack of historic knowledge of natural enemies of *P. minor*, several factors suggest that biological control plays an important role in regulating mealybug numbers. Ants have been observed feeding on the honeydew excretions of mealybugs (Kairo *et al.*, 2008). Although some ants may be predaceous, others are known to protect this important food source from predators. Mealybug populations closely associated with ants tend to be larger than non-tended populations of the same species (Lamb, 1974; Buckley and Gullan, 1991; Franco *et al.*, 2004). In studies of mealybugs – probably *P. minor* – infesting passion fruit, the destruction of natural enemies by pesticides increased mealybug numbers (Williams, 1991). As with other potential or secondary pests, problems with *P. minor* may be induced by pesticides. Table 16.1 lists known predators and parasitoids of *P. minor*.

Predators

As many as 47 mealybug predators are found in diverse insect orders and families such as Coleoptera (coccinellids), Diptera (cecidomyiids), Neuroptera (chrysopids and hemerobiids), Lepidoptera (lycanids) and Hemiptera (Moore, 1988). One of the most important predators of

Table 16.1. Reported natural enemies of *Planococcus minor*.

	Family	Species	Reference
Predators	Anthocoridae	*Calliodis* sp.	(Francis, 2011)
	Cecidomyiidae	*Diadiplosis coccidarum* Cockerell[3]	(Kairo *et al.*, 2008; Francis 2011; Stocks and Roda, 2011)
	Coccinellidae	*Brumoides suturalis* (Fabricius)	(Chandrababu *et al.*, 1997)
		Cryptolaemus affinis Crotch	(Szent-Ivany and Stevens, 1966)
		Cryptognatha nodiceps Marshall	(Francis, 2011)
		Tenuisvalvae bisquinquepustulata Fabricius	(Francis, 2011)
		Diomus sp.	(Francis, 2011)
		Diomus robert Gordon	(Francis, 2011)
	Syrphidae	*Ocyptamus stenogaster*	(Francis, 2011)
Parasitoids	Encyrtidae	*Leptomastix dactylopii* Howard	(Nagarkatti *et al.*, 1992; Kairo *et al.*, 2008; Francis, 2011)
		Aenasius advena Compere	(Bhuiya *et al.*, 2000)
		Coccidoxenoides perminutus Girault[1]	(Kairo *et al.*, 2008; Francis, 2011)
		Gahaniella tertia Kerrich[2]	(Kairo *et al.*, 2008; Francis, 2011)
		Coccidoctonus trinidadensis Crawford[2]	(Kairo *et al.*, 2008; Francis, 2011)
	Signiphoridae	*Signiphora* n. sp. (Woolley) mexicanus group	(Kairo *et al.*, 2008; Francis, 2011)

P. minor is *Cryptolaemus montrouzieri* Mulsant, a generalist feeder, which has been utilized extensively against many mealybugs and scale insects (Smith and Armitage, 1931; Reddy and Seetharama, 1997; Mani and Krishnamoorthy, 2008). *C. affinis* Crotch was also reported to be effective against *P. minor* in Papua New Guinea (Szent-Ivany and Stevens, 1966). *Brumoides suturalis* (Fabricius) has also been investigated in some detail as a potential control agent for a number of mealybug pests including *P. minor* (Chandrababu et al., 1997, 1999). In recent studies conducted in Trinidad, populations of *P. minor* were found to be very low and attacked by a complex of natural enemies including several Coccinellid species and the gall midge, *Diadiplosis coccidarum* (Cecidomyidae) (Kairo et al., 2008). Additionally, *D. coccidarum* was found attacking *P. minor* in South Florida (Stocks and Roda, 2011).

Fig. 16.3 *Coccidoxenoides perminutus* adult female parasitizing mealybugs.

Parasitoids

Important hymenopteran parasitoids of *Planococcus* spp. belong to the family Encyrtidae and include the solitary endoparasitoids *Leptomastix dactylopii* Howard, *Leptomastidea abnormis* (Girault), *Anagyrus pseudococci* (Girault) and *Coccidoxenoides perminutus* Girault (Bartlett, 1961; Berlinger, 1977; Noyes and Hayat, 1994) (Fig 16.3). Other reported genera that have been reared from *Planococcus* spp. include *Aenasuis*, *Gyranusoidea*, *Pseudaphycus* and *Pativana* (Ben-Dov et al., 2011). However, in biological control programs against *P. citri* in particular, two of the most widely used of these encyrtid wasps have been *L. dactylopii* and *C. perminutus* (Noyes and Hayat, 1994).

16.6.4 Ant control

Ant species often engage in facultative mutualisms with pest Hemiptera. Large outbreaks of sometimes seemingly inconspicuous hemipterans are correlated to the presence of attendant ants likely because they can disrupt the activity of natural enemies (Buckley and Gullan, 1991; Franco et al., 2004; Daane et al., 2007; Mgocheki and Addison, 2009). Therefore, biological control could be enhanced by disrupting the activity of ants. Chemical tactics available to manage ant populations include insecticide-treated baits, ground, trunk or foliar treatments or placing insecticide-treated bands around trunks (Franco et al., 2004). Blocking the ants' path to the mealybugs can also be achieved by placing sticky bands around the tree trunk. Flood irrigation and soil disturbance such as plowing under cover crops can also be used to disrupt ant populations.

16.6.5 Mating disruption, mass trapping, and lure and kill

The identification of the sex pheromone of *P. minor* combined with techniques to synthesize the active component to stereospecific purity has opened up new opportunities to improve monitoring techniques and control tactics (mass trapping, mating disruption, and lure and kill). The existence of facultative parthenogenesis *P. minor* would limit the use of pheromones for pest management. Studies would need to be conducted to verify if *P. minor* is an obligate amphimictic species, similar to what was found for *P. citri* (Borges da Silva et al., 2010). Additionally, little has been done on using mealybug pheromones as a management tactic (Franco et al., 2004; Daane et al., 2006; Walton et al., 2006). A 2-year study of mass trapping of *P. citri* males conducted in small citrus plots showed that mass trapping could significantly reduce the number of males; however, the male reduction obtained was not enough to significantly reduce fruit infestation. Therefore, the pheromone trapping system employed could not reduce the number of attracted males effectively, probably because many of the trapped males originated from outside the experimental plots. Therefore, more work is needed on the design of trapping systems before mass trapping can become a viable option for mealybug suppression (Howse et al., 1998). Mating disruption was found not to affect *P. ficus* populations in heavily infested vineyards, possibly due to the fact that at high mealybug densities, adult males would emerge in close proximity to females (Daane et al., 2006; Walton et al., 2006). However, Daane et al. (2006) consistently found higher parasitism rates of the exposed mealybugs in the mating disruption plots, suggesting the encyrtid parasitoid *Anagyrus* may cue in on the mealybug pheromone, and either remain in the vineyard aggressively searching for mealybug hosts, or be pulled in from nearby vineyards.

16.6.6 Cultural, physical, mechanical

Specific cultural management practices for *P. minor* have not been reported. However, common strategies to manage other mealybugs would likely impact *P. minor*. Proper sanitation practices are very important in managing the spread of mealybugs that can be transported on farm equipment, plant parts and clothing of workers (Buss and Turner, 2006). To reduce the spread of these mealybugs, farm equipment and harvesting supplies should be cleaned of all plant parts prior to movement to an uninfested area. Plants should be inspected for signs of mealybug infestation before purchase or installation. All infested material should be destroyed, and the area thoroughly cleaned (especially important in greenhouses and nurseries). When infestations are low, mealybugs could be removed by rubbing, or picking them from affected plants. Additionally, mealybugs can be removed mechanically by spraying a steady stream of water at reasonably high pressure on the host plant. Once on the ground, the mealybugs will be vulnerable to ground predators. In citrus, pruning is used also to open 'windows' in the tree crown in order to expose cryptic mealybug populations inside the tree crown to light, thus changing the microclimate and ensuring greater exposure to natural enemies (Franco et al., 2004). Mealybugs often thrive in warm, humid environments, so an increase in air flow or decrease in plant density in the area can make conditions less conducive. Soil fertility can play both a positive and a negative role in mealybug management. Scale insects often lay more eggs and survive better on plants receiving excess nitrogen, so avoiding over-fertilizing plants may help reduce the growth of mealybug populations. However, improved plant nutrition of cassava resulted in the production of larger cassava mealybugs, which in turn resulted in a higher proportion of female *Apoanagyrus lopezi* parasitic wasps with higher fertility levels (Schulthess et al., 1997). Improved fertilization of cassava also enhanced the antibiotic properties of cassava against mealybug infestations (Neuenschwander, 2003).

16.6.7 Quarantine methods

Using quarantines to contain a pest such as *P. minor* would be difficult because the insect has a very large host range and could easily escape detection. In the USA, common quarantine action includes prohibiting movement of all host material from the infested area, unless an effective control treatment is available. The treatment for mealybugs usually entails a chemical spray or drench. The plant material will also normally require a phytosanitary certificate issued by a regulatory agency, saying that the material was treated according to the requirements, and based on visual inspection, has been found to be free of pests.

16.6.8 Host plant resistance

Host resistance has not been reported for *P. minor*. However, plant host susceptibility to *P. minor* varies widely (Venette and Davis, 2004) and the mealybug has shown distinct preferences to certain species (Cox, 1989). Additionally, there are highly susceptible citrus varieties for the similar species *P. citri* (Franco et al., 2004). This suggests that there may be plant-resistant mechanisms available that could limit the impact of the pest.

16.7 Conclusions

P. minor has characteristics that indicate that the mealybug could become a serious economic pest. These include its wide host range, global distribution, potential for vectoring viruses and cryptic nature, which makes it possible for the pest to escape detection during inspections. Additionally, *P. minor*'s morphological similarity to other *Planococcus* species may allow the pest to escape detection during routine field surveys until the mealybug has become established. Fortunately, the recent developments in molecular markers and the identification and synthesis of the sex pheromone have provided tools to help with the timely and accurate detection of the pest, so that measures can be taken to mitigate economic damage. Once established, the vast host range of *P. minor* makes wide-scale chemical management unrealistic. However, the recent discovery of *Leptomastix dactylopii* and *Coccidoxenoides perminutus* attacking *P. minor*, as well as several predators, suggests that these natural enemies may suppress populations of the pest so that insecticide use maybe unnecessary in the landscape. Integrated pest management strategies developed for other pest *Planococcus* species will also help to reduce the impact of the pest in production systems, where management practices may disrupt the effectiveness of natural enemies. Note: Since the time of the orginal writing, *P. minor* was confirmed in the U.S. Populations were found not to have increased after 2 years of monitoring male numbers with pheromone traps and colonies with visual surveys. Natural enemies were also found attacking the pests. As a result, the U.S. down regulated the pest from "actionable" to "non-actionable" at ports of entry.

References

Arnett, R.H. (1993) Homoptera (Cicadas, leafhoppers, aphids, scale insects, and allies). In: *American Insects: A Handbook of the Insects of America North of Mexico*. Sandhill Crane Press, Gainesville, Florida, 232–243.

Bartlett, B.R. (1961) The influence of ants upon parasites, predators and scale insects. *Annals of the Entomological Society of America* 54, 543–551.

Batra, R.C., Brar, S.S., Khangura, J.S. and Dhillon, W.S. (1987) A new record of *Planococcus pacificus* Cox (Pseudoccidae: Hemiptera) as a pest of grapevine in India. *Punjab Horticultural Journal* 27, 250–251.

Ben-Dov, Y. (1994) *A Systematic Catalogue of the Mealybugs of the World (Insecta: Homoptera: Coccoidea: Pseudococcidae and Putoidae).* Intercept Ltd, Andover, UK.

Ben-Dov, Y., Miller, D.R. and Gibson, G.A.P. (2011) ScaleNet.

Berlinger, M.J. (1977) The Mediterranean vine mealybug and its natural enemies in southern Israel. *Phytoparasitica* 5, 3–14.

Bierl-Leonhardt, B.A., Moreno, D.S., Schwarz, M., Fargerlund, J. and Plimmer, J.R. (1981) Isolation, identification and synthesis of the sex pheromone of the citrus mealybug, *Planococcus citri* (Risso). *Tetrahedron Letters* 22, 389–392.

Biswas, J. and Ghosh, A.B. (2000) Biology of the mealybug, *Planococcus minor* (Maskell) on various host plants. *Environment & Ecology* 18, 929–932.

Borges da Silva, E., Mendel, Z. and Franco, J. (2010) Can facultative parthenogenesis occur in biparental mealybug species? *Phytoparasitica* 38, 19–21.

Buckley, R. and Gullan, P.J. (1991) More aggressive ant species (Hymenoptera: Formicidae) provide better protection for soft scales and mealybugs (Homoptera: Coccidae, Pseudococcidae). *Biotropica* 23, 282–286.

Buss, E.A. and Turner, J.C. (2006) *Scale Insects and Mealybugs on Ornamental Plants. Feature Creatures.* University of Florida, Florida Cooperative Extension Service, Institute of Food and Agricultural Sciences, Gainesville, Florida.

Chandrababu, A., Gautam, R.D., and Garg, A.K. (1997) Feeding potential and associated behaviour of predatory beetle, *Brumoides suturalis* (Fabricius). *Annals of Plant Protection Sciences* 5, 53–60.

Chandrababu, A., Gautam, R.D., and Garg, A.K. (1999) Biology of ladybird beetle, *Brumoides suturalis* (Fabricus) on aphids and mealbugs. *Annals of Plant Protection Sciences* 7, 13–18.

Cid, M., Pereiro, S., Cabaleiro, C. and Segura, A. (2010) Citrus mealybug (Hemiptera: Pseudococcidae) movement and population dynamics in an arbor-trained vineyard. *Journal of Economic Entomology* 103, 619–630.

Cox, J.M. (1981) Identification of *Planococcus citri* (Homoptera: Pseudococcidae) and the description of a new species. *Systematic Entomology* 6, 47–53.

Cox, J.M. (1983) An experimental study of morphological variation in mealbugs (Homoptera: Coccoidea: Pseudococcidae). *Systematic Entomology* 8, 361–382.

Cox, J.M. (1989) The mealybug genus *Planococcus* (Homoptera: Pseudococcidae). *Bulletin of the British Museum (Natural History)* 58, 1–78.

Cox, J.M. and Freeston, A.C. (1985) Identification of mealybugs of the genus *Planococcus*

(Homoptera: Pseudococcidae) occurring on cacao throughout the world. *Journal of Natural History* 19, 719–728.

Cox, J.M. and Wetton, M.N. (1988) Identification of the mealybug *Planococcus halli* Ezzat & McConnell (Hemiptera: Pseudococcidae) commonly occurring on yams (*Dioscorea* spp.) in Africa and the West Indies. *Bulletin of Entomological Research* 78, 561–571.

Daane, K.M., Bentley, W.J., V., W., Malakar-Kuenen, R., Millar, J., Ingels, C., Weber, E. and Gispert, C. (2006) New controls in vestigated for vine mealybug. *California Agriculture* 31–38.

Daane, K.M., Sime, K.R., Fallon, J. and Cooper, M.L. (2007) Impacts of Argentine ants on mealybugs and their natural enemies in California's coastal vine yards. *Ecological Entomology* 32, 583–596.

Francis, A. (2011) Investigation of bio-ecological factors influencing infestation by the passion-vine mealybug (*Planococcus minor*) (Maskell) (Hemiptera: Pseudococcidae) in Trinidad for application towards its management. PhD thesis, Florida Agriculture and Mechanical University/ University of Florida, Tallahassee, Florida.

Franco, J., Suma, P., da Silva, E., Blumberg, D. and Mendel, Z. (2004) Management strategies of mealybug pests of citrus in Mediterranean countries. *Phytoparasitica* 32, 507–522.

Franco, J.C., Zada, A. and Mendel, Z. (eds) (2009) *Novel Approaches for the Management of Mealybug Pests*. Springer Science, New York, New York.

Geiger, C.A. and Daane, K.M. (2001) Seasonal movement and distribution of the grape mealybug (Homoptera:Pseudococcidae):Developing a sampling program for San Joaquin valley vineyards. *Journal of Economic Entomology* 94, 291–301.

Gill, R. (2004) *Guide to the Identification of Common Adult Male Mealybugs*. Plant Pest Diagnostics Center, California Dept. of Food and Agriculture, Sacramento, California.

Gullan, P.J. and Kosztarab, M. (1997) Adaptations in scale insects. *Annual Review of Entomology* 42, 23–50.

Hinkens, D.M., McElfresh, J.S. and Millar, J.G. (2001) Identification and synthesis of the sex pheromone of the vine mealybug, *Planococcus ficus*. *Tetrahedron Letters* 42, 1619–1621.

Ho, H.Y., Hung, C.C., Chuang, T.H. and Wang, W.L. (2007) Identification and synthesis of the sex pheromone of the passion vine mealybug, *Planococcus minor* (Maskell). *Journal of Chemical Ecology* 33, 1986–1996.

Kairo, M.T.K., Francis, A. and Roda, A.L. (2008) Developing strategic research for biological control of new pest threats: the passion vine mealybug, *Planococcus minor* a case study. *Proceedings of the Caribbean Food Crop Society* 44, 118–123.

Krishnamoorthy, A. and Singh, S. (1987) Biological control of citrus mealybug, *Planococcus citri*, with an introduced parasite, *Leptomastix dactylopii* in India. *BioControl* 32, 143–148.

Lamb, K.P. (1974) *Hemiptera*. Academic Press, London.

Maity, D.K., Sahoo, A.K. and Mandal, S.K. (1998) Evaluation of laboratory hosts for rearing and mass multiplication of *Planococcus minor* (Maskell) (Pseudococcidae: Hemiptera). *Environment and Ecology* 16, 530–532.

Malausa, T., Fenis, A., Warot, S., Germain, J.F., Ris, N., Prado, E., Botton, M., et al. (2010) DNA markers to disentangle complexes of cryptic taxa in mealybugs (Hemiptera: Pseudococcidae). *Journal of Applied Entomology* 135, 142–155.

Mani, M. and Krishnamoorthy, A. (2008) Biological suppression of the mealybugs *Planococcus citri* (Risso), *Ferrisia virgata* (Cockerell) and *Nipaecoccus viridis* (Newstead) on pummelo with *Cryptolaemus montrouzieri* Mulsant in India. *Journal of Biological Control* 22, 169–172.

Martinez, M. and Suris, M. (1998) Biology of *Planococcus minor* Maskell (Homoptera: Pseudococcidae) under laboratory conditions / Biología de *Planococcus minor* Maskell (Homoptera: Pseudococcidae) en condiciones de laboratorio. *Revista de Proteccion Vegetal* 13, 199–201.

Meyerdirk, D.E.R., Warkentin, R., Atta vian, B., Gersabeck, E., Francis, A. and Francis, G. (1998) *Biological Control of Pink Hibiscus Mealybug Project Manual*. US Department of Agriculture, Animal Plant Health Inspection Service, Plant Protection and Quarantine, Riverdale, Maryland.

Mgocheki, N. and Addison, P. (2009) Interference of ants (Hymenoptera: Formicidae) with biological control of the vine mealybug *Planococcus ficus* (Signoret) (Hemiptera: Pseudococcidae). *Biological Control* 49, 180–185.

Millar, J.G. (2008) Stereospecific synthesis of the sex pheromone of the passion vine mealybug, *Planococcus minor*. *Tetrahedron Letters* 49, 315–317.

Millar, J.G., Daane, K.M., McElfresh, J.S., Moreira, J.A., Malakar-Kuenen, R., Guillen, M. and Bentley, W.J. (2002) Development and optimization of methods for using sex pheromone for monitoring the mealybug *Planococcus ficus*

(Homoptera: Pseudococcidae) in California vineyards. *Journal of Economic Entomology* 95, 706–714.

Miller, D.R. and Kosztarab, M. (1979) Recent advances in the study of scale insects. *Annual Review of Entomology* 24, 1–27.

Mittler, T.E. and Douglas, A.E. (2003) Honeydew. In: Resh, V.H. and Cardé, R.T. (eds), *Encyclopedia of Insects*. Academic, Amsterdam, The Netherlands.

Moore, D. (1988) Agents used for biological control of mealybugs (Pseudococcidae). *Biocontrol News and Information* 9, 209–225.

Myers, L.E. (1932) Two economic greenhouse mealybugs of Mississippi. *Journal of Economic Entomology* 25, 891–896.

Neuenschwander, P. (2003) Biological control of cassava and mango mealybugs. In: Neuenschwander, P., Borgemeister, C. and J., L. (eds) *Biological Control in IPM Systems in Africa*. CAB International, Wallingford, UK, pp. 45–59.

Noyes, J.S. and Hayat, M. (1994) *Oriental Mealybug Parasitoids of the Anagyrini (Hymenoptera: Encyrtidae)*. CAB International, Wallingford, UK.

Panis, A. (1969) Observations faunistiques et biologiques sur quelques Pseudococcidae (Homoptera, Coccoidea) vivant dans le midi de la France. *Annales de Zoologie, Écologie Animale* 1, 211–244.

Ravuiwasa, K.T., Lu, K.-H., Shen, T.-C. and Hwang, S.-Y. (2009) Effects of irradiation on *Planococcus minor* (Hemiptera: Pseudococcidae). *Journal of Economic Entomology* 102, 1774–1780.

Reddy, K.B. and Seetharama, H.G. (1997) Integrated management of mealybugs in coffee. *Indian Coffee* 61, 26–28.

Rung, A., Scheffer, S.J., Evans, G. and Miller, D. (2008) Molecular identification of two closely related species of mealybugs of the genus *Planococcus* (Homoptera: Pseudococcidae). *Annals of the Entomological Society of America* 101, 525–532.

Rung, A., Miller, D.R. and Scheffer, S.J. (2009) Polymerase chain reaction-restriction fragment length polymorphism method to distinguish three mealybug groups within the *Planococcus citri – P. minor* species complex (Hemiptera: Coccoidea: Pseudococcidae). *Journal of Economic Entomology* 102, 8–12.

Sahoo, A.K. and Ghosh, A.B. (2001) Descriptions of all instars of the mealybug *Planococcus minor* (Maskell) (Homoptera, Pseudococcidae). *Environment and Ecology* 19, 436–445.

Sahoo, A.K., Ghosh, A.B., Mandal, S.K. and Maiti, D.K. (1999) Study on the biology of the mealybug, *Planococcus minor* (Maskell) Pseudococcidae: Hemiptera. *Journal of Interacademicia* 3, 41–48.

Santa Cecilia, L.V.C., Reis, P.R. and Souza, J.C. (2002) About the nomenclature of coffee mealybug species in Minas Gerais and Espirito Santo States, Brazil. *Neotropical Entomology* 31, 333–334.

Schulthess, F., Neuenschwander, P. and Gounou, S. (1997) Multi-trophic interactions in cassava, *Manihot esculenta*, cropping systems in the subhumid tropics of West Africa. *Agriculture, Ecosystems & Environment* 66, 211–222.

Shukla, R.P. and Tandon, P.L. (1984) India-insect pests on custard apple. *Plant Protection Bulletin, FAO* 32, 31–31.

Smith, H.S. and Armitage, H.M. (1931) The biological control of mealybugs attacking citrus. *California University Agricultural Experiment Station Bulletin* 509, 74.

Stocks, I. and Roda, A.L. (2011) The Passionvine mealybug, *Planococcus minor* (Maskell), a New Exotic Mealybug in South Florida (Hemiptera: Pseudococcidae). *PestAlert*. Division of Plant Industry, Florida Department of Agriculture and Consumer Services.

Szent-Ivany, J.J.H. and Stevens, R.M. (1966) Insects associated with *Coffea arabica* and some other crops in the Wau-Bulolo area of New Guinea. *Papua and New Guinea Agricultural Journal* 18, 101–119.

Tandon, P.L. and Verghese, A. (1987) New insect pests of certain fruit crops. *Indian Journal of Horticulture* 44, 121–122.

Thirumurugan, A. and Gautam, R.D. (2001) Relative toxicity of some insecticides to mealybug, *Planococcus pacificus* (Pseudococcidae, Hemiptera). *Annals of Plant Protection Sciences* 9, 135–136.

Tsai, C.W., Chau, J., Fernandez, L., Bosco, D., Daane, K.M. and Almeida, R.P.P. (2008) Transmission of Grapevine leafroll-associated virus 3 by the Vine Mealybug (*Planococcus ficus*). *Phytopathology* 98, 1093–1098.

USDA (2010) *Port Information Network (PIN-309): Quarantine Status Database*. US Department of Agriculture, Animal Plant Health Inspection Service, Plant Protection and Quarantine, Riverdale, Maryland.

Venette, R.C. and Davis, E.E. (2004) *Mini Risk Assessment, Passionvine Mealybug: Planococcus minor (Maskell) (Pseudococcidae: Hemiptera)*. National Cooperative Agricultural Pest Survey Target Pests Pest Risk Assessment. US Department of Agriculture, Animal Plant Health Inspection Service, Plant Protection and Quarantine, Riverdale, Maryland.

Walton, V.M., Daane, K.M., Bentley, W.J., Millar, J.G., Larsen, T.E. and Malakar-Kuenen, R. (2006) Pheromone-based mating disruption of *Planococcus ficus* (Hemiptera: Pseudococcidae) in California vineyards. *Journal of Economic Entomology* 99, 1280–1290.

Watson, G.W. and Chandler, L.R. (2000) *Identification of Mealybugs Important in the Caribbean Region*. Commonwealth Science Council and CAB International, Egham, UK.

Williams, D.J. (1982) The distribution of the mealybug genus *Planococcus* (Hemiptera: Pseudococcidae) in Melanesia, Polynesia and Kiribati. *Bulletin of Entomological Research* 72, 441–455.

Williams, D.J. (1985) *Australian mealybugs*. British Museum (Natural History), London.

Williams, D.J. (1991) Superfamily Coccoidea. In: Naumann, I.D., Came, P.B., Lawrence, J.F., Nielsen, E.S., Spradbery, J.P., Taylor, R.W., Whitten, M.J. *et al.* (eds), *The Insects of Australia. A textbook for Students and Research Workers*. Melbourne University Press, Melbourne, pp. 457–464.

Williams, D.J. (2004) *Mealybugs of Southern Asia*. Natural History Museum, London.

Williams, D.J. and Granara de Willink, M.C. (1992) *Mealybugs of Central and South America*. CAB International, Wallingford, UK.

Williams, D.J. and Watson, G.W. (1988) *The Scale Insects of the Tropical South Pacific Region, Part 2: The Mealybugs (Pseudococcidae)*. CAB International, Wallingford, UK.

17 The Citrus Orthezia, *Praelongorthezia praelonga* (Douglas) (Hemiptera: Ortheziidae), a Potential Invasive Species

Takumasa Kondo,[1] Ana Lucia Peronti,[2] Ferenc Kozár[3] and Éva Szita[3]
[1]*Corporación Colombiana de Investigación Agropecuaria, Corpoica, Colombia;*
[2]*Departamento de Ecologia e Biologia Evolutiva, Universidade Federal de São Carlos (UFSCar), São Carlos/SP, Brazil;* [3]*Plant Protection Institute, Hungarian Academy of Sciences, Budapest, Hungary*

The citrus orthezia, *Praelongorthezia praelonga* (Douglas) (Hemiptera: Ortheziidae), is a highly polyphagous scale insect that causes plant damage both directly by its feeding and indirectly due to its associated sooty molds. This Neotropical species currently is largely confined to Central and South America and the Caribbean Region, but has the potential to be invasive if accidentally introduced into other climatically suitable parts of the world. The citrus orthezia was recently introduced into the Afro-tropical region where it has become a pest. This chapter provides a brief summary of the vast literature on the citrus orthezia, which is often difficult to access, including its taxonomy, biology, host records, economic importance, world distribution, integrated pest management (including chemical, mechanical, cultural, physical and biological control strategies) and quarantine methods. The scale insect can have multiple generations per year and has a lengthy life cycle lasting between 40 and 200 days.

17.1 Introduction

The scale insects are sap-sucking hemipteran insects that include all members of the superfamily Coccoidea. These are closely related to aphids (Aphidoidea), whiteflies (Aleyrodoidea) and jumping plant lice (Psylloidea), which make up the suborder Sternorrhyncha (Gullan and Martin, 2009). Currently, there are up to 32 extant scale insect families recognized, depending on the taxonomic authority (Gullan and Cook, 2007). Scale insects are generally divided into two informal groups, the archaeococcoids and the neococcoids. Except for one family (the Putoidae), the archaeococcoids generally can be recognized by the presence of 2–8 pairs of abdominal spiracles, which are absent in the neococcoids. The Ortheziidae are commonly known as ensign scales and belong to the archaeococcoid group.

Until the 1990s, the Ortheziidae included about 80 species in six genera, with most species described in the genus *Orthezia* Bosc d'Antic (Williams and Watson, 1990). The family was revised by Kozár (2004) and this work includes ten new genera, 198 species (including 94 new species and seven fossil species), four subfamilies, nine tribes and 20 valid genera (including three fossils). Recently, three new species in leaf litter have been described from Japan, namely *Newsteadia yanbaruensis* Tanaka and Amano (2005), *Ortheziola mizushimai* Tanaka and Amano (2007), and

Ortheziolamameti maeharai Tanaka and Amano (2007). Currently there are 202 (194 extant and eight fossil) species of ensign scales (family Ortheziidae) described (Kozár, 2004; Ben-Dov et al., 2011). The extant taxa are distributed in 18 genera as follows: *Acropygorthezia* LaPolla and Miller (1 sp.), *Arctorthezia* Cockerell (4 spp.), *Graminorthezia* Kozár (11 spp.), *Insignorthezia* Kozár (10 spp.), *Jermycoccus* Kozár and Konczné Benedicty (1 sp.), *Matileortheziola* Kozár and Foldi (1 sp.), *Mixorthezia* Morrison (18 spp.), *Neomixorthezia* Kozár (3 spp.), *Neonipponorthezia* Kozár (2 spp.), *Newsteadia* Green (57 spp.), *Nipponorthezia* Kuwana (6 spp.), *Nipponorthezinella* Kozár (2 spp.), *Orthezia* Bosc (23 spp.), *Orthezinella* Silvestri (1 sp.), *Ortheziola* Šulc (9 spp.), *Ortheziolacoccus* Kozár (17 spp.), *Ortheziolamameti* Kozár (5 spp.) and *Praelongorthezia* Kozár (23 spp.). Known fossils include: *Arctorthezia antiqua* Koteja and Zak-Ogaza, 1988a; *Creorthezia hammanaica* Koteja and Azar, 2008; *Jersicoccus kurthi* Koteja, 2000; *Newsteadia succini* Koteja and Zak-Ogaza, 1988b; *Ochyrocoris electrina* Menge, 1856; *Orthezia* sp. (Menge, 1856), *Palaeonewsteadia huaniae* Koteja, 1987a; and *Protorthezia aurea* Koteja, 1987b.

Two ortheziid species, namely the citrus orthezia *P. praelonga* (Douglas) and the lantana bug *Insignorthezia insignis* (Browne) are highly polyphagous, and are considered as pests wherever they occur. *I. insignis* is already distributed in all zoogeographic regions of the world, whereas *P. praelonga* is of Neotropical origin, and its distribution was restricted to the Caribbean Region and Central and South America until recently. Although not reported in the scientific literature, *P. praelonga* was introduced to the tropical areas of West Africa in the early 2000s, namely to the Republic of the Congo (Brazzaville) and the Democratic Republic of the Congo (A. Kiyindou, Brazzaville, Congo, pers. comm. 19 January 2011).

According to Koteja (2004), most species of ensign scales live in soil litter in humid habitats and feed on fungi, moss and roots of different plants. The true hosts or host preferences of most species are unknown. Some specialization can be found in the tribe Ortheziini; some species feed on grasses, others on woody plants, although rarely on conifers, and many others feed on herbaceous plants, mostly in the Nearctic and Neotropical regions. Unlike most of the species in the Ortheziidae, the citrus orthezia is highly polyphagous and can cause severe infestations associated with sooty molds and dieback of branches of its hosts. Most biological information on this insect has been published in Spanish and Portuguese in journals and reports that might not be readily available outside South America. Due to its wide host range, *P. praelonga* may be considered an insect pest with a potential for becoming an invasive species outside its area of current distribution. This possibility could be accelerated by climate change, which can support the establishment, spread and outbreaks of scale insect pests in new areas (Kozár, 1997; Miller et al., 2002, 2005; Miller and Miller, 2003; Wang et al., 2010).

17.2 Economic Importance

The citrus orthezia, *P. praelonga*, is known mainly due to the damage it causes to citrus trees in several countries of the Neotropical region. According to Matile-Ferrero and Étienne (2006), it is consistently harmful to *Malpighia* and *Bougainvillea* in Guadeloupe. Very large populations were observed also on *Spathodea*, causing mortality of trees, *Codiaeum* and *Plumeria*. Initial attacks in Guadeloupe were first noted on *Citrus* in 2004, and in 2006 outbreaks were observed causing the death of trees. No effective natural enemy could be detected except for a species of Drosophilidae (Diptera), tentatively identified as *Rhinoleucophenga brasiliensis*, which is present sporadically. The drosophilid develops on the surface of the ensign scale, and it is itself parasitized by *Aprostocetus* sp. (Hymenoptera: Eulophidae) (Matile-Ferrero and Étienne, 2006).

The citrus orthezia has been recorded as a pest on fruit trees since 1973 in Colombia, in the state of Antioquia (ICA, 1973). In 1975 it was recorded in Palmira, in the state of Valle del Cauca, attacking citrus trees and causing severe infestations associated with sooty molds and dieback of branches (ICA, 1975). García-Roa (1995) reported an increasing number of hosts of the citrus orthezia, including *Citrus* spp., figs and coffee, but it preferred ornamental plants and weeds. In Colombia, *P. praelonga* is a common pest of citrus (Fig. 17.1) and is a cause of sooty mold which grows on the honeydew and of dieback of branches. In severe cases the tree is killed. In the state of Valle del Cauca, ornamental croton (*Codiaeum variegatum*) (Fig. 17.2A) and *Bougainvillea* spp. (Fig. 17.2B) are the most important reservoirs of

Fig. 17.1 Three species of citrus heavily infested with *P. praelonga* showing symptoms of sooty mold. Photos taken in Colombia, by T. Kondo.

Fig. 17.2 *P. praelonga*. (A). On *Codiaeum variegatum*. (B). On *Bougainvillea* sp., causing deformation of young leaves. (C). On *Schefflera* sp. (D). On mistletoe (Loranthaceae). (E). On *Anthurium* sp. (F). On *Begonia* sp. Photos by T. Kondo.

this insect, and they can become a source for permanent infestations. The wide range of plant hosts, high reproductive capacity and poor biological control of *P. praelonga* have contributed to this insect's reputation as a severe pest.

When *P. praelonga* attains high population densities, it will readily deplete its hosts of sap, weakening the plants, and causing dieback of branches, eventually killing its host. Severe infestations of *P. praelonga* on citrus plants are often associated with the lack of natural enemies, and this is usually a direct consequence of pesticide use. The practice of chemical control for *P. praelonga*, besides being inefficient, often destroys the natural balance in the ecosystem, causing the appearance of new insect pests and the resurgence of others (García-Roa, 1995). In the Republic of the Congo and the Democratic Republic of the Congo, *P. praelonga* has been reported to cause the complete defoliation of trees (A. Kiyindou, Brazzaville, Congo, pers. comm. 19 January 2011).

In Brazil, *P. praelonga* is widely distributed: Acre (Lima, 1981); Amapá (Kogan, 1964; Silva and Jordão, 2005); Bahia (Carvalho, 2006; Silva et al., 1968; Teixeira et al., 2001); Ceará (Vieira et al., 1976); Espírito Santo (Cassino and Menezes, 1984; Culik et al., 2007; Lima, 1981); Goiás (Garcia, 1999; Lima, 1981); Maranhão (Lima, 1981); Minas Gerais (Cassino et al., 1991); Pará (Cockerell, 1900; Autuori and Fonseca, 1932; Costa Lima, 1936; Lepage, 1938; Teixeira et al. 2001; Magalhães and Silva, 2008); Paraíba (Lopes et al., 2007); Paraná

(Albuquerque et al., 2002; Vernalha, 1970); Pernambuco (Barbosa et al., 2007; Carvalho and Carvalho, 1941; Kogan, 1964; Lago, 1981; Pyenson, 1938; Teixeira et al., 2001); Piauí (Lima, 1981); Rio Grande do Sul (Silva et al., 1968; Teixeira et al. 2001); Rio de Janeiro (Robbs, 1947, 1951; Berry, 1957; Giacometti, 1962; Gonçalves, 1962; Kogan, 1964; Lima and Cassino, 1974; Lima, 1981; Cassino et al., 1991); São Paulo (Hempel, 1900; Autuori and Fonseca, 1932; Lepage, 1938; Gonçalves and Cassino, 1978; Prates, 1989; Cassino et al., 1991; Peronti et al., 2001); Sergipe (Robbs, 1974; Silva et al., 1979).

In Brazil, this species is one of the most important insects associated with *Citrus* spp., causing serious losses (Cassino et al., 1991). Benvenga et al. (2001) reported that *P. praelonga* became a key pest of citrus due to the difficulty in controlling it. The insects prefer to feed on the underside of leaves except in very high populations, when they occur on both leaf surfaces, and on twigs, flowers and trunks. Sooty molds generally caused by a fungus, *Capnodium* sp., develop on the honeydew produced by the insects, causing a reduction in photosynthesis. There is a reduction in fruit yield, and dieback of highly infested plants (García-Roa, 1995).

17.3 Host Range

Although most species of ensign scales can be considered either monophagous or oligophagous, *P. praelonga* is highly polyphagous and has been recorded on more than 182 host plant species in 50 plant families (Douglas, 1891; Hempel, 1900; Bourne, 1923; Lizer y Trelles, 1942; Robbs, 1951; Morrison, 1925, 1952; Kogan, 1964; Lima and Cassino, 1974; Gonçalves and Gonçalves, 1976; Cruz and Oliveira, 1979; Cabrita et al., 1980; Medeiros et al., 1980; Silva and Gravena, 1980; Lago, 1981; Lima, 1981; Cassino and Menezes, 1984; Laurence, 1991; Cassino et al., 1991, 1993; Cassino et al., 1991, 1993; Albuquerque et al., 2002; Garcia, 2004; Carvalho, 2006; Culik et al., 2007; Kondo, 2009; Ben-Dov et al., 2011; and in this present study).

17.3.1 Host plants

The source(s) of each record is/are given in parentheses after each plant name. Acanthaceae: *Graptophyllum* sp., *Hemigraphis colourata* (Lima and Cassino, 1974), *Pseuderanthemum atropurpureum* (Lago, 1981), *Sanchezia* sp. (Carvalho, 2006; Hempel, 1900; Morrison, 1925), *S. nobilis* (Douglas, 1891), *Thunbergia* sp. (Morrison, 1952), *T. speciosa* (Lima, 1981). Amaranthaceae: *Achyranthes* sp. (Morrison, 1952), *Alternanthera dentata rubiginosa* (Lago, 1981), *Amaranthus* sp. (Cabrita et al., 1980). Anacardiaceae: *Anacardium occidentale* (Lima and Cassino, 1974; Lago, 1981), *Mangifera indica* (Lago, 1981; Kondo, 2009), *Mangifera* sp. (Morrison, 1952). Annonaceae: *Annona muricata* (Lago, 1981). Apiaceae: *Petroselinum crispum* (Matile-Ferrero and Étienne, 2006), *Pimpinella anisum* (Lima and Cassino, 1974). Apocynaceae: *Cryptostegia madagascariensis* (Cassino et al., 1991), *Plumeria alba* (Lima and Cassino, 1974; Matile-Ferrero and Étienne, 2006), *P. rubra* (Matile-Ferrero and Étienne, 2006; Lima and Cassino, 1974). Araceae: *Anthurium andreanum* (Medeiros et al., 1980), *Anthurium* sp. (Medeiros et al., 1980; the present study) (Fig. 17.2E), *Philodendron* sp. (Lima and Cassino, 1974). Araliaceae: *Hedera helix* (Lima and Cassino, 1974), *Schefflera arboricola* (Cassino and Menezes, 1984), *Schefflera* sp. (Culik et al., 2007; the present study) (Fig. 17.2C). Arecaceae: *Cocos nucifera* (Lima and Cassino, 1974). Asparagaceae: *Dracaena* sp. (Lima, 1981). Asteraceae: *Ageratum conyzoides* (Cassino et al., 1993), *Baccharis* sp. (Morrison, 1952), *Bidens pilosa* (Medeiros et al., 1980; Barbosa et al., 2007), *Chrysanthemum morifolium*, *Coreopsis grandiflora* (Cassino et al., 1991), *Conyza* sp. (Barbosa et al., 2007); *Dahlia* sp. (Lima, 1981), *Eupatorium* sp. (Lima and Cassino, 1974), *Sphagneticola trilobata* (Carvalho, 2006; Lago, 1981), *Sphagneticola* sp. (Lago, 1981), *Vernonia cinerea* (Cassino et al., 1993), *V. squamulosa* (Lizer y Trelles, 1942), *Vernonia* sp. (Morrison, 1952); *Wedelia paludosa* (Garcia, 2004); *Wedelia* sp. (Lago, 1981). Bignoniaceae: *Spathodea campanulata* (Lima and Cassino, 1974; Garcia, 1999; Matile-Ferrero and Étienne, 2006), *Tabebuia* sp. (Culik et al., 2007), *Tecoma speciosa* (Lima, 1981). Begoniaceae: *Begonia* sp. (the present study) (Fig.17.2F). Boraginaceae: *Cordia corymbosa* (Cassino et al., 1993). Bromeliaceae: *Ananas comosus* var. *comosus* (Culik et al., 2009), *Ananas sativus* (Cruz and Oliveira, 1979), *Tillandsia aeranthos* (Cassino et al., 1993). Cactaceae: *Cactus* sp. (Cabrita et al., 1980). Caprifoliaceae: *Lonicera* sp. (Morrison, 1925). Caricaceae: *Carica papaya* (Robbs, 1951), *Carica* sp. (Morrison, 1925).

Combretaceae: *Terminalia catappa* (Medeiros *et al.*, 1980). Commelinaceae: *Commelina benghalensis* (Barbosa *et al.*, 2007). Convolvulaceae: *Merremia dissecta, Ipomoea carnea* (Matile-Ferrero and Étienne, 2006), *I. fistulosa* (Cassino *et al.*, 1981). Cucurbitaceae: *Cucurbita pepo* (Lima and Cassino, 1974; Teixeira *et al.*, 2004), *C. moschata* (the present study), *C. charantia* (Cassino *et al.*, 1993). Euphorbiaceae: *Acalypha wilkesiana* (Cassino *et al.*, 1993), *Acalypha* sp. (Lago, 1981; Morrison, 1952), *Codiaeum variegatum* (Lago, 1981; Matile-Ferrero and Étienne, 2006; the present study) (Fig. 17. 2A), *C.* sp. (Morrison, 1925), *Croton* sp. (Hempel, 1900; Kogan, 1964), *Euphorbia tirucalli* (Lima and Cassino, 1974), *Euphorbia* sp. (Morrison, 1925), *Jatropha integerrima* (Matile-Ferrero and Étienne, 2006), *Manihot esculenta* (Lago, 1981), *M. utilissima* (Lima and Cassino, 1974), *Sapium* sp. (Morrison, 1925). Fabaceae: *Bauhinia alba, B. variegata* (Garcia, 1999), *B.* sp. (Medeiros *et al.*, 1980), *B. monandra* (Lima, 1981; Matile-Ferrero and Étienne, 2006), *Caesalpinia peltophoroides* (Lima and Cassino, 1974), *Cajanus indicus* (Lima, 1981), *Cajanus* sp. (Morrison, 1952), *Cassia* sp. (Kogan, 1964), *Centrosema virginianum* (Cassino, 1993), *Gliricidia sepium* (Matile-Ferrero and Étienne, 2006), *Gliricidia* sp. (Laurence, 1991), *Haematoxylum campechianum* (Matile-Ferrero and Étienne, 2006), *Haematoxylum* sp. (Morrison, 1925), *Indigofera hirsuta* (Cassino, 1993), *Macroptilium* sp. (Medeiros *et al.*, 1980), *Mimosa pudica* (Barbosa *et al.*, 2007), *Pterocarpus violaceus* (Lima, 1981). Gesneriaceae: *Besleria* sp. (Morrison, 1952). Goodeniaceae: *Scaevola plumieri* (Matile-Ferrero and Étienne, 2006). Lamiaceae: *Aegiphila pernambucensis* (Kogan, 1964), *Coleus blumei* (Lago, 1981), *Coleus* sp. (Robbs, 1947), *Hyptis* sp. (Hempel, 1900; Morrison, 1925), *Leonotis nepetifolia* (Lima and Cassino, 1974), *Mentha piperita* (Lago, 1981). Lythraceae: *Lawsonia inermis* (Lago, 1981). Loranthaceae: *Loranthus* sp. (Morrison, 1925; the present study) (Fig. 172D). Malpighiaceae: *Byrsonima sericea* (Lago, 1981), *Malpighia emarginata* (Matile-Ferrero and Étienne, 2006), *M. glabra* (Bourne 1923; Albuquerque *et al.*, 2002), *Malpighia* sp. (Barbosa *et al.*, 2007; Morrison, 1925). Malvaceae: *Dombeya acutangula* (Lima, 1981), *Gossypium* sp. (Morrison, 1925), *Hibiscus rosa-sinensis* (Lima and Cassino, 1974), *H. syriacus* (Lima, 1981), *H. tiliaceus* (Lago, 1981; Lima and Cassino, 1974), *Malvastrum coromandelianum, Malvastrum* sp. (Lima and Cassino, 1974), *Malvaviscus* sp. (Morrison, 1952), *Sida urens* (Cassino, 1993), *S. rhombifolia* (Cassino, 1993), *Sida* sp. (Lima and Cassino, 1974), *Theobroma cacao* (Kirkpatrick, 1957), *Triumfetta semitriloba* (Cassino, 1993). Moraceae: *Artocarpus altilis* (Lago, 1981), *A. heterophyllus* (Lago, 1981), *Ficus canonii* (Medeiros *et al.*, 1980). Myrtaceae: *Eugenia jambos* (Lima and Cassino, 1974), *E. uniflora* (Lago, 1981), *Psidium guajava* (Cassino *et al.*, 1981), *P. araca* (Lago, 1981), *P.* sp. (Kogan, 1964). Nyctaginaceae: *Bougainvillea spectabilis* (Carvalho and Carvalho, 1941; Kogan, 1964), *Bougainvillea* sp. (Matile-Ferrero and Étienne, 2006; Morrison, 1952; the present study) (Fig. 172B), *Mirabilis jalapa* (Medeiros *et al.*, 1980), *Pisonia* sp. (Morrison, 1952; Garcia, 2004). Passifloraceae: *Passiflora edulis, P. quadrangularis* (Lago, 1981). Phyllanthaceae: *Breynia nivosa* (Lago, 1981), *Phyllanthus distichus* (Lima and Cassino, 1974), *P. corcovadensis* (Medeiros *et al.*, 1980), *Phyllanthus* sp. (Morrison, 1952). Piperaceae: *Piper marginatum, P. nigrum, P.* sp. (Lago, 1981). Plumbaginaceae: *Plumbago coerulea* (Lago, 1981). Poaceae: *Brachiaria purpurascens* (Medeiros *et al.*, 1980), *Digitaria insularis* (Barbosa *et al.*, 2007), *Panicum plantagineum* (Cabrita *et al.*, 1980), *P.* sp. (Medeiros *et al.*, 1980), *Saccharum* sp. (Morrison, 1925). Polypodiaceae: *Polypodium vacciniifolium* (Cassino *et al.*, 1993). Polygonaceae: *Coccoloba uvifera* (Lima, 1981), *Coccoloba* sp. (Morrison, 1925, 1952), *Triplaris felipensis* (Lima, 1981), *T. surinamensis, Triplaris* sp. (Medeiros *et al.*, 1980). Portulacaceae: *Portulaca* sp. (Cabrita *et al.*, 1980). Rosaceae: *Eriobotrya japonica* (Medeiros *et al.*, 1980), *Rosa* sp. (Morrison, 1925). Rubiaceae: *Coffea arabica* (Lago, 1981; Lima and Cassino, 1974), *C. canephora* (Culik *et al.*, 2007), *Coffea* sp. (Morrison, 1925), *Gardenia florida* (Lago, 1981), *G. jasminoides* (Cassino *et al.*, 1981), *Ixora coccinea* (Cassino *et al.*, 1981), *I.* sp. (Matile-Ferrero and Étienne, 2006), *Paederia* sp. (Morrison, 1925), *Pentas* sp. (Morrison, 1952). Rutaceae: *Citrus aurantium* (Lago, 1981), *C. latifolia* (Silva and Gravena, 1980), *C. limonia* (Robbs, 1947), *C. limetta* (Cockerell, 1900), *C. paradisi* (Matile-Ferrero and Étienne, 2006), *C. reticulata* (Lima and Cassino, 1974), *C. sinensis* (Robbs, 1947; Silva and Jordão, 2005), *C. sinensis* varieties: folha-murcha, Natal (Oliveira *et al.*, 1979), Pera, Ponkan, Valencia (Cabrita *et al.*, 1980), *Citrus* sp. (Matile-Ferrero and Étienne, 2006; Morrison, 1925), *Fortunella* sp. (Morrison, 1952). Sapindaceae: *Talisia esculenta* (Garcia, 2004). Sapotaceae: *Achras sapota*

(Garcia, 2004). Solanaceae: *Brunfelsia* sp. (Medeiros et al., 1980), *Capsicum frutescens* (Lago, 1981), *Capsicum* sp. (Douglas, 1891; Hempel, 1900), *Solanum asperum* (Lago, 1981), *S. balbisii* (Kogan, 1964), *S. tuberosum* (Kogan, 1964). Verbenaceae: *Durante repens* (the present study). Violaceae: *Viola* sp. (Gonçalves and Gonçalves, 1976).

17.4 Geographical Origin of *P. praelonga* (Douglas)

The citrus orthezia occurs in the Caribbean and in Central and South America (Fig. 17.3) and, although of Neotropical origin, its exact native range is unknown.

17.4.1 World distribution

Nearctic: Mexico. Neotropical: Antigua and Barbuda; Argentina; Barbados; Bolivia; Brazil; British Virgin Islands; Colombia; Dominica; Ecuador; Grenada; Guadeloupe; Guyana; Jamaica; Marie-Galante; Marinique; Panama; Peru; Puerto Rico and Vieques Island; Saint Croix; St. Barthélemy; St. Martin; Trinidad and Tobago; US Virgin Islands; Venezuela (Ben-Dov et al. 2011; Matile-Ferrero and Étienne, 2006). Afro-tropical region: Republic of the Congo and the Democratic Republic of the Congo.

17.5 Classification and Taxonomy

Kingdom: Animalia
Class: Insecta
Order: Hemiptera
Family: Ortheziidae
Genus: *Praelongorthezia*
Species: *Praelongorthezia praelonga* (Douglas)

Common names: English: citrus orthezia (Ebeling, 1959); croton bug (Bourne, 1923); horned lamellated scale (Cockerell, 1896). Spanish: piojo blanco harinoso de los cítricos (García Roa, 1995); ortezia de los cítricos (Kondo, 2009). Portuguese: piolho branco, ortézia (Cesnik and Ferraz, 2003), piolho branco dos citros) (Rodrigues, 2004).

According to Williams and Watson (1990), members of the Ortheziidae are defined by distinctive waxy symmetrical plates on the dorsum and margin. The adult female usually secretes a long ovisac from a spine band on the venter. The ovisac remains attached to the body when the insect is mobile. Slide-mounted specimens can be diagnosed by having a broadly oval body and

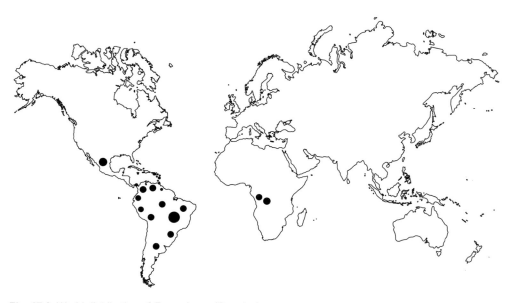

Fig. 17.3 World distribution of *P. praelonga* (Douglas).

antennae with 3–8 segments, often with a hair-like or thick apical spine on the last antennal segment. The legs are well developed and each is usually long and slender, with the trochanter and femur fused, and tibia and tarsus sometimes fused. The anal ring is located at the apex of the abdomen and possesses 6 setae and a conspicuous pore band. Abdominal spiracles are present, totalling 4–8 pairs. Spines are grouped into definite bands and clusters especially on the dorsum. An ovisac band is present on the venter of the abdomen, is usually wide, and continuous from just posterior to the hind coxae and around the margin; this band of pores secretes the characteristic ovisac. Setae are slender, with basal collars. Pores are normally quadrilocular, but sometimes other types are also present. Adult males are usually large, with dusky wings, and a pair of large compound eyes and well-developed legs and antennae.

17.6 External Morphology

According to Szita *et al.* (2010), the shape, number and arrangement of the wax plates of live (or dead) ensign scales have been useful for genus and in some cases for species identification. Digital photos of the venter and dorsum of ensign scales can be used for the quick separation of species.

The dorsal wax plates consist of a medial wax plate, submedial wax plate and marginal wax plates, and the ventral wax plates consists of midcoxal and postcoxal wax plates (Fig. 17.4). The arrangement of the wax plates of three common species of the tribe Ortheziini are shown in Fig. 17.5. It should be noted that the shape and structure of the wax plates in live insects differ significantly from the spine groups seen on microscopic slides in the same species (e.g., Figs 17.5C and 17.5F versus Fig. 17.7).

17.6.1 Practical key for separation of genera in the tribe Ortheziini (Adapted from Szita et al., 2010)

1 – Generally found on Poaceae   *Graminorthezia*
          (not illustrated)
  – Generally found on plants other than Poaceae   2
2 – Wax plates around coxae poorly developed, midcoxal wax plate absent, with only a few small wax protrusions on mid thorax
          Insignorthezia (Fig. 17.5 A, D)
  – Wax plates on venter well developed   3
3 – Submedial wax plates on dorsum well developed   *Orthezia* (Fig. 17.5 B, E)
  – Submedial wax plates on dorsum poorly developed or absent, when present composed of a loose mealy wax
          Praelongorthezia (Fig. 17.5 C, F)

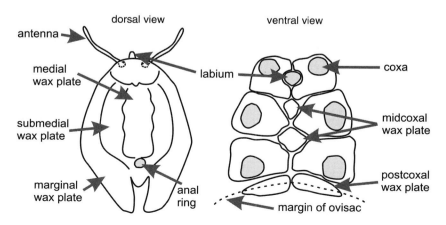

Fig. 17.4 Dorsal and ventral external morphology of an ensign scale showing position of wax plates. Illustration by É. Szita.

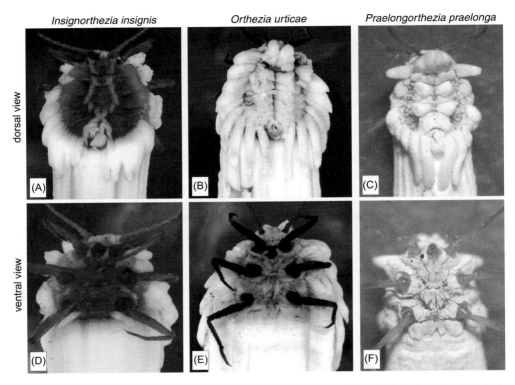

Fig. 17.5 Dorsal and ventral view of wax plates of some ensign scales. (A) and (D). *Insignorthezia insignis*. (B) and (E). *Orthezia urticae*. (C) and (F). *P. praelonga*. Photos (A), (B), (D) and (E) by É. Szita; (C) and (F) by T. Kondo.

17.7 Genus *Praelongorthezia* Kozár, 2004

Type species: *Orthezia praelonga* Douglas, 1891

The genus *Praelongorthezia* was described by Kozár (2004) and includes species with the following combination of features: dorsum of the adult female only partly covered by wax plates, with bare narrow bands. Slide-mounted specimens with 7- or 8-segmented antennae, with variable number, shape, size and type of setae; with flagellate sensory setae sometimes present on segments VII and VIII. Eye stalks protruding, thumb-like, not fused with bases of antennae. Legs well developed, leg setae robust, spine-like or hair-like; trochanter and femur fused, tibia and tarsus not fused; claw digitules mostly spine-like, claw with a denticle. Labium 1-segmented, with many setae. Stylet loop short, usually as long as labium. Atrium of thoracic spiracles without pores. With bands or rows of wax plates within ovisac band. Head of dorsum with sclerotized cephalic plates of different forms. Anal ring present in fold of derm on dorsal surface, with six setae. Discoidal pores with one or four loculi (openings), scattered over surface. Abdominal spiracles located mostly on dorsum, numbering seven or eight pairs. *Praelongorthezia* belongs to the tribe Ortheziini, subfamily Ortheziinae, a group that also includes two other genera, namely *Insignorthezia* Kozár and *Graminorthezia* Kozár (Kozár, 2004).

The genus *Praelongorthezia* includes 24 species (Ben-Dov et al., 2011) which are distributed in the Nearctic and Neotropical regions. *P. acapulcoa* is widely distributed in Central America, and only *P. praelonga* is more widely distributed in different parts of Central and South America (Kozár, 2004). Recently, *P. praelonga* has become a pest of citrus and ornamental plants in the Republic of the Congo and the Democratic Republic of the Congo, this being the first report of *P. praelonga* from the

African continent. Species belonging to this genus show variable host plant preferences. Several live on Poaceae, or Asteraceae. Some other species are polyphagous.

17.7.1 *Praelongorthezia praelonga* (Douglas)

Orthezia praelonga Douglas, 1891: 246–247.
Praelongorthezia praelonga (Douglas): Kozár, 2004: 420 (change of combination).

Description

Eggs oval in shape, smooth, initially whitish, turning greenish in colour when near eclosion. Infertile eggs brownish. 1st- to 3rd-instar nymphs similar in shape. 1st-instar nymph (crawler) when leaving ovisac, already with a whitish wax that covers its body; wax increasing in volume after feeding begins. The basic difference from one instar to the next is the increase in size and the molting of exuviae. In the 4th instar, the males have a light bluish colour with the body longer than that of female. The hind wings and the long, slender legs begin to develop in the 4th-instar (pupal) males. The 4th-instar female is the adult stage that produces the ovisac (Figs 17.6 and 17.7). The adult males have a pair of wings and an elongated white tail, composed of waxy filaments (Fig. 17.8B, inset). The female has three nymphal instars and the male has four (Cesnik and Ferraz, 2003; Garcia-Roa, 1995).

UNMOUNTED ADULT FEMALE (FIG. 17.6). Insects nearly 2 mm long, and 1.25 mm wide, ovisac sometimes as much as 6 mm (Kozar, 2004). Body completely covered with very fragile white secretion dorsally, showing a more or less distinct, but at most narrow, bare streak near each margin, separating the dorsal and marginal plates, with the secretion arranged in the usual lateral and dorsal tufts (Morrison, 1925).

The adult female has an ovisac that can reach up to five times its body length. The dorsum has two rows of waxy secretions that run parallel from the head to the apex of the abdomen, giving the dorsum a ribbed appearance; on the apex of the head there are two curved waxy tufts that touch at their tips; on each side of the pair of anterior tufts there is a longer pointed tuft

Fig. 17.6 Adult female, nymphs and exuviae of *P. praelonga* (Douglas). Photo by T. Kondo.

protruding sideways and slightly anteriorly; just posterior to these there are about six shorter tufts or waxy plates that point outwards and slightly in a posterior direction; on both sides of the abdominal apex there are two longer plates, each composed of two waxy tufts that point posteriorly, and between these two tails there is a shorter median pointed tuft. The long posterior plates are found above the ovisac. The area between the median rows of waxy plates and the marginal plates is bare and has a dark yellowish-green colour (T. Kondo, Palmirea, Colombia, pers. obs., 6 September 2010) (Figs 17.4 and 17.6). The waxy plates of the venter are well developed, but fragmented (Fig. 17.5F).

MOUNTED ADULT FEMALE (FIG. 17.7). Description adapted from Kozár (2004). Insect body 1.3–1.6 mm long, 1.0–1.3 mm wide, oval. Length of antennal segments (in µm): I 102; II 87, III 107, IV 92, V 102, VI 102, VII 97, VIII 209 long; apical segment weakly club shaped; apical seta 19 µm long, sensory seta on segment VIII 22 µm long. Antennal segment VII with one sensory seta. Sensory pore present on each of antennal segments II and III. Microseta near apex of antenna absent; all segments of antennae covered with

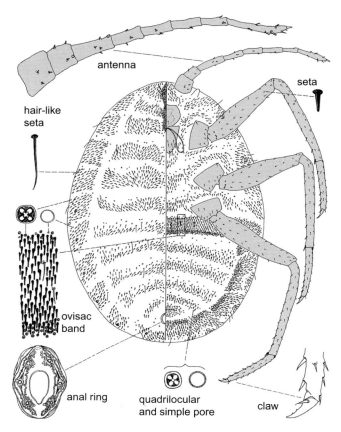

Fig. 17.7 *P. praelonga* (Douglas), adult female (Adapted from drawing by Zsuzsanna Konczné-Benedicty, from Kozár, 2004).

small number of spine-like, straight setae, longest seta 7 µm long. Eyestalk elongate conical.

VENTER. Labium 202 µm long, with apical setae quite long, robust. Stylet loop as long as labium. Legs (length in µm): front coxa 115, middle 120, hind 134; front trochanter + femur 459, middle 464, hind 536; front tibia 469, middle 515, hind 546; front tarsus 286, middle 286, hind 288; base of tarsus with a sensory pore; hind claw 62 µm long, with 2 denticles; claw digitules spine-like, 15 µm long; legs with rows of setae: longest on trochanter + femur 7–12 µm long; trochanter with four sensory pores on each surface. Well-developed wax plate bands present in marginal areas of head and thorax, with small groups surrounding each thoracic spiracle and with wax plates in front of each coxa; with four loose bands of spines within ovisac band. Atrium of thoracic spiracles surrounded by wax spines without quadrilocular pores associated with spiracular opening; diameter of anterior thoracic spiracles 38 µm. Setae few, scattered on margin and within ovisac band, several more associated with posterior quadrilocular pores surrounding vulva. Ovisac band well developed, with one row of quadrilocular and simple pores on anterior and with a band of quadrilocular pores on posterior margin of ovisac band. Quadrilocular pores 4 µm in diameter with a central loculus; predominant near anterior edges of spine bands. Simple pores, 3–5 µm in diameter, present with quadrilocular pores.

DORSUM. Membranous, except for a narrow, chitinized median stripe, running backward from anterior margin. Short wax plate bands covering mid-dorsum, divided medially. Spines at margin of abdominal wax plates 18 µm long and apically

Fig. 17.8 (A). Tree trunk with male pupae of *P. praelonga*. (B). Dry leaf on soil surface with male puparia. Inset showing adult male. Photos taken in Colombia, by T. Kondo.

capitate, up to 38 μm long. Hair-like setae present in marginal wax plates, also present in very small numbers on other wax plates. Pores with four loculi; present in marginal areas of abdomen, between wax plate bands, some present near anal ring. Several quadrilocular pores forming groups near marginal wax plates, associated with simple pores of two sizes. Anal ring 96 μm wide, and 130 μm long; with 1–2 inner and outer rows of round pores and spinulae present on each pore; longest anal ring seta 148 μm long. Abdominal spiracles numbering seven pairs on each side of body, each spiracle with sclerotized vestibule.

Biology

The nymphs and adult females of *P. praelonga* move about the entire host plant in search of a suitable feeding site. The 1st two male instars are similar to those of the female; however, after completing the 2nd instar, the males stop feeding and move to the tree trunk (Figs 17.8A and 17.8B), where they remain segregated until they develop into adults (inset, Fig. 17.8B).

According to a study conducted on citrus in São Paulo, Brazil, populations of *P. praelonga* are highest in colder and drier months of the year, when there is low precipitation and low relative humidity (Prates and Pinto, 1987). The optimal temperature for the development of *P. praelonga* is c. 25°C, with the maximum temperature limit at 38°C, and the minimum limit at 15°C. The insects have a lifespan of 40–200 days. In the field, males are common flying at dusk, and copulating pairs are found at this time on infested plants. Newly hatched nymphs form aggregations and feed around their adults (Lima, 1981).

The duration of the life cycle of *P. praelonga* is affected by the hosts they feed on and by temperature. Lima (1981) studied the biology of *P. praelonga* under laboratory conditions (temperature 27.5 ± 1°C and relative humidity 73.8 ± 0.4 and a photoperiod of 14 h). The life cycle was completed in 81.5 ± 7.3 days on *Solanum tuberosum* and 60.9 ± 2.5 days on *Alternanthera dentata* f. *rubiginosa*. The type of diet influenced the oviposition period and mean number of laid eggs, which were significantly higher in adults reared on sprouted potato tubers. The longevity of females was directly correlated with the length of the oviposition period and the number of laid eggs. Males, after emerging as adults, showed an average lifespan of 5.0 ± 0.7 days. Copulation took place within the first 7.2 ± 0.5 h after emergence. The number of matings per male was 8 ± 1 to 24 ± 2 copulations when one or more females were available. The duration of each copulation ranged from 1.4 ± 0.7 to 11.2 min, with an average of 6.4 ± 2.5 min. The sex ratio was 0.67, indicating an approximate ratio of one male to two females.

Neves et al. (2010) studied the temperature requirements and voltinism of *P. praelonga* reared on *Citrus limonia* in Brazil, and determined that the number of generations of the species per year varied from 5.0 to 6.6.

Restrepo et al. (1991) studied the life cycle of *P. praelonga* under field conditions in Colombia. The eggs hatched about 7 days after they are laid, and the three female nymphal stages, i.e., 1st-instar nymph or crawler, 2nd- and 3rd-instar nymph, lasted for 31, 35 and 64 days, respectively. The male prepupal stage lasted for an average of 32 days and the male pupal stage for 4 or 5 days. The winged adult males lived for up to 8 days and were quite active fliers. The adult females lived for up to 90 days. Each adult female laid 85–106 eggs. The 1st-instar nymphs, which aggregated around their mothers, started dispersing onto the foliage soon after molting to the 2nd instar.

NOTE ON MALE FLIGHT ACTIVITY. Lima (1981) reported that males of *P. praelonga* fly at dusk in Brazil, but in Colombia, they have been observed flying in the morning (T. Kondo, Palmira, Colombia, pers. obs., 13 December 2008).

Ant (Hymenoptera: Formicidae) associations

There are few reports of associations of *P. praelonga* with ants. Robbs (1951) reported a symbiotic relationship of *Solenopsis saevissima* (Smith) with the citrus orthezia. More recently Zinger et al. (2006) observed two species of *Camponotus*, *Cephalotes* sp., *Solenopsis* sp. and *Brachymyrmex* sp. tending *P. praelonga*, but they determined by the frequency of the ants visiting the scale insects that these were facultative associations and that the species *Camponotus* sp. was the most frequent visitor.

Detection methods

Very few species of scale insects can be identified from their external morphology. Most species need careful study of slide-mounted specimens under a high-powered microscope. Thus, well-prepared specimens are essential for identification. In life, the citrus orthezia generally is easy to recognize due to its characteristic waxy secretions compared to many other scale insects. Male prepupa and pupae of *P. praelonga* are generally found forming groups at the bases of the tree trunks, in bark crevices and sheltered areas of the host.

Control methods

CHEMICAL. Carvalho (2006) discussed the use of chemical control as follows. In Brazil, there are no registered chemical products that can be used to control *P. praelonga* on most fruits; however, in future, when products become available, preference should be given to products that are less harmful to humans and the environment. During drought periods, in the months of July through December, in the State of Bahia, there is an increase in populations of *P. praelonga*, making it necessary to control focal points of infestation. The use of chemical products should be targeted at focal points in the orchard, and applications to the entire orchard should be avoided in order to preserve the natural enemies that keep *P. praelonga* under an ecological balance.

BIOLOGICAL CONTROL. In Colombia, Velásquez et al. (1992) studied beneficial agents for control of the citrus orthezia and found that predators (see below), although usually present in low numbers, can help reduce populations. In Brazil and Colombia, bacteria and fungi have been evaluated as control agents, with promising results. It is necessary to increase and diversify the beneficial fauna of the citrus orthezia in order to replace the use of insecticides. The current practice of chemical control is not recommended because it is ineffective, risky, harmful and costly.

The Colombian Agriculture Institute (ICA) and the Corporación Colombiana de Investigación Agropecuaria (Corpoica) conducted research on pest management strategies to control this important citrus pest. In its native range, at least in Colombia, the citrus orthezia appears to be kept under control by various natural enemies, including an unidentified species of Hyperaspidinae (Coleoptera: Coccinellidae) (Fig. 17.9B); *Ambracius dufouri* and *Proba vittiscutis* (Hemiptera: Miridae); *Chrysopa* sp. (Neuroptera: Chrysopidae); and a fly species identified as Drosophilidae, of which the larvae feed on the ortheziid eggs (Velásquez et al., 1992). The fungus *Colletotrichum* sp. also has been found causing natural death of the insect (Garcia-Roa, 1995).

Fig. 17.9 Predators of *P. praelonga*. (A). *Ambracius* sp. (Hemiptera: Miridae). (B). Hyperaspidinae (Coleoptera: Coccinellidae). (C). *Harmonia axiridis* (Coleoptera: Coccinellidae). Photos by T. Kondo.

However, the common use of chemical control often results in the breakdown of the ecological balance between *P. praelonga* and its natural enemies, resulting in outbreaks of this pest. For the control of the citrus orthezia, an integrated management strategy that combines the use of insecticidal soaps, oils, cultural and biological control methods is recommended (Garcia-Roa, 1995).

For the State of Rio de Janeiro, Brazil, Gonçalves (1962) provided a list of natural enemies, including the wasp *Cales noacki* (Hymenoptera: Aphelinidae), and the flies *Gitona brasiliensis* and *G. fluminensis* (Diptera: Drosophilidae). In another study, Gonçalves (1963) noted that *Pentilia egena* and *Scymnus* sp. (Coleoptera: Coccinellidae) prey upon ortheziid eggs inside their ovisac, and also listed several pathogenic fungi: *Verticillium lecanii*, *Cladosporium* sp. and *Fusarium* sp. Known natural enemies of *P. praelonga* are listed in Table 17.1.

PHYSICAL CONTROL. Alternative methods for the control of the citrus orthezia on citrus and other host plants include the use of high-pressure water sprays and the use of soaps and oils that affect the spiracles and tracheal system (Garcia-Roa, 1995). Garcia *et al.* (1992) reported that the combined action of water, soaps, oils, predators and entomopathogenic fungi and cultural practices can help to substantially lower high citrus orthezia populations on citrus.

CULTURAL CONTROL. Well-managed groves where procedures include good weed control, fertilization, irrigation and pruning, are a good strategy for keeping healthy trees, which is essential for successful integrated pest management (IPM). Periodical inspection is important in detecting early infestations and focal points of the citrus orthezia (Garcia-Roa, 1995).

Among the IPM strategies, cultural control can reduce and in some cases eradicate the citrus orthezia if the infestation is in its early stages (Carvalho, 2006). Recommended cultural control practices for *P. praelonga* include: (i) removal of weeds around infested trees; (ii) pruning and destruction of infested twigs and dead plant material; (iii) fertilizing plants with minerals and organic fertilizers to increase plant resistance; (iv) elimination of plants and weeds that will serve as reservoirs of the pest in or around the citrus areas; and (v) harvesting infested fruit only after healthy fruit has been harvested, to avoid further dispersal of the pest.

MECHANICAL CONTROL. When small sources of infestations of the citrus orthezia are found, these can be destroyed manually. In order to achieve a good control with this method, frequent and thorough inspections must be carried out in the plantation, by carefully inspecting twigs and leaves for incipient infestations of the pest and then destroying them. By conducting this manual inspection periodically, it is possible to prevent the dispersion of the pest in the orchard.

QUARANTINE

Detection and identification. As for other scale insects, the citrus orthezia may be difficult to detect at quarantine inspections because of its relatively small size (especially the nymphs) and sessile nature. *P. praelonga* may be confused with other species of ensign scales, including *P. molinarii* (Morrison) and the polyphagous lantana bug, *Insignorthezia insignis*. Taxonomic keys and illustrations for Ortheziidae are provided by Kozár (2004) and Morrison (1925, 1952).

Table 17.1. Natural enemies of *P. praelonga* (Douglas).

INSECTA

ORDER COLEOPTERA
    Family Coccinellidae (11 spp.)
    Azya luteipes (Lago, 1981; Cassino *et al.*, 1991; Fernandes *et al.*, 2005)
    Cladis sp. (Wolcott, 1960)
    Harmonia axyridis (the present study) (Fig. 17.9C)
    Harmonia sp. (Garcia-Roa, 1995)
    Hyperaspis anexa (Bobadilla *et al.*, 1999, in Garcia, 2004)
    H. notata (Cassino *et al.*, 1991)
    Hyperaspis sp. (Velásquez *et al.*, 1992; Garcia-Roa, 1995)
    Hyperaspidinae (the present study) (Fig. 17.9B)
    Olla abdominalis (Lima, 1981)
    Pentilia egena (Gonçalves, 1963)
    Scymnus sp. (Gonçalves, 1963; Fernandes *et al.*, 2005)
    S. limbativentris (Cassino *et al.*, 1991)
ORDER DIPTERA
    Family Anthomyzidae (1 sp.)
    Species not determined (Fulmek, 1943)
    Family Chamaemyiidae (1 sp.)
    Melaleucopis sp. (Bobadilla *et al.*, 1999, in Garcia, 2004)
    Family Drosophilidae (4 spp.)
    Gitona brasiliensis (Gonçalves, 1962; Lago, 1981; Lopes and Maia, 2008)
    G. fluminensis (Gonçalves, 1962)
    Species not determined (Velásquez *et al.*, 1992)
    Rhinoleucophenga brasiliensis? (Matile-Ferrero and Étienne, 2006)
    Familiy Syrphidae (2 spp.)
    Ocyptamus sp. (Cassino *et al.*, 1991)
    Salpingogaster conopida (Cassino *et al.*, 1991)
ORDER HEMIPTERA
    Family Anthocoridae (1 sp.)
    Species not determined (Garcia-Roa, 1995)
    Familiy Miridae (2 spp.)
    Ambracius dufouri (Cassino *et al.*, 1991; Velásquez *et al.*, 1992)
    Ambracius sp. (the present study) (Fig. 17.9A)
    Familiy Reduviidae (2 spp.)
    Heza insignis (Cassino *et al.*, 1991; Lago, 1981)
    Species not determined (Garcia-Roa, 1995)
ORDER HYMENOPTERA
    Familiy Aphelinidae (1 sp.)
    Cales noacki (Gonçalves, 1962)
    Family not determined (1 sp.)
    Parasitoid wasp (Garcia-Roa, 1995)
ORDER NEUROPTERA
    Familiy Chrysopidae (4 spp.)
    Chrysopodes sp. (Benvenga *et al.*, 2004)
    Ceraeochrysa cubana (Benvenga *et al.*, 2004)
    Chrysopa sp. (Cassino *et al.*, 1991; Velásquez *et al.*, 1992)
    Species not determined (Garcia-Roa, 1995; Fernandes *et al.*, 2005)

MOLLUSCA: GASTROPODA (1 sp.)
    Oxystyla pulchella (Cruz *et al.*, 1999, in Garcia, 2004)

FUNGI (13 spp.)
    Aschersonia sp. (Cassino *et al.*, 1991)
    Beauveria sp. (Cassino *et al.*, 1991)

Continued

Table 17.1. Continued.

FUNGI

Beauveria bassiana (Viegas *et al.*, 1995)
Beauveria bassiana (Viegas *et al.*, 1995)
Cladosporium sp. (Gonçalves, 1963; Prates and Novo, 1979; Pinto and Prates, 1980; Lago, 1981)
Cladosporium herbarum var. *aphidicola* (Prates, 1980; Cassino *et al.*, 1991)
Colletotrichum gloeosporioides (Cesnik and Bettiol, 1998; Cesnik and Ferraz, 2003; Teixeira *et al.*, 2001; Viegas *et al.*, 1995)
Colletotrichum sp. (Garcia-Roa, 1995)
Fusarium sp. (Benvenga *et al.*, 2004; Gonçalves, 1963)
Lecanicillium longisporium (Alves *et al.*, 2004; Molina, 2007)
Metarhizium anisopliae (Viegas *et al.*, 1995)
Syngliocladium sp. (Alves *et al.*, 2004; Molina, 2007)
Verticilium lecanii (Gonçalves, 1963; Garcia, 2004)

Discussion

P. praelonga is highly polyphagous, but it appears to prefer certain crops, such as citrus (various species and cultivars) and some common ornamental and medicinal plants. The plant families with the highest number species that are hosts to *P. paraelonga* are: Fabaceae (17 spp.), Malvaceae (13 spp.), Euphorbiaceae (11 spp.), Rubiaceae (9 spp.), Rutaceae (9 spp.), Acanthaceae (7 spp.), Lamiaceae (6 spp.), Solanaceae (6 spp.), Myrtaceae (5 spp.) and Polygonaceae (5 spp.).

Despite its potential for becoming an invasive species in the tropical and subtropical climates of the USA and other countries, *P. praelonga* has not become established in these areas to date. In contrast to many invasive scale insect species, *P. praelonga* has a lengthy life cycle and is a relatively large insect (body about 1.5 mm long, with an ovisac up to 6 mm long), making it easier to detect at quarantine inspections. Also, it reproduces sexually, which means that single nymphs or unmated females are unable to start new infestations. In Europe several scale insect species, such as *Pseudaulacaspis pentagona* (Targioni Tozzetti) have become established in new localities, and their distribution is altering because of climate change (Kozár, 1997). In the USA, a high number of invasive species of Pseudococcidae and Coccidae were listed by Miller *et al.* (2002) and Miller and Miller (2003). The only introduced ortheziid pest listed for the USA is *I. insignis* (Miller *et al.*, 2005). The recent detection of *P. praelonga* in Africa shows that this insect has the potential to be introduced into and become established into new areas.

Acknowledgements

The work of Drs F. Kozár and Éva Szita was supported by OTKA Grant No 75889. Many thanks to Dr Antoine Kiyindou for information on the distribution of *P. praelonga* in Africa. Thanks to Mr Guillermo Gonzalez for the identification of Coccinellidae. Special thanks to Drs Penny Gullan and Douglas Williams for reviewing an earlier draft of this chapter, and for useful comments.

References

Albuquerque, F.A., Pattaro, F.C., Borges, L.M., Lima, R.S. and Zabini, A.V . (2002) Insetos associados à cultura da aceroleira (*Malpighia glabra* L.) na região de Mar ingá, Estado do Paraná. *Acta Scientiarum* 24, 1245–1249.

Alves, S.B., Humber, R., Lopes , R.B., Tersi, F. and Padulla, F.L. (2004) Ocorrência da doença salmão-da-ortézia causada por um hifomiceto entomopatogênico. Proceedings XX Congresso Brasileiro de Entomologia, Gramado, Brazil, pp. 279.

Autuori, M. and Fonseca, J. (1932) Lista dos principais insetos que atacam plantas cítr icas no Brasil. *Revista de Entomologia* 2, 202–216.

Barbosa, F.R., Gonzaga Neto, L., Lima, G.K. and Carvalho, R.S. (2007) Manejo e Controle da Cochonilha Ortézia (*Orthezia praelonga*), em Plantios Irrigados de Acerola, no Submédio São Francisco. Petrolina: Embrapa Semi-Árido, Circular Técnica, Petrolina PE, Br azil, 83, pp. 1–8.

Ben-Dov, Y., Miller, D.R. and Gibson, G.A.P. (2011) ScaleNet: a database of the scale insects of the world. Available from: www.sel.barc.usda. gov/scalenet/scalenet.htm, accessed 6 August 2012.

Benvenga, S.R., Araújo Júnior, N. de and Gravena, S. (2001) Cochonilha or tézia. *Informativo do Manejo Ecológico de Pragas, Jaboticabal* 25, p. 280.

Benvenga, S.R., Gravena, S., Silva, J.L., Araujo Jr, N. and Amorim, L.C.S. (2004) Manejo prático da Cochonilha Ortézia em pomares de citros . *Laranja, Cordeirópolis*, 25, 291–312.

Berry, P.A. (1957) Relatór io das obser vações sobre o controle biológico da *Orthezia praelonga* no Br asil. Relatório datilografado depositado nos arquivos da Seção de Entomologia e P arasitologia do Instituto de Ecologia Experimentação Agrícola do Rio de Janeiro, Rio de Janeiro, Brazil, pp. 7.

Bourne, B.A. (1923) Repor t on the Depar tment of Agriculture, Barbados, 1922–1923. Report (Barbados Department of Agriculture), Christ Church, Barbados, pp. 25.

Cabrita, J.R.M., Pinto, W.B. de S ., Prates, H.S. and Novo, P.S. (1980) Constatação da cochonilha *Orthezia praelonga* (Douglas, 1981) (Hom., Or theziidae) em pomares cítricos dos municípios de Severínea, Monte Azul Paulista, Araraquara, Bebedoura e Pitangueiras, Estado de São P aulo. Proceedings of the Br azilian Congress of Entomology, São Paulo, Brazil, pp. 68.

Carvalho, M.B. and Carvalho, R.F. (1941) Segunda contribuição para um catálogo dos insetos de Pernambuco. *Arquivos do Instituto de Pesquisas Agronômica* 3, 13–24.

Carvalho, R.S. (2006) Controle integ rado da Or tézia em pomares e hortos comerciais. Circular Tecnica 82, Embrapa, Cruz das Almas, Brazil, pp. 6.

Cassino, P.C.R. and Menez es, E.B. (1984) Notas bionômicas sobre *Orthezia praelonga* Douglas, 1891 (Hom., Ortheziidae), no Estado do Espírito Santo. Proceedings of the 9th Brazilian Congress of Entomology, Londrina, Brazil, pp. 86.

Cassino, P.C.R., Lima, A.F ., Peixoto, A.L. and Medeiros, L.S.A. (1981) Hospedeiros de *Orthezia praelonga* Douglas, 1891 (Hom., Ortheziidae) no Estado do Rio de J aneiro. *Anais do Congresso Brasileiro de Fruticultura* 6, 1363–1368.

Cassino, P.C.R., Lima, A.F.D. and Racca Filho , F. (1991) *Orthezia praelonga* Douglas, 1891 em plantas citricas no Br asil (Homoptera, Ortheziidae). *Arquivos da Universidade Federal Rural (Rio de Janeiro)* 14, 35–57.

Cassino, P.C.R., Perruso, J.D. and Nascimento , F.N.D. (1993) [Contr ibution to kno wledge of bioecology interactions between white flies (Homoptera, Aleyrodidae) and *Orthezia praelonga* Douglas, 1891 (Homoptera, Ortheziidae) in the citr us environment.] (In P ortuguese; Summary In English). *Anais da Sociedade Entomologica do Brasil* 22, 209–212.

Cesnik, R. and Bettiol, W. (1998) P otencial fitopatogênico de *Colletotrichum gloeosporioides*, agente de controle biológico de *Orthezia praelonga. Laranja, Cordeirópolis* 19, 261–268.

Cesnik, R. and F erraz, G.J.M. (2003) Biologia e controle biológico de *Orthezia praelonga. Manejo Integrado de Plagas y Agroecología (Costa Rica)*, 70, 90–96.

Cockerell, T.D.A. (1896) Coccidae or scale insects . VIII. *Bulletin of the Botanical Department, Jamaica* 3, 8–10.

Cockerell, T.D.A. (1900) Notas sobre coccidas Brazileiras. *Revista do Museu Paulista, São Paulo* 4, 363–364.

Costa Lima, A. (1936) *Terceiro Catálogo dos Insetos que Vivem nas Plantas do Brasil.* Directoria de Estatistica da Producção, Rio de Janeiro, Brazil.

Cruz, C. de A. da and Oliv eira, A.M. de (1979) Occurrence of *Orthezia praelonga* on pineapple. *Comunicado Técnicos, Empresa Pesquisa Agropecuário, Estado do Rio de Janeiro* 10, 1–2.

Cruz, J.D., Marques, O.M. and Nascimento , A.S. (1999) Consumo de *Orthezia praelonga* Douglas, 1891 (Insecta: Ortheziidae) por *Oxystila pulchella* Spix, 1827 (Gastropoda: Bulimullidae) em laboratorio. *Sitientibus, Feira de Santana* 20, 81–88.

Culik, M.P., Martins, D.S., Ventura, J.A., Peronti, A.L.B.G., Gullan, P.J. and K ondo, T. (2007) Coccidae , Pseudococcidae, Ortheziidae, and Monophlebidae (Hemiptera: Coccoidea) of Espérito Santo. *Biota Neotropica* 7, 61–65.

Culik, M.P., Ventura, J.A. and dos S . Martins, D. (2009) Scale insects (Hemiptera: Coccidae) of Pineapple in the state of Espírito Santo, Brazil. *Acta Horticulturae* 822, 215–218.

Douglas, J.W. (1891) Notes on some Br itish and exotic Coccidae (No . 21). *Entomologist's Monthly Magazine* 27, 244–247.

Ebeling, W. (1959) *Subtropical Fruit Pests.* Division of Agricultural Sciences, University of California, Los Angeles, California.

Fernandes, D.P.F., Rodrigues, W.C., Cassino, P.C.R., Zinger, K. and Spolidoro, M.V. (2005) Flutuação populacional de *Orthezia praelonga* (Sternorryncha, Ortheziidae) e seus inimigos naturais em tanger ina cv. Poncã em cultiv o orgânico em Seropédica, RJ. CD-ROM. Anais da XV Jor nada de Iniciação Científica da

Universidade Federal Rural do Rio de Janeiro, Rio de Janeiro, Brazil.
Fulmek, L. (1943) Wirtsindex der Aleyrodiden- und Cocciden- Parasiten. *Entomologische Beihefte* 10, 1–100.
Garcia, A.H. (1999) Levantamento, identificação e avaliação dos danos de insetos em árvores ornamentais na área urbana de Goiânia (GO). *Pesquisa Agropecuária Tropical* 29, 77–81.
Garcia, M.O. (2004) Utilização de Fungos Entomopatogênicos para o controle de *Orthezia praelonga* (Sternorryncha: Ortheziidae). MSc thesis, Universidade de São Paulo, Piracicaba, Brazil.
García-Roa, F. (1995) Manejo de *Orthezia praelonga*, plaga de cítricos. Programa de choque tecnológico. Produmedios, Pasto, Colombia, pp. 9.
García-Roa, F., Nuñez, B.L., Varón de Agudelo, F. and Reyes, E. (1992) Avances sobre el manejo de *Orthezia praelonga* Douglas en cítricos. Proceedings XIX Congreso Socolen, Produmedios, Manizales, Colombia, pp. 31.
Giacometti, D.C. (1962) Áreas citrícolas Brasileiras e a ocorrência de Orthezia spp. Instituto de Ecologia e Experimentação Agrícolas, Rio de Janeiro. *Anais do primeiro Simpósio Brasileiro sobre combate Biológico. Boletim Técnico* 21, 61–64.
Gonçalves, C.R. (1962) Perspectivas de combate biológico das principais pragas das plantas cultivadas na baixada Fluminense. Boletim do Instituto de Ecologia e Experimentação Agrícolas, Rio de Janeiro. *Anais do primeiro Simpósio Brasileiro sobre Combate Biológico*, 21, 73–76.
Gonçalves, C.R. (1963) Procedimento da *Orthezia praelonga* na Baixada Fluminense e seu combate racional. *Boletim do Campo* 19, 12–15.
Gonçalves, C.R. and Cassino, P.C.R. (1978) Problems caused by *Orthezia praelonga* in citrus cultivation. *Encontro Nacional Citricultura* 5, 1–5.
Gonçalves, C.R. and Gonçalves, A.J.L. (1976) Observações sobre as moscas da família Syrphidae predadora de homópteros. *Anais da Sociedade Entomológica do Brasil* 5, 3–10.
Gullan, P.J. and Cook, L.G. (2007) Phylogeny and higher classification of the scale insects (Hemiptera: Sternorrhyncha: Coccoidea). In: Zhang, Z.-Q. and Shear, W.A. (eds) *Linnaeus Tercentenary: Progress in Invertebrate Taxonomy. Zootaxa* 1668, 413–425.
Gullan, P.J. and Martin, J.H. (2009) Sternorrhyncha (jumping plant-lice, whiteflies, aphids and scale insects). In: Resh, V.H. and Cardé, R.T. (eds) *Encyclopedia of Insects.* 2nd edn. Elsevier, San Diego, California, pp. 957–967.

Hempel, A. (1900) As coccidas Brasileiras. *Revista do Museu Paulista* 4, 365–537.
ICA (Instituto Colombiano Agropecuario) (1973) Programa de Entomología. Notas y Noticias Entomológicas (material mimeografiado), ICA, Palmira, Colombia.
ICA (Instituto Colombiano Agropecuario) (1975) Programa de Entomología. Notas y Noticias Entomológicas (material mimeografiado), ICA, Palmira, Colombia.
Kirkpatrick, T.W. (1957) *Insect Life in the Tropics.* Longmans, Green & Co., London.
Kogan, M. (1964) Nota sobre as espécies do gênero Orthezia Bosq d'Antic, 1784, de importância econômica e que ocorrem no Brasil. *Agronomia* 22, 134–144.
Kondo, T. (2009) *Los Insectos Escama (Hemiptera: Coccoidea) del Mango, Mangifera indica L. (Anacardiaceae) en Colombia.* Novedades Técnicas, Revista Regional. Corpoica, Centro de Investigación Palmira, Colombia 10, 41–44.
Koteja, J. (1987a) *Palaeonewsteadiea huaniae* gen. et sp. n. (Homoptera, Coccinea, Ortheziidae) from Baltic amber. *Polskie Pismo Entomologiczne* 57, 235–240.
Koteja, J. (1987b) *Protorthezia aurea* gen. et sp. n. (Homoptera, Coccinea, Oretheziidae) from Baltic amber. *Polskie Pismo Entomologiczne* 30, 241–249.
Koteja, J. (2000) Scale insects (Homoptera, Coccinea) from Upper Cretaceous New Jersey amber. In: Grimaldi, D. (ed.) *Studies on Fossils in Amber, with Particular Reference to the Cretaceous of New Jersey.* Backhuys Publishers, Leiden, The Netherlands, pp. 147–229.
Koteja, J. (2004) Scale insects (Hemiptera: Coccinea) from Cretaceous Myanmar (Burmese) amber. *Journal of Systematic Palaeontology* 2, 109–114.
Koteja, J. and Azar, D. (2008) Scale insects from Lower Cretaceous amber of Lebanon (Hemiptera: Sternorrhyncha: Coccinea). *Alavesia* 2, 133–167.
Koteja, J. and Zak-Ogaza, B. (1988a) *Arctorthezia antiqua* sp. n. (Homoptera, Coccinea) from Baltic amber. *Annales Zoologici Polska Akademia Nauk, Instytut Zoologiczny* 41, 1–8.
Koteja, J. and Zak-Ogaza, B. (1988b) *Newsteadia succini* sp. n. (Homoptera, Coccinea) from Baltic amber. *Annales Zoologici Polska Akademia Nauk, Instytut Zoologiczny* 41, 9–14.
Kozár, F. (1997) Insects in a changing world. *Acta Phytopathologica et Entomologica Hungarica* 32, 129–139.

Kozár, F. (2004) *Ortheziidae of the World.* Plant Protection Institute, Hungarian Academy of Sciences, Budapest.

Lago, I.C.S. (1981) Biologia de *Orthezia praelonga* Douglas, 1891 (Homopter a - Or theziidae), distribuição geográfica, plantas hospedeir as e ocorrência de inimigos natur ais, em Pernambuco. MSc thesis, Universidade Federal Rural de P ernambuco, Recife, Brazil.

Laurence, G.A. (1991) The scale insects of Trinidad and Tobago. *Journal of the Agricultural Society, Trinidad* 1 19–23, 26–29, 32–36.

Lepage, H.S. (1938) Catálogo dos coccídeos do Brasil. *Revista do Museu Paulista* 23, 432.

Lima, A.F. (1981) Bioecologia de *Orthezia praelonga* Douglas, 1891 (Homoptera, Ortheziidae). MSc thesis, Escola Super ior de Ag ricultura "Luiz de Queiroz", Universidade de São Paulo, Piracicaba, Brazil.

Lima, A.F. de, and Cassino, P.C.R. (1974) No vos hospedeiros de *Orthezia praelonga* Douglas, 1891 (Homoptera, Ortheziidae) no Estado do Rio de J aneiro. *Arquivos da Universidade Federal Rural do Rio de Janeiro* 4, 73–74.

Lizer y Trelles, C.A. (1942) La colección coccidológica de Pedro Jorgensen. *Notas del Museo de la Plata, Zoologia* 7, 69–80.

Lopes, E.B., Albuquerque, I.C. and Moura, F.T. (2007) Perfil da citricultura de Matinhas, PB, visando o mercado nacional. *Revista Tecnologia & Ciência Agropecuária, João Pessoa* 1, 1–7.

Lopes, L.S. and Maia, W.J.M.S. (2008) Estudo da bioecologia de *Gitona brasiliensis* sobre *Orthezia praelonga* em *Citrus* spp. Anais do VI Seminário de iniciação cientifica da UFRA e XII Seminário de iniciação cientifica da EMBRAPA Amazônia Oriental. No 27. http://anaispibic2008.cpatu.embrapa.br/index2.html, accessed November 2008.

Magalhães, T.L.M. and Silva, W.J.M. (2008) Estudo da biologia do par asitóide *Gitona brasiliensis* (Lima, 1950) (Dipter a: Drosophilidae), tendo como hospedeiro *Orthezia praelonga* (Douglas, 1981) (Hemiptera: Ortheziidae). Anais do VI Seminário de iniciação cientifica da UFRA e XII Seminário de iniciação cientifica da EMBRAPA Amazônia Oriental, No 50. http://anaispibic2008.cpatu.embrapa.br/index2.html, accessed November 2008.

Matile-Ferrero, D. and Étienne, J. (2006) Cochenilles des Antilles françaises et quelques autres îles des Caraïbes [Hemiptera, Coccoidea]. *Revue Française d'Entomologie (N.S.)* 28, 161–190.

Medeiros, L.S.A., Junger, C.M., Rodrigues Filho, I.L., Lima, A.F. and Cassino, P.C.R. (1980) No vos hospedeiros de *Orthezia praelonga* Douglas, 1891 (Hom., Ortheziidae) no Estado do Rio de Janeiro. 7th Br azilian Congress of Zoology. Abstract, Mossoró, Brazil, pp. 72–73.

Menge, F.A. (1856) *Lebenszeichen Vorweltlicher, im Bernstein Eingeschlossener Thiere.* [Sign of life of prehistoric animals enclosed in amber] Programm der Öffentlichen Prüfung der Schüler der Petrischule, Danzig, Kafemann. pp. 32.

Miller, D.R., Miller, G.L. and Watson, G.W. (2002) Invasive species of mealyb ugs (Hemiptera: Pseudococcidae) and their threat to U.S. agriculture. *Proceedings of the Entomological Society of Washington* 104, 825–836.

Miller, D.R., Miller, G.L., Hodges, G.S. and Davidson. J.A. (2005) Introduced (Hemiptera: Coccoidea) of the United States and their impact on US Agriculture. *Proceedings of the Entomological Society of Washington* 107, 123–159.

Miller, G.L. and Miller, D.R. (2003) Invasive soft scales (Hemiptera: Coccidae) and their threat to U.S. agriculture. *Proceedings of the Entomological Society of Washington* 105, 832–846.

Molina, J.H.G. (2007) Efeito do Controle Microbiano em insetos sugadores em três sistemas de manejo de pr aga de citros. PhD thesis, University of São Paulo, São Paulo, Brazil.

Morrison, H. (1925) Classification of scale insects of the subf amily Ortheziinae. *Journal of Agricultural Research* 30, 97–154.

Morrison, H. (1952) *Classification of the Ortheziidae.* Supplement to *Classification of Scale Insects of the Subfamily Ortheziinae.* United States Department of Ag riculture Technical Bulletin 1052, Beltsville, Maryland, pp. 1–80.

Neves, A.D., Haddad, M. de Lar a, Zério, N.G. and Postali Parra, J.R. (2010) Exigências témicas e estimativas do número de gerações de ortézia dos citros cr iadas em limão-cr avo. *Pesquisa Agropecuária Brasileira, Brasília* 45, 791–796.

Oliveira, A.M., Cr uz, C.A., Lima, A.F. and Vasconcellos, H.O. (1979) Obser vações preliminares sobre infestação de *Orthezia praelonga* em laranja Natal e f olha-murcha, no Estado do Rio de J aneiro. PESAGRO-RIO, Rio de Janeiro, Brazil, pp. 2.

Peronti, A.L.B.G., Miller, D.R. and Sousa-Silva, C.R. (2001) Scale insects (Hemipter a: Coccoidea) of ornamental plants from São P aulo, Brazil. *Insecta Mundi* 15, 247–255.

Pinto, W.B and Pr ates, H.S. (1980) Inimigos naturais da cochonilha *Orthezia praelonga* Douglas, 1891 em pomares de citros no Estado de São P aulo. Proceedings of the 4th Brazilian Congress of Entomology, Campinas, Brazil, pp. 193.

Prates, H.S. (1980) Piolho br anco, a pr aga dos citros. *Agroquímica Ciba-Geigy* 12, 11–13.

Prates, H.S. (1989) Coccídeos: pragas em potencial para a citricultura. *Revista Agrotécnica CIBA-Geigy, São Paulo* 5, 16–22.

Prates, H.S. and Novo, J.P.S. (1979) *Orthezia praelonga* Douglas, 1891 – "Piolho branco" ou "ortézia", praga dos pomares cítricos. Secretaria de Agricultura, Coordenadoria de Assistência Técnica e Integral, Centro de Orientação Técnica, São Paulo, Brazil, pp. 8.

Prates, H.S. and Pinto, W.B.S. (1987). Orthezia: uma praga potencial. *Casa da Agricultura* 9, 16–19.

Pyenson, L. (1938) The problems of applied entomology in Pernambuco, Brasil. *Revista de Entomologia, Rio de Janeiro* 69, 16–31.

Restrepo, G.H., Ochoa L.P., León, M.G. and de la Cruz, T. (1991) Ciclo de vida y hábitos de *Orthezia* sp., plaga de cítricos. In: Resumenes VIII Congreso Socolen, Santa Fe de Bogotá, Colombia, p. 23.

Robbs, C.F. (1947) O "piolho branco" da laranjeira, uma ameaça da citricultura do Distrito Federal. *Boletim do Campo, Rio de Janeiro* 3, 1–4.

Robbs, C.F. (1951) Principais cochonilhas das plantas cítricas na Baixada Fluminense. *Boletim do Campo, Rio de Janeiro* 7, 5–13.

Robbs, C.F. (1974) Ortézia: descrição e combate. EEB, Sergipe, Brazil, pp. 6.

Rodrigues, W.C. (2004) Inimigos naturais de pragas de plantas cítricas no Estado do Rio de Janeiro. *Info Insetos, Informativo dos Entomologistas do Brasil* 1, 1–7.

Silva, A.G.A., Gonçalves, C.R., Galvão, D.M., Gonçalves, A.J.L., Gomes, J., Silva, M.N. and Simoni, L. (1968) *Quarto Catálogo dos Insetos que Vivem nas Plantas do Brasil: Seus Parasitos e Predadores*. Ministério da Agricultura, Laboratório Central de Patologia Vegetal, Rio de Janeiro, Brazil.

Silva, L.M.S. and Gravena, S. (1980) *Salpingogaster conopida* (Phillipi, 1865) Diptera, Syrphidae, novo predador de *Orthezia praelonga* Douglas 1891 (Homoptera, Orthezidae). *Revista Brasileira de Fruticultura, Campinas* 2, 79–80.

Silva, L.M.S., Vieira, G., Emidio Filho, J.E. and Trindade, J. (1979) Aldicarbe, uma nova opção para o controle da "Orthezia" em citros. Anais do 5th Congresso Brasileiro de Fruticultura, Pelotas, Brazil, pp. 43–46.

Silva, R.A. and Jordão, A.L. (2005) Pragas dos citros no Estado do Amapá. *Revista Científica Eletrônica de Agronomia* 4, 277–280.

Szita, É., Kozár, F. and Konczné Benedicty, Z. (2010) Study of some new macromorphological characters in Ortheziidae (Hemiptera, Coccoidea). Poster. XIIth International Symposium on Scale Insect Studies, 6–9 April, Chania, Greece.

Tanaka, H. and Amano, H. (2005) *Newsteadia yanbaruensis*, a new species from Okinawa Is., Japan (Hemiptera, Ortheziidae). *Japanese Journal of Systematic Entomology* 11, 283–286.

Tanaka, H. and Amano, H. (2007) First records of the subfamily Ortheziolinae (Hemiptera: Ortheziidae) in Japan, with descriptions of two new species. *Zootaxa* 1516, 31–37.

Teixeira, M.A., Bettiol, W. and Cesnik, R. (2001) Patogenicidade do fungo *Colletotrichum gloeosporioides*, patógeno de *Orthezia praelonga*, para folhas, frutos e flores cítricas. *Summa Phytopathologica* 27, 352–357.

Teixeira, M.A., Bettiol, W., Cesnik, R. and Vieira, R.F. (2004) Patogenicidade do fungo *Colletotrichum gloeosporioides*, patógeno da cochonilha *Orthezia praelonga*, a diversos frutos e a plântulas de abobrinha. *Revista Brasileira de Fruticultura* 26, 356–358.

Velásquez V.H., Nuñez, B. and García, R.F. (1992) Avances en el reconocimiento y evaluación de agentes benéficos de *Orthezia praelonga* Douglas. Proceedings XIX Congress of Socolen, Produmedios, Manizales, Colombia, pp.15.

Vernalha, M.M. (1970) Uma nova praga da Cafeicultura paranaense. *Arquivo de Biologia e Tecnologia* 13, 29–33.

Viegas, E.C., Sampaio, H.N., Carvalho, P.O.L., Perruso, J.C. and Cassino, P.C.R. (1995) Controle alternativo de *Orthezia praelonga* Douglas, 1891 (Hemiptera, Ortheziidae) em laboratório. Proceedings 15th Congress of Entomology, Caxambú. Anais da Sociedade Entomológica do Brazil, Caxambú, Brazil, p. 333.

Vieira F.V., Pontes A.A. and Santos, J.H.R. (1976) Ocorrência de pragas hortícolas em Fortaleza. *Ciência Agronômica* 6, 99–103.

Wang, Y.P., Watson, G.W. and Zhang, R. (2010) The potential distribution of an invasive mealybug *Phenacoccus solenopsis* and its threat to cotton in Asia. *Agricultural and Forest Entomology* 12, 403–416.

Williams, D.J. and Watson, G.W. (1990) *The Scale Insects of the Tropical South Pacific Region. Pt. 3: The Soft Scales (Coccidae) and Other Families*. CAB International, Wallingford, UK.

Wolcott, G.N. (1960) Efficiency of ladybeetles (Coccinellidae: Coleoptera) in insect control. *Journal of Agriculture of the University of Puerto Rico* 44, 166–172.

Zinger, K., Spolidoro, M.V., Rodrigues, W.C. and Cassino, P.C.R. (2006) Atendimento de Formigas em Cochonilhas em Cultivo Orgânico de Tangerina cv. Poncã (*Citrus reticulata* Blanco). CD-ROM. Proceedings of the XXI Brazilian Congress of Entomology (Abstracts), Recife, Brazil.

18 Potential Invasive Species of Scale Insects for the USA and Caribbean Basin

Gregory A. Evans[1] and John W. Dooley[2]

[1]*USDA/APHIS/PPQ/National Identification Service, 10300 Baltimore Avenue, BARC-West, Bldg. 005, Rm 09A, Beltsville, Maryland 20705, USA;* [2]*USDA/APHIS/PPQ 389 Oyster Point Blvd, Suite 2A, South San Francisco, California 94080, USA*

18.1 Introduction

History has shown that when an exotic pest enters and establishes in a country outside its native range, it often takes only a little time for the species to spread to other countries in the region. Therefore, it is mutually beneficial for those working in quarantine and crop protection in all regions of the world to work together to stop, or at least to slow down, the movement of pests. This chapter deals with potential invasive scale insect pests for the USA and the Caribbean Basin.

There are approximately 7500 known species of scale insects (Coccoidea) belonging to 45 families (extant and fossil); however, the most common species, and the ones that are usually intercepted at US ports of entry on plant material, belong to one of the following eight families: Asterolecaniidae, Coccidae, Dactylopiidae, Diaspididae, Monophlebidae (formerly part of Margarodidae), Ortheziidae, Pseudococcidae and Stictococcidae. The number of species in these families that are known worldwide is shown in Table 18.1. In addition, the number of species known to occur in the USA and in the following seven zoogeographic regions: Nearctic (NA), Neotropical (NT), Afrotropical (AF), Palearctic (PA), Oriental (OR), Australasian (AU) and Antarctica (AN) is provided, with the percentage of the total number of species represented in the USA and the respective zoogeographic regions indicated beside the number of species (Table 18.1). These data were extracted from ScaleNet (2011), an online database that includes all published information on scale insect species. A few subspecies might be included as part of the total number of species; in addition, the creators of ScaleNet place the northern part of Mexico in the NA, whereas southern Mexico is in the NT region. Northern mainland China is included in the PA, whereas southern China is in the OR region. This differs from the interception data presented herein, in that all of Mexico is included in the NT region; all of mainland China, Japan and Korea are placed in the Eastern PA region; and the records from the AU region are separated into those from Australia and those from the PI.

All scale insects are obligate plant feeders, and therefore are almost always found closely associated with plants, commonly being intercepted on imported plant material. In its native area, an ecological balance usually exists between the population of the scale insect and its natural enemies, which maintains the population of the scale insect below economic injury levels. However, when a species is separated from its natural enemies, primarily due to human mediation

Table 18.1. Number of scale insect species of selected families by zoogeographic region, and the percentage of the total number of known species (ScaleNet, 2011).

Region	Asterolecaniidae		Coccidae		Diaspididae		Pseudococcidae	
	Total species	% total species	Total species	% total species	Total species	% total species	Total species	% total species
USA	19	7.9	106	9.4	329	13.3	361	15.9
NA	29	12.1	158	14.0	419	17.0	445	19.5
NT	51	21.3	298	26.3	365	14.8	285	12.5
PA	74	30.8	289	25.6	643	26.1	721	31.7
AF	25	10.4	274	24.2	573	23.2	297	13.0
OR	104	43.3	196	17.3	913	37.0	429	18.8
AU	35	14.6	172	15.2	420	17.0	451	19.8
AN	0	0	0	0	1	<1.0	0	0
WW	240	100	1131	100	2465	100	2277	100

Region	Dactylopiidae		Monophlebidae		Ortheziidae		Stictococcidae	
	Total species	% total species	Total species	% total species	Total species	% total species	Total species	% total species
USA	4	40.0	10	4.1	31	15.7	0	0
NA	6	60.0	27	11.0	42	21.2	0	0
NT	9	90.0	42	17.1	64	32.3	0	0
PA	3	30.0	31	12.6	22	11.1	0	0
AF	6	60.0	63	25.6	39	19.7	17	100
OR	2	20.0	73	29.7	27	13.6	0	0
AU	5	50.0	41	16.7	22	11.1	0	0
AN	0	0	0	0	0	0	0	0
WW	10	100	246	100	198	100	17	100

NA, Nearctic; NT, Neotropical; AF, Afrotropical; PA, Palearctic; OR, Oriental; AU, Australasian; AN, Antarctica; WW, Worldwide.

(transported to a new area or by the non-selective use of chemical pesticides), then high infestations of the scale often occur, resulting in economic damage. Therefore, any scale insect species that has been separated from its natural enemies and survives the environmental conditions in its new environment, has the potential to become a pest species in its non-native habitat. Virtually all of the major scale insect pests in the USA and Caribbean basin are non-native species.

The introduction of an exotic species has often been followed by large population outbreaks and subsequent economic plant damage. The cycad aulacaspis scale, *Aulacaspis yasumatsui*, was an obscure species in South-East Asia until it arrived in Florida in 1996; thereafter, its population exploded and has now spread to many other countries. However, not all invasive species become major pest species. In fact, Miller and Miller (1990) reported that only about 8% of the known species of armored scales had been reported as pests somewhere in the world. Nevertheless, since it is difficult to predict the impact an invasive species might have, all invading scale species are of interest to plant-protection workers.

Regardless of the efforts by plant-protection officers, discovery of an invasive scale often happens after the scale population has become established and population outbreaks have occurred; however, some species have been discovered shortly after they entered the country, and measures were taken to eradicate the species before it had a chance to become established. Examples of invasive species of armored scales that have recently been introduced and have been established in a region include: *Aulcaspis yasumatsui* Takagi and *Poliaspis cycadis* Comstock on cycads in Florida, and *Parlagena bennetti* Williams, a pest of palms that is now known to occur in Trinidad, St Andres Island (Colombia) and the Bay Islands (Honduras).

In Table 18.1, the number of potential invasive species for a country or region was determined by subtracting the number of species known to occur in the country or region from the total number of species known worldwide. This number does not take into account the number of cryptic and other species that have not been described.

For example, currently 329 species of armored scales are known to occur in the USA, representing 13.3% of the total number of species (2465) known worldwide. Therefore, 2136 (86.7%) species are not known to occur in the USA but could potentially invade and establish there and perhaps become economic pests. The potential of an exotic species to enter a region and become an economic pest is a function of the characteristics of the species, and the probability that it will arrive, gain entrance and establish in the country and/or region. Some of the important characteristics of a scale-insect species are: (i) ability to adapt to a new habitat; (ii) reproductive rate; (iii) life cycle (number of generations per annum); (iv) natural enemies (native and exotics); (v) host plants (economic crops and alternate hosts); and (vi) the part(s) of the plant on which it feeds (fruit, leaves, roots). Information on its known plant hosts, distribution and other information can be gathered quickly from ScaleNet (2011). Knowing which scale insect species are associated with each imported plant species, and their distribution, may provide important clues to plant protection and quarantine workers with respect to possible presence of a scale species on the imported hosts and its identity, and information to determine the threat that the species poses and the appropriate action that must be taken.

Other factors that influence the ability of a scale insect to arrive and gain entrance into the region include: (i) its pest status in the exporting country; (ii) its plant hosts, including imported and alternate hosts; (iii) the frequency and volume of the host that is imported; and (iv) pest frequency on the imported plant material at the ports of entry.

The increase in the diversity and volume of agricultural products being shipped throughout the world and their rapid movement on the world market has also increased the movement of these pests and the likelihood of their introduction and establishment in the region. Plant material is being imported into the USA from nearly every country in the world. In addition, there is much greater diversity and a much higher volume of plant species being imported than ever before. Exporting countries are bringing new lands into production, which sometimes results in native species moving from their usual hosts onto the crop (an alternate host). In addition, the management of these species is sometimes complicated by the lack of taxonomic expertise to identity the species, and insufficient knowledge about the fauna of many regions of the world.

The Animal and Plant Health Inspection Service (APHIS/USDA) and the Department of National Homeland Security, US Customs and Border Protection (NHS/CBP) are responsible for preventing the introduction of exotic species into the USA. Currently, there are several US inspection stations distributed throughout the USA and its territories at airports, seaports and land border ports. APHIS has for many years maintained the AQAS (Agricultural Quarantine Activity Systems) database, which is an internal database of records of insects, mites, mollusks and other pests intercepted at US ports of entry (Table 18.2). These data include the pest species, host plant, the part of the host plant on which it was found and the country from which the shipment originated. The data are limited, however, because some pests are not identified to the species level, usually because the identifiable stage of the species was not intercepted, or the condition of the specimen was unsatisfactory for identification. The true origin of the interception is sometimes questionable, especially when a specimen is intercepted in a passenger's baggage, and to a lesser extent in cargo, even though certification of origin is usually provided.

When species were intercepted on imported plant material, this indicates that they are known to feed on at least one plant species, and because they were intercepted at a port of entry this increases the probability that they may gain entrance into the country and region.

The number of interceptions of scale insects has varied by zoogeographic region (Table 18.2). In this table, the NA region includes only Canada and the USA; all records from Mexico are included

Table 18.2. Interceptions of soft scales (Coccidae), armored scales (Diaspididae) and mealybugs (Pseudococcidae) at US ports of entry (2007–2010) by zoogeographic region (USDA/APHIS, 2012).

Region	Coccidae		Diaspididae		Pseudococcidae	
	Total interceptions	% Total	Total interceptions	% Total	Total interceptions	% Total
NA	14	0.4	164	0.4	95	0.4
NT	1,607	49.2	29,849	74.6	16,518	65.9
WPA	169	5.2	3,704	9.3	1,178	4.7
AF	54	1.6	640	1.6	380	1.5
EPA	37	1.1	557	1.4	431	1.7
OR	273	8.3	2,080	5.2	3,502	14.0
Australian	17	0.5	147	0.4	143	0.6
PI	939	28.7	1,004	2.5	1,794	7.2
UNK	163	5.0	1,855	4.6	1,028	4.0
Total	3,273	100	40,000	100	25,064	100

NA, Nearctic; NT, Neotropical; AF, Afrotropical; OR, Oriental; AU, Australasian; WPA, Western Palearctic; EPA, Eastern Palearctic; PI, Pacific Islands; UNK, Unknown

in the NT region. China (entire), Japan and South Korea are in the Eastern PA region, and the records from the AU region are separated into those from Australia and those from the PI.

For scales, armored scales and mealybugs, there were many more interceptions made from the NT countries, which accounted for 49.2%, 74.6% and 65.9% of the interceptions of soft scales, armored scales and mealybugs, respectively. Many of the species that were intercepted from this region were already present in the Caribbean basin region. The next highest number of interceptions of soft scales was from the PI region which had 28.7% of the total interceptions, with the majority of those from Hawai'i. The Western PA had the next highest number of interceptions of armored scales (9.3%); and the OR region had the next highest number of interceptions of mealybugs (14.0%).

The ability of an exotic scale insect species to gain entry and establish in the USA and Caribbean basin region is also influenced by its host and the part of the host that is imported (Table 18.3). Scale insects, especially armored scales, that have been found on imported fresh fruit are believed to be less likely to survive and establish in the new habitat than those imported on leaves or material for propagation. Possible exceptions must be made for soft scales and mealybugs, since most species in these families are mobile in their nymphal and adult stages. The number of interceptions of soft scales, armored scales and mealybugs from the top 12 host plants of each of the three families is shown in Table 18.3. For soft scales, *Gardenia* (19.6%) was the most common host, followed by *Schlefflera* (11.2%). Both of these hosts are ornamental species that are imported as leaves and/or propagated material. For armored scales, *Citrus* (50.4%), *Mangifera indica* (18.0%), *Persea americana* (15.9%) and *Psidium guajava* (1.5%) were the most common hosts. Together they accounted for 85.8% of all interceptions of armored scales. Material of all of these fruit species is imported as fruit and the probability of a species being found on these hosts gaining entry and establishing is less than for pests on hosts that are imported as leaves. Only 9.2% of the armored scales were found on hosts other than those listed in Table 18.3. For mealybugs, *Musa* (22.6%), *Ananas comosus* (5.9%), *Annona* spp. (5.7%) and *Nephelium* (4.6%) were the most common hosts. All of these hosts are imported as fruit.

The numbers of different scale insect species intercepted at US ports of entry between 1 January 2007 and 31 December 2010 are categorized by their relative distribution in the Caribbean basin region as being either: (i) widespread; (ii) limited; or (iii) not known to occur in the Caribbean basin region (Table 18.4).

If the species is known to occur in the USA, Mexico and at least one of the countries of Central America and/or a Caribbean island, it is designated as being widespread; if the species

Table 18.3. Interceptions (Total int.) of soft scales (Coccidae), armored scales (Diaspididae) and mealybugs (Pseudococcidae) at US ports of entry (2007–2010) by their top 12 imported hosts (USDA/APHIS, 2012).

Coccidae			Diaspididae			Pseudococcidae		
Host genus	Total int.	% Total	Host genus	Total int.	% Total	Host genus	Total int.	% Total
Gardenia (L)	640	19.6	Citrus (F)	20,172	50.4	Musa (F)	5,669	22.6
Schlefflera (L)	366	11.2	Mangifera (F)	7,212	18.0	Ananas (F)	1,476	5.9
Citrus (F)	212	6.5	Persea (F)	6,343	15.9	Annona (F)	1,432	5.7
Nephelium (F)	108	3.3	Psidium (F)	602	1.5	Nephelium (F)	1,141	4.6
Mangifera (F)	105	3.2	Chamaedorea (L)	413	1.0	Ocimum (L)	860	3.4
Alyxia (L)	105	3.2	Musa (F)	269	0.7	Garcinia (F)	858	3.4
Laurus (L)	75	2.3	Areca (L)	232	0.6	Citrus (F)	858	3.4
Psidium (F)	73	2.2	Annona (F)	229	0.6	Alpinia (L)	826	3.3
Annona (F)	70	2.1	Cocos (F)	227	0.6	Psidium (F)	655	2.6
Alpinia (L)	68	2.1	Draecaena (L)	219	0.5	Salvia (L)	475	1.9
Carica (F)	59	1.8	Dimocarpus (F)	208	0.5	Carica (F)	415	1.7
Musa (F)	58	1.8	Murraya (L)	200	0.5	Sechium(F)	346	1.4
Other	1,334	40.7	Other	3,674	9.2	Other	10,053	40.1
Total	3,273	100	–	40,000	100	–	25,064	100

F, imported fresh fruit; L, leaves

has not been known to occur in at least one of the countries of these four areas, then it is designated as being limited in distribution. For armored scales (Diaspididae), 71 (50.7%) of the 140 species that have been intercepted are widespread in the Caribbean Basin. Since they are already present in all or many of the countries in the region, they are not considered as potential invasive species, and therefore are not included in the discussion of each family. Of the remaining species that have been intercepted, 41 species (29.3%) are limited in distribution and 28 species (20%) are not known to occur in the region. These are considered to be potential invasive species for the region or for other countries within the region.

The number of interceptions of different scale insect species at US ports of entry between 2007 and 2010 has been categorized by species that are widespread, limited or not known to occur in the Caribbean basin region (Table 18.5). For example, for armored scales (Diaspididae), 20,063 (50.2%) of all interceptions (40,000) were of species that are widely distributed in the Caribbean basin region, 16,048 (40.1%) of the interceptions were of species that occur in the region but are limited in their distribution, and 190 (0.5%) were interceptions of species that are not known to occur in the region.

Specimens identified at the family or genus level represented 3699 (9.2%) of the interceptions. Although the 28 intercepted species that are not known to occur in the region only represented 190 (0.5%) of all interceptions, their potential impact should not be underestimated. The entry and establishment of any of these species in the region could have major repercussions (e.g., the introduction and establishment of *Aulacaspis yasumatsui*).

The number of interception records of scale-insect species that are not known to occur in the Caribbean basin or have limited distribution in the region, their region of origin, their distribution in the Caribbean basin region, if the species is known to occur in the USA, the country and host plant on which the species was intercepted and other notes are shown in Table 18.6. Since the focus of this chapter is on the potential invasive scale-insect species to the Caribbean basin region, we have excluded those species that are widely distributed in the region. Species of the eight families of scale insects are discussed separately by family.

The species of scale insects that have been intercepted at US ports between 2007 and 2010 are discussed below. An asterisk (*) indicates those species that are of particular interest as potential pest species.

Table 18.4. Number of species (# sp.) and percentage of total number of scale insect species (% sp.) intercepted at US ports of entry (2007–2010) by family and their distribution in the Caribbean basin region (USDA/APHIS, 2012).

Distribution in Caribbean basin	Asterolecanidaeae # sp./(%) sp.	Coccidae # sp./(%) sp.	Diaspididae # sp./(%) sp.	Pseudococcidae # sp./(%) sp.	Dactylopidae # sp./(%) sp.	Monophlebidae # sp./(%) sp.	Ortheziidae # sp./(%) sp.	Strictococcidae # sp./(%) sp.
Widespread in region	4 / 50%	24 / 61.5%	71 / 50.7%	33 / 34.0%	2 / 66.6%	1 / 12.5%	2 / 50.0%	0 / 0%
Limited in region	2 / 25%	7 / 17.9%	41 / 29.3%	30 / 30.9%	1 / 33.3%	5 / 62.5%	1 / 25.0%	0 / 0%
Not known in region	2 / 25%	8 / 20.5%	28 / 20.0%	34 / 35.1%	0 / 0%	2 / 25.0%	1 / 25.0%	3 / 100%
Total # species intercepted	8 / 100%	39 / 100%	140 / 100%	97 / 100%	3 / 100%	8 / 100%	4 / 100%	3 / 100%

Table 18.5. Interceptions of different scale insect species (# int.) at US ports of entry (2007–2010), and the percentage of total number of species (% sp.) intercepted by distribution in the Caribbean basin region (USDA/APHIS, 2012).

Distribution in Caribbean basin	Asterolecanidae # int./ (% int.)	Coccidae # int./ (% int.)	Diaspididae # int./ (% int.)	Pseudococcidae # int./ (% int.)	Dactylopidae # int./ (% int.)	Monophlebidae # int./ (% int.)	Ortheziidae # int./ (% int.)	Strictococcidae # int./ (% int.)
Widespread in region	4 / 4%	1,984 / 60.6%	20,063 / 50.2%	6,715 / 26.9%	82 / 36.9%	64 / 16.4%	56 / 25.9%	0 / 0%
Limited in region	52 / 52%	19 / 0.6%	16,048 / 40.1%	1,879 / 7.5%	0 / 0%	27 / 6.9%	13 / 6.0%	0 / 0%
Not known in region	6 / 6%	44 / 1.3%	190 / 0.5%	419 / 1.7%	0 / 0%	5 / 1.3%	1 / 0.5%	25 / 78.1%
Unidentified to genus or species	38 / 38%	1,226 / 37.5%	3,699 / 9.2%	15,920 / 63.9%	140 / 63.1%	292 / 74.9%	146 / 67.6%	7 / 21.9%
Total # species intercepted	100 / 100%	3,273 / 100%	40,000 / 100%	24,933 / 100%	222 / 100%	390 / 100%	216 / 100%	32 / 100%

Table 18.6. Interceptions of scale insects of limited distribution or not known to occur in Caribbean basin (USDA/APHIS, 2012).

Species	Total int.	Region of orig.	Region of dist.	In USA	Interceptions and notes
Asterolecaniidae (pit scales)					
Asterolecanium ungulatum	2	OR	N	no	Thailand on *Durio*
Palmaspis inlabefactum	50	NT	L	no	Mexico on *Chamaedorea*
*Palmaspis phoenicis	4	WPA	N	no	India, Jordan, Pakistan, Turkey on *Phoenix*
Total = 3 species	100	–	–	–	–
Coccidae (soft scales)					
*Ceroplastes japonicus	7	EPA	N	no	Armenia, France, Georgia, Germany, Russia on *Laurus, Prunus*; pest of kiwi and persimmon
Coccus moestus	4	AF	L	no	Ecuador, Honduras, Puerto Rico on *Nephelium, Prunus* and palms
Drepanococcus chiton	16	OR	N	no	Asia on *Dimocarpus, Garcinia, Lansium, Nephelium*
Kilifia deltoides	1	AF	N	no	India on *Mangifera*
*Lichtensia viburni	1	WPA	N	no	Ireland on *Hedera*
*Philephedra broadwayi	1	NT	L	no	Grenada on *Annona*; also in South America and Caribbean on *Mangifera, Spondias, Theobroma*
Protopulvinaria longivalvata	7	OR	L	no	Costa Rica, Hawai'i, Jamaica, Puerto Rico on *Cinnamomum, Morinda, Piper, Plumeria, Schefflera*
*Taiwansaissetia formicarii	8	OR	N	no	India, Thailand, Vietnam on *Areca, Durio, Lansium, Mangifera*; also on *Garcinia, Lagerstroemia, Persea*
Udinia catori	8	AF	N	no	Cameroon, Gambia, Nigeria, Senegal on *Cola, Mangifera*; also on *Citrus, Ficus, Persea, Psidium*
Udinia farquharsoni	1	AF	N	no	Kenya on *Mangifera*; also on *Citrus, Coffea, Durio, Ficus, Gardenia, Theobroma*
Total = 10 species	3,273	–	–	–	–
Diaspididae (armored scales)					
Abgrallaspis aguacatae	4,691	NT	L	no	Mexico on *Persea*
Abgrallaspis mendax	10	NT	L	no	Guatemala, Mexico on orchids
Acutaspis scutiformis	15	NT	L	yes	Costa Rica on *Aglaonema*
Acutaspis umbonifera	15	NT	L	yes	Mexico, Central and South America; various plants
Aonidiella comperei	16	OR	L	no	Brazil, India, Korea, Sri Lanka; in Central America and Caribbean on *Carica, Citrus*
Aspidiella hartii	102	NT	L	no	Africa, Barbados, Haiti, Thailand, Vietnam on roots of *Dioscorea* and *Zingiber*; pest of yams
Aspidiotus capensis	1	AF	N	no	South Africa on *Encephalartos*, cycads
*Aulacaspis crawii	2	EPA	L	no	China on *Citrus, Cymbidium*; also on *Hibiscus, Murraya, Poncirus*, orchids; Asia, El Salvador

Continued

Table 18.6. Continued.

Species	Total int.	Region of orig.	Region of dist.	In USA	Interceptions and notes
Aulacaspis ima	1	EPA	N	no	India on Cinnamomum
Aulacaspis rosarum	1	EPA	L	no	Mexico on Rosa; also in EPA, OR and PI region
Aulacaspis tubercularis	6,245	OR	L	yes	Widespread; many hosts but most common on Mangifera (mango); USA (Florida)
*Chrysomphalus pinnulifer	13	OR	L	no	Australia, Asia, Italy, Colombia, Dominican Republic on Citrus, Garcinia, Mangifera, Persea, pest of tea
Chrysomphalus propsimus	5	OR	N	no	Philippines, Vietnam on Annona, Cocos, Garcinia
Duplachionaspis natalensis	3	AF	L	no	Peru, Virgin Islands on Cymbopogon; also in Brazil
Fiorinia coronata	2	PI	N	no	Hawai'i, Vietnam on Chrysalidocarpus, Cocos, palms
*Fiorinia proboscidaria	7	OR	N	no	China, Hawai'i, Indonesia, Vietnam on Citrus, roses, palms
Greenaspis sp.	1	Asia	L	no	Dominican Republic, Grenada, Puerto Rico on Cymbopogon
Leonardianna sp.	58	NT	L	no	Dominican Republic, Trinidad on Canella, Guatteria
Lepidosaphes cornuta	3	OR	N	no	Singapore, Thailand on Michelia, Piper
*Lepidosaphes laterochitinosa	679	WPA	L	no	Asia, Australia, Costa Rica on many hosts including Citrus, Mangifera and ornamentals
*Lepidosaphes malicola	4	WPA	L	no	Dominican Republic, Italy on Malus, Prunus, Pyrus
*Lepidosaphes pistaciae	68	WPA	N	no	India, Egypt, Iran, Jordan, Syria, Tunisia, Turkey on pistacios; also on Malus, Prunus, Pyrus and others
Lepidosaphes rubrovittata	7	OR	L	no	Asia, Jamaica, Puerto Rico, South Africa on cycads, Psidium, Murraya, Nepenthes and others
Lepidosaphes similis	15	PI	N	no	Asia, PI, Brazil on Areca, Cocos, Dracaena, Musa, palms
*Lepidosaphes tubulorum	1	OR	N	no	Asia, Guam; ornamental pest
*Leucapis riccae	1	WPA	N	no	Europe on Olea
Lindingaspis misrae	6	OR	N	no	India on Cycas, Mangifera, Murraya, Tamarindus
Pseudoparlatoria mammata	30	NT	L	no	Mexico on Chamaedorea, palms; also in Panama
Melanaspis calura	5	NT	L	no	Costa Rica, Ecuador, Guatemala, Honduras, Mexico on Spondias; also on Crataegus, Mammea
Melanaspis inopinata	26	WPA	N	no	Middle East on pistacios, Dioscorea; Malus, Prunus
Parlatoreopsis chinensis	1	EPA	N	no	Taiwan on Acer; also Hibicus, Malus, Prunus, Pyrus
Parlatoria citri	6	OR	N	no	China, Indonesia, Vietnam on Citrus, Pyrus
*Parlatoria crypta	17	OR	N	no	India, Pakistan on Mangifera, Murraya

Continued

Table 18.6. Continued.

Species	Total int.	Region of orig.	Region of dist.	In USA	Interceptions and notes
Parlatoria fluggeae	1	AF	N	no	Germany on Adenia; also on Ficus, Hibiscus; Hawai'i
Parlatoria multipora	1	EPA	N	no	South Korea on Cornus
*Parlatoria pseudaspidiotus	101	OR	L	no	India, Mexico on Mangifera; Caribbean on orchids
*Parlatoria ziziphi	9,864	WPA	L	no	widespread on Citrus; eradicated from Florida, USA
*Pinnaspis theae	1	OR	L	no	Brazil on Thea; Colombia, Guadeloupe, Martinique
Pseudaonidia corbetti	1	OR	N	no	China on Garcinia, ornamentals
Pseudaonidia curculiginis	1	OR	N	no	Philippines on Pholidota; also on Gardenia, palms; Malaysia, Thailand, Vietnam
Pseudaulacapis brimblecombei	2	AUS	N	no	Australia, New Zealand on Macadamia, Proteaceae
Pseudaulacaspis eugeniae	8	AUS	N	no	Australia on Proteaceae (ornamentals); also reported on Prunus and ornamentals from Asia
Pseudaulcaspis barberi	5	OR	N	no	India on Mangifera
Rolaspis leucadendri	2	AF	N	no	South Africa on Leucadendron, Proteaceae
*Salicicola archangelskyae	1	WPA	N	no	Central Asia on Malus, Pyrus; pest in Central Asia
Selenaspidopsis browni	2	NT	L	no	Mexico on Chamaedorea (palms)
Silvestraspis uberifera	3	OR	N	no	Vietnam on Cinnamomum, Machilus, Pimenta
Thysanofiorinia leei	1	OR	N	no	Hawai'i, Korea, India, Taiwan on Dimocarpus, Nephelium, Litchi
*Unaspis yanonensis	192	EPA	L	no	Asia, Australia, China, Greece, Hawai'i, Japan; citrus pest, Araceae, Alyxis, Citrullus, Fortunella
Total = 46 species	40,000	–	–	–	–
Dactylopiidae					
Dactylopius coccus	4	NT	L	no	on Opuntia from Mexico
Total = 1 species	140	–	–	–	–
Monophlebidae (giant scales)					
*Crypticerya genistae	14	NT	L	yes	Grenada, Puerto Rico, Trinidad, Virgin Is. on Cajanus, Caesalpina, Origanum, Spondias; in Florida, USA
*Drosicha sp.	3	OR	N	no	India, Pakistan, Philippines; on Annona, Psidium
*Icerya seychellarum	9	OR	L	no	Cook Islands, Philippines, Taiwan on Alyxia, Lycium, Piper, Rosa; reported from Guadeloupe
Total = 3 species	390	–	–	–	–
Ortheziidae (ensign scales)					
Arctorthezia cataphracta	1	?	N	yes	UK, on Cantharellus; in Alaska, Canada
Insignorthezia pseudoinsignis	4	NA	W	yes	El Salvador, Guatemala, Mexico; on Citrus, Tillandsia, Thymus; in Louisiana and Texas, USA

Continued

Table 18.6. Continued.

Species	Total int.	Region of orig.	Region of dist.	In USA	Interceptions and notes
*Praelongorthezia praelonga	13	NT	L	no	Dominica, Jamaica, Puerto Rico, Virgin Islands; on Cajanus, Croton, Kalanchoe, Mentha, Rosmarinus
Total = 3 species	216	–	–	–	–
Pseudococcidae (mealybugs)					
*Crisicoccus matsumotoi	3	EPA	N	no	Japan, South Korea on Pyrus
Delottococcus proteae	3	AF	N	no	South Africa on Brunia, Leucadendron, Protea
Dysmicoccus hypogaeus	1	AU	N	no	Australia on Leucospermum
*Dysmicoccus lepelleyi	26	OR	N	no	China, Indonesia, Taiwan, Thailand, Vietnam on Garcinia, Lansium, Musa, Nephelium, Punica
Dysmicoccus queenslandianus	1	AU	N	no	Australia on Allocasuarina
Exallomochlus camur	2	OR	N	no	Philippines on Annona, Lansium
*Exallomochlus hispidus	57	OR	N	no	Asia, on Dimocarpus, Garcinia, Lansium, Murraya, Nephelium
Exallomochlus philippinensis	5	OR	N	no	Philippines on Dimocarpus, Durio, Lansium
Ferrisia terani	3	NT	L	no	Guatemala, Mexico on Rosmarinus, Citrus
Formicococcus robustus	7	OR	N	no	India on Mangifera; also on Citrus, Ficus, Vitis
Hordeolicoccus heterotrichus	4	OR	N	no	Thailand, Vietnam on Nephelium
Hordeolicoccus nephelii	1	OR	N	no	Vietnam on Garcinia; also on Artocarpus, Nephelium
Laminicoccus pandani	1	PI	N	no	Hawai'i on Pandanus
*Maconellicoccus multipori	11	OR	N	no	Philippines, Singapore, Vietnam on Garcinia, Lansium, Nephelium
Nipaecoccus annonae	1	NT	L	no	Puerto Rico on Annona; also Cocos, Persea, Psidium
Nipaecoccus jonmartini	3	NT	L	no	Costa Rica, Puerto Rico on Calathea, Musa, Nephelium
*Nipaecoccus viridis	11	OR	L	yes	widespread; in Costa Rica; on many hosts; recently found in Florida, USA
Palmicultor browni	7	PI	L	no	Australia, Bahamas, Colombia, Puerto Rico on Cocos
Palmicultor palmarum	23	PI	L	no	Widespread on palms; in Mexico and Bahamas
*Paracoccus burnerae	5	AF	N	no	China, India, South Africa on Citrus, Nephelium, Pyrus
Paracoccus ferrisi	12	NT	L	no	El Salvador, Mexico on Citrus, Eriobotrya, Fernaldia, Ocimum, Origanum, Rosmarinus
Paracoccus herreni	2	NT	L	no	El Salvador, Mexico on Annona, Fernaldia
*Paracoccus interceptus	81	OR	N	no	widespread in Asia on Dimocarpus, Garcinia, Lansium, Litchi, Nephelium
Paracoccus mexicanus	9	NT	L	no	Mexico on Ocimum, Rosmarinus

Continued

Table 18.6. Continued.

Species	Total int.	Region of orig.	Region of dist.	In USA	Interceptions and notes
Paracoccus solani	19	NT	L	no	Mexico on *Agave, Ananas, Leucaena, Mentha, Salvia*
Paraputo odontomachi	4	OR	N	no	Cambodia, Vietnam on *Garcinia, Nephelium*
*Phenacoccus franseriae	254	NT	L	no	Mexico, common on *Salvia*; also intercepted on *Ocimum, Origanum, Rosmarinus, Thymus* (herbs)
Phenacoccus graminicola	1	WPA	L	yes	New Zealand on *Feijoa*; in South Africa and California, USA
Phenacoccus hakeae	8	AU	N	no	Australia on *Banksia, Leptospermum, Persoonia*
Phenacoccus stelli	6	AF	N	no	South Africa on *Brunia, Leucadendron, Symphoricarpos*
Planococcus halli	75	NT	L	no	AF common on *Dioscorea*; widespread in NT
*Planococcus kenyae	1	AF	N	no	Kenya on *Musa*; pest of coffee
*Planococcus lilacinus	88	OR	L	no	Asia on *Annona, Dimocarpus, Garcinia, Lansium, Litchi, Nephelium*; also intercepted Costa Rica, Peru
Planococcus litchi	5	OR	N	no	China, Philippines, Thailand on *Garcinia, Litchi, Nephelium*
Pseudococcus aurantiacus	11	OR	N	no	Asia on *Garcinia, Nephelium*
*Pseudococcus baliteus	165	OR	N	no	Asia on *Garcinia, Lansium, Nephelium, Litchi*
*Pseudococcus cryptus	47	NT	L	no	widespread on *Garcinia, Lansium, Litchi, Nephelium*
Pseudococcus dendrobiorum	3	AU	N	no	Taiwan, Thailand on orchids
Pseudococcus elisae	1,219	NT	L	no	NT, common on *Musa*
Pseudococcus eucalypticus	1	AU	N	no	Australia on *Chamelaucium*
Pseudococcus landoi	21	NT	L	no	Costa Rica, Haiti, Honduras, Mexico, South America on *Citrus, Musa, Nephelium, Sechium*
Pseudococcus zamiae	1	AU	N	no	Australia on *Macrozamia*
Puto barberi	22	NT	L	no	NT on *Cajanus, Hibiscus, Origanum*
Rastrococcus expeditionis	1	OR	N	no	Vietnam on *Dimocarpus*
Rastrococcus iceryoides	6	OR	N	no	India, Thailand on *Ficus, Murraya, Psidium*
*Rastrococcus invadens	1	OR	N	no	Philippines on *Psidium*
Rastrococcus spinosus	1	OR	N	no	Vietnam on *Artocarpus*
Rastrococcus tropicasiaticus	3	OR	N	no	India, Philippines, Vietnam on *Murraya, Nephelium*
Rhizoecus amorphophalli	4	OR	L	no	Thailand on roots of *Curcuma, Zingiber, Typhonium*; in Guadeloupe
Spinococcus convolvuli	1	WPA	N	no	Israel on *Salvia*
Vryburgia distincta	1	AF	N	no	AF on *Echeveria*
Vryburgia viator	1	AF	N	no	AF on *Leucospermum*
Total interceptions = 52 species	24,933	–	–	–	–

Continued

Table 18.6. Continued.

Species	Total int.	Region of orig.	Region of dist.	In USA	Interceptions and notes
Stictococcidae					
Parastictococcus multispinosus	2	AF	N	no	Nigeria on *Dennettia*
Stictococcus intermedius	6	AF	N	no	Cameroon, Nigeria on Annonaceae, *Cola*, *Persea*
Stictococcus sjostedti	17				Cameroon, Nigeria on Annonaceae, *Cola*, *Persea*, *Piper*, *Uvaria*
Stictococcus sp.	7	AF	N	no	Nigeria on Annonaceae, *Bactris*, *Dennetiia*, *Phaseolus*, *Solanum*
Total interceptions = 3 species	32	–	–	–	–

Total int., total intercepted; Region of orig., region of origination; Region of dist., region of distribution; NA, Nearctic; NT, Neotropical; AF, Afrotropical; OR, Oriental; AU, Australasian; WPA, Western Palearctic; EPA, Eastern Palearctic; PI, Pacific Islands; W, widespread; L, limited; N, not known to occur in region; *Species of particular interest as potential invasive species based on their pest status, known hosts and distribution, and the number of times they have been intercepted.

18.2 Family Asterolecaniidae (Pit Scales)

Asterolecaniids are a relatively small family of scale insects with 240 species known worldwide; of these, 19 (7.9%) and 51 (21.3%) species are known to occur in the USA and the NT region, respectively. Russell (1941) reported that species in this family are destructive and some are potential enemies of economic plants, but their impact is probably underestimated because of their inconspicuousness. *Asterolecanium coffeae* Newstead was reported to cause heavy damage to coffee in Kenya. Of the eight species of asterolecaniids that have been intercepted, four are widespread in the region and represent 61.5% of the interceptions. Two species of limited distribution in the region have been intercepted; *Palmaspis inlabefactum* (Russell) was intercepted on *Chamaedorea* palms from Mexico and is only known to occur in Mexico, and *Planchonia stentae* (Brain) was intercepted on *Ocimum* from Antigua and is known to occur in the USA (California and Florida), Puerto Rico and Martinique, and on a wide variety of hosts. Two species are not known to occur in the region. *A. ungulatum* Russell was intercepted on durian *Durio zibethanus* from Thailand, and *Palmaspis phoenicis* (Ramachandra Rao) on *Phoenix* sp. palms from Pakistan and the Middle East. All of these are tropical or subtropical species that could become established in the southern USA, Mexico, Central America and Caribbean islands. *Palmaspis phoenicis* is reported to be a pest of date palms (*Phoenix dactylifera*) in Israel, Iraq, Egypt and elsewhere in North Africa (Howard, 2001).

18.3 Family Coccidae (Soft Scales)

Soft scales are the third largest family of scale insects, with 1131 species known worldwide, of which 106 (9.4%) are known to occur in the USA. Many species are known to attack agricultural, ornamental and greenhouse crops throughout the world. They damage plants by their direct feeding that removes the plant sap and also by the secretion of large amounts of honeydew. This serves as a medium for sooty mold fungus, which not only inhibits phytosynthesis but also greatly reduces the aesthetic qualities of the plants. Most of the soft scales of economic importance in the USA are *Ceroplastes*, *Coccus* and *Saissetia* species (Hamon and Williams, 1984).

Of the 29 species of soft scales that have been intercepted, 24 species (61.5%), representing 60.6% of all interceptions of soft scales, are widely distributed in the region. Seven species (17.9%) are limited in distribution and represent 0.6% of the interceptions. Of these, *Coccus capparidis*, *Eulecanium tiliae*, *Kilifia americana* and

Parthenolecanium corni are known to occur in the USA. The following three species are limited in distribution in the region, and are not known to occur in the USA or to be pest species. *Coccus moestus*, an AF species that is present in Central America and the Caribbean region, was intercepted on *Nephelium* spp., *Prunus* spp. and palms; *Philephedra broadwayi*, an NT species, was intercepted from Grenada on an *Annona* sp. and occurs in other Caribbean countries on such economic hosts as *Mangifera indica*, *Spondias* spp. and *Theobroma cacao*. *Protopulvinaria longivalvata*, an OR species that is known to occur in Central America and on some of the Caribbean Islands, was intercepted on *Cinnamomum*, *Morinda*, *Piper*, *Plumeria* and *Schefflera*. The following seven species (20.5%) are not known to occur in the region and represent 20.5% of the interceptions: *Ceroplastes japonicus*, *Drepanococcus chiton*, *Kilifia deltoides*, *Lichtensia viburni*, *Taiwansaissetia formicarii*, *Udinia catori* and *Udinia farquharsoni*. Of these, Miller and Miller (2003) listed *Ceroplastes japonicus* as being a major pest or threat to the USA. This species was described from Japan but occurs throughout much of Europe on a wide range of ornamental and fruit species and is reported to be a pest of kiwi fruit (*Actinidia deliciosa*) and persimmons (*Diospyros kaki*). It is Holarctic (i.e. Nearctic and Paleactic) in distribution, and may not be well adapted to the Caribbean basin region, but poses a threat to the central and northern parts of the USA. Miller and Miller (2003) list *Drepanococcus chiton*, *Lichtensia viburni* and *Taiwansaissetia formicarii* as potential threats to the USA. The first is an Asian species known to attack several tropical fruit species of economic importance that are grown in the Caribbean basin region; *Lichtensia viburni* is a European species that is a pest of ornamentals but also attacks *Prunus*; and *Taiwansaissetia formicarii* is an Asian species that attacks several ornamental and fruit species that could become an important pest species in the Caribbean basin region. *Udinia catori* and *U. farquharsoni* are AF species that attack several ornamental and fruit species and could become important pest species in the Caribbean basin region.

18.4 Family Diaspididae (Armored Scales)

Armored scales (Hemiptera: Diaspididae) comprise the largest and most diverse group of scale insects and are found on terrestrial plants on every continent (including AN). There are 2465 species known worldwide; of these, 329 (13.3%) are known to occur in the USA and 365 (14.8%) in the NT region. They include many economically important pest species of forest, fruit and ornamental crops throughout the world, and are among the most common species found on imported plant products. A review of the armored species intercepted at US ports of entry between 2007 and 2010 (Table 18.3) showed that 71 (50.7%) of the identified species were widely distributed in the Caribbean basin region and comprised 50.2% of all interceptions of armored scales; 41 species (29.3%) were limited in distribution and comprised 40.1% of the interceptions; 28 species (20.0%) were not known to occur in the region and comprised 0.5% of the interceptions; unidentified genera and species accounted for 9.2% of the interceptions. Three species, *Parlatoria ziziphi*, *Aulcaspis tubercularis* and *Abgrallaspis aguacatae* were exceptionally common, accounting for 20,800 (52%) of the total interceptions. The species are discussed by their zoogeographic region of origin.

18.4.1 Neotropical species (8)

1. *Abgrallaspis aguacatae* Evans, Watson and Miller was recently discovered on avocados imported from Mexico (Evans et al., 2009). Currently, it is only known to occur on avocados in Mexico. The impact that this scale would have on agriculture in the USA and elsewhere in the region is unknown, but it does not appear to cause significant damage to avocados in Mexico.
2. *Abgrallaspis mendax* (McKenzie) was described on orchids from Guatemala and was intercepted on orchids from Mexico.
3. *Acutaspis scutiformis* (Cockerell) was described on avocados from Mexico. In the Caribbean basin region it is known to occur in Texas (USA), Mexico and Guatemala, but has not been reported from the Caribbean islands. It was intercepted on plants of *Aglaonema* exported from Costa Rica for propagation.
4. *Acutaspis umbonifera* (Newstead) was described on *Lecythis* from Guyana and is reported from other South American countries and New York (USA) on *Anthurium*, *Attalea*, *Pereskia* and

Heliconia. It has been intercepted on various plants for propagation from Mexico and Central and South America.

5. *Aspidiella hartii* (Cockerell), the yam scale, was described from Trinidad and occurs throughout much of the Caribbean basin region. It is reported as a pest of yams (*Dioscorea*), especially in storage, and also occurs on *Curcuma* (tumeric) and *Zingiber* (ginger) roots. It is a phloem feeder; when large populations are present, yellowing, defoliation and reduction in fruit set and loss in plant vigour may occur. Stored yams encrusted with scales become dry and fibrous. It was intercepted on *Zingiber* and *Curcuma* from Asia, on *Dioscorea* from Africa and Haiti and on *Zingiber* from Barbados.

6. *Hemiberlesia gliwicensis* (Komosinka) was described on Billbergia from Poland. It was intercepted on Tillandsia from Brazil.

7. *Pseudoparlatoria mammata* (Ferris) was intercepted on *Chamaedorea* palms from Mexico. It is not reported as a pest and its potential impact on agricultural crops in other countries in the region is unknown.

8. *Selenaspidopsis browni* Nakahara was intercepted on *Chamaedorea* palms from Mexico. It is not reported as a pest and its potential impact on agricultural crops in other countries in the region is unknown.

18.4.2 Afrotropical species (4)

1. *Aspidiotus capensis* Newstead was described from South Africa on cycads and is only known to attack cycads. It was intercepted *Encephalartos* sp. from South Africa in 2006 and 2007. Although it has not been reported as a pest, the absence of known natural enemies indicates that it has the potential to become a pest of cycads outside its native region.

2. *Duplachionaspis natalensis* (Maskell) is an AF species that was intercepted on lemongrass (*Cymbopogon*) from the Caribbean basin region, and may pose a threat to other grass species in the region.

3. *Parlatoria fluggeae* Hall was described on *Fluggea* from Zimbabwe. It occurs in Algeria, Hawai'i, Morocco and Taiwan. Its reported hosts include *Ficus* and *Hibiscus*, and it was intercepted on *Adenia* from Germany.

4. *Rolaspis leucadendri* (Brain) was described from South Africa and is only known to attack Proteaceae. It was intercepted on *Berzelia* and *Leucadendron* (Proteaceae) from South Africa.

18.4.3 Western Palearctic species (5)

1. *Lepidosaphes malicola* Borchsenius was described from Armenia, and is also reported in other European and Middle Eastern countries, China and India. It was intercepted on apples from Denmark and Italy and on *Prunus* from the Dominican Republic. It should be considered as a potential economic pest since it attacks apples, pears, roses and other hosts, and Miller and Davidson (1990) list it as a pest.

2. *Lepidosaphes pistaciae* Archangelskaia was described from Turkmenistan on pistachios (*Pistacia vera*), and is present in other Central Asian countries, the Middle East and Pakistan. It was intercepted on pistachios from India, Egypt, Iran, Jordan, Kuwait, Lebanon, Syria, Tunisia and Turkey. It is not reported as a pest, but hosts include important plant species such as *Malus*, *Prunus*, *Pyrus*, pistachios and azaleas.

3. *Leucaspis riccae* (Targioni–Tozzetti) was described from Italy and reported in other European and Central Asian countries and in Argentina. Its hosts include ornamentals (*Nerium, Erica, Ephedra*), figs (*Ficus*) and olives (*Olea*). It was intercepted on olives from Italy. Miller and Davidson (1990) list it as a pest species.

4. *Melanaspis inopinata* (Leonardi) was described from Italy on pears and 'mandorlo' and is also reported from other European countries and Pakistan. Its hosts include *Malus*, *Pistacia*, *Prunus* and ornamentals. It was intercepted on pistachios from Iran, Jordan, Lebanon and Syria, and on *Dioscorea* from Nigeria. It is a pest of fruit trees, mainly of Rosaceae, in southern Europe, the Middle East and Armenia, and therefore should be considered a potential threat to the region.

5. *Salicicola archangelskyae* (Lindinger) was described on pears (*Pyrus communis*) from Uzbekistan. Its host also include *Cydonia, Malus, Punica, Olea* and ornamental species. It was intercepted on a plant, and The Netherlands was listed as the origin.

18.4.4 Eastern Palearctic species (6)

1. *Aulacaspis crawii* (Cockerell) is an Asian species, known to attack citrus, mango, orchids and other ornamentals. It was intercepted on mango fruit in baggage from El Salvador in 2001; and on *Citrus* fruit, and leaves of *Aglaia, Buxus, Cymbidium, Pelea* and *Syringa* from Asia and Hawai'i. It is not reported as a pest, but should be considered as a potential threat since it attacks several economic hosts.
2. *Aulacaspis ima* Scott was described from China on *Lindera* and intercepted on *Cinnamomum* from India in 2007.
3. *Aulacaspis rosarum* Borchsenius is an Asian species also known to occur in the PI. Its hosts include *Rosa, Rubus* and *Ficus*. It was intercepted on roses from Mexico in 1986 and 2009.
4. *Parlatoreopsis chinensis* (Marlatt) was described on crab apple (*Malus*) from China. It is also known to occur in the USA (California, Florida, Missouri), Iran and Japan; its hosts also include *Hibiscus, Prunus* and *Pyrus*. It was intercepted on *Acer* from Taiwan.
5. *Parlatoria multipora* McKenzie was described from China, and intercepted on *Cornus* from South Korea.
6. *Unaspis yanonensis* (Kuwana), the arrowhead scale, was described from Japan where it is a major of pest of *Citrus*. It is of economic importance in China. Heavy infestations of the scale cause dieback of terminal branches, and if severe, may kill the tree. Old World interceptions of the scale include *Citrus* (fruit) from Asia, Australia, Greece, Hawai'i; Araceae (leaves) from Australia; *Alyxia* (leaves) from the Cook Islands; *Citrullus* (fruit) from Japan; and *Fortunella* (fruit) from China and Japan. New World interceptions include: *Citrus* (fruit) from Panama in 2001; *Mangifera indica* (fruit) from Nicaragua in 1992; and *Pouteria* (fruit) from Mexico in 2007. Since all of the New World interceptions were taken from baggage, it is uncertain whether *U. yanonensis* is established in the New World, and must be confirmed by collections made on plants in these countries or that are known to have originated there.

18.4.5 Oriental species (23)

1. *Aonidiella comperei* McKenzie is an Asian species, known to occur in the Caribbean basin region and Central America on *Carica papaya*, *Citrus, Cocos nucifera* and other economic crop species, but is not reported to be a pest.
2. *Aulcaspis tubercularis* Newstead, the white mango scale, was described from Java on *Cinnamomum*. It is intercepted primarily on mango and occurs throughout the region. A moderate infestation of this species was discovered in Miami, Florida, in 2002, but it has not been reported in Mexico or Central America. It injures leaves and fruits and is reported to cause significant damage to mangoes in South Africa (Colyn and Schaffer, 1993; Watson, 2002).
3. *Chrysomphalus pinnulifer* (Maskell) is an Asian species reported as a pest of tea in Kenya and India, and a potential invasive pest of *Citrus*. Among its economic hosts are *Annona* spp., *Citrus, Cocos nucifera, Cycas* spp., *Garcinia mangostana, Mangifera indica, Musa, Prunus* spp. and *Vitis vinifera*. It is widespread in the Old World and was intercepted on *Citrus* from the Dominican Republic in 2007 and on *Persea americana* from Guatemala in 2006. It should be considered as a potential threat since it is reported as reaching pest status and has many economic hosts.
4. *Chrysomphalus propsimus* Banks is an Asian species. It is also reported from Hawai'i and usually is found on leaves of palms and *Pandanus*. It was intercepted on *Cocos nucifera* and *Garcinia mangostana* from Vietnam and on *Annona* spp. from the Philippines.
5. *Fiorinia coronata* Williams and Watson was described from palms from the Solomon Islands, and intercepted on *Cocos nucifera* from Vietnam and on palms from Hawai'i. It is not reported as a pest.
6. *Fiorinia proboscidaria* Green was described from Sri Lanka, and is also reported in Hawai'i and other PI and Asia. It was intercepted on *Citrus* from Indonesia and Vietnam and on *Areca* palms from Hawai'i. Its hosts include *Mangifera indica, Citrus*, roses, palms and other species. It is not reported as a pest, but its hosts include several economically important species.
7. *Greenaspis* sp. This species is in an Asian genus. It was intercepted on lemongrass (*Cymbopogon*) from the Caribbean basin region, and may pose a threat to other grass species in the region.
8. *Lepidosaphes cornuta* Ramakrishna Ayyar was described from India on *Piper nigrum*. It was intercepted on *Michelia* from Thailand, and is not reported as a pest.

9. *Lepidosaphes laterochitinosa* was described from a garden in the UK, but is probably native to Asia. It has been intercepted from Asia, Australia and Costa Rica on many hosts including *Citrus*, *Mangifera indica*, orchids, palms and various ornamental species. It has not been reported as an economic pest, but since its hosts include many economic plants and very little is known of its natural enemies, it could pose a threat to agriculture in the region.

10. *Lepidosaphes rubrovittata* Cockerell is an Asian species; it is known to attack *Anthurium*, palms, cacti, cycads, *Ficus*, *Heliconia* and other ornamental species. It is known to occur in Puerto Rico, and was intercepted on *Alpinia* from Costa Rica.

11. *Lepidosaphes similis* Beardsley was described from Micronesia, and is present in other PI. Its hosts include *Areca*, *Coco nucifera* and other palms and *Musa*. It was intercepted on palms from Asia and the PI, and on *Musa* from the Philippines.

12. *Lepidosaphes tubulorum* Ferris was described on *Sapium* from Taiwan. Its hosts include *Malus*, *Prunus* spp., *Vitis vinifera* and various ornamental species. Miller and Davidson (1990) list it as a pest species.

13. *Lindingaspis misrae* (Laing) was described on *Tamarindus* from India. Its reported hosts also include *Capparis* and cycads. It was intercepted on *Azadirachta*, *Mangifera indica* and *Murraya* from India and is not reported to be a pest species.

14. *Parlatoria citri* McKenzie was described from Indonesia and is also reported in Nigeria, India and Thailand. It was intercepted on *Citrus* from Indonesia and Vietnam, and on *Pyrus* from China.

15. *Parlatoria crypta* McKenzie was described from India, and is also reported in the Comoros, Iran, Iraq, Saudi Arabia and Pakistan; it is a pest species. It was intercepted on *Mangifera indica* from India and Pakistan, and on *Murraya* from India.

16. *Parlatoria pseudaspidiotus* Lindinger is an Asian species that is most often associated with orchids but is also known to occur on *Mangifera indica* and other hosts. Miller and Davidson (1990) list it as a pest species. It was intercepted on *M. indica* fruit from India and Mexico and on orchids from the Caribbean islands.

17. *Parlatoria ziziphi* (Lucas), the black parlatoria scale, is by far the most frequently intercepted species. It is almost always intercepted on *Citrus* fruit, and on shipments originating in all the regions where *Citrus* is grown. An infestation was discovered in 1995 on *Citrus* trees in Miami, Florida, but apparently the species was eradicated before it could become established. It is a pest species and considered to be a threat to agriculture in the USA.

18. *Pinnaspis theae* (Maskell) is an Asian species known to occur in Guadeloupe and Martinique. It was intercepted on *Thea* from Brazil, but its hosts also include *Camellia*, *Cordyline* and *Punica*. Miller and Davidson (1990) list it as a pest species.

19. *Pseudaonidia corbetti* Hall and Williams was described from Malaysia. It primarily attacks ornamentals, and was intercepted on *Garcinia mangostana* from China.

20. *Pseudaonidia* (=*Stringaspidiotus*) *curculiginis* (Green) was described from Indonesia, and is also known from Malaysia, Philippines, Vietnam and Thailand. Its hosts include *Gardenia* and palms; it was intercepted on *Pholidota* from the Philippines.

21. *Pseudaulacaspis barberi* (Green) was described from Sri Lanka. Its hosts include *Mangifera indica* and ornamentals. It was intercepted on mango fruit from India.

22. *Silvestraspis uberifera* (Lindinger) was described from the Philippines on *Mallotus*; it is also known from China, Taiwan and Hong Kong. Its reported hosts also include *Artocarpus*, *Cinnamomum*, *Machilus* and *Syzygium* and it was intercepted on *Pimenta* from Vietnam.

23. *Thysanofiorinia leei* Williams was described from Hong Kong on *Nephelium*, and is also reported in Hawai'i, India, Taiwan on *Litchi chinensis* and *Nephelium lappaceum*. It was intercepted on *Dimocarpus longan* from South Korea.

18.4.6 Australian species (2)

1. *Pseudaulacaspis brimblecombei* Williams was described from Australia on *Macadamia integrifolia*. It has only been reported to occur in Australia, and was intercepted on *Protea* and *Telopea* from New Zealand.

2. *Pseudaulacaspis eugeniae* (Maskell) was described from Australia. Its reported hosts include *Eucalyptus*, *Melaleuca*, *Michelia*, palms and other ornamentals. It was intercepted on *Leucospermum*, *Protea* (Proteaceae) and *Scaevola* from Australia.

18.5 Family Dactylopiidae

This is a small family of scale insects comprising one genus and ten species that are found on cacti; four (40%) and nine (90%) species are known to occur in the USA and NT, respectively. Of the 140 interceptions of this family, *Dactylopius opuntiae* (Cockerell) was the most common species, followed by *D. confusus* (Cockerell). Both of these species occur in the USA. *Dactylopius coccus* Costa is not known to occur in the USA. Species in this family are not considered to be an important potential threat to agriculture in the region because of their specificity to cacti and because several of the species are already present in the region.

18.6 Family Monophlebidae (Giant Scales)

Until recently, the species in the family Monophlebidae were part of the family Margarodidae. Currently, it comprises 246 species with ten (4.1%) and 42 (17.1%) of the species known to occur in the USA and the NT region, respectively.

1. The most notable species in this family is the cottony-cushion scale, *Icerya purchasi* (Maskell), a major pest worldwide, which represented 64 (41.3%) of the 155 interceptions of monophlebids that were identified to species.
2. *Crypticerya genistae* (Hempel) was described from Brazil on *Genista*. It has a very wide host range that includes palms, strawberries, grasses and legumes. It is known to occur in Guadeloupe and in Florida (USA). It is important to stop the movement of this pest, to prevent its spread to other countries in the region.
3. The genus *Drosicha* comprises 24 species that are primarily Asian in distribution. It also occurs in the PI region, but is not known to occur in the New World. Specimens were intercepted on *Annona* and *Psidium guajava* from India, Pakistan and the Philippines.
4. *Icerya seychellarum* (Westwood) was described from a palm imported from the Seychelles Islands. It is a very widespread species with a very large and diverse host range. It is a common pest on various host plants in parts of Asia, several PI, Japan and Africa. In the Caribbean basin region, it has been reported from Guadeloupe (Matile-Fererro and Étienne, 2006). It was intercepted from the Cook Islands, Philippines and Taiwan on *Alyxia, Lycium, Piper* and *Rosa*; it poses a potential threat to the Caribbean basin region.

18.7 Family Ortheziidae (Ensign Scales)

The family Ortheziidae comprises 198 species, of which 31 (15.7%) and 64 (32.3%) are known to occur in the USA and the NT region, respectively. Of the 216 interceptions of this family, 146 were identified to family level. *Insignorthezia insignis* (Browne) was the most common species intercepted; this is a very widespread species and has been found on many hosts.

1. *Arctorthezia cataphracta* (Olafsen) is known to occur in several European countries, Canada and Alaska (USA) and was intercepted on *Cantharellus* from the UK. It appears that it may be able to establish in the northern USA, but may not be able to survive in the NT region.
2. *Insignorthezia pseudoinsignis* (Morrison) is known to occur in the USA (Louisiana, Texas), Mexico, El Salvador and Guatemala on *Citrus, Duranta, Gardenia, Solanum* and *Tillandsia*.
3. *Praelongorthezia praelonga* (Douglas) is probably the most important potential pest species of this family in the southern USA and Caribbean basin region. Ebeling (1959) reported it as a pest of *Citrus* in South America and tropical North America (Mexico). It occurs in Mexico and in most of the Caribbean islands, and in Panama.

18.8 Family Pseudococcidae (Mealybugs)

Mealybugs are the second largest family of scale insects, comprising 2277 species, of which 361 (15.9%) and 285 (19.5%) are known to occur in the USA and NT region, respectively. Many species of the family are major pests of fruit and ornamental crops. The species are discussed by their zoogeographic region of origin.

18.8.1 Neotropical species (13)

1. *Ferrisia terani* Williams and Granara de Willink was described from Argentina on *Citrus* and has

also been reported from Guatemala. It was intercepted on *Rosmarinus* from Mexico.

2. *Nipaecoccus annonae* Williams and Granara de Willink was described on *Annona* from St Vincent Island. In the Caribbean basin region, it has also been reported from Guadeloupe, Martinique and San Marino and on *Cocos nucifera*, *Persea americana* and *Psidium guajava*. It was intercepted from Puerto Rico on *Annona*.

3. *Nipaecoccus jonmartini* Williams and Granara de Willink was described from Panama and is reported in the Caribbean basin region from Mexico, Costa Rica and El Salvador on *Citrus*, *Nephelium lappaceum*, *Persea americana*, *Psidium guajava* and other hosts. It was intercepted from Costa Rica and Puerto Rico on *Calathea*, *Musa*, and *N. lappaceum*.

4. *Paracoccus ferrisi* Ezzat and McConnell was described from Mexico on *Gardenia* and is not reported to occur in other countries. It was intercepted from El Salvador and Mexico on *Citrus*, *Eriobotrya japonica*, *Fernaldia*, *Ocimum*, *Origanum* and *Rosmarinus*.

5. *Paracoccus herreni* Williams and Granara de Willink was described from Mexico on *Manihot esculentum* and is also reported from Nicaragua. It was intercepted from El Salvador and Mexico on *Annona* and *Fernaldia*.

6. *Paracoccus mexicanus* Ezzat and McConnell was described from Mexico on *Sedum* and was intercepted from Mexico on *Ocimum* and *Rosmarinus*.

7. *Paracoccus solani* Ezzat and McConnell was described from Arizona (USA) on *Erigeron*. It is also reported from Mexico and Costa Rica. It was intercepted from Mexico on *Agave*, *Ananas comosus*, *Leucaena*, *Mentha* and *Salvia*.

8. *Phenacoccus franseriae* Ferris was described from Mexico on *Franseria* (=*Ambrosia*). It is often intercepted from Mexico on *Salvia*, but has also been intercepted on *Ocimum*, *Origanum*, *Rosmarinus* and *Thymus* (herbs).

9. *Planococcus halli* Ezzat and McConnell was described from St Kitts on yams (*Dioscorea*). It has been reported from Central America and the Caribbean islands. It is usually intercepted on yams from the AF region.

10. *Pseudococcus cryptus* Hempel was described from Brazil on the roots of coffee. It has a wide host range and is known to occur in the Caribbean basin region in Central America and the Caribbean islands. It was intercepted on *Garcinia mangostana*, *Lansium domesticum*, *Litchi chinensis* and *Nephelium lappaceum*.

11. *Pseudococcus elisae* Borchsenius was described on imported bananas from Colombia. It is known to occur in Central and South America and the Caribbean islands. Its hosts include *Annona*, *Citrus* and *Codiaeum*, but it is most commonly found and intercepted on bananas.

12. *Pseudococcus landoi* (Balachowsky) was described from Colombia and is known to occur in Mexico, Central America and the Caribbean islands. It was intercepted from Costa Rica, Haiti, Honduras, Mexico, and South America on *Citrus*, *Musa*, *N. lappaceum* and *Sechium*.

13. *Puto barberi* (Cockerell) was described from Antigua and is known to occur in South America and in the Caribbean islands. It has many hosts and was intercepted on *Cajanus*, *Hibiscus*, *Morinda* and *Origanum*.

18.8.2 Western Palearctic species (2)

1. *Phenaccoccus graminicola* Leonardi was described from Italy on a grass. In the New World, it is known to occur in Argentina and California (USA). It was intercepted on *Feijoa* from New Zealand.

2. *Spinococcus convolvuli* Ezzat was described on *Convolvulus* from Egypt. It was intercepted on *Salvia* from Israel.

18.8.3 Afrotropical species (6)

1. *Delottococcus proteae* (Hall) was described on *Protea* from Zimbabwe. It is only known to attack plants in the Proteaceae. It was intercepted from South Africa on *Brunia*, *Leucadendron* and *Protea*.

2. *Paracoccus burnerae* (Brain) was described from *Passiflora* in South Africa. It has a wide host range and is reported as a citrus pest in South Africa where it is among the three most important mealybug species. It was intercepted from China, India and South Africa on *Citrus*, *N. lappaceum* and *Pyrus*.

3. *Phenacoccus stelli* (Brain) was described on *Borbonia* in South Africa, but is more common on Proteaceae species. It was intercepted from South Africa on *Brunia*, *Leucadendron* and *Symphoricarpos*.

4. *Planococcus kenyae* (Le Pelley), the coffee mealybug, was described on coffee in Kenya, where it is a major pest of coffee. It could become an important pest of coffee in the Caribbean basin region. It was intercepted on bananas from Kenya.
5. *Vryburgia distincta* (De Lotto) was described on *Galenia* in South Africa. It was intercepted on *Echeveria* from Africa.
6. *Vryburgia viator* (De Lotto) was described on *Pyrus* in Kenya. It was intercepted on *Leucospermum* from Africa.

18.8.4 Eastern Palearctic species (1)

1. *Crisicoccus matsumotoi* (Siraiwa) was described on various plants from Japan, and is reported to occur on several hosts including *Citrus*, *Malus* and *Pyrus*. It was intercepted on *Pyrus* from Japan and South Korea. It is not reported as a pest species but could become a pest of apples and pears in the northern USA.

18.8.5 Oriental species (20)

1. *Dysmicoccus lepelleyi* (Betrem) was described on *Annona* in Indonesia. Its hosts include several economic crops such as *Annona*, *Citrus*, *Coffea arabica*, *Garcinia mangostana*, *Mangifera indica* and *Musa*, which indicates that it could be a potential pest for the Caribbean basin region. It was intercepted from China, Indonesia, Taiwan, Thailand and Vietnam on *G. mangostana*, *Lansium domesticum*, *Musa*, *Nephelium lappaceum* and *Punica*.
2. *Exallomochlus camur* Williams was described from the Philippines and in South-East Asia and South Korea. Its hosts include several economic species. It was intercepted from the Philippines on *Annona* and *L. domesticum*.
3. *Exallomochlus hispidus* (Morrison) was described from Singapore on *Gordonia*. Its hosts include several economic species. It was intercepted on *Dimocarpus longan*, *G. mangostana*, *L. domesticum*, *Murraya* and *N. lappaceum* from Asia.
4. *Exallomochlus philippinensis* Williams was described from the Philippines on *L. domesticum*. It is only known to occur in the Philippines where it was intercepted on *D. longan*, *Durio zibethanus* and *L. domesticum*.

5. *Formicococcus robustus* (Ezzat and McConnell) was described from India on mangoes and has also been observed on *Citrus*, *Ficus* and *Vitis vinifera*. It was intercepted from India on mango fruit.
6. *Hordeolicoccus heterotrichus* Williams was described on *N. lappaceum* from Malaysia. It has also been reported on *G. mangostana* and other hosts from Singapore, Thailand and Vietnam. It was intercepted from Thailand and Vietnam on *N. lappaceum*.
7. *Hordeolicoccus nephelii* (Takahashi) was described on *N. lappaceum* from Malaysia. It was intercepted from Vietnam on *G. mangostana*.
8. *Maconellicoccus multipori* (Takahashi) was described in Malaysia. It was intercepted on *G. mangostana*, *L. domesticum* and *N. lappaceum* from the Philippines, Singapore and Vietnam.
9. *Nipaecoccus viridis* (Newstead) was described on *Hygrophila* in India. It has a wide host range. It was intercepted on a plant from Costa Rica and was recently found in Florida (USA).
10. *Paracoccus interceptus* Lit was described on *L. domesticum* from the Philippines. It is widespread in Asia and found on several economic crops such as *Annona*, *Citrus*, *G. mangostana* and *M. indica*. It is periodically intercepted on *D. longan*, *G. mangostana*, *L. domesticum*, *Litchi chinensis*, and *N. lappaceum* from Asia.
11. *Paraputo odontomachi* (Takahashi) was described on *Elaeocarpus* in Malaysia and was intercepted on *G. mangostana* and *N. lappaceum* from Cambodia and Vietnam.
12. *Planococcus lilacinus* (Cockerell) was described on *Citrus* in the Philippines. It has a very wide host and geographic range and has been reported in Central America and the Caribbean islands. It was intercepted from Asia on *Annona*, *D. longan*, *G. mangostana*, *L. domesticum*, *L. chinenesis* and *N. lappaceum*; and on *Polyscias* from Costa Rica and *Solanum* from Peru. It has also been reported as a pest of cacao (*Theobroma cacao*).
13. *Planococcus litchi* Cox was described on *L. chinensis* from Hong Kong. It was intercepted from China, Philippines and Thailand on *G. mangostana*, *L. chinensis* and *N. lappaceum*.
14. *Pseudococcus aurantiacus* Williams was described on *Callophyllum* in Brunei. It was intercepted on *G. mangostana* and *N. lappaceum* from Asia.
15. *Pseudococcus baliteus* Lit was described on *Ficus* from the Philippines. It is frequently intercepted on *G. mangostana*, *L. domesticum*, *N. lappaceum* and *L. chinensis* fruit from South-East Asia.

16. *Rastrococcus expeditionis* Williams was described from Indonesia and was intercepted on *D. longan* from Vietnam.
17. *Rastrococcus iceryoides* (Green) was described on *Capparis* in India. It is widely distributed in the OR region and has a large host range, but is particularly common on mango fruit. It was intercepted from India and Thailand on *Ficus*, *Murraya* and *Psidium guajava*.
*18. *Rastrococcus invadens* Williams was described on mango fruit in Pakistan. It is widely distributed in the Orient and occurs on a wide range of hosts. It has been reported to be a pest of mangoes and was also intercepted on *P. guajava* from the Philippines.
19. *Rastrococcus tropicasiaticus* Williams was described on *Ficus* in Malaysia and is also reported on *Citrus*, *M. indica* and *N. lappaceum*. It was intercepted from India, Philippines and Vietnam on *Murraya* and *N. lappaceum*.
20. *Rhizoecus amorphophalli* Betrem was described on the roots of *Amorphophallus* in Java. Species in this genus feed on the roots of plants. This species is known to occur in Guadeloupe. It was intercepted from Thailand on roots of *Curcuma*, *Zingiber* and *Typhonium*, and is also known to occur on yams (*Dioscorea*). The potential of this species to become an economic pest of these crops is unclear.

18.8.6 Australian species (6)

1. *Dysmicoccus hypogaeus* Williams was described from Australia and was intercepted from Australia on *Leucospermum*.
2. *Dysmicoccus queenslandianus* Williams was described on *Macadamia integrifolia* in Australia and was intercepted from Australia on *Allocasuarina*.
3. *Phenacoccus hakeae* Williams was described on Proteaceae in Australia and is only known to attack species of that family. It was intercepted from Australia on *Banksia*, *Leptospermum* and *Persoonia*.
4. *Pseudococcus dendrobiorum* Williams was described on *Dendrobium* in Australia and is only known to attack orchids. It has also been reported in India, Indonesia, Malaysia, Philippines, Sri Lanka, South Korea and Thailand. It was intercepted from Taiwan and Thailand on orchids.
5. *Pseudococcus eucalypticus* Williams was described on *Eucalyptus* in Australia where it is only known to occur. It was intercepted from Australia on *Chamelaucium*.
6. *Pseudococcus zamiae* (Lucas) was described from *Zamia* that had originated in Australia. It is only known to occur on Arecaceae and Zamaiceae species in Australia. It was intercepted on a *Macrozamia* from Australia.

18.8.7 Pacific Island species (3)

1. *Laminicoccus pandani* (Cockerell) was described on *Pandanus* from the Marquesas Islands. It is also reported on *Cocos nucifera*, *Musa* and sugarcane on several of the PI. It was intercepted on *Pandanus* from Hawai'i.
2. *Palmicultor browni* Williams was described on coconut in the Solomon Islands. It was intercepted on *C. nucifera* from Australia, Bahamas, Colombia and Puerto Rico.
3. *Palmicultor palmarum* (Ehrhorn) was described on palms in Hawai'i. It is usually found on palms and is known to occur in the Caribbean basin region in Florida (USA), Mexico and the Caribbean islands (but not yet reported from Central America). It is reported to cause little damage to adult palm trees but may kill germinating palms. It was intercepted on palms from Mexico and the Bahamas.

18.9 Family Stictiococcidae

This is a small family consisting of three genera and 17 species, all of which are only known to occur in the AF region. Of the 32 interceptions of this family, *Stictococcus sjostedti* Cockerell was the most common species. It was intercepted from Cameroon and Nigeria on Annonaceae, *Cola*, *Persea americana*, *Piper nigrum* and *Uvaria*; *Stictococcus intermedius* Newstead was intercepted from Cameroon and Nigeria on Annonaceae, *Cola* and *P. americana*; and *Parstictococcus multispinosus* (Newstead) was intercepted from Nigeria on *Dennettia*, but is also known to occur on *Annona*, *Cola*, *Hibiscus* and *Theobroma cacao*. Most species in this family are not reported to be important pest species, except for *Stictococcus sojostedti* which is reported as a pest of *T. cacao* in Western Africa; however, they could possibly become pests in the NT countries if they are separated from their natural enemies.

18.10 Conclusions

The approach that has sometimes been taken to address the introduction of an invasive species into a country or region is to take control measures after the species has established and its population has exploded to the point where it is causing economic damage to plants. Sooner or later, the species is likely to spread to other areas of a country and to other countries in the region. A more effective approach involves implementation of preventive measures that will prevent a species from entering and establishing in the country or region. Borrowing an analogy from the game of basketball, the best way to keep your opponent from scoring is to not allow him to have the ball. Similarly, a species that is not allowed to enter a country, cannot cause damage. The introduction and establishment of an invasive scale insect species in any country in the Caribbean basin region is a threat to all of the countries in the region. Therefore, it is mutually beneficial for those working in quarantine and crop protection in each of the countries in the region to work together to stop, or at least to slow down, the movement of pests that threaten crops in their region. This will require adequate resources to be allocated and personnel in each of the countries who are trained to inspect, detect, identify and assess the threat that a scale insect and other organism poses to agriculture in their country. Training in detection, sample and specimen processing and identification, along with identification manuals and other identification materials, need to be made available. Plant protection workers throughout the region need to be in constant communication with each other to alert others to the discovery of a species that was intercepted on plant material or found within the country; only by working together will we be able stop the establishment of an invasive pest in the region.

References

Colyn, J. and Schaffer, B. (1993) The South African mango industry. Fourth International Mango Symposium, Miami, Florida, 5–10 July 1992. *Acta Horticulturae* 341, 60–68.

Ebeling, W. (1959) *Subtropical Fruit Pests*. Division of Agricultural Sciences, University of California, Los Angeles, California, pp. 436.

Evans, G.A., Watson, G.W. and Miller, D.R. (2009) A new species of armored scale (Hemiptera: Coccoidea: Diaspididae) found on a vocado fruit from Mexico and a key to the species of armored scales found on a vocado worldwide. *Zootaxa* 1991, 57–68.

Hamon, A.B. and Williams, M.L. (1984) *The Soft Scales of Florida (Homoptera: Coccoidea: Coccidae). Arthropods of Florida and Neighboring Areas, Volume 11*. Division of Plant Industry, Florida Department of Consumer Services, Gainesville, Florida, pp. 194.

Howard, F.W. (2001) Sap-feeders on palm. Hemiptera: Sternorrhyncha. In: Howard, F.W., Moore, D., Giblin-Davis, R.M. and Abad, R.G. (eds), *Insects on Palms*. CAB International, Wallingford, UK, pp. 161–227.

Matile-Ferrero, D. and Étienne, J. (2006) Cochenilles des Antilles françaises et de quelques autres îles Caraïbes [Hemiptera, Coccoidea]. (Summary in English). *Revue Française d'Entomologie* 28, 161–190.

Miller, D.R. and Davidson, J.A. (1990) A list of the armored scale insect pests. In: Rosen, D. (ed.) *Armored Scale Insects, Their Biology, Natural Enemies and Control, World Crop Pests*, Vol. 4B. Elsevier, Amsterdam, The Netherlands, pp. 299–306.

Miller, G.L. and Miller, D.R. (2003) Invasive soft scales (Hemiptera: Coccidae) and their threat to U.S. agriculture. *Proceedings of the Entomological Society of Washington* 105, 832–846.

Russell, L.M. (1941) *A Classification of the Scale Insect Genus Asterolecanium*. USDA Miscellaneous Publication 424, Washington, DC, pp. 322.

ScaleNet (2011) ScaleNet query index. www.sel.barc.usda.gov/scalenet/query.htm, accessed February 2011.

USDA/APHIS (2012) AQAS (Agricultural Quarantine Activity Systems) internal database. https://mokcs14.aphis.usda.gov/aqas, accessed February 2011.

Watson, G.W. (2002) Arthropods of economic importance. Diaspididae. http://nlbif.eti.uva.nl/bis/daspididae, accessed February 2011.

19 Recent Adventive Scale Insects (Hemiptera: Coccoidea) and Whiteflies (Hemiptera: Aleyrodidae) in Florida and the Caribbean Region

Ian Stocks

Division of Plant Industry, Florida Department of Agriculture & Consumer Services, 1911 SW 34th Street, PO Box 147100, Gainesville, Florida 32614-7100, USA

19.1 Introduction

The flora and fauna of Florida and the Caribbean region are in continual flux. Underlying this flux is the dynamic and complex international trade of plant material involving the USA, the countries of the Caribbean region, South America, Europe and Asia. Vast quantities of fruits and vegetables, and propagative and non-propagative plants, are distributed through ports in the Caribbean region and Florida, and inspection of this material results in a constant struggle against the threat of adventive species. Among the most prominent of potentially and actually adventive animal species are hemipterous insects in the family Aleyrodidae (the whiteflies) and numerous families in the superfamily Coccoidea, such as Diaspididae (armored scales), Pseudococcidae (mealybugs) and soft scales (Coccidae). Several factors conspire to make insects in these groups some of the most noxious of plant pests. First, they are typically quite small, and are therefore more likely to pass unnoticed through inspection sites. Second, many species are parthenogenic. Therefore, it is possible that a new infestation could be started from the introduction on plant material of a single juvenile female, or even a single egg mass. Third, all of these insects are plant parasites, and are therefore almost always in close association with their host plant; thus, they are likely to be introduced already attached to a suitable host. The ecological and economic impact of adventive scales and whiteflies is enhanced primarily by the fact that the natural complement of predators and parasitoids is often absent in the new environment, but also by the fact that, given an expanded diversity of novel potential host plant species, the insects become far more polyphagous than they are in their native environment.

Complicating still further attempts to control the flow of adventive scales and whiteflies is the fact that, in many cases, we have very little information about the species beyond the name, known host(s) and collection locale (which may or may not be the original provenance). Also, it is not uncommon that undescribed species are discovered because they are adventive and have become, at least initially, prominent members of the plant–pest insect community.

This chapter will focus on recent invasions of scales and whiteflies as a result of international trade among Florida and the countries and territories that comprise the Caribbean region. The data presented in table format are derived primarily from the extensive database maintained by the Florida Department of Agriculture, Division of Plant Industry (FDACS-DPI), and ScaleNet, an online database that endeavours to maintain an exhaustive compendium of literature pertaining to scale insects.[i] ScaleNet offers numerous database search options that enable the user to 'data-mine' under various parameters, such as distribution, host and literature citation information. Other information presented is derived from the primary literature, especially Miller et al. (2005), which is an excellent synopsis of the adventive Coccoidea fauna of the USA.

The length of time under consideration, and therefore the list of relevant taxa, is somewhat arbitrary, since the number of adventive taxa in Florida and the Caribbean region has been growing inexorably since record-keeping began. Also, the list is biased toward a description of those taxa introduced into Florida. This is primarily due to the fact that FDACS-DPI has a long history of detailed record-keeping, but also to the intensive sampling and surveying that takes place within Florida. Thus, Florida overall has a better-known fauna than the islands in the Caribbean region. The taxa considered in this review are presented in Table 19.1.

To place these recent developments in perspective, of ~7400 described species of Coccoidea (Miller et al., 2005), there are c. 1300 species of Coccoidea in the Neotropical (NT) region, and c. 320 species of this total are recorded from at least one territory in the Caribbean region (75 species of Pseudococcidae; 132 species of Diaspididae; 55 species of Coccidae; 54 species in assorted families). By comparison, ScaleNet retrieves ~1020 species of Coccoidea in the continental USA, of which ~340 species are recorded from Florida (67 species of Pseudococcidae; 165 species of Diaspididae, 45 species of Coccidae, 61 species in assorted families). A comparison of the respective lists of species indicates that while

Table 19.1. Taxa discussed in this chapter. The year of first report in Florida is given, and the known ranges as reported by ScaleNet and Evans (2008).

Family	Scientific Name	First Recorded in Florida	Range
Pseudococcidae	*Nipaecoccus viridis* (Newstead)		
	Pseudococcus dendrobiorum Williams	2009	AF, AU, OR, PA
	Palmicultor lumpurensis (Takahashi)	2009	AU, OR, PA
	Palmicultor palmarum (Ehrhorn)	2003	AU, NE, OR, PA
	Chaetococcus bambusae (Maskell)	1999	AU, NE, NT, OR, PA
	Paracoccus marginatus Williams and Granara de Willink	1998	AF, AU, NA, NT, OR, PA
	Maconellicoccus hirsutus (Green)	1998	AU, AF, NE, NT
	Planococcus minor (Maskell)	2002	AF, AU, OR, PA, NT, NA
		2010	AF, AU, OR, NA, NT
Kerriidae	*Paratachardina pseudolobata* Kondo and Gullan	1999	NA
	Tarchardiella mexicana (Comstock)	1985	NA
Coccidae	*Pulvinaria psidii* Maskell	1985	AU, AF, NA, NT, OR, PA
	Protopulvinaria longivalvata Green	NR, intercepted	AF, AU, NT, OR
	Phalacrococcus howertoni Hodges & Hodgson	2008	NA, NT
Diaspididae	*Poliaspis cycadis* Comstock	2007	NA, OR, PA
	Unachionaspis tenuis (Maskell)	2007	OR, PA
	Aulacaspis yasumatsui Takagi	1996	AU, NA, NT, OR, PA
	Aulacaspis tubercularis Newstead	2002	AF, AU, OR, PA, NT, NA
	Duplachionaspis divergens (Green)	2000	AU, OR, PA, NA, NT
Monophlebidae	*Crypticerya genistae* (Hempel)	2006	NT, NA
Aleyrodidae: Aleurodicinae	*Aleurodicus dugesii* Cockerell	1996	NT, AU, OR
	Aleurodicus rugioperculatus Martin	2009	NT, NA
Aleyrodidae: Aleyrodinae	*Siphoninus phillyreae* (Haliday)	2010	NA, NT, AU, AF
	Dialeurodes schefflerae Hodges and Dooley	1986	NA, NT, AU, OR

OR, Oriental; PA, Palearctic; AU, Australasian; NT, Neotropical; AF, Afrotropical; NR, not recorded in Florida.

there is substantial similarity, there are nonetheless significant differences. It is these differences that are of concern, because they are the primary source of species that are available for transport between Florida and the Caribbean region. Given the diversity in numerous characteristics of the territories and countries that comprise the countries and territories in the Caribbean region, it is not surprising that species distribution across the Caribbean region is uneven; some countries are relatively well sampled and support a large number of species, while others, for various reasons, are less well sampled or support a lower diversity. Even many common and polyphagous species are unrecorded from many Caribbean region countries. For instance, the seriously pestiferous mealybug *Maconellicoccus hirsutus* (Green) (Pink hibiscus mealybug), is officially recorded from only ten of the 27 territories in the Caribbean region; *Coccus viridis* (Green) (green scale) from 11 territories; *Ceroplastes floridensis* Comstock (Florida wax scale) from 13 territories; *Coccus hesperidum* L. (brown soft scale) from 12 territories; *Pinnaspis strachani* (Cooley) (lesser snow scale) from 15 territories.

Compilation of detailed distribution and species-richness data for Aleyrodidae is complicated by the lack of a whitefly database similar to ScaleNet; however, Evans (2008) enumerated species richness for the biogeographic regions of the world, so rough numbers are available. The distributional and host data given for the species discussed in this review are derived from that source.

Worldwide, there are 1560 species of whitefly in 161 genera (Martin and Mound, 2007). In the NT region, there are 15 genera with c. 120 species in the subfamily Aleurodicinae, and 42 genera with c. 210 species in the subfamily Aleyrodinae. The FDACS-DPI database contains 72 species recorded from Florida, but, as with the scales, this number increases almost yearly. No data are readily available for the number of whitefly species in the Caribbean region, but as the number of monographic systematic works increases, and our knowledge of the distribution of species improves (Martin, 2008), compiling those data will become easier. For instance, Martin (2008) lists 14 species from the Caribbean region in the mainly NT genus *Aleurodicus* (Aleurodicinae), and there are three from Florida, including *A. dispersus* Russell and the recently introduced and established species *A. dugesii* and *A. rugioperculatus*; however, although many of the Caribbean records pertain to Trinidad, whose fauna has a distinctly South American character (Peck *et al.*, 2002), there are almost certainly more species of *Aleurodicus* more widely distributed in the Caribbean region, but that are as yet officially unrecorded.

Several species will be excluded from consideration in this chapter because they will be discussed elsewhere in this book: *Planococcus minor* (Maskell) (Pseudococcidae), *Praelongorthezia praelonga* (Douglas) (Ortheziidae) and *Singhiella simplex* (Singh) (Aleyrodidae).

19.2 Mealybugs: Pseudococcidae

Some of the most seriously noxious Coccoidea are mealybugs, attacking not just plants grown and sold for decoration and aesthetic purposes, but numerous fruit and vegetable crops. In Florida, a large proportion of the known species are adventive. Two particularly pestiferous species have been introduced in recent years, the papaya mealybug (*Paracoccus marginatus*) and the pink hibiscus mealybug (*Maconellicoccus hirsutus*), but each appears to be under significant natural control by parasitoid wasps and predatory lady beetles. Given the very recent arrival in Florida of two additional species of mealybugs, the lebbeck mealybug (*Nipaecoccus viridis*) and the passionvine mealybug (*Planococcus minor*), it is too early to know what effect these two well-known pest species will have on Florida agriculture and horticulture.

19.2.1 *Nipaecoccus viridis* (Newstead)

Common names: lebbeck mealybug, spherical mealybug, karoo thorn mealybug. Distribution: Australasian (AU), Afrotropical (AF), Oriental (OR), Palearctic (PA), Nearctic (NE) (USA: Florida). ScaleNet reports that it is present in Mexico and the Bahamas, but these records are apparently based on misidentifications (Meyerdirk *et al.*, 1988; Dr G. Evans and Dr D. Miller, pers. comm.).

Five other species of *Nipaecoccus* occur throughout the Caribbean region: *N. annonae*

Willams and Granara de Willink (Guadeloupe and Martinique), *N. filamentosus* (Cockerell) (Turks and Caicos, Puerto Rico, Haiti), *N. neogaeus* Willams and Granara de Willink (Trinidad) and *N. pitkini* Willams and Granara de Willink (Trinidad and Tobago). *Nipaecoccus floridensis* Beardsley, which is possibly native to Florida, appears to be restricted to Florida. Only the potentially pestiferous species *N. nipae* (Maskell) is widespread in the Caribbean region, occurring on 17 islands. In general, specimens of *Nipaecoccus* require slide mounting for positive identification, but there are some recognizable differences between species in the colour and shape of the wax produced.

Lebbeck mealybug was first recorded in Florida in late 2009 from Palm Beach County attacking dodder (*Cuscuta exaltata* Engelm; Cuscutaceae) growing in a natural area, and its discovery prompted the release of a pest alert (Stocks and Hodges, 2010). A ground survey to determine the extent of the infestation was initiated, which revealed that it had become widespread in the natural area, feeding on a variety of host plants, and was possibly responsible for the demise of the dodder (K. Griffiths and A. Derksen, pers. comm.); however, it was of only limited distribution in the surrounding suburban areas, where it was found to be feeding most frequently on *Citrus* sp. and *Gardenia jasminoides* Ellis (Rubiaceae). As of February 2011, the host list had grown to 14 plant species, with *Citrus* spp., *G. jasminoides* and dodder seeming to be preferred hosts (FDACS-DPI database). In citrus-growing areas of Jordan, where lebbeck mealybug had been a pest prior to the introduction of biological controls, it caused such extensive damage that orchards were burned in an effort to eradicate it. Through their feeding, developing fruits are aborted and drop prematurely, and toxins in the saliva kill the terminal branch tip near feeding sites, causing die-back (Meyerdirk *et al.*, 1988).

Almost contemporaneous with the discovery of the mealybug in Florida was the discovery of predators and parasitoids associated with *N. viridis* populations. Adult flies reared from masses of mealybug ovisacs were identified as *Leucopis* sp. (Chamaemyiidae), known predators of mealybugs. Also reared from sequestered infested plant material was the parasitoid wasp *Pachyneuron eros* Girault (Pteromalidae). Curiously, it is not clear whether these wasps are parasitoids of the mealybugs, the predatory flies or perhaps both, as both mealybugs and flies are listed as known hosts of *Pachyneuron* wasps. If they are parasitoids of the mealybugs, they would join the other wasp fauna that might exert some measure of biological control, such as *Anagyrus kamali* Moursi (Encyrtidae). After it was reared and released for the control of pink hibiscus mealybug (PHM), *A. kamali* can now be found in both Florida and the Caribbean region (Hoy *et al.*, 2006; Division of Plant Industry, 2004). Overall, some 65 species of Hymenoptera primary parasitoids have been recorded from *N. viridis* (Noyes, 2003), with *A. indicus* (Subba Rao) having an exceptional record of control in Guam and Jordan (Meyerdirk *et al.*, 1988).

19.2.2 *Pseudococcus dendrobiorum* Williams

Common name: dendrobium mealybug. Distribution: AU, OR, PA, NE (USA: Florida, Hawai'i).

This species was reported from Florida in 2009 from a population of *Phalaenopsis* orchids maintained in a greenhouse on the University of Florida campus. Although it prompted the publication of a pest alert (Hodges and Buss, 2009), it has not been collected since, and the population may have been eradicated. Other than distribution information and host data, nothing has been published on the biology, ecology or economic impact of this species. Hosts are exclusively Orchidaceae, including the genera *Ascoglossum*, *Cymbidium*, *Dendrobium*, *Phalaenopsis*, *Pholidota* and *Promatocalpum*. Records from USDA-ARS report 16 interceptions in the past 25 years, all but one from orchid shipments, and all originating from Asia (15 interceptions) and Australia (one interception).

19.2.3 *Palmicultor browni* (Williams)

No common name. Distribution: AU, NE (USA: Florida).

Nothing was known about the biology or ecology of this host prior to its discovery in Florida in 2002, and very little has been learned since. In the literature, it is recorded from coconut, oil palm (*Elaeis guineensis* Jacq.), *Howea forsteriana*

(F. Muell. ex H. Wendl.) Becc. and *Veitchia* spp. In Florida, it has a broad palm-host range, with coconut (*Cocos nucifera* L.), *Phoenix roebelenii* O'Brien and *Adonidia merrillii* (Becc.) Becc. (Arecaceae) accounting for 60% of records. The majority of records (58%) are from Miami-Dade County, but it is also recorded from Broward, Collier, Monroe and Orange counties. Heavy infestations of this species may cause dieback (Dr G. Hodges, pers. comm.).

19.2.4   *Palmicultor palmarum* (Ehrhorn)

Common name: Ehrhorn's palm mealybug. Distribution: AU, PA, OR, NT (Bahamas, Bermuda, Guadeloupe, Jamaica, St Barthélémy, St Martin, St Croix), NE (Mexico, USA: Florida).

This pest was described from Hawai'i from specimens feeding on palms. Ali (1987), citing Beardsley (1966), speculated that it may be indigenous to the Pacific region, and Williams and Martin (2003) speculated that the natural hosts are Arecaceae and possibly Pandanaceae (*Freycinetia* sp.). ScaleNet also records Fabaceae (*Kentia* sp.) and Poaceae (*Phyllostachys* sp.) as additional hosts, but these are such aberrant records that they could be erroneous. Very little is known about the biology or ecology of this species, but it does not seem to be a large threat to palms in general. It was first recorded in the New World in Nassau, Bahamas, in 1980 (Williams, 1981), and has been recorded from Florida 26 times since its first collection in 1999. In Florida, it has been recorded from the following counties: Broward, Flagler, Miami-Dade, Monroe, Palm Beach, Polk and Sarasota, although most frequently from Miami-Dade. Collection records indicate that numerous species of palms are suitable hosts, but *Phoenix roebelenii* O'Brien and coconut are preferred.

19.2.5   *Palmicultor lumpurensis* (Takahashi) (=*Trionymus lumpurensis*)

No common name. Distribution: AU, OR, PA, NE (USA: Florida).

This mealybug is presumably endemic to the Old World, possibly China or Malaysia. Hosts are exclusively Poaceae, especially *Bambusa* spp. Very little has been published about this species, so presumably it is not of economic concern in its native range, nor are there reports of its status in Australia, where it has also been recorded. In Florida, it has been collected 65 times, primarily from *Bambusa* spp., but also *Arundinaria* sp., *Dendrocalamus hamiltonii* Nees and Arn. ex Munro, *Gigantochloa atroviolacea* Widjaja, *Guadua angustifolia* Kunth and *Phyllostachys nigra* (Lodd. ex Lindl.) Munro. It appears to have been introduced into Florida via the international trade in exotic bamboo, in particular into the landscaping of an amusement park (Hodges, 2004a; Hodges and Hodges, 2004). It is now widely distributed across Florida, but does not appear to be having a large negative impact.

19.2.6   *Chaetococcus bambusae* (Maskell)

Common name: giant bamboo scale. Distribution: AF, AU, OR, PA, NE (USA: Florida), NT (Bermuda, Guadeloupe, Jamaica, Puerto Rico, Trinidad, US Virgin Islands).

Giant bamboo mealybug was first introduced into Florida in 1956, but was thought to have been eradicated (Hodges and Hodges, 2004). It was subsequently re-collected in Miami-Dade County 1998, and has been infrequently and sporadically collected around the state. Hosts are exclusively species of grasses, primarily bamboo, including *Bambusa* spp., *Phyllostachys* sp., *Dendrocalamus latiflorus* Munro and *D. asper* Backer ex K. Heyne (FDACS-DPI database), and *Gigantochloa* spp., *Lingnania chungii* (McClure) McClure, *Miscanthus* sp. and *Schizostachyum* sp. (ScaleNet). Hodges and Hodges (2004) reported that since the species was not causing economic damage, the initial infestation in 1998 was not discovered until older leaf sheaths were peeled back from the bamboo for cosmetic reasons.

19.2.7   *Paracoccus marginatus* Williams and Granara de Willink

Common name: papaya mealybug, la cochenille du papayer. Distribution: AU, AF; NE (Mexico, Florida); NT (Antigua, Cuba, Dominican Republic, Guadeloupe, Haiti, St Kitts, Martinique, Puerto Rico, St Martin and St Barthélémy, British Virgin Islands, US Virgin Islands).

Papaya mealybug is an excellent example of an insect species moving from the relative obscurity of the newly described, to an increasingly wide distribution as a pest species as it moves globally through international trade, and back to a status as pest of negligible importance. As such, it is a testament to the power of classical biological control. Papaya mealybug is thought to be a native of Mexico or Central America (Walker et al., 2006), and was described relatively recently (Williams and Granara de Willink, 1992; Miller et al., 1999; Miller and Miller, 2002); however, it moved rapidly through the Caribbean region, and by 2006 was present in at least 12 Caribbean islands. It was first recorded in Florida in 1998 and spread rapidly throughout the state. ScaleNet records papaya mealybug from 45 species in 20 families, but in Florida, hibiscus accounts for 67% of host records, with *Jatropha integerrima* Jacq. and *Acalypha wilkesiana* Muell. Arg. (Euphorbiaceae), and *Carica papaya* L. (Caricaceae) also common hosts. Twenty-five percent of records are from Mimi-Dade County.

The wide host range, attraction to certain hosts (such as hibiscus and papaya) and rapid worldwide spread made papaya mealybug an ideal candidate for classical biological control. Thus, parasitoids collected from mealybug populations in Mexico were artificially reared as part of various USDA programs (Meyerdirk et al., 2004; Muniappan et al., 2006; Amarasekare et al., 2009). The parasitoids *Acerophagus papayae* Noyes and Schauff, *Anagyrus loecki* Moyes and *Pseudoleptomastix mexicana* Noyes and Schauff (Encyrtidae) were released almost contemporaneously in Florida (2003), Guam (2002) and Palau (2003–2004) from cultures maintained in Puerto Rico. Amarasekare et al. (2009) reported that, of the three parasitoid species reared and released in Florida, only *A. papayae* and *A. loecki* were recovered from experimental plots, with *A. papayae* accounting for the highest mealybug mortality. Papaya mealybug population reduction in Florida, as measured by the number of samples submitted to FDACS-DPI, has been dramatic, with a decline from a peak in 2004 of 230 samples, to 28 samples in 2010. In quantitative studies of field effectiveness of parasitoid controls, a reduction in mealybugs of as much as 97% has been observed (cited in Meyerdirk et al., 2004).

19.2.8 *Maconellicoccus hirsutus* (Green)

Common names: pink hibiscus mealybug, grape mealybug, La cochenille de l'Hibiscus, la cochinilla rosada del hibisco. Distribution: AU, AF, OR, PA, NE (USA: California, Florida, Louisiana, Texas), NT (Cuba, Guadeloupe, Grenada, Haiti, St Kitts and Nevis, Martinique, Puerto Rico, St Barthélémy and St Martin, Trinidad and Tobago).

Pink hibiscus mealybug (PHM) is one of the most recognizable names in the pantheon of pest insect species, familiar throughout much of the tropical to sub-tropical world to agricultural workers and the ornamental growing public alike. This pest of worldwide distribution was first recorded from Florida in 2002 (Hodges, 2002), and its invasion into Florida from the Caribbean region, where it had been moving from island to island since the mid-1990s, proved to be inevitable despite robust quarantine measures. PHM is one of the most polyphagous species of all known Coccoidea, having been recorded from over 340 plant species in 75 families, with Euphorbiaceae, Fagaceae and Malvaceae appearing to be the most preferred of host families (ScaleNet). In Florida, it has been recorded from 175 host species, but by far the majority of records (90%) are from *Hibiscus rosa-sinensis*. Other common hosts are *Viburnum odoratissimum* Ker-Gawl (Caprifoliaceae), *Trema micrantha* (L.) Blume (Ulmaceae), *Talipariti tiliaceum* (L.) Fryxell (Malvaceae), *Senna polyphylla* (Jacq.) Irwin and Barneby (Fabaceae) and *Calophyllum* spp. (Clusiaceae). Of c. 3600 Florida records since 2002, 25% are from Miami-Dade County and 15% from Broward County.

In Florida, although PHM is particularly destructive to hibiscus, various cultivars differ in their degree of susceptibility and degree of damage incurred through feeding (Vitullo et al., 2009). With hibiscus, PHM feeding typically induces abnormalities in the length of the internode, resulting in a gall-like condition known as 'bunchy-top' or 'rosetting', which negatively affects the aesthetic value of the plant (Meyerdirk et al., 2001). Prolonged feeding results in defoliation and eventually death of the plant.

In anticipation of its eventual arrival into the NT region, the USDA, in conjunction with state and university cooperators, and the agricultural agencies of several countries in the Caribbean region, initiated biocontrol programs to help reduce the economic impact of PHM on horticulture and agriculture in the Caribbean region and Florida. Economic losses in the Caribbean region following introduction were significant, with Grenada reporting losses of US$3.5–10 million over the 1996–1997 season, and Trinidad and Tobago anticipated losses in excess of US$100 million per annum (Meyerdirk et al., 2001). PHM was found in Louisiana in 2006 (Hodges et al., 2007), Texas in 2007 (Anonymous, 2011a) and Georgia in 2008 (Anonymous, 2011b), each infestation presumably the result of shipments of infested nursery stock from Florida, even though FDACS-DPI quarantines shipments of stock positive for PHM.

Of the 30 species of parasitoid wasps listed by the Universal Chalcidoid Database as documented primary host for PHM, the two encyrtid species *Anagyrus kamali* Moursi and *Gyranusoidea indica* Shafee, Alam and Agarwal were deemed most suitable for importation, rearing and release. Each parasitoid has a preferred host stage for the deposition of the egg and development of the larva, with *Anagyrus kamali* females preferring 3rd-instar PHM larvae, and *G. indica* females preferring late 2nd-instar larvae (Sagarra and Vincent, 1999; Ahmed et al., 2007; Roltsch, 2007). Additionally, two coccinellid predators (*Cryptolaemus montrouzieri* Mulsant and *Scymnus coccivora* Ayyar) were released, but only *C. montrouzieri*, the mealybug destroyer, appears to be effective in Florida. Through these biocontrol efforts, the reduction in PHM pest load has been dramatic, with countries in the Caribbean region experiencing persistent reductions of 90–95% (Osborne et al., 2011). In Florida, PHM is a quarantinable pest, which helps reduce the overall pest load in the environment. However, it remains a common landscaping pest.

19.2.9 *Planococcus lilacicus* (Cockerell)

Common names: coffee mealybug, oriental cacao mealybug. Distribution: AF, AU, OR, PA, NT (Dominican Republic, Haiti).

ScaleNet lists 89 recorded host plants in 36 families, but this mealybug is most widely known in the NT region as a pest of coffee (*Coffea arabica* L.; Rubiaceae) and cacao (*Theobroma cacao* L.; Sterculiaceae). However, Fernando and Kanagaratnam (1987) reported that in Sri Lanka, *P. lilacinus* was found feeding on the peduncle and stalk of the inflorescence of coconut, causing it to dry up, and Waite and Martinez Barrera (2002) reported that it was a minor pest of avocado (*Persea americana* Mill; Lauraceae) in the Philippines, where it occasionally caused early fruit drop. In a study to determine which biocontrol agents would be most suitable for control of *P. lilacinus* in various crop systems in India, Mani (1995) found that several predators and parasitoids from Java and the Philippines were associated with the mealybug, and were possibly suitable for release in India. The species *Tetracnemoidea indica* Ayyar and *Leptomastix dactylopii* Howard (Encyrtidae) and *Aprostocetus purpureus* Cameron (Eulophidae), and the predators *Spalgis epius* (Westwood) (Lycaenidae), *Brumus* sp., *S. coccivora* and *Cryptolaemus montrouzieri* Mulsant (Coccinellidae), *Triommata coccidivora* (Felt) (Cecidomyiidae), and *Cacoxenus perspicax* (Knab.) (Drosophilidae) were recovered from infested crops, but *T. indica*, *A. purpureus* and *S. epius* appeared to be the most efficient at reducing mealybug populations, with *L. dactylopii* having little effect. In a separate study, Mani and Krishnamoorthy (1990a) found that *C. montrouzieri* eliminated *P. lilacinus* populations on pomegranate (*Punica granatum* L.; Lythraceae) fruits in India. In the laboratory, Mani and Krishnamoorthy (1990b) found that the predatory larvae of the lacewing *Mallada desjardinsi* (Navás) (=*Mallada boninensis* (Okamato)) (Chrysopidae) was also effective against passionvine mealybugs.

19.3 Soft Scales: Coccidae

Numerous pestiferous soft-scale species are found in Florida and the Caribbean region, including the recently described croton scale, *Phalacrococcus howertoni* Hodges and Hodgson, and there are fears that several other pest species could be introduced. The soft wax scale, *Ceroplastes destructor* Newstead, is a major pest of *Citrus* in parts of its introduced range, but thus far is not known from the New World. *Vinsonia stellifera* (Westwood) (=*Ceroplastes stellifera*), believed in 1954 to have been eradicated from

Florida (Hodges, 2004b), is being collected more frequently, predominantly in Miami-Dade County, and occurring most commonly on *Schefflera* spp. and *Ixora* sp. Also, in 2007, it was newly reported from New Providence, Bahamas, on *Melaleuca quinquenervia* (Cav.) S.F. Blake (Myrtaceae), a previously unreported host (Blackwood and Pratt, 2007).

19.3.1 *Protopulvinaria longivalvata* Green

No common name. Distribution: AF, AU, OR, NT (South America, Guadeloupe, Haiti, Martinique, Puerto Rico, US Virgin Islands), NE (USA: Florida, interceptions only).

This presumably Old World species began appearing in the Caribbean region in the 1950s (Guadeloupe). It is frequently intercepted by the United States Department of Agriculture, with 200 quarantine interceptions from 1985 to 2010 from shipments or passengers originating in South and Central America and the Caribbean region, including Antigua and Barbuda, Dominica, Jamaica, St Vincent and the Grenadines, and Trinidad and Tobago. The most frequent interceptions are from the Dominican Republic (23%), Puerto Rico (18%) and Jamaica (12%) (USDA-ARS database). This scale has been intercepted in quarantine in Florida by FDACS-DPI inspectors seven times in the past 2 years, all from *Schefflera arboricola* (Hayata) Merr. (Araliaceae) plants originating in Costa Rica. ScaleNet lists 21 host plant species in 12 families, including *Mangifera indica* L. (Anacardiaceae), *Psidium guajava* L. (Myrtaceae), *Gardenia* spp. and *Citrus* spp. The most frequent host interceptions by the USDA were on *Citrus* spp. (29%), *Schefflera* sp. (15%) and *M. indica* (12%). *Protopulvinaria longivalvata* is very similar to *P. pyriformis* Cockerell (pyriform scale), and needs to be slide mounted for accurate identification. There is relatively little information on *P. longivalvata*, but thus far it does not appear to be as pestiferous as pyriform scale, which has a host list of 68 species in 34 families (but does not include *S. arboricola* as a known host). No data are available regarding natural enemies, but observations on submitted samples with parasitoid emergence holes indicate that there may be at least one species of parasitoid wasp attacking it (Anonymous, 2009).

19.3.2 *Pulvinaria psidii* Maskell (=*Chloropulvinaria psidii* Borchsenius)

Common names: green shield scale, guava mealy scale, guava scale. Distribution: AF, AU, NE (USA: Alabama, DC, Florida, Georgia, Missouri, Mississippi, New York, Pennsylvania); NT (South America, Central America, Antigua, Bahamas, Bermuda, Cuba, Dominican Republic, Guadeloupe, Haiti, Jamaica, St Kitts, Montserrat, Martinique, Puerto Rico, Trinidad and Tobago, St Vincent and the Grenadines, US Virgin Islands).

Green shield scale was first recorded in Florida in 1985 (FDACS-DPI database), but it has been moving slowly through the Caribbean islands since the late 1890s (Jamaica 1895, Cuba 1926, Guadeloupe 1957), being present in 17 Caribbean countries by 1993 (ScaleNet, 2011). ScaleNet lists 207 host species in 62 families, with Moraceae, Myrtaceae, Rubiaceae and Rutaceae the most commonly reported families. The FDACS-DPI database lists 118 host species, with a distinct preference for *Schinus terebinthifolius* Raddi (Anacardiaceae), *Ficus* spp. (Moraceae), *Ixora* sp. and *Gardenia* spp. (Rubiaceae), and agreeing in general with that reported by ScaleNet. Nada *et al.* (1990) reported that *P. psidii* (as *Cribropulvinaria psidii*) was one of the three most important pests of mango in Egypt. El-Minshawy and Moursi (1976) reported that *P. psidii* was also a serious pest of guava in Egypt, but was also found throughout the year attacking *S. terebinthifolius*, *Meyerta sinclairii* Seem. (Araliaceae) and *Jasminum humile* L. (Oleaceae). In Bangalore, India, Puttarudriah and ChannaBasavanna (1957) reported that *P. psidii* was attacked by *Cryptolaemus montrouzieri*, Mulsant (Coleoptera: Coccinellidae) (mealybug destroyer). In Bermuda, *Microterys kotinskyi* (Fullaway) and *Coccophagus* (=*Aniseristus*) *ceroplastae* (Howard) were released as biocontrol agents, but *M. kotinskyi* was primarily responsible for control (Bennett *et al.*, 1976). The predators *C. montrouzieri* and *Azya luteipes* (Mulsant) (Coccinellidae) were also released and became established. In combination with the parasitoid wasps, green shield scale was reduced to noneconomic levels in Bermuda. Similarly, in South Africa, Annecke and Moran (1982) found that coccinellid predators were common and that populations of green shield scale were parasitized so heavily that the scale is no longer of economic importance (cited in de Villiers, 2001).

19.3.3 *Phalacrococcus howertoni* Hodges and Hodgson

Common name: croton scale. Distribution: NE (USA: Florida); US Virgin Islands (St Thomas, St Croix; FDACS-DPI database); Figs 19.1 and 19.2.

Croton scale was first detected in Florida in 2008 and soon thereafter was determined to be an undescribed species (Hodges, 2008). The description followed in 2010, by which time croton scale was recorded from 72 host plant species in 34 families. As of February 2011, the host list had grown to 90 species in 36 families, with 60% of the host records from *Codiaeum variegatum* (L.) A. Juss. (Euphorbiaceae; croton).

Fig. 19.1 Croton scale *Phalacrococcus howertoni* Hodges and Hodgson (Coccidae) adult female on croton (*Codiaeum variegatum*). Photograph, Ian Stocks.

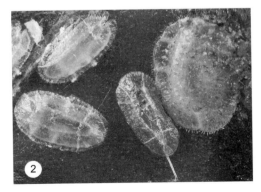

Fig. 19.2 Croton scale immature male puparia (left), adult male under puparium (center), immature female (right). Photograph, Ian Stocks.

Bursera simaruba L. (Burseraceae; gumbo-limbo) is the next most common host, with 7% of records, but *Ficus* spp. (Moraceae) and *P. guajava* L. are also common hosts. Thus far, naturally occurring infestations are limited to the southern-most regions of Florida, with 56% and 10% of records from Miami-Dade County and Broward County, respectively; however, due to the transport of infested nursery stock, FDACS-DPI has records of croton scale from 27 counties.

At this time, relatively little is known regarding the biology of this pest species, but some life history details have been discovered (Hodges and Hodgson, 2010; C. Mannion and S. Brown, unpublished data). In primary areas of infestation in south Florida, year-round warm weather allows for overlapping generations throughout the year, with females taking roughly 30 days to complete development. Female fecundity is high, with egg production of up to 400 eggs per female. Natural predators have been associated with croton scale infestations, including *C. montrouzieri* Mulsant, *Azya orbigera* Mulsant (Coccinellidae) (H. Liere, pers. comm.) and *Laetilia coccidivora* Comstock (as *Laelilla* in Hodges and Hodgson, 2010). The commercially available parasitoid wasp *Metaphycus flavus* (Howard) (Hymenoptera: Encyrtidae) has also been collected in the field, but is apparently at such low levels that it has little effect on population size.

Infestations can build rapidly and can cover almost all above-ground parts of the plant (especially croton). Females prefer the stems and petioles, while males prefer the adaxial leaf surface (personal observation). A heavily infested croton becomes highly stressed through loss of nutrients and the production of honeydew, and can eventually succumb. At Sanibel Island (Lee County, Florida), conservation staff have observed this scale on both native and non-native plants in a wildlife refuge. They report that gumbo limbo and strangler fig (*Ficus aurea* Nutt.) are showing signs of stress, with branch and twig die-back attributed to heavy scale infestations (J. Evans and S. Brown, pers. comm.). They report that, in addition to gumbo-limbo and strangler fig, firebush (*Hamelia patens* Jacq.), wild coffee (*Psychotria nervosa* Sw.), and Bahama wild coffee (*P. ligustrifolia* (Northrop) Millsp.) (Rubiaceae), paradisetree (*Simarouba glauca* DC; Simaroubaceae), and marlberry (*Ardisia escallonoides* Schiede and Deppe ex Schltdl. and Cham.)

and myrsine (*Myrsine cubana* DC) (Myrsinaceae) are also heavily affected. However, horticultural experience with this scale is still so limited that it is too early to tell what effect it will have on the natural landscape in the long term.

19.4 Lac Scales: Kerriidae

Two lac scales are known from Florida, both of which are adventive, and either or both of which could easily become established widely in the Caribbean region. *Paratachardina pseudolobata* Kondo and Gullan is seriously pestiferous, whereas the pest status of *Tachardiella mexicana* (Comstock) is still under investigation. Although an additional 17 species of Kerriidae are found in the NT region, only *Austrotachardiella gemmifera* (Cockerell) (=*Tachardia gemmifera*), which is endemic to Jamaica on *Chrysobalanus* spp. (Chrysobalanaceae), and *P. pseudolobata*, which is adventive, are currently found in the Caribbean region (ScaleNet, 2011). Other than the report by Cockerell (cited in ScaleNet, 2011) that *A. gemmifera* was destroying *Chrysobalanus* spp. trees in Kingston, no life-history data are published for this species.

19.4.1 *Paratachardina pseudolobata* Kondo and Gullan

Common name: lobate lac scale, escama lobada de laca. Distribution: AU, NE (USA: Florida), NT (Bahamas: Andros, Grand Bahamas, New Providence).

Confusion as to the identity of this scale was finally clarified by Kondo and Gullan (2007) in their revision of the genus *Paratachardina*. The lac insect christened lobate lac scale was originally identified as *P. l. lobata* (Chamberlin), but was found by Kondo and Gullan (2007) upon a more detailed analysis to be a distinct and undescribed species. Thus, the literature on the lobate lac scale is confused, with some publications referring to what is actually *P. l. lobata*, and others referring to the new species *P. pseudolobata*; this should be borne in mind when consulting the literature. For instance, it was originally claimed that the lobate lac scale (as *P. l. lobata*) was native to India and Sri Lanka (Howard and Pemberton, 2003). In fact, lobate lac scale (as *P. pseudolobata*) has an unknown provenance, which hinders the task of finding parasitoids for biocontrol considerably.

Lobate lac scale was first discovered in Florida in 1999, but subsequent examination of slides in the Florida State Collection of Arthropods (FSCA) revealed that the scale has been in several Bahamian islands since at least 1992. In Florida, by 2004 the scale had spread to six counties and was by that time considered to be a serious pest (Anonymous, 2004; Mannion *et al.*, 2005), prompting the publication of several factsheets (Hamon and Hodges, 2004; Mannion *et al.*, 2005; Howard *et al.*, 2009). The host list for this species is extensive, with up to 307 plant species enumerated in a report by Howard *et al.* (2006). While many plants from the list of known host plants are variable in their suitability as hosts or susceptibility once infested, there are a number of species that are clearly preferred hosts. Howard and Pemberton (2003) list wax myrtle (*Morella cerifera* (L.) Small = *Myrica cerifera*; Myricaceae), *Chrysobalanus icaco* L. (Chrysobalanaceae), *Conocarpus erectus* L. (Combretaceae), *Myrsine guianensis* (Aublet) Kuntze, *Psychotria* spp., *Annona* spp. (Annonaceae), *Averrhoa carambola* L. (Oxalidaceae), *Litchi chinensis* Sonn. (Sapindaceae), *Ficus* spp., *H. rosa-sinensis* and *M. indica* as preferred hosts. Wax myrtle is so susceptible that several plants in one survey plot in south Florida died within 1 year after becoming heavily infested (Howard *et al.*, 2006), and another experimental host had to be found because the wax myrtles died soon after experimental infestation.

Controlling this pest scale has proved challenging. The hard resinous test is extremely effective at preventing topically applied insecticides from reaching the scale. Soil drenches of imidacloprid were effective at suppressing scale populations (Howard and Pemberton, 2003), but long-term suppression will require finding, rearing and releasing parasitoid biocontrol agents. Initial reports of effective biocontrol agents are now known to refer to *P. lobata lobata*, and are therefore unsuitable candidates; however, field populations of lobate lac scale monitored in south Florida yielded *Metaphycus* sp. and *Ammonoencyrtus* sp. (Encyrtidae), but the numbers of these parasitoids relative to the number of potential hosts was very low (Howard and Pemberton, 2003).

19.4.2 *Tachardiella mexicana* (Comstock) (also including *T. texana* as a junior synonym)

No common name. Distribution: NE (Mexico, USA: Texas, Florida).

Native to the south-west of the USA and to Mexico, this lac scale was first discovered in Florida in 1985 infesting *Acacia* sp. at an amusement park in Orange County, and a second disjunct population was discovered at a nursery in Seminole County in 1987 infesting Texas ebony (*Ebenopsis ebano* (Berl.) Barneby and Grimes = *Pithecellobium flexicaule*). Since then, this scale has been found in other locations and on other hosts, including *Lysiloma latisiliquum* L. (Benth.), *L. sabicu* Benth., *Acacia pinetorum* F.J. Herm. and *A. cornigera* (L.) Willd. Overall, this lac scale shows a marked preference for Fabaceae, including *Mimosa* sp. in its native range. Thus, it is disconcerting that another population of this scale was discovered in 2010 infesting wax myrtle at the same amusement park.

Nothing is known about the biology of this scale in its native environment. In Florida, it can have profound consequences for the tree or shrub, leading to defoliation and death of the plant (S. Brown, unpublished data). In Lee County, massive populations killed several wild tamarind (*L. latisiliquum*) trees, and numerous small wax myrtle shrubs were dead or dying in Orange County (pers. obs.). Massive quantities of honeydew are secreted, leading to a proliferation of sooty mold on the plant, surrounding plants and ground. Although initially each scale starts growth in isolation, in heavy infestations larger females will coalesce into a hard resinous mass that completely envelops the branch. Thus far, no parasitoids are known, and control efforts using horticultural oils are only minimally effective, even at low infestation levels.

19.5 Armored Scales: Diaspididae

Armored scales are notorious tramp species, primarily due to their small size and the attendant difficulty in finding them during inspections. Also, because of the protective cover that they manufacture (which substantially protects them from topical insecticides), and the fact that they feed directly from plant cells as opposed to xylem or phloem (which reduces their exposure to systemic insecticides), they are notoriously hard to control once established. Five adventive species have recently been detected in Florida: *Unachionaspis tenuis* (Maskell), *Aulacaspis yasumatsui* Takagi, *Poliaspis cycadis* (Comstock), *Aulacaspis tubercularis* Newstead and *Duplachionaspis divergens* (Green). Unfortunately, other than host and distribution information, no information is available about the biology of either *P. cycadis* or *U. tenuis*.

19.5.1 *Unachionaspis tenuis* (Maskell)

No common name. Distribution: PA, OR, NE (USA: Florida, South Carolina).

Hosts are grasses (Poaceae), including: *Bambusa* sp., *Phyllostachys* spp., *Pleioblastus* spp., *Sasa* sp. and *Shibataea kumasaca* (Zoll. ex Steud.) Makino. In Nassau County, Florida, it has been taken exclusively from *Bambusa* sp. There is one additional unpublished record of this species from one collection in South Carolina in a botanical garden, also on *Bambusa* sp.

19.5.2 *Aulacaspis yasumatsui* Takagi

Common names: cycad aulacaspis scale, cycad scale, sago palm scale. Distribution: AU, OR, PA, NE (USA: Florida, Texas, Louisiana), NT (Cayman Islands, Puerto Rico, US Virgin Islands: St John, Barbados); Figs 19.3–19.4.

Thought to be native to Thailand, cycad scale has spread widely across the globe and

Fig. 19.3 Cycad scale, *Aulacaspis yasumatsui* Takagi (Diaspididae) adult female scale covers on adaxial leaflet surface. Photograph, Ian Stocks.

Fig. 19.4 Cycad scale, primarily male puparia on adaxial leaflet surface. Photograph, Ian Stocks.

can now be found wherever *Cycas* spp. (Cycadales: Cycadaceae) and *Zamia* spp. (Cycadales: Zamiaceae) grow naturally or are cultivated (Hodgson and Martin, 2001; Germain and Hodges, 2007). Cycad scale was first collected in Florida in 1996, presumably having been introduced via the international trade in ornamental cycads (Weissling and Howard, 1999). Other regions of the southern US that are suitable for the outdoor cultivation of cycads (Louisiana and Texas) also have the scale (Germain and Hodges, 2007), but thus far in California it remains unrecorded outside of nursery settings (G. Watson, pers. comm.). ScaleNet reports that the Caribbean region distribution includes Cayman Islands, Puerto Rico and the US Virgin Islands (St John). The Puerto Rico records are derived from the FDACS-DPI sample submission database, but the Cayman Islands and US Virgin Islands reports were unverified at the time of their publication (Howard and Weissling, 1999). As reported by the Barbados Ministry of Agriculture and Rural Development website, it has been present there since 2003 (Lavine, 2010). Reports of the scale's presence in Martinique since 2005 were cited by Germain and Hodges (2007), and it is now considered established; however, given the rapidity of the worldwide spread of this pest, and the size of the populations that can build up, it is very unlikely that the Caribbean region in general will escape the establishment of this scale. For a general discussion of the biology of cycad scale, the reader is referred to Howard *et al.* (1999), and the synopsis available at Ben-Dov *et al.* (2011).

While the commercially popular *Cycas revoluta* is by far the most common host in Florida, and may in fact be the primary native host for *A. yasumatsui*, it has also been collected from six other commercially available species, *Microcycas calocoma* (Miq.) DC (native to Cuba and endangered; Vovides *et al.*, 1997), the Zamiaceae hosts *Dioon* spp., *Encephalartos* spp. and *Macrozamia moorei* F.J. Muell. and *Stangeria eriopus* (Kunze) (Stangeriaceae). A heavily infested plant will have scales on all parts of the plant, including the roots (Howard *et al.*, 1999).

In its native range, naturally occurring parasitoids and predators help maintain populations below levels that cause significant damage to the plants (Hodgson and Martin, 2001). However, outside its native range, pest load can cause from 70% to 100% mortality of infested *C. revoluta* (Hodgson and Martin, 2001). Howard *et al.* (1999) reported that leaves of recently infested experimental *C. revoluta* plants were necrotic within 112 days of exposure, with the death of the plant occurring in 1 year post exposure. In control experiments, levels of pesticide application (imidacloprid) exceeding that practicable by home gardeners were only marginally successful, perhaps because the plants were rapidly reinfested from the root populations untouched by the pesticide application (Howard *et al.*, 1999). To date, the only pesticide products with known efficacy are certain formulations of horticultural oils, the primary benefit being a reduction in the crawler stage (Howard and Weissling, 1999).

Natural enemies, which often provide the best and most efficient long-term control, have been discovered and have provided partial control. In Florida, the scale predator *Cybocepahlus nipponicus* Endroudy-Younga (Nitidulidae, originally identified as *C. binotatus*) and parasitoid *Coccobius fulvus* (Compere and Annecke) (Encyrtidae) were bred for a release program. These are now established throughout the cycad scale infestation range, and can eliminate a scale population (Howard and Weissling, 1999). Recently, other predators and parasitoids have been recovered from cycad scale-infested areas. In south Texas, where ornamental cycads are grown commercially, the predator *Rhizobius lophanthae* Blaisdell (Coccinellidae) and *Aphytis* sp. *lingnanensis* group (Aphelinidae) appear to be keeping the cycad scale at relatively low levels (Flores and Carlson, 2009).

Field recognition of this species is relatively straightforward, although slide-mounting adult female specimens is necessary for positive identification. The adult female scale is white to off-white, tear-drop to roughly circular in outline and 1–2 mm along the longest axis (Plate 2). Beneath the scale cover, the insect is orange in colour. Females preferentially colonize the adaxial surface of the leaflet, with relatively lower density on the rachis and abaxial surface; however, the male scale covers, which are <1 mm long and parallel sided, preferentially colonize the abaxial leaflet surface and both sides of the rachis. In heavily infested plants, almost the entire surface of the leaf will be covered. Care should be taken in identifying cycad scale in the field, because the white magnolia scale (= false oleander scale), *Pseudaulacaspis cockerelli* (Cooley) has a similar scale cover with a similarly shaped body; however, white magnolia scale adult females and their eggs are yellow, and they typically do not build up large populations on the leaflets.

19.5.3 *Poliaspis cycadis* Comstock

No common name. Distribution: OR, PA, NE (USA: California, Florida, DC).

Hosts include *Cycas* spp. (Cycadaceae), *Dioon edule* (Zamiaceae), *Gaultheria* spp. (Ericaceae) and *Microsemia* sp. (Brassicaceae). UK and Scottish records are presumably from hothouse plants, and in fact the type series is from Washington DC on *Cycas revoluta* Thunb., *Dioon edule* Lindl. and *Microsemia* sp. plants grown in a conservatory. In Florida, it has been collected in Miami-Dade County from *Dioon* sp., but appears to be either not established or at very low levels in the environment (FDACS-DPI database; Hodges and Dixon, 2007).

19.5.4 *Aulacaspis tubercularis* Newstead

Common names. cinnamon scale, mango scale, white mango scale, escama del mango, escama blanca del mango. Distribution: PA, OR, AF, AU, NE (USA: Florida), NT (Antigua, Aruba, Barbados, Bermuda, British Virgin Islands, Dominican Republic, Grenada, Guadeloupe, Jamaica, Martinique, Puerto Rico, Saint Croix, St Lucia, Trinidad and Tobago, US Virgin Islands).

This scale was first observed in Florida in 2002, and has since become a common, though not very significant, pest of mango (FDACS-DPI database). ScaleNet lists 44 host species records in 17 families, with a heavy bias to members of the Lauraceae; however, thus far, Lauraceae of Florida appear to have not become common host plants, with one record each from *Cinnamomum zeylanicum* Garcin ex Blume and *Persea palustris* (Raf.) Sarg. Seventy-five percent of records are from Miami-Dade County, but we have records from five other primarily southern Florida counties.

19.5.5 *Duplachionaspis divergens* (Green)

No common name. Distribution: AU, OR, PA, NT, NE (USA: Florida, intercepted from Alabama and Texas, possibly established in Texas).

This species was first discovered in 2002 on *Miscanthus* sp., but an examination of slides of unidentified scales from *Miscanthus* revealed that it has been in Florida since at least 2000 (Hodges, 2004c). Evans and Hodges (2007) published a brief article on this species in which they provide a synopsis of the known ecology, economic impact and biological control issues. This species is now widely distributed around Florida, known from 35 counties. In 2010, FDACS-DPI received a sample from St Croix, US Virgin Islands, on *Bothriochloa pertusa* (L.) A. Camus. It has been collected exclusively from Poaceae, including *Andropogon* spp., *Cimbopogon citratus* (DC ex Nees) Stapf, *Eustachys* sp., *Miscanthus sinensis* Andersson, *Pennisetum* sp., *Saccharum* spp., *Stenotaphrum secundatum* (Walt.) O. Kuntze, *Tripsacum dactyloides* (L.) L. and *T. floridanum* Porter ex Vasey. In regions where sugarcane is grown, it has caused minor economic impact, but parasitoids appear to suppress the scale to below economic levels. In Florida, *Aphytis* sp. *lingnanensis* group and *Encarsia citrine* (Craw) (Aphelinidae) have been reared from populations of *D. divergens*.

19.6 Cushiony Scales: Monophlebidae

Nine species of monophlebid scales are known from the Caribbean. Perhaps the most seriously pestiferous is *Icerya p. purchasi* Maskell (citrus fluted scale, cottony cushion scale), which is present on 14 Caribbean islands, but *Crypticerya genistae* (Hempel), known from Guadeloupe, also causes economic injury. *Icerya s. seychellarum* (Westwood) has recently been recorded from Guadeloupe on *M. indica* and *Citrus* sp., and from Martinique on *P. guajava* (Matile-Ferrero and Étienne, 2006), but nothing is known yet of its biology on these islands. Williams and Butcher (1987) reported that this species is capable of killing *Citrus* trees, and Beardsley (1966) reported that it was a severe pest of breadfruit (*Artocarpus altilis* (Parkinson) Fosberg).

19.6.1 *Crypticerya genistae* (Hempel) (=*Icerya genistae*)

Distribution: NT (Barbados, Guadeloupe), NE (USA: Florida).

There is no accepted common name for this species, but images of it are returned after an internet search using the name 'white partridge pea bug'. Little has changed regarding our knowledge of this species since a Pest Alert was created in 2006 (Hodges, 2006) and short technical note written in 2008 (Hodges *et al.*, 2008). This species was first collected in Florida in Broward County in 2006, and has since spread throughout Broward County and into Miami-Dade and Palm Beach Counties. The host list, which is quite broad at over 50 species, continues to grow. There is little evidence of host preference pattern other than a slight bias to Asteraceae, Fabaceae and Euphorbiaceae (especially *Chamaesyce* spp.), an observation in Florida that was also made for this species in Guadeloupe (Étienne and Matile-Ferrero, 2008). Hodges *et al.* (2008) reported that it is a pest of several vegetable crops in Barbados, where the Ministry of Agriculture is currently surveying for this pest and searching for biocontrol agents. Étienne and Matile-Ferrero (2008) also reported that the coccinellid beetle *Rodolia cardinalis* (Mulsant), the vedalia beetle, was found associated with the mealybugs. In Guadeloupe, Gagné and Étienne (2009) found the presumably predatory midge *Pectinodiplosis erratica* (Felt) (Cecidomyiidae) along with remains of *C. genistae* feeding on *Mimosa* sp. (Fabaceae).

19.7 Whiteflies: Aleyrodidae

This family contains many well-travelled species, several of which have had profound and long-lasting impacts on agriculture, namely *Dialeurodes citri* (Ashmead) (citrus whitefly), *Aleurocanthus woglumi* Ashby (citrus blackfly) and *Aleurodicus dispersus* Russell (spiraling whitefly). In recent years, several species of whiteflies have become established either in Florida, the Caribbean region or both. Four species are of interest for this review, at least three of which either could be, or are currently, found in the Caribbean region.

19.7.1 *Dialeurodes schefflerae* Hodges and Dooley (Aleyrodinae)

Common name: schefflera whitefly. Distribution: Indonesia, USA (Florida, Hawai'i); NT (Puerto Rico).

Dialeurodes schefflerae was described in 2007, but a systematic review of unidentified slide specimens collected from *Schefflera* species in preparation for the published description indicated that it has been in Hawai'i since c. 1960, California since 1988–1990 and Florida since 1986 (Hodges and Dooley, 2007). Various lines of evidence, such as host-plant affiliation and affinity with other *Dialeurodes* species, suggest that it is native to Asia. Collection data indicate that it is polyphagous within the genus *Schefflera*, but has a marked preference for the popular landscape and potted-plant species *S. arboricola* (Hayata) Merr. Hodges and Dooley (2007) noted that no serious damage occurs to the plant, even though the whiteflies can reach quite high densities on the leaf (pers. obs.). The current distribution in Florida includes 16 counties, with greatest abundance in two of the southernmost counties, Broward and Miami-Dade. The records for San Juan, Puerto Rico are based on pre-departure quarantine interceptions of infested propagative *Schefflera* sp. plants (Hodges and Dooley, 2007).

19.7.2 *Siphoninus phillyreae* (Haliday) (Aleyrodinae)

Common names: ash whitefly, pomegranate whitefly, la mosca blanca del Fresno, mosca blanca del granado. Distribution: Western PA; AU; NT

(Mexico, Central America, USA: California, Florida, Georgia, Hawai'i, North Carolina, Virginia; also reported from Arizona, New Mexico and Nevada).

Throughout much of its range, ash whitefly has historically been a severe pest on a wide variety of commercial and non-commercial plant species, especially pomegranate and *Citrus* spp. (Abd-Rabou, 2006). Ash whitefly was discovered in California in 1988, where it quickly became a severe pest that resulted in millions of dollars of economic loss. A similar situation ensued when ash whitefly appeared on ornamental pear trees (*Pyrus calleryana* Decne. 'Bradford'; Rosaceae) in North Carolina in 1993. Taking advantage of the results of biological control programs developed in the Middle East, whitefly parasitoid *E. inaron* Walker (=*E. partenopea* Masi) (Aphelinidae) was imported into California and reared for eventual release (Pickett *et al.*, 1996). Two years after release, the parasitoid had effected a nearly complete suppression of ash whitefly on both major hosts, ash (*Fraxinus* sp.) and pomegranate. Additionally, several other parasitoids and the beetle predator *Clitostethus arcuatus* (Risso) (Coccinellidae) have been used in biological control programs in other parts of the ash whitefly's range. In Mexico, Myartseva (2006) discovered *Eretmocerus* sp. (Aphelinidae) and the hyperparasite *Signiphora aleyrodis* Ashmead (Signiphoridae) attacking ash whitefly, and she concluded that the parasitoids had been introduced along with the whitefly. Overall, 23 Hymenoptera parasitoids or associates of ash whitefly are reported in the Universal Chalcidoid Database.

In Florida in 2010, a small population of ash whitefly was discovered on containerized pomegranate trees in the nursery grounds of an amusement park in Orange County, the discovery of which prompted the release of a FDACS-DPI Pest Alert (Stocks and Hodges, 2010b). Later in 2010, a second population was discovered on *Citrus* trees on private property in Bay County, and is presumed to be the result of a separate introduction. The parasitoid *E. inaron* was reared from the whiteflies from Orange County and an unidentified parasitoid was reared from the whiteflies in Bay County.

Giant whitefly was first discovered in the USA in Texas in 1991, and by 1996 was established in California, Louisiana and Florida. This whitefly is a highly polyphagous species that can build to severe infestation levels, producing such large quantities of long stringy wax that the plant itself becomes obscured (Hodges, 2004d; Smith and Fox, 2004; Martin, 2008). The FDACS-DPI database lists 51 host plant species, but 66% of the samples received were from *H. rosa-sinensis*; however, there appears in recent years to be a substantial decrease in the populations, reflected in a decrease in samples submitted to FDACS-DPI. The reasons for this are not entirely clear, but the decrease may be due to suppression by parasitoids. Shortly after the giant whitefly's discovery in Texas, *Entedononecremnus krauteri* Zolnerowich and Rose (Eulophidae) was collected from puparia, and subsequently reared and released in California. The other parasitoids associated with giant whitefly are: *Encarsia guadeloupe* Viggiani, *E. brasiliensis* Hempel, *E. meritoria* Gahan, *E. noyesi* Hayat (=*Encarsiella noyesi*) (Eulophidae) and *Idioporus affinis* LaSalle and Polaszek (Pteromalidae) (Noyes, 2003). The beetle *Delphastus catalinae* Horn (Coccinellidae), a generalist whitefly predator, has also been recovered from giant whitefly infestations (Evans, 2008). This beetle is present in Florida (FSCA museum records), but several related *Delphastus* species are found throughout the Caribbean region, including *D. chapini* Gordon (Trinidad), *D. guiniculus* Gordon (Dominican Republic), *D. nebulosus* Chapin (Puerto Rico), *D. pallidus* (LeConte) (Bahamas, Cuba, Dominican Republic, Virgin Islands), *D. pusillus* (LeConte) (Cuba, Jamaica) (Gordon, 1970; FSCA museum data).

In Florida, release of *E. krauteri* was initiated in 1997 in Seminole, Indian River, St Lucie and Volusia Counties. In 1998, *E. noyesi* and *Encarsia* sp. were added to the biocontrol program. Survey data from 2001 found that *E. krauteri* and *E. noyesi* had become established and appeared to be having a significant effect on giant whitefly populations.

19.7.3 *Aleurodicus dugesii* Cockerell (Aleurodicinae)

Common name: giant whitefly. Distribution: NE (USA: Arizona, California, Florida, Hawai'i, Louisiana, Texas; Mexico) NT; Pakistan; Java.

19.7.4 *Aleurodicus rugioperculatus* Martin

Proposed common name: rugose spiraling whitefly. Distribution: USA (Florida); Belize, Guatemala, Mexico (Martin, 2008) (Figs 19.5–19.6).

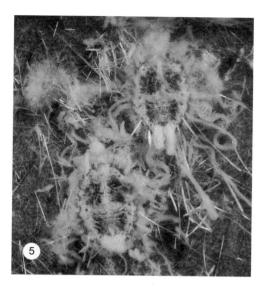

Fig. 19.5 Puparia of rugose spiraling whitefly, *Aleurodicus rugioperculatus* Martin. Photograph, Lyle Buss.

Fig. 19.6 Rugose spiraling whitefly, adult female. Photograph, Lyle Buss.

Rugose spiraling whitefly was described from Belize in 2004 from *Cocos nucifera* L. (Martin, 2004), with subsequent records from Mexico and Guatemala on *Caryocar amygdaliferum* Mutis (Caryocaraceae), *P. guajava*, *P. americana* Mill., *Musa* sp. (Musaceae) and *Melia* sp. (Meliaceae). No additional information about the biology or ecology of this whitefly was published, so all that is currently known about rugose spiraling whitefly is derived from limited experience with it in Florida. The first specimens received by FDACS-DPI were collected in March 2009, from Miami-Dade County, and by December 2010, FDACS-DPI had accumulated 87 records. In response, two factsheets were released to inform homeowners, landscapers and plant regulatory officials (Mannion, 2010; Stocks and Hodges, 2010c).

The majority of records are from Miami-Dade County, but through the spread of infested nursery stock, it is now recorded from two additional counties. In Miami-Dade County, rugose spiraling whitefly has spread rapidly through the environment on its own to infest an ever-increasing area. Furthermore, the host list, which is currently at 37 species, keeps expanding as more collections are made. *Bursera simaruba* (L.) Sarg. (Burseraceae) is the most common, with 23% of the host records, followed by coconut with 10%, *Bucida buceras* L. (Combretaceae) with 9% and *P. americana* with 7% of records. While coconut is a preferred palm host, at least five other palm species are suitable hosts.

Rugose spiraling whitefly is, at present, considered to be seriously pestiferous, causing extensive aesthetic damage to the host plants and physical damage by the production of prodigious quantities of honeydew. The time since infestations were noted has been too brief to determine if it causes significant health problems to the tree, but it seems likely that it will have severe negative consequences. In many cases, the leaves of *Bursera simaruba* and palms have been entirely covered with puparia and the sticky flocculent wax they produce. The honeydew promotes the growth of sooty mold, further compromising the aesthetic quality of the tree and hindering normal tree physiology. Whether rugose spiraling whitefly continues to increase its range and effect in the landscape will almost certainly depend on the role of parasitoids and predators. Martin (pers. comm.) noted parasitoid emergence holes in puparia from Belize, and at least one species of parasitoid wasp (*Aleuroctonus vittatus* Dozier; Eulophidae) has been associated with infestations in Florida, often at high levels, and samples received by FDACS-DPI frequently show signs of parasitism (I. Stocks, pers. obs.).

Acknowledgements

The author would like to thank several people for their assistance during the preparation of this chapter. Beverly Pope (FDACS-DPI) helped to locate several obscure articles; Dr Greg Evans (USDA-APHIS) extracted interception data from USDA databases; Dr Catharine Mannion (UF-IFAS), Dr Greg Hodges (FDACS-DPI), Jenny Evans (Sanibel-Captiva Conservation Foundation) and Stephen Brown (UF-IFAS); Gillian Watson (CDFA) provided unpublished information on several species of scales and whiteflies discussed in this article. Lyle Buss, UF, kindly allowed reproduction of photographs.

Note

i www.sel.barc.usda.gov/ScaleNet/ScaleNet.htm, accessed 8 August 2010.

References

Abd-Rabou, S. (2006) Biological control of the pomegranate whitefly, Siphoninus phillyreae (Homoptera: Sternorrhyncha: Aleyrodidae) by using the bioagent, Clitostethus arcuatus (Coleoptera: Coccinellidae). Journal of Entomology (Faisalabad) 3, 331–335.

Ahmed, S.A., El-Saedi, A.A, Ahmed, H.M and Sayed, A.M. (2007) Host-par asitoid relationship between the par asitoid, Anagyrus kamali Mourse (Hymenopter a: Encyrtidae) and the pink hibiscus mealyb ug, Maconellicoccus hirsutus (Green), (Homopter a: Pseudococcidae). Egyptian Journal of Biological Pest Control 17, 107–113.

Ali, M. (1987) P alm mealybug, Palmicultor palmarum (Homoptera: Pseudococcidae), new to the Indian subcontinent. Annals of the Entomological Society of America 80, 501.

Amarasekare, K.G., Mannion, C .M. and Epsky, N.D. (2009) Efficiency and estab lishment of three introduced parasitoids of the mealyb ug Paracoccus marginatus (Hemiptera: Pseudococcidae). Biological Control 51, 91–95.

Annecke, D.P. and Mor an, V.C. (1982) Insects and Mites of Cultivated Plants in South Africa. Butterworths, Durban, South Afr ica, pp. 383.

Anonymous (2004) Lobate lac scale and Melaleuca: a de vastating insect aided b y an invasive tree. Available at http://tame .ifas.ufl. edu/pdfs/publications/Lobate.pdf.

Anonymous (2009) Protopulvinaria longivalvata. Available at www.freshfromflorida.com/pi/enpp/ triology/4802/triology_4802_entomology.html, accessed 9 August 2012.

Anonymous (2011a) Pink Hibiscus Mealyb ug in Texas. Available at http://etipm.tam u.edu/ insect_alerts/ph_mealybug.cfm.

Anonymous (2011b) Pink Hibiscus Mealybug Found in North Georgia. Available at www.plantmanagementnetwork.org/pub/php/news/2008/ PHMealybug, accessed 9 August 2012.

Beardsley, J.W. (1966) Insects of Micronesia, Homoptera: Coccoidea. Insects of Micronesia 6, 377–562.

Ben-Dov, Y., Miller, D.R. and Gibson, G.A.P. (2011) Remarks Concerning the biology of a scale query results. www.sel.barc.usda.gov/scalecgi/ bionote.exe?Family=Diaspididae&genus=aula caspis&species=yasumatsui&subspecies=&r mktype=biology, accessed 7 August 2012.

Bennett, F.D., Rosen, D., Cochreau, P. and Wood, B.J. (1976) Biological control of pests of tropical fruits and n uts. In: Huffaker, C.B and Messenger, P.S. (eds) Theory and Practice of Biological Control. Academic Press , New York, New York, pp. 359–395.

Blackwood, J.S. and Pr att, J.D. (2007) Ne w host and expanded geographic range of stellate scale (Hemiptera: Coccidae: Ceroplastinae). Florida Entomologist 90, 413–414.

de Villiers, E.A. (2001) Pulvinaria psidii (Cockerell). Guava scale. In: van den Berg, M.A., de Villiers, E.A. and Joubert, P.H. (eds) Pests and Beneficial Arthropods of Tropical and Noncitrus Subtropical Crops in South Africa. ARC-Institute for Tropical and Subtropical Crops , Nelspruit, South Africa, pp. 174.

Division of Plant Industr y (2004) Pink hibiscus mealybug biological control. Available at www. freshfromflorida.com/pi/methods/phm.html.

El-Minshawy, A.M. and Moursi, K. (1976) Biological studies on some soft scale-insects (Hom., Coccidae) attacking guava trees in Egypt. Zeitschrift fur Angewandte Entomologie 81, 363–371.

Étienne, J. and Matile-Ferrero, D. (2008). Crypticerya genistae (Hempel), nouv eau danger en Guadeloupe (Hemiptera, Coccoidea, Monophlebidae). Entomologique d'Egypte 113, 517–520.

Evans, G.A. (2008) The whiteflies (Hemipter a: Aleyrodidae) of the and their host plants and natural enemies. Available at www.sel.barc.usda. gov:8080/1WF/World-Whitefly-Catalog.pdf.

Evans, G.A and Hodges , G.S. (2007) Duplachionaspis divergens (Hemiptera: Diaspididae), a new exotic pest of sugarcane and other g rasses in Flor ida. Florida Entomologist 90, 392–393.

Fernando, L.C.P. and Kanagar atnam, P. (1987) New records of some pests of the cocon ut inflorescence and developing fruit and their natural enemies in Sri Lanka. *Cocos* 5, 39–42.

Flores, D. and Carlson, J. (2009) Fortuitous establishment of *Rhizobius lophanthae* (Coleoptera: Coccinellidae) and *Aphytis lingnanensis* (Hymenoptera: Encyrtidae) in south Texas on the cycad aulacaspis scale, *Aulacaspis yasumatsui* (Hemiptera: Diaspididae). *Southwestern Naturalist* 34, 489–491.

Gagné, R.J. and Étienne, J. (2009) Note on the Cecidomyiidae from Guadeloupe (West Indies) with description of a new species of *Paracalmonia* (Diptera). *Bulletin de la Société Entomologique de France* 114, 337–350.

Germain, J.F. and Hodges, G.S. (2007) First report of *Aulacaspis yasumatsui* (Hemiptera: Diaspididae) in Africa (Ivory Coast), and update on distribution. *Florida Entomologist* 90, 755–756.

Gordh, G. and Headrick, D.H. (2001) *A Dictionary of Entomology*. CABI Publishing, Wallingford, UK.

Gordon, R.D. (1970) A review of the genus *Delphastus* Caset (Coleoptera: Coccinellidae). *Proceedings of the Entomological Society of Washington* 72, 356–369.

Hamon, A.B. and Hodges, G.S. (2004) Lobate lac scale, *Paratachardina pseudolobata* Kondo & Gullan (Hemiptera: Kerriidae). Available at www.freshfromflorida.com/pi/enpp/ento/paratachardina.html.

Hodges, A., Roda, A. and Hodges, G.S. (2007) Pink hibiscus mealybug. Available at www.plantmanagementnetwork.org/proceedings/npdn/2007/posters/6.asp, accessed 8 August 2012.

Hodges, G. and Hodgson, C.J. (2010) *Phalacrococcus howertoni*, a new genus and species of soft scale (Hemiptera: Coccidae) from Florida. *Florida Entomologist* 93, 8–23.

Hodges, G.S. (2002) Pink hibiscus mealybug, *Maconellicoccus hirsutus* (Green). Available at www.freshfromflorida.com/pi/pest-alerts/maconellicoccus-hirsutus.html, accessed 8 August 2012.

Hodges, G.S. (2004a) A bamboo mealybug *Palmicultor lumpurensis* (Takahashi) Coccoidea: Pseudococcidae. Available at www.freshfromflorida.com/pi/enpp/ento/t-lumpurensis.html.

Hodges, G.S. (2004b) Stellate scale, *Vinsonia stellifera* (Westwood) Coccoidea: Coccidae. Available at www.freshfromflorida.com/pi/enpp/ento/v.stellifera.html.

Hodges, G.S. (2004c) *Duplachionaspis divergens* (Green) - A new pest of sugarcane and other grasses in Florida (Hemiptera: Diaspididae). Available at www.freshfromflorida.com/pi/enpp/ento/d.divergens.html.

Hodges, G.S. (2004d) Giant whitefly, *Aleurodicus dugesii* Cockerell, in Florida. Available at www.freshfromflorida.com/pi/pest-alerts/aleurodicus-dugesii.html, accessed 8 August 2012.

Hodges, G.S. (2006) *Crypticerya genistae* (Hemiptera: Margarodidae). A new exotic scale insect for Florida. Available at www.freshfromflorida.com/pi/enpp/ento/c.genistae.html

Hodges, G.S. (2008) A new exotic soft scale insect on croton in south Florida (Hemiptera: Coccoidea: Coccidae). Available at www.freshfromflorida.com/pi/enpp/ento/coccidea_coccidae.html.

Hodges, G.S. and Buss, L.J. (2009) An orchid mealybug, *Pseudococcus dendrobiorum* Williams (Hemiptera: Pseudococcidae. Available at www.freshfromflorida.com/pi/pest-alerts/pseudococcus-dendrobiorum.html, accessed 8 August 2012.

Hodges, G.S. and Dixon, W.N. (2007) The poliaspis cycad scale *Poliaspis cycadis* Comstock (Hemiptera: Diaspididae): A new exotic scale insect for Florida. Available at www.freshfromflorida.com/pi/enpp/ento/poliapsis_cycadis.html.

Hodges, G.S. and Dooley, J.W. (2007) A new species of *Dialeurodes* Cockerell (Hemiptera: Aleyrodidae) on *Schefflera* Forst and Forst in Florida. *Insecta Mundi* 16, 1–5.

Hodges, G.S. and Hodges, A.C. (2004) New invasive species of mealybugs, *Palmicultor lumpurensis* and *Chaetococcus bambusae* (Hemiptera: Coccoidea: Pseudococcidae), on bamboo in Florida. *Insecta Mundi* 87, 396–397.

Hodgson, C.J. and Martin, J.H. (2001) Three noteworthy scale insects (Hemiptera: Coccoidea) from Hong Kong and Singapore, including *Cribropulvinaria tailungensis*, new genus and species (Coccidae), and the status of the cycad-feeding *Aulacaspis yasumatsui* (Diaspididae). *Raffles Bulletin of Zoology* 49, 227–250.

Hodges, G.S., Hodges, A.C. and Unruh, C.M. (2008) A new exotic pest from Florida's natural areas: *Crypticerya genistae* (Hemiptera: Monophlebidae). *Florida Entomologist* 91, 335–337.

Howard, F.W. and Pemberton, R.W. (2003) The lobate lac scale, a new pest of trees and shrubs in Florida: implications for the Caribbean region. *Proceedings of the Caribbean Food Crops Society* 39, 91–94.

Howard, F.W. and Weissling, T.J. (1999) Questions and answers about cycad aulacaspis scale insect. *Proceedings of the Florida State Horticultural Society* 112, 243–245.

Howard, F.W., Hamon, A., McLaughlin, M., Weissling, T. and Yang, S. (1999) *Aulacaspis yasumatsui* (Hemiptera: Sternorrhyncha: Diaspididae), a scale insect pest of cycads recently introduced into Florida. *Florida Entomologist* 82, 14–27.

Howard, F.W., Pemberton, R.W., Hodges, G.S., Steinberg, B., McLean, D. and Liu, H.(2006) Host plant range of lobate lac scale, *Paratachardina lobata*, in Florida. *Proceedings of the Florida State Horticultural Society* 119, 398–408.

Howard, F.W, Pemberton, R., Hamon, A., Hodges, G.S., Steinberg, B., Mannion, C.M., McLean, D. et al. (2009) Lobate lac scale, *Paratachardina lobata lobata* (Chamberlin) (Hemiptera: Sternorrhyncha: Coccoidea: Kerriidae). Available at http://edis.ifas.ufl.edu/in471, accessed 8 August 2012.

Hoy, M., Hamon, A. and Nguyen, R. (2006) Pink hibiscus mealybug, *Maconellicoccus hirsutus* (Green) (Insecta: Hemiptera: Pseudococcidae) Available at http://entnemdept.ufl.edu/creatures/orn/mealybug/mealybug.htm, accessed 8 August 2012.

Kondo, T. and Gullan, P.J. (2007) Taxonomic review of the lac insect genus Paratachardina Balachowsky (Hemiptera: Coccoidea: Kerriidae), with a revised key to genera of Kerriidae and description of two new species. *Zootaxa* 1617, 1–41.

Lavine, G. (2010) Identifying and controlling the Sago Palm Scale (*Aulacaspis yasumatsui*). Available at www.agriculture.gov.bb.

Mani, M. (1995) Studies on the natural enemies of oriental mealybug, *Planococcus lilacinus* (Ckll.) (Homoptera: Pseudococcidae) in India.*Journal of Entomological Research* 19, 61–70.

Mani, M. and Krishnamoorthy, A. (1990a) Outbreak of mealybugs and record of their natural enemies on pomegranate. *Journal of Biological Control* 4, 61–62.

Mani, M. and Krishnamoorthy, A. (1990b) Predation of *Mallada boninensis* on *Ferrisia virgata, Planococcus citri* and *P. lilacinus. Journal of Biological Control* 4, 122–123.

Mannion, C. (2010) Gumbo limbo spiraling whitefly, a new whitefly in south Florida. Available at http://miami-dade.ifas.ufl.edu/documents/GumboLimboSpiralingWhitefly.pdf, accessed 8 August 2012.

Mannion, C., Howard, F., Hodges, G. and Hodges, A. (2005) Lobate lac scale, *Paratachardina lobata* (Chamberlin). Regional IPM center pest alert. Available at www.ncipmc.org/alerts/lobatelacscale/lobate_lac_scale.pdf.

Martin, J.H. (2004) Whiteflies of Belize (Hemiptera: Aleyrodidae). Part 1- introduction and account of the subfamily Aleurodicinae Quaintance & Baker. *Zootaxa* 681, 1–119.

Martin, J.H. (2008) A revision of *Aleurodicus* Douglas (Sternorrhyncha, Aleyrodidae), with two new genera proposed for palaeotropical natives and an identification guide to world genera of Aleurodicinae. *Zootaxa* 1835, pp. 100.

Martin, J.H. and Mound, L. (2007) An annotated check list of the world's whiteflies (Insecta: Hemiptera: Aleyrodidae). *Zootaxa* 1492, 1–84.

Matile-Ferrero, D. and Étienne, J. (2006) Cochenilles des Antilles françaises et de quelques autres îles Caraïbes (Hemiptera, Coccoidea). *Revue Française d'Entomologie* 28, 161–190.

Meyerdirk, D.E., Khashimuddin, S. and Bashir, M. (1988) Importation, colonization and establishment of *Anagyrus indicus* (Hym: Encyrtidae) on *Nipaecoccus viridis* (Hom: Pseudococcidae) in Jordan. *Entomophaga* 33, 229–237.

Meyerdirk, D.E., Warkentin, R., Attavian, B., Gersabeck, E., Francis, A., Adams, M. and Francis, G. (2001) Biological control of pink hibiscus mealybug project manual. United States Department of Agriculture, Animal and Plant Health Inspection Service, Plant Protection and Quarantine, Available at www.aphis.usda.gov/import_export/plants/manuals/domestic/downloads/phm.pdf, pp. 194.

Meyerdirk, D.E., Warkentin, R., Attavian, B., Gersabeck, A., Francis, A., Adams, M. and Francis, G. (2002) Biological control of pink hibiscus mealybug project manual. Available at www.aphis.usda.gov/import_export/plants/manuals/domestic/downloads/phm.pdf.

Meyerdirk, D.E., Muniappan, R., Warkentin, R., Maba, J. and Reddy, G.V.P. (2004) Biological control of papaya mealybug, *Paracoccus marginatus* (Hemiptera: Pseudococcidae) in Guam. *Plant Protection Quarterly* 19, 110–114.

Miller, D.R. and Miller, G.L. (2002) Redescription of *Paracoccus marginatus* Williams and Granara de Willink (Hemiptera: Coccoidea: Pseudococcidae), including descriptions of the immature stages and adult male. *Proceedings of the Entomological Society of Washington* 104, 1–23.

Miller, D.R., Williams, D.J. and Hamon, A.B. (1999) Notes on a new mealybug pest in Florida and the Caribbean: the papaya mealybug,

Paracoccus marginatus Williams and Granara de Willink. Insecta Mundi 13, 179–181.

Miller, D.R., Miller, G.L. and Watson, G.W. (2005a) Invasive species of mealybugs (Hemiptera: Pseudococcidae) and their threat to U.S. agriculture. Proceedings of the Entomological Society of Washington 104, 825–836.

Miller, D.R., Miller, G.L, Hodges, G.S and Davidson, J.A. (2005b) Introduced scale insects (Hemiptera: Coccoidea) of the United Sates and their impact on U.S. agriculture. Proceedings of the Entomological Society of Washington 107, 123–158.

Muniappan, R., Meyerdirk, D.E., Sengebau, F.M., Berringer, D.D. and Reddy, G.V.P. (2006) Classical biological control of the papaya mealybug, Paracoccus marginatus (Hemiptera: Pseudococcidae) in the republic of Palau. Florida Entomologist 89, 212–217.

Myartseva, S.N. (2006) Siphoninus phillyreae (Haliday) (Hemiptera: Sternorrhyncha: Aleyrodidae) and its parasitoid, Encarsia inaron (Walker) (Hymenoptera: Encyrtidae): two new records of insects for Mexico. Entomological News 117, 451–453.

Nada, S., Abd Rabo, S. and El Deen Hussein, G. (1990) Scale insects infesting mango trees in Egypt (Homoptera: Coccoidea). In: Koteja, J. (ed.) Proceedings of the Sixth International Symposium of Scale Insects Studies, Part II. Cracow, pp. 133–134.

Noyes, J.S. (2003) Universal Chalcidoidea Database. Available at: www.nhm.ac.uk/entomology/chalcidoids/index.html, accessed 8 August 2012.

Osborne, L., Mannion, C., Griffiths, K., Hornby, P., Roda, A. and Meyerdirk, D. (2011) Pink hibiscus mealybug (PHM) Maconellicoccus hirsutus (Green). Available at www.aphis.usda.gov/plant_health/plant_pest_info/phmb/downloads/project-information.pdf, accessed 8 August 2012.

Peck, S.B., Cook, J. and Hardy Jr, J.D., (2002) Beetle fauna of the island of Tobago, Trinidad and Tobago, West Indies. Insecta Mundi 16, 9–23.

Pickett, C.H., Ball, J.C., Casanave, K.C., Klonsky, K.M., Jetter, K.M., Bezark, L.G. and Schoenig, S.E. (1996) Establishment of the ash whitefly parasitoid Encarsia inaron (Walker) and its economic benefit to ornamental street trees in California. Biological Control 6, 260–272.

Puttarudriah, M. and ChannaBasavanna, G.P. (1957) Notes on some predators of mealybugs (Coccidae: Hemiptera). Mysore Agricultural Journal 32, 4–19.

Roltsch, W. (2007) Insectary rearing of the pink hibiscus mealybug and its parasitoids in a desert environment. Available at www.cdfa.ca.gov/phpps/ipc/biocontrol/pdf/insects/phm_parasite-production.pdf, accessed 8 August 2012.

Sagarra, L. and Vincent, C. (1999) Influence of host stage on oviposition, development, sex ratio, and survival of Anagyrus kamali Moursi (Hymenoptera: Encyrtidae), a parasitoid of the hibiscus mealybug, Maconellicoccus hirsutus Green (Homoptera: Pseudococcidae). Biological Control 15, 51–56.

Smith, T. and Fox, A.J. (2004) Biological control of giant whitefly, Aleurodicus dugesii Cockerell, in Florida. Available at www.freshfromflorida.com/pi/methods/giant-whitefly-bc.html, accessed 8 August 2012.

Stocks, I.C. and Hodges, G.S. (2010a) Nipaecoccus viridis (Newstead), a new exotic mealybug in South Florida (Coccoidea: Pseudococcidae). Available at www.freshfromflorida.com/pi/pest_alerts/pdf/nipaecoccus-viridis-pest-alert.pdf.

Stocks, I.C. and Hodges, G.S. (2010b) Ash whitefly, Siphoninus phillyreae (Haliday), a new exotic whitefly (Hemiptera: Aleyrodidae) in central Florida, and Encarsia inaron, its parasitoid (Hymenoptera: Aphelinidae). Available at www.freshfromflorida.com/pi/pest_alerts/pdf/ash-whitefly-pest-alert.pdf.

Stocks, I.C. and Hodges, G.S. (2010c) Aleurodicus rugioperculatus Martin, a new exotic whitefly in South Florida (Hemiptera: Aleyrodidae). Available at www.freshfromflorida.com/pi/pest_alerts/pdf/aleurodicus-rugioperculatus-pest-alert.pdf.

Vitullo, J., Zhang, A., Mannion, C. and Bergh, J.C. (2009) Expression of feeding symptoms from pink hibiscus mealybug (Hemiptera: Pseudococcidae) by commercially important cultivars of hibiscus. Florida Entomologist 92, 248–254.

Vovides, A.P., Ogata, N., Sosa, V. and Peña-García, E. (1997) Pollination of endangered Cuban cycad Microcycas calocoma (Miq.) A.DC. Botanical Journal of the Linnean Society 125, 201–210.

Waite, G.K. and Martinez Barrera, R. (2002) Insect and mite pests. In: Whiley, A.W., Schaffer, B. and Wolstenholme, B.N. (eds) The Avocado. Botany, Production and Uses. CAB International, Wallingford, UK.

Walker, A., Hoy, M. and Meyerdirk, D. (2006) Paracoccus marginatus. Available at http://entnemdept.ufl.edu/creatures/fruit/mealybugs/papaya_mealybug.htm, accessed 8 August 2012.

Weissling, T.J. and Howard, F.W. (1999) *Aulacaspis yasumatsui*. Available at http://entnemdept.ufl.edu/creatures/orn/palms/cycad_scale.htm.

Williams, D.J. (1981) New records of some important mealybugs (Hemiptera: Pseudococcidae). *Bulletin of Entomological Research* 71, 243–245.

Williams, D.J. and Butcher, C.F. (1987) Scale insects (Hemiptera: Coccoidea) of Vanuatu. *New Zealand Entomologist* 9, 88–99.

Williams, D.J. and Granara de Willink, M.C. (1992) *Mealybugs of Central and South America*. CAB International, Wallingford, UK, pp. 635.

Williams, D.J. and Martin, J.H. (2003) A palm mealybug, *Palmicultor palmarum* (Ehrhorn) (Hem. Pseudococcidae), now found in the Canary Islands. *Entomologist's Monthly Magazine* 139, 178.

20 Biology, Ecology and Control of the Ficus Whitefly, *Singhiella simplex* (Hemiptera: Aleyrodidae)

Jesusa Crisostomo Legaspi,[1] Catharine Mannion,[2] Divina Amalin[2] and Benjamin C. Legaspi, Jr[3]

[1]*US Department of Agriculture, Agricultural Research Service, CMAVE/Florida A&M University - Center for Biological Control, 6383 Mahan Dr., Tallahassee, Florida 32308, USA;* [2]*Tropical Research and Education Center, University of Florida, 18905 SW 280th Street, Homestead, Florida 33031, USA;* [3]*Florida Public Commission, Tallahassee, Florida 32399, USA*

20.1 Economic Importance

Whiteflies are small Homopteran insects that cause crop damage by extracting phloem sap, excreting honeydew that serves as a medium for fungi, or acting as vectors of economically important viral pathogens (Byrne and Bellows, 1991). Crop loss can exceed 50% yield reduction as their importance as economic pests appears to increase continually. The ficus whitefly, *Singhiella simplex* (Singh) (Hemiptera: Aleyrodidae), is an economic pest of *Ficus* plant species in India, Burma and China (Hodges, 2007). The whitefly has been most commonly found infesting weeping fig (*Ficus benjamina* L.) (Moraceae) (Fig. 20.1). However, it has also been reported on *F. altissima* Blume (lofty fig, false banyan tree), *F. bengalensis* L. ('banyan tree'), *F. microcarpa* L. f. (Cuban laurel), *F. aurea* Nutt. (strangler fig), *F. lyrata* Warb. (fiddle-leaf fig), *F. racemosa* L. (Cluster Fig, Indian Fig) and *F. maclellandii* King (banana-leaf fig) (Mannion *et al.*, 2008). When disturbed, small clouds of the tiny gnat-like insects emerge from whitefly-infested foliage. Severe infestations result in leaf dropping or shedding and defoliation. In common with other whiteflies, the ficus whitefly can cause serious injury to host plants by sucking sap, resulting in wilting, yellowing, stunting, defoliation or plant death (Osborne, 2008).

20.2 Geographic Distribution

Although *S. simplex* has historically been known as a pest of *Ficus* in India, Burma and China, its arrival in the continental US is relatively recent. Recently, the whitefly was intercepted at entry points on *Ficus* plants imported into Korea from China (Suh *et al.*, 2008). Possibly the earliest record is that of the Florida Department of Agriculture and Consumer Services, Division of Plant Industry (FDACS-DPI) in South Florida on 3 August 2007 on *F. benjamina* (Hodges, 2007). A similar report was made by the Miami-Dade County Extension, University of Florida – Institute of Food and Agricultural Sciences. Initial reports of ficus whitefly in 2007 were confined to Miami and surrounding areas such as Coral Gables, Davie, Opa-laka and Hialeah. By 2011, FDACS-DPI surveys reported expansion from as far south as Key West, to as far north as St Augustine, although Miami continued to account for a significant percentage of records (44.3%; 128 reports from the city of Miami out of 289 total reports) (Fig. 20.2). Clusters of reports were found in the west coast of Florida from Naples up to St Petersburg and Tampa. The preferred host plant for the ficus whitefly was *Ficus benjamina* with 80.3% of

Fig. 20.1 *Ficus* defoliation caused by ficus whitefly (photo by H. Glenn, University of Florida).

records (232/289) (Table 20.1). Overall, plants in the genus *Ficus* comprised 95.2% of records (275/289), with isolated reports on host plants such as *Hibiscus* and mango.

20.3 Descriptive Biology

Very little is known about the biology and life history of the ficus whitefly. Eggs are usually laid on leaf undersides (Fig. 20.3) and hatch into crawlers. The crawlers are mobile and begin to feed. Early nymphal stages can be very difficult to detect. The nymphs become immobile feeders, usually oval and flat in shape (Mannion et al., 2008). During the pupal stage, the nymphs turn tan to light green with red eyes, and measure about 1.3 mm in length. The adult whitefly is yellow, and the wings are white with a faint grey band towards the middle (Hodges 2007; Mannion et al., 2008).

20.4 Reproductive Biology and Life Table Analysis

Like other whiteflies in its genus, *S. simplex* is assumed to have at least three generations per year in Florida (Hodges, 2007) with a life cycle completed within about 1 month (Mannion et al., 2008). Detailed reproductive biology and life table studies at five different constant temperatures were performed by Legaspi et al. (2011). In the laboratory, development rates (reciprocal of duration times) were studied at 15, 20, 25, 27, 30

and 35°C on leaf cuttings of *F. benjamina*. No insects survived the 35°C treatment. Total duration of immature stages varied from 97.11 d at 15°C to 25.23 d at 30°C (Table 20.2). Within each immature lifestage, development rates increased linearly with temperature and were described using linear equations. For the combined immature stages (eggs to pupae), the effect of temperature on development was described adequately using both linear regressions and a nonlinear Briere model:

$$r(T) = aT(T - T_0)\sqrt{T_L - T} \quad (20.1)$$

where a is an empirical constant, r is development rate, T is temperature, T_0 is the lower developmental threshold, and T_L is lethal temperature (Briere et al., 1999) (Fig. 20.4). The linear model estimated the lower developmental threshold temperature (T_0) to be 10.6°C. By comparison, the Briere model estimated T_0 of 7.3°C and upper lethal temperature of 45.9°C. The thermal requirement for development from eggs to pupae was calculated to be 487.8 degree days. Life table parameters for the whitefly at each temperature are shown in Table 20.3. Ficus whitefly reproduction was highest at 27°C: R_0, GRR, T, r, λ and DT were 23.114 ♀/♀, 24.25 ♀/♀, 31.413 d, 0.0999 ♀/♀/d, 1.105 ♀/♀/d and 6.93 d, respectively. The calculations assumed a 1:1 sex ratio which may have underestimated actual reproductive potential, because the sex ratio of immatures that successfully emerged was female-biased (79.4%; 15♂: 58♀).

The combined effects of temperature and adult female age were analyzed using the nonlinear regression model of Enkegaard (1993): eggmean = $(p+qT)$ d $\exp(-wTd)$; where T is temperature. The Enkegaard model did not provide a very good fit to the observed data (Fig. 20.5), possibly because of high variability in fecundity and paucity of data points. Female adult survivorship was plotted on a linear scale (Fig. 20.6). Duration of adulthood was significantly longer at 15°C compared to all other temperatures tested, averaging 8.0 d, compared to 4.2, 2.8 and 2.5 at 25, 27, and 30°C, respectively.

Temperature was not found to affect lifetime fecundity significantly. At 15, 25, 27 and 30°C, lifetime fecundity per female averaged 27.0, 37.9, 46.2 and 27.7 eggs, respectively. The temperature effect was not significant, probably

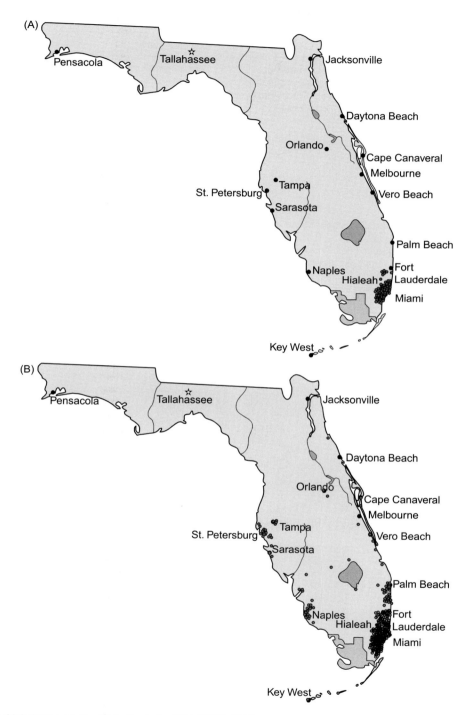

Fig. 20.2 Ficus whitefly distribution in the USA (2007–2011). (A) Initial distribution reported in 2007; (B) distribution from 2007 to February 2011. Based on surveys by the Florida Department of Agriculture and Consumer Services, Division of Plant Industry. In 2007, reports were confined to Miami and surrounding areas. By 2011, reports included St Petersburg and Tampa, as well as one from St Augustine (Hodges, 2007).

Table 20.1. Host plants for ficus whitefly.

Host plant	Common name	Number of records
Callicarpa americana L. (Lamiales: Lamiaceae)	American Beautyberry	1
Citrus × paradisi Mcfayden (Sapindales: Rutaceae)	Grapefruit	1
Coccoloba uvifera L. (Caryophyllales: Polygonaceae)	Seagrape, baygrape	1
Ficus altissima Blume (Rosales: Moraceae)	Council tree	1
F. aurea Nutt.	Florida strangler fig	9
F. benghalensis L.	Indian banyan	4
F. benjamina L.	Weeping fig	232
F. lyrata Warb.	Fiddleleaf fig	1
F. maclellandii King	Alii fig, banyan leaf fig	7
F. microcarpa L. f.	Chinese banyan	17
F. salicaria Berg	Willow leaf fig	1
Ficus sp.		3
Hibiscus rosa-sinensis L. (Malvales: Malvaceae)	Chinese hibiscus, China rose, show flower	1
Mangifera indica L. (Sapindales: Anacardiaceae)	Mango	1
Manilkara zapota (L.) P. Royen (Ebenales: Sapotaceae)	Sapodilla	1
Ruellia brittoniana Morong (Scrophulariales: Acanthaceae)	Britton's wild petunia	8

Fig. 20.3 Life stages of the ficus whitefly. (A) Eggs; (B) red-eyed nymph (pupa); (C) adult (photos by H. Glenn, University of Florida).

Table 20.2. Effect of temperature on immature survival (days) of the ficus whitefly, *Singhiella simplex* (means ± SE) (Legaspi et al., 2011).

Life stage	Temperature (°C)				
	15	20	25	27	30
Eggs	36.00 ± 0.82a	21.00 ± 0.26b	11.00 ± 0.00c	10.21 ± 0.14cd	8.00 ± 0.00e
First	1.56 ± 0.18ab	1.00 ± 0.00b	2.19 ± 0.32a	1.00 ± 0.00b	1.00 ± 0.00b
Second	18.78 ± 1.34a	11.12 ± 0.18b	6.06 ± 0.28c	6.05 ± 0.26c	4.38 ± 0.33c
Third	11.67 ± 0.5a	6.18 ± 0.30b	3.06 ± 0.25c	6.21 ± 0.36b	3.38 ± 0.21c
Fourth	22.56 ± 0.62a	12.88 ± 0.34b	8.56 ± 0.39c	4.05 ± 0.38d	5.54 ± 0.37d
Pre-pupae	3.67 ± 0.64a	1.75 ± 0.25b	1.00 ± 0.00c	1.05 ± 0.05b	1.23 ± 0.12b
Pupae	3.67 ± 0.71a	2.38 ± 0.39b	1.06 ± 0.06c	1.10 ± 0.07c	1.23 ± 0.17c
Total immatures	97.11 ± 2.11a (9)	56.31 ± 0.69b (16)	32.81 ± 0.68c (16)	29.74 ± 0.26d (19)	25.23 ± 0.17e (13)

Insects in 35°C treatment did not survive and are excluded from analysis; each stage analyzed separately for effects of temperature on life stage duration (means ± SE); numbers in parentheses indicates sample size; within each row, means followed by different letters are significantly different Tukey HSD; $P < 0.05$).

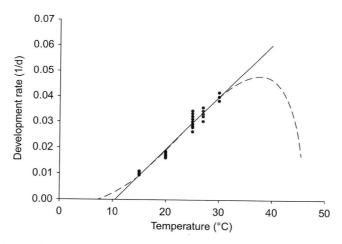

Fig. 20.4 Effect of temperature on development rate (egg to pupae): linear model $r(T) = -0.0216421 + 0.0020499T$ (SE of estimates = 0.0010686 and 0.0000435, respectively; $F = 2216.8$; $df = 1, 71$; $P < 0.01$; $R^2 = 0.97$). Briere model (dashed line) estimates: a = 0.0000146, $T_0 = 7.3120084$; $T_L = 45.9512202$ (ASE = 0.0000020, 1.0080836 and 3.3177988, respectively; $R^2 = 0.97$) (from Legaspi et al., 2011).

due to high variability. Also, lower daily fecundity at lower temperatures may have been compensated by longer ovipositional periods.

20.5 Controlling the Ficus Whitefly

20.5.1 Chemical control

Mannion *et al.* (2008) recommend drenching soil around the bases of trees or hedges with neonicotinoid compounds such as imidacloprid or clothianidin. These insecticides are widely known to attack the insect central nervous system while displaying reduced toxicity to mammals. When applied properly, neonicotinoids should provide adequate whitefly control for 4–8 months, although monitoring after 3 months is suggested, with possible spot treatments where needed. Although soil application is the preferred control, foliar treatments may be necessary during extreme infestations. In such cases,

Table 20.3. Life history parameters for S. simplex (Legaspi et al., 2011).

Parameter	Temperature (°C)			
	15	25	27	30
Net reproductive rate (R_0)[a]	13.055	21.464	23.114	13.834
Gross reproductive rate (GRR)[b]	19.495	29.415	24.25	16.62.0
Generation time (T)[c]	99.3	35.48	31.413	26.86
Intrinsic rate of increase (r)[d]	0.0258	0.0864	0.0999	0.0978
Finite rate of increase (λ)[e]	1.0262	1.0902	1.105	1.103
Doubling time (DT)[f]	26.87	8.022	6.93	7.087

[a] $R_0 = \Sigma l_x m_x$ summation of survivorship multiplied by age-specific fecundity per female, expressed in units of ♀/♀; egg numbers divided by 2 assuming 1:1 sex ratio
[b] $GRR = \Sigma m_x$; summation of age-specific fecundity per female in ♀/♀
[c] $T = (\Sigma x l_x m_x) / R_0$ in days
[d] $r \approx (\ln R_0) / T$ in ♀/♀/day
[e] $\lambda = e^r$ in ♀/♀/day
[f] $DT = \ln(2)/r$ in days

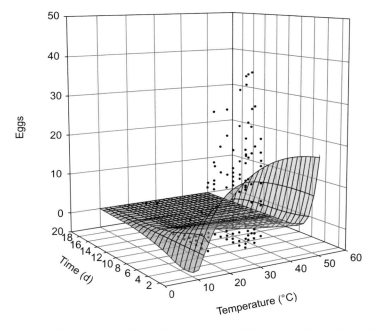

Fig. 20.5 Enkegaard surface showing effects of temperature and time on numbers of eggs laid. Model was: eggs = $(-30.2105187+2.6177368T)\ d\exp(-0.0342074Td)$ (ASE = 8.83, 0.556, 0.0038, respectively) $R^2 = 0.16$; where T is temperature and d is time (Enkegaard, 1993; Legaspi et al., 2011).

recommended foliar insecticides include flonicamid (novel insecticide), abamectin (also an acaricide/nematicide), azadirachtin (insect growth regulator), *Beauveria bassiana* (entomopathogenic fungus), pyriproxyfen (juvenile hormone analogue), pymetrozine (novel antifeedant), endosulfan (organochlorine), spiromesifen (lipid biosynthesis inhibitor), buprofezin (chitin synthesis inhibitor), bifenthrin (pyrethroid) and acetamiprid (neonicotinoid). To prevent the development of resistance, insecticides should be rotated based on differing modes of action.

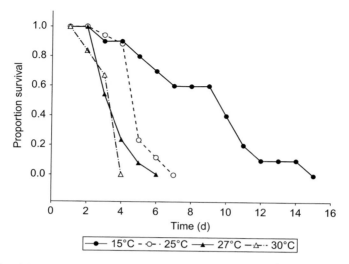

Fig. 20.6 Female adult survivorship (plotted on linear scale) (Legaspi et al., 2011).

20.5.2 Biological control

Biological control agents used against whiteflies typically include parasitic Hymenoptera such as *Encarsia formosa* Gahan (Aphelinidae) (Hoddle et al., 1998) or *Eretmocerus* spp. (van Lenteren and Martin, 2006), predatory coccinellid beetles (Obrycki and Kring, 1998; Arnó et al., 2010), entomopathogenic fungi such as *Beauveria bassiana* (Balsamo) Vuillemin and *Paecilomyces fumosoroseus* (Wize) Brown and Smith (Poprawski and Jones, 2001), or combinations of these agents (Labbé et al., 2009; Bardin et al., 2008).

Whitefly-infested *Ficus* plants in Miami that were sampled for natural enemies (G. Hodges, C. Mannion, unpublished data) yielded several species of parasitic wasps and predatory beetles (Coleoptera: Coccinellidae). The wasps were identified to be *Encarsia protransvena* Viggiani (Hymenoptera: Aphelinidae) and *Amitus benetti* Viggiani and Evans (Hymenoptera: Platygasteridae) (Fig. 20.7). *Encarsia protransvena* is distributed worldwide and attacks whitefly hosts such as the greenhouse whitefly, *Trialeurodes vaporariorum* (Westwood) (Homoptera: Aleyrodidae) (Giorgini, 2001). *Amitus benetti* is a parasitoid of the silverleaf whitefly, *Bemisia argentifolii* Bellows and Perring (Homoptera: Aleyrodidae) (Ryckewaert and Alauzet, 2002). The beetles were the multicolored Asian lady beetle, *Harmonia axyridis* Pallas; the ashy gray lady beetle, *Olla v-nigrum* (Mulsant); *Exochomus childreni* Mulsant; *Chilocorus nigritus* (F.); and the dark blue lady beetle, *Curinus coeruleus* (Mulsant). *Harmonia axyridis* has been used as a biological control agent against aphids and other species of Hemipteran pests (Koch, 2003; Yoon et al., 2010). *Olla v-nigrum* is indigenous in arboreal habitats throughout much of the USA, and is known to be a natural enemy of aphids and psyllids (Michaud, 2001). *Exochomus childreni* has been evaluated as a predator of mites in citrus (Villanueva et al., 2004). *Chilocorus nigritus* has been used successfully as a biological control agent of scale pests throughout the tropical and tropical environments (Ponsonby, 2009). *Curinus coeruleus* is a predator of citrus psyllids (Soemargono et al., 2008).

The generalist predator *Delphastus catalinae* (Horn) (Coleoptera: Coccinellidae) was evaluated as a potential biological control agent of ficus whitefly (J.C. Legaspi, unpublished data). *Delphastus* is widely known as a whitefly predator (e.g., Heinz et al., 1999; Legaspi et al., 2006). Female predators were starved for a 24-h period, then allowed to feed for a 24-h period on ficus whiteflies of different stages: eggs, small and large larvae. Initial prey numbered 200 eggs, 100 small larvae and 50 large larvae. Within the 24-h feeding period, *Delphastus* adults consumed about 150 eggs or 40 small larvae (Fig. 20.8). Predation on large larvae was minimal.

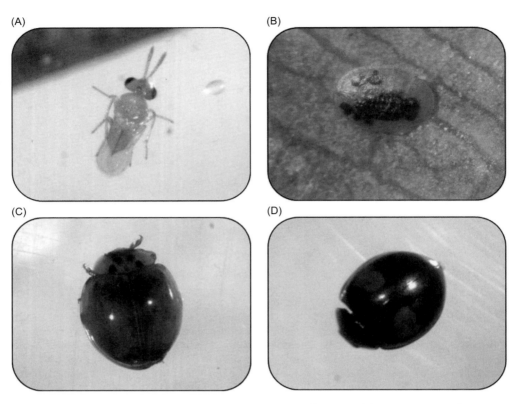

Fig. 20.7 *Encarsia protransvena* adult (A) and parasitized whitefly nymph (B). *Harmonia axyridis* (C) and *Olla v-nigrum* (D) adults (photos by H. Glenn, University of Florida).

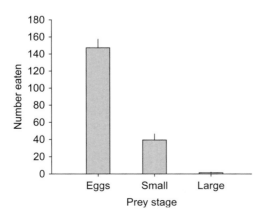

Fig. 20.8 Predation by *Delphastus catalinae* on different life stages of the ficus whitefly (JCL unpublished data). Female predators were starved for 24 h, then given a feeding period of 24 h.

Therefore, *Delphastus* may be a useful control agent when the whiteflies are in the egg or small larval stages, but unlikely to be successful against large larvae.

Acknowledgments

Technical assistance was provided by Neil Miller (USDA, ARS, CMAVE), Keith Marshall, Jr: and William Zeigler (Florida A&M University). Luis Bradshaw and Phellicia Perez (University of Florida) assisted in the field collection of the ficus whitefly and natural enemies. We thank Ru Nguyen, Greg Hodges, Ian Stocks, Trevor Smith (Division of Plant Industry, Gainesville, Florida) and Jesse A. Logan (US Forest Service, retired) for assistance and helpful discussions.

This chapter presents the results of research only. The use of trade, firm or corporation names in this publication is for the information and convenience of the reader. Such use does not constitute an official endorsement or approval by the US Department of Agriculture or the Agricultural Research Service of any product or service to the exclusion of others that may be suitable.

References

Arnó, J.R., Gabarra, R., Liu, T.-X., Simmons, A.M. and Gerling, D. (2010) Natural enemies of *Bemisia tabaci*: predators and parasitoids. In: Stansly, P.A. and Naranjo, S.E. (eds) *Bemisia: Bionomics and Management of a Global Pest*. Springer, Amsterdam, The Netherlands, pp. 385–421.

Bardin, M., Fargues, J. and Nicot, P.C. (2008) Compatibility between biopesticides used to control grey mould, powdery mildew and whitefly on tomato. *Biological Control* 46, 476–483.

Briere, J.-F., Pracros, P., Le Roux, A.-Y. and Pierre, J.-S. (1999) A novel rate model of temperature-dependent development for arthropods. *Environmental Entomology* 28, 22–29.

Byrne, D.N. and Bellows, T.S. (1991) Whitefly biology. *Annual Review of Entomology* 36, 431–457.

Enkegaard, A. (1993) The poinsettia strain of the cotton whitefly, *Bemisia tabaci* (Homoptera: Aleyrodidae), biological and demographic parameters on poinsettia (*Euphorbia pulcherrima*) in relation to temperature. *Bulletin of Entomological Research* 83, 535–546.

Giorgini, M. (2001) Induction of males in thelytokous populations of *Encarsia meritoria* and *Encarsia protransvena*: A systematic tool. *BioControl* 46, 427–438.

Heinz, K.M., Brazzle, J.R., Parella, M.P. and Pickett, C.H. (1999) Field evaluations of augmentative releases of *Delphastus catalinae* (Horn) (Coleoptera: Coccinellidae) for suppression of *Bemisia argentifolii* Bellows & Perring (Homoptera: Aleyrodidae) infesting cotton. *Biological Control* 16, 241–251.

Hoddle, M.S., Van Driesche, R.G. and Sanderson, J.P. (1998) Biology and use of the whitefly parasitoid *Encarsia formosa*. *Annual Review of Entomology* 43, 645–669.

Hodges, G. (2007) The fig whitefly *Singhiella simplex* (Singh) (Hemiptera: Aleyrodidae): A new exotic whitefly found on ficus species in south Florida. Division of Plant Industry, Florida Department of Agriculture and Consumer services, http://www.freshfromflorida.com/pi/pest-alerts/singhiella-simplex.html, accessed 10 August 2012.

Koch, R.L. (2003) The multicolored Asian lady beetle, *Harmonia axyridis*: A review of its biology, uses in biological control, and non-target impacts. *Journal of Insect Science* 3, 1–16.

Labbé, R.M., Gillespie, D.R., Cloutier, C. and Brodeur, J. (2009) Compatibility of an entomopathogenic fungus with a predator and a parasitoid in the biological control of greenhouse whitefly. *Biocontrol Science and Technology* 19, 429–446.

Legaspi, J.C., Simmons, A.M. and Legaspi Jr, B.C. (2006) Prey preference by *Delphastus catalinae* (Coleoptera: Coccinellidae) on *Bemisia argentifolii* (Homoptera: Aleyrodidae): effects of plant species and prey stages. *Florida Entomologist* 89, 218–222.

Legaspi, J.C., Mannion, C., Amalin, D. and Legaspi Jr, B.C. (2011) Life table analysis and development of *Singhiella simplex* (Hemiptera: Aleyrodidae) under different constant temperatures. *Annals of the Entomological Society of America* 104, 451–458.

Mannion, C., Osborne, L., Hunsberger, A., Mayer, H. and Hodges, G. (2008) *Ficus Whitefly: A New Pest in South Florida*. Institute of Food and Agricultural Sciences, University of Florida, Gainesville, Florida.

Michaud, J.P. (2001) Numerical response of *Olla v-nigrum* (Coleoptera: Coccinellidae) to infestations of Asian citrus psyllid, (Hemiptera: Psyllidae) in Florida. *Florida Entomologist* 4, 608–612.

Obrycki, J.J. and Kring, T.J. (1998) Predaceous Coccinellidae in biological control. *Annual Review of Entomology* 43, 295–321.

Osborne, L. (2008) The 'Ficus Whitefly': A new pest in south Florida. Invasive Arthropod Working Group. University of Florida, www.mrec.ifas.ufl.edu/lso/IAWG/FIG/default.asp, accessed 10 August 2012.

Ponsonby, D.J. (2009) Factors affecting utility of *Chilocorus nigritus* (F.) (Coleoptera: Coccinellidae) as a biocontrol agent. *CAB Reviews: Perspectives in Agriculture, Veterinary Science, Nutrition and Natural Resources* 4, 1–20.

Poprawski, T.J. and Jones, W.A. (2001) Host plant effects on activity of the mitosporic fungi *Beauveria bassiana* and *Paecilomyces fumosoroseus* against two populations of *Bemisia* whiteflies (Homoptera: Aleyrodidae). *Mycopathologia* 151, 11–20.

Ryckewaert, P. and Alauzet, C. (2002) The natural enemies of *Bemisia argentifolii* in Martinique. *BioControl* 47, 115–126.

Soemargono, A., Ibrahim, Y.B., Ibrahim, R. and Osman, M.S. (2008) Life table and demographic parameters of the metallic blue ladybeetle, *Curinus coeruleus* Mulsant, fed with the Asian citrus psyllid, *Diaphorina citri* Kuwayama. *Pertanika Journal of Tropical Agricultural Science* 31, 1–10.

Suh, S.-J., Evans, G.A. and Oh, S.-M. (2008) A checklist of intercepted whiteflies (Hemiptera: Aleyrodidae) at the Republic of Korea ports of entry. *Journal of Asia-Pacific Entomology* 11, 37–43.

Van Lenteren, J.C. and Martin, N.A. (2006) Biological control of whiteflies. In: Albajes, R., Gullino, M.L. and van Lenteren, J.C. (eds) *Integrated Pest and Disease Management in Greenhouse Crops*. Kluwer, Dordrecht, The Netherlands, pp. 202–216.

Villanueva, R.T., Michaud, J.P. and Childers, C.C. (2004) Ladybeetles as predators of pest and predacious mites in citrus. *Journal of Entomological Science* 39, 23–29.

Yoon, C., Seo, D.-K., Yanga, J.-O., Kang, S.-H. and Kim, G.-H. (2010) Attraction of the predator, *Harmonia axyridis* (Coleoptera: Coccinellidae), to its prey, *Myzus persicae* (Hemiptera: Aphididae), feeding on Chinese cabbage. *Journal of Asia-Pacific Entomology* 4, 255–260.

21 Invasion of Exotic Arthropods in South America's Biodiversity Hotspots and Agro-Production Systems

K.A.G. Wyckhuys,[1] **T. Kondo,**[2] **B.V. Herrera,**[1] **D.R. Miller,**[3] **N. Naranjo**[4] **and G. Hyman**[1]

[1]*International Center for Tropical Agriculture CIAT, Recta Palmira-Cali, Cali, Valle del Cauca, Colombia;* [2]*Corporación Colombiana de Investigación Agropecuaria, Corpoica, Colombia;* [3]*Systematic Entomology Laboratory, Agricultural Research Service, U.S. Department of Agriculture, Beltsville, Maryland 20705, USA;* [4]*Horticulture Research Center CIAA, Universidad Jorge Tadeo Lozano, Bogota, Colombia*

21.1 Introduction

Worldwide, exotic invasive species are of increased concern, and are responsible for environmental and economic problems. Although some exotic species are efficiently kept at bay through both biotic and abiotic processes, several lack effective natural enemies and undergo explosive population increases and geographic spread. Such species commonly transform and negatively affect (native) ecosystems, threaten biotic integrity and contribute to the disappearance of endangered species (Reid and Miller, 1989; IUCN, 1991; Luken and Thieret, 1997; Sheley and Petroff, 1999; Mack, 2000; Clavero and Garcia-Berthou, 2005). Alien species affect native species and communities in a direct fashion through herbivory, predation or parasitism. In a more complex, indirect fashion, invasive species have been documented to cause apparent competition, indirect mutualism/commensalism, exploitative competition, and trophic cascades (White *et al.*, 2006; Wagner and Van Driesche, 2010). Positive interactions among invaders can facilitate or accelerate further invasion, eventually leading to so-called 'invasional meltdown' (e.g., Simberloff and Von Holle, 1999; O'Dowd *et al.*, 2003). Through this myriad of mechanisms, some species can inflict severe damage on crops, livestock and native ecosystems (Pimentel, 1986; Simberloff, 1989; Worner, 1994). Invasive species are also known to interfere with ecosystems, (i.e., pollination or (natural) biological control) (Hooper *et al.*, 2005). Although invasive species greatly affect ecosystems, many of these effects remain unexplored (Kenis *et al.*, 2009).

Different species can bring about different kinds of impacts. On a guild level, invasive herbivores not only directly interfere with growth and reproduction of endemic plant species, but also retard ecosystem recovery after disturbance, facilitate the invasion of plant species and alter the broader composition of local plant communities (Nuñez *et al.*, 2010). Invasive (generalist) predators or parasitoids not only impact native herbivores but can also weaken herbivore suppression by acting upon (native) key natural enemies (Snyder and Evans, 2006; Crowder and Snyder, 2010). Invasive pollinators or flowering plants interfere with pollination (Ghazoul, 2004; Aizen *et al.*, 2008). Invasive species such as *Aedes aegypti, Culex pipiens* or *A. albopictus* establish in urban areas as natural ecosystems and vector agents of human diseases (Lounibos, 2002). On a taxa level, exotic taxa are thought to be more invasive and cause far larger impacts than native ones (Strauss *et al.*, 2006; Wilson and Holway, 2010).

Both negative and positive relationships exist between native biodiversity and the invasion of exotic species. Fridley *et al.* (2007) described how rich native ecosystems are likely to harbor a wide variety of invasive species. On the other hand, Elton's 'biodiversity-invasibility hypothesis' describes how high biodiversity increases the competitive environment of communities and makes them more difficult to invade (Elton, 1958). South America includes multiple biodiversity hotspots, under three defining criteria: endemism, species richness and threatened status (Orme *et al.*, 2005). Its undeniable value for global biodiversity makes this continent particularly appealing from both angles of the invasion paradox; it could either be very vulnerable to invasion or exceptionally resilient to arthropod invaders.

In a compilation of invasive species records, Pimentel *et al.* (2001) documented >120,000 non-native species of plants, animals and microbes in the USA, UK, Australia, South Africa, India and Brazil. In these countries, non-native species are estimated to cause damage at >US$314 billion per annum. In Europe, Australia and North America, arthropod invasions are relatively well documented, while in other parts of the world, information regarding arthropod invasions is scarce and dispersed (e.g., Farji-Brener and Corley, 1998; Matthews and Brand, 2004). In many South American countries, comprehensive invasive species inventories are lacking and records only seem to exist for conspicuous invaders that affect key agricultural commodities. In this region, an exception can be made for the Galapagos Islands, where a long-standing presence of ecologists has permitted the exhaustive documenting of 463 alien insect species (Causton *et al.*, 2006). A good, regional appreciation of nature and distribution of invasive arthropods could aid in the design of management schemes, develop early warning systems and ultimately help prevent future invasions (Kenis *et al.*, 2007).

In this study, we have attempted to document the most significant arthropod invasions in mainland South America, with particular emphasis on scales (e.g., Coccidae, Diaspididae and Pseudococcidae). We have developed a regional species inventory, based upon (gray) literature revision, and employed niche modelling to visualize the geographic distribution of (potentially) some of the most worrisome species. In no way is our invasive species inventory for this region intended to be complete.

21.2 Materials and Methods

A broad literature revision was carried out to screen invasive species records in mainland South America. Global (e.g., ISI Web of Knowledge, CAB Abstracts) and regional scientific literature databases (e.g., Scielo) were consulted and queries were formulated for invasive or exotic arthropods. For invasive scales, expert knowledge, e.g., TK, DRM or ALP and specialized databases such as ScaleNet were consulted. For a few countries, we accessed lists of invasive arthropods, generally in reports from national environmental institutes. For each species, we noted species name, family, presumed region of origin and habitat associations in the invaded range. We also documented the initial record of occurrence in South America, diet type and breadth, and eventual broader impacts of a given species.

From the entire list of invasive species, we selected nine species that were termed 'top invaders'. Top invaders were designated based upon three different criteria:

1. host range;
2. geographic distribution as reflected by number of country records; and
3. a broad measure of impact. Impact was determined in three possible ways: (i) economic impact for agricultural crops; (ii) number of habitat types with reported species presence; and (iii) reported impacts on native species or communities. As an illustrative example of an unsuccessful invasion, we describe the situation of *Brontispa longissima* (Coleoptera: Hispidae), a key invasive pest in South-East Asia that was reported in 1936 in Uruguay but apparently did not establish successfully.

We added a tenth species, the coconut hispid, *Brontispa longissima* (Coleoptera: Hispidae). Although this species was recorded in 1936 in Flores, Uruguay, it apparently did not successfully establish and spread.

For each of the ten species, we employed climate-based niche modelling to describe their geographical distribution in South America. Climate matching between native and potential areas is often used to evaluate the risk of establishment and subsequent spread of exotic species (Baker, 2002; Wyckhuys et al., 2009). As a first step for niche modelling, detailed occurrence records were obtained from entomological collections at the International Center for Tropical Agriculture (CIAT), the Global Biodiversity Information Facility (GBIF) and scientific literature. These occurrence records from the region of origin are used to train niche models for a given species.

For construction of the environmental layers of subsequent niche models, we used climate data from the global WorldClim database[i] (Hijmans et al., 2005). A total of 20 climatic variables were used as environmental layers. These were composed of 19 bioclimatic variables, based on the climate profile of a given location. Within this dataset, monthly temperature and rainfall values were included and represent annual trends, seasonality and extreme or limiting environmental factors (Busby, 1991). Additionally, we incorporated the number of consecutive dry months.

For a given species, occurrence records were entered in Maxent software (i.e., Maximum Entropy Species Distribution Modelling). This modelling package estimated probability distribution of maximum entropy that satisfies a certain set of constraints, expressed as a function of the environmental variables (Phillips et al., 2006). For Maxent training, a minimum of 16 records was used for each species. Logistic output gives an estimate of probability of species occurrence as a function of the selected environmental variables (Phillips and Dudík, 2008). For each species, ten different models were run and median measures were used to plot the final distribution. Results were processed in ArcGis 9.3. After transformation from Ascii to Grid format, the floating points grid files were converted to integer grids and classified in five categories using an equal interval from 0 to 100.

21.3 Results

Detailed records were obtained for a total of 273 arthropod species (Table 21.1). Thanks to an in-depth assessment of invasive species records within the order Hemiptera, we could report the largest numbers of invasives in the families Coccidae (34 spp.), Diaspididae (82 spp.) and Pseudococcidae (29 spp.). Other families that included a fair number of invasive species were Formicidae (13 spp.), Aphididae (8 spp.), Asterolecaniidae (8 spp.) and Nymphalidae (5 spp.). The vast majority of invasive species were herbivores. Additionally, we recorded a large number of pests of stored grains, including Chrysomelidae, Curculionidae and Pyralidae, some hematophages and a restricted set of natural enemies (i.e., parasitoids and predators). The dominance of herbivores undoubtedly was skewed through the large number of records of Diaspididae and Coccidae.

Table 21.1. Number of occurrence records and source of respective information for the ten invasive species. Occurrence records were used used either for Maxent model training or validation.

Species	Source	Training	Test	Total
Aspidiotus destructor	Literature review	31	4	35
Bemisia tabaci	CIAT cassava entomology laboratory	92	11	103
Brontispa longissima	Literature review	18	3	21
Ceratitis capitata	Literature review and GBIF	90	11	101
Coccus viridis	Literature review	18	3	22
Milviscutulus mangiferae	Literature review	16	2	18
Pheidole megacephala	Literature review and GBIF	27	4	31
Pseudococcus longispinus	GBIF and literature review	18	2	20
Toxoptera citricidus	Literature review	16	2	18
Xylosandrus compactus	Literature review	18	2	20

Among all species, 285 of them were associated with agricultural habitats, 173 with natural habitats, 48 with urban habitats and 35 with forestry systems. Most species could be considered habitat generalists, being reported from more than one habitat type.

Of the 211 species for which credible data were found regarding their first record in South America, as few as 15 species were reported prior to 1900. On the other hand, as many as 54 species were reported during the past 20 years.

Based upon host range, number of country reports in South America and estimated impact, we designated the following nine species as 'top invaders' in the region:

Bemisa tabaci (Hemiptera: Aleyrodidae)
Aspidiotus destructor (Hemiptera: Diaspididae)
Ceratitis capitata (Diptera: Tephritidae)
Coccus viridis (Hemiptera: Coccidae)
Milviscutulus mangiferae (Hemiptera: Coccidae)
Pheidole megacephala (Hymenoptera: Formicidae)
Pseudococcus longispinus (Hemiptera: Pseudococcidae)
Toxoptera citricidus (Hemiptera: Aphididae)
Xylosandrus compactus (Coleoptera: Scolytidae)

For each of the above species, geographical distribution was computed using climate-based niche modelling   each species, a different set of occurrence records was used to train Maxent models (Table 21.2).

21.4 Discussion

The low number of reported invasive species in South America could be ascribed to numerous factors. First, the low level of economic development and associated international trade may make this region considerably less prone to biological invasions (Nuñez and Pauchard, 2009). Throughout the world, geographic and taxonomic patterns of biological invasions are determined by trends in human trade and transport (Perrings et al., 2005; Meyerson and Mooney, 2007). Consequently, lower levels of trade mean fewer introductions from fewer sources to fewer locations, and a lowered probability of establishment and spread of new species (Wilson et al., 2009).

Second, establishment and further geographic spread of introduced species has been linked to many biotic and abiotic patterns and processes. Phenotypic similarity between invaders and native communities is thought to greatly reduce invasion success, which could either be explained through direct overlap in resource use (e.g., Darwin's natural-release hypothesis), or sharing of common natural enemies (e.g., natural enemy release) (Strong et al., 1984; Strauss et al., 2006). Thus, in the highly biodiverse parts of South America, invaders may be more likely to overlap in resource use with endemic species, or encounter effective local natural enemies. The latter is exemplified by the effective biological control of the invasive leafminer *Liriomyza huidobrensis* by endemic natural enemies in Indonesia (Rauf et al., 2000; Hidrayani et al., 2005). Several years after its initial arrival in Java, *L. huidobrensis* was effectively controlled by local parasitoids in small-scale potato fields. In this case, field size, level of agricultural intensification and high diversity of locally available natural enemies possibly all contributed in keeping an invasive pest at bay. The same may be true for certain invasive species in both natural and agricultural habitats in biodiverse parts of South America.

A third factor that may explain the low number of invasive species in South America could be the critical lack of trained taxonomists and ecologists. As for our study, the meticulous work of a Latin America-based scale taxonomist, i.e., T.K., with the help of DM and AP, provided valuable information regarding a total of 165 invasive hemipterans. If similar information could be compiled for other arthropod orders, a much more comprehensive and larger species list could easily be generated.

Many of the recorded invasive species appear to be accidental introductions; however, some species (i.e., *Harmonia axyridis*, *Megarhyssa nortoni*, *Rodolia cardinalis*, *M. praecellens* and *Psyllaephagus bliteus*) were intentionally introduced for biological control purposes. The first of these, *H. axyridis*, definitely did not receive the necessary risk assessment attention prior to its release in this continent (Koch et al., 2006). For other species, exceptionally shallow pre-release risk assessments were carried out. As stressed by Lavandero et al. (2006), a much more profound assessment should be conducted prior to releasing exotic natural enemies in South America. This will not only substantially increase the odds of success of classical biological control, but prevent this valuable practice from being conducted in a

Table 21.2. Country records, dietary specifications and habitat associations of selected arthropod invaders in (mainland) South America.

Order/Family	Species	Origin	Ecosyst[a]	Diet[b]	Impacts	Country reports[c]	First record
Araneae							
Araneidae	*Cyrtophora citricola*	S. Europe	U	Pred	Natural enemy	Co, Br	1996
Filistatidae	*Kukulcania hibernalis*	S.W. USA	U	Fruct	–	Pa, Br	–
Theridiidae	*Latrodectus geometricus*	Zimbab	U	Pred	Human health	Co, Ar, Ve	–
Blattodea							
Blattellidae	*Blattella germanica*	Africa	U, A	Scav	Household pest	Ve, Pa, Ec	–
Blattidae	*Periplaneta americana*	Africa	U, A	Scav	Household pest	Ve	–
	Periplaneta australasiae	Africa	U, A	Scav	Household pest	Ve	–
Coleoptera							
Anobiidae	*Lasioderma serricorne*	India, Zimbab.	A, U	Herb	Sweet pepper, tobacco	Ve	–
Anthribidae	*Araecerus fasciculatus*	Africa	A	Herb	Stored coffee, cacao, maize, rice, etc.	All	–
Bostrichidae	*Dinoderus minutus*	India, Phillip.	A, N	Herb	Bamboo, orchids	Ve, Co, Br, Ar	–
	Rhyzopertha dominica	–	A, U	Herb	Stored grain, cereals	Ve, Ch, Ar, Co, Ec	1980
Cerambycidae	*Phoracantha recurva*	Australia	F	Herb	*Eucalyptus* spp.	Ch	1997
Chrysomelidae	*Brontispa longissima*	S-E. Asia	F	Herb	Coconut and other palms	Ur	1936
	Acanthoscelides obtectus	Africa	A, U	Herb	Stored beans, peas, lentils, etc.	All	–
	Callosobruchus chinensis	China	A, U	Herb	Bean, cowpea, chick pea, soy	Ve	–
	Callosobruchus maculatus	Africa	A, U	Herb	Cowpea, chickpea and pea	Ve, Br	–
	Zabrotes subfasciatus	Mexico	A, U	Herb	Stored bean, cowpea, lentils, soy	All	–
Coccinellidae	*Harmonia axyridis*	E. Asia	A, N, F, U	Pred	Predator of aphids and scale, can displace native coccinellids	Ar, Br	1990
	Rodolia cardinalis	Australia	A, N	Pred	Predator of mealybugs	Ve, Ec	–
Curculionidae	*Anthonomus grandis*	Central America	A, F, C	Herb	Cotton, Malvaceae	All	1949
	Gonipterus scutellatus	Australia	F	Herb	*Eucalyptus* spp.	Ch	1998
	Hylastes ater	Europe	N, F	Herb	*Pinus* spp., vector plant-pathogenic fungi	Ch	1980
	Sitophilus granarius	Europe	A, U	Herb	Stored grains	All	–
	Sitophilus oryzae	India	A, U	Herb	Stored grains	All	–
	Sitophilus zeamais	India	A, U	Herb	Stored grains	Br, Co, Ve, Pe, Ch	–

Continued

Table 21.2. Continued

Order/Family	Species	Origin	Ecosyst[a]	Diet[b]	Impacts	Country reports[c]	First record
Dermestidae	*Trogoderma granarium*	India	A, U	Herb	Stored grains and oilseeds	Ve, Ur	–
	Trogoderma ornatum	Europe	A, U	Herb	Stored grains	Ve, Pe	–
Lagriidae	*Lagria villosa*	Africa	A	Herb	Soybean, fruits, coffee	Ar, Br	1976
Scarabaeidae	*Digitonthophagus gazella*	Indo-Africa	A	Copr	Dung beetle	Co, Ch, Br, Bo	1995
Scolytidae	*Xylosandrus compactus*	S.-E. Asia	A, N, F	Herb	Cacao, citrus, avocado, coffee, etc.	Br	–
	Orthotomicus erosus	Africa, Europe	N, F	Herb	*Pinus*, *Picea*, *Abies* spp.	Ch	1980s
	Hypothenemus hampei	Ethiopia	A	Herb	Coffee	Co, Br, Pe, Ec, Ve	1926
Tenebrionidae	*Tribolium castaneum*	Europe	A, U	Herb	Grains	All	–
	Tribolium confusum	Europe	A, U	Herb	Grains	All	–
Diptera							
Culicidae	*Aedes albopictus*	S.-E. Asia	C, U	Hema	Human disease vector	All	1980
	Aedes aegypti	Ethiopia	C, U	Hema	Vector of dengue	Ve, Ur, Ar, Pe, Br, Co	1919
Drosophilidae	*Zaprionus indianus*	Africa, India	A	Fruct	Commercial fruits	Br, Ur	1999
	Drosophila suzukii	Asia	A	Fruct	Commercial fruits	Ec	2005
	Scaptodrosophila latifasciaeformis	Africa	A	Fruct	Commercial fruits	Br	1943
	Drosophila malerkotliana	S.-E. Asia	A, N	Fruct	Fruit pest, competes with native *Drosophila* spp.	Br	1976
Muscidae	*Haematobia irritans*	Europe	A	Hema	Ectoparasite of cattle	Br, Ar, Ve, Co	1937
Richardiidae	*Melanoloma viatrix*	–	A	Herb	Pineapple	Ve, Co	1991
Tephritidae	*Bactrocera carambolae*	S.-E. Asia	A	Herb	Commercial fruits	Br, Su, Gu	1986
	Ceratitis capitata	Mediterr.	A	Herb	Commercial fruits	Ve, Co, Br, Ar, Ec, Pe, Ch	1900
Hemiptera							
Aclerdidae	*Aclerda takahashii*	Oriental	A	Herb	Sugarcane	Br	1932
Aleyrodidae	*Bemisia tabaci*	India	A, U	Herb	Cotton, beans, cassava, etc.	All	1980s
	Dialeurolonga sp.	Asia, Africa	A	Herb	*Citrus*, Myrtaceae	Ch	2005
Aphididae	*Cinara cupressi*	Mediterr.	N, F	Herb	Conifers	Ar, Ch	2003
	Greenidea ficifola	Asia	A, N	Herb	*Ficus* sp., guava	Co, Br	2005

Family	Species	Origin	Code	Habit	Host/Feeding	Country	Year
	Macrosiphum euphorbiae	Nearctic	A	Herb	Potato, bean, tomato; virus vector	Ve, Ch, Co	–
	Myzocallis boerneri	Europe	F	Herb	Quercus spp.	Ch	2005
	Myzus persicae	Asia	A	Herb	Brassica spp., flowers	All	–
	Rhopalosiphum rufiabdominalis	E. Asia	A	Herb	Attacks roots of rice, wheat	Ve, Ar, Pe	1978
	Sipha flava	S.-E. Asia	A	Herb	Sugarcane, Poaceae	Ve, Co, Br, Ar, Ch	1998
	Toxoptera citricidus	E. Asia	A	Herb	Citrus, other Rutaceae; virus vector	All	1930s
Asterolecaniidae	Asterodiaspis variolosa	Palearctic	F	Herb	Quercus sp.	Ar, Br, Ch	1922
	Asterolecanium puteanum	Nearctic	F, N	Herb	Ilex spp., Persea sp.	Br	2006
	Bambusaspis bambusae	–	A	Herb	Polyphagous	Ar, Br, Co, Gu, Su, Ve	1917
	Bambusaspis bambusicola	Palearctic/ Asia	A	Herb	Polyphagous	Ar, Br, Co, Gu, Su, Ve	1923
	Bambusaspis caudata	Asia	A	Herb	Poaceae	Br	1930
	Bambusaspis miliaris	–	A	Herb	Poaceae	Br, Co, Gu, Su, Ve	1938
	Bambusaspis scirrosis	Asia	A	Herb	Poaceae, bamboo	Br	1941
Cicadellidae	Planchonia stentae	Africa	A	Herb	Polyphagous	Co	2001
	Homalodisca vitripennis	N.E. Mexico	A, N	Herb	Almond, grape, stone fruit	Ch	2005
Coccidae	Ceroplastes actiniformis	Asia	A, N	Herb	Polyphagous	Br, Co	1968
	Ceroplastes ceriferus	Asia	A, N	Herb	Polyphagous	Ch	1828
	Ceroplastes floridensis	–	A, N	Herb	Polyphagous	Br, Co, Ec, Gu, Pe	1914
	Ceroplastes rubens	Africa	A, N	Herb	Polyphagous	Co	2008
	Ceroplastes rusci	Africa	A, N	Herb	Polyphagous	Ar, Br, Gu	1917
	Ceroplastes stellifer	Asia	A, N	Herb	Polyphagous	Br, Co, Gu	1904
	Coccus alpinus	Africa	A, N	Herb	Polyphagous	Br	2010
	Coccus celatus	Africa	A, N	Herb	Polyphagous	Br	2010
	Coccus hesperidum	Asia	A, N	Herb	Polyphagous	Ar, Br, Ch, Co, Ec, Gu	1902
	Coccus longulus	Asia	A, N	Herb	Polyphagous	Br, Co, Ec	1989
	Coccus moestus	Australia	A, N	Herb	Polyphagous	Gu	2010
	Coccus viridis	Africa	A, N	Herb	Polyphagous	Bo, Br, Co, Gu, Pe, Ve	1917

Continued

Table 21.2. Continued

Order/Family	Species	Origin	Ecosyst[a]	Diet[b]	Impacts	Country reports[c]	First record
	Cryptinglisia lounsburyi	Africa	A	Herb	*Geranium, Pelargonium* spp.	Ar	1999
	Ericerus pela	Asia	A, N	Herb	Oligophagous	Br	1848
	Kilifia acuminata	Asia	A, N	Herb	Polyphagous	Ar, Br, Co, Su	1989
	Mesolecanium nigrofasciatum	Nearctic	A, N	Herb	Polyphagous	Ar	1999
	Milviscutulus mangiferae	Asia	A, N	Herb	Polyphagous	Br, Co, Ec, Gu, Ve	1917
	Parasaissetia nigra	Africa, Asia	A, N	Herb	Polyphagous	Ar, Br, Co, Ec, Gu, Pe	1917
	Parthenolecanium corni	Palearctic	A, N	Herb	Polyphagous	Ar, Br, Ch	1999
	Parthenolecanium perlatum	Palearctic	A, N	Herb	Polyphagous	Ar, Br, Pa, Ur	1972
	Parthenolecanium persicae	Asia	A, N	Herb	Polyphagous	Ar, Br, Ch	1922
	Protopulvinaria longivalvata	Asia	A, N	Herb	Polyphagous	Br, Co	1989
	Protopulvinaria pyriformis	Asia	A, N	Herb	Polyphagous	Ar, Br, Ch, Co, Gu, Pe	1999
	Pulvinaria elongata	Africa	A, N	Herb	Grasses	Co, Gu, Ve	1917
	Pulvinaria flavicans	Australia	A, N	Herb	Grasses	Gu	1917
	Pulvinaria floccifera	Palearctic	A, N	Herb	Polyphagous	Ar	1943
	Pulvinaria iceryi	Africa	A, N	Herb	Grasses	Co	1942
	Pulvinaria psidii	—	A, N	Herb	Polyphagous	Br, Co, Gu, Ve	1930
	Pulvinaria vitis	Palearctic	A, N	Herb	Polyphagous	Ar, Br	1999
	Pulvinariella mesembryanthemi	—	—	Herb	—	Ar	1999
	Saissetia coffeae	Africa	A, N	Herb	Polyphagous	Ar, Br, Ch, Co, Gu, Pe	1897
	Saissetia miranda	Africa	A, N	Herb	Polyphagous	Ar	1999
	Saissetia neglecta	Africa	A, N	Herb	Polyphagous	Ar, Co, Ve	1971
	Saissetia oleae	Africa	A, N	Herb	Polyphagous	Ar, Br, Ch, Co, Pe	1941

Diaspididae	Andaspis hawaiiensis	Australia	A, N	Herb	Polyphagous	Co	2001
	Aonidia lauri	Palearctic	A, N	Herb	Polyphagous	Br	1900
	Aonidiella aurantii	Asia	A, N	Herb	Polyphagous	Ar, Br, Ch	1937
	Aonidiella citrina	Asia	A, N	Herb	Polyphagous	Ar, Ch	1968
	Aonidiella inonatae	Asia	A, N	Herb	Polyphagous	Ec	1982
	Aonidiella orientalis	Asia	A, N	Herb	Polyphagous	Br, Co	1938
	Aonidiella taxus	Asia	A, N	Herb	Polyphagous	Ar, Br	1937
	Aspidiella sacchari	Asia	A, N	Herb	Grasses	Co, Gu	1959
	Aspidiotus destructor	Australia	A, N	Herb	Polyphagous	Br, Co, Ec, Gu, Pe	1893
	Aspidiotus excisus	Asia	A, N	Herb	Polyphagous	Co, Ec, Gu, Su, Ve	1982
	Aspidiotus nerii	Australia	A, N	Herb	Polyphagous	Ar, Br, Ch, Co	1938
	Aulacaspis rosae	Africa	A, N	Herb	Polyphagous	All	1917
	Aulacaspis tubercularis	Asia	A, N	Herb	Polyphagous, mango	Br, Co, Gu, Su, Ve	1942
	Carulaspis juniperi	Palearctic	F	Herb	Junipers	Ar, Br, Ch	2005
	Carulaspis minima	Palearctic	F	Herb	Junipers and cypresses	Ar, Br, Ch, Co, Ur	1973
	Chionaspis pinifoliae	Nearctic	F	Herb	Conifers (*Pinus* spp.)	Ch	2002
	Chrysomphalus aonidum	Asia	A, N	Herb	Polyphagous	Br, Ch, Co, Gu	1937
	Chrysomphalus dictyospermi	Asia	A, N	Herb	Polyphagous	Ar, Br, Ch, Co, Gu	1889
	Chrysomphalus pinnulifer	Asia	A, N	Herb	Polyphagous	Ar, Br	1916
	Dactylaspis lobata	Nearctic	A, N, F	Herb	Fabaceae and Ulmaceae	Br	1971
	Diaspidiotus ancylus	Nearctic	A, N	Herb	Polyphagous	Ar, Br, Ch	1938
	Diaspidiotus ostreaeformis	Palearctic	A, N	Herb	Polyphagous	Ar	1936
	Diaspidiotus perniciosus	Palearctic	A, N	Herb	Polyphagous	All	1938
	Diaspidiotus uvae	Nearctic	A, N	Herb	Polyphagous	Ar, Br	1938
	Duplachionaspis natalensis	Africa	A, N, F	Herb	Cypresses, Asparagaceae and Poaceae	Br	1936
	Dynaspidiotus britannicus	Palearctic	A, N	Herb	Polyphagous	Br	1938
	Epidiaspis leperii	Palearctic	A, N	Herb	Polyphagous	Ar, Ch, Ur	1905
	Fiorinia fioriniae	—	A, N	Herb	Polyphagous	Ar, Br, Pe	1902

Continued

Table 21.2. Continued.

Order/Family	Species	Origin	Ecosyst[a]	Diet[b]	Impacts	Country reports[c]	First record
	Fiorinia theae	Asia	A, N	Herb	Polyphagous	Ar	2002
	Froggattiella penicillata	Asia	A, N	Herb	Poaceae, bamboo	Gu	1988
	Furchadaspis zamiae	–	N	Herb	Cycads	Ar, Ch	1938
	Hemiberlesia camarana	Africa	A, N	Herb	Musaceae, Myrtaceae, Vitaceae	Br	1938
	Hemiberlesia lataniae	–	A, N	Herb	Polyphagous	All	1900
	Hemiberlesia palmae	–	A, N	Herb	Polyphagous	All	1956
	Hemiberlesia rapax	Palearctic	A, N	Herb	Polyphagous	All	1900
	Howardia biclavis	Africa	A, N	Herb	Polyphagous	All	1914
	Ischnaspis longirostris	Africa	A, N	Herb	Polyphagous	Ar, Br, Co, Ec, Gu, Su	1889
	Kuwanaspis bambusicola	–	A, N	Herb	Poaceae, bamboo	Br	1954
	Kuwanaspis linearis	–	A, N	Herb	Poaceae, bamboo	Co	1976
	Kuwanaspis vermiformis	Asia	A, N	Herb	Poaceae, bamboo	Co	1976
	Lepidosaphes beckii	Asia	A, N	Herb	Polyphagous	All	1902
	Lepidosaphes conchiformis	Palearctic	A, N	Herb	Polyphagous	Ar, Ch	1912
	Lepidosaphes gloverii	Asia	A, N	Herb	Polyphagous	Ar, Bo, Br, Co, Ec	1952
	Lepidosaphes lasianthi	–	A, N	Herb	Oligophagous	Gu	1922
	Lepidosaphes pinnaeformis	Asia	A, N	Herb	Orchidaceae	Ar	1907
	Lepidosaphes ulmi	Palearctic	A, N	Herb	Polyphagous	Ar, Br, Ch	1900
	Leucaspis pusilla	Palearctic	N, F	Herb	Pinus spp.	Ar	1938
	Leucaspis riccae	Palearctic	A, N	Herb	Polyphagous, pest of olives	Ar	1937
	Lindingaspis rossi	Australia	A, N	Herb	Polyphagous	Ar, Br, Ch, Pe	1930
	Lopholeucaspis cockerelli	Europe	A, N	Herb	Polyphagous	Br, Co, Ec, Pe, Ur	1938
	Lopholeucaspis japonica	Palearctic	A, N	Herb	Polyphagous	Br	1923
	Melanaspis smilacis	Nearctic	A, N	Herb	Polyphagous	Br	1951
	Morganella longispina	–	A, N	Herb	Polyphagous	Br, Gu	1889
	Oceanaspidiotus spinosus	–	A, N	Herb	Polyphagous	Ar, Br, Co, Pe, Ur	1982

Odonaspis greenii	—	A, N	Herb	Poaceae, bamboo	Gu, Su	1988
Odonaspis paucipora	Asia	A, N	Herb	Poaceae, *Bambusa tulda*	Gu	1988
Odonaspis ruthae	—	A, N	Herb	Poaceae, grasses	Ar, Bo, Br, Ch, Pe	1938
Odonaspis saccharicaulis	Asia	A, N	Herb	Poaceae	Br, Ve	1900
Parlatoria blanchardi	Palearctic	A, N	Herb	Polyphagous, Arecaceae	Ar, Bo, Br	1938
Paraselenaspidus madagascariensis	Africa	A, N	Herb	Polyphagous	Gu	1958
Parlatoria camelliae	Asia	A, N	Herb	Polyphagous	Ar, Ch	1939
Parlatoria cinerea	—	A, N	Herb	Polyphagous	Ar, Br, Co, Gu, Su	1934
Parlatoria crotonis	—	A, N	Herb	Polyphagous	Br, Gu	1939
Parlatoria oleae	Asia	A, N	Herb	Polyphagous	Ar, Br, Bo	1938
Parlatoria pergandii	Asia	A, N	Herb	Polyphagous	Ar, Br, Co, Ec, Gu	1902
Parlatoria proteus	Europe	A, N	Herb	Polyphagous	All	1901
Parlatoria pseudaspidiotus	—	A, N	Herb	Polyphagous	Co, Gu, Su	1982
Parlatoria ziziphi	—	A, N	Herb	Polyphagous, *Citrus* spp.	All	1938
Pinnaspis aspidistrae	Asia	A, N	Herb	Polyphagous	Ar, Br, Ch, Co, Gu	1869
Pinnaspis buxi	Asia	A, N	Herb	Polyphagous	All	1897
Pinnaspis mussaendae	—	A, N	Herb	Oligophagous	Br	1937
Pinnaspis strachani	Asia	A, N	Herb	Polyphagous, *Hibiscus* sp.	Br, Ch, Co, Ec, Gu	1923
Pinnaspis theae	Asia	A, N	Herb	Polyphagous	Co	1959
Pinnaspis yamamotoi	Europe	A, N	Herb	*Dracaena* sp.	Ve	1965
Pseudaonidia trilobitiformis	Palearctic	A, N	Herb	Polyphagous	All	1900
Pseudaulacaspis pentagona	Palearctic	A, N	Herb	Polyphagous, Rosaceae	All	1902
Pseudotargionia glandulosa	Africa	A, N	Herb	Oligophagous, *Acacia* spp.	Br	1938
Rutherfordia major	Asia	A, N	Herb	Polyphagous, lychee	Co, Ve	1945
Selenaspidus articulatus	Africa	A, N	Herb	Polyphagous	All	1900
Thysanofiorinia nephelii	Asia	A, N	Herb	Oligophagous, lychee, logan	Br	1938
Unaspis citri	Asia	A, N	Herb	Polyphagous, *Citrus* spp.	All	1896
Unaspis euonymi	Palearctic	A, N	Herb	Polyphagous	Ar, Bo	1942

Continued

Table 21.2. Continued.

Order/Family	Species	Origin	Ecosyst[a]	Diet[b]	Impacts	Country reports[c]	First record
Eriococcidae	*Acanthococcus araucariae*	Australia	A, N	Herb	Oligophagous, *Araucaria* spp.	Ar, Br, Ch, Co, Ur, Ve	1963
	Acanthococcus coccineus	Nearctic	A, N	Herb	Cactaceae	Br	1963
	Acanthococcus dubius	–	A, N	Herb	Polyphagous	Br	2007
Kerriidae	*Kerria lacca*	Asia	A, N	Herb	Polyphagous	Gu	1893
Lecanodiaspididae	*Lecanodiaspis rufescens*	–	A, N	Herb	Polyphagous	Ch	1972
Monophlebidae	*Icerya purchasi*	Australia	A, N	Herb	Polyphagous	All	1918
	Icerya seychellarum	Asia	A, N	Herb	Poaceae	Co, Gu	1946
Ortheziidae	*Graminorthezia monticola*	Nearctic	A, N	Herb	Fagaceae, Poaceae, orchids, bromeliads	Br	2004
	Newsteadia trisegmentalis	–	A, N	Herb	Arecaceae	Ve	2001
Phoenicococcidae	*Phoenicococcus marlatti*	Palearctic	A, N	Herb	Arecaceae	Br	1938
Psyllidae	*Diaphorina citri*	Asia	A	Herb	Rutaceae, mainly *Citrus* spp.; vector of greening disease	Br, Ar, Ur, Pa, Ve, Co	1942
Pseudococcidae	*Glycaspis brimblecombei*	Australia	F	Herb	*Eucalyptus* spp.	Ch	2002
	Amonostherium lichtensioides	Nearctic	A, N	Herb	Asteraceae, Lamiaceae, Poaceae, Verbenaceae	Ve	2006
	Antonina graminis	Asia	A, N	Herb	Cyperaceae, Poaceae	All	1964
	Brevennia rehi	Asia	A, N	Herb	Oligophagous, Poaceae	Br	2010
	Chaetococcus bambusae	Palearctic	A, N	Herb	Poaceae	Br, Su	1898
	Chorizococcus rostellum	–	A, N	Herb	Polyphagous	Ar, Br, Pe	1972
	Dysmicoccus boninsis	Palearctic	A, N	Herb	Polyphagous	All	1950
	Hypogeococcus spinosus	Nearctic	A, N	Herb	Cactaceae, Euphorbiaceae	Ar, Br, Ve	1992
	Maconellicoccus hirsutus	Asia	A, N	Herb	Polyphagous	Co, Gu, Ve	2000
	Paludicoccus distichlium	Nearctic	A, N	Herb	*Distichlis* spp.	Ar	1953
	Phenacoccus artemisiae	Nearctic	A, N	Herb	Asteraceae, Poaceae	Ar	2007
	Phenacoccus hurdi	Nearctic	A, N	Herb	Oligophagous	Br	2007
	Phenacoccus solani	–	A, N	Herb	Polyphagous	Br, Co, Ec, Pe, Ve	1940
	Phenacoccus solenopsis	Nearctic	A, N	Herb	Polyphagous	Ar, Br, Ch, Co, Ec	1992
	Planococcus citri	Asia	A, N	Herb	Polyphagous	All	1900
	Planococcus ficus	Palearctic	A, N	Herb	Polyphagous	Ar, Br, Ch, Ur	1901

	Species	Origin	Status	Feeding guild	Impact	Country	Year
	Planococcus halli	–	A, N	Herb	Polyphagous	Br, Co, Gu	1989
	Planococcus lilacinus	Asia	A, N	Herb	Polyphagous	Co	1989
	Planococcus minor	–	A, N	Herb	Polyphagous	Ar, Br, Co, Gu	1991
	Planococcus vovae	Palearctic	A, N	Herb	Polyphagous	Br	1935
	Pseudococcus calceolariae	Australia	A, N	Herb	Polyphagous	Br, Ch, Co	1992
	Pseudococcus comstocki	Palearctic	A, N	Herb	Polyphagous	Ar, Br, Co	1952
	Pseudococcus cryptus	Asia	A, N	Herb	Polyphagous	Ar, Br, Pa	1918
	Pseudococcus longispinus	Australia	A, N	Herb	Polyphagous	All	1952
	Pseudococcus maritimus	Nearctic	A, N	Herb	Polyphagous	Ar, Br, Ch, Gu	1950
	Pseudococcus sorghiellus	Nearctic	A, N	Herb	Polyphagous	Ar	1992
	Pseudococcus viburni	–	A, N	Herb	Polyphagous	All	1983
	Rhizoecus falcifer	Palearctic	A, N	Herb	Poaceae, sugarcane	Su	1933
	Saccharicoccus sacchari	–	A, N	Herb	Polyphagous, Cactaceae	All	1950
	Spilococcus mamillariae	Nearctic	A, N	Herb	Polyphagous	Br	1992
Putoidae	*Puto yuccae*	Nearctic	A, N	Herb	Polyphagous	Co	1989
Hymenoptera							
Apidae	*Apis mellifera scutellata*	Africa	A, N, U	Pol	Competition with native pollinators	Pe, Br, Ar	1957
	Bombus ruderatus	Europe	N	Pol	Competition with native pollinators	Ar, Ch	1996
Encyrtidae	*Psyllaephagus bliteus*	Mexico	F	Par	Parasitoid of *Glycaspis brimblecombei*	Ch	2003
Formicidae	*Anoplolepis gracilipes*	S.-E. Asia	A, C, N, F, U	Herb	Cinnamon, *Citrus* sp., coffee and coconut	Ch	1859
	Anoplolepis gracilipes	S.-E. Asia	A, C, N, F, U	Omni	Polyphagous	Bo, Ch	1859
	Cardiocondyla wroughtonii	Palearctic	U	–	–	Ve	–
	Monomorium destructor	India	F, U	Omni	Displaces native ants, attacks arthropods	Br, Ec	1999
	Monomorium floricola	E. Africa	U	Omni	Household pest, tends homopterans	Ve	–
	Monomorium pharaonis	E. Africa	U, F	Omni	Household pest	Ec, Ve	–
	Nylanderia fulva	Brazil	N, A, F	Omni	Coffee, maize, sugarcane, fruits; tends homopterans	Co, Ar, Ch, Bo, Ve	1971
	Paratrechina longicornis	SE Asia	A, U, N, F	Omni	Displaces native ants, vectors pathogens, household pest	Br, Ch, Ec, Ve	1859

Continued

Table 21.2. Continued.

Order/Family	Species	Origin	Ecosyst[a]	Diet[b]	Impacts	Country reports[c]	First record
	Pheidole megacephala	S Africa	A, C, N, F, U	Omni	Household pest, displaces native ants	Br	–
	Strumigenys emmae	Africa	N	Omni	–	Ve	–
	Tapinoma melanocephalum	Africa	A, F, U	Omni	Household pest	Br, Ec, Ve	–
	Tetramorium caespitum	Mediterr.	U	Omni	Stored grains, attacks small vertebrates	Ve, Br, Ch, Ar	–
	Tetramorium simillimum	Africa	U	Omni	–	Ve, Ec, Br	–
Ichneumonidae	Megarhyssa nortoni	New Zealand	F	Par	Parasitoid of Sirex noctilio	Ch	2005
	Megarhyssa praecellens	China	F	Par	Parasitoid of Tremex fuscicornis	Ch	2005
Megachilidae	Anthidium manicatum	Europe	N	Pol	Pollinator	–	–
Mymaridae	Anaphes nitens	S. Africa	F	Par	Introduced for control of Gonipterus scutellatus	Ch	1998
Siricidae	Sirex noctilio	Europe	F, N	Herb	Pinus, Abies, Larix, Picea, Pseudotsuga spp.	Ar, Br, Ur	1980
	Tremex fuscicornis	Asia	F	Herb	Salix, Populus spp.	Ch	2000
Vespidae	Polistes versicolor	–	A, F, N	Pred	Direct predation on arthropods	Ec – i.e., Galapagos	–
	Vespula germanica	Europe	A, F, U	Pred	Direct predation on arthropods and vertebrates	Ar, Ch	1980
Isopoda							
Platyarthridae	Niambia squamata	South Africa	–	–	–	Br	–
Isoptera							
Kalotermitidae	Cryptotermes brevis	India	U	Xil	Pest	Ve, Ch	–
Lepidoptera							
Bombycidae	Bombyx mori	China	U	Herb	–	Ve	–
Galleridae	Corcyra cephalonica	Europe	U	–	Pest	Ve	–
Gelechiidae	Hypolimnas misippus	Africa, Indomal.	C	–	–	Ve	–
	Pectinophora gossypiella	Polynesia	A	–	–	Ve	–
	Sitotroga cerealella	Europe	U	–	Pest	Ve	–
Gracillariidae	Phyllocnistis citrella	Asia	A	Herb	Citrus pest	Ve	–
Nymphalidae	Cynthia cardui	Africa, Europe, Asia	C	–	–	Ve	–
	Ephestia elutella	Europe	U	–	Pest	Ve	–

Family	Species	Origin	Ecosystem[a]	Diet[b]	Impact	Country[c]	Year
	Galleria mellonella	Europe	U	–	Competition, Pest	Ve	–
	Tineola bisselliella	Europe	U	–	Pest	Ve	–
	Vanessa atalanta rubria	N. America, Bermuda and Antilles	C	–	–	Ve	–
Pyralidae	Cadra cautella	Europe	U	–	Pest	Ve	–
	Corcyra cephalonica	Europe	U	–	Pest	Ve	–
	Ephestia kuehniella	Europe	U	–	Pest	Ve	–
	Plodia interpunctella	India	U	–	Pest	Ve	–
Orthoptera							
Acrididae	Rhammatocerus schistocercoides	–	–	–	–	Co	–
	Schistocerca pallens	Africa	U	–	Pest	Ve	–
Thysanoptera							
Thripidae	Frankliniella brevicaulis	–	–	–	–	Ve, Br	–
	Fulmekiola serrata	W. India	A	–	Pest	Ve	–
	Retithrips syriacus	N.E. Australia	A	–	Pest	Ve	–
	Stenchaetothrips biformis	Europe	A	–	Pest	Ve	–
	Thrips palmi	S-E. Asia	A	Herb	Cucurbitaceae pest	Co, Ve	1985
Trombidiformes							
Eriophyidae	Phyllocoptruta oleivora	–	–	–	Pest	Ar	–
Tetranychidae	Bryobia rubrioculus	–	A	Herb	Apple, Pear, Peach pest	Ar	–
	Raoiella indica	India	A	Herb	Palm trees, Orchids, Plantain	Ve	2007
	Schizotetranychus hindustanicus	India	A	Herb	*Citrus* pest	Br, Ve, Co	2004

[a]Ecosystem types: U: Urban; A: Agricultural; N: Natural; F: Forest; C: Coastal/riverine;
[b]Diet types: Pred: predator; Par: parasitoid; Pol: pollinator; Fruct: fructivore; Herb: herbivore; Omni: ominivore; Scav: scavenger; Hema: hematophage; Copr: coprofagous; Xil: xilofago.
[c]Country reports: Ar: Argentina; Bo: Bolivia; Br: Brazil; Co: Colombia; Ch: Chile; Ec: Ecuador; Gu: Guayana; Pa: Paraguay; Pe: Peru; Su: Suriname; Ur: Uruguay; Ve: Venezuela. All: throughout South America.

hit-and-miss manner. In the meantime, it will help to stop the intentional release of species that could imperil native fauna and flora.

The bulk of invasive species was reported from agro-production systems; however, many of these species remain unreported from natural systems, while undoubtedly affecting their composition and functioning in some fashion. For example, Kaiser et al. (2008) described how an invasive coffee pest in Mauritius significantly reduced the reproductive success of a threatened native plant. The effect of invasive pests on natural ecosystems, however, may be easily overlooked.

A large proportion of invasive species originates from the broader Asian region, which somehow contrasts with past human colonization of the continent, which was primarily from Europe and the Mediterranean region. Intensive trade with Asia in the horticultural and agricultural sector seems to explain such a pattern. For example, the use of exotic *Ficus* species as ornamental trees and increased marketing with Asian countries may explain the regional occurrence of *Greenidea ficicola* (Sousa-Silva et al., 2005). Another example is the recent joint establishment of *Diaphorina citri* with citrus greening disease, which has become a major concern for the Latin American citrus industry (Bonani et al., 2009). The design and enforcement of strict quarantine measures, orchestrated at a regional instead of a national level, could at least slow down the arrival of invasive pests in South America. Such an approach, aimed at invasive species prevention, is the most environmentally desirable strategy that South American nations can adopt (Hulme, 2006). It also provides a valuable and particularly appealing alternative to the ineffective 'firefighting' strategy that many South American nations are implementing.

Invasive species inventories can be a step towards targeted management actions, which could include containment, eradication and control actions for certain prioritized species. Without question, many more species than our nine 'top invaders' are impacting South America's agricultural or natural ecosystems. Some of these species may already be well incorporated into local ecosystems, and their removal could do more harm than good (Zavaleta et al., 2001). Food web and functional role analyses should therefore be incorporated in any prioritizing exercise. For control of already well-established species, different management options could be considered. Highly selective insecticides with low mammalian and non-target toxicity could prove an option for species with restricted distribution (Gentz, 2009). In sensitive, natural ecosystems and for species with wide distribution, classical biological control could constitute a powerful strategy (Hoddle, 2004; Van Driesche et al., 2010); however, as highlighted above, thorough pre-release ecological risk assessments should precede the selection of any species for release. In the meantime, Messing and Wright (2006) have stressed the importance of a rational, efficient and transparent regulatory framework, which should ideally operate at the regional level instead of at a national scale. We recognize that these kinds of frameworks may take some effort to establish in some parts of the developing world. Nevertheless, they are crucial to take full advantage of biological control as a practice that benefits agriculture as well as environmental protection. Last but not least, in case invasive species have affected native biota to a great extent, one could consider re-establishing native species and/or restoring ecosystem functions (Hulme, 2006).

In many South American countries, containment and eradication of invasive species may be less feasible options, as a well-developed scouting network and well-trained taxonomists are frequently not present. Focusing monitoring activities on potential ports of entry could be an effective and relatively low-cost option for early detection of invasive species that enter through cargo or passenger transport. In this effort, pest risk maps could help inform pest management decisions (i.e., trade restrictions, pest surveys and domestic quarantines) (Venette et al., 2010). For example, our niche modelling indicates that the coconut hispid *B. longissima* encounters high levels of climatic suitability along the Ecuadorian coastline. Quarantine measures for selected South-East Asian imports could help prevent the arrival of this serious pest insect in this part of South America.

Acknowledgements

We would like to thank Alex Wild (University of Illinois) for invaluable input on invasive ants. Alejandro Sosa (USDA-ARS) and Hernan

Norambuena (INIA Carillanca) provided information regarding invasive species in Argentina. Many thanks to Ana Lucia B.G. Peronti (Universidade Federal de São Carlos) for her help on invasive Coccoidea of Brazil. We are grateful to the Humboldt Institute in Bogota, Colombia, for providing access to invasive species databases for Colombia.

Note

[i] www.worldclim.org, accessed 13 August 2012.

References

Aizen, M.A., Morales, C.L. and Morales, J.M. (2008) Invasive mutualists erode native pollination webs. *PLoS Biology* 6, 396–403.

Albuquerque, C.M., Melo-Santos, M.A., Bezerra, M.A., Barbosa, R.M., Silva, D.F. and Silva, E.D. (2000) Primeiro registro de *Aedes albopictus* em área da Mata Atlântica, Recife, PE, Brasil. *Revista de Saúde Pública* 34, 314–315.

Anonymous (2010a) Global Invasive Species Database. IUCN/SSC Invasive Species Specialist Group. www.issg.org, accessed 13 August 2012.

Anonymous (2010b) Invasive Species Database. Instituto Alexander von Humboldt, Colombia. http://ef.humboldt.org.co, accessed 17 March 2010.

Anonymous (2010c) Invasive Species Database. I3N-Argentina, Universidad Nacional del Sur, Argentina. In: www.inbiar.org.ar, consulted on 17 March 2010.

Anonymous (2010d) Invasive Species Database. I3N-Paraguay, Asociación Guyra Paraguay. www.i3n.org.py, accessed 19 March 2010.

Anonymous (2010e) Invasive Species Database. I3N-Brasil, Instituto Hórus. http://institutohorus.org.br, accessed 19 March 2010.

Aragón, S., Cantor, F., Cure, J.R. and Rodriguez, D. (2007) Capacidad parasítica de *Praon* pos. *occidentale* (Hymenoptera: Braconidae) sobre *Macrosiphum euphorbiae* (Hemiptera: Aphididae) en condiciones de laboratorio. *Agronomía Colombiana* 25, 142–148.

Ashburner, M., Golic, K. and Hawley, S.H. (2005) *Drosophila: A Laboratory Handbook*. Cold Spring Harbor Laboratory Press, New York, New York.

Autran, E. (1907) Las Cochinillas Argentinas. *Boletín de la Argentina Ministerio de Agricultura* 7, 145–200.

Azevedo, A. (1923) Insectos nocivos ás principaes culturas do Estado da Bahia. *Correio Agricola* 1, 151–155.

Balachowsky, A.S. (1948) Les cochenilles de France, d'Europe, du nord de l'Afrique et du bassin Méditerranéen. IV. Monographie des Coccoidea, classification - Diaspidinae (Premiere partie). (1er partie). *Actualités Scientifiques et Industrielles* 1054, 243–394.

Balachowsky, A.S. (1950) Les cochenilles de France, d'Europe, du Nord de l'Afrique et du Bassin Méditerranéen. V. - monographie des Coccoidea; Diaspidinae (deuxième partie) Aspidiotini. *Entomologique Applicata Actualités Sciences et Industrielles* 1087, 397–557.

Balachowsky, A.S. (1951) Les cochenilles de France, d'Europe, du Nord de l'Afrique et du bassin Méditerranéen. VI. - monographie des Coccoidea; Diaspidinae (Troisième partie) Aspidiotini (fin). (3e partie). *Entomologique Applicata Actualités Sciences et Industrielles* 1127, 561–720.

Balachowsky, A.S. (1953) Les cochenilles de France d'Europe, du Nord de l'Afrique, et du bassin Méditerranéen. VII. - Monographie des Coccoidea; Diaspidinae-IV, Odonaspidini-Parlatorini. *Actualités Scientifiques et Industrielles* 1202, 725–929.

Balachowsky, A.S. (1954) *Les Cochenilles Paléarctiques de la Tribu des Diaspidini*. Memoires Scientifiques de l'Institut Pasteur, Paris, pp. 450.

Balachowsky, A.S. (1958) Les cochenilles du Sahara Francais. 1ère partie. *Travaux de l'Institut de Recherches Sahariennes* 3, 31–53.

Balachowsky, A.S. (1959) Otras cochinillas nuevas de Colombia. *Revista de la Academia Colombiana de Ciencias Exactas, Físicas y Naturales* 10, 362–366.

Ballou, C.H. (1945) Notas sobre insectos daninos observados en Venezuela - 1938–43 - Coccidae. *Proc. 3d Conf. Inter-Amer. Agr. Caracas* 34, 1–151.

Bartlett, B.R. (1978) Coccidae. In: Clausen, C.P. (ed.) *Introduced Parasites and Predators of Arthropod Pests and Weeds: A World Review*. Agricultural Research Service, United States Department of Agriculture, Washington, DC, pp. 57–74.

Bartra, P., Urrelo, G. and Rodriguez, S. (1982) Biología de la broca del café *Hypothenemus hampei* Ferr. (Coleoptera: Ipidae), en Tingo Maria - Perú. *Tropicultura* 2, 17–31.

Ben-Dov, Y. (1974) A revision of *Ischnaspis* Douglas with a description of a new allied genus (Homoptera: Diaspididae). *Journal of Entomology* 43, 19–32.

Ben-Dov, Y. (1988) *A Taxonomic Analysis of the Armored Scale Tribe Odonaspidini of the World (Homoptera: Coccoidea: Diaspididae).* United States Department of Agriculture Technical Bulletin No. 1723, pp. 142.

BirdLife International (2000) *Threatened Birds of the World.* Lynx Editions and BirdLife International, Barcelona, Spain and Cambridge, UK.

Bodkin, G.E. (1914) The scale insects of British Guiana. A preliminary list with an account of their host plants, natural enemies, and controlling agencies. *Journal of the Board of Agriculture of British Guiana* 7, 106–124.

Bodkin, G.E. (1917) Notes on the Coccidae of British Guiana. *Bulletin of Entomological Research* 8, 103–109.

Bodkin, G.E. (1922) The scale insects of British Guiana. *Journal of the Board of Agriculture of British Guiana* 15, 56–63.

Bonani, J.P., Fereres, A., Garzo, E., Miranda, M.P., Appezzato-Da-Gloria, B. and Lopez, J.R.S. (2009) Characterization of electrical penetration graphs of the Asian citrus psyllid, *Diaphorina citri*, in sweet orange seedlings. *Entomologia Experimentalis et Applicata* 134, 35–49.

Bondar, G. (1914) Praga das laranjeiras e outras auranciaceas. *Boletim da Agricultura (São Paulo)* 15, 1064–1106.

Boscán, N., Rosales, C.J. and Godoy, F. (2000) La mosca del fruto de la piña *Melanoloma viatrix* (Diptera: Richardiidae) nuevo insecto plaga en Venezuela. *Agronomía Tropical* 50, 135–140.

Brockerhoff, E.G., Jones, D.J., Kimberley, M.O., Suckling, D.M. and Donaldson, T. (2006) Nationwide survey for invasive wood-boring and bark beetles (Coleoptera) using traps baited with pheromones and kairomones. *Forest Ecology and Management* 228, 234–240.

Busby, J.R. (1991) BIOCLIM: a bioclimate analysis and prediction system. *Plant Protection Quarterly* 6, 8–9.

Bustillo, A.E. (2006) Una revisión sobre la broca del cafe, *Hypothenemus hampei* (Coleoptera: Curculionidae: Scolytinae), en Colombia. *Revista Colombiana de Entomologia* 32.

Calabria, G., Máca, J., Bächli, G., Serra, L. and Pascual, M. (2011) First records of the potential pest species *Drosophila suzukii* (Diptera: Drosophilidae) in Europe. *Journal of Applied Entomology*, doi: 10.1111/j.1439–0418.2010.01583.x.

Callaway, R.M., Thelen, G.C., Rodriguez, A. and Holben, W.E. (2004) Soil biota and exotic plant invasions. *Nature* 427, 731–733.

Campbell, F.T. (2004) Mediterranean pine engraver beetle *Orthotomicus erosus* Wollaston. The Nature Conservancy: Potential Exotic Pest Threats to North American Forests. http://tncweeds.ucdavis.edu, revised on 26 March 2010.

Carrera, A. and Cermeli, M. (2001) Fluctuación e identificación de áfidos en tres localidades productoras de papa (*Solanum tuberosum* L.) en el estado Monagas, durante los años 1987 – 1990. *Entomotropica* 16, 67–72.

Carver, M. (1978) The Black citrus aphids, *Toxoptera citricidus* (Kirkaldy) and *T. aurantii* (Boyer de Fonscolombe) (Homoptera: Aphididae). *Australian Journal of Entomology* 17, 263–210.

Causton, C.E., Lincango, M.P. and Poulsom, T.G.A. (2004) Feeding range studies of *Rodolia cardinalis* (Mulsant), a candidate biological control agent of *Icerya purchasi* Maskell in the Galápagos Islands. *Biological Control* 29, 315–325.

Causton, C.E., Peck, S.B., Sinclair, B.J., Roque-Albelo, L., Hodgson, C.J. and Landry, B. (2006) Alien insects: threats and implications for conservation of Galápagos Islands. *Annals of the Entomological Society of America* 99, 121–143.

Cermeli, M., Morales, P. and Godoy, F. (2000) Presencia del psílido asiático de los cítricos *Diaphorina citri* Kuwayama (Hemiptera: Psyllidae) en Venezuela. *Boletin Entomologia Venezolana* 15, 235–243.

Chada, H.L. and Wood, E.A. (1960) Biology and control of the rhodes-grass scale. *United States Department of Agriculture Technical Bulletin* 1221, 1–21.

Charlin, C.R. (1973) Coccoidea de Isla de Pascua (Homoptera). (In Spanish). *Revista Chilena de Entomología* 7, 111–114.

Chavannes, A. (1848) Notice sur deux Coccus cériféres du Brésil. (In French). *Bulletin de la Societe Entomologique de France (ser.2)* 6, 139–145.

Chiesa Molinari, O. (1937) Coccidae (cochinillas) que atacan el olivo en Argentina. *Boletín de Agricultura-Industrial San Juan, Argentina* 3, 166–173, 230–246.

Clavero, M. and Garcia-Berthou, E. (2005) Invasive species are leading cause of extinctions. *Trends in Ecology and Evolution* 20, 110.

Cockerell, T.D.A. (1893a) Coccidae, or scale insects - II. *Bulletin of the Botanical Department, Jamaica* 40, 7–9.

Cockerell, T.D.A. (1893b) A new lac-insect from Jamaica. *Canadian Entomologist* 25, 181–183.

Cockerell, T.D.A. (1896) Coccidae or scale insects. - IX. *Bulletin of the Botanical Department, Jamaica (n.s.)* 3, 256–259.

Cockerell, T.D.A. (1897) Coccidae or scale insects. - XI. *Bulletin of the Botanical Department, Jamaica* 4, 149–151.

Cockerell, T.D.A. (1898) Mais algumas Coccidae, colligidos pelo Dr. F. Noack. *Revista do Museu Paulista. São Paulo* 3, 501–503.

Cockerell, T.D.A. (1902) A catalogue of the Coccidae of South America. *Revista Chilena de Historia Natural* 6, 250–257.

Cockerell, T.D.A. (1905) Three new South American Coccidae. *Entomological News* 16, 161–163.

Collingwood, C.A., Tigar, B.J. and Agosti, D. (1997) Introduced ants in the United Arab Emirates. *Journal of Arid Environments* 37, 505–512.

Corley, J. and Villacide, J. (2008) Dispersal and dynamics of the woodwasp *Sirex noctilio* in Argentina. *USDA Research Forum on Invasive Species*, 13–15.

Corn, M.L., Buck, E.H., Rawson, J. and Fischer, E. (1999) *Harmful Non-Native Species: Issues for Congress*. Congressional Research Service, Library of Congress, Washington, DC.

Corseuil, E. and Silva, T.L. da (1971) A tribo Diaspidini no Rio Grande do Sul (Homoptera, Diaspididae). *Arquivos do Museu Nacional, Rio de Janeiro* 54, 109–112.

Costa Lima, A. (1934) Sobre alguns coccideos. *Archivos do Instituto de Biologia Vegetal* 1, 131–138.

Costa Lima, A. (1936) *Terceiro catálogo dos insectos que vivem nas plantas do Brasil*. Directoria de Estatistica da Produccao, Rio de Janeiro, pp. 460.

Cox, J.M. (1989) The mealybug genus *Planococcus* (Homoptera: Pseudococcidae). *Bulletin British Museum (Natural History)* 58, 1–78.

Crowder, D.W. and Snyder, W.E. (2010) Eating their way to the top? Mechanisms underlying the success of invasive insect generalist predators. *Biological Invasions* 12, 2857–2876.

David, R.D., Quiroz, J.A., Yepez, F. and Smith, A.H. (2009) Nota técnica nuevo registro de *Greenidea ficicola* Takahashi (Hemiptera: Sternorrhyncha: Aphididae) en gua yabo *Psidium guajava* (Myrtaceae) en Antioquia, Colombia. *Revista Facultad Nacional de Agronomia Medellín* 62, 4999–5002.

Davis-Merlen, G. (1998) New introductions and special law for Galapagos. *Aliens* 7, 10–11.

de Almeida, L.M. and Borges, V. (2002) Primeiro registro de *Harmonia axyridis* (Palias) (Coleoptera, Coccinellidae): um coccinelídeo originário da região Paleártica. *Revista Brasileira de Zoologia* 19, 941–944.

de Almeida, W.R., Carneiro, T.C., Roda, S.A. and de Sá, M. (2009) *Contextualização Sobre Espécies Exóticas Invasoras: Dossiê Pernambuco*. Concervação Internacional, Brasil. 63p.

De Lotto, G. (1971) A preliminary note on the black scales (Homoptera, Coccidae) of North and Central America. *Bulletin of Entomological Research* 61, 325–326.

De Oliveira, C.R.F., Faroni, L.R.D.A., Guedes, R.N.C. and Araujo, A.P.A. (2006) Sobrevivência do acaro *Acarophenax lacunatus* (Cross & Krantz) (Prostigmata: Acarophenacidae) na ausência de alimento. *Neotropical Entomology* 35, 506–510.

Delabie, J.H.C., Do Nascimento, I.C., Pacheco, P. and Casimiro, A.B. (1995) Community structure of house-infesting ants (Hymenoptera: Formicidae) in Southern Bahia, Brazil. *Florida Entomologist* 78, 264–270.

Donzé, G. and Guérin, P.M. (1994) Behavioral attributes and parental care of Varroa mites parasitizing honey bee brood. *Behavioral Ecology and Sociobiology* 34, 305–319.

Durán, M.L. and Cortés, P.R. (1941) Otro enemigo natural de la *Saissetia oleae* (Bern.) nuevo para Chile. *Agricultura Técnica* (Chile) 5, 98–99.

Eiswerth, M.E. and Johnson, W.S. (2002) Managing nonindigenous invasive species: insights from dynamic analysis. *Environmental and Resource Economics* 23, 319–342.

Elton, C.S. (1958) *The Ecology of Invasion by Animals and Plants*. Chapman and Hall, London.

Enserink, M. (2008) A mosquito goes global. *Science* 32, 864–866.

Espinoza, J.C. and Medina, T. (2002) Plagas de granos almacenados en zonas calidas de la provincia de Huancayo. In: *Proceedings of the XLIV National Entomology Meeting*, Sociedad Entomológica del Perú, Lima, pp. 164.

FAO (Food and Agriculture Organization) (2004) *Report of the Expert Consultation on Coconut Beetle Outbreak in APPPC Member Countries*. RAP Publication 2004/29, 26–27 October 2004, Bangkok.

Farji-Brener, A.G. and Corley, J.C. (1998) Successful invasions of hymenopteran insects into NW Patagonia. *Ecologia Austral* 8, 237–249.

Fernandez, F. and Cordero, J. (2007) Biología de la broca del café *Hypothenemus hampei* en condiciones de laboratorio. *Bioagro* 19, 35–40.

Ferris, G.F. (1937) *Atlas of the Scale Insects of North America*. Stanford University Press, Palo Alto, California, pp. S1–S136.

Ferris, G.F. (1941) The genus *Aspidiotus* (Homoptera; Coccoidea; Diaspididae) (Contribution no. 28). *Microentomology* 6, 33–69.

Ferris, G.F. (1950) *Atlas of the Scale Insects of North America. (ser. 5) [v. 5]. The Pseudococcidae (Part I)*. Stanford University Press, Palo Alto, California, pp. 278.

Ferris, G.F. (1953) *Atlas of the Scale Insects of North America, v. 6, The Pseudococcidae (Part II)*. Stanford University Press, Palo Alto, California, pp. 506.

Ferris, G.F. and Rao, V.P. (1947) The genus *Pinnaspis* Cockerell (Homoptera: Coccoidea: Diaspididae). (Contribution No. 54). *Microentomology* 12, 25–58.

Figueroa Potes, A. (1946) Catalogación inicial de las cochinillas del Valle del Cauca (Homoptera - Coccoidea). *Revista Facultad de Agronomía, Montevideo Universidad* 6, 196–220.

Figueroa Potes, A. (1952) Catálogos de los atropodos de las clases Arachnida e Insecta encontrados en el hombre, los animales y las plantas de la República de Colombia-II. *Acta Agronomica* 2, 199–223.

Foldi, I. and Kozar, F. (2005) New species of *Cataenococcus* and *Puto* from Brazil and Venezuela, with data on other species (Hemiptera, Coccoidea). *Nouvelle Revue d'Entomologie* 22, 305–312.

Foldi, I. and Kozar, F. (2007) New species and new records of *Eriococcus* (Hemiptera, Coccoidea, Eriococcidae) from South America. *Zootaxa* 1573, 51–64.

Fonseca, J.P. da (1972) A cochonilha *Lecanium deltae* (Lizer, 1917) em citrus, no Brasil. *O Biologico, Sao Paulo* 38, 213–215.

Fridley, J.D., Stachowicz, J.J., Naeem, S., Sax, D.F., Seabloom, E.W., Smith, M.D., Stohlgren, T.J. et al. (2007) The invasion paradox: reconciling pattern and process in species invasions. *Ecology* 88, 3–17.

Garcia, A.C.L., Gottschalk, M.S., Audino, G.F., Rohde, C., Valiati, V.H. and Valente, V.L.S. (2005) First evidence of *Drosophila malerkotliana* in the extreme South of Brazil (Porto Alegre, Rio Grande do Sul, Brazil). *Drosophila Information Service* 88, 28–30.

Gentz, M.C. (2009) A review of chemical control options for invasive social insects in island ecosystems. *Journal of Applied Entomology* 133, 229–235.

Ghazoul, J. (2004) Alien abduction: disruption of native plant–pollinator interactions by invasive species. *Biotropica* 36, 156–164.

Gill, R.J. (1988) *The Scale Insects of California: Part 1. The Soft Scales (Homoptera: Coccoidea: Coccidae)*. California Department of Food and Agriculture, Sacramento, California, pp. 132.

Gill, R.J. (1993) *The Scale Insects of California: Part 2. The Minor Families (Homoptera: Coccoidea)*. California Department of Food and Agriculture, Sacramento, California, pp. 241.

Gill, R.J. (1997) *The Scale Insects of California: Part 3. The Armored Scales (Homoptera: Diaspididae)*. California Department of Food and Agriculture, Sacramento, California, pp. 307.

Gill, R.J., Nakahara, S. and Williams, M.L. (1977) A review of the genus *Coccus* Linnaeus in America north of Panama (Homoptera: Coccoidea: Coccidae). *Occasional Papers in Entomology, State of California, Department of Food and Agriculture* 24, 44.

González, R.H. (1983) Manejo de plagas de la vid. *Publicaciones en Ciencia Agricultura, (Universidad de Chile, Facultad Ciencia Agrarias, Veterinarias, y Forestales)* 13, 1–115.

González, R.H. and Charlin, R. (1968) Nota preliminar sobre los insectos cóccoideos de Chile. *Revista Chilena de Entomología* 6, 109–113.

Gonzalez, W.L., Fuentes-Contreras, E. and Niemeyer, H.M. (1998) Una nueva especie de afido (Hemiptera: Aphididae) detectada en Chile: *Sipha flava* (Forbes). *Revista Chilena de Entomología* 25, 87–90.

Granara de Willink, M.C. (1991) Cochinillas harinosas de importancia economica encontradas en la Argentina: actualizacion sistematica y nueva lista de hospederos. *Boletin de la Academia Nacional de Ciencias* (Cordoba, Argentina) 59, 259–271.

Granara de Willink, M.C. (1999) Las cochinillas blandas de la República Argentina (Homoptera: Coccoidea: Coccidae). *Contributions on Entomology International* 3, 1–183.

Granara de Willink, M.C. and Szumik, C. (2007) Phenacoccinae de Centro y Sudamérica (Hemiptera: Coccoidea: Pseudococcidae): Sistemática y Filogenia. *Revista de la Sociedad Entomológica Argentina* 66, 29–129.

Granara de Willink, M.C., Pirovani, V.D. and Ferreira, P.S.F. (2010) *Coccus* species affecting *Coffea arabica* in Brazil (Coccoideae: Coccidae) and the redescription of two species. *Neotropical Entomology* 39, 391–399.

Gray, E.J. (1828) *Spicilegia Zoologica; or Original Figures and Short Systematic Descriptions of New and Unfigured Animals. Part 1.* Treuttel, Wurtz & Co., London, pp. 12.

Green, E.E. (1930a) Notes on some Coccidae collected by Dr. Julius Melzer, at São Paulo, Brazil (Rhynch.). *Stettiner Entomologische Zeitung* 91, 214–219.

Green, E.E. (1930b) Fauna Sumatrensis (Bijdrag Nr. 65). Coccidae. *Tijdschrift voor Entomologie* 73, 279–297.

Green, E.E. (1933) Notes on some Coccidae from Surinam, Dutch Guiana, with descriptions of new species. *Stylops* 2, 49–58.

Green, E.E. (1937) An annotated list of the Coccidae of Ceylon, with emendations and additions to date. *Ceylon Journal of Science Section B. Zoology and Geology* 20, 277–341.

Hambleton, E.J. (1935) Notas sobre Pseudococcinae de importáncia economica no Brasil com a descripçao de quatro especies novas. *Archivos do Instituto Biologico, São Paulo* 6, 105–120.

Hambleton, E.J. (1976) A revision of the new world mealybugs of the genus *Rhizoecus* (Homoptera: Pseudococcidae). *United States Department of Agriculture Technical Bulletin* 1522, 1–88.

Hempel, A. (1900) As coccidas Brasileiras. *Revista do Museu Paulista, São Paulo* 4, 365–537.

Hempel, A. (1901) On some new Brazilian Hemiptera. Coccidae. *Annals and Magazine of Natural History* 8, 388–391.

Hempel, A. (1904) [The result of the study of several collections of scales sent to the Instituto Agronomico and Sr. Carlos Moreira, from Museu Nacional, Rio de Janeiro.] (In Portuguese). *Boletim da Agricultura (São Paulo)* 1, 311–323.

Hempel, A. (1918) Descripção de sete novas espécies de coccidas. *Revista do Museu Paulista, São Paulo* 10, 193–208.

Hempel, A. (1932) Descripção de vinte a duas espécies novas de coccideos (Hemiptera – Homoptera. *Revista de Entomologia* 2, 310–339.

Hendricks, H.J. and Kosztarab, M. (1999) *Revision of the Tribe Serrolecaniini (Homoptera: Pseudococcidae).* Walter de Gruyter, New York, New York.

Hidrayani, P., Rauf, A., Ridland, P.M. and Hoffmann, A.A. (2005) Pesticide applications on Java potato fields are ineffective in controlling leafminers, and have antagonistic effects on natural enemies of leafminers. *International Journal of Pest Management* 51, 181–187.

Hijmans, R.J., Cameron, S.E., Parra, J.L., Jones, P.G. and Jarvis, A. (2005) Very high resolution interpolated climate surfaces for global land areas. *International Journal of Climatology* 25, 1967–1978.

Hincapie, C.A., Lopera, A. and Ceballos, M. (2008) Actividad insecticida de extractos de semilla de *Annona muricata* (Anonaceae) sobre *Sitophilus zeamais* (Coleoptera: Curculionidae). *Revista Colombiana de Entomología* 34, 76–82.

Hoddle, M.S. (2004) Restoring balance: using exotic species to control invasive exotic species. *Conservation Biology* 18, 38–49.

Hooper, D.U., Chapin, F.S., Ewel, J.J., Hector, A., Inchausti, P., Lavorel, S., Lawton, J.H., *et al.* (2005) Effects of biodiversity on ecosystem functioning: a consensus of current knowledge. *Ecological Monographs* 75, 3–35.

Howell, J.O. and Kosztarab, M. (1972) Morphology and systematics of the adult females of the genus *Lecanodiaspis* (Homoptera: Coccoidea: Lecanodiaspididae). *Research Division Bulletin Virginia Polytechnic Institute and State University* 70, 1–248.

Hulme, P.E. (2006) Beyond control: wider implications for the management of biological invasions. *Journal of Applied Ecology* 43, 835–847.

Hurley, B.P., Slippers, B. and Wingfield, M.J. (2007) A comparison of the control results for the alien invasive woodwasp, *Sirex noctilio*, in the southern hemisphere. *Agricultural and Forest Entomology* 9, 159–171.

ICA (2009) *Mosca del Mediterráneo (Ceratitis capitata) en Colombia.* Boletin Epidemiologico, Instituto Colombiano Agropecuario ICA, Colombia.

Ihering, H. von (1897) Os piolhos vegetaes (*Phytophthires*) do Brazil. *Revista do Museu Paulista* 2, 385–420.

Imwinkelried, J.M., Fava, F.D. and Trumper, E.V. (2004) Pulgones que atacan al cultivo de trigo. Boletín N° 7.

Issa, S. (2000) A checklist of the termites from Venezuela (Isoptera: Kalotermitidae, Rhinotermitidae, Termitidae). *Florida Entomologist* 83, 379–382.

IUCN (International Union for the Conservation of Nature and Natural Resources) (1991) *Caring for the Earth: A Strategy for Sustainable Living.* IUCN, Gland, Switzerland.

IUCN (2010) SSC Invasive Species Specialist Group (ISSG) www.issg.org/database, revised on 5 March 2010.

Kaiser, C.N., Hansen, D.M. and Muller, C.B. (2008) Exotic pest insects: another perspective on coffee and conservation. *Oryx* 42, 143–146.

Kaplan, J.K. (2004) What's buzzing with africanized honey bees? *Agriculture Research* 52, 4–8.

Kenis, M., Rabitsch, W., Auger-Rozenberg, M.A. and Roques, A. (2007) How can alien species inventories and interception data help us prevent insect invasions? *Bulletin of Entomological Research* 97, 489–502.

Kenis, M., Auger-Rozenberg, M.A., Roques, A., Timms, L., Péré, C., Cock, M.J.W., Settele, J. et al. (2009) Ecological effects of invasive alien insects. *Biological Invasions* 11, 21–45.

Kirkaldy, G.W. (1902) Hemiptera. *Fauna Hawaiiensis* 3, 93–174.

Koch, R.L. (2003) The Multicolored Asian Lady Beetle, *Harmonia axyridis*: a review of its biology, uses in biological control, and non-target impacts. *Journal of Insect Science* 3, 32–48.

Koch, R.L., Venette, R.C. and Hutchinson, W.D. (2006) Invasions by *Harmonia axyridis* (Pallas) (Coleoptera: Coccinellidae) in the Western Hemisphere: Implications for South America. *Neotropical Entomology* 35, 421–434.

Kondo, T. (2001) Las cochinillas de Colombia (Hemiptera: Coccoidea). *Biota Colombiana* 2, 31–48.

Kondo, T. (2008) *Ceroplastes rubens* Maskell (Hemiptera: Coccidae), a new coccid record for Colombia. *Boletin del Museo de Entomologia de la Universidad del Valle* 9, 66–68.

Kozár, F. (2004) *Ortheziidae of the World*. Plant Protection Institute, Hungarian Academy of Sciences, Budapest, pp. 525.

Kozár, F. and Konczné Benedicty, Z. (2001) Revision of *Newsteadia* (Homoptera: Coccoidea) of the Nearctic and Neotropic regions, with descriptions of new species. *Acta Phytopathologica et Entomologica Hungarica* 36, 123–142.

Kuwana, S.I. (1923) On the genus *Leucaspis* in Japan. *Dobutsugaku Zasshi* 35, 321–324.

Lavandero, B., Munoz, C. and Barros, W. (2006) The Achilles heel for biological control: a new vision for your success. *AgroCiencia* 22, 111–123.

Lazzarotto, C.M. and Noemberg, SM. (2005) Análise faunística de afídeos (Hemiptera, Aphididae) na Serra do Mar, Paraná, Brasil. *Revista Brasileira de Entomologia* 49, 270–274.

Lepage, H.S. (1938) Catálogo dos coccídeos do Brasil. *Revista do Museu Paulista* 23, 327–491.

Lima, A.M. (1942) *Insectos do Brazil; Homopteros 3*. Imprenso National, Rio de Janeiro, Brazil, pp. 101.

Lima Filho, M.G., Monteiro, A.G., Anjos, M.D., Garcia, A.C., Valente, V.L. and Rohde, C. (2008) Ecologia e genética da espécie in vasora *Zaprionus indianus* na região nordeste do Brasil. Resumos do 54º Congresso Brasileiro de Genética.

Lizer y Trelles, C.A. (1916) Un cóccido asiático nuevo para la República Argentina: '*Chrysomphalus dictyospermi pinnulifera*' (Mask.) (Hem. Hom.). *Physis* 2, 177.

Lizer y Trelles, C.A. (1922) Tres coccidos nuevos para la fauna Argentina. *Physis* 6, 99–100.

Lizer y Trelles, C.A. (1936) Algunas cochinillas nuevas para la fauna de la República Argentina. *Physis* 12, 113–116.

Lizer y Trelles, C.A. (1937) Cochinillas exóticas introducidas en la República Argentina y daños que causan. *Jornadas Agronómicas y Veterinarias* 1937, 341–362.

Lizer y Trelles, C.A. (1942) La colección coccidológica de Pedro Jorgensen. *Notas del Museo de la Plata Zoologia* 7, 69–80.

Lopez, F., Hiriart, D. and Vargas, M. (2005) *Informe de Biodiversidad. Proyecto planta de vinificación. Viña ocho tierras, Limarí, Ovalle.* Sociedad Agrícola Limarí Poniente Ltda, Limarí, Chile.

Lounibos, L.P. (2002) Invasions by insect vectors of human disease. *Annual Review of Entomology* 47, 233–266.

Luken, J.O. and Thieret, J.W. (1997) *Assessment and Management of Plant Invasions*. Springer-Verlag, New York, New York.

Mack, R.N. (2003) Phylogenetic constraint, absent life forms and pre-adapted alien plants: a prescription for biological invasions. *International Journal of Plant Sciences* 164, S185–S196.

Mack, R.N. and Lonsdale, W.M. (2001) Humans as global plant dispersers: Getting more than we bargained for. *Bioscience* 51, 95–102.

Mack, R.N., Simberloff, D., Lonsdale, W.M., Evans, H., Clout, M. and Bazzaz, F. (2000) Biotic invasions: causes, epidemiology, global consequences, and control. *Ecological Applications* 10, 689–710.

Malcolm, T., Sanford, H. and Hall, G. (1991) *African Honey Bee: What You Need to Know*. IFAS Extension Bulletin, University of Florida.

Mamet, J.R. (1943) A revised list of the Coccoidea of the islands of the western Indian Ocean, south of the equator. *Mauritius Institute Bulletin* 2, 137–170.

Mamet, J.R. (1958) The *Selenaspidus* complex (Homoptera, Coccoidea). *Annales du Musée Royal du Congo Belge. Zoologiques, Miscellanea Zoologica, Tervuren* 4, 359–429.

Mariategui, P.G., Speicys, C. and Urretabizkaya, N. (2004) Evaluación de la dinámica poblacional de *Philonthus flavolimbatus* (Coleoptera, Staphylinidae), en mater ia fecal bovina: su análisis como potencial biocontrolador de *Haematobia irritans* (Diptera, Muscidae) en campos de la cuenca del Río Salado, Buenos Aires, República Argentina. *Boletin de la Sociedad Entomologica Argentina* 34, 233–236.

Martin, S.J. (1994) Ontogenesis of the mite *Varroa jacobsoni* in w orker brood of the hone y bee *Apis mellifera* under natur al conditions. *Experimental and Applied Acarology* 18, 87–100.

Martin, S.J. and Kr yger, P. (2001) Reproduction of *Varroa destructor* in South African honey bees: does cell space influence Varroa male sur vivorship? *Apidologie* 33, 51–61.

Martinez, J.E. (2009) Detección serológica y molecular de virus en áfidos asociados a cultivos de tomate de árbol con síntomas de virosis en Antioquia, Cundinamarca y Nar iño. MSc thesis, Universidad Nacional Medellin, Medellin, Colombia.

Matile-Ferrero, D., Étienne, J. and Tiego, G. (2000) Introduction de deux r avageurs d'importance pour la Guy ane française: *Maconellicoccus hirsutus* et *Paracoccus marginatus* (Hem., Coccoidea, Pseudococcidae). *Bulletin de la Société Entomologique de France* 105, 485–486.

Matthews, S. and Brand, K. (2004) *Africa Invaded: The Growing Danger of Invasive Alien Species*. Global Invasive Species Prog ramme, Cape Town, South Africa, pp. 79.

Maxwell-Lefroy, H. (1902) Scale insects of the West Indies. *West Indian Bulletin* 3, 295–319.

McGlynn, T.P. (1999) The worldwide transfer of ants: geographical distribution and ecological invasions. *Journal of Biogeography* 26, 535–548.

McIntire, S.J. (1889) Fur ther notes upon some remarkable Coccidae from Br itish Guiana. *Journal of the Quekett Microscopical Club Ser. II* 4, 22–25.

McKenzie, H.L. (1937) Mor phological differences distinguishing California red scale , yellow scale, and related species (Homopter a, Diaspididae). *University of California Publications in Entomology* 6, 323–335.

McKenzie, H.L. (1938) The genus *Aonidiella* (Homoptera; Coccoidea: Diaspididae). *Microentomology* 3, 1–36.

McKenzie, H.L. (1939) A re vision of the gen us *Chrysomphalus* and supplementar y notes on the genus *Aonidiella* (Homoptera: Coccoidea: Diaspididae). *Microentomology* 4, 51–77.

McKenzie, H.L. (1945) A revision of *Parlatoria* and closely allied genera (Homoptera: Coccoidea: Diaspididae). *Microentomology* 10, 47–121.

McKenzie, H.L. (1953) Two new *Selenaspidus* scales infesting *Euphorbia* in Calif ornia. (Homoptera; Coccoidea; Diaspididae). Scale studies - P art XII. *Bulletin of the California Department of Agriculture* 42, 53–58.

McKenzie, H.L. (1967) *Mealybugs of California with Taxonomy, Biology, and Control of North American Species (Homoptera: Coccoidea: Pseudococcidae)*. University of Calif ornia Press, Berkeley, California, pp. 526.

Merrill, G.B. and Chaffin, J. (1923) Scale insects of Florida. *Quarterly Bulletin of the Florida State Plant Board* 7, 177–298.

Messing, R.H. and Wright, M.G. (2006) Biological control of invasive species: solution or pollution? *Frontiers in Ecology and the Environment* 4, 132–140.

Meyerson, L.A. and Mooney, H.A. (2007) Invasive alien species in an er a of globalization. *Frontiers in Ecology and the Environment* 5, 199–208.

Miller, D.R. (1975) A re vision of the gen us *Heterococcus* Ferris with a diagnosis of *Brevennia* Goux (Homopter a: Coccoidea: Pseudococcidae). *United States Department of Agriculture Technical Bulletin* 1497, 1–61.

Miller, D.R., Miller, G.L., Hodges, G.S. and Davidson, J.A. (2005) Introduced scale insects (Hemiptera: Coccoidea) of the United States and their impact on U .S. agriculture. *Proceedings of the Entomological Society of Washington* 107, 123–158.

Miller, R.R., Williams, J.D. and Williams, J.E. (1989) Extinctions of Nor th American fishes dur ing the past century. *Fisheries* 14, 22–38.

Montilla, R., García, J.L., Lacruz, L. and Durán, D. (2007) *Spalangia drosophilae* Ashmead (Hymenoptera: Pteromalidae) par asitoide de pupas de la mosca de la piña *Melanoloma viatrix* Hendel (Dipter a: Richardiidae) en Trujillo, Venezuela. *Agronomía Tropical* 57, 107–112.

Montilla, R., Lacr uz, L. and Dur án, D. (2008) Distribucion geografica de *Melanoloma viatrix* Hendel (Diptera: Richardiidae) en Trujillo, Venezuela. *Agronomía Tropical* 58, 403–407.

Morgan, A.C.F. (1889) Obser vations on Coccidae (No. 5). *Entomologist's Monthly Magazine* 25, 349–353.

Morrison, H. (1939) Taxonomy of some scale insects of the genus *Parlatoria* encountered in plant quarantine inspection w ork. *United States Department of Agriculture, Miscellaneous Publications* 344, 1–34.

Mosquera, P.F. (1976) *Escamas Protegidas Más Frecuentes en Colombia.* Boletín Técnico, Ministerio de Ag rícola Instituto Colombiano Agropecuario, División de Sanidad Vegetal, 38, 1–103.

Munir, B. and Sailer, R.I. (1985) Population dynamics of the tea scale, Fiorinia theae (Homoptera: Diaspididae), with biology and lif e tables. *Environmental Entomology* 14, 742–748.

Munroe, L. (2001) Er adicating the car ambola fruit fly, *Bactrocera carambolae*, from Guyana. IV Seminario Científico Internacional de Sanidad Vegetal, 11–15 J une 2001, Varadero, Cuba.

Nakahara, S. (1982) *Checklist of the Armored Scales (Homoptera: Diaspididae) of the Conterminous United States.* United States Department of Ag riculture, Animal and Plant Health Inspection Service, pp. 110.

Natal, D., Urbinatti, P., Taipe-Lagos, C., Cereti-Junior, W., Diederichsen, A. and Souza, R. (1997) Encontro de *Aedes albopictus* (Skuse) em Bromeliaceae na per iferia de São P aulo, SP, Brasil. *Revista de Saudade Pública* 31, 517–518.

Navarro, I.L., Roman, A.K., Gomez, F .H. and Perez, H.A. (2009) Primer registro de *Digitonthophagus gazella* (Fabricius, 1787) par a el departamento de Sucre , Colombia. *Revista Colombiana de Ciencia Animal* 1, 60–64.

Navia, D. and Marsado, A.L. (2010) First repor t of the citrus hindu mite, *Schizotetranychus hindustanicus* (Hirst) (Prostigmata:Tetranychidae) in Brazil. *Neotropical Entomology* 39, 140–143.

Newstead, R. (1893) Obser vations of Coccidae (No. 5). *Entomologist's Monthly Magazine* 29, 185–188.

Newstead, R. (1901) *Monograph of the Coccidae of the British Isles.* Ray Society, London, pp. 220.

Newstead, R. (1914) Notes on scale-insects (Coccidae). Part II. *Bulletin of Entomological Research* 4, 301–311.

Newstead, R. (1917a) Obser vations on scale-insects (Coccidae) - IV . *Bulletin of Entomological Research* 8, 1–34.

Newstead, R. (1917b) Obser vations on scale-insects (Coccidae) -V. *Bulletin of Entomological Research* 8, 125–134.

Niño, L., Cer meli, M., Becerr a, F. and Flores , M. (2001) Fluctuación pob lacional de afidos alados en dos localidades productor as de papa en el Estado Mérida, Venezuela. *Revista Latinoamericana de la Papa* 12, 57–71.

Noriega, J.A. (2002) First report of the presence of the genus *Digitonthophagus* (Coleoptera: Scarabaeidae) in Colombia. *Caldasia* 24, 213–215.

Nuñez, M.A. and P auchard, A. (2009) Biological invasions in developing and developed countries: does one model fit all? *Biological Invasions* 12, 707–714.

Nuñez, M.A., Baile y, J.K. and Schw eitzer, J.A. (2010) Population, community and ecosystem effects of e xotic herbivores: a g rowing global concern. *Biological Invasions* 12, 297–301.

Ochoa, L.P. (1989) *Lista de Insectos Dañinos y Otras Plagas en Colombia.* Boletín Técnico 43, Ministerio de Agricultura, Instituto Colombiano Agropecuario, pp. 1–662.

O'Dowd, D.J., Green, P.T. and Lak e, P.S. (2003) Invasional meltdown on an oceanic island. *Ecology Letters* 6, 812–817.

Ojasti, J. (2001a) Estr ategia Regional de Biodiversidad para los P aíses del Trópico Andino. Especies exóticas invasoras. Convenio de cooperación técnica A TN/JF-5887-RG CAN - BID.

Ojasti, J. (2001b) Estudio sobre el estado actual de las especies exóticas. Secretaría General de la Comunidad Andina. www.comunidadandina. org, revised on 10 March 2010.

Ojasti, J. (2001c) *Informe Sobre las Especies Exóticas en Venezuela.* Ministerio del Ambiente y de los Recursos Natur ales, Caracas, pp. 205.

Oliveira, C.M., Flechtmann, C.A.H. and Frizzas, M.R. (2007) First record of *Xylosandrus compactus* (Eichhoff) (Coleoptera: Curculionidae: Scolytinae) on soursop, *Annona muricata* L. (Annonaceae) in Br azil, with a list of host plants. *Coleopterists Bulletin* 62, 45–48.

Olovacha, I.V. (2008) El prob lema social y de distribución del agua de r iego en la parroquia Pasa, Canton Ambato , Provincia del Tungurahua. BSc thesis, Universidad Estatal de Bolivar, 180 p.

Orme, C.D.L., Davies, R.G., Burgess , M., Eigenbrod, F., Pickup, N., Olson, V.A., Webster, A.J. *et al.* (2005) Global hotspots of species r ichness are not cong ruent with endemism or threat. *Nature* 436, 1016–1019.

Pellizzari, G. and Camporese , P. (1994) The *Ceroplastes* species (Homopter a: Coccoidea) of the Mediterr anean basin with emphasis on *C. japonicus* Green. *Annales de la Société Entomologique de France (N.S)* 30, 175–192.

Penella, J.S. (1942) El cultiv o del mango . *El Agricultor Venezolano* 6, 8–13.

Perrings, C., Dehnen-Schmutz, K., Touza, J. and Williamson, M. (2005) How to manage biological invasions under globalization. *Trends in Ecology & Evolution* 20, 212–215.

Phillips, S.J. and Dudík, M. (2008) Modeling of species distributions with Maxent: new extension for comprehensive evaluation. *Ecography* 31, 161–175.

Phillips, S.J., Anderson, R.P. and Schapire, R.E. (2006) Maximum entropy modeling of species geographic distributions. *Ecological Modelling* 190, 231–259.

Pimentel, D. (1986) Biological invasions of plants and animals in agriculture and forestry. In: Mooney, H.A. and Drake, J.A. (eds) *Ecology of Biological Invasions of North America and Hawaii.* Springer Verlag, New York, New York, pp. 149–162.

Pimentel, D., McNair, S., Janecka, J., Wightman, J., Simmonds, C., O'Connell, C., Wong, E. et al. (2001) Economic and environmental threats of alien plant, animal, and microbe invasions. *Agriculture, Ecosystems and Environment* 84, 1–20.

Pons, L. and Bliss, R.M. (2007) A tiny menace island-hops Caribbean. *USAD/ARS Agricultural Research Magazine* 55, 4–6.

Ponzoni, F.J. (1996) Dados TM/Landsat na identificação do ataque da vespa-damadeira em plantios de Pinus sp. Anais VIII Simpósio Brasileiro de Sensoriamento Remoto, Salvador, Brasil, 14–19 Abril 1996, INPE, pp. 557–566.

Porter, C.E. (1912) Notas para la zoología ecónomica de Chile. III. Adiciones a la lista de los cóccidos. *Revista Chilena de Historia Natural* 16, 22–23.

Qin, T.K. and Gullan, P.J. (1994) Taxonomy of the wax scales (Hemiptera: Coccidae: Ceroplastinae) in Australia. *Invertebrate Taxonomy* 8, 923–959.

Qin, T.K., Gullan, P.J., Beattie, G.A.C., Trueman, J.W.H., Cranston, P.S., Fletcher, M.J. and Sands, D.P.A. (1994) The current distribution and geographical origin of the scale insect pest *Ceroplastes sinensis* (Hemiptera: Coccidae). *Bulletin of Entomological Research* 84, 541–549.

Qin, T.K., Gullan, P.J. and Beattie, A.C. (1998) Biogeography of the wax scales (Insecta: Hemiptera: Coccidae: Ceroplastinae). *Journal of Biogeography* 25, 37–45.

Quijada, T., Marchán, V., Carucí, P., Jiménez, M. and García, M. (2002) Efecto del control químico sobre *Haematobia irritans* (Diptera: Muscidae) durante un año en bovinos de la Parroquia Moroturo, municipio Urdaneta del estado Lara, Venezuela. *Revista Científica* 12, 601–603.

Rauf, A., Shepard, B.M. and Johnson, M.W. (2000) Leafminers in vegetables, ornamental plants and weeds in Indonesia: surveys of host crops, species composition and parasitoids. *International Journal of Pest Mangement* 46, 257–266.

Reid, W.V. and Miller, K.R. (1989) *Keeping Options Alive: The Scientific Basis for Conserving Biodiversity.* World Resources Institute, Washington, DC, pp. 128.

Rodriguez, J.P. (2001) La amenaza de las especies exoticas para la conservacion de la biodiversidad Suramericana. *Interciencia* 26, 479–483.

Rojas-Gil, Y. and Brochero, H. (2008) Hallazgo de *Aedes aegypti* (Linnaeus 1762), en el casco urbano del corregimiento de La Pedrera, Amazonas, Colombia. *Biomédica* 28, 587–596.

Roque-Albelo, L., Berg, M. and Galarza, M. (2006) *Polizontes peligrosos: dispersión de insectos entre las Islas Galapagos en barcos de turismo.* Informe de investigación. Fundación Charles Darwin para las Islas Galapagos A.I.S.B.L., pp. 23.

Rosen, D. and DeBach, P. (1978) Diaspididae. In: Clausen, C.P. (ed.) *Introduced Parasites and Predators of Arthropod Pests and Weeds: a World Review.* United States Department of Agriculture, Agricultural Research Service, Washington, DC, pp. 78–128.

Rossi, G.C., Pascual, N.T. and Kristicevic, F.J. (1999) First record of *Aedes albopictus* (Skuse) from Argentina. *Journal of the American Mosquito Control Association* 15, 422.

Ruiz, C. and Lanfranco, D. (2008) Los escarabajos de corteza en Chile: una revisión de la situación actual e implicancias en el comercio internacional. *Bosque* 29, 109–114.

Russell, L.M. (1941) *A Classification of the Scale Insect Genus Asterolecanium.* United States Department of Agriculture, Miscellaneous Publications 424, pp. 1–319.

Sáenz, A., Montoya, J.A. and Tistl, M. (2004) Artrópodos asociados con *Guadua angustifolia* almacenada en Pereira, Colombia. *Manejo Integrado de Plagas y Agroecología* 71, 59–63.

SAG (2005) Informativo Fitosanitario Forestal, Servicio Agrícola y Ganadero, Gobierno de Chile. www.sag.gob.cl, accessed 29 December 2010.

Sailer, R.J. (1983) History of insect introductions. In: Wilson, C.L. and Graham, C.L. (eds) *Exotic Plant Pests and North American Agriculture.* Academic Press, New York, New York, pp. 15–38.

Salazar, T.J. (1972) Contribución al conocimiento de los Pseudococcidae del Perú. *Revista Peruana de Entomología* 15, 277–303.

Sant'Ana, A.L. (1996) Primeiro encontro de *Aedes albopictus* (Skuse) no Estado do P araná, Brasil. *Revista de Saúde Pública* 30, 392–393.

Sauers-Muller, A. (2005) Host plants of the car ambola fruit fly, *Bactrocera carambolae* Drew & Hancock (Diptera: Tephritidae), in Sur iname, South America. *Neotropical Entomology* 34, 203–214.

Schmale, K. (1968) Biological control of the bean weevil, *Acanthoscelides obtectus* by a hymenopteran parasitoid, as par t of an IPM system. PhD thesis , Rheinisch-Westfälische Technische Hochschule, Aachen, Germany.

Schüttler, E. and Karez, C.S. (2009) Especies exóticas invasoras en las Reservas de Biosfera de América Latina y el Car ibe. UNESCO, Montevideo.

Sereno, F.T.P.S. and Sereno, J.R.B. (2000) Estudio comparativo de la atracción de la *Haematobia irritans* a las mater ias fecales de bo vinos y búfalos en el Pantanal Brasileño. *Arch. Zootec.* 49, 285–290.

Sheley, R.L. and P etroff, J.K. (1999) *Biology and Management of Noxious Rangeland Weeds*. Oregon State Univ ersity Press, Corvallis, Oregon.

Signoret, V. (1869) Essai sur les cochenilles ou gallinsectes (Homoptères - Coccides), 5e partie. *Annales de la Société Entomologique de France* 9, 431–452.

Silva d'Araujo, G.A., Goncalv es, C.R., Galvao, G.M. and Goncalv es, D.M. (1968) 132–200 *Fourth Catalog of Insects That Live in Brazil. Parte II. Insects, Hosts and Natural Enemies.* Vol. 1, Ministerio da Cultura, Rio de J aneiro, pp. 622.

Simberloff, D. (1986) Introduced insects: a biogeographic and systematic perspectiv e. In: Mooney, H.A. and Drake, J.A. (eds) *Ecology of Biological Invasions of North America and Hawaii.* Springer Verlag, New York, New York, pp. 3–26.

Simberloff, D. (1989) Which insect introductions succeed and which f ail? In: Drake, J.A., Mooney, H.A., Di Castr i, F., Groves, R.H., Kruger, F.J., Rejmanek, M. and Williamson, M. (eds) *Biological Invasions: A Global Perspective.* Wiley, New York, New York, pp. 61–75.

Simberloff, D. and Von Holle, B. (1999) P ositive interactions of nonindigenous species: Invasional meltdown? *Biological Invasions* 1, 21–32.

Snyder, W.E. and Ev ans, E.W. (2006) Ecological effects of in vasive arthropod generalist predators. *Annual Review of Ecology, Evolution, and Systematics* 37, 95–122.

Sousa-Silva, C.R., Brombal, J.C. and Ilharco, F.A. (2005) *Greenidea ficicola* Takahashi (Hemiptera: Greenideidae), a ne w aphid in Brazil. *Neotropical Entomology* 34, 1023–1024.

Sponangel, K.W. (1994) *La Broca del Café Hypothenemus hampei en Plantaciones de Café Robusta en la Amazonía Ecuatoriana*. Wissenschaftlicher Fachverlag, Giessen, Germany, pp. 185.

Stein, M., Inés , G. and Almirón, W.R. (2002) Principales criaderos para *Aedes aegypti* y culícidos asociados, Argentina. *Revista do Saúde Pública* 36, 627–630.

Steinweden, J.B. (1946) The identity of cer tain common American species of *Pulvinaria* (Homoptera: Coccoidea: Coccidae). (Contribution no. 49). *Microentomology* 11, 1–28.

Stohlgren, T.J., Binkley, D., Chong, G.W ., Kalkhan, M.A., Schell, L.D., Bull, K.A., Otsuki,Y. *et al.* (1999) Exotic plant species in vade hot spots of nativ e plant div ersity. *Ecological Monographs* 69, 25–46.

Strauss, S.Y., Webb, C.O. and Salamin, N. (2006) Exotic taxa less related to nativ e species are more in vasive. *Proceedings of the National Academy of Sciences* 103, 5841–5845.

Strong, D.R., Lawton, J.H. and Southwood, T.R.E. (1984) *Insects on Plants: Community Patterns and Mechanisms*. Harvard University Press, Cambridge, Massachusetts.

Stumpf, C.F. and Lambdin, P.L. (2000) Distribution and known host records f or *Planchonia stentae* (Hemiptera: Coccoidea: Asterolecaniidae). *Florida Entomologist* 83, 368–369.

Stumpf, C.F. and Lambdin, P .L. (2006) *Pit scales (Sternorrhyncha: Coccoidea) of North and South America*. Tennessee Agricultural Experiment Station, Univ ersity of Tennessee Institute of Ag riculture, Knoxville, Tennessee, pp. 231.

Suarez, V.H. and Busetti, M.R. (1996) Variación estacional y efecto de la mosca de los cuernos en novillos de invernada en la región semiárida pampeana. *Veterinaria Argentina*, XIII 129, 654–660.

Tabatadze, E.S. and Yasnosh, V.A. (1999) Population dynamics and biocontrol of the J apanese scale, *Lopholeucaspis japonica* (Cockerell) in Georgia. *Entomologica* 33, 429–434.

Tambasco, F.J., Sá, L.A.N., Nardo, E.A.B. de, and Tavares, M.T. (2000) The pink mealyb ug,

Maconellicoccus hirsutus (Green): an imported pest now of quarantinable importance encountered in British Guiana (Guyana). *Seminario de Atualidades em Protecao Florestal: Incenios, Pragas e Doencas* 30, 85–93.

Thadeu, A., Guglielmone, A.A. and Martins, J.R. (2002) *Mosca de los Cuernos (*Haematobia irritans*): Control Sustentable y Resistencia a los Insecticidas*. Corpoica, Colombia.

Thuiller, W., Richardson, D.M., Pysek, P., Midgley, G.F., Hughes, G.O. and Rouget, M. (2005) Niche-based modeling as a tool for predicting the risk of alien plant invasions at a global scale. *Global Change Biology* 11, 2234–2250.

Van Driesche, R.G., Carruthers, R.I., Center, T., Hoddle, M.S., Hough-Goldstein, J., Morin, L., Smith, L. *et al.* (2010) Classical biological control for the protection of natural ecosystems. *Biological Control* 54, S2–S33.

Vasquez, C., Quiros, M., Aponte, O. and Sandoval, D.M. (2008) First report of *Raoiella indica* Hirst (Acari: Tenuipalpidae) in South America. *Neotropical Entomologist* 37, 739–740.

Velarde, N., Díaz, A., Sedano, J. and Castro, J. (2002) Plagas de granos almacenados en zonas calidas de la provincial de Huancayo. In: Proceedings of the XLIV National Entomology Meeting, Sociedad Entomológica del Perú, Lima.

Venette, R.C., Kriticos, D.J., Magarey, R.D., Koch, F.H., Baker, R.H.A., Worner, S.P., Gomez Raboteaux, N.N. *et al.* (2010) Pest risk maps for invasive alien species: a roadmap for improvement. *BioScience* 60, 349–362.

Vidaurre, T., Noriega, J.A. and Ledezma, M.J. (2008) First report on the distribution of *Digitonophagus gazelle* (Coleoptera: Scarabaeidae) in Bolivia. *Acta Zoologica Mexicana* 24, 217–220.

Villanueva, R. and Roubik, D. (2004) Why are African honey bees and not European bees invasive? Pollen diet diversity in community experiments. *Apidologie* 35, 481–491.

Vitousek, P.M., D'Antonio, C.M., Loope, L.L., Rejmanek, M. and Westbrooks, R. (1997) Introduced species: a significant component of human-caused global change. *New Zealand Journal of Ecology* 21, 1–16.

Wagner, D.L. and Van Driesche, R.G. (2010) Threats posed to rare or endangered insects by invasions of nonnative species. *Annual Review of Entomology* 55, 547–568.

Wang, Y., Watson, G.W. and Zhang, R. (2010) The potential distribution of an invasive mealybug *Phenacoccus solenopsis* and its threat to cotton in Asia. *Agricultural and Forest Entomology* 12, 403–416.

Watson, G.W. (2002) *Arthropods of Economic Importance: Diaspididae of the World*. World Biodiversity Database, ETI Information Services (Expert Center for Taxonomic Identification), Berkshire, UK.

Watson, G.W., Voegtlin, D.J., Murphy, S.T. and Foottit, R.G. (1999) Biogeography of the *Cinara cupressi* complex (Hemiptera: Aphididae) on Cupressaceae, with description of a pest species introduced into Africa. *Bulletin of Entomological Research* 89, 271–283.

Welbourn, C. (2007) Pest Alert: Red palm mite *Raoiella indica* Hirst (Acari: Tenuipalpidae). Florida Department of Agriculture. http://www.freshfromflorida.com/pi/pest-alerts/raoiella-indica.html, accessed 13 August 2012.

White, E.M., Wilson, J.C. and Clarke, A.R. (2006) Biotic indirect effects: a neglected concept in invasion biology. *Diversity and Distributions* 12, 443–455.

Wille, J.E. (1940) Übersicht der landwirtschaftlich wichtigen Insekten von Peru (Südamerika). *Zeitschrift für Pflanzenbrau und Pflanzenschutz* 50, 24–388.

Williams, D.J. and Watson, G.W. (1988) *The Scale Insects of the Tropical South Pacific Region. Part 1. The Armoured Scales (Diaspididae)*. CAB International, Wallingford, UK, pp. 290.

Williams, D.J. and Watson, G.W. (1990) *The Scale Insects of the Tropical South Pacific Region. Part 3. The Soft Scales (Coccidae) and Other Families*. CAB Publishing, Wallingford, UK, pp. 267.

Wilson, E.E. and Holway, D.A. (2010) Multiple mechanisms underlie displacement of solitary Hawaiian Hymenoptera by an invasive social wasp. *Ecology* 91, 3294–3302.

Wilson, J.R.U., Dormontt, E.E., Prentis, P.J., Lowe, A.J. and Richardson, D.M. (2009) Something in the way you move: dispersal pathways affect invasion success. *Trends in Ecology & Evolution* 24, 136–144.

Worner, S.P. (1994) Predicting the establishment of exotic pests in relation to climate. In: Sharp, J.L. and Hallman, G.J. (eds) *Quarantine Treatments for Pests of Food Plants*. Westview Press, Boulder, Colorado, pp. 11–32.

Wyckhuys, K.A.G., Koch, R.L., Kula, R.R. and Heimpel, G.E. (2009) Potential exposure of a classical biological control agent of the soybean aphid, *Aphis glycines*, on non-target aphids in North America. *Biological Invasions* 11, 857–871.

Yust, H.R. and Cevallos, M.A. (1956) Lista preliminar de plagas de la agricultura del Ecuador. *Revista Ecuatoriana de Entomología y Parasitología* 2, 425–442.

Zarzuela, M.F., Ribeiro, M.C. and Campos-Farinha, A.E. (2002) Distribuição de formigas urbanas em um hospital da região sudeste do Brasil. *Arquivos do Instituto Biologico* 69, 85–87.

Zavala, J.A., Casteel, C.L., DeLucia, E.H. and Berenbaum, M.R. (2008) Anthropogenic increase in carbon dioxide compromises plant defense against invasive insects. *Proceedings of the National Academy of Sciences PNAS* 105, 5129–5133.

Zavaleta, E.S., Hobbs, R.J. and Mooney, H.A. (2001) Viewing invasive species removal in a whole-ecosystem context. *Trends in Ecology and Evolution* 8, 454–459.

Ziller, S.R., Reaser, J.K., Neville, L.E. and Brandt, K. (2005) *Invasive Alien Species in South America: National Reports & Directory of Resources.* Global Invasive Species Program, Cape Town, South Africa, pp. 114.

22 Likelihood of Dispersal of the Armored Scale, *Aonidiella orientalis* (Hemiptera: Diaspididae), to Avocado Trees from Infested Fruit Discarded on the Ground, and Observations on Spread by Handlers

M.K. Hennessey,[1] J.E. Peña,[2] M. Zlotina[1] and K. Santos[2]

[1]*USDA-APHIS-PPQ, 1730 Varsity Dr., Suite 300, Raleigh, North Carolina 27606, USA;* [2]*Tropical Research and Education Center, University of Florida/IFAS, Homestead, Florida 33031, USA*

We investigated the likelihood of infestation of orchard trees by crawlers of oriental red scale, *Aonidiella orientalis* (Hemiptera: Diaspididae), originating from artificially infested fruit discarded into an orchard. In a favorable climate, the percentage of crawlers settling on a tree from very heavily infested fruit discarded when crawlers were emerging, from fruit with a long shelf life, and from fruit discarded near a tree, was low. Infestation was higher when fruit was in contact with the tree than when it was placed 2 m away. It is concluded that establishment via the pathway of commercially produced fruit for consumption is low, because such fruit has not been observed to be as heavily infested, and is not as likely to be discarded in an orchard as was the study fruit. A second part of the study investigated if fruit handlers could become infested with crawlers and be a pathway for establishment. Fruit handlers did receive a low percentage of crawlers on their clothes when they engaged in brushing crawlers from heavily (>5000 crawlers per fruit) infested fruit in the laboratory. Importantly, the pathway from the handlers' clothes to trees in an orchard has not been demonstrated. Good sanitation and safeguarding practices in the packing houses and the orchards can further ensure the prevention of armored scale introduction via the commodity intended for human consumption.

22.1 Introduction

Certain groups of quarantine pests may be introduced into the USA with fresh fruit for consumption imported from foreign countries (Mau and Kessing, 1992; CDFA, 2003). Introduction refers to the entry of a pest resulting in its establishment (IPPC, 2010). It is a responsibility of Plant Protection and Quarantine (PPQ) of the US Department of Agriculture's Animal and Plant Health Inspection Service (USDA APHIS) to conduct risk assessments determining which quarantine pest species may be introduced with a specific fruit commodity from a foreign country, and the likelihood and consequences of such introductions. If the risk is unacceptable, then risk management is applied to reduce the risk to an acceptable level.

While fruit flies and other arthropods with a high capability for dispersal via natural- or human-mediated means are certainly an obvious threat, quarantine species of armored scales have not been considered by PPQ as likely to be introduced with the importation of fresh fruit for consumption. Most stages of the scales are sedentary and attached to a piece of fruit during most of their lives, thus being unable to leave the host and establish new populations. The only stage that is mobile and likely to disperse is the 1st-instar nymph, also known as the crawler stage; however, crawlers have been considered to be unlikely colonizers of new hosts in their new environment because they are minute, fragile, unable to fly and ambulate slowly; the life stage is very short-lived and crawlers tend to settle on the natal plant close to their mother (Oda, 1963; Greathead, 1972; Nielsen and Johnson, 1973; Koteja, 1990; Singh and Singh Ojha, 2005).

The crawler's wandering behavior generally serves to disperse it away from the mother onto new growth of the same host (Bennett and Brown, 1958; Brown, 1958). Crawlers are considered to be pre-adapted to behavior that favors dispersal within a single tree, because survival tends to decrease if movement between host trees could occur (Hanks and Denno, 1994). Settling of crawlers on the same tree on which they were born appears to be part of a greater genetic adaptation and could be manifested in increased fecundity (McClure, 1990). Armored scales seldom spread from plant to plant unless the crowns of the plants are in contact (Bodenheimer, 1951; Beardsley and Gonzalez, 1975). Long-range dispersal is influenced by environmental conditions (wind velocity, light, temperature and relative humidity), the innate behavior to initiate wandering and settling, and availability of acceptable settling sites (Hulley, 1962; Timlin, 1964; Willard, 1974). Human activities are also known to facilitate long-distance dispersal and therefore to increase risk of introduction of crawlers into new areas. Peairs and Merril (1916) considered that there was a low likelihood of the crawlers of San Jose scale, *Diaspidiotus perniciosus*, carried from tree to tree by persons working in the orchards or by horses and cattle. Magsig-Castillo et al. (2010) observed phoresy of crawlers on other insects in the laboratory.

Invading species entering a new region generally have rather poor prospects for survival and colonizing new areas (Mack et al., 2000). The risk associated with the introduction of a species is a product of the likelihood that it will be introduced into the area and the consequences of the pest's introduction. While the presumed likelihood that a species will be introduced can remain low in specific pathways, the consequences of that introduction could be high for certain species–host combinations. Currently, there are 2465 species of armored scales that have been described worldwide (G. Evans, Beltsville, Maryland, pers. comm., 2010). Miller and Davidson (1990) reported that 199 species of armored scales were considered pests in at least some part of the world, which represented about 8% of the total number of described species known at that time. Although this appears to be a relatively small number, the economic impact of some species has been significant.

A low probability (or likelihood) of the introduction of armored scales on imported commodities for consumption has been previously concluded by Miller (1985) and APHIS (2007). This conclusion was based on the assumption that the fruit would be commercially produced and took into account the sedentary nature of scale insects, their inability to disperse actively over long distances, and other conditions to be met before establishment could occur. Commercially grown imported fruit for consumption that is infested with armored scales is permitted by PPQ to enter the USA, following inspection and reporting of any species found, but is not subject to further mitigation measures.

The assertion that armored scales have a low likelihood of introduction was challenged in 2007 by the California Avocado Commission after the California Department of Food and Agriculture (CDFA) detected a new species of armored scale, subsequently described by Evans et al. (2009) as *Abgrallaspis aguacatae*, on shipments of commercially grown avocado fruit (*Persea americana* Mill.) 'Hass' imported from Mexico. It was suspected that the species could enter and establish later on California avocados.

One of the scenarios that was suggested as to how the species might become established on California avocados was by the discarding of imported Mexican fruit in orchards near California packing houses where avocados are re-packed. Another scenario suggested the possibility of the transfer of crawlers from infested fruits to workers during re-packing, and from there to a nearby orchard when those workers pick fruit or tend orchards.

To address the concerns about the introduction of exotic armored scales species on imported Mexican avocados, the current study was designed to answer the following questions:

1. How likely are armored scales on infested fruit that has been discarded in an orchard to survive and establish in the area?
2. What is the likelihood that crawlers can be spread by workers from an infested fruit in a packing house to the trees outside? We designed experiments assuming the 'worst case scenario' using fruit that was very heavily infested in the laboratory with scales and artificially discarding it near an orchard tree, or handling it at the time when crawlers were most numerous on it. We did not evaluate whether the settling of crawlers on the avocado trees led to a reproducing population long-term that would indicate actual establishment.

22.2 Materials and Methods

22.2.1 Stock

The experimental insect species was the oriental red scale, *Aonidiella orientalis* (Newstead), a pest species that is established in the USA (Miller and Davidson, 2005). A culture at the University of Florida was maintained on butternut squash (*Cucurbita moschata* Duchesne) in the laboratory at c. 21°C, 40% relative humidity, and 12:12 l:d. For our tests, fresh store-bought test fruit were washed and allowed to dry, and placed in contact with squash from the culture, so that scale crawlers could move onto the new host. The infestation on the test fruit was allowed to increase until the first-generation females were ready to produce or were producing crawlers. At that point, test fruit with high levels of infestation were selected for the experiment. At weekly intervals, the number of crawlers on the fruit placed in the field was estimated. Elder and Smith (1995) reported that it required c. 10 days for *A. orientalis* crawlers to settle on a new host. It was therefore assumed that crawlers were continuously emerging from week to week, and that the same crawler was not counted 2 weeks after the previous count, because crawlers would not take as long as 2 weeks to settle after emerging. The total number of crawlers emerged on a fruit over a 30-day period was estimated from the total count for all fruit from day 0 to that for day 14.

22.2.2 Site

An isolated grassy (grass c. 10 cm tall) fallow orchard, located at the University of Florida Tropical Research and Education Center in Homestead, was the study site from 2007 to 2009. The nearest other trees in the orchard were avocados c. 50 m south and coconuts c. 150 m east. Twelve 'Hass' avocado trees c. 1.5 m tall were obtained from a local nursery before each experiment. The trees had leaves but no flowers. All trees underwent 100% visual inspection for the presence of armored scales before the experiment, and any scales were removed. The trees were then planted at a spacing of 10 m from each other. Each tree was used for only one replicate. At weekly intervals, trees were inspected for settled scales and their numbers recorded as either on the leaf or stem. We used data from the Florida Automated Weather Network (FAWN) System to record wind direction, temperature and rainfall.

22.2.3 Experiments

Fruit surface area

To calculate fruit surface area, a procedure similar to that used by Schroeder (1950) with 'Hass' avocado was followed. Small sections of skin were carefully removed, placed on a tracing paper, their outline traced and the unknown irregular shapes weighed. A piece of the same paper 1 cm^2 in size was cut and weighed and the unknown area of the fruit calculated (Bigelow, 1958). The surface area of orange (*Citrus sinensis* (L.) Osbeck was approximately 165.9 cm^2, tangerines (*Citrus* sp., × *C. reticulata* Blanco) were 112.5 cm^2, avocados (*Persea Americana* Mill.) were 120.5 cm^2 and squash was 341 cm^2.

Experiment 1

Infested squash fruit were placed on the ground 2 m from the trunk of a selected tree. To account for wind direction, a fruit was placed at each of the following points relative to the tree: north (N), south (S), east (E), west (W), north-east (NE), north-west (NW), south-east (SE) and south-west (SW). Three trees were the untreated controls (C1, C2, C3). The treatments were designed to test if the crawlers would ambulate or be blown by the wind or otherwise move at least 2 m to a tree and settle on it. We sampled the number of crawlers on each squash on different areas of the fruit by making ten observations of a lensfield area of a hand-held lens. The equivalent of each lensfield was 1 cm^2. We calculated the number of crawlers on each fruit based on the surface area of the fruit, and assumed a uniform distribution. We recorded the number of crawlers on each fruit on day 0 (day before treatment – DBT) and, following the same procedure as above, on days 7, 14, 21 and 30 after placement of the fruit in the field (days after the treatment – DAT). We inspected for scale infestations on avocado trees using a 10×-hand lens on day 0 (DBT) and 7, 14, 21 and 30 DAT, recording the number of scales that settled on the upper side and underside of the leaves and the stems of each tree. There was one replicate.

Experiments 2–6

The experiments were designed taking into account the following fruit sizes: orange, c. 165.9 cm^2; tangerines c. 112.5cm^2, avocado c. 120.5 cm^2. We placed individual crawler-infested fruit at a distance of 2 m from the tree in N, S, E, W, NE, NW, SE and SW wind directions. Three fruits were also placed in direct contact with the tree (T1, T2 and T3). Only one tree was as an untreated control (C). The numbers of crawlers produced on the fruit and new scales established on each avocado tree were evaluated in the same manner as in the experiment with the squash. There was one replicate for avocado, two for tangerine and two for orange.

Experiment 7

Crawler dispersal by workers. Two persons brushed crawlers from a scale-infested squash onto an un-infested squash in the laboratory for c. 10 min. Immediately afterwards, each worker pressed 15 one-sided adhesive tape segments (10 × 2 cm) onto the other person's clothing. The tape segments were then removed and placed on index cards. A dissecting microscope was used for counting crawlers on the segments. The experiment was replicated 20 times using the same two persons.

22.3 Results

22.3.1 Experiment 1

Squash

The average minimum and maximum temperatures between March and the end of April, 2008, were 13°C and 29°C, respectively. Rainfall ranged from 0.0025 to 1.5 cm per day. The wind direction ranged from 86 to 233 degrees. The numbers of crawlers per infested fruit ranged between 136 and 1125 on day 0. More than 12,000 crawlers emerged (Table 22.1). The production of crawlers on the infested squash within 30 days of their placement in the field varied among individual fruit. Crawler infestation dropped on most squash during the first 7 days after placement of the infested fruit in the field. In general, crawler production was reduced by 60–100% by the end of the experiment (Table 22.1). Fire ants (*Solenopsis* sp.), a general predator, were observed foraging in high numbers on some of the squashes immediately after placement in the field.

Thirty-six crawlers out of over 12,000 counted on all squashes settled on a new

Table 22.1. Total number of crawlers per infested squash over time (March–April 2008).

Treatment	0 DBT	7 DAT	14 DAT	21 DAT	30 DAT
C1	0	0	0	0	0
C2	0	0	0	0	0
C3	0	0	0	0	0
E	954	306	1636	545	0
N	1091	443	559	375	68
S	954	204	375	136	0
W	1125	136	0	0	0
NE	1670	204	784	238	102
NW	136	0	34	0	0
SE	988	647	0	0	0
SW	920	5285	1739	716	306

host, which was 0.3% of those that emerged (Table 22.2). Seven of the 11 trees were infested. Scales were detected on two of the control trees 10 m from a squash. Wind direction seemed to have no effect on which tree was infested. Tree contact with the fruit was not included in the squash experiment.

Experiment 2

Tangerine (June–July 2008), first replicate, no table. Scale-infested tangerines were placed in the field on 24 June 2008. The initial number of crawlers per tangerine ranged from 180 to 360. No crawlers were observed on the tangerines on 3 July and 18 July 2008. At the end of the experiment, the tangerines were completely decayed or missing from the experimental site. We presume that wildlife removed the tangerines. No crawlers settled on the trees.

Experiment 3

Orange (September–October 2008), first replicate. Temperature between 16 September and 7 October 2008 ranged between 21°C and 31°C, with the rainfall ranging from 0.03 to 0.7 cm per day. Direction of the wind ranged between 66 and 94 degrees. The number of crawlers per fruit at the beginning of the experiment varied between 49 and 248, and a total of over 1300 crawlers emerged (Table 22.3). While crawler production was observed on a fruit placed in the W direction throughout the 30 days, crawlers were not present on most of the fruit by day 14. Despite the relatively high production of crawlers on the fruit located W of a tree 30 days after its placement in the field, no crawlers settled on that tree. No crawlers settled on the control tree.

Thirty six individuals of >1300 (2.8%) estimated crawlers produced settled on five out of the 12 trees, 30 of which were on trees in contact with the fruit (Table 22.4). No scales were found on the control tree. Wind direction did not seem to affect which tree was colonized.

Experiment 4

Orange (December 2008–January 2009), second replicate. The daily temperature between 31 December 2008 and 22 January 2009 fluctuated between 9°C and 27°C; rainfall ranged from

Table 22.3. Total number of crawlers per infested orange over time (September–October 2008).

Treatment	0 DBT	7 DAT	14 DAT	21 DAT	30 DAT
C	0	0	0	0	0
T1	83	16	0	0	0
T2	50	0	0	0	0
T3	116	0	0	0	0
E	99	215	0	0	0
N	99	0	0	0	0
S	49	16	0	0	0
W	99	165	464	429	398
NE	248	0	0	0	0
NW	66	33	0	0	0
SE	99	50	0	0	0
SW	83	99	33	0	0

Table 22.2. Total number of scales dispersing from an infested squash fruit and settling on avocado trees (March–April 2008).

Treatment	Scales/ leaf	Scales/ stem	Total number of scales per tree
C1	1	0	1
C2	0	0	0
C3	4	0	4
E	0	0	0
N	1	0	1
S	1	15	16
W	0	0	0
NE	0	0	0
NW	2	1	3
SE	0	10	10
SW	1	0	1

Table 22.4. Total number of scales dispersing from an infested orange fruit and settling on avocado trees (September–October 2008).

Treatment	Scales/ leaf	Scales/ stem	Total number of scales per tree
C	0	0	0
T1	1	0	1
T2	2	1	3
T3	0	26	26
E	0	0	0
N	0	0	0
S	0	0	0
W	0	0	0
NE	0	0	0
NW	0	0	0
SE	2	0	2
SW	4	0	4

0 to 0.1 cm per day. Wind direction ranged between 116 and 145 degrees. The number of crawlers emerging from fruit at the beginning of the experiment ranged from 99 to 841, and >3000 emerged. Crawlers were observed on the infested fruit for up to 14 days after the beginning of the experiment. No crawlers were recorded on the fruit 30 days after placing the oranges in the field (Table 22.5).

Forty-eight of >3000 (1.6%) crawlers had settled on two out of 12 trees (Table 22.6). Forty-seven (98%) of the settlers were on a tree in contact with a fruit. No scales were observed on the control trees.

Experiment 5

Tangerine (September–October 2009), second replicate. The average daily temperature between 16 September and 23 October 2008 fluctuated between 15°C and 29°C, and rainfall ranged between 0 and 0.4 cm per day. Wind direction fluctuated between 69 and 123 degrees. The number of crawlers per fruit at the beginning of the experiment ranged from 37 to 179, and a total of >600 emerged. Viable populations of crawlers remained on the fruit for up to 14 days after placement of the infested fruit in the field, but only on tangerines located in the N and SE direction to the tree (Table 22.7). Fruit were completely decomposed after 14 days.

Wind did not seem to affect which tree was infested. Crawlers remained 14 days on the fruit placed in the N and SE direction; however, no crawlers had settled on the trees located in those directions. Out of over 600 crawlers that emerged, 16 (2.6%) settled, with most on trees that were in contact with the infested fruit (Table 22.8). The crawlers settled on three trees out of 12.

Experiment 6

'Hass' Avocados, one replicate. The experiment was begun on 30 April 2009. Average daily temperature was 25°C, rainfall was 2.3 cm, and wind direction was 35 degrees. Crawlers per infested avocado fruit introduced into the field ranged from 337 to 650, and totaled over 5000 for all fruit on day 0. No scales were observed on the fruit from day 7 to the end of the experiment (day 28) (Table 22.9).

Table 22.6. Total number of scales dispersing from an infested orange fruit and settling on avocado trees (December 2008–January 2009).

Treatment	Scales/ leaf	Scales/ stem	Total number of scales per tree
C	0	0	0
T1	0	47	47
T2	0	0	0
T3	0	0	0
E	0	0	0
N	0	0	0
S	0	0	0
W	0	1	1
NE	0	0	0
NW	0	0	0
SE	0	0	0
SW	0	0	0

Table 22.5. Total number of crawlers per infested orange over time (December 2008–January 2009).

Treatment	0 DBT	7 DAT	14 DAT	30 DAT
C	0	0	0	0
T1	214	0	66	0
T2	132	33	16	0
T3	478	33	0	0
E	115	33	0	0
N	99	33	0	0
S	181	0	0	0
W	841	0	0	0
NE	660	0	0	0
NW	165	0	0	0
SE	132	33	0	0
SW	561	16	16	0

Table 22.7. Total number of crawlers per infested tangerine over time (September–October 2008).

Treatment	0 DBT	7 DAT	14 DAT
C	0	0	0
T1	67	0	0
T2	78	0	0
T3	37	0	0
E	90	0	0
N	67	11	11
S	56	22	0
W	45	0	0
NE	56	0	0
NW	179	0	0
SE	145	22	22
SW	56	0	0

Table 22.8. Total number of scales dispersing from an infested tangerine fruit and settling on avocado trees (September–October 2008).

Treatment	Scales/leaf	Scales/stem	Total number of scales per tree
C	0	0	0
T1	0	0	0
T2	0	0	0
T3	0	10	10
E	1	0	1
N	0	0	0
S	0	0	0
W	0	0	0
NE	5	0	5
NW	0	0	0
SE	0	0	0
SW	0	0	0

Table 22.9. Total number of crawlers per infested avocado over time (April–May 2009).

Treatment	0 DBT	7 DAT	14 DAT	21 DAT	28 DAT
C	0	0	0	0	0
T1	504	0	0	0	0
T2	360	0	0	0	0
T3	433	0	0	0	0
E	408	0	0	0	0
N	554	0	0	0	0
S	410	0	0	0	0
W	530	0	0	0	0
NE	361	0	0	0	0
NW	650	0	0	0	0
SE	554	0	0	0	0
SW	337	0	0	0	0

Scales settled only on those trees where infested avocado fruit were in contact with the tree (Table 22.10). A total of three of the 12 trees were infested with 15 settlers (0.3%) of the >5000 crawlers emerged. The control was not infested.

Experiment 7

Crawlers dispersed by workers. The mean (*n*=20 trials) number of crawlers on a handler's clothes following brushing activity was 15. Each squash had an estimated 5000 emerging crawlers at the time of brushing.

Table 22.10. Total number of scales dispersing from infested avocado fruit and settling on avocado trees (April–May 2009).

Treatment	Scales/leaf	Scales/stem	Total number of scales per tree
C	0	0	0
T1	0	8	8
T2	0	5	5
T3	0	2	2
E	0	0	0
N	0	0	0
S	0	0	0
W	0	0	0
NE	0	0	0
NW	0	0	0
SE	0	0	0
SW	0	0	0

22.4 Discussion

We observed no obvious relationships in our experiments between the wind direction and the location of the infested trees. We did not study the effect of long-distance dispersal of crawlers from outside the study area; however, it is possible that crawlers were blown in from the surrounding trees. As for the active short-range movement of crawlers, there is considerable evidence that diaspidid crawlers can move across sand or bare soil only for very short distances (Mathis, 1947; Quayle, 1911; Schweig and Grunberg, 1936; Singh and Singh Ojha, 2005). This was confirmed in the present study, in that a very low percentage of emerged crawlers colonized trees 2 m away.

We did not investigate phoresy of crawlers on other insects, and it was not observed to occur during these studies. This phenomenon has been observed in the laboratory (Magsig-Castillo *et al.*, 2010); however, its importance for diaspidid dispersal in the field remains to be determined.

Female scales were able to produce crawlers on all fruit for at least 7 days after placing them in the field. The highest numbers of crawlers were observed for all fruits at the very beginning of the experiment (day 0), after which scale production generally declined. Between two and 12 times more crawlers emerged on squash than on avocado, oranges and tangerines. Large fruit (i.e., squash) that is heavily infested with scales continues producing crawlers for c. 14–30 days in the field, while on a less-infested fruit (i.e., avocado), crawler production lasted c. 7 days. Small fruits (i.e., oranges, avocado and tangerines) were, in general, not able to sustain heavy scale infestation for >7 days. It seems that fruit with a short shelf-life become unsuitable for survival of scales sooner than fruit with a long shelf-life. Generally, more settlers occurred on the new host when the average number of crawlers per fruit discarded nearby was 5000 than when the number of crawlers was only 554 (e.g., squash and avocado, respectively).

Schweig and Grunberg (1936) found that infested fruit were of little or no importance in spreading Florida red scale, Chrysomphalus ficus Ashmead [= C. aonidum (L.)], in Citrus orchards. They regarded infested seedlings or infested leaves as the principal means of spread of the scale to new areas, and that if crawlers emerged on a fruit, they had a tendency to settle on the same fruit. The insects that settled on picked fruits died after the first month due to decay of the fruit. Adult females survived under field conditions on the tree for 50–70 days compared with the 3–4 weeks of the same generation on picked fruits (Schweig and Grunberg, 1936). In the present study, fruit decomposition was an important factor in crawler emergence; it appears that the duration of crawler production correlates with the shelf-life of the fruit in general. For example, crawler emergence ceased sooner on fruit which decomposed quickly (i.e., avocados and tangerines), compared to squash and oranges.

Tree contact by an infested fruit is of great importance for colonization of a new host. Schweig and Grunberg (1936) showed that when infested fruit were piled under clean trees for 1–2 months, no infestation by Chrysomphalus aonidum was observed except where branches were low enough to sweep the ground and come into immediate contact with the infested fruits. Similarly, Melis (1943), after testing the ability of crawlers to cross exposed soil, concluded that infested fruits were of no importance in spreading San Jose scale, Diaspidiotus perniciosus (Comstock), unless the fruit was placed in direct contact with susceptible hosts. While colonization of trees within 2 m from a heavily infested fruit took place, most infestations occurred when a fruit with crawlers was in contact with the tree. In our study, 20 of 58 trees became infested with new scales. In the experiments with direct contact (47 tested trees), eight out of 13 infested trees were in contact with the infested fruit. For avocado, settlers were only recorded on the trees that were in contact with the infested fruit.

Approximately 1.5% (0.3–2.8%; mean of four types of fruit, as reported above) of the total number of crawlers emerged in the field from heavily infested hosts settled on a new host. The scale infestation rate on imported fruit entering the USA is very likely to be many times lower than those rates observed in the current study.

About 0.3% of crawlers that emerged on 'Hass' avocados settled on trees in the present study. Morse et al. (2009) found 5247 live, sessile diaspidid scales and 2511 live eggs or crawlers in 140 cartons of imported Mexican 'Hass' avocados (Table 22.11). The 140 cartons contained 7343 avocado fruit. Assuming a uniform distribution of all scales, crawlers and eggs (7758 individuals) among all fruit, we can conclude that each of 7343 avocados was infested by an average of 1.05 individuals; however, only crawlers (or eggs that will produce crawlers) are the dispersing life stage of the armored scales. Therefore, given 2511 live eggs or crawlers on 7343 avocados, each fruit would be infested with 0.34 individuals capable of dispersal. From those, only 0.3% would be expected to settle on a new host, based on the avocado results above, thus representing 0.001% colonizers out of 7343 fruit. Such a low percentage is concluded to be unlikely to lead to introduction of a new diaspidid species into the USA via commercially produced fruit for consumption.

Uncertainty remains regarding the total effect of the environment on the outcome of the experiment (weather, temperature, wind, predators, etc.) and concerning ability of the settled scales to establish a self-perpetuating population. Peairs and Merrill (1916) considered that there was a low likelihood of the crawlers of San Jose scale, Diaspidiotus perniciosus, being carried from tree to tree by persons working in the orchards or by horses and cattle. Armored scales seldom spread from plant to plant unless the crowns of the plants are in contact (Bodenheimer, 1951; Beardsley and Gonzalez, 1975). In the present study, brushing heavily infested squash did result in a small percentage of the brushed crawlers landing on the handlers' clothes. In addition to the fruit-to-clothes pathway, a pathway from the clothes to trees in an orchard remains to be demonstrated and quantified in order to conclude that it is viable.

Table 22.11. Infestation rate of crawlers from different fruit and number settled on a new host.

Fruit	Studies	Number of crawlers per infested fruit initially (Min/Max)	Total number of scales per new hosts in all replicates	Number of infested trees/ Total trees
Avocado	Morse et al., 2009	0.34 (average)	Not determined	Not determined
Avocado	Current study	463/650	15	3/12
Tangerine (Exp. 5)	Current study	79/145	16	3/12
Orange (Exp. 3)	Current study	325/841	48	2/11
Orange (Exp. 4)	Current study	79/248	36	5/12
Squash (Exp. 1)	Current study	979/1670	36	7/11

22.5 Conclusions

The following conclusions about armored scale introduction are extrapolated from the present study:

- Even with heavily infested discarded fruit, the likelihood of infesting trees in an orchard appears unlikely.
- The likelihood of infestation is higher if the infested fruit is in a direct contact with the tree.
- There appears to be a positive relationship between the number of crawlers per fruit and the number of settlers per tree.
- Shelf life of the discarded infested commodity also seems to be positively related to subsequent settling on new hosts, with longer shelf-life favoring more infestation.
- Real-world conditions, under which commercial fruit is produced, imported and might be discarded, are unlikely to be as favorable to settling as the conditions in the present study (i.e., heavily infested fruit being discarded near a host).
- Based on the infestation levels found on commercially produced fruit for consumption imported into the USA, it is unlikely that the discard pathway is a viable pathway that would lead to the introduction of an exotic armored scale pest into the environment.
- As long as the imported commodity remains within the designated pathway (including usual household disposal practices associated with the fruit for consumption, i.e., landfills), the likelihood of an introduction will remain low.

It is possible that fruit handlers could get crawlers on their clothes if they handled very heavily infested fruit from which crawlers were emerging and while conducting artificial laboratory exercises. It would be less likely to occur under more realistic conditions, including a low infestation rate and real packing-house handling activities. Importantly, the pathway from the handlers' clothes to live trees in an orchard has not been demonstrated. Good sanitation and safeguarding practices in the packing houses and the orchards can further ensure the prevention of armored scale introduction via the commodity destined for human consumption.

Acknowledgements

This research was funded by USDA-APHIS and conducted in collaboration with the University of Florida. We thank E. Jones (USDA-APHIS) for the advice on experimental design, G. Evans (USDA-APHIS) for the identifications of armored scales and review of the manuscript, and S. Robertson and R. Ahern (USDA-APHIS) for their reviews of the manuscript and suggestions for its improvement.

References

APHIS (Animal and Plant Health Inspection Service) (2007) *Phytosanitary Risks Associated with Armoured Scales in Commercial Shipments of Fruit for Consumption to the United States (8 June 2007 Revision).* Agency report, USDA-APHIS-PPQ, Center f or Plant Health and Technology, Raleigh, North Carolina.

Beardsley, J.W. and Gonzalez, R.H. (1975) The biology and ecology of armored scales. *Annual Review of Entomology* 20, 47–73.

Bennett, F.D. and Brown, S. (1958) Life history and sex determination in the diaspine scale *Pseudaulacaspis pentagona* (Tar.). *Canadian Entomologist* 90, 317–325.

Bigelow, P. (1958) How to accurately measure irregular 2-D surface areas. *Scientific American Magazine*. August, 107.

Bodenheimer, F.S. (1951) *Citrus Entomology in the Middle East with Special References to Egypt, Iran, Iraq, Palestine, Syria, Turkey*. Junk, The Hague.

Brown, C.E. (1958) Dispersal of the pine needle scale, *Phenacaspis pinifoliae* (Fitch). *Canadian Entomologist* 90, 685–690.

CDFA (California Department of Food and Agriculture) (2003) *Preventing Biological Pollution: The Mediterranean Fruit Fly Exclusion Program*. Report to the California Legislature. CDFA Medfly Exclusion Program.

Elder, R.J. and Smith, D. (1995) Mass rearing of *Aonidiella orientalis* (Newstead) (Hemiptera: Diaspididae) on butternut gramma. *Journal of the Australian Entomological Society* 34, 253–254.

Evans, G.A., Watson, G.W. and Miller, D.R. (2009) A new species of armored scale (Hemiptera: Coccoidea: Diaspididae) found on a vocado fruit from Mexico and a key to armored scale species found on avocado worldwide. *Zootaxa* 1191, 57–68.

Greathead, D.J. (1972) Dispersal of the sugar-cane *Aulacaspis tegalensis* (Zhnt.) by air currents. *Bulletin of Entomological Research* 61, 547–558.

Hanks, L.M. and Denno, R. (1994) Local adaptation in the armored scale insect *Pseudaulascapis pentagona* (Homoptera: Diaspididae). *Ecology* 78, 2301–2310.

Hulley, P. (1962) On the behavior of the crawlers of the citrus mussel scale, *Lepidosaphes beckii* (Newm.). *Journal of Entomological Society of South Africa* 25, 56–72.

IPPC (International Plant Protection Convention) (2010) *Glossary of Phytosanitary Terms*. International Standards for Phytosanitary Measures publication no. 5. Secretariat of the International Plant Protection Convention, United Nations, FAO, Rome.

Koteja, J. (1990) Life history. In: Rosen, D. (ed.) *Armored Scale Insects. Their Biology, Natural Enemies and Control. Volume A*. Elsevier, Amsterdam, The Netherlands, pp. 243–254.

Mack, R.N., Simberloff, D., Lonsdale, W.M., Evans, H., Clout, M. and Bazzaz, F.A. (2000) Biotic invasions: causes, epidemiology, global consequences, and control. *Ecological Applications* 10, 689–710.

Magsig-Castillo, J., Morse, J.G., Walker, G.P., Bi, J.L., Rugman-Jones, P.F., and Stouthamer, R. (2010) Phoretic dispersal of armored scale crawlers (Hemiptera: Diaspididae). *Journal of Economic Entomology* 103, 1172–1179.

Mathis, J. (1947) Biology of the Florida red scale in Florida. *Florida Entomologist* 29, 13–35.

Mau, R.F. and Kessing, J.L.M. (1992) *Ceratitis capitata* (Wiedemann). *Crop Knowledge Master*. On-line database. University of Hawaii, College of Tropical Agriculture and Human Resources, Hawaii Department of Agriculture.

McClure, M.S. (1990) Coevolution of armored scale insects and their host plants: speciation processes. In: Rosen, D. (ed.) *Armored Scale Insects. Their Biology, Natural Enemies and Control. Volume A*. Elsevier, Amsterdam, The Netherlands, pp. 165–168.

Melis, A. (1943) Contributo alla conoceza dell' *Aspidiotus perniciosus* Comst. *Redia* 29, 1–170 (in Italian).

Miller, D.R. (1985) *Pest Risk Assessment of Armored Scales on Certain Fruit*. Agency report, USDA-ARS, Beltsville, Maryland.

Miller, D.R. and Davidson, J.A. (1990) A list of the armored scale insect pests. In: Rosen, D. (ed.) *Armored Scale Insects. Their Biology, Natural Enemies and Control. Volume A*. Elsevier, Amsterdam, The Netherlands, pp. 299–306.

Miller, D.R. and Davidson, J.A. (2005) *Armored Scale Insect Pests of Trees and Shrubs (Hemiptera: Diaspididae)*. Comstock Publishing Associates, Ithaca, New York.

Morse, J.G., Rugman-Jones, P.F., Watson, G.W., Robinson, L.J., Bi, J.L and Stouthamer, R. (2009) High levels of exotic armored scales on imported avocados raise concerns regarding USDA–APHIS' phytosanitary risk assessment. *Journal of Economic Entomology* 102, 855–867.

Nielsen, D. and Johnson, N.E. (1973) Contribution to the life history and dynamics of the pine needle scale, *Phenacaspis pinifoliae*, in central New York. *Annals of Entomological Society* 66, 34–43.

Oda, T. (1963) Studies on the dispersion of the mulberry scale *Pseudaulacaspis pentagona*. *Japanese Journal of Ecology* 13, 41–46 (In Japanese, English summary).

Peairs, L.M. and Merrill, J.H. (1916) *The San-Jose Scale (Aspidiotus perniciosus Comstock)*. Agricultural Experiment Station Bulletin no. 214. Kansas State Agricultural College, Manhattan, Kansas.

Quayle, H.J. (1911) Locomotion of certain young scale insects. *Journal of Economic Entomology* 4, 301–306.

Schroeder, C.A. (1950) The structure of the skin or rind of the avocado. *California Avocado Society Yearbook* 34, 169–176.

Schweig, C. and Grunberg, A. (1936) The problem of black scale, *Chrysomphalus ficus* (Ashm.) in Palestine. *Bulletin of Entomological Research* 27, 677–714.

Singh, K. and Singh Ojha, R. (2005) Observations on the behaviour of crawler (first stage larva) of *Aonidiella orientalis* (Newsted) (Homoptera: Coccoidea: Diaspididae). *Himalayan Journal of Environment and Zoology* 19, 209–211.

Timlin, J.S. (1964) The biology, bionomics, and control of *Parlatoria pittospori* Mask. (Hemiptera, Diaspidiade): a pest on apples in New Zealand. *New Zealand Journal of Agricultural Research* 7, 536–550.

Willard, J.R. (1974) Horizontal and vertical dispersal of California red scale, *Aonidiella aurantii* (Mask.), (Homoptera: Diaspididae) in the field. *Australian Journal of Zoology* 22, 531–548.

23 Insect Life Cycle Modelling (ILCYM) Software – A New Tool for Regional and Global Insect Pest Risk Assessments under Current and Future Climate Change Scenarios

Marc Sporleder, Henri E.Z. Tonnang, Pablo Carhuapoma, Juan C. Gonzales, Henry Juarez and Jürgen Kroschel

International Potato Center (CIP), Apartado 1558, Lima 12, Peru

23.1 Introduction

The relationship between aspects of an insect's life-history such as development, survival and reproduction, and environmental variables (e.g., temperature) can be well described by process-based phenology models. These models can be used to identify environments where insects might persist, and they are realistic and preferable tools to predict the risks of establishment and population growth potential of invasive insect species. This chapter describes a software tool, Insect Life Cycle Modelling (ILCYM), to support the development of process-based temperature-driven and age-stage structured insect phenology models, and to apply these models for insect species distribution and risk mapping. ILCYM is an open-source computer-aided tool built on R-codes with a user-friendly interface in Java computing language. It is linked to an uDig platform, which is a geographic information system (GIS), an application that contains basic tools for managing and mapping geographic information. The software has three main modules: the model builder, the validation and simulation module, and the potential population distribution and risk mapping module. For the demonstration of the various modelling steps and outputs, we have used the phenology of the potato tuber moth *Phthorimaea operculella* (Zeller) (Lepidoptera: Gelechiidae) as an example. This chapter discusses the positioning of ILCYM among available process-based insect modelling computer-aided packages, some special features of ILCYM, the software limitations and future prospects.

Interest in models to predict the environmental suitability for invasive insect pest species has grown radically in the last two decades. In particular, the need to understand the impact of climate change on the potential distribution of pests has accelerated the demand for new tools to estimate the potential risk of their invading new environments and agricultural regions. For this purpose, risk maps are good communication tools that use different spatial scales, from regional to worldwide maps, to visualize the potential pest distribution and the economic damage it may inflict on crops. Thus, risk maps are used to inform policy and management in this field to aid in strategic pest-management decisions such as restrictions on the importation of certain crops in international trade, implementation of quarantine measures and the design of pest surveys (Baker, 1996; Braasch *et al.*, 1996; Baker *et al.*, 2000; McKenney *et al.*, 2003).

There are two distinct approaches prevalent in the modelling of insect pests and the risk of establishment and expansion; these can be described as (i) 'inductive' and (ii) 'deductive'. The 'inductive' approach combines through statistical or machine-learning methods, the known occurrence records of insect species with digital layers of environmental variables. It uses minimal data sets and simple functions to describe the species' response to temperature and other climatic factors. Generally, presence/absence data or occurrence data only, from different locations, are sufficient for creating risk maps. The combination of occurrence records and environmental variables can be performed through the application of climate-match functions that seek out the establishment potential of an invasive species to new areas, by comparing the long-term meteorological data for each selected location where the species is absent, with the same data for the location of origin or locations where the species prevails (Sutherst et al., 2000; Sutherst and Maywald, 1991). For applying this approach, computer-aided tools such as CLIMEX (Peacock and Worner, 2006; Vanhanen et al., 2008a, b; Wilmot Senaratne et al., 2006) and BIOCLIM (Kohlmann et al., 1988; Steinbauer et al., 2002) have been developed and used to predict insect species' demography for pest risk analysis (Rafoss, 2003; Sutherst, 1991; Zalucki and Furlong, 2005) and possible climate change effects (Sutherst and Maywald, 1990).

The 'inductive' modelling approach has made considerable advances and a great number of computer programs, including BioMOD, GARP and HABITAT, have been developed (reviewed by Venette et al., 2010). This modelling approach showed advantages where detailed information about insect species is not available; however, critical limitations include the failure to consider the species' biological characteristics in the modelling framework (Venette et al., 2010). Hence, resulting risk maps may inform about potential establishment but they do not provide information on the species' population growth and damage potential, or temporal population change within a cropping season or year, in a given region.

By contrast, the 'deductive' approach uses a process-based climatic response model – a phenology model – for a particular insect species of interest. Phenology models are analytical tools for the evaluation, understanding and prediction of the dynamics of insect populations in ecosystems under a variety of environmental conditions and management practices, and more recently they have also been used in phytosanitary risk assessments (Baker, 1991; Jarvis and Baker, 2001a, b). The development of insects, as in other ectothermic organisms, depends on the ambient temperature. This temperature dependency can be applied in a process-oriented framework; forecasting the potential distribution of insect species is completely independent of observed occurrences, and this approach is therefore referred to as 'deductive'.

The difference between the 'inductive' and 'deductive' modelling approach is the level of abstraction, which is higher in the 'inductive' or 'climate match' approach, in which the mathematical methods employed lead to a greater generality. Instead, process-based phenology models are either detailed or simplified mathematical models, which describe the basic physiological principles of the insect species' growth, namely its development, survival and reproduction; the complexity of these models can range from simple models with no age structure and limited environmental inputs to age-stage structured or multi-species models with complex environmental drivers. The two approaches do not necessarily compete, but may also be used to complement each other.

Degree-day models are often used to describe the linear development of insects using the accumulation of temperature above the minimum temperature threshold (Allen, 1976); see Nietschke et al., 2007). However, due to the non-linearity of the development curve, especially when temperature deviates from the intrinsic optimal temperature of a species, degree-day models are poor predictors of insect development. This method works well for intermediate temperatures, but produces errors (i.e., significant deviations from the real development) when daily temperature fluctuates to extremes (Stinner et al., 1974; Worner, 1992). Modern, more progressive models use non-linear functions of higher biological significance (e.g., Logan et al., 1976; Sharpe and DeMichele, 1977), and include stochastic functions for variability in development times among individuals within a population (Sharpe et al., 1981; Wagner et al., 1984).

Computer-aided modelling packages such as DYMEX (Kriticos et al., 2003), NAPPFAST (Nietschke et al., 2008), ECAMON (Trnka et al.,

2007) or ILCYM (Sporleder et al., 2007, 2009) support the development of process-oriented temperature-driven and age-stage structured insect phenology/population models. The latter, ILCYM (Insect Life Cycle Modelling software, version 2.1), has recently been developed and made freely available by the International Potato Center (CIP), Lima, Peru.[i] This chapter describes the application of ILCYM software, which supports the development of process-oriented temperature-driven and age-stage structured insect phenology/population models. ILCYM interactively leads the user through the steps for developing insect phenology models, for conducting simulations, and for producing potential population distribution and risk mapping under current or future temperature (climate change) scenarios. The phenology model developed for the potato tuber moth *Phthorimaea operculella* (Zeller) (Lepidoptera: Gelechiidae) is used to demonstrate the resulting modelling outputs.

23.2 The ILCYM modelling concept

23.2.1 Theory

Exothermic organisms such as insects cannot internally regulate their own temperature. Their body temperature depends on the temperature to which they are exposed in the environment and can change abruptly, for example, from night-time to day-time temperatures. Therefore, they are heterothermic organisms (with fluctuating body temperatures) in contrast to the homeothermy of endothermic organisms which regulate their body temperature by generating endogenous heat. Insects require a certain amount of heat to develop from one developmental stage to another in their life cycle (Andrewartha and Birch, 1955; Uvarov, 1931). Their developmental rate, which is a reciprocal value of the developmental time, is determined by the metabolic rate, which reflects the velocity of the energy-supplying biochemical processes in the organism. Besides development, other processes that determine an insect species' life history, such as survival and reproduction, are also strongly influenced by temperature. Every insect species is adapted to a specific temperature range. Development and reproduction, and hence population increase, occur at physiological temperatures that are delimited by an upper and lower developmental threshold. Within this range, there is an intrinsic optimal temperature for optimal development, highest survival and highest reproduction. Temperature-dependent development of insects from one stage to another does not follow a linear relationship. Several biochemical or mathematical models have been developed that describe well the temperature-dependent development rates for insects (Janisch, 1932; Stinner et al., 1975; Sharpe and DeMichele, 1977; Lactin et al., 1995; Brière et al., 1999). Even if functions for temperature-dependent mortality and reproduction are not addressed to any great extent in the literature, it is possible to describe temperature-dependent mortality and reproduction adequately by distinct non-linear equations. In ILCYM a set of functions is established that describes the effect of temperature on development, immature mortality, adult senescence, reproduction and, if required, on female rates in the progeny. In addition, distribution models are included for describing variation in development between individuals within a single insect species population and alternating age-dependent daily oviposition frequencies. These functions are compiled to an overall phenology model that makes it possible to generate life table parameters for a given range of fluctuating temperatures and allows temporal and special simulations of population parameters according to real or interpolated temperature data. One example for experimentally establishing all required functions that describe the insect's life history (i.e., development, mortality and reproduction) by temperature is the model established for the potato tuber moth (Sporleder et al., 2004). Similar experiments have been carried out and published for many insect species, including pest species of global proportions, including *Bemisia tabaci* Gennadius (Muñiz and Nombela, 2001), the psocid *Liposcelis tricolor* Badonnel (Dong et al., 2007), *Liriomyza sativae* Blanchard (Haghani et al., 2007) and *Helicoverpa armigera* Hübner (Mironidis and Savopoulou-Soultani, 2008); and important parasitoids used as biological control agents, such as *Lysiphlebia mirzai* Shuja-Uddin, a parasitoid of *Toxoptera citricida* Kirkaldy (Liu and Tsai, 2002), *Anagyrus pseudococci* Girault attacking the vine mealybug *Planococcus ficus* Signoret (Daane, 2004) and *Diadegma anurum* Thomson attacking *Plutella xylostella* L. (Golizadeh *et al.*, 2008). Full knowledge of their phenology has not been applied for simulating the species' temporal population growth potential or the species' spatial potential distribution according to temperature.

23.2.2 Main features

ILCYM is an open-source computer-aided tool built on R and Java codes and linked to uDig platform, which is a geographic information system (GIS), an application that contains basic tools for mapping and managing geographic information. The software package consists of three modules (see Fig. 23.1 for ILCYM's main menu):

1. The 'Model builder', which facilitates the development of insect phenology models based on life table data derived from constant-temperature studies;

2. The 'Validation and simulation module', which provides tools for validating the developed phenology model using insect life tables established at fluctuating temperature conditions. It also utilizes the obtained model for deterministic and stochastic simulations.

3. The 'Potential population distribution and risk mapping module', in which the phenology models are implemented in a GIS environment that allows for spatial – global or regional – simulations of species activities and mapping.

23.2.3 Experimental data required

ILCYM requires either life table or single-stage cohort study data collected over a range of constant temperatures in which the insect species develops. In life table experiments a group of individuals of the same age is observed from the beginning of their egg stage until the death of all adults; phenological events (i.e., development to the next stage, mortality, adult emergence, longevity and fecundity) are recorded at constant intervals (generally one day). In single-stage cohort studies, individuals of the same age for each developmental stage, such as fresh eggs, emerging larvae or prepupa, are exposed to different constant temperatures. The phenological

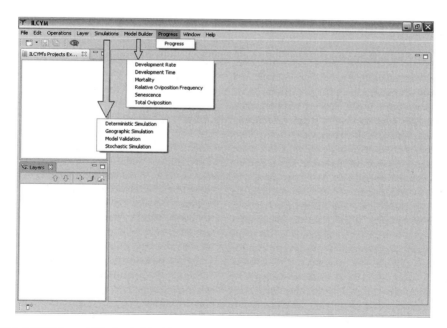

Fig. 23.1 ILCYM (Insect Life Cycle Modelling) software main menu indicating different features of the package, the model builder and model validation and simulation tools. The status of models under development can be seen in the 'Progress' pop-up window.

events are monitored until the development to the next stage or death of all individuals in the cohort. Data on daily reproduction per female at each temperature is required for constructing the oviposition sub-model. If the rate of females and males in the progeny is expected to be variable with temperature, the proportion of females in the progeny needs to be assessed, in addition to the daily female oviposition.

23.2.4 Model development using the 'model builder'

ILCYM provides a list of non-linear functions which are adequate for describing the temperature dependency of the different processes in the species life history (i.e., development, survival and reproduction). For each specific temperature process the function that fits the experimental data best can be selected (see Fig. 23.2 for outputs produced by ILCYM). When a function for each specific temperature-dependent process in the insect's life history has been identified, the program compiles the overall phenology model automatically.

23.2.5 Validation of established models using stochastic simulation

The validation tool in ILCYM allows the user to evaluate the ability of the developed phenology model to reproduce the insect species' behavior under fluctuating temperature conditions. This is achieved by comparing experimental life table data obtained from fluctuating temperature studies with model outputs produced by using the same temperature records as input data (Fig. 23.3). ILCYM stochastically simulates a user-defined number of life tables, each with a user-defined number of individuals, through rate summation and random determination for each individual's survival, development to the next stage and sex. Stochasticity in reproduction is calculated according to the variance observed in the data on total oviposition per female used for developing the model. For each life table analysis, the life table parameters are calculated using standard methods (Birch, 1948; Southwood and Henderson, 2000).

The standard errors for the simulated parameters are used for statistical comparison between simulated outputs and the experimental life table parameters established at fluctuating temperatures. When a considerable gap is found between both simulated and experimentally determined results, the user should revise the plausibility of all established functions used in the overall model. If the overall fit of a particular sub-model or the precision of its best-fitting parameters is unsatisfactory, parameters can be improved using the modelling tools provided in the 'model builder' or, if improvement of the model fails, another model may be more appropriate. ILCYM calculates various criteria, including AIC, for comparing and choosing models. If no model can be fitted successfully, for instance due to large scatter in the data or missing data at extreme temperatures, it is recommended to generate more data to increase the precision of estimated parameters. In particular, more data at extreme temperatures might be helpful for increasing the precision of model parameters that determine temperature limits for development and survival.

23.2.6 Deterministic simulation tools

The 'deterministic simulation' is another feature of ILCYM which simulates populations using a rate summation and cohort up-dating approach over a long-term period (one or more years), with multiple overlapping generations, for a specific location, based on minimum and maximum daily temperatures, and which visualizes the potential population increase. At the present stage of ILCYM development, this simulation considers only the growth process as an unbounded process in which the population grows without limit if uncontrolled (Fig. 23.4).

23.2.7 Potential population distribution and risk mapping

Temperature data used

For spatial simulations, ILCYM uses temperature data for present and future climate change scenarios. For present scenarios, WorldClim[ii] is used.

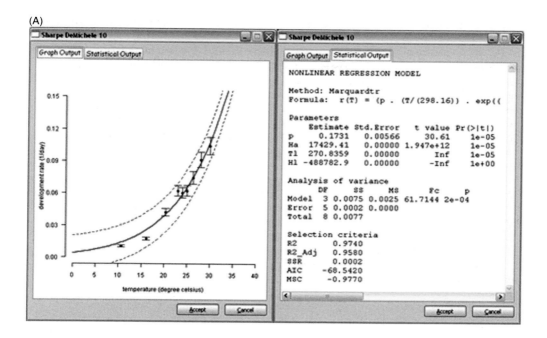

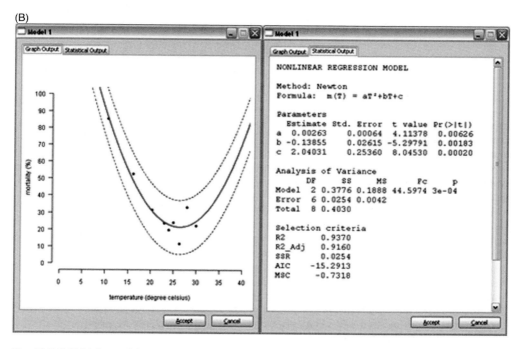

Fig. 23.2 ILCYM (Insect Life Cycle Modelling) software outputs showing functions and parameter values used to describe simulation of *P. operculella* larva development (A) and mortality (B); data shown are from Sporleder *et al.* (2004). The output on the right side presents the equation fitted, model parameter values and their standard errors, ANOVA table, and selection criteria for comparing the model fit with other models.

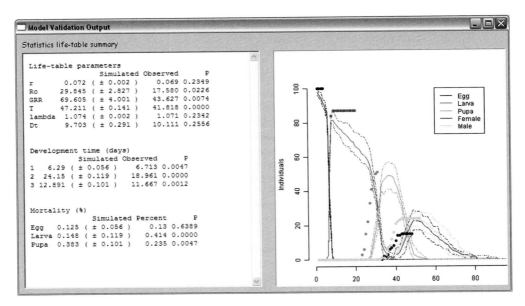

Fig. 23.3 Parameter values and graphical display of *P. operculella* model validation. Left: average values obtained from four life-stage simulations compared to the value obtained from the experimental life table; P indicated the probability that the value from the experimental life table is within the range of values from simulated life table (student *t*-test). Right: observed and simulated stage-specific survival; the dots (•) are observed data and lines are the results from stochastically simulated life tables (Tmin = 15°C and Tmax 25°C) for *P. operculella* stages egg, larva, pupa, female and male, respectively. Full lines: average survival; scattered lines: maximum and minimum survival obtained from the four life tables simulated.

This is a global source of climate surface data at four different spatial resolutions (30 arc-second; 2, 5 and 10 arc min) encompassing monthly aggregated climate variables. The data are freely available and were described by Hijmans et al. (2005). The data were compiled from monthly averages of climate as measured at weather stations from a large number of global, regional, national and local sources, mostly from the 1950–2000 period (Hijmans et al., 2005). For climate change scenarios, ILCYM uses the emission scenario SRES-A1B projections for the year 2050 according to the IPCC assessment report (IPCC, 2007b); however, other climate change scenarios (for the years 2020, 2030, 2040, 2060, 2080 and 2100) can be used. The SRES-A1B scenario incorporated in ILCYM assumes a future world of very rapid economic growth, global population that peaks in mid-century and declines thereafter, and rapid introduction of new and more efficient technologies with balanced emphasis on all energy sources (IPCC, 2007b).

For spatial simulation of risk indices, ILCYM uses monthly minimum and maximum temperature data directly; however, because this aggregation might raise problems of temporal scale, daily minimum and maximum temperatures can be interpolated for simulating within-year variation of the population growth parameters. While more computationally complex than interpolating point phenology results, this method is considered to provide more flexible and robust results.

23.2.8 Spatial simulation

The option is devoted to spatial simulation of the population distribution. Populations are simulated grid-based within a defined area according to grid-specific daily temperatures interpolated from available databases. If the study insect is a pest, the tool can plot 'risk' indices on a map, based on simulation results for visualizing the establishment risk, the spread and damage potential of that pest species.

Model calculations are based on daily minimum and maximum air temperatures; either real point-based weather station records or

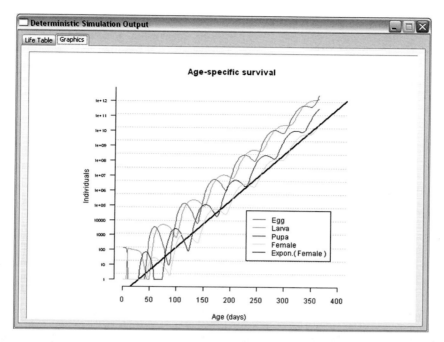

Fig. 23.4 Potential population increase of *P. operculella* over a period of 1 year with multiple overlapping generations using the model established by Sporleder *et al.* (2004) for a constant temperature of 15°C (minimum) and 25°C (maximum), respectively. Model started with 100 eggs; seven overlapping generations were produced but the stable age-stage distribution was not reached after 1 year.

interpolated grid-point data. A 15-min time step length was chosen for the model. Temperature in each 15-min time step is calculated using a cosine function for half-day temperature predictions. The equation used for the first half-day is:

$$T_i = \left(\frac{Max - Min}{2}\right) \times \cos\left(\frac{\pi \times (i+0.5)}{48}\right) + \frac{Max - Min}{2}$$

where T_i is the temperature (in °C) of time step i (i = 1, 2, 3, …48), and *Min* and *Max* are daily minimum and maximum temperatures. The calculation is then repeated to obtain T_i for the second half-day employing the minimum temperature, *Min*, of the following day in the equation.

In order to reduce the simulation computation time, and further, due to the availability of data sets (which are often monthly), ILCYM replaces the *Min* and *Max* daily temperatures in the equation by monthly average minimum and maximum temperatures, respectively. With the assumption T_i retains a constant value for all days of the month with the exception of the last day, which uses the temperature of the second half-day of the first day of next month to calculate T_i.

The phenological simulation is executed in two phases. First, the program calculated stage-specific daily (monthly) development rates (d_k) and mortality rates (m_k) for each immature development stage, k, adult senescence rates for females and males (s_F and s_M) and total fecundity per female (R), according to temperature starting from Julian day 1 (1 January) to Julian day 365 (31 December) of the year. From these results the following (life table) parameter estimates are generated for each Julian day or each month:

1. Generation length in days (T) = $(1/d_1) + (1/d_2)$ +…..+ $(1/d_K) + (1/s_F) \times TR_{50\%}$

where d_k is the median development rate for the immature life stages, k, s_F is the median senescence time for female adults and $TR_{50\%}$ is the normalized age of females until median oviposition.

2. Net reproduction rate (R_0) = $f \times IS \times FR$

where f is the fecundity per female, IS is the immature survival rate (= $\Pi\, 1-m_k$), and FR is the female rate in the progeny

3. Intrinsic rate of increase (r_m) = $\ln(R_0) / T$

4. Finite rate of increase (λ) = $\exp(r_m)$
5. Doubling time (Dt) = $\ln(2) / r_m$

From these results three indices are calculated:

1. Establishment index (survival risk) = $(1-x_{Egg}) \times (1-x_{Larva})(1-x_{Pupa})$

where x_k is the number of days in which all individuals of a given immature life stage k would not survive (i.e., the number of days in which the stage specific mortality, m_k, equals 1). The index is 1 when a certain proportion of all immature life stages of the pest survive throughout the year.

2. Generation index = $\dfrac{\Sigma\, 365\ /\, T_x}{365}$

where Tx is the generation time in days calculated for each Julian day x (x = 1, 2, 3, ..., 365). The generation index estimates the mean number of generations that may be produced within a given year. Development times decline with increasing temperatures, thereby increasing the speed of population build-up. However, because immature survival and oviposition also decline at extremely high temperatures, the generation numbers per annum may not correspond with the real population growth rates at extreme temperatures.

3. Activity index = $\log \Pi\, \lambda_x$

This index is explicitly related to the finite rate of population increase, λ, which takes the whole life history of the pest into consideration. For example, an index value of 4 would illustrate a potential population increase by a factor of 10,000 within 1 year if all other population-limiting factors, including food availability, are neglected.

These three indices are visualized in maps to evaluate the potential distribution area and population growth potential in space (see Fig. 23.5 for demonstration using the generation index in maps). For further details, the life table parameters

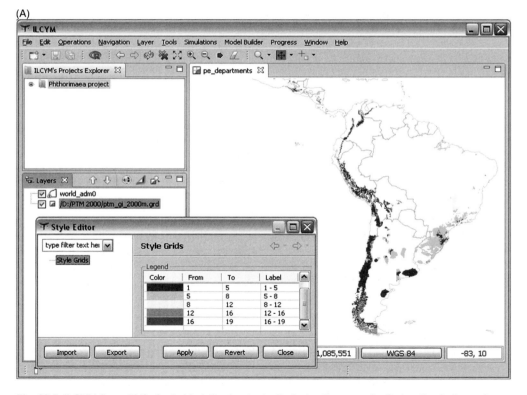

Fig. 23.5 ILCYM (Insect Life Cycle Modelling) outputs displaying the maps for Generation indices of *P. operculella* simulated using monthly average minimum and maximum temperature data for the (A) current scenario (year 2000); and (B) future scenario (year 2050) for the potato production areas in South America.

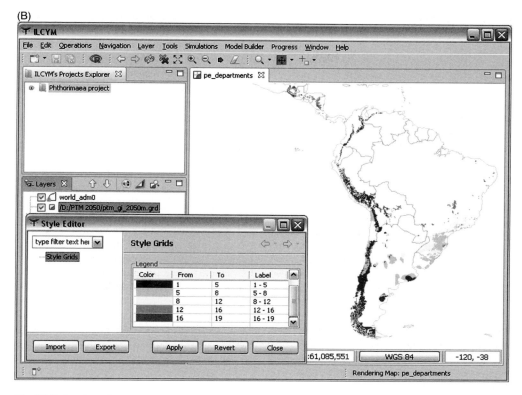

Fig. 23.5 Continued.

for a selected area (grid) in the map can be plotted against Julian day or month to visualize the within-year variability of population development due to seasonal climate variation.

23.3 Positioning ILCYM within process-based modelling tools

23.3.1 'Inductive' versus 'deductive' modelling approaches

ILCYM software uses a 'deductive' process-based climatic modelling approach. This approach is also exclusively applied in the DYMEX software (Kriticos *et al.*, 2003). Other software also include elements of 'inductive' approaches, as in ECAMON (Trnka *et al.*, 2007), CLIMEX (Sutherst and Maywald, 1985), and NAPPFAST (Nietschke *et al.*, 2008).

DYMEX is a tool with a graphical interface for modelling population dynamics of organisms through assembling and linking of modular objects. This package uses a discrete time model that tracks the status of development of individuals within a cohort. At each time step of model building a new cohort is created, and the constituent individuals enter a new life stage and are independently tracked (Yonow *et al.*, 2004).

ECAMON stands for 'environmental change assessment models for the European corn borer, *Ostrinia nubilalis* (Hübner)'. This is a specific modelling tool that integrates the phenology of this pest measured in degree days and occurrence data into one single framework and produces maps via ArcInfo (Trnka *et al.*, 2007). In a comparative study, Kocmánková *et al.* (2008) demonstrated that ECAMON predicted pest occurrence and distribution superior to the output obtained from the application of CLIMEX. This may stimulate the application of this modelling approach to other insect pests.

The NAPPFAST system is a web-based graphic user interface that links climatic and geographic databases with upper temperature thresholds of a pest, as well as the degree-day requirements for each phenological stage (Nietschke et al., 2008). It employs development time by degree-day accumulation as the only phenological event.

CLIMEX software is probably the best-known and used computer-aided tool for modelling the environmental suitability for invasive insect pest species (Sutherst and Maywald, 1985). It makes use of the pest species responses to a series of stress indices (cold, hot, dry and wet) that define the species' particular environmental limits for a sustained growth and development of the population. These stress indices are combined to the species' growth indices to yield an 'eco-climatic index' that describes the total favorability of geographic locations for the species (Rafoss, 2003; Sutherst, 1991; Zalucki and Furlong, 2005). Through the inclusion of an 'eco-climatic index' that is added to the simple inbuilt climate-match function, the CLIMEX software uses partly 'deductive' and partly 'inductive' modelling approaches.

This dual property of 'deductive' and 'inductive' is also shared by ECAMON and NAPPFAST. Within the 'deductive' modelling approach, one can differentiate between 'inferred deduction' and 'pure deduction' mechanisms. 'Inferred deduction' refers to situations in which the process-oriented mechanisms introduced into the modelling framework mimic the real situation, while 'pure deduction' captures direct measurement of the relationships between aspects of life-history (development, survival, productivity, etc.) of individual and environmental variables. An example of application is the 'compare locations' function in CLIMEX, encompassing the eco-climatic index that infers linear expressions representing, for example, the species' growth, survival and reproduction, and the 'match climate' function that compares the meteorological data from different places, with no reference to the preferences of a given species. Within 'pure deduction', detailed knowledge of climatic factors (such as temperature preferences, obtained from detailed laboratory studies on the species' development) is required, in order to predict where the species might occur and populations might develop. This mechanism is only and purely applied in the ILCYM software.

23.3.2 Linear versus non-linear model methods

In the pure biological sense, equations used in process-oriented models can be based on empirical and non-linear models; however, most computer-aided tools presented here (CLIMEX, ECAMON and NAPPFAST) apply linear response models to temperature, but in contradiction to non-linear responses observed in insect species. This linear degree-day modelling method may produce errors, especially at extreme temperatures where the development time curtails. Generally, linear models do not account properly for within-day temperature fluctuations, especially when daily temperatures fluctuate frequently between extreme temperature values, which might cause significant errors in predicting daily population growth rates (see Kaufmann effect in Worner, 1992). The non-linearity is properly captured in ILCYM for expressing the complete species life history, which is mainly determined by the development time from one stage to another, mortality, longevity and reproduction. In ILCYM the establishment of a phenology model for a particular species is based on detailed laboratory assessments for parameter determination that produces life-table parameters, and allows the simulation of populations according to real or interpolated temperature data for a given region and time.

23.3.3 Climate change analysis and mapping

Climate change is inevitably causing serious challenges for the development and interpretations of pest risk maps. In this context, proper incorporation of climate change analysis within tools for insect risk mapping is crucial. Key challenges for the development of computer-aided tools are the type of resolution to be used in mapping to forecast impacts of climate change on species and the selection of an appropriate SRES emission scenario. For ILCYM it was opined that the coarse resolution of 100 or 200 km (or even more) of the original general circulation model (GCM) projections is simply not realistic or practical enough for assessing the distribution of agricultural pests on a landscape level, particularly in the tropics, where orographic and climatic conditions vary significantly across relatively small distances. Additionally, Ramirez and Jarvis (2010) stated that changes in topography and climate variables are not the only factors accounting for variability; soils and socioeconomic drivers, which vary over small distances, also influence ecosystems and increase uncertainties, making the forecasting and assessment models more inaccurate and complicated to calibrate. For these reasons the ILCYM software applies freely available, downscaled temperature-change data sets produced by Ramirez and Jarvis (2010) http://gisweb.ciat.cgiar.org/ rather than direct outputs from GCM. These authors applied the thin plate spline spatial interpolation of anomalies (delta) of original GCM outputs for the downscaling. Further, ILCYM opted by default to the SRES-A1 emission scenario projections of the year 2050 as per the IPCC (2007a) assessment report. Additionally, the software provides options to analyze any downscale data sets (with downscale methodology different from that of thin plate spline) as well as other SRES emission scenarios (A2, B1 and B2) of the years 2020, 2030, 2080, 2100, and so on, provided the user uploads the data sets for the corresponding scenarios into the software package.

23.3.4 Applications for classical biocontrol programs

Many classical biological control programs have involved introductions and releases of exotic species using the 'climate matching' tools as useful guide to find climate similarities and differences between the species' native range of distribution and intended release areas (van Lenteren et al., 2006). This approach, however, is not a reliable basis for predicting survival and establishment specifically in the context of climate change. In tackling this aspect, ILCYM has a component that allows users to study the synchronization of the phenology between pests and natural enemies (parasitoids). This is performed by estimating the timing and numbering of individuals in the different development stages. ILCYM allows the development of an independent parasitoid phenology model that can be linked with the pest (host) model through predicted development times, parasitoid fertility and parasitism rates for simulations of host population growth and development under constant or varying temperatures. Further, the parasitoid phenology model can be directly connected to GIS mapping, to identify those regions in which the parasitoid can potentially establish (establishment index) and to indicate regionally its potential efficacy (activity index) to control its host. Overall, such information will be vital in guiding understanding, and projecting how the biological control agent may adapt well or fail to adapt in the future with potential change in climate temperature.

23.3.5 Considering sex ratios in population growth models

A possible concern with the potential change in climate is that an increase in extreme events may occur. Extreme events such as heat and cold waves have direct impacts on temperature values that affect species development and progenies. For insects and other organisms, it is well known that temperatures below a certain value favor the development of males only, whereas temperatures surpassing a certain value may favor only females. Single-sex offspring are then typical in a single nest, but an equilibrium population sex ratio is usually achieved by within-habitat microclimatic differences among nests and dispersal among populations (West et al., 2002). Thus, changes in the incidence of extreme temperatures (high or low) through time could result in highly skewed population sex ratios. As an attempt to understand the effects of extreme temperature events on the overall population of species, ILCYM introduced a sub-function to describe species sex ratios and female rate variability in dependence of temperature.

23.4 Outlook

23.4.1 Special features of ILCYM

The primary feature that distinguishes ILCYM from other software is the relative ease and speed with which the overall phenology model of a particular species can be developed when data from constant temperature experiments are available.

Model development is based on statistical analysis that determines the precision (standard errors) of established model parameters. The model-builder environment of this package makes programming and coding for such models unnecessary, and hence makes the model development more readily accessible to researchers in the field. ILCYM has a graphic user interface, and a library of appropriate functions for the different processes that determine an insect species' life history. Selection is based on analytic and statistical tools for choosing appropriate models through the fitting of sub-functions to experimental data. Different models for a particular process can be compared using different criteria of model selections such as the AIC. ILCYM provides users with summary output tables showing statistical analysis and high resolution graphics. The ILCYM 'model builder' places strong emphasis on 'parameter determination' for models. Parameters are stored in an inbuilt database, and the model under development is compiled step by step as well as evaluated (using validation and sensitivity analysis of the model) and, if necessary, modified using several tools provided within the software. Fitting a model remains a complicated task and should be done carefully. There is no standard method to choose between competing models; that is, the decision on choosing a model should not be left to a computer program, and the user needs experience and skills to identify an appropriate model that best delivers biologically meaningful and statistically significant parameters. When all required models (sub-models) that describe the species life history are determined, the overall phenology model is compiled automatically.

A common aspect of the above-mentioned available computer-aided tools is their ability only to map species-suitable environmental distribution, with no reference to risk assessments. ILCYM generates an activity index (AI) offering the possibility of estimating the population's temporarily potential growth within an area of interest. It also offers users the possibility of calculating (via stochastic or deterministic simulations at constant or fluctuating temperature) the complete life-table parameters for a species under investigation. These parameters include the net reproduction rate, mean generation time, intrinsic rate of natural increase, finite rate of increase and the doubling time.

To date, the majority of tools mentioned in this text are only available to a small number of researchers. Software packages such as DYMEX and CLIMEX are not open-source products and their costs (thousands of US dollars) probably make them unaffordable to research institutions in developing countries. Instead, ILCYM software is freely available and can be downloaded from the International Potato Center (CIP), Lima, Peru (see Endnote (i) for web address).

23.4.2 Limitations and future prospects of ILCYM

ILCYM requires input data in appropriate quantity and quality. Collecting this is quite demanding and time consuming; this might constitute an important factor for the massive use of the software. When using ILCYM for determining species' potential environmental suitability, the program specifically guides the user to understand the temperature effects that limit the geographical distribution of the species and to identify the temperature optima that favor population growth. The approach therefore takes temperature as the only influencing variable on insect phenology, because the effects of this climate factor are well assessed through laboratory experiments. However, further effects of other abiotic factors such as rainfall and humidity, as well as species-specific limiting factors and environmental stresses, should be quantified and included in the concept for better accuracy.

23.5 Conclusions

Many modelling approaches, tools and computer packages are available for analyzing insect species potential ranges. Each of the approaches (either 'inductive' or 'deductive') used in the different packages and models presents advantages and disadvantages. Referring to the fact that errors and uncertainties in pest risk mapping are unavoidable, the focus and emphasis should be on combining several tools and methods in conjunction with expert opinions, for better prediction of establishment risk and population growth potential of a species of interest.

Notes

i www.cipotato.org/ilcym, accessed 14 August 2012.
ii www.worldclim.org, accessed 14 August 2012.

References

Allen, J.C. (1976) A modified sine wave method for calculating degree days. *Environmental Entomology* 5, 388–396.

Andrewartha, H. and Birch, L. (1955) *The Distribution and Abundance of Animals*. University of Chicago Press, Chicago, Illinois.

Baker, C.R.B. (1991) The validation and use of a life-cycle simulation model for risk assessment of insect pests. *Bulletin OEPP* 21, 615–622.

Baker, R.H.A. (1996) Developing a European pest risk mapping system 1. *EPPO Bulletin* 26, 485–494.

Baker, R.H.A., Sansford, C.E., Jarvis, C.H., Cannon, R.J.C., MacLeod, A. and Walters, K.F.A. (2000) The role of climatic mapping in predicting the potential geographical distribution of non-indigenous pests under current and future climates. *Agriculture, Ecosystems & Environment*, 82, 57–71.

Birch, L.C. (1948) The intrinsic rate of natural increase of an insect population. *Journal of Animal Ecology* 17, 15–26.

Braasch, H., Wittchen, U. and Unger, J.G. (1996) Establishment potential and damage probability of *Meloidogyne chitwoodi* in Germany 1. *EPPO Bulletin* 26, 495–509.

Brière, J.F., Pracros, P., le Roux, A.Y. and Pierre, J.S. (1999) A novel rate model of temperature-dependent development for arthropods. *Environmental Entomology* 28, 22–29.

Daane, K. (2004) Temperature-dependent development of *Anagyrus pseudococci* (Hymenoptera: Encyrtidae) as a parasitoid of the vine mealybug, *Planococcus ficus* (Homoptera: Pseudococcidae). *Biological Control* 31, 123–132.

Dong, P., Wang, J.-J., Jia, F.-X. and Hu, F. (2007) Development and reproduction of the Psocid *Liposcelis tricolor* (Psocoptera: Liposcelididae) as a function of temperature. *Annals of the Entomological Society of America* 100, 228–235.

Golizadeh, A., Kamali, K., Fathipour, Y. and Abbasipour, H. (2008) Life table and temperature-dependent development of *Diadegma anurum* (Hymenoptera: Ichneumonidae) on its host *Plutella xylostella* (Lepidoptera: Plutellidae). *Environmental Entomology* 37, 38–44.

Haghani, M., Fathipour, Y., Talebi, A.A. and Baniameri, V. (2007) Thermal requirement and development of *Liriomyza sativae* (Diptera: Agromyzidae) on cucumber. *Journal of Economic Entomology* 100, 350–365.

Hijmans, R.J., Cameron, S.E., Parra, J.L., Jones, P.G. and Jarvis, A. (2005) Very high resolution interpolated climate surfaces for global land areas. *International Journal of Climatology* 25, 1965–1978.

IPCC (2007a) *Climate Change 2007: Impacts, Adaptation and Vulnerability*. Contribution of Working Group II to the Fourth Assessment Report of the Intergovernmental Panel on Climate Change, Cambridge University Press, Cambridge, UK.

IPCC (2007b) *Fourth Assessment Report (AR4). Climate Change 2007: Synthesis Report*. Intergovernmental Panel on Climate Change, Geneva, Switzerland, pp. 104.

Janisch, E. (1932) The influence of temperature on the life history of insects. *Transactions of the Entomological Society of London* 80, 137–168.

Jarvis, C.H. and Baker, R.H.A. (2001a) Risk assessment for nonindigenous pests: 1. Mapping the outputs of phenology models to assess the likelihood of establishment. *Diversity and Distributions* 7, 223–235.

Jarvis, C.H. and Baker, R.H.A. (2001b) Risk assessment for nonindigenous pests: 2. Accounting for interyear climate variability. *Diversity and Distributions* 7, 237–248.

Kocmánková, E., Trnka, M., Žalud, Z., Semerádová, D., Dubrovský, M., Muška, F. and Možný, M. (2008) Comparison of two mapping methods of potential distribution of pests under present and changed climate. *Plant Protection Science* 44, 49–56.

Kohlmann, B., Nix, H. and Shaw, D.D. (1988) Environmental predictions and distributional limits of chromosomal taxa in the Australian grasshopper *Caledia captiva* (F.). *Oecologia* 75, 483–493.

Kriticos, D.J., Brown, J.R., Maywald, G.F., Radford, I.D., Nicholas, D.M., Sutherst, R.W. and Adkins, S.W. (2003) SPAnDX: a process-based population dynamics model to explore management and climate change impacts on an invasive alien plant, *Acacia nilotica*. *Ecological Modelling* 163, 187–208.

Lactin, D.J., Holliday, N.J., Johnson, D.L. and Craigen, R. (1995) Improved rate model of temperature-dependent development by arthropods. *Environmental Entomology* 24, 68–75.

Liu, Y.H. and Tsai, J.H. (2002) Effect of temperature on development, survivorship, and fecundity of *Lysiphlebia mirzai* (Hymenoptera: Aphidiidae), a parasitoid of *Toxoptera citricida* (Homoptera: Aphididae). *Environmental Entomology* 31, 418–424.

Logan, J.A., Wollkind, D.J., Hoyt, S.C. and Tanigoshi, L.K. (1976) An analytic model for description of temperature dependent rate phenomena in arthropods. *Environmental Entomology* 5, 1133–1140.

McKenney, D.W., Hopkin, A.A., Campbell, K.L., Mackey, B.G. and Foottit, R. (2003) Opportunities for improved risk assessments of exotic species in Canada using bioclimatic modeling. *Environmental Monitoring and Assessment* 88, 445–461.

Mironidis, G.K. and Savopoulou-Soultani, M. (2008) Development, survivorship, and reproduction of *Helicoverpa armigera* (Lepidoptera: Noctuidae) under constant and alternating temperatures. *Environmental Entomology* 37, 16–28.

Muñiz, M. and Nombela, G. (2001) Differential variation in development of the B- and Q-biotypes of *Bemisia tabaci* (Homoptera: Aleyrodidae) on sweet pepper at constant temperatures. *Environmental Entomology* 30, 720–727.

Nietschke, B.S., Magarey, R.D., Borchert, D.M., Calvin, D.D. and Jones, E. (2007) A developmental database to support insect phenology models. *Crop Protection* 26, 1444–1448.

Nietschke, B.S., Borchert, D.M., Magarey, R.D. and Ciomperlik, M.A. (2008) Climatological potential for *Scirtothrips dorsalis* (Thysanoptera: Thripidae) establishment in the United States. *Florida Entomologist* 91, 79–86.

Peacock, L. and Worner, S. (2006) Using analogous climates and global insect pest distribution data to identify potential sources of new invasive insect pests in New Zealand. *New Zealand Journal of Zoology* 33, 141–145.

Rafoss, T. (2003) Spatial stochastic simulation offers potential as a quantitative method for pest risk analysis. *Risk Analysis* 23, 651–661.

Ramirez, J. and Jarvis, A. (2010) *Downscaling Global Circulation Model Outputs: The Delta Method Decision and Policy Analysis*. Working Paper 1, International Center for Tropical Agriculture (CIAT), Cali, Colombia.

Sharpe, P.J.H. and DeMichele, D.W. (1977) Reaction kinetics of poikilotherm development. *Journal of Theoretical Biology* 64, 649–670.

Sharpe, P.J.H., Schoolfield, R.M. and Butler Jr, G.D. (1981) Distribution model of *Heliothis zea* (Lepidoptera: Noctuidae) development times. *Canadian Entomologist* 113, 845–856.

Southwood, T.R.E. and Henderson, P.A. (2000) *Ecological Methods*. Blackwell Science Ltd, Oxford, UK.

Sporleder, M., Kroschel, J., Gutierrez Quispe, M.R. and Lagnaoui, A. (2004) A temperature-based simulation model for the potato tuberworm, *Phthorimaea operculella* Zeller (Lepidoptera; Gelechiidae). *Environmental Entomology* 33, 477–486.

Sporleder, M., Kroschel, J. and Simon, R. (2007) Potential changes in the distributions of the potato tuber moth, *Phthorimaea operculella* Zeller, in response to climate change by using a temperature-driven phenology model linked with geographic information systems (GIS). In XVI International Plant Protection Congress, Congress Proceedings Vol. 1, BCPC, Glasgow, pp. 360–361.

Sporleder, M., Simon, R., Gonzales, J., Carhuapoma, P., Juarez, H., De Mendiburu, F. and Kroschel, J. (2009) *ILCYM - Insect Life Cycle Modeling. A Software Package for Developing Temperature-Based Insect Phenology Models with Applications for Regional and Global Pest Risk Assessments and Mapping*. User manual. International Potato Center, Lima.

Steinbauer, M.J., Yonow, T., Reid, I.A. and Cant, R. (2002) Ecological biogeography of species of *Gelonus*, *Acantholybas* and *Amorbus* in Australia. *Austral Ecology* 27, 1–25.

Stinner, R.E., Gutierrez, A.P. and Butler Jr, G.D. (1974) An algorithm for temperature-dependent growth rate simulation. *Canadian Entomologist* 106, 519–524.

Stinner, R.E., Butler Jr, G.D., Bacheler, J.S. and Tuttle, C. (1975) Simulation of temperature-dependent development in population-dynamics models. *Canadian Entomologist* 107, 1167–1174.

Sutherst, R.W. (1991) Pest risk analysis and the greenhouse effect. *Review of Agricultural Entomology* 79, 1177–1187.

Sutherst, R.W. and Maywald, G.F. (1985) A computerised system for matching climates in ecology. *Agriculture, Ecosystems & Environment* 13, 281–299.

Sutherst, R.W. and Maywald, G.F. (1990) Impact of climate change on pests and diseases in Australasia. *Search (Sydney)* 21, 230–232.

Sutherst, R.W. and Maywald, G.F. (1991) Climate modelling and pest establishment. Climate-matching for quarantine, using CLIMEX. *Plant Protection Quarterly* 6, 3–7.

Sutherst, R.W., Collyer, B.S. and Yonow, T. (2000) The vulnerability of Australian horticulture to the Queensland fruit fly, *Bactrocera (Dacus) tryoni*, under climate change. *Australian Journal of Agricultural Research* 51, 467–480.

Trnka, M., Muška, F., Semerádová, D., Dubrovský, M., Kocmánková, E. and Zalud, Z. (2007) European corn borer life stage model: regional estimates of pest development and spatial distribution under present and future climate. *Ecological Modelling* 207, 61–84.

Uvarov, B.P. (1931) Insects and climate *Transactions of the Entomological Society of London* 79, 1–247.

Vanhanen, H., Veteli, T. and Niemelä, P. (2008a) Potential distribution ranges in Europe for *Ips hauseri*, *Ips subelongatus* and *Scolytus morawitzi*, a CLIMEX analysis. *EPPO Bulletin* 38, 249–258.

Vanhanen, H., Veteli, T.O. and Niemelä, P. (2008b) Potential distribution ranges in Europe for *Aeolesthes sarta*, *Tetropium gracilicorne* and *Xylotrechus altaicus*, a CLIMEX analysis. *EPPO Bulletin* 38, 239–248.

van Lenteren, J.C., Bale, J., Bigler, F., Hokkanen, H.M.T. and Loomans, A.J.M. (2006) Assessing risks of releasing exotic biological control agents of arthropod pests. *Annual Review Entomology* 51, 609–634.

Venette, R.C., Kriticos, D.J., Magarey, R.D., Koch, F.H., Baker, R.H.A., Worner, S.P., Gómez Raboteaux, N.N. *et al.* (2010) Pest risk maps for invasive alien species: a roadmap for improvement. *BioScience* 60, 349–362.

Wagner, T.L., Wu, H.I., Sharpe, P.J.H. and Coulson, R.N. (1984) Modeling distributions of insect development time: a literature review and application of the Weibull function. *Annals of the Entomological Society of America* 77, 475–487.

West, S.A., Reece, S.E. and Sheldon, B.C. (2002) Sex ratios. *Heredity* 88, 117–124.

Wilmot Senaratne, K.A.D., Palmer, W.A. and Sutherst, R.W. (2006) Use of CLIMEX modelling to identify prospective areas for exploration to find new biological control agents for prickly acacia. *Australian Journal of Entomology* 45, 298–302.

Worner, S.P. (1992) Performance of phenological models under variable temperature regimes: consequences of the Kaufmann or rate summation effect. *Environmental Entomology* 21, 689–699.

Yonow, T., Zalucki, M.P., Sutherst, R.W., Dominiak, B.C., Maywald, G.F., Maelzer, D.A. and Kriticos, D.J. (2004) Modelling the population dynamics of the Queensland fruit fly, *Bactrocera (Dacus) tryoni*: a cohort-based approach incorporating the effects of weather. *Ecological Modelling* 173, 9–30.

Zalucki, M.P. and Furlong, M.J. (2005) Forecasting Helicoverpa populations in Australia: a comparison of regression based models and a bioclimatic based modelling approach. *Insect Science* 12, 45–56.

Index

Abgrallaspis
    A. aguacatae 333
    A. mendax 333
Aceria litchii see Lychee erinose mite (LEM)
Aceria tosichella see Wheat curl mite (WCM)
Acutaspis
    A. scutiformis 333
    A. umbonifera 333–334
Adventive scales insects
    Aleyrodidae *see* Aleyrodidae
    Coccidae *see* Coccidae
    database 342–343
    Diaspididae *see* Diaspididae
    distribution and species-richness data 344
    Florida 343
    hemipterous insects 342
    Kerriidae *see* Kerriidae
    Monophlebidae 354–355
    Pseudococcidae *see* Pseudococcidae
Afrotropical species
    Diaspididae
        Aspidiotus capensis 334
        Duplachionaspis natalensis 334
        Parlatoria fluggeae 334
        Rolaspis leucadendri 334
    Pseudococcidae
        Delottococcus proteae 338
        Paracoccus burnerae 338
        Phenacoccus stelli 338
        Planococcus kenyae 339
        Vryburgia see Vryburgia
Agro-production systems *see* Exotic arthropods
Aleurodicus
    A. dugesii 356
    A. rugioperculatus 356–357

Aleyrodidae
    Aleurodicus dugesii 356
    Aleurodicus rugioperculatus 356–357
    Dialeurodes schefflerae 355
    Siphoninus phillyreae 355–356
Anastrepha grandis
    behavior *see* Behavior
    distribution and hosts 197–198
    fruits/vegetables 197
    life cycle 198
    morphology *see* Morphology
    population dynamics 198–199
    synonymy 198
Anastrepha ludens
    behavior 194
    distribution and hosts 192–193
    life cycle 194
    mating 195
    morphology *see* Morphology
    population dynamics 195
    synonymy 193
Anastrepha obliqua
    behavior *see* Behavior
    distribution and hosts 195
    life cycle 196
    morphology *see* Morphology
    population dynamics 1997
    synonymy 195–196
Anastrepha suspensa 228–230
Aonidiella orientalis
    calculation, fruit surface area 403
    Chrysomphalus aonidum 408
    crawlers per infestation
        avocado over time 406, 407
        orange over time 405, 406

Aonidiella orientalis (Continued)
    squash over time 404
    tangerine over time 406
  crawler's wandering behavior 402
  decomposition, fruit 408
  Diaspidiotus perniciosus 408
  economic impact 402
  favorable climate 401
  FAWN 403
  female scales 407
  fruit sizes 404
  heavily infested discarded fruit 409
  human activities 402
  imported Mexican avocados 403
  infestation rate, crawlers 408, 409
  infested seedlings/infested leaves 408
  infested squash fruit 404
  minimum and maximum temperatures 404
  natural / human-mediated 402
  PPQ 401
  quarantine pest species 401
  real-world conditions 409
  scales dispersing infestation
    avocado fruit 406, 407
    orange fruit 405, 406
    squash fruit 405
    tangerine fruit 406, 407
  shelf-life 409
  stock 403
  wind direction 407
Armored scales *see Aonidiella orientalis*; Diaspididae
Aulacaspis
  A. crawii 335
  A. ima 335
  A. rosarum 335
  A. tubercularis 354
  A. yasumatsui
    cycad scale 352, 353
    distribution 352
    field recognition 353–354
    natural enemies 353
Australian species
  Diaspididae
    Pseudaulacaspis brimblecombei 336
    Pseudaulacaspis eugeniae 336
  Pseudococcidae
    Dysmicoccus see Dysmicoccus
    Phenacoccus hakeae 340
    Pseudococcus see Pseudococcus
Avocado trees *see Aonidiella orientalis*
Avocado Weevils
  seed borer
    Heilipus lauri see Heilipus lauri
    Heilipus pittieri 41
    Heilipus trifasciatus 41–42
  stem borers
    Heilipus albopictus 42

    Heilipus apiatus 42–44
    Heilipus catagraphus 44
    Heilipus elegans 44
    Heilipus rufipes 44–45

Bactrocera carambolae
  adult feeding 216–217
  economic importance 218–219
  funding and coordination 216
  host range and geographic distribution 217–219
  life stages 217
  male annihilation technique 219–220
  mating 216
  OFF 215–216
  oviposition 216
  potential risk, Florida 220
  protein baiting technique 220
  sampling and monitoring techniques 217
  South America 216
  Suriname and Indonesia 216
  taxonomic status 216
Bactrocera invadens
  in Africa 220–221
  B. cucurbitae and *B. zonata* 221
  control tactics 224
  development 223
  distribution 222–223
  economic importance 224
  host list 223
  interspecific competition 223–224
  invasion history 221
  taxonomic status 221
  temporal and spatial abundance 223
Bactrocera species
  anthropogenic dispersion 214–215
  B. invadens see Bactrocera carambolae
  CFF *see Bactrocera carambolae*
  commercial fruit species 214
  family members 214
  host range and colonization ability 215
  large and rapid rate, population growth 214
  Medfly Program in California 215
  natural ability, dispersion 214
  Tephritidae 214
Behavior
  A. grandis
    mating 198
    oviposition 198
  A. obliqua
    mating 196–197
    oviposition 196
Behavioral control, *Copitarsia* spp. 176
Biological control
  Copitarsia spp. 176
  Diabrotica speciosa 79
  Heilipus lauri 45

Index

Neoleucinodes elegantalis 150–153
Singhiella simplex 369–370
Tecia solanivora Povolny 130–131
Tuta absoluta
    entomopathogens 112–113
    parasitoids 105–109
    predators 109–111

Carambola fruit fly (CFF) *see Bactrocera carambolae*
Ceratitis capitata 228
Chemical control
    Copitarsia spp. 176
    Diabrotica speciosa 78
    Heilipus lauri 45
    Neoleucinodes elegantalis 150
    Singhiella simplex 367–368
    Tecia solanivora Povolny 130
    Tuta absoluta 116–117
Chrysomphalus
    C. pinnulifer 335
    C. propsimus 335
Citrus infestation
    Anastrepha grandis and *A. suspensa* 228
    caribbean fruit fly 228
    C. capitata 228
    chemical peaks 233
    Diachasmimorpha longicaudata 236–237
    D-limonene and β-ocimene 232
    economic impact, tephritid pests 229
    eggs and first instar larvae 231
    evaluation, tephritid 237
    first-instar larvae 233, 235
    gas chromatography 229
    headspace volatiles 229
    high resolution GC analysis 233, 234
    'Marsh' grapefruits 233
    normalized GC peak area 231, 232
    oviposition 233, 235
    ripening process 233
    sample preparation 229–230
    sensitive chemical detection technology 237
    single mated female 229
    statistical analysis 231
    tephritid-infested grapefruit 233
    tropical tephritid fruit flies 228
    ultra-fast GC analysis 236
    visual inspections 229
    volatile collections and chemical analysis 230–231
Citrus nest-webbing mite (CNWM) 252
Citrus orthezia
    Afro-tropical region 301
    classification and taxonomy 306–307
    climate change 302
    description 301
    economic importance 302–304
    ensign scales 302, 304
    extant taxa 302
    external morphology 307–308
    genus *Praelongorthezia* 308–309
    geographical origin 306
    host plants 304–306
    Newsteadia yanbaruensis 301
    Ortheziidae 301
    Ortheziid species 302
    Ortheziolamameti maeharai 302
    scale insects 301
Climate-based niche model 375
CLIMEX software 422
CNWM *see* Citrus nest-webbing mite (CNWM)
Coccidae
    description 348–349
    Phalacrococcus howertoni 350–351
    Protopulvinaria longivalvata 349
    Pulvinaria psidii 349
COI *see* Cytochrome oxidase-1 (COI)
Conotrachelus 35
Control methods, *P. praelonga*
    biological 312–313, 314–315
    chemical 312
    cultural 313
    invasive species, tropical and subtropical climates 315
    mechanical 313
    ornamental and medicinal plants 315
    physical 313
    quarantine 313
Copitarsia spp.
    control
        behavioral 176
        biological 176
        chemical 176
        cultural 176
        regulatory 174–175
    description 160
    economic status and frequency 171, 172
    host plants
        acreage of crops 169
        crop species 165–169
        literature lists 165
        records 165
    interception records 171, 174–175
    life cycle
        adult *C. corruda* 163
        eggs 162
        fecundity 162
        geographic distribution 164–165
        geographic ranges 163
        growth 162–163
        larval instars 163
        pupal stage 163–164
        rearing 164

Copitarsia spp. (*Continued*)
  reproduction rates 162
  temperature and diet 163
 monitoring
  ratio, acetate *vs*. alcohol 174
  (Z)-9-tetradecenyl acetate (Z9D14:Ac)
   and (Z)-9-tetradecenol
   (Z9D14:OH) 171, 174
 pathway analysis
  biosecurity procedures 177
  escape of potential mated females 177–178
  estimated risk, importation 176–177
  uncertainty, climate simulation 178
 taxonomy
  geographic information 161–162
  list of species 160–161
 type and extent of damage
  damages caused in crop plants 170
  pest density 169
  quinoa and artichoke 169
Cultural control
 Copitarsia spp. 176
 Stenoma catenifer 92
 Tecia solanivora Povolny 131–132
Cushiony scales *see* Monophlebidae
Cytochrome oxidase-1 (COI) 290

Darna pallivitta see also Nettle caterpillar
 feeding list 186, 187–188
 larval duration 185
 life stages 184
 pupal duration 185
Deductive modelling approach 413
Degree-day models 413
Diabrotica speciosa
 biological aspects and life cycle 75–76
 corn root system 76–77
 egg laying and fecundity 75–76
 geographical distribution 76
 host plants and damage
  adults 77
  behavioral control 78
  biological control 79
  chemical control 78
  larvae 76–77
 insecticides 78
 larval stage 75
 longevity 75–76
 origin 75
 plant resistance 78–79
 rearing techniques
  cages for adults maintenance 79–80
  eggs collection 80
  larval and pupal development 80–81
  larval food 80
  substrates and containers for 80

Diaspididae
 afrotropical species *see* Afrotropical species
 armored scales 333
 Aulacaspis tubercularis 354
 Aulacaspis yasumatsui 352–354
 Australian species *see* Australian species
 Caribbean basin 333
 description 352
 Duplachionaspis divergens 354
 Eastern Palearctic species *see* Eastern Palearctic species
 neotropical species *see* Neotropical species
 oriental species *see* Oriental species
 Poliaspis cycadis 354
 Unachionaspis tenuis 352
 Western Palearctic species *see* Western Palearctic species
Dracaena massangeana 188
DYMEX software 421
Dysmicoccus
 D. hypogaeus 340
 D. queenslandianus 340

Eastern Palearctic species
 Diaspididae
  Aulacaspis see Aulacaspis
  Parlatoreopsis chinensis 335
  Parlatoria multipora 335
  Unaspis yanonensis 335
 Pseudococcidae 339
EF-1α *see* Elongation factor 1α (EF-1α)
Elongation factor 1α (EF-1α) 290
Entomopathogens 112–113
Environmental change assessment models for the European corn borer, *Ostrinia nubilalis* (ECAMON) 421
Exallomochlus
 E. camur 339
 E. hispidus 339
 E. philippinensis 339
Exotic arthropods
 agro-production systems 388
 'biodiversity-invasibility hypothesis' 374
 biotic and abiotic processes 373, 376
 compilation 374
 dietary specifications and habitat associations 376–387
 economic development and associated international trade 376
 food web and functional role 388
 geographical distribution 376
 'invasional meltdown' 373
 materials and methods
  climate-based niche model 375
  coconut hispid 375
  environmental layers 375

Maxent software 375
ScaleNet 374
'top invaders' 374
natural ecosystems and vector agents 374
occurrence records, Maxent model training or validation 375
pest risk maps 388
pollination 373
targeted management actions 388
trained taxonomists and ecologists 376

FAWN *see* Florida Automated Weather Network (FAWN)
FDACS-DPI *see* Florida Department of Agriculture and Consumer Services, Division of Plant Industry (FDACS-DPI)
Ficus whitefly
    biological control 369–370
    chemical control 367–368
    economic importance 363, 364
    geographic distribution
        FDACS-DPI surveys 363, 365
        host plants 364, 366
    life stages and biology 364, 366
    reproductive biology and life table analysis
        description 364
        eggs, red-eyed nymph and adult 364, 366
        Enkegaard model 364, 368
        female adult survivorship 364, 369
        immature survival, temperature 364, 367
        parameters 364, 368
        temperature, development rate 364, 367
Florida Automated Weather Network (FAWN) 403
Florida Department of Agriculture and Consumer Services, Division of Plant Industry (FDACS-DPI) 363, 365
Fruit flies
    Antillean/mango *see* Anastrepha obliqua
    bait stations and male annihilation 200
    biological control 201–202
    chemical control 200–201
    chitosan polymers 204
    climatological and biological data 199
    cultural and physical practices 200
    GPS-GIS information 204–205
    habitat manipulation *see* Habitat manipulation, Anastrepha
    management strategies 204
    Mexican *see* Anastrepha ludens
    objectives 204
    pestiferous *Anastrepha* species 192
    phenological data 199
    plant resistance 203
    postharvest treatment *see* Postharvest treatment, Anastrepha
    regulatory control 204
    sampling and damage evaluation 199
    SIT *see* Sterile insect technique (SIT)
    South American cucurbit *see* Anastrepha grandis
    trapping techniques 199–200

Gall midges
    biology and life cycle 241
    Contarinia spp. 240–241
    control tactics
        biological 243
        chemical 243
        infested material 243
        insecticides 243
        quarantine *see* Quarantine control, gall midge
    damage 242
    distribution 241
    HPGM 240
    Prodiplosis longifila *see* Prodiplosis longifila
    rearing method 243
    sampling and monitoring techniques 242–243
    seasonality 241–242
    USDA-APHIS 240
Gas chromatography (GC)
    GC-MS *see* GC-mass spectrometry (GC-MS)
    grapefruit volatiles 231, 232
    headspace volatiles 234
    rapid method 233
    separation 229, 233
    Trace™ 230
    ultra-fast unit 236, 237
GC-mass spectrometry (GC-MS)
    high resolution analysis, broad peak 231–232
    α-pinene, sabinene and β-myrcene 233, 234
    replicate fruits per treatment 231
    SPME collections 231
Geographic information systems (GIS) 15
GIS *see* Geographic information systems (GIS)

Habitat manipulation, *Anastrepha*
    entire geographic region 202
    interception, traps 203
    push-pull strategy 203
    trap cropping 203
Heilipus albopictus
    biology 42
    damage 42
    distribution 42
    host plants 42
    length and wide 42
Heilipus apiatus 42–44
Heilipus catagraphus 44
Heilipus elegans
    damage 44
    distribution 44
    host plants 44

Heilipus elegans (Continued)
    length 44
    morphology 44
Heilipus lauri
    biological control 46
    biology
        adult 41
        egg 39, 41
        larva 41
        pupa 41
    characteristics 36
    chemical control 46
    damage assessment 45
    developmental stages 36–38
        adults 38, 39
        egg 36
        larva 36–37
        pupa 37–38
    distribution 38
    economic thresholds 45
    host plants 38–39
    life cycle 40
    origin 36
    sampling and monitoring techniques 45
Heilipus pittieri
    damage 41
    vs. *H. lauri* 41
    host plants 41
    length and rostrum 41
Heilipus rufipes 44–45
Heilipus trifasciatus Fabricius 41–42
Hindustan citrus mite (HCM)
    control tactics
        biological 255
        chemical 255
        regulatory 255
    description 252
    economic impact 252–253
    host plants 252
    morphological and biological aspects 253–254
    restricted distribution 252
    spread over areas 252
    symptoms 254–255
Hordeolicoccus
    H. heterotrichus 339
    H. nephelii 339
Hot pepper gall midge (HPGM)
    Contarinia lycopersici 240–241
    gall midges *see* Gall midges
    in Jamaica 241
    larvae feed 242
    multivoltine 241
    pest management 243
    post harvest handling 243–244
    Prodiplosis longifila 240–241
    quarantine control measurement 244
    undetermined species, *Contarinia* spp. 241

Icerya
    I. purchasi 337
    I. seychellarum 337
Inductive modelling approach 413
Insect life cycle modelling (ILCYM) software
    activity index 424
    deterministic simulation tools 416, 419
    future prospects 424
    life table and single-stage cohort study
        data 415–416
    limitations 424
    main features 412, 415
    model builder 416, 424
    population distribution and risk mapping
        spatial simulation 418–421
        temperature data 416, 418
    process-based modelling tools
        classical biological control programs 423
        climate change 422–423
        inductive *vs.* deductive approach 421–422
        linear *vs.* non-linear model 422
        species sex ratios 423
    temperature-dependent insects
        development 414
    validation tool 416, 418
Integrated pest management (IPM)
    chemicals used, Saudi Arabia 17
    mass trapping-based operation 17
    pheromone based area-wide programs 16
    RPW-IPM *see* RPW-IPM programs
Interception records, *Copitarsia* spp.
    APHIS AdHoc-309 database 171
    commodities reported 171, 175
    country of origin 171, 174
    various ports in USA 171, 174
IPM *see* Integrated pest management (IPM)

Kerriidae
    description 351
    Paratachardina pseudolobata 351
    Tachardiella mexicana 352

Lac scales *see* Kerriidae
Larva
    Diabrotica speciosa 75, 80–81
    Heilipus lauri 36–37, 41
    Neoleucinodes elegantalis 144–146
    Neoleucinodes silvaniae 141
Lepidopteran pest
    avocado fruit
        Antaeotricha nictitans 94
        Cryptaspasma sp. nr *lugubris* 93–94
        evolutionary range 87
        importations 86
        pest fauna 87

risk assessment reports preparation 87
Stenoma catenifer see Stenoma catenifer
Tortricidae 93
Lepidosaphes
L. *cornuta* 335
L. *laterochitinosa* 336
L. *rubrovittata* 336
L. *similis* 336
L. *tubulorum* 336
Life cycle
Anastrepha grandis 198
Anastrepha ludens 194
Anastrepha obliqua 196
Copitarsia spp. 162–164
Diabrotica speciosa 75–76
heilipus lauri 40
Neoleucinodes elegantalis 144
Singhiella simplex 364
Tecia solanivora Povolny 127–129
Tuta absoluta 100–101
Lychee erinose mite (LEM)
control tactics
biological 259
chemical 258–259
cultural 258
description 255
dissemination 257
distribution 255–256
economic impact 256
Eriophyoidea mites 255
foliage 258
fruit crop 256
hand lens/stereoscope microscope 258
morphological and biological aspects 256–257
orchard 256
symptoms 257–258

Mass trapping 113
Mating disruption 113–114
Maxent model training or validation 375, 376
Mealybugs *see* Pseudococcidae
Mite invasions in South America
acarological history 251
arachnids 251
Hindustan citrus mite *see* Hindustan citrus mite (HCM)
lychee erinose mite *see* Lychee erinose mite (LEM)
panicle rice mite *see* Panicle rice mite (PRM)
phytophagous 251
red palm mite *see* Red palm mite (RPM)
socio-economic importance 251–252
wheat curl mite *see* Wheat curl mite (WCM)
Monophlebidae
Crypticerya genistae 337, 355
Drosicha genus 337

Icerya see Icerya
Icerya p. purchasi 354–355
Morphology
A. *grandis*
adults 198
egg and larvae 198
A. *ludens*
adult 194
egg stage 193
larvae 193
puparium 193
A. *obliqua*
adults 196
egg stage 196
larvae 196

NAPPFAST system 421–422
Neoleucinodes dissolvens 143
Neoleucinodes elegantalis
adult 146–147
control strategies
biological control 150–153
chemical control 150
cultural and physical control 154
plant resistance 153–154
quarantine methods 154
economic losses 139
eggs 144
host plant 138
insect behavior and damage 150
larva 144–146
life cycle 144
monitoring and sampling techniques 148–149
origin and distribution 139–140
oviposition sites 147
pupa 146
pupation sites 147
rearing techniques 147–148
taxonomy 137–138
Neoleucinodes imperialis 142
Neoleucinodes prophetica
adult 141
distribution 142
female genitalia 141–142
male genitalia 141
Neoleucinodes silvaniae
adult 140
distribution 141
female genitalia 141
larva 141
male genitalia 140
Neoleucinodes torvis
adult 142
distribution 143
female genitalia 142–143
male genitalia 142

Neotropical species
    Diaspididae
        Abgrallaspis see *Abgrallaspis*
        Acutaspis see *Acutaspis*
        Aspidiella hartii 334
        Hemiberlesia gliwicensis 334
        Malleolaspis mammata 334
        Selenaspidopsis browni 334
    Pseudococcidae
        Ferrisia terani 337–338
        Nipaecoccus see *Nipaecoccus*
        Paracoccus see *Paracoccus*
        Phenacoccus franseriae 338
        Planococcus halli 338
        Pseudococcus see *Pseudococcus*
        Puto barberi 338
Nettle caterpillar
    classical biological control 190
    control strategies 186
    emergence times 188
    field observations 186, 187–188
    larval duration 184, 185
    life stages 183–184
    limacodids 190
    no-choice host range tests 186–188, 189
    plant species 188
    pupal duration 184, 185
    starfruit trees 188
Nipaecoccus
    N. annonae 338
    N. jonmartini 338

Oriental species
    Diaspididae
        Aonidiella comperei 335
        Aulcaspis tubercularis 335
        Chrysomphalus see *Chrysomphalus*
        Fiorinia coronata 335
        Fiorinia proboscidaria 335
        Greenaspis sp. 335
        Lepidosaphes see *Lepidosaphes*
        Lindingaspis misrae 336
        Parlatoria see *Parlatoria*
        Pinnaspis theae 336
        Pseudaonidia corbetti 336
        Pseudaonidia curculiginis 336
        Pseudaulacaspis barberi 336
        Silvestraspis uberifera 336
        Thysanofiorinia leei 336
    Pseudococcidae
        Dysmicoccus lepelleyi 339
        Exallomochlus see *Exallomochlus*
        Formicococcus robustus 339
        Hordeolicoccus see *Hordeolicoccus*
        Maconellicoccus multipori 339
        Nipaecoccus viridis 339

        Paracoccus interceptus 339
        Paraputo odontomachi 339
        Planococcus see *Planococcus*
        Pseudococcus see *Pseudococcus*
        Rastrococcus see *Rastrococcus*
        Rhizoecus amorphophalli 340
    Ortheziidae
        Arctorthezia cataphracta 337
        Insignorthezia insignis 337
        Insignorthezia pseudoinsignis 337
        Praelongorthezia praelonga 337

Pacific Island species
    Laminicoccus pandani 340
    Palmicultor browni 340
    Palmicultor palmarum 340
Palmicultor
    P. browni 345–346
    P. lumpurensis 346
    P. palmarum 346
'Palm weevils' 2
Panicle rice mite (PRM)
    in Asian rice crops 271
    continental Central America 271
    control tactics
        biological 274
        chemical 274
        cultural 275
        host plant resistance 274–275
    damage and economic impact 271–272
    detection 273
    dissemination 273
    morphological and biological aspects 272–273
    rice 270–271
    tarsonemid mite 270
    tropical Asia 271
Paracoccus
    P. ferrisi 338
    P. herreni 338
    P. mexicanus 338
    P. solani 338
Parasitoids 105–109
Parlatoria
    P. citri 336
    P. crypta 336
    P. pseudaspidiotus 336
Planococcus
    P. lilacinus 339
    P. litchi 339
Planococcus minor
    adult female mealybug 290
    Coccidoxenoides perminutus 297
    control tactics
        ant 295
        biological 294–295

chemical 293
regulatory 293–294
cultural, physical and mechanical
management 296
damage and economic thresholds 292–293
description 288
host plant range 288–289
host plant resistance 296
identification, species 289
indirect economic impact 289
Leptomastix dactylopii 297
life cycle 290–291
mating disruption, mass trapping 295–296
mealybug 289
misidentification 289
molecular identification 290
Nymphs 290
quarantine methods 296
rearing techniques 291–292
sampling and monitoring techniques 292
species with origins in Old World 289
viruses and cryptic nature 296–297
Plant Protection and Quarantine (PPQ) 401
Plant resistance
diabrotica speciosa 78–79
Neoleucinodes elegantalis 153–154
Tecia solanivora Povolny 131
tuta absoluta 114–116
Postharvest treatment, *Anastrepha*
fruit irradiation 204
fruit temperature treatment 203–204
PPQ *see* Plant Protection and Quarantine (PPQ)
Praelongorthezia praelonga see also Citrus orthezia
ant associations 312
biology 311–312
control methods *see* Control methods,
P. praelonga
description 309
detection methods 312
dorsum 310–311
mounted adult female 309–310
4th-instar female 309
unmounted adult female 309
venter 310
Predators 109–111
Pritchardia hillebrandei 188, 190
Prodiplosis longifila
biological control 247–249
biology and life cycle 244–245
chemical control 247
and *Contarinia lycopersici* 240–241
control tactics 247, 248
damage 245–246
distribution 244
horticultural crops 240
hosts 244
sampling and monitoring techniques 246–247

seasonality 245
taxonomy 244
Pseudaulacaspis
P. brimblecombei 336
P. eugeniae 336
Pseudococcidae
Afrotropical species *see* Afrotropical species
Australian species *see* Australian species
Chaetococcus bambusae 346
description 344
Eastern Palearctic species 339
Maconellicoccus hirsutus 347–348
Mealybugs 337
neotropical species *see* Neotropical species
Nipaecoccus viridis 344–345
oriental species *see* Oriental species
Pacific Island species *see* Pacific Island species
Palmicultor browni 345–346
Palmicultor lumpurensis 346
Palmicultor palmarum 346
Paracoccus marginatus 346–347
Planococcus lilacinus 348
Pseudococcus dendrobiorum 345
Western Palearctic species *see* Western
Palearctic species
Pseudococcus
P. aurantiacus 339
P. baliteus 339
P. cryptus 338
P. dendrobiorum 340
P. elisae 338
P. eucalypticus 340
P. landoi 338
P. zamiae 340
Pupa
Diabrotica speciosa 80–81
Heilipus lauri 37–38, 41
Neoleucinodes elegantalis 146

Quarantine control, gall midge
grower registration and trace-back system 243
in Jamaica 243, 244
pest management 243
port inspection 244
post harvest handling 243–244

Raoiella indica see Red palm mite (RPM)
Rastrococcus
R. expeditionis 340
R. iceryoides 340
R. invadens 340
R. tropicasiaticus 340
Rearing techniques
Copitarsia spp. 164
diabrotica speciosa

Rearing techniques (*Continued*)
    cages for adults maintenance 79–80
    eggs collection 80
    larval and pupal development 80–81
    larval food 80
    substrates and containers for 80
    Neoleucinodes elegantalis 147–148
    Tuta absoluta 101–102
Red palm mite (RPM)
    the Americas 260
    Brazilian host plants 261
    control tactics
        biological 265–266
        chemical 265
        description 264–265
        host plant resistance 266
    damage and economic impact 261
    description 260
    detection 264
    dissemination 263
    field surveys 260
    morphological and biological aspects 262–263
    plant species 261
    population dynamics 263
    rearing techniques 263–264
Red palm weevil (RPW)
    adults, *Rhynchophorus* species
        and *Dynamis* diagnostic traits 3
        heads of males, *R. quadrangulus* and *R. cruentatus* 3, 5, 6
        R. bilineatus and *R. ferrugineus* 3, 8
        R. palmarum 3, 4
        scutellum, *R. phoenicis* 3, 6
    biology, detection and distribution
        galleries 5–7
        larval stages 7
        RPW females 5
        RPW males 4
        symbiotic relationship 7–8
    infestation 1
    local, regional, national and European regulations
        eradication 22, 23
        failure, ermadication program 22
        measures 21
        research activities 21–22
    North Africa 25–27
    Northern Mediterranean basin and Canary islands
        adult *Phoenix dactylifera* palms 19, 20
        palm locations, EU 18
        Spanish autonomous communities 19
        year of detection, RPW 20, 21
    research 27
    Rhynchophorus aggregation pheromones 9–11
    Rhynchophorus nematode symbionts 8–9
    strategies, Spain
        monitoring, early detection 25
        plantations 25
        preventive treatments 22–24
        pruning 24
        removal and destruction, affected palms 25
        sanitation 24, 25
        trapping 25
    trapping *see* RPW trapping
Red ring disease (RRD) 8
Red ring nematode (RRN) 8
Regulatory control, *Copitarsia* spp. 174–175
Rhynchophorus bilineatus
    heads of males 3, 4
    ovaries 9
Rhynchophorus cruentatus
    genital capsule 8
    heads of males 3, 5
Rhynchophorus ferrugineus see Red palm weevil (RPW)
Rhynchophorus palmarum
    oil palm 16
    RPW 8
    trapping 11–12
RPW *see* Red palm weevil (RPW)
RPW-IPM programs
    date palms 17–18
    pheromone-based 15–17
RPW trapping
    definition 11
    density 15, 16
    food baits 13–14
    insecticide, traps 14–15
    monitoring, mass trapping
        ferrugineol-based pheromone traps 16
        infestation 17
        management 16–17
        pheromone traps, GIS 15
    protocols
        design 11
        pheromone lure efficiency, release rate and longevity 12–13
    trap placement 15
RRD *see* Red ring disease (RRD)
RRN *see* Red ring nematode (RRN)

SAW *see* Surface acoustic wave (SAW)
Scale insects, the USA and Caribbean Basin
    ability 322
    APHIS/USDA 322
    AQAS 322
    armored scales 323, 324
    armored scales and mealybugs 322–323, 324
    Asterolecaniidae 332
    characteristics 322
    Coccidae 332–333
    Dactylopiidae 337
    description 320
    Diaspididae *see* Diaspididae
    diversity and volume, agricultural products 322

economic damage, plants 341
exotic pest enters 320
exotic species 321
exporting countries 322
interception 322
invasive species 321–322
limited distribution 324, 327–332
Monophlebidae 337
natural enemies 320–321
NHS/CBP 322
number of interceptions 323, 326
number of species and percentage 323, 325
Ortheziidae 337
Pseudococcidae *see* Pseudococcidae
quarantine and crop protection 341
selected families 320, 321
specimens identification, family/genus 324
Stictiococcidae 340
training 341
Schizotetranychus hindustanicus see Hindustan citrus mite (HCM)
Singhiella simplex see Ficus whitefly
SIT *see* Sterile insect technique (SIT)
Soft scales *see* Coccidae
Solanum tuberosum L. *see Tecia solanivora*
Solid phase microextraction (SPME)
    GC analysis, grapefruits 231, 232, 235
    headspace volatiles 234
    tephritid-infested citrus 230
SPME *see* Solid phase microextraction (SPME)
Steneotarsonemus spinki see Panicle rice mite (PRM)
Stenoma catenifer
    biology and ecology 89–90
    common names 88
    control measures
        cultural control 92
        insecticides 92
        natural enemies 91–92
        sex pheromone 90–91
        sterile insect technique 92
    developmental and reproductive biology 90
    distribution 88–89
    invasion potential 92–93
    taxonomy 88–89
Sterile insect technique (SIT) 202
Surface acoustic wave (SAW) 231
Synthetic pheromone 103–104

Tecia solanivora
    action thresholds 129
    autocidal control 130
    biological control 130–131
    biology and life history 127
    chemical control 130
    cultural control 131–132
    damage 129

host plant resistance 131
life cycle 127–129
mating 128
regulatory quarantine methods 132
related genera and distribution 126–127
sampling and monitoring techniques 129
Tephritidae 214
Trapping *see* RPW trapping
Tuta absoluta
    biological control
        entomopathogens 112–113
        parasitoids 105–109
        predators 109–111
    biology 101
    chemical control 116–117
    cultural practices 116
    distribution 99
    life cycle 100–101
    life history parameters 101
    pheromone traps
        monitoring 104
        plant sampling 105
        synthetic pheromone 103–104
    plant hosts and damage 102–103
    plant resistance 114–116
    rearing techniques 101–102
    semiochemical management
        attract and kill 114
        mass trapping 113
        mating disruption 113–114
    spread 99–100
    thermal requirements 102

The United States Department of Agriculture-Animal and Plant Health Inspection Service (USDA-APHIS) 240

Vryburgia
    V. distincta 339
    V. viator 339

Western Palearctic species
    Diaspididae
        Lepidosaphes malicola 334
        Lepidosaphes pistaciae 334
        Leucaspis riccae 334
        Melanaspis inopinata 334
        Salicicola archangelskyae 334
    Pseudococcidae
        Phenacoccus graminicola 338
        Spinococcus convolvuli 338
Wheat curl mite (WCM)
    biological and morphological aspects 268–269

Wheat curl mite (WCM) (*Continued*)
    biological, morphological and molecular studies 266
    control tactics
        Aceria tosichella colonies 268, 269–270
        chemical 270
        cultural 270
        host plant resistance 270
        regulatory 270
    damage and economic impact 268
    description 266
    dissemination 269
    Eriophyidae mite 266
    eriophyid mites 267
    production areas 266–267
    quite often surveys 267
    subsequent surveys 267
    symptoms 269
    in Uruguay and Argentina 267
    WCM populations 266
Whiteflies *see* Aleyrodidae